Supergravity

Supergravity, together with string theory, is one of the most significant developments in theoretical physics. Although there are many books on string theory, this is the first-ever authoritative and systematic account of supergravity.

Written by two of the most respected workers in the field, it provides a solid introduction to the fundamentals of supergravity. The book starts by reviewing aspects of relativistic field theory in Minkowski spacetime. After introducing the relevant ingredients of differential geometry and gravity, some basic supergravity theories ($D = 4$ and $D = 11$) and the main gauge theory tools are explained. The second half of the book is more advanced: complex geometry and $N = 1$ and $N = 2$ supergravity theories are covered. Classical solutions and a chapter on anti-de Sitter/conformal field theory (AdS/CFT) correspondence complete the text.

Numerous exercises and examples make it ideal for Ph.D. students, and with applications to model building, cosmology, and solutions of supergravity theories, this text is an invaluable resource for researchers. A website hosted by the authors, featuring solutions to some exercises and additional reading material, can be found at www.cambridge.org/supergravity.

Daniel Z. Freedman is Professor of Applied Mathematics and Physics at the Massachusetts Institute of Technology. He has made many research contributions to supersymmetry and supergravity: he was a co-discoverer of the first supergravity theory in 1976. This discovery has been recognized by the award of the Dirac Medal and Prize in 1993, and the Dannie Heineman Prize of the American Physical Society in 2006.

Antoine Van Proeyen is Head of the Theoretical Physics Section at the KU Leuven, Belgium. Since 1979, he has been involved in the construction of various supergravity theories, the resulting special geometries, and their applications to phenomenology and cosmology.

Our conventions

The metric is 'mostly plus', i.e. $- + \ldots +$. The curvature is

$$R_{\mu\nu\rho\sigma} = g_{\rho\rho'}(\partial_\mu \Gamma^{\rho'}_{\nu\sigma} - \partial_\nu \Gamma^{\rho'}_{\mu\sigma} + \Gamma^{\rho'}_{\mu\tau}\Gamma^{\tau}_{\nu\sigma} - \Gamma^{\rho'}_{\nu\tau}\Gamma^{\tau}_{\mu\sigma})$$
$$= e^a_\rho e^b_\sigma \left(\partial_\mu \omega_{\nu ab} - \partial_\nu \omega_{\mu ab} + \omega_{\mu ac}\omega_\nu{}^c{}_b - \omega_{\nu ac}\omega_\mu{}^c{}_b\right).$$

Ricci tensor and energy–momentum tensors are defined by

$$R_{\mu\nu} = R^\rho{}_{\nu\rho\mu}, \qquad R = g^{\mu\nu} R_{\mu\nu},$$
$$R_{\mu\nu} - \tfrac{1}{2} g_{\mu\nu} R = \kappa^2 T_{\mu\nu}.$$

Covariant derivatives involving the spin connection are, for vectors and spinors,

$$D_\mu V^a = \partial_\mu V^a + \omega_\mu{}^{ab} V_b, \qquad D_\mu \lambda = \partial_\mu \lambda + \tfrac{1}{4}\omega_\mu{}^{ab}\gamma_{ab}\lambda.$$

We use (anti)symmetrization of indices with 'weight 1', i.e.

$$A_{[ab]} = \tfrac{1}{2}(A_{ab} - A_{ba}) \qquad \text{and} \qquad A_{(ab)} = \tfrac{1}{2}(A_{ab} + A_{ba}).$$

The Levi-Civita tensor is

$$\varepsilon_{0123} = 1, \qquad \varepsilon^{0123} = -1.$$

The dual, self-dual, and anti-self-dual of antisymmetric tensors are defined by

$$\tilde{H}^{ab} \equiv -\tfrac{1}{2}\mathrm{i}\varepsilon^{abcd} H_{cd}, \qquad H^{\pm}_{ab} = \tfrac{1}{2}(H_{ab} \pm \tilde{H}_{ab}), \qquad H^{\pm}_{ab} = \left(H^{\mp}_{ab}\right)^*.$$

Structure constants are defined by

$$[T_A, T_B] = f_{AB}{}^C T_C.$$

The Clifford algebra is

$$\gamma_\mu\gamma_\nu + \gamma_\nu\gamma_\mu = 2g_{\mu\nu}, \qquad \gamma_{\mu\nu} = \gamma_{[\mu}\gamma_{\nu]}, \ldots$$
$$(\gamma^\mu)^\dagger = \gamma^0\gamma^\mu\gamma^0,$$
$$\gamma_* = (-\mathrm{i})^{(D/2)+1}\gamma_0\gamma_1\cdots\gamma_{D-1};$$

in four dimensions:

$$\gamma_* = \mathrm{i}\gamma_0\gamma_1\gamma_2\gamma_3, \qquad \varepsilon_{abcd}\gamma^d = \mathrm{i}\gamma_*\gamma_{abc}.$$

The Majorana and Dirac conjugates are

$$\bar{\lambda} = \lambda^T C, \qquad \bar{\lambda} = \mathrm{i}\lambda^\dagger\gamma^0.$$

We mostly use the former. For Majorana fermions the two are equal.

The main SUSY commutator is

$$[\delta(\epsilon_1), \delta(\epsilon_2)] = \tfrac{1}{2}\bar{\epsilon}_2\gamma^\mu\epsilon_1\partial_\mu.$$

p-form components are defined by

$$\phi_p = \frac{1}{p!}\phi_{\mu_1\cdots\mu_p}\mathrm{d}x^{\mu_1}\wedge\cdots\wedge\mathrm{d}x^{\mu_p}.$$

The differential acts from the left:

$$\mathrm{d}A = \partial_\nu A_\mu\,\mathrm{d}x^\nu\wedge\mathrm{d}x^\mu, \qquad A = A_\mu\mathrm{d}x^\mu.$$

Supergravity

DANIEL Z. FREEDMAN
Massachusetts Institute of Technology, USA

and

ANTOINE VAN PROEYEN
KU Leuven, Belgium

CAMBRIDGE
UNIVERSITY PRESS

University Printing House, Cambridge CB2 8BS, United Kingdom

Cambridge University Press is part of the University of Cambridge.

It furthers the University's mission by disseminating knowledge in the pursuit of education, learning and research at the highest international levels of excellence.

www.cambridge.org
Information on this title: www.cambridge.org/9780521194013

First published 2012
Reprinted 2013

A catalogue record for this publication is available from the British Library

Library of Congress Cataloguing in Publication data
Freedman, Daniel Z.
Supergravity / Daniel Z. Freedman and Antoine Van Proeyen.
p. cm.
ISBN 978-0-521-19401-3 (hardback)
1. Supergravity. I. Van Proeyen, Antoine. II. Title.
QC174.17.S9F735 2012
530.14′23–dc23
2011053360

ISBN 978-0-521-19401-3 Hardback

Additional resources for this publication at www.cambridge.org/supergravity

Contents

Preface

The main purpose of this book is to explore the structure of supergravity theories at the classical level. Where appropriate we take a general D-dimensional viewpoint, usually with special emphasis on $D = 4$. Readers can consult the Contents for a detailed list of the topics treated, so we limit ourselves here to a few comments to guide readers. We have tried to organize the material so that readers of varying educational backgrounds can begin to read at a point appropriate to their background. Part I should be accessible to readers who have studied relativistic field theory enough to appreciate the importance of Lagrangians, actions, and their symmetries. Part II describes the differential geometric background and some basic physics of the general theory of relativity. The basic supergravity theories are presented in Part III using techniques developed in earlier chapters. In Part IV we discuss complex geometry and apply it to matter couplings in global $\mathcal{N} = 1$ supersymmetry. In Part V we begin a systematic derivation of $\mathcal{N} = 1$ matter-coupled supergravity using the conformal compensator method. The going can get tough on this subject. For this reason we present the final physical action and transformation rules and some basic applications in two separate short chapters in Part VI. Part VII is devoted to a systematic discussion of $\mathcal{N} = 2$ supergravity, including a short chapter with the results needed for applications. Two major applications of supergravity, classical solutions and the AdS/CFT correspondence, are discussed in Part VIII in considerable detail. It should be possible to understand these chapters without full study of earlier parts of the book.

Many interesting aspects of supergravity, some of them subjects of current research, could not be covered in this book. These include theories in spacetime dimensions $D < 4$, higher derivative actions, embedding tensors, infinite Lie algebra symmetries, and the positive energy theorem.

Like many other subjects in theoretical physics, supersymmetry and supergravity are best learned by readers who are willing to 'get their hands dirty'. This means actively working out problems that reinforce the material under discussion. To facilitate this aspect of the learning process, many exercises for readers appear within each chapter. We give a rough indication of the level of each exercise as follows:

Level 1. The result of this exercise will be used later in the book.
Level 2. This exercise is intended to illuminate the subject under discussion, but it is not needed in the rest of the book.
Level 3. This exercise is meant to challenge readers, but is not essential.

These levels are indicated respectively by single, double or triple gray bars in the outside margin.

A website featuring solutions to some exercises, errata and additional reading material, can be found at www.cambridge.org/supergravity.

Dan Freedman
Toine Van Proeyen
October 2011

Acknowledgements

We thank Eric Bergshoeff, Paul Chesler, Bernard de Wit, Eric D'Hoker, Henriette Elvang, John Estes, Gary Gibbons, Joaquim Gomis, Renata Kallosh, Hong Liu, Marián Ledó, Samir Mathur, John McGreevy, Michael Peskin, Leonardo Rastelli, Kostas Skenderis, Stefan Vandoren, Bert Vercnocke and Giovanni Villadoro. We thank the students in various courses (Leuven advanced field theory course, Doctoral schools in Paris, Barcelona, Hamburg), and also Frederik Coomans, Serge Dendas, Daniel Harlow, Andrew Larkoski, Jonathan Maltz, Thomas Rube, Walter Van Herck and Bert Van Pol for their input in the preparation of this text and their critical remarks.

Our home institutions have supported the writing of this book over a period of years, and we are grateful. We also thank the Galileo Galilei Institute in Florence and the Department of Applied Mathematics and Theoretical Physics in Cambridge for support during extended visits, and the Stanford Institute for Theoretical Physics for support and hospitality, indeed a home away from home, during multiple visits when we worked closely together.

A.V.P. wil in het bijzonder zijn moeder bedanken voor de sterkte en voortdurende steun die hij van haar gekregen heeft. He also thanks Marleen and Laura for the strong support during the work on this book. D.Z.F. thanks his wife Miriam for her encouragement to start this project and continuous support as it evolved.

Introduction

Two developments in the late 1960s and early 1970s set the stage for supergravity. First the standard model took shape and was decisively confirmed by experiments. The key theoretical concept underlying this progress was *gauge symmetry*, the idea that symmetry transformations act independently at each point of spacetime. In the standard model these are internal symmetries, whose parameters are Lorentz scalars $\theta^A(x)$ that are *arbitrary functions* of the spacetime point x. These parameters are coordinates of the compact Lie group $SU(3) \otimes SU(2) \otimes U(1)$. Scalar, spinor, and vector fields of the theory are each classified in representations of this group, and the Lagrangian is invariant under group transformations. The special dynamics associated with the non-abelian gauge principle allows different realizations of the symmetry in the particle spectrum and interactions that would be observed in experiments. For example, part of the gauge symmetry may be 'spontaneously broken'. In the standard model this produces the 'unification' of weak and electromagnetic interactions. The observed strength and range of these forces are very different, yet the gauge symmetry gives them a common origin.

The other development was global (also called rigid) supersymmetry [1, 2, 3]. It is the unique framework that allows fields and particles of different spin to be unified in representations of an algebraic system called a superalgebra. The symmetry parameters are spinors ϵ_α that are *constant*, independent of x. The simplest $\mathcal{N} = 1$ superalgebra contains a spinor supercharge Q_α and the energy–momentum operator P_a. The anti-commutator of two supercharges is a translation in spacetime. The $\mathcal{N} = 1$ supersymmetry algebra has representations containing massless particles of spins $(s, s - 1/2)$ for $s = 1/2, 1, \ldots$ and somewhat larger representations containing particles with a common non-vanishing mass. Thus supersymmetry always unites bosons, integer spin, with fermions, half-integer spin. The focus of early work was interacting field theories of the $(1/2, 0)$ and $(1, 1/2)$ multiplets. It was found that the ultraviolet divergences of supersymmetric theories are less severe than in the standard model due to the cancelation between bosons and fermions in loop diagrams.

Unbroken supersymmetry requires a spectrum of particles in equal-mass boson–fermion pairs. This is decidedly not what is observed in experiments. So if supersymmetry is realized in Nature, it must appear as a broken symmetry. Through the years much theoretical effort has been devoted to the construction of extensions of the standard model with broken supersymmetry. It is hypothesized that the as yet unseen superpartners of the known particles will be produced at the Large Hadron Collider (LHC) accelerator, thus confirming a supersymmetric version of the standard model. The advantages of supersymmetric models include the following:

- Milder ultraviolet divergences permit an improved and more predictive treatment of electroweak symmetry breaking.
- When extrapolated using the renormalization group, the three distinct gauge couplings of the standard model approach a common value at high energy. The unification of couplings is a major success.
- Supersymmetry provides natural candidates for the particles of cosmological cold dark matter.

The role of gauge symmetry in the standard model suggested that a gauged form of supersymmetry would be interesting and perhaps more powerful than the global form. Such a theory would contain gauge fields for both spacetime translations P_a and SUSY transformations generated by Q_α. Thus gauged supersymmetry was expected to be an extension of general relativity in which the graviton acquires a fermionic partner called the gravitino. The name supergravity is certainly appropriate and was used even before the theory was actually found. It was reasonable to think that the gauge fields of the theory would be the vierbein, $e^a_\mu(x)$, needed to describe gravity coupled to fermions, and a vector–spinor field, $\psi_{\mu\alpha}(x)$, for the gravitino. The graviton and gravitino belong to the $(2, 3/2)$ representation of the algebra. A Lagrangian field theory of supergravity was formulated in the spring of 1976 in [4]. The approach taken was to modify the known free field Lagrangian for $\psi_{\mu\alpha}$ to agree with gravitational gauge symmetry and then find, by a systematic procedure, the additional terms necessary for invariance under supersymmetry transformations with arbitrary $\epsilon_\alpha(x)$. Soon an alternative approach appeared [5] in which the most complicated calculation required in [4] is avoided.

Research in supergravity became a very intense activity in the years following its discovery. One early direction was the construction of Lagrangian field theories in which the spin-(2, 3/2) gravity multiplet is coupled to the $(1/2, 0)$ and $(1, 1/2)$ multiplets of global supersymmetry. This is the framework of matter-coupled supergravity. It shares the positive features of global symmetry listed above. In addition supergravity provides new scenarios for the breaking of supersymmetry. In particular, the structure of the supergravity Lagrangians allows SUSY breaking with vanishing vacuum energy and thus vanishing cosmological constant. This feature is not available without the coupling of matter fields to supergravity. Matter-coupled supergravity theories typically contain scalar fields, which can be useful in constructing phenomenological models of inflationary cosmology.

A spin-3/2 particle is the key prediction of supergravity. SUSY breaking gives it a mass whose magnitude depends on the breaking mechanism. Unfortunately it appears difficult to detect it at the LHC because it is coupled to matter with the feeble strength of quantum gravity. However, gravitinos can be copiously produced in the ultra-high-temperature environment at or near the big bang. Gravitino production leads to important constraints on early universe cosmology.

A second direction of research involves the construction of theories with several supercharges $Q_{i\alpha}$, $i = 1, 2, \dots, \mathcal{N}$. Such extended supergravity theories can be constructed up to the limit $\mathcal{N} = 8$ in spacetime dimension $D = 4$. Beyond that the superalgebra representations necessarily contain particles of spin $s \geq 5/2$, for which no consistent interactions exist. Many of the ultraviolet divergences expected in a field theory containing gravity are known to cancel in the maximal $\mathcal{N} = 8$ theory, and some theorists speculate that it is ultraviolet finite to all orders in perturbation theory.

Supergravity theories in spacetime dimensions $D > 4$ have been constructed up to the bound $D = 11$ (which is again due to the higher spin consistency problem). Two 10-dimensional supergravity theories, known as the Type IIA and Type IIB theories, are related to the superstring theories that carry the same names. Roughly speaking, supergravity appears as the low-energy limit of superstring theory. This means that the dynamics of the lowest-energy modes of the superstring are described by supergravity. But these statements do not do justice to the intimate and rich relation of these two theoretical frameworks.

The very important anti-de Sitter/conformal field theory (AdS/CFT) correspondence provides one example of this relation. It was based on the remarkable conjecture that Type IIB string theory on the product manifold $\text{AdS}_5 \otimes S^5$ is equivalent to the maximal $\mathcal{N} = 4$ global supersymmetric gauge theory. However, concrete tests and predictions of AdS/CFT usually involve working in the limit in which classical supergravity is a valid approximation to string theory.

The scope of supergravity is broad. There is a supergravity-inspired approach to positive energy and stability in gravitational theories. Many classical solutions of supergravity have the special Bogomol'nyi–Prasad–Sommerfield (BPS) property and therefore satisfy tractable first order field equations. The scalar sectors of supergravity theories involve non-linear σ-models on complex manifolds with new geometries of interest in both physics and mathematics.

To summarize: supergravity is based on the gauge principle of local supersymmetry and is thus connected to fundamental ideas in theoretical physics. Supergravity effects may turn out to be observable at the LHC. Further there is important input from cosmology. This real side of the subject is far from confirmation, but it must be taken seriously. In addition there are several more theoretical applications such as BPS solutions and AdS/CFT. Active research continues on most branches of supergravity although 35 years have passed since it was first formulated.

PART I

RELATIVISTIC FIELD THEORY IN MINKOWSKI SPACETIME

1 Scalar field theory and its symmetries

The major purpose of the early chapters of this book is to review the basic notions of relativistic field theory that underlie our treatment of supergravity. In this chapter we discuss the implementation of internal and spacetime symmetries using the model of a system of free scalar fields as an example. The general Noether formalism for symmetries is also discussed. Our book largely involves classical field theory. However, we adopt conventions for symmetries that are compatible with implementation at the quantum level.

Our treatment is not designed to teach the material to readers who are encountering it for the first time. Rather we try to gather the ideas (and the formulas!) that are useful background for later chapters. Supersymmetry and supergravity are based on symmetries such as the spacetime symmetry of the Poincaré group and much more!

As in much of this book, we assume general spacetime dimension D, with special emphasis on the case $D = 4$.

1.1 The scalar field system

We consider a system of n real scalar fields $\phi^i(x)$, $i = 1, \dots, n$, that propagate in a flat spacetime whose metric tensor

$$\eta_{\mu\nu} = \eta^{\mu\nu} = \mathrm{diag}(-, +, \dots, +) \tag{1.1}$$

describes one time and $D - 1$ space dimensions. This is Minkowski spacetime, in which we use Cartesian coordinates x^μ, $\mu = 0, 1, \dots, D - 1$, with time coordinate $x^0 = t$ (with velocity of light $c = 1$).

Practicing physicists and mathematicians are largely concerned with fields that satisfy nonlinear equations. However, linear wave equations, which describe free relativistic particles, have much to teach about the basic ideas. We therefore assume that our fields satisfy the Klein–Gordon equation

$$\Box\phi^i(x) = m^2\phi^i(x)\,, \tag{1.2}$$

where $\Box = \eta^{\mu\nu}\partial_\mu\partial_\nu$ is the Lorentz invariant d'Alembertian wave operator.

The equation has plane wave solutions $e^{\pm i(\vec{p}\cdot\vec{x}-Et)}$, which provide the wave functions for particles of spatial momentum $\vec{p}$, with spatial components p^i, and energy $E = p^0 = \sqrt{\vec{p}^2 + m^2}$. The general solution of the equation is the sum of a positive frequency part, which can be expressed as the $(D-1)$-dimensional Fourier transform in the plane waves $e^{i(\vec{p}\cdot\vec{x}-Et)}$, plus a negative frequency part, which is the Fourier transform in the $e^{-i(\vec{p}\cdot\vec{x}-Et)}$,

$$\begin{aligned} \phi^i(x) &= \phi^i_+(x) + \phi^i_-(x)\,, \\ \phi^i_+(x) &= \int \frac{d^{D-1}\vec{p}}{(2\pi)^{(D-1)}2E} e^{i(\vec{p}\cdot\vec{x}-Et)} a^i(\vec{p})\,, \\ \phi^i_-(x) &= \int \frac{d^{D-1}\vec{p}}{(2\pi)^{(D-1)}2E} e^{-i(\vec{p}\cdot\vec{x}-Et)} a^{i*}(\vec{p}). \end{aligned} \tag{1.3}$$

In the classical theory the quantities $a^i(\vec{p})$, $a^{i*}(\vec{p})$ are simply complex valued functions of the spatial momentum $\vec{p}$. After quantization one arrives at the true quantum field theory[1] in which $\mathbf{a}^i(\vec{p})$, $\mathbf{a}^{i*}(\vec{p})$ are annihilation and creation operators[2] for the particles described by the field operator $\phi^i(\vec{x})$.

The Klein–Gordon equation (1.2) is the variational derivative $\delta S/\delta\phi^i(x)$ of the action

$$S = \int d^D x\, \mathcal{L}(x) = -\tfrac{1}{2}\int d^D x \left[\eta^{\mu\nu}\partial_\mu\phi^i\partial_\nu\phi^i + m^2\phi^i\phi^i\right]. \tag{1.4}$$

The repeated index i is summed. The action is a *functional* of the fields $\phi^i(x)$. It is a real number that depends on the configuration of the fields throughout spacetime.

1.2 Symmetries of the system

Consider a set of fields such as the $\phi^i(x)$ that satisfy equations of motion such as (1.2). A general symmetry of the system is a mapping of the configuration space, $\phi^i(x) \to \phi'^i(x)$, with the property that if the original field configuration $\phi^i(x)$ is a solution of the equations of motion, then the transformed configuration $\phi'^i(x)$ is also a solution. For scalar fields and for most other systems of interest in this book, we can restrict attention to symmetry transformations that leave the action invariant. Thus we require that the mapping has the property[3,4]

$$S[\phi^i] = S[\phi'^i]. \tag{1.5}$$

Here is an example.

[1] When desirable for clarity we use bold face to indicate the operator in the quantum theory that corresponds to a given classical quantity.

[2] In the conventions above, creation and annihilation operators are normalized in the quantum theory by $[\mathbf{a}(\vec{p}),\, \mathbf{a}^*(\vec{p}\,')] = (2\pi)^{D-1} 2E\, \delta(\vec{p} - \vec{p}\,')$.

[3] Such mappings must also respect the boundary conditions. This requirement can be non-trivial, e.g. Neumann and Dirichlet boundary conditions for the bosonic string lead to different spacetime symmetry groups. We will mostly assume that field configurations vanish at large spacetime distances.

[4] One important exception is the electromagnetic duality symmetry, which is discussed in Sec. 4.2.

Exercise 1.1 *Verify that the map $\phi^i(x) \to \phi'^i(x) = \phi^i(x+a)$ satisfies (1.5) if a^μ is a constant vector. This symmetry is called a global spacetime translation.*

We consider both spacetime symmetries, which involve a motion in Minkowski spacetime such as the global translation of the exercise, and internal symmetries, which do not. Internal symmetries are simpler to describe, so we start with them.

1.2.1 SO(n) internal symmetry

Let $R^i{}_j$ be a matrix of the orthogonal group SO(n). This means that it is an $n \times n$ matrix that satisfies

$$R^i{}_k \delta_{ij} R^j{}_\ell = \delta_{k\ell}\,, \qquad \det R = 1. \tag{1.6}$$

It is quite obvious that the linear map,

$$\phi^i(x) \to \phi'^i(x) = R^i{}_j \phi^j(x)\,, \tag{1.7}$$

satisfies (1.5) and is an internal symmetry of the action (1.4). This symmetry is called a continuous symmetry because a matrix of SO(n) depends continuously on $\frac{1}{2}n(n-1)$ independent group parameters. We will discuss one useful choice of parameters shortly. We also call the symmetry a global symmetry because the parameters are constants. In Ch. 4 we will consider local or gauged internal symmetries in which the group parameters are arbitrary functions of x^μ.

It is worth stating the intuitive picture of this symmetry. One may consider the field ϕ^i as an n-dimensional vector, that is an element of $\mathbb{R}^n$. The transformation $\phi^i \to R^i{}_j\phi^j$ is a rotation in this internal space. Such a rotation preserves the usual norm $\phi^i\delta_{ij}\phi^j$.

We now introduce the Lie algebra of the group SO(n). To first order in the small parameter ϵ, we write the infinitesimal transformation

$$R^i{}_j = \delta^i_j - \epsilon r^i{}_j. \tag{1.8}$$

This satisfies (1.6) if $r^i{}_j = -r^j{}_i$. Any antisymmetric matrix $r^i{}_j$ is called a generator of SO(n). The *Lie algebra* is the linear space spanned by the $\frac{1}{2}n(n-1)$ independent generators, with the commutator product

$$[r, r'] = r\,r' - r'r. \tag{1.9}$$

Note that matrices are multiplied[5] as $r^i{}_k r'^k{}_j$.

A useful basis for the Lie algebra is to choose generators that act in each of the $\frac{1}{2}n(n-1)$ independent 2-planes of $\mathbb{R}^n$. For the 2-plane in the directions $\hat{i}\,\hat{j}$ this generator is given by

$$r_{[\hat{i}\hat{j}]}{}^i{}_j \equiv \delta^i_{\hat{i}}\delta_{\hat{j}j} - \delta^i_{\hat{j}}\delta_{\hat{i}j} = -r_{[\hat{j}\hat{i}]}{}^i{}_j. \tag{1.10}$$

[5] Some mathematical readers may initially be perturbed by the indices used to express many equations in this book. We will follow the standard conventions used in physics. Unless ambiguity arises we use the Einstein summation convention for repeated indices, usually one downstairs and one upstairs. The summation convention incorporates the standard rules of matrix multiplication.

Note the distinction between the coordinate plane labels in brackets with hatted indices and the row and column indices. The commutators of the generators defined in (1.10) are

$$[r_{[\hat{i}\hat{j}]}, r_{[\hat{k}\hat{l}]}] = \delta_{\hat{j}\hat{k}} r_{[\hat{i}\hat{l}]} - \delta_{\hat{i}\hat{k}} r_{[\hat{j}\hat{l}]} - \delta_{\hat{j}\hat{l}} r_{[\hat{i}\hat{k}]} + \delta_{\hat{i}\hat{l}} r_{[\hat{j}\hat{k}]} . \tag{1.11}$$

The row and column indices are suppressed in this equation, and this will be our practice when it causes no ambiguity. The equation implicitly specifies the structure constants of the Lie algebra in the basis of (1.10).

In this basis, a finite transformation of SO(n) is determined by a set of $\frac{1}{2}n(n-1)$ real parameters $\theta^{\hat{i}\hat{j}}$ which specify the angles of rotation in the independent 2-planes. A general element of (the connected component) of the group can be written as an exponential

$$R = \mathrm{e}^{-\frac{1}{2}\theta^{\hat{i}\hat{j}} r_{[\hat{i}\hat{j}]}} . \tag{1.12}$$

1.2.2 General internal symmetry

It will be useful to establish the notation for the general situation of linearly realized internal symmetry under an arbitrary connected Lie group G, usually a compact group, of dimension $\dim G$. We will be interested in an n-dimensional representation of G in which the generators of its Lie algebra are a set of $n \times n$ matrices $(t_A)^i{}_j$, $A = 1, 2, \ldots, \dim G$. Their commutation relations are[6]

$$[t_A, t_B] = f_{AB}{}^C t_C , \tag{1.13}$$

and the $f_{AB}{}^C$ are structure constants of the Lie algebra. The representative of a general element of the Lie algebra is a matrix Θ that is a superposition of the generators with real parameters θ^A, i.e.

$$\Theta = \theta^A t_A . \tag{1.14}$$

An element of the group is represented by the matrix exponential

$$U(\Theta) = \mathrm{e}^{-\Theta} = \mathrm{e}^{-\theta^A t_A} . \tag{1.15}$$

We consider a set of scalar fields $\phi^i(x)$ which transforms in the representation just described. The fields may be real or complex. If complex, the complex conjugate of every element is also included in the set. A group transformation acts by matrix multiplication on the fields:

$$\phi^i(x) \to \phi'^i(x) \equiv U(\Theta)^i{}_j \phi^j(x) . \tag{1.16}$$

[6] Although it is common in the physics literature to insert the imaginary i in the commutation rule, we do not do this in order to eliminate 'i's in most of the formulas of the book. This means that compact generators t_A in this book are anti-hermitian matrices.

If the system is complex, the representation of G will typically be reducible, and the full set $\{\phi^i\}$ splits into equal numbers of fields and conjugates on which $U(\Theta)$ acts independently. It is not necessary to distinguish the real and complex cases explicitly in our notation.

We assume that the equations of motion of the system are obtained from an action

$$S[\phi^i] = \int \mathrm{d}^D x\, \mathcal{L}(\phi^i, \partial_\mu \phi^i)\,, \tag{1.17}$$

which is invariant under (1.16). In this chapter we consider internal symmetry with the property that the Lagrangian density is invariant, that is

$$\mathcal{L}(\phi^i, \partial_\mu \phi^i) = \mathcal{L}(\phi'^i, \partial_\mu \phi'^i). \tag{1.18}$$

This property is stronger than (1.5).

Exercise 1.2 *Verify that Lagrangian density (1.4) is invariant under the* SO(n) *symmetry of Sec. 1.2.1, but not under the spacetime translation of Ex. 1.1.*

An infinitesimal transformation of the group is defined as the truncation of the exponential power series in (1.15) to first order in Θ. This gives the field variation (matrix and vector indices suppressed)

$$\delta\phi = -\Theta\phi\,, \tag{1.19}$$

which defines the action of a Lie algebra element on the fields.

It is important to define iterated Lie algebra variations carefully. The definition we make below may seem unfamiliar. However, we show that it does give a representation of the algebra. Later, in Sec. 1.4, we will see that the definition is compatible with implementation of symmetry transformations by Poisson brackets of their conserved Noether charges at the classical level and (in Sec. 1.5) by unitary transformation after quantization.

The action of a transformation δ_2 with Lie algebra element Θ_2 followed by a transformation δ_1 with element Θ_1 is defined by

$$\delta_1\delta_2\phi \equiv -\Theta_2\delta_1\phi = \Theta_2\Theta_1\phi. \tag{1.20}$$

The second variation acts only on the dynamical variables of the system, the fields ϕ^i, and is not affected by the matrix Θ_2 that multiplies ϕ^i. In detail,

$$\delta_1\delta_2\phi = \theta_1^A\theta_2^B t_B t_A \phi. \tag{1.21}$$

The commutator of two symmetry variations is then

$$\begin{aligned}[]
[\delta_1, \delta_2]\phi &= -[\Theta_1, \Theta_2]\phi \equiv \delta_3\phi\,,\\
\Theta_3 &\equiv [\Theta_1, \Theta_2] = f_{AB}{}^C\theta_1^A\theta_2^B t_C.
\end{aligned} \tag{1.22}$$

The commutator of two Lie algebra transformations is again an algebra transformation by the element $\Theta_3 = [\Theta_1, \Theta_2]$.

It follows that finite group transformations compose as

$$\phi \xrightarrow{\theta_2} \phi' = U(\Theta_2)\phi \xrightarrow{\theta_1} \phi'' = U(\Theta_2)U(\Theta_1)\phi. \tag{1.23}$$

We show in Sec. 1.5 that this agrees with the composition of the unitary transformations which implement the symmetry in the quantum theory.

1.2.3 Spacetime symmetries – the Lorentz and Poincaré groups

The Lorentz group is defined as the group of homogeneous linear transformations of coordinates in D-dimensional Minkowski spacetime that preserve the Minkowski norm of any vector. We write

$$x^\mu = \Lambda^\mu{}_\nu x'^\nu \qquad \text{or} \qquad x'^\mu = \Lambda^{-1\mu}{}_\nu x^\nu\,, \tag{1.24}$$

and require that

$$x^\mu \eta_{\mu\nu} x^\nu = x'^\mu \eta_{\mu\nu} x'^\nu. \tag{1.25}$$

The Poincaré group is defined by adjoining global translations and considering

$$x'^\mu = \Lambda^{-1\mu}{}_\nu \left(x^\nu - a^\nu\right). \tag{1.26}$$

In this section we review the properties of these groups, their Lie algebras and the group action on fields such as $\phi^i(x)$.

If (1.25) holds for any vector x^μ, it follows that

$$\Lambda^\mu{}_\rho \eta_{\mu\nu} \Lambda^\nu{}_\sigma = \eta_{\rho\sigma}. \tag{1.27}$$

This is the defining property of the Λ matrices. If the Minkowski metric were replaced by the Kronecker delta, $\delta_{\mu\nu}$, these conditions would define the orthogonal group $\mathrm{O}(D)$, but here we have the pseudo-orthogonal group $\mathrm{O}(D-1,1)$. For most purposes we need only the connected component of this group, which we call the proper Lorentz group.

The metric tensor (and its inverse) are used to lower (or raise) vector indices. Thus one has, for example, $x_\mu = \eta_{\mu\rho} x^\rho$ and $\Lambda_{\mu\nu} = \eta_{\mu\rho} \Lambda^\rho{}_\nu$. Upper or lower indices are called contravariant or covariant, respectively.

Exercise 1.3 *Show that (1.24) and (1.27) imply*

$$\Lambda_{\mu\nu} = (\Lambda^{-1})_{\nu\mu}\,, \qquad \Lambda^\mu{}_\nu = (\Lambda^{-1})_\nu{}^\mu\,,$$
$$x'_\mu = (\Lambda^{-1})_\mu{}^\nu x_\nu = x_\nu \Lambda^\nu{}_\mu. \tag{1.28}$$

The first relation of the exercise resembles the standard matrix orthogonality property, but it holds for Lorentz when both indices are down (or both up), which is not the correct position for their action as linear transformations. Indeed, Lorentz matrices must be multiplied with indices in up–down position, viz. $\Lambda^\mu{}_\rho \Lambda'^\rho{}_\nu$.

We now introduce the Lie algebra of the Lorentz group, proceeding in parallel[7] to the discussion of the group SO(n) in Sec. 1.2.1. For a small parameter ϵ, we write the infinitesimal transformation

$$\Lambda^\mu{}_\nu = \delta^\mu_\nu + \epsilon m^\mu{}_\nu + \cdots . \qquad (1.29)$$

It is straightforward to see that (1.29) satisfies (1.27) to first order in ϵ as long as the generator (m with two lower indices) is antisymmetric, viz.

$$m_{\mu\nu} \equiv \eta_{\mu\rho} m^\rho{}_\nu = -m_{\nu\mu} . \qquad (1.30)$$

The *Lie algebra* is the real linear space spanned by the $\frac{1}{2}D(D-1)$ independent generators, with the commutator product $[m, m'] = m\,m' - m'm$. These matrices must also be multiplied as $m^\mu{}_\rho m'^\rho{}_\nu$, but the forms with both indices down, as in (1.30), (or both up) are often convenient.

A useful basis for the Lie algebra is to choose generators that act in each of the $\frac{1}{2}D(D-1)$ coordinate 2-planes. For the 2-plane in the directions ρ, σ this generator is given by

$$m_{[\rho\sigma]}{}^\mu{}_\nu \equiv \delta^\mu_\rho \eta_{\nu\sigma} - \delta^\mu_\sigma \eta_{\rho\nu} = -m_{[\sigma\rho]}{}^\mu{}_\nu . \qquad (1.31)$$

Note the distinction between the coordinate plane labels in brackets and the row and column indices. In this basis, a finite proper Lorentz transformation is specified by a set of $\frac{1}{2}D(D-1)$ real parameters $\lambda^{\rho\sigma} = -\lambda^{\sigma\rho}$ and takes exponential form

$$\Lambda = \mathrm{e}^{\frac{1}{2}\lambda^{\rho\sigma} m_{[\rho\sigma]}} . \qquad (1.32)$$

When matrix indices are restored, we have, with the representation (1.31), the series[8]

$$\Lambda^\mu{}_\nu = \delta^\mu{}_\nu + \lambda^\mu{}_\nu + \tfrac{1}{2}\lambda^\mu{}_\rho \lambda^\rho{}_\nu + \cdots . \qquad (1.33)$$

The commutators of the generators defined in (1.31) are

$$[m_{[\mu\nu]}, m_{[\rho\sigma]}] = \eta_{\nu\rho} m_{[\mu\sigma]} - \eta_{\mu\rho} m_{[\nu\sigma]} - \eta_{\nu\sigma} m_{[\mu\rho]} + \eta_{\mu\sigma} m_{[\nu\rho]} . \qquad (1.34)$$

These equations specify the structure constants of the Lie algebra, which may be written as

$$f_{[\mu\nu][\rho\sigma]}{}^{[\kappa\tau]} = 8\eta_{[\rho[\nu}\delta^{[\kappa}_{\mu]}\delta^{\tau]}_{\sigma]} . \qquad (1.35)$$

Note that antisymmetrization is always done with 'weight 1', see (A.8), such that the right-hand side can be written as eight terms with coefficients ± 1.

Exercise 1.4 *Check that (1.34) leads to (1.35). To compare with (1.13), you have to replace each of the indices A, B, C by antisymmetric combinations, e.g. $A \rightarrow [\mu\nu]$. Moreover, you have to insert a factor $\frac{1}{2}$ each time that you sum over such a combined index to avoid double counting, as e.g. the factor $\frac{1}{2}$ in (1.32). Therefore we rewrite (1.34) as*

[7] Specifically, it is Λ^{-1} which is the analogue of R in (1.12) and of $U(\theta)$ in (1.15).

[8] Thus, we now replace in (1.29) $\epsilon m^\mu{}_\nu$ by $\frac{1}{2}\lambda^{\rho\sigma} m_{[\rho\sigma]}{}^\mu{}_\nu$.

$$[m_A, m_B] = f_{AB}{}^C m_C \rightarrow [m_{[\mu\nu]}, m_{[\rho\sigma]}] = \tfrac{1}{2} f_{[\mu\nu][\rho\sigma]}{}^{[\kappa\tau]} m_{[\kappa\tau]}. \tag{1.36}$$

Under a symmetry, each group element is mapped to a transformation of the configuration space of the dynamical fields. This map must give a group homomorphism. For the Lorentz matrix Λ, the transformation of the scalar fields is defined as

$$\phi^i(x) \xrightarrow{\Lambda} \phi'^i(x) \equiv \phi^i(\Lambda x). \tag{1.37}$$

Using (1.24), we find that $\phi'^i(x') = \phi^i(x)$.

Exercise 1.5 *Show that the action (1.4) is invariant under the transformation (1.37).*

We now define differential operators which implement the coordinate change due to an infinitesimal transformation. A transformation in the $[\rho\sigma]$ 2-plane is generated by

$$L_{[\rho\sigma]} \equiv x_\rho \partial_\sigma - x_\sigma \partial_\rho. \tag{1.38}$$

The commutator algebra of these operators is isomorphic to (1.34), and we thus have a realization of the Lie algebra acting as differential operators on functions.

Exercise 1.6 *Compute the commutators $[L_{[\mu\nu]}, L_{[\rho\sigma]}]$ and show that they agree with that of (1.34) for matrix generators. Show that to first order in $\lambda^{\rho\sigma}$*

$$\phi^i(x^\mu) - \tfrac{1}{2}\lambda^{\rho\sigma} L_{[\rho\sigma]} \phi^i(x^\mu) = \phi^i(x^\mu + \lambda^{\mu\nu} x_\nu). \tag{1.39}$$

We then define the differential operator

$$U(\Lambda) \equiv \mathrm{e}^{-\frac{1}{2}\lambda^{\rho\sigma} L_{[\rho\sigma]}}. \tag{1.40}$$

Using this operator, the mapping (1.37) which defines the action of finite Lorentz transformations on scalar fields can then be written as

$$\phi^i(x) \rightarrow \phi'^i(x) = U(\Lambda)\phi^i(x) = \phi^i(\Lambda x). \tag{1.41}$$

Box 1.1 **Symmetries**

Symmetries are implemented by transformations that act on fields.

We see that both Lorentz and internal symmetries (see (1.16)) are implemented by linear operators acting on the classical fields, a differential operator for Lorentz and a matrix operator for internal. Both operators depend on group parameters in the same way.

By expanding the exponential, this defines the infinitesimal Lorentz transformation of the scalars:

$$\delta(\lambda)\phi^i(x) = \phi'^i(x) - \phi^i(x) = \phi^i(\Lambda x) - \phi^i(x) = \lambda^\mu{}_\nu x^\nu \partial_\mu \phi^i(x) = -\tfrac{1}{2}\lambda^{\mu\nu} L_{[\mu\nu]}\phi^i(x). \tag{1.42}$$

This definition does give a homomorphism of the group; namely the product of maps, first with Λ_2, then with Λ_1, produces the compound map associated with the product $\Lambda_1\Lambda_2$. This can be seen from the sequence of steps

$$\begin{aligned} \phi^i(x) \xrightarrow{\Lambda_2} \phi'^i(x) = U(\Lambda_2)\phi^i(x) \xrightarrow{\Lambda_1} \phi''^i(x) &= U(\Lambda_2)U(\Lambda_1)\phi^i(x) \\ &= U(\Lambda_2)\phi^i(\Lambda_1 x) \\ &= \phi^i(\Lambda_1\Lambda_2 x). \end{aligned} \tag{1.43}$$

As mentioned above, a symmetry transformation acts *directly on the fields*. This convention determines the order of operations in the first line. In the second and third lines, we use the action of the differential operators to arrive at a transformation with matrix product $\Lambda_1\Lambda_2$. This is exactly the same for internal symmetries in (1.23), where the action of U is obtained by matrix multiplication.

Exercise 1.7 *It is instructive to check (1.43) for Lorentz transformations which are close to the identity. Specifically, use the definition (1.40) to show that the order $\lambda_1\lambda_2$ terms in the product $U(\Lambda_2)U(\Lambda_1)\phi^i$ of differential operators acting on ϕ^i agrees with terms of the same order in $\phi^i(\Lambda_1\Lambda_2 x)$.*

Calculate the infinitesimal commutator $[U(\Lambda_2), U(\Lambda_1)]$ to order $\lambda_1\lambda_2$. Show that the commutator is a Lorentz transformation with (matrix) parameters $[\lambda_1, \lambda_2]$. Note that the second transformation acts on $\phi^i(x)$, and not on the x-dependent factor in (1.42).

It is important to extend the Lorentz transformation rules to covariant and contravariant vector fields, which are, respectively, sections of the cotangent and tangent bundles of Minkowski spacetime. The transformation of a general covariant vector field $W_\mu(x)$ can be modeled on that of the gradient of a scalar $\partial_\mu\phi(x)$. From (1.37) we find

$$\partial_\mu\phi(x) \to \partial_\mu\phi'(x) = \frac{\partial}{\partial x^\mu}\phi(\Lambda x) = \Lambda^{-1}{}_\mu{}^\nu(\partial_\nu\phi)(\Lambda x), \tag{1.44}$$

where we have used the chain rule and (1.28) in the last step. Thus we define the transformation of a general covariant field as

$$W_\mu(x) \to W'_\mu(x) \equiv \Lambda^{-1}{}_\mu{}^\nu W_\nu(\Lambda x). \tag{1.45}$$

For contravariant vectors we assume a transformation of the form

$$V^\mu(x) \to B^\mu{}_\sigma V^\sigma(\Lambda x). \tag{1.46}$$

The matrix can determined by requiring that the inner product $V^\mu(x)W_\mu(x)$ transforms as a scalar. This fixes $B^\mu{}_\sigma = \Lambda^{-1\mu}{}_\sigma$, and we have the transformation rule

$$V^\mu(x) \to V'^\mu(x) \equiv \Lambda^{-1\mu}{}_\nu V^\nu(\Lambda x). \tag{1.47}$$

Next we define generators of Lorentz transformations appropriate to the covariant and contravariant vector representations. They are combined differential/matrix operators given by

$$J_{[\rho\sigma]}V^\mu(x) \equiv (L_{[\rho\sigma]}\delta^\mu{}_\nu + m_{[\rho\sigma]}{}^\mu{}_\nu)V^\nu(x),$$
$$J_{[\rho\sigma]}W_\nu(x) \equiv (L_{[\rho\sigma]}\delta_\nu{}^\mu + m_{[\rho\sigma]\nu}{}^\mu)W_\mu(x). \tag{1.48}$$

We avoid an overly decorative notation by suppressing row and column indices on $J_{[\rho\sigma]}$. For each case a finite Lorentz transformation is implemented by the operator

$$U(\Lambda) = \mathrm{e}^{-\frac{1}{2}\lambda^{\rho\sigma}J_{[\rho\sigma]}}. \tag{1.49}$$

Exercise 1.8 *Verify that* $J_{[\rho\sigma]}(V^\mu W_\mu) = L_{[\rho\sigma]}(V^\mu W_\mu)$.

The Lorentz group has many representations, both higher rank tensor and spinor representations (to be discussed in Ch. 2) and combinations thereof. Let $\psi^i(x)$ denote a set of fields where i is now an index of the components of a general representation. There is a corresponding Lie algebra representation with matrices $m_{[\rho\sigma]}$ which act on the indices and differential/matrix generators

$$J_{[\rho\sigma]} = L_{[\rho\sigma]}\mathbb{1} + m_{[\rho\sigma]}, \tag{1.50}$$

in which $\mathbb{1}^i{}_j = \delta^i_j$ is the unit matrix. A finite Lorentz transformation is then the mapping (suppressing the index i for simplicity of the notation)

$$\psi(x) \to \psi'(x) = U(\Lambda)\psi(x) = \mathrm{e}^{-\frac{1}{2}\lambda^{\rho\sigma}m_{[\rho\sigma]}}\psi(\Lambda x). \tag{1.51}$$

The precise forms of the operators $m_{[\rho\sigma]}$, $J_{[\rho\sigma]}$ and $U(\Lambda)$ depend on the representation under study.

Spacetime translations $x^\mu \to x'^\mu = x^\mu - a^\mu$ are much simpler because they are implemented in the same way in all representations of the Lorentz group, namely by the mapping

$$\psi^i(x) \to \psi'^i(x) = \psi^i(x+a) = U(a)\psi^i(x), \tag{1.52}$$
$$U(a) = \mathrm{e}^{a^\mu P_\mu}, \tag{1.53}$$
$$P_\mu = \partial_\mu = \frac{\partial}{\partial x^\mu}. \tag{1.54}$$

In (1.52) we have defined the generator P_μ and the finite translation operator $U(a)$ which are differential operators. Finite transformations of the Poincaré group are implemented by the operator $U(a,\Lambda) \equiv U(\Lambda)U(a)$, which acts as follows:

$$\begin{aligned}\psi(x) \rightarrow \psi'(x) \equiv U(a, \Lambda)\psi(x) &= U(\Lambda)U(a)\psi(x) \\ &= e^{-\frac{1}{2}\lambda^{\rho\sigma} m_{[\rho\sigma]}}\psi(\Lambda x + a). \end{aligned} \quad (1.55)$$

Exercise 1.9 *Prove that $U((\Lambda)^{-1}a)U(\Lambda) = U(\Lambda)U(a)$. Verify for operators which are close to the identity that*

$$\begin{aligned}U(a)\phi(\Lambda' x + b) &= \phi(\Lambda' x + \Lambda' a + b), \\ U(\Lambda)\phi(\Lambda' x + b) &= \phi(\Lambda'\Lambda x + b). \end{aligned} \quad (1.56)$$

The Lie algebra of the Poincaré group contains the $D(D+1)/2$ generators $J_{[\mu\nu]}$, P_μ. The following commutation rules complete the specification of the Lie algebra:

$$\begin{aligned}[J_{[\mu\nu]}, J_{[\rho\sigma]}] &= \eta_{\nu\rho} J_{[\mu\sigma]} - \eta_{\mu\rho} J_{[\nu\sigma]} - \eta_{\nu\sigma} J_{[\mu\rho]} + \eta_{\mu\sigma} J_{[\nu\rho]}, \\ [J_{[\rho\sigma]}, P_\mu] &= P_\rho \eta_{\sigma\mu} - P_\sigma \eta_{\rho\mu}, \\ [P_\mu, P_\nu] &= 0. \end{aligned} \quad (1.57)$$

Exercise 1.10 *Verify (1.57).*

We now discuss the implementation of this Lie algebra on fields. The treatment is parallel to the discussion of internal symmetry at the end of Sec. 1.2.2. The infinitesimal variation of the fields $\psi^i(x)$ is defined as the first order truncation of the exponential in (1.55):

$$\delta\psi = [a^\mu P_\mu - \tfrac{1}{2}\lambda^{\rho\sigma} J_{[\rho\sigma]}]\psi. \quad (1.58)$$

The $J_{[\rho\sigma]}$ operator appropriate to the representation is used, and representation indices are suppressed for simplicity. Consider now the transformation δ_2 with group parameters (a_2, λ_2) followed by transformation δ_1 with parameters (a_1, λ_1). The result is defined as

$$\begin{aligned}\delta_1\delta_2\psi &= \delta_1[a_2^\mu P_\mu - \tfrac{1}{2}\lambda_2^{\rho\sigma} J_{[\rho\sigma]}]\psi \\ &= [a_2^\mu P_\mu - \tfrac{1}{2}\lambda_2^{\rho\sigma} J_{[\rho\sigma]}][a_1^\nu P_\nu - \tfrac{1}{2}\lambda_1^{\kappa\tau} J_{[\kappa\tau]}]\psi. \end{aligned} \quad (1.59)$$

The second transformation acts only on the field variable. With some care one can calculate the commutator of two variations, which yields a third transformation with parameters (a_3, λ_3), that is

$$\begin{aligned}[\delta_1, \delta_2]\psi &= \delta_3\psi, \\ \lambda_3^{\rho\sigma} &= [\lambda_1, \lambda_2]^{\rho\sigma} = \tfrac{1}{4} f_{[\mu\nu][\kappa\tau]}{}^{[\rho\sigma]}\lambda_1^{\mu\nu}\lambda_2^{\kappa\tau}, \\ a_3^\mu &= a_{1\rho}\lambda_2^{\rho\mu} - a_{2\rho}\lambda_1^{\rho\mu} = \tfrac{1}{2} f_{\nu[\rho\sigma]}{}^\mu \left(a_1^\nu\lambda_2^{\rho\sigma} - a_2^\nu\lambda_1^{\rho\sigma}\right) \end{aligned} \quad (1.60)$$

(remember Ex. 1.4). Note that the parameters of infinitesimal transformations compose with the structure constants of the Lie algebra, exactly as in the case of internal symmetry.

Exercise 1.11 *Verify the rule of combination of parameters in (1.60).*

As in the discussion of internal symmetry at the end of Sec. 1.2.2, the product of two finite transformations of the Poincaré group, the first with parameters $(a_2,\ \lambda_2)$ followed by the second with parameters $(a_1,\ \lambda_1)$, is given by

$$\psi(x) \rightarrow \psi''(x) = U(a_2,\ \lambda_2)U(a_1,\ \lambda_1)\psi(x). \tag{1.61}$$

The compound map is a representation, as we discuss at the end of Sec. 1.5 below.

1.3 Noether currents and charges

It is well known that, using the Noether method, one can construct a conserved current for every continuous global symmetry of the action of a classical field theory. Integrals of the time component of the currents are conserved charges, and an infinitesimal symmetry transformation is implemented on fields by the Poisson bracket of charge and field. Poisson brackets and the correspondence principle provide a useful bridge to the quantum theory in which finite group transformations are implemented through unitary transformations. In this section and the next we review this formalism in order to show that the conventions for symmetries used are compatible with readers' previous study of symmetries in quantum field theory. The Noether formalism has other important applications in supersymmetry and supergravity which enter in later chapters of this book.

We assume that we are dealing with a system of scalar fields $\phi^i(x)$, $i = 1, \ldots, n$, whose Lagrangian density is a function of the fields and their first derivatives, as described by the action (1.17). The Euler–Lagrange equations of motion are

$$\frac{\partial}{\partial x^\mu}\frac{\delta \mathcal{L}}{\delta \partial_\mu \phi^i(x)} - \frac{\delta \mathcal{L}}{\delta \phi^i(x)} = 0. \tag{1.62}$$

We define a generic infinitesimal symmetry variation of the fields by

$$\delta\phi^i(x) \equiv \epsilon^A \Delta_A \phi^i(x), \tag{1.63}$$

in which the ϵ^A are constant parameters. This formula includes as special cases the various internal and spacetime symmetries discussed in Sec. 1.2. In these cases each $\Delta_A\phi^i(x)$ is linear in ϕ^i and given by a matrix or differential operator applied to ϕ^i. Specifically,

$$\text{internal} \quad \epsilon^A \Delta_A \phi^i \rightarrow -\theta^A (t_A)^i{}_j \phi^j, \tag{1.64}$$

$$\text{spacetime} \quad \epsilon^A \Delta_A \phi^i \rightarrow \left[a^\mu \partial_\mu - \tfrac{1}{2}\lambda^{\rho\sigma}(x_\rho \partial_\sigma - x_\sigma \partial_\rho)\right]\phi^i. \tag{1.65}$$

If (1.63) is a symmetry of the theory, then the action is invariant, and the variation of the Lagrangian density is an explicit total derivative, i.e. $\delta\mathcal{L} = \epsilon^A \partial_\mu K^\mu_A$. This must hold for *all* field configurations, not merely those which satisfy the equations of motion (1.62). In detail, the variation of the Lagrangian density is

$$\delta\mathcal{L} \equiv \epsilon^A \left[\frac{\delta\mathcal{L}}{\delta\partial_\mu\phi^i}\partial_\mu\Delta_A\phi^i + \frac{\delta\mathcal{L}}{\delta\phi^i}\Delta_A\phi^i \right] = \epsilon^A\partial_\mu K^\mu_A. \tag{1.66}$$

Using (1.62) we can rearrange (1.66) to read $\partial_\mu J^\mu_A = 0$, where J^μ_A is the Noether current

$$J^\mu{}_A = -\frac{\delta\mathcal{L}}{\delta\partial_\mu\phi^i}\Delta_A\phi^i + K^\mu_A. \tag{1.67}$$

This is a conserved current, by which we mean that $\partial_\mu J^\mu_A = 0$ for all solutions of the equations of motion of the system.

We temporarily assume that the symmetry parameters are arbitrary functions $\epsilon^A(x)$. In this case the variation of the action is

$$\begin{aligned}\delta S &= \int \mathrm{d}^D x \left[\frac{\delta\mathcal{L}}{\delta\partial_\mu\phi^i}\partial_\mu(\epsilon^A\Delta_A\phi^i) + \frac{\delta\mathcal{L}}{\delta\phi^i}\epsilon^A\Delta_A\phi^i \right] \\ &= \int \mathrm{d}^D x \left[\epsilon^A\partial_\mu K^\mu_A + (\partial_\mu\epsilon^A)\frac{\delta\mathcal{L}}{\delta\partial_\mu\phi^i}\Delta_A\phi^i \right] \\ &= -\int \mathrm{d}^D x\, (\partial_\mu\epsilon^A) J^\mu_A. \end{aligned} \tag{1.68}$$

The use of varying parameters $\epsilon^A(x)$ is usually an efficient way to obtain the conserved Noether current. Note that surface terms from partial integrations in the manipulations above have been neglected because the field configurations are assumed to vanish at large spacetime distances.

For each conserved current one can define an integrated Noether charge, which is a constant of the motion, i.e. independent of time. Suppose that we have a foliation of Minkowski spacetime by a family of space-like $(D-1)$-dimensional surfaces $\Sigma(\tau)$. A space-like surface has a time-like normal vector n^μ at every point.[9] Then for each Noether current there is an integrated charge

$$Q_A = \int_{\Sigma(\tau)} \mathrm{d}\Sigma_\mu\, J^\mu{}_A(x)\,, \tag{1.69}$$

which is conserved, that is independent of τ for all solutions that are suitably damped at infinity. The simplest foliation is given by the family of equal-time surfaces $\Sigma(t)$, which are flat $(D-1)$-dimensional hyperplanes with fixed $x^0 = t$. In this case

$$Q_A = \int \mathrm{d}^{D-1}\vec{x}\, J^0{}_A(\vec{x}, t). \tag{1.70}$$

We now discuss the specific Noether currents for the linear internal and spacetime transformations of this chapter. To simplify the discussion we restrict attention to systems with conventional scalar kinetic term, so the Lagrangian density is

[9] The Minkowski space norm of any vector v^μ is $v^\mu\eta_{\mu\nu}v^\nu$. A vector is called space-like if its norm is positive, time-like if the norm is negative, and null for vanishing norm. $\mathrm{d}\Sigma_\mu$ is a vector proportional to n_μ which represents a surface element orthogonal to $\Sigma(\tau)$.

$$\mathcal{L} = -\tfrac{1}{2}\partial_\mu \phi^i \partial^\mu \phi^i - V(\phi^i). \tag{1.71}$$

For internal symmetry, substitution of the first line of (1.64) in (1.67) gives

$$J^\mu{}_A = -\partial^\mu \phi t_A \phi, \tag{1.72}$$

provided that the potential V is invariant, that is, $\delta V = \partial_i V \delta\phi^i = -\partial_i V (t_A)^i{}_j \phi^j = 0$. For spacetime translations, the index A of the generic current (1.67) is replaced by the vector index ν, and the Noether current obtained from (1.65) and (1.67) is the conventional stress tensor (or energy–momentum tensor)

$$T^\mu{}_\nu = \partial^\mu \phi\, \partial_\nu \phi + \delta^\mu_\nu \mathcal{L}. \tag{1.73}$$

For Lorentz transformations, the index A becomes the antisymmetric pair $\rho\sigma$, and (1.65) and (1.67) give the current

$$M^\mu{}_{[\rho\sigma]} = -x_\rho T^\mu{}_\sigma + x_\sigma T^\mu{}_\rho. \tag{1.74}$$

The conserved charges for internal, translations, and Lorentz transformations are denoted by T_A, P_μ, and $M_{[\rho\sigma]}$, respectively. They are given by

$$\begin{aligned} T_A &= \int \mathrm{d}^{D-1}\vec{x}\, J^0{}_A, \\ P_\mu &= \int \mathrm{d}^{D-1}\vec{x}\, T^0{}_\mu, \\ M_{[\rho\sigma]} &= \int \mathrm{d}^{D-1}\vec{x}\, M^0{}_{[\rho\sigma]}. \end{aligned} \tag{1.75}$$

Note that one does not need the detailed form of the stress tensor (1.73) to show that the current (1.74) is conserved. The situation is indeed simpler. A current of the form (1.74) is conserved if $T_{\mu\nu}$ is both conserved and symmetric, $T_{\mu\nu} = T_{\nu\mu}$. For many systems of fields, such as the Dirac field discussed in the next chapter, the stress tensor given by the Noether procedure is conserved but not symmetric. In all cases one can modify $T_{\mu\nu}$ to restore symmetry.

In general the symmetry currents obtained by the Noether procedure are not unique. They can be modified by adding terms of the form $\Delta J^\mu{}_A \equiv \partial_\rho S^{\rho\mu}{}_A$, where $S^{\rho\mu}{}_A = -S^{\mu\rho}{}_A$. The added term is identically conserved, and the Noether charges are not changed by the addition since $\Delta J^0{}_A$ involves total spatial derivatives. It is frequently the case that the Noether currents of spacetime symmetries need to be 'improved' by adding such terms in order to satisfy all desiderata, such as symmetry of $T_{\mu\nu}$. Another example is the stress tensor of the electromagnetic field which we discuss in Ch. 4. The Noether stress tensor is conserved but neither gauge invariant nor symmetric. It can be made gauge invariant and symmetric by improvement, and the improved stress tensor is naturally selected by the coupling of the electromagnetic field to gravity.

1.4 Symmetries in the canonical formalism

In this section we discuss the implementation of symmetries in the canonical formalism at the classical and quantum level. In much modern work in quantum field theory the canonical formalism has been superseded by the use of Feynman path integrals, but canonical methods provide a quick pedagogical treatment of the issues of immediate concern. For scalar fields ϕ^i, the canonical coordinates at fixed time $t = 0$ are the field variables $\phi(\vec{x}, 0)$ at each point $\vec{x}$ of space, and the canonical momenta are given by $\pi(\vec{x}, 0) = \delta S/\delta\partial_t\phi(\vec{x}, 0)$. For the action (1.71), the canonical momentum is $\pi_i = \partial_0\phi^i = -\partial^0\phi^i$.

We consider explicitly the special cases of internal symmetry, space translations and rotations in which the vector K_A^μ of (1.66) has vanishing time component. In these cases the formula (1.70) for the Noether charge simplifies to

$$\begin{aligned} Q_A &= -\int \mathrm{d}^{D-1}\vec{x}\, \frac{\delta\mathcal{L}}{\delta\partial_0\phi^i}\Delta_A\phi^i \\ &= -\int \mathrm{d}^{D-1}\vec{x}\, \pi_i\Delta_A\phi^i . \end{aligned} \tag{1.76}$$

We work in this generic notation and ask readers to verify the results using (1.64) for internal symmetry and (1.65) for space translations and rotations. The following results are also valid for time translations and Lorentz boosts, although the manipulations needed are a little more complicated.

We remind readers that the basic (equal-time) Poisson bracket is $\{\phi^i(\vec{x}), \pi_j(\vec{y})\} = \delta^i_j\delta^{D-1}(\vec{x}-\vec{y})$. The Poisson bracket of two observables $A(\phi, \pi)$ and $B(\phi, \pi)$ is

$$\{A, B\} \equiv \int \mathrm{d}^{D-1}\vec{x}\left(\frac{\delta A}{\delta\phi^i(\vec{x})}\frac{\delta B}{\delta\pi_i(\vec{x})} - \frac{\delta A}{\delta\pi_i(\vec{x})}\frac{\delta B}{\delta\phi^i(\vec{x})}\right). \tag{1.77}$$

Poisson brackets $\{A, \{B, C\}\}$ obey the Jacobi identity

$$\{A, \{B, C\}\} + \{B, \{C, A\}\} + \{C, \{A, B\}\} = 0. \tag{1.78}$$

It is now easy to see that the infinitesimal symmetry transformation $\Delta_A\phi^i$ is generated by its Poisson bracket with the Noether charge Q_A. In detail

$$\Delta_A\phi^i(x) = \{Q_A, \phi^i(x)\} = -\int \mathrm{d}^{D-1}\vec{y}\,\{\pi_j(\vec{y})\Delta_A\phi^j(\vec{y}), \phi^i(x)\}. \tag{1.79}$$

Further, Poisson brackets of the conserved charges obey the Lie algebra of the symmetry group,

$$\{Q_A, Q_B\} = f_{AB}{}^C Q_C. \tag{1.80}$$

Exercise 1.12 *Readers are invited to verify (1.80) for internal symmetry, spatial translations and rotations using the Noether charges given in (1.75) and the structure constants of the subalgebra of Poincaré transformations that do not change the time coordinate.*

In the Poisson bracket formalism an iterated symmetry variation, δ_2 with parameters ϵ_2^B followed by δ_1 with parameters ϵ_1^A, is given by

$$\delta_1\delta_2\phi^i = \epsilon_1^A\epsilon_2^B\{Q_A, \{Q_B, \phi^i\}\}. \tag{1.81}$$

Using the Jacobi identity (1.78) one easily obtains the commutator

$$[\delta_1, \delta_2]\,\phi^i = f_{AB}{}^C\epsilon_1^A\epsilon_2^B\{Q_C, \phi^i\}. \tag{1.82}$$

Note that the symmetry parameters compose exactly as in (1.22) and (1.60).

Exercise 1.13 *Derive from (1.79) that* $[\Delta_A, \Delta_B] = f_{AB}{}^C\Delta_C$.

When Poisson brackets are available, we define symmetry operators as in (1.79). However, in practice we streamline the notation by omitting explicit Poisson brackets and simply use the notation $\Delta_A\phi^i$ to indicate the transformation rules. In the three cases of interest we replace $\Delta_A\phi$ by $T_A\phi$, $P_\mu\phi$ and $M_{\mu\nu}\phi$ and write the explicit transformation rules as

$$\begin{aligned} T_A\phi^i &= -(t_A)^i{}_j\phi^j, \\ P_\mu\phi^i &= \partial_\mu\phi^i, \\ M_{[\mu\nu]}\phi^i &= -J_{[\mu\nu]}\phi^i. \end{aligned} \tag{1.83}$$

In the rest of the book, the symmetries described are produced by these operators. The operators satisfy the algebra (1.80) (with Q replaced by T), and (1.57) (with J replaced by M). The minus signs are necessary due to the steps in (1.20)–(1.22) and (1.58)–(1.60), which change the order of the operators when they act on fields.

1.5 Quantum operators

In the quantum theory each classical observable becomes an operator[10] in Hilbert space, which we denote by bold-faced type, e.g. $A(\phi, \pi) \to \mathbf{A}(\mathbf{\Phi}, \mathbf{\Pi})$. The correspondence principle states that, if the Poisson bracket of two observables gives a third observable, i.e. $\{A, B\} = C$, then the commutator of the corresponding operators is $[\mathbf{A}, \mathbf{B}]_{\text{qu}} = \text{i}\mathbf{C}$. Note that we use $\hbar = 1$.

After quantization the symmetry operators become the operator commutators

$$\begin{aligned} \Delta_A\mathbf{\Phi}^i &= -\text{i}\left[\mathbf{Q}_A, \mathbf{\Phi}^i\right]_{\text{qu}}, \\ [\mathbf{Q}_A, \mathbf{Q}_B]_{\text{qu}} &= \text{i}\, f_{AB}{}^C\mathbf{Q}_C. \end{aligned} \tag{1.84}$$

[10] Subtleties such as operator ordering in the definition of $\mathbf{A}(\mathbf{\Phi}, \mathbf{\Pi})$ are ignored because they are not relevant for the questions of interest to us.

The first relation implies that a finite group transformation with parameters ϵ^A is implemented by the unitary transformation

$$\mathbf{\Phi}^i(x) \to \mathrm{e}^{-\mathrm{i}\epsilon^A \mathbf{Q}_A}\mathbf{\Phi}^i(x)\mathrm{e}^{\mathrm{i}\epsilon^A \mathbf{Q}_A} = U(\epsilon)\mathbf{\Phi}^i(x). \tag{1.85}$$

Here $U(\epsilon)$ is a generic notation for a finite group transformation. More specifically, for internal symmetry $U(\epsilon) \to U(\Theta)$ of (1.15), for translations $U(\epsilon) \to U(a)$ of (1.53), and for Lorentz $U(\epsilon) \to U(\Lambda)$ of (1.49). For finite transformations of an internal symmetry group G or the Poincaré group, (1.85) reads

$$\begin{aligned}\mathrm{e}^{-\mathrm{i}\theta^A \mathbf{T}_A}\Phi^i(x)\mathrm{e}^{\mathrm{i}\theta^A \mathbf{T}_A} &= \mathrm{e}^{-\Theta}\Phi^i(x)\,,\\ \mathrm{e}^{-\mathrm{i}[a^\mu \mathbf{P}_\mu + \frac{1}{2}\lambda^{\rho\sigma}\mathbf{M}_{[\rho\sigma]}]}\Phi^i(x)\mathrm{e}^{\mathrm{i}[a^\mu \mathbf{P}_\mu + \frac{1}{2}\lambda^{\rho\sigma}\mathbf{M}_{[\rho\sigma]}]} &= \Phi^i(\Lambda x + a).\end{aligned} \tag{1.86}$$

Exercise 1.14 *Verify the corresponding quantum operator relation*

$$[\delta_1, \delta_2]\,\Phi^i = -\mathrm{i} f_{AB}{}^C \epsilon_1^A \epsilon_2^B \left[\mathbf{Q}_C, \Phi^i\right]_{\mathrm{qu}}. \tag{1.87}$$

It is also useful to verify the composition of finite group transformations. A transformation with parameters ϵ_2^A followed by another one with parameters ϵ_1^A is found by applying (1.85) twice. One obtains

$$\begin{aligned}\mathrm{e}^{-\mathrm{i}\epsilon_1^A \mathbf{Q}_A}\mathrm{e}^{-\mathrm{i}\epsilon_2^B \mathbf{Q}_B}\mathbf{\Phi}^i(x)\mathrm{e}^{\mathrm{i}\epsilon_2^B \mathbf{Q}_B}\mathrm{e}^{\mathrm{i}\epsilon_1^A \mathbf{Q}_A} &= U(\epsilon_2)\mathrm{e}^{-\mathrm{i}\epsilon_1^A \mathbf{Q}_A}\mathbf{\Phi}^i(x)\mathrm{e}^{\mathrm{i}\epsilon_1^A \mathbf{Q}_A}\\ &= U(\epsilon_2)U(\epsilon_1)\mathbf{\Phi}^i(x).\end{aligned} \tag{1.88}$$

This agrees with (1.23) for internal symmetry and its analogue for spacetime transformations. Furthermore the group composition law for the product $\mathrm{e}^{\mathrm{i}\epsilon_2^B \mathbf{Q}_B}\mathrm{e}^{\mathrm{i}\epsilon_1^A \mathbf{Q}_A}$ of unitary operators is the same as for the classical operators $U(\epsilon_2)U(\epsilon_1)$, so we do get a consistent representation of the symmetry group.

Exercise 1.15 *For a free scalar field, use (1.75) and (1.73) to express the Hamiltonian $H = P^0$ in terms of the canonical momenta and coordinates*

$$H = \tfrac{1}{2}\int \mathrm{d}^{D-1}\vec{x}\left[\pi^2 + (\vec{\partial}\phi)^2\right]. \tag{1.89}$$

Check that this leads, using (1.3), to the quantum commutation relation

$$\begin{aligned}[H, \Phi]_{\mathrm{qu}} &= -\mathrm{i}\pi = -\mathrm{i}\partial_0\phi\\ &= \int \frac{\mathrm{d}^{D-1}\vec{p}}{(2\pi)^{(D-1)}2E}E\left(-\mathrm{e}^{\mathrm{i}(\vec{p}\cdot\vec{x}-Et)}a(\vec{p}) + \mathrm{e}^{-\mathrm{i}(\vec{p}\cdot\vec{x}-Et)}a^*(\vec{p})\right).\end{aligned} \tag{1.90}$$

Express the Hamiltonian in terms of $a(\vec{p})$ and $a^*(\vec{p})$:

$$H = \tfrac{1}{2}\int \frac{\mathrm{d}^{D-1}\vec{p}\,E}{(2\pi)^{(D-1)}2E}\left[a^*(\vec{p})a(\vec{p}) + a(\vec{p})a^*(\vec{p})\right]. \tag{1.91}$$

Using

$$\left[a(\vec{p}), a^*(\vec{p}\,')\right]_{\mathrm{qu}} = (2\pi)^3 2E(\vec{p})\delta^3(\vec{p} - \vec{p}\,'). \tag{1.92}$$

you can then reobtain (1.90).

1.6 The Lorentz group for $D = 4$

In Sec. 1.2.3 we introduced fields transforming in a general finite-dimensional representation of the D-dimensional Lorentz group $\mathrm{SO}(D-1,1)$ without giving a detailed description of these representations. There are several special features of the case $D = 4$, which reduce the description of the finite-dimensional representations of SO(3, 1) to those of the familiar representations of SU(2), as we now review.

First we note that the proper subgroup of O(3, 1) is characterized by the two conditions $\det(\Lambda) = 1$ and $\Lambda^0{}_0 \geq 1$. The latter means that the sign of the time coordinate of any point is preserved. There are three disconnected components, which contain the product of the discrete transformations P, T, and PT, describing inversion in space and/or time, with a proper transformation. Lorentz transformations in the disconnected components satisfy either $\det(\Lambda) = -1$ or $\Lambda^0{}_0 \leq -1$ or both.

Let $m_{[\mu\nu]}$ denote the matrices of a representation of the Lie algebra (1.34) for $D = 4$. The six independent matrices consist of three spatial rotations $J_i = -\frac{1}{2}\varepsilon_{ijk} m_{[jk]}$ (where ε_{ijk} is the alternating symbol with $\varepsilon_{123} = 1$) and three boosts $K_i = m_{[0i]}$. It is a straightforward and important exercise to show, using (1.34), that the linear combinations

$$\begin{aligned} I_k &= \tfrac{1}{2}(J_k - \mathrm{i}K_k), \qquad k = 1,2,3, \\ I'_k &= \tfrac{1}{2}(J_k + \mathrm{i}K_k), \end{aligned} \tag{1.93}$$

satisfy the commutation relations of two independent copies of the Lie algebra $\mathfrak{su}(2)$, viz.

$$\begin{aligned} [I_i, I_j] &= \varepsilon_{ijk} I_k, \\ [I'_i, I'_j] &= \varepsilon_{ijk} I'_k, \\ [I_i, I'_j] &= 0. \end{aligned} \tag{1.94}$$

Note that the operators (1.93) are defined for the complexified algebra. The complexified Lie algebra of $\mathfrak{so}(3,1)$ is thus related to $\mathfrak{su}(2) \oplus \mathfrak{su}(2)$. As such all finite-dimensional irreducible representations of $\mathfrak{so}(3,1)$ are obtained[11] from products of two representations of $\mathfrak{su}(2)$ and thus classified by the pair of non-negative integers or half-integers (j, j'). The (j, j') representation has dimension $(2j+1)(2j'+1)$. The representations (j, j') and (j', j) are inequivalent representations when $j \neq j'$. The four-dimensional vector representation of (1.31) for $D = 4$ is denoted by $(\frac{1}{2}, \frac{1}{2})$.

Exercise 1.16 *Verify (1.94).*

[11] To be a representation of the group SO(3, 1) the sum $j + j'$ must be an integer. But in fact we are interested in the covering group $\mathrm{SL}(2,\mathbb{C})$, since this allows the representations for fermions.

2 The Dirac field

The Dirac equation is based on special representations of the Lorentz group, called spinor representations. They were discovered by Élie Cartan in 1913 [6, 7]. These representations are very different from the D-dimensional defining representation discussed in Ch. 1 (and from tensor products of the defining representation). It is remarkable and profound that spinor representations are realized in Nature and are not just mathematical curiosities. They describe fermionic particles such as the electron and quarks, and they are required for supersymmetry.

Our treatment of the Dirac equation should be viewed as part of our review of the basic notions of relativistic field theory needed to move ahead in this book. Many readers will already be familiar with the Dirac field. We advise them to skim this chapter to learn our conventions and then move on to Ch. 3 in which the Clifford algebra and Majorana spinors are discussed. It is this material that is really essential in applications to SUSY and supergravity later in the book.

2.1 The homomorphism of $\mathrm{SL}(2,\mathbb{C}) \rightarrow \mathrm{SO}(3,1)$

Spinor representations exist for all spacetime dimensions D. We introduce them for $D = 4$. In the notation of Sec. 1.6, a general spinor representation is labeled $(j,\ j')$ with j a half-integer and j' an integer, or vice versa. The fundamental spinor representations are $(\frac{1}{2}, 0)$ and its complex conjugate $(0, \frac{1}{2})$. We now discuss the important $2:1$ homomorphism of the group $\mathrm{SL}(2,\mathbb{C})$ of unimodular 2×2 complex matrices onto the connected component of $\mathrm{O}(3,1)$. It will lead to an explicit description of the $(\frac{1}{2}, 0)$ and $(0, \frac{1}{2})$ representations, and it is central to the treatment of fermions in quantum field theory.

First we note that a general 2×2 hermitian matrix can be parametrized as

$$\mathbf{x} = \begin{pmatrix} x^0 + x^3 & x^1 - \mathrm{i}x^2 \\ x^1 + \mathrm{i}x^2 & x^0 - x^3 \end{pmatrix} \tag{2.1}$$

and that $\det \mathbf{x} = -x^\mu \eta_{\mu\nu} x^\nu$, which is the negative of the Minkowski norm of the 4-vector x^μ. This suggests a close relation between the linear space of hermitian 2×2 matrices and four-dimensional Minkowski space. Indeed, there is an isomorphism between these spaces, which we now elucidate.

For this purpose we introduce two complete sets of 2×2 matrices

$$\sigma_\mu = (-\mathbb{1}, \sigma_i), \qquad \bar{\sigma}_\mu = \sigma^\mu = (\mathbb{1}, \sigma_i), \tag{2.2}$$

where $\mathbb{1}$ is the unit matrix, and the three Pauli matrices are

$$\sigma_1 = \begin{pmatrix} 0 & 1 \\ 1 & 0 \end{pmatrix}, \qquad \sigma_2 = \begin{pmatrix} 0 & -i \\ i & 0 \end{pmatrix}, \qquad \sigma_3 = \begin{pmatrix} 1 & 0 \\ 0 & -1 \end{pmatrix}. \tag{2.3}$$

The index μ on the matrices of (2.2) is a Lorentz vector index, suggesting that the matrices are 4-vectors. We will make this precise shortly, and, in anticipation, we will raise and lower these indices using the Minkowski metric.

Exercise 2.1 *Show that*

$$\sigma_\mu \bar{\sigma}_\nu + \sigma_\nu \bar{\sigma}_\mu = 2\eta_{\mu\nu}\mathbb{1}, \tag{2.4}$$
$$\mathrm{tr}(\sigma^\mu \bar{\sigma}_\nu) = 2\delta^\mu{}_\nu. \tag{2.5}$$

Using the matrices σ_μ and $\bar{\sigma}_\mu$ and (2.5), we easily find

$$\mathbf{x} = \bar{\sigma}_\mu x^\mu, \qquad x^\mu = \tfrac{1}{2}\,\mathrm{tr}(\sigma^\mu \mathbf{x}), \tag{2.6}$$

which gives the explicit form of the isomorphism. Given the 4-vector x^μ, one can construct the associated matrix $\mathbf{x}$ from the first equation of (2.6), and one can obtain x^μ from $\mathbf{x}$ using the second equation.

Let A be a matrix of SL(2, $\mathbb{C}$), and consider the linear map

$$\mathbf{x} \to \mathbf{x}' \equiv A\mathbf{x}A^\dagger. \tag{2.7}$$

The associated 4-vectors are also linearly related, i.e. $x'^\mu \equiv \phi(A)^\mu{}_\nu x^\nu$, and (2.6) can be used to obtain the explicit form of the matrix $\phi(A)$,

$$\phi(A)^\mu{}_\nu = \tfrac{1}{2}\mathrm{tr}(\sigma^\mu A \bar{\sigma}_\nu A^\dagger). \tag{2.8}$$

Since the transformation (2.7) preserves $\det \mathbf{x}$, the Minkowski norm $x^\mu \eta_{\mu\nu} x^\nu$ is invariant. This means that the matrix $\phi(A)$ satisfies (1.27); it must be a Lorentz transformation, and we can write, using (1.24),

$$\Lambda^{-1\mu}{}_\nu = \phi(A)^\mu{}_\nu. \tag{2.9}$$

Since the group SL(2, $\mathbb{C}$) is connected [8], Λ lies in the connected component of O(3, 1), i.e. the proper Lorentz group.

Here are some exercises to help familiarize readers with this important homomorphism.

Exercise 2.2 *Verify that (2.8) is a group homomorphism by showing that* $\phi(AB) = \phi(A)\phi(B)$.

Exercise 2.3 *Show that the kernel of the homomorphism consists of the matrices* $(\mathbb{1}, -\mathbb{1})$.

Exercise 2.4 *Show that* $A\bar{\sigma}_\mu A^\dagger = \bar{\sigma}_\nu \Lambda^{-1\nu}{}_\mu$ *and* $A^\dagger \sigma_\mu A = \sigma_\nu \Lambda^\nu{}_\mu$*. This gives precise meaning to the statement that the matrices* $\bar{\sigma}_\mu$ *and* σ_ν *are 4-vectors.*

Let us introduce two sets of matrices that will turn out to be generators of the Lie algebra $\mathfrak{so}(3,1)$ in the $(\frac{1}{2},0)$ and $(0,\frac{1}{2})$ representations. We define them in terms of the matrices $\sigma_\mu, \bar\sigma_\nu$ as

$$\sigma_{\mu\nu} = \tfrac{1}{4}(\sigma_\mu\bar\sigma_\nu - \sigma_\nu\bar\sigma_\mu)\,,$$
$$\bar\sigma_{\mu\nu} = \tfrac{1}{4}(\bar\sigma_\mu\sigma_\nu - \bar\sigma_\nu\sigma_\mu)\,. \tag{2.10}$$

Note that $\sigma^{\mu\nu\dagger} = -\bar\sigma^{\mu\nu}$. The finite Lorentz transformation (1.32) is then represented as

$$L(\lambda) = \mathrm{e}^{\frac{1}{2}\lambda^{\mu\nu}\sigma_{\mu\nu}}\,, \tag{2.11}$$
$$\bar L(\lambda) = \mathrm{e}^{\frac{1}{2}\lambda^{\mu\nu}\bar\sigma_{\mu\nu}}\,. \tag{2.12}$$

Exercise 2.5 *Show that*

$$L(\lambda)^\dagger = \bar L(-\lambda) = \bar L(\lambda)^{-1}\,. \tag{2.13}$$

Exercise 2.6 *Use (2.4) to show that the commutator algebras of $\sigma^{[\mu\nu]}$ and $\bar\sigma^{[\mu\nu]}$ are isomorphic to (1.34). According to (1.93) and (1.94), the commutators of the representatives $I_k = -\frac{1}{2}(\frac{1}{2}\varepsilon_{ijk}\sigma_{ij} + \mathrm{i}\sigma_{0k})$ and $I'_k = -\frac{1}{2}(\frac{1}{2}\varepsilon_{ijk}\sigma_{ij} - \mathrm{i}\sigma_{0k})$ should satisfy (1.94). Check that in this case $I_k = 0$ and I'_k indeed satisfy these commutation relations. This means that the matrices $\sigma_{\mu\nu}$ are the generators of the $(0,\frac{1}{2})$ representation.*

The representation matrices L and $\bar L$ of (2.11) and (2.12) are directly related to the SL(2, ℂ) map (2.7). For a Lorentz transformation Λ^{-1} related to SL(2, ℂ) matrix A by (2.9), we can identify $A = L^{-1}$ and $A^\dagger = \bar L$.

We now argue that one can reach any proper Lorentz transformation from SL(2, ℂ) via the homomorphism (2.7). We show explicitly that the SL(2, ℂ) matrix $A = L(\lambda)^{-1}$ with $\lambda_{03} = -\lambda_{30} = -\rho$ and other $\lambda_{\mu\nu} = 0$ maps to a Lorentz boost in the 3-direction under (2.7). The parameter ρ is conventionally called the rapidity. We use $A^\dagger\sigma_\mu A = \sigma_\nu\Lambda^\nu{}_\mu$, which was proven in Ex. 2.4, to obtain the corresponding Lorentz transformation. Using $A = \mathrm{e}^{-\frac{1}{2}\rho\sigma_3}$ we compute

$$A^\dagger \mathbb{1} A = \mathbb{1}\cosh\rho - \sigma_3\sinh\rho\,,$$
$$A^\dagger\sigma_3 A = -\mathbb{1}\sinh\rho + \sigma_3\cosh\rho\,,$$
$$A^\dagger\sigma_{1,2}A = \sigma_{1,2}\,. \tag{2.14}$$

On the right-hand side we recognize the matrix of the Lorentz boost

$$\Lambda^{-1} = \begin{pmatrix} \cosh\rho & 0 & 0 & -\sinh\rho \\ 0 & 1 & 0 & 0 \\ 0 & 0 & 1 & 0 \\ -\sinh\rho & 0 & 0 & \cosh\rho \end{pmatrix}. \tag{2.15}$$

Similarly, one can consider the SL(2, ℂ) matrices $A = L(\lambda)^{-1}$ with non-vanishing λ_{ij} only, and show that (2.7) maps them to rotations of the spatial σ_i with any desired rotation angle.

Exercise 2.7 *Show that (2.7) works as claimed for spatial rotations about the* 1, 2, 3 *axes.*

Since any proper Lorentz transformation can be expressed as a product of rotations and a boost in the 3-direction, our discussion shows that the homomorphism is 'onto'.

In any even spacetime dimension $D = 2m$ the situation is similar to what we have described in detail in four dimensions. The Lie algebra $\mathfrak{so}(2m-1, 1)$ has a pair of inequivalent fundamental spinor representations of dimension 2^{m-1}. The products of exponentials of matrices of either representation satisfy the composition rules of a Lie group that is called $\mathrm{Spin}(2m-1, 1)$. The latter is the universal covering group of the connected component of $\mathrm{O}(2m-1, 1)$, related by a $2:1$ homomorphism.

For odd $D = 2m + 1$ the situation of spinor representations is somewhat different and will be described in Ch. 3.

2.2 The Dirac equation

We follow, at least roughly, the historical development, and introduce the Dirac equation as a relativistic wave equation describing a particle with internal structure. Such a particle is described by a multi-component field, e.g. $\Phi^M(x)$. The indices M label the components of a column vector that transforms under some finite-dimensional representation of the Lorentz group. For the particular case of the Dirac field the representation is closely related to the fundamental spinor representations we have just discussed. We again work for general D, with special emphasis on $D = 4$.

Dirac postulated that the electron is described by a complex valued multi-component field $\Psi(x)$ called a spinor field, which satisfies the first order wave equation

$$\not{\partial}\Psi(x) \equiv \gamma^\mu \partial_\mu \Psi(x) = m\Psi(x)\,. \tag{2.16}$$

The quantities γ^μ, $\mu = 0, 1, \ldots, D-1$, are a set of square matrices, which act on the indices of the spinor field Ψ. Applying the Dirac operator again, one finds

$$\begin{aligned}\not{\partial}^2\Psi &= m^2\Psi\,,\\ \tfrac{1}{2}\{\gamma^\mu\gamma^\nu + \gamma^\nu\gamma^\mu\}\partial_\mu\partial_\nu\Psi &= m^2\Psi\,.\end{aligned} \tag{2.17}$$

If we require that the second order differential operator on the left is equal to the d'Alembertian, then we fulfill the physical requirement of plane-wave solutions discussed in Ch. 1. This means that the matrices must satisfy

$$\{\gamma^\mu, \gamma^\nu\} \equiv \gamma^\mu\gamma^\nu + \gamma^\nu\gamma^\mu = 2\,\eta^{\mu\nu}\,\mathbb{1}\,, \tag{2.18}$$

where $\mathbb{1}$ is the identity matrix in the spinor indices.

The condition (2.18) is the defining relation of the Clifford algebra associated with the Lorentz group. The D matrices γ^μ are the generating elements of the Clifford algebra, and a basis of the algebra consists of the unit matrix and all independent products of the generators. The structure of the Clifford algebra is important, and it is discussed systematically for general D in Sec. 3.1. For immediate purposes, we note that there is an irreducible representation by square matrices of dimension $2^{[D/2]}$, where $[x]$ is the largest integer less than or equal to x. The representation is unique up to equivalence for even $D = 2m$, and there are two inequivalent representations for odd dimensions. It is always possible to choose a representation in which the spatial γ-matrices γ^i, $i = 1, 2, \ldots, D-1$, are hermitian and γ^0 is anti-hermitian. We will always work in such a representation, which we call a hermitian representation.

For generic D, the matrices γ^μ are necessarily complex, so the spinor field must have complex components. After Dirac's work, Majorana discovered that there are real representations in $D = 4$ dimensions, and it is now known that such Majorana representations exist in dimensions $D = 2, 3, 4 \bmod 8$. In these dimensions, one may impose the constraint that the field $\Psi(x)$ is real. The special case of Majorana spinors is very important for supersymmetry and supergravity, and it will be discussed in Sec. 3.3. In this chapter we assume that $\Psi(x)$ is complex.

There are various levels of interpretation of the components of $\Psi(x)$. In the first quantized formalism and in many classical applications, they are simply complex numbers. However, when second quantization is introduced through the fermionic path integral, the components of $\Psi(x)$ are anti-commuting Grassmann variables, which satisfy $\{\Psi_\alpha(x), \Psi_\beta(y)\} = 0$. Finally in the second quantized operator formalism, they are operators in Hilbert space. The equations we write are valid in both of the first two situations. Although the quantized theory appears rarely in this book, our basic formulas are compatible with the canonical formalism, positive Hilbert space metric and positive energy.

Equivalent representations of the Clifford algebra describe equivalent physics. Therefore the physical implications of the Dirac formalism should be independent of the choice of representation. Indeed much of the physics can be deduced in a representation independent fashion, but an explicit representation is convenient for some purposes. We display one useful representation for $D = 4$, namely a Weyl representation in which the 4×4 γ^μ have the 2×2 Weyl matrices of (2.2) in off-diagonal blocks:

$$\gamma^\mu = \begin{pmatrix} 0 & \sigma^\mu \\ \bar{\sigma}^\mu & 0 \end{pmatrix}. \tag{2.19}$$

There are 'block off-diagonal' representations of this type in all even dimensions. This is shown in Ex. 3.11 of Ch. 3.

One important fact about the Clifford algebra in general dimension D is that the commutators

$$\Sigma^{\mu\nu} \equiv \tfrac{1}{4}\left[\gamma^\mu, \gamma^\nu\right] \tag{2.20}$$

are generators of a $2^{[D/2]}$-dimensional representation of the Lie algebra of SO$(D-1,1)$. It is a straightforward exercise to show, using only the Clifford property (2.18), that the commutator algebra of the matrices $\Sigma^{\mu\nu}$ is isomorphic to (1.34). An explicit representation is not needed.

Exercise 2.8 *Do this straightforward exercise mentioned above.*

Exercise 2.9 *Show, using only (2.18), that* $[\Sigma^{\mu\nu},\gamma^\rho] = 2\gamma^{[\mu}\eta^{\nu]\rho} = \gamma^\mu\eta^{\nu\rho} - \gamma^\nu\eta^{\mu\rho}$.

In the Weyl representation given above, one sees that the matrices $\Sigma^{\mu\nu}$ are block diagonal with the 2-component $\sigma^{\mu\nu}$ and $\bar{\sigma}^{\mu\nu}$ of (2.11) and (2.12) as the diagonal entries. The four-dimensional spinor representation of $\mathfrak{so}(3,1)$ is therefore reducible. Indeed, it is the direct sum of the irreducible $(\frac{1}{2},0)$ and $(0,\frac{1}{2})$ representations discussed in Sec. 2.1. Of course, reducibility of the representation $\Sigma^{\mu\nu}$ holds for any choice of the γ^μ, since the matrices $\Sigma^{\mu\nu}$ are equivalent to those of the Weyl representation. In all even dimensions, there is an analogous reduction to the direct sum of a $\frac{1}{2}2^{[D/2]}$-dimensional irreducible representation of $\mathfrak{so}(D-1,1)$ plus the conjugate representation (see Sec. 3.1.6). In odd spacetime dimension the representation given by the $\Sigma^{\mu\nu}$ is irreducible.

Finite proper Lorentz transformations are represented by the matrices

$$L(\lambda) = \mathrm{e}^{\frac{1}{2}\lambda^{\mu\nu}\Sigma_{\mu\nu}}\,, \tag{2.21}$$

and we have

$$L(\lambda)\gamma^\rho L(\lambda)^{-1} = \gamma^\sigma \Lambda(\lambda)_\sigma{}^\rho\,. \tag{2.22}$$

From this one can deduce the important Lorentz covariance property of the Dirac equation; given *any solution* $\Psi(x)$, then

$$\Psi'(x) = L(\lambda)^{-1}\Psi(\Lambda(\lambda)x) \tag{2.23}$$

is also a solution.

Exercise 2.10 *Use the result of Ex. 2.9 and (2.21) to prove (2.22) and (2.23).*

Since the Dirac field is also a solution of the Klein–Gordon equation, the general solution is again the sum of positive and negative frequency parts in analogy with (1.3). This Fourier expansion reads

$$\begin{aligned}
\Psi(x) &= \Psi_+(x) + \Psi_-(x)\,,\\
\Psi_+(x) &= \int \frac{\mathrm{d}^{(D-1)}\vec{p}}{(2\pi)^{D-1}2E}\mathrm{e}^{\mathrm{i}(\vec{p}\cdot\vec{x}-Et)}\sum_s u(\vec{p},s)c(\vec{p},s)\,,\\
\Psi_-(x) &= \int \frac{\mathrm{d}^{(D-1)}\vec{p}}{(2\pi)^{D-1}2E}\mathrm{e}^{-\mathrm{i}(\vec{p}\cdot\vec{x}-Et)}\sum_s v(\vec{p},s)d(\vec{p},s)^*\,.
\end{aligned} \tag{2.24}$$

The * indicates complex conjugation in the classical theory and an operator adjoint after quantization.

The new features in the spinor case are the momentum space wave functions[1] $u(\vec{p}, s)$ and $v(\vec{p}, s)$, which are column vectors (with the same number of components as the field Ψ, namely $2^{[D/2]}$ components), and s is a discrete label with $\frac{1}{2}2^{[D/2]}$ values. These describe the various 'spin states' of the Dirac particle. For $m \neq 0$ these states transform in an irreducible representation of SO$(D-1)$, which is the subgroup of SO$(D-1, 1)$ that fixes the time-like energy–momentum vector $(E, \vec{p})$.

The analogous expansions of the Dirac adjoint field $\bar{\Psi}$, defined in the next section, involve the conjugate quantities c^* and d. In the second quantized theory of the complex spinor field, $c(\vec{p}, s)$, $c(\vec{p}, s)^*$, $d(\vec{p}, s)$ and $d(\vec{p}, s)^*$ are the annihilation and creation operators for particles and anti-particles of momentum $\vec{p}$ and spin quantum number s. For a Majorana spinor field, anti-particles are not distinct from particles and we replace $d, d^* \to c, c^*$ in these expansions. Furthermore, v is related to u^*, which the reader can check later in Ex. 3.38.

2.3 Dirac adjoint and bilinear form

Our task in this section is to find a suitable Lorentz invariant bilinear form for the Dirac field that can be used to construct the Lagrangian density and other fundamental quantities. Under an infinitesimal Lorentz transformation in the $[\mu\nu]$-plane, the variations of Ψ and its adjoint $\Psi^\dagger$ are

$$\begin{aligned}\delta\Psi(x) &= -\tfrac{1}{2}\lambda^{\mu\nu}(\Sigma_{\mu\nu} + L_{[\mu\nu]})\Psi(x) = -\tfrac{1}{2}\lambda^{\mu\nu}\Sigma_{\mu\nu}\Psi(x) + \lambda^{\mu}{}_{\nu}x^{\nu}\partial_{\mu}\Psi(x)\,,\\ \delta\Psi^\dagger(x) &= -\tfrac{1}{2}\lambda^{\mu\nu}\Psi^\dagger\Sigma_{\mu\nu}{}^\dagger + \lambda^{\mu}{}_{\nu}x^{\nu}\partial_{\mu}\Psi(x)^\dagger\,.\end{aligned} \tag{2.25}$$

This can be compared with (1.58) with $J_{[\mu\nu]} = \Sigma_{[\mu\nu]} + L_{[\mu\nu]}$. We suppose that our Lorentz invariant non-degenerate bilinear form may be written as

$$\Psi^\dagger\beta\Psi\,, \tag{2.26}$$

where β is some square, invertible matrix that we want to find. Lorentz invariance requires that

$$\Sigma^\dagger_{\mu\nu}\beta + \beta\Sigma_{\mu\nu} = 0\,. \tag{2.27}$$

We look for a real bilinear form,[2] so we will choose a hermitian matrix β.

The generators of spatial rotations Σ_{ij} are anti-hermitian but those of boosts Σ_{0i} are hermitian. Thus we cannot just choose β to be the identity. This may be understood as follows. If the Lorentz group were compact, it would have finite-dimensional unitary representations, in which the representatives of its Lie algebra would be anti-hermitian, satisfying $\Sigma^\dagger_{\mu\nu} = -\Sigma_{\mu\nu}$, and $\Psi^\dagger\Psi$ would be the desired Lorentz scalar. However, the Lorentz group is non-compact, and it has no finite-dimensional unitary representations and the required anti-hermitian property holds only for spatial rotations but not for boosts.

[1] For $D = 4$, these wave functions will be discussed in detail in Sec. 2.5.

[2] In this book the adjoint operation on a pair of quantities is defined to include both conjugation and reversal of order, i.e. $(AB)^\dagger \equiv B^\dagger A^\dagger$.

Nevertheless, it is easy to check that (2.27) is satisfied if we take β to be any multiple of γ^0. This form is valid in any hermitian representation and is unique. It gives

$$\beta\gamma_\mu\beta^{-1} = -\gamma_\mu^\dagger\,,$$
$$\beta\Sigma_{\mu\nu}\beta^{-1} = -\Sigma_{\mu\nu}{}^\dagger\,. \tag{2.28}$$

The last relation is a rewriting of (2.27), which shows that the Dirac spinor representation of the Lorentz group is equivalent to the transposed, complex conjugate representation.

It is convenient to make the specific choice

$$\beta = \mathrm{i}\gamma^0\,. \tag{2.29}$$

We then define the Dirac adjoint, a row vector, by

$$\bar\Psi = \Psi^\dagger\beta = \Psi^\dagger\mathrm{i}\gamma^0\,, \tag{2.30}$$

and write our invariant bilinear form as

$$\bar\Psi\Psi\,. \tag{2.31}$$

Exercise 2.11 *Show that the bilinear form $\bar\Psi\Psi$ has signature $(2,2)$ in four dimensions.*

Exercise 2.12 *Show that $(\bar\Psi_1\Psi_2)^\dagger = \bar\Psi_2\Psi_1$ for any pair of Dirac spinors and is valid for both commuting and anti-commuting (Grassmann-valued) components.*

2.4 Dirac action

We can now define the action of the free Dirac field,

$$S[\bar\Psi,\Psi] = -\int \mathrm{d}^Dx\,\bar\Psi[\gamma^\mu\partial_\mu - m]\Psi(x)\,. \tag{2.32}$$

The condition that it is stationary reads (including integration by parts in the second term)

$$\delta S[\bar\Psi,\Psi] = -\int \mathrm{d}^Dx\{\overline{\delta\Psi}[\gamma^\mu\partial_\mu - m]\Psi \;-\; \bar\Psi[\gamma^\mu\overleftarrow{\partial}_\mu + m]\delta\Psi\} = 0\,. \tag{2.33}$$

The variations $\delta\Psi$ and $\overline{\delta\Psi}$ are arbitrary infinitesimal quantities related by conjugation as in (2.30). Given one choice, one may consider another with $\delta'\Psi = \mathrm{i}\delta\Psi$ and $\overline{\delta'\Psi} = -\mathrm{i}\overline{\delta\Psi}$. With the second choice, the relative sign between the two terms in (2.33) changes. Thus the coefficients of $\delta\Psi$ and $\overline{\delta\Psi}$ in (2.33) must vanish independently. The Euler–Lagrange variational process therefore yields the Dirac equation (2.16) and its conjugate

$$\bar\Psi[\gamma^\mu\overleftarrow{\partial}_\mu + m] = 0\,. \tag{2.34}$$

Note that, in the discussion above, we assumed that the components of Ψ were ordinary complex numbers. In later work we will want to take them to be either anti-commuting

Grassmann numbers or operators on Hilbert space. Since the manipulations we carried out did not require a change of the order of Ψ and $\delta\Psi$ they remain valid in that more general situation.

2.5 The spinors $u(\vec{p}, s)$ and $v(\vec{p}, s)$ for $D = 4$

In this section we construct the $\vec{p}$-space spinors $u(\vec{p}, s)$ and $v(\vec{p}, s)$, which appear in the plane wave expansion (2.24) of the free Dirac field. We treat the case $D = 4$ explicitly, but the ideas involved are similar in all spacetime dimensions. The details are not essential to the understanding of supergravity, but it is useful to complete the story of the expansion (2.24). Furthermore, the spinors we find for the Dirac field are used to form the momentum space wave functions for the gravitino in Ch. 5.

Since $\Psi_\pm(x)$ in (2.24) satisfy the Dirac equation (2.16) and the plane waves $\mathrm{e}^{\mathrm{i}p\cdot x}$ for different 4-vectors p^μ are linearly independent, it follows that the spinors satisfy the algebraic equations

$$\begin{aligned} \gamma \cdot p\, u(\vec{p}, s) &= -\mathrm{i}m\, u(\vec{p}, s)\,, \\ \gamma \cdot p\, v(\vec{p}, s) &= +\mathrm{i}m\, v(\vec{p}, s)\,. \end{aligned} \tag{2.35}$$

We will solve these equations using the Weyl representation (2.19) in which the equation for $u(\vec{p}, s)$ in (2.35) becomes

$$\begin{pmatrix} 0 & \sigma \cdot p \\ \bar{\sigma} \cdot p & 0 \end{pmatrix} \begin{pmatrix} w_1 \\ w_2 \end{pmatrix} = -\mathrm{i}m \begin{pmatrix} w_1 \\ w_2 \end{pmatrix}. \tag{2.36}$$

Here w_1 and w_2 are a temporary notation for the upper and lower components of $u(\vec{p}, s)$. The equation for $v(\vec{p}, s)$ is similar, differing only in the sign of the right-hand side.

We assume that $m > 0$ and take a direct approach to the solution of (2.36), suggested by the treatment of Sec. 3.3 of [9]. We note that the matrices $-\sigma \cdot p$ and $\bar{\sigma} \cdot p$ each have the two positive eigenvalues $E \pm |\vec{p}|$, and they therefore have matrix square roots $\sqrt{-\sigma \cdot p}$ and $\sqrt{\bar{\sigma} \cdot p}$ defined by taking the positive root in each eigenspace. We also have the relations $-\sigma \cdot p\, \bar{\sigma} \cdot p = -\bar{\sigma} \cdot p\, \sigma \cdot p = m^2$. Using this information it is easy to see that

$$u(p) = \begin{pmatrix} \sqrt{-\sigma \cdot p}\, \xi \\ \mathrm{i}\sqrt{\bar{\sigma} \cdot p}\, \xi \end{pmatrix} \tag{2.37}$$

is a solution of (2.36) for any two-component spinor ξ. Similarly

$$v(p) = \begin{pmatrix} \sqrt{-\sigma \cdot p}\, \eta \\ -\mathrm{i}\sqrt{\bar{\sigma} \cdot p}\, \eta \end{pmatrix} \tag{2.38}$$

is a solution of the corresponding equation for $v(p)$ for any two-component η. Note that we have omitted the index s, which describes the spin state of the particle because that information is determined by the choice of ξ and η, to which we now turn.

It is convenient to choose spin states which are eigenstates of the helicity, the component of angular momentum in the direction of motion of the particle. Therefore we define spinors $\xi(\vec{p}, \pm)$ that satisfy

$$\vec{\sigma} \cdot \vec{p}\, \xi(\vec{p}, \pm) = \pm |\vec{p}|\, \xi(\vec{p}, \pm) . \tag{2.39}$$

Note that $\vec{\sigma} \cdot \vec{p} \equiv \sigma^i p^i$ is summed over the spatial components only. We assume these spinors to be normalized:

$$\xi(\vec{p}, \pm)^\dagger \xi(\vec{p}, \pm) = 1 , \qquad \xi(\vec{p}, \pm)^\dagger \xi(\vec{p}, \mp) = 0 . \tag{2.40}$$

Since the angular momentum operator is $\vec{J} = \frac{1}{2}\vec{\sigma}$, the spinor $\xi(\vec{p}, \pm)$ is an eigenstate of $\vec{p} \cdot \vec{J}/|\vec{p}|$ with eigenvalue $\pm 1/2$. We also choose

$$\eta(\vec{p}, \pm) = -\sigma_2\, \xi(\vec{p}, \pm)^* , \tag{2.41}$$

which is then normalized in the same way as ξ in (2.40).

Exercise 2.13 *Assume that* $(p^1, p^2, p^3) = |\vec{p}|(\sin\beta\cos\alpha, \sin\beta\sin\alpha, \cos\beta)$. *Find* $\xi(\vec{p}, \pm)$ *explicitly.*

Exercise 2.14 *Show that* $\vec{\sigma} \cdot \vec{p}\, \eta(\vec{p}, \pm) = \mp |\vec{p}|\, \eta(\vec{p}, \pm)$.

We may now specify the precise spinors to be inserted in the Fourier expansion (2.24) as

$$u(\vec{p}, \pm) = \begin{pmatrix} \sqrt{E \mp |\vec{p}|}\, \xi(\vec{p}, \pm) \\ \mathrm{i}\sqrt{E \pm |\vec{p}|}\, \xi(\vec{p}, \pm) \end{pmatrix} , \tag{2.42}$$

$$v(\vec{p}, \pm) = \begin{pmatrix} \sqrt{E \pm |\vec{p}|}\, \eta(\vec{p}, \pm) \\ -\mathrm{i}\sqrt{E \mp |\vec{p}|}\, \eta(\vec{p}, \pm) \end{pmatrix} . \tag{2.43}$$

Let us note that the spinors (2.42) and (2.43) have a smooth massless limit and satisfy the *massless* Dirac equation in this limit. The limit simplifies because of the relations $(E \pm \vec{\sigma} \cdot \vec{p})\xi(\vec{p}, \pm) = 2E\xi(\vec{p}, \pm)$ and $(E \pm \vec{\sigma} \cdot \vec{p})\xi(\vec{p}, \mp) = 0$, where $E = |\vec{p}|$ for a massless particle. Thus the massless u spinors take the simple form[3]

$$u(\vec{p}, -) = \sqrt{2E} \begin{pmatrix} \xi(\vec{p}, -) \\ 0 \end{pmatrix} , \qquad u(\vec{p}, +) = \sqrt{2E} \begin{pmatrix} 0 \\ \mathrm{i}\xi(\vec{p}, +) \end{pmatrix} . \tag{2.44}$$

Similarly, one finds for massless v spinors

$$v(\vec{p}, -) = \sqrt{2E} \begin{pmatrix} 0 \\ -\mathrm{i}\eta(\vec{p}, -) \end{pmatrix} , \qquad v(\vec{p}, +) = \sqrt{2E} \begin{pmatrix} \eta(\vec{p}, +) \\ 0 \end{pmatrix} . \tag{2.45}$$

Exercise 2.15 *The following properties of bilinears of the* u, v *spinors are frequently useful. Derive them.*

$$\begin{aligned} \bar{u}(\vec{p}, s)\, u(\vec{p}, s') &= -\bar{v}(\vec{p}, s) v(\vec{p}, s') = -2m\delta_{ss'} , \\ \bar{u}(\vec{p}, s) v(\vec{p}, s') &= \bar{v}(\vec{p}, s) u(\vec{p}, s') = 0 , \\ \bar{u}(\vec{p}, s)\gamma^\mu u(\vec{p}, s) &= \bar{v}(\vec{p}, s)\gamma^\mu v(\vec{p}, s) = -2\mathrm{i}p^\mu . \end{aligned} \tag{2.46}$$

[3] Since Lorentz transformations cannot change the helicity of a massless particle, the relative phase of positive and negative helicity spinors is arbitrary, and the relative phase of the corresponding spinors is arbitrary. One can redefine $u(\vec{p}, +) \to \alpha u(\vec{p}, +)$ and $v(\vec{p}, +) \to \alpha^* v(\vec{p}, +)$ where α is an arbitrary phase; $\alpha = -\mathrm{i}$ is particularly convenient.

2.6 Weyl spinor fields in even spacetime dimension

Let's return to the case of even dimension $D = 2m$. We saw in Sec. 2.2 that the Dirac representation of the Lorentz group is reducible if $D = 4$. Since there is a 'block off-diagonal' representation for any even dimension, the same is true for all $D = 2m$ with two irreducible subrepresentations, each of dimension $2^{(m-1)}$. This suggests that a Dirac spinor $\Psi(x)$ is not the simplest Lorentz covariant field. In a sense this is correct. One can define a Weyl field $\psi(x)$ with $2^{(m-1)}$ components, which is defined to transform as

$$\psi(x) \to \psi'(x) = L(\lambda)^{-1}\psi(\Lambda(\lambda)x), \tag{2.47}$$

where $L(\lambda)$ is defined as in (2.11), but now for any even dimension. One can also define a field $\bar{\chi}(x)$ that transforms in the conjugate representation, namely as

$$\bar{\chi}(x) \to \bar{\chi}'(x) = \bar{L}(\lambda)^{-1}\bar{\chi}(\Lambda(\lambda)x), \tag{2.48}$$

with $\bar{L}(\lambda)$ defined as in (2.12). There are Lorentz invariant wave equations for these fields:

$$\bar{\sigma}^\mu \partial_\mu \psi(x) = 0, \tag{2.49}$$

$$\sigma^\mu \partial_\mu \bar{\chi}(x) = 0. \tag{2.50}$$

Exercise 2.16 *Extend the discussion of Lorentz transformations in Sec. 2.2 to any even dimension and show that the wave equations are indeed Lorentz invariant. Show that (2.49) and (2.50) imply that* $\Box\psi(x) = 0$ *and* $\Box\bar{\chi}(x) = 0$.

The Weyl equations thus admit plane wave solutions $\mathrm{e}^{\pm \mathrm{i}p\cdot x}$ with $p^0 = |\vec{p}|$ and therefore describe *massless* particles. For $D = 4$ the fields ψ and $\bar{\chi}$ each describe particles of a single helicity value, namely negative for ψ and positive for $\bar{\chi}$, and their anti-particles of opposite helicity. The plane wave expansion of $\psi(x)$ is

$$\psi(x) = \int \frac{\mathrm{d}^{(D-1)}\vec{p}}{(2\pi)^{\frac{1}{2}(D-1)}\sqrt{2E}} \left[\mathrm{e}^{\mathrm{i}p\cdot x} c(\vec{p}, -) + \mathrm{e}^{-\mathrm{i}p\cdot x} d(\vec{p}, +)^*\right] \xi(\vec{p}, -). \tag{2.51}$$

It follows from (2.49) that one can choose the same momentum space spinor $\xi(\vec{p}, -)$ in the positive and negative frequency terms. The expansion for $\bar{\chi}(x)$ is similar.

For $D = 2m$ the expansion contains a sum over 2^{m-2} values of the 'spin label' s, and the spin states of the particle transform in an irreducible spinor representation of $\mathrm{SO}(D-2)$.

To write a kinetic action for the Weyl field we need both $\psi(x)$ and its adjoint $\psi(x)^\dagger$. From these we can construct a bilinear form that transforms as a vector.

Exercise 2.17 *Use (2.47) and the results of Exs. 2.4 and 2.5 to show that, under the Lorentz transformation* Λ,

$$\psi(x)^\dagger \bar{\sigma}^\mu \psi(x) \to \Lambda^{-1\mu}{}_\nu \psi(\Lambda x)^\dagger \bar{\sigma}^\nu \psi(\Lambda x). \tag{2.52}$$

The Lorentz invariant hermitian action is then

$$S[\psi, \bar{\psi}] = -\int \mathrm{d}^D x\, \mathrm{i}\psi^\dagger \bar{\sigma}^\mu \partial_\mu \psi. \tag{2.53}$$

There is an analogous action for the field $\bar{\chi}(x)$ and its adjoint.

It is intriguing that, with a single Weyl field (either ψ or $\bar{\chi}$), there is no way to introduce a mass. The candidate wave equation

$$\bar{\sigma}^{\mu}\partial_{\mu}\psi(x) = m\psi(x) \tag{2.54}$$

is *not* Lorentz invariant.

Exercise 2.18 *Why not?*

One can describe massive particles using both $\psi(x)$ and $\bar{\chi}(x)$. In fact this is the secret content of a single Dirac field in any even dimension, and this can be exhibited using a Weyl representation of the γ-matrices.

Exercise 2.19 *Show this! Write the Dirac field as the column*

$$\Psi(x) = \begin{pmatrix} \psi(x) \\ \bar{\chi}(x) \end{pmatrix} \tag{2.55}$$

and show that the Dirac equation (2.16) in the representation (2.19) is equivalent to the pair of equations

$$\bar{\sigma}^{\mu}\partial_{\mu}\psi(x) = m\bar{\chi}(x)\,, \qquad \sigma^{\mu}\partial_{\mu}\bar{\chi}(x) = m\psi(x)\,. \tag{2.56}$$

Exercise 2.20 *Show that the Dirac Lagrangian in (2.32) can be rewritten in terms of* ψ, $\bar{\chi}$ *and their adjoints as*

$$\mathcal{L} = \mathrm{i}\Big[-\psi^{\dagger}\,\bar{\sigma}\cdot\partial\,\psi + \bar{\chi}^{\dagger}\,\sigma\cdot\partial\,\bar{\chi} - m\,\bar{\chi}^{\dagger}\,\psi + m\,\psi^{\dagger}\,\bar{\chi}\Big]\,. \tag{2.57}$$

Show that each of the four terms is a Lorentz scalar. Note that the result in (2.13) holds for all even $D = 2m$.

2.7 Conserved currents

2.7.1 Conserved U(1) current

In this section we discuss the global U(1) symmetry property of the Dirac field in any spacetime dimension. The free Dirac action (2.32) is invariant under the global U(1) phase transformation $\Psi(x) \to \Psi'(x) \equiv \mathrm{e}^{\mathrm{i}\theta}\Psi(x)$. The conserved Noether current for this symmetry is

$$J^{\mu} = \mathrm{i}\bar{\Psi}\gamma^{\mu}\Psi\,. \tag{2.58}$$

The time component is positive in all Lorentz frames for commuting complex spinors,

$$J^{0} = \Psi^{\dagger}\Psi > 0\,. \tag{2.59}$$

Thus, the vector J^{μ} is generically future-directed time-like.

One of Dirac's original motivations for his famous equation was that, unlike the Klein–Gordon equation, the quantity J^{0} could be regarded as a *positive* probability density. This is important in the first quantized version of the theory that Dirac considered, but it ceases to be an issue in the second quantized quantum field theory.

2.7.2 Energy–momentum tensors for the Dirac field

The action (2.32) of a complex spinor field is also invariant under spacetime translations and Lorentz transformations. For translations the Noether current (1.67) obtained from the action (2.32) is

$$T_{\mu\nu} = \bar{\Psi}\gamma_\mu\partial_\nu\Psi + \eta_{\mu\nu}\mathcal{L}\,, \tag{2.60}$$

where the Lagrangian density $\mathcal{L}$ is the integrand of (2.32). This tensor, which is sometimes called the canonical energy–momentum tensor, is conserved on the first index only and non-symmetric, which is different from the scalar stress tensor in (1.72). Since symmetry fails, the simple form (1.74) of the Noether current for Lorentz transformations does not hold if $T_{\mu\nu}$ is used. For these reasons the canonical stress tensor needs to be improved, and we now guide the reader through some exercises that accomplish this.

Exercise 2.21 *It is well known that the Lagrangian density of a field theory can be changed by adding a total divergence $\partial_\mu B^\mu$, since the Euler–Lagrange equations are unaffected. Show that the addition of $\frac{1}{2}\partial_\mu(\bar{\Psi}\gamma^\mu\Psi)$ brings the action to the form*

$$S' = -\int \mathrm{d}^D x \left[\tfrac{1}{2}\bar{\Psi}\gamma^\mu \overleftrightarrow{\partial}_\mu\Psi - m\bar{\Psi}\Psi\right]. \tag{2.61}$$

Note that the antisymmetric derivative is defined as

$$A\overleftrightarrow{\partial}_\mu B \equiv A(\partial_\mu B) - (\partial_\mu A)B\,. \tag{2.62}$$

The advantage of the form (2.61) of the Dirac theory is that the Lagrangian density $\mathcal{L}'$ is hermitian as an operator in Hilbert space.

Exercise 2.22 *Show that the Noether stress tensor obtained using $\mathcal{L}'$ is*

$$T'_{\mu\nu} = \tfrac{1}{2}\bar{\Psi}\gamma_\mu\overleftrightarrow{\partial}_\nu\Psi + \eta_{\mu\nu}\mathcal{L}'\,. \tag{2.63}$$

Show that $T'_{\mu\nu} - T_{\mu\nu} = \partial^\rho S_{\rho\mu\nu}$ where the tensor $S_{\rho\mu\nu}$ satisfies $S_{\rho\mu\nu} = -S_{\mu\rho\nu}$, as in the discussion of improved Noether currents in Sec. 1.3.

Exercise 2.23 *Show that the addition of $\Delta T_{\mu\nu} = \frac{1}{4}\partial^\rho\left(\bar{\Psi}\{\Sigma_{\rho\mu}, \gamma_\nu\}\Psi\right)$ to $T'_{\mu\nu}$ produces the symmetric energy–momentum tensor*

$$\Theta_{\mu\nu} = \tfrac{1}{4}\bar{\Psi}(\gamma_\mu\overleftrightarrow{\partial}_\nu + \gamma_\nu\overleftrightarrow{\partial}_\mu)\Psi + \eta_{\mu\nu}\mathcal{L}'. \tag{2.64}$$

Note that symmetry currents are evaluated 'on-shell', i.e. one should assume that Ψ and $\bar{\Psi}$ satisfy the Dirac equation. The last term $\mathcal{L}'$ then vanishes.

Exercise 2.24 *Consider the variation $\delta(\lambda)\Psi = -\frac{1}{2}\lambda^{\rho\sigma}(\Sigma_{\rho\sigma} + L_{[\rho\sigma]})\Psi$, where $L_{[\rho\sigma]}$ was defined in (1.38), under infinitesimal Lorentz transformations and the analogue for $\delta\bar{\Psi}$. These variations may be obtained using (2.21) and (2.23). Show that the Noether current and Lorentz generators $M_{[\rho\sigma]}$ can be written in the form of (1.74), but using the symmetric stress tensor (2.64).*

3 Clifford algebras and spinors

The Dirac equation is a relativistic wave equation that is first order in space and time derivatives. The key to this remarkable property is the set of γ-matrices, which satisfy the anti-commutation relations (2.18):

$$\gamma^\mu \gamma^\nu + \gamma^\nu \gamma^\mu = 2\eta^{\mu\nu} \mathbb{1} . \tag{3.1}$$

These matrices are the generating elements of a Clifford algebra which plays an important role in supersymmetric and supergravity theories. In the first part of this chapter we discuss the structure of this Clifford algebra for general spacetime dimension D. For a generic value of D, the γ-matrices are intrinsically complex. This is why we assumed that the Dirac field is complex in the previous chapter. In certain spacetime dimensions the representation of the Clifford algebra is real, which means that the γ-matrices are conjugate to real matrices. In this case the basic spinor field may be taken to be real and is called a Majorana spinor field. Since the Majorana field has a smaller number of independent components, it is fair to say that, when it exists, it is more fundamental than the Dirac field. For this reason Majorana spinors are selected in supersymmetry and supergravity. We study the special properties of Majorana spinors in the second part of this chapter.

In the body of the chapter we take a practical approach, intended as a guide to the applications needed later in the book. Further supporting arguments are collected in Appendix 3A at the end of the chapter.

3.1 The Clifford algebra in general dimension

3.1.1 The generating γ-matrices

The main purpose of this section is to discuss the Clifford algebra associated with the Lorentz group in D dimensions. To be concrete, we start with a general and explicit construction of the generating γ-matrices. It is simplest first to construct Euclidean γ-matrices, which satisfy (3.1) with Minkowski metric $\eta_{\mu\nu}$ replaced by $\delta_{\mu\nu}$:

$$\begin{aligned} \gamma^1 &= \sigma_1 \otimes \mathbb{1} \otimes \mathbb{1} \otimes \ldots , \\ \gamma^2 &= \sigma_2 \otimes \mathbb{1} \otimes \mathbb{1} \otimes \ldots , \end{aligned}$$

$$\begin{aligned}
\gamma^3 &= \sigma_3 \otimes \sigma_1 \otimes \mathbb{1} \otimes \ldots, \\
\gamma^4 &= \sigma_3 \otimes \sigma_2 \otimes \mathbb{1} \otimes \ldots, \\
\gamma^5 &= \sigma_3 \otimes \sigma_3 \otimes \sigma_1 \otimes \ldots, \\
\ldots &= \ldots .
\end{aligned} \tag{3.2}$$

These matrices are all hermitian with squares equal to $\mathbb{1}$, and they mutually anti-commute. Suppose that $D = 2m$ is even. Then we need m factors in the construction (3.2) to obtain $\gamma^\mu, 1 \leq \mu \leq D = 2m$. Thus we obtain a representation of dimension $2^{D/2}$.

For odd $D = 2m + 1$ we need one additional matrix, and we take γ^{2m+1} from the list above, but we keep only the first m factors, i.e. deleting a σ_1. Thus there is no increase in the dimension of the representation in going from $D = 2m$ to $D = 2m + 1$, and we can say in general that the construction (3.2) gives a representation of dimension $2^{[D/2]}$, where $[D/2]$ means the integer part of $D/2$.

Euclidean γ-matrices do have physical applications, but we need Lorentzian γ for the subject matter of this book. To obtain these, all we need to do is pick any single matrix from the Euclidean construction, multiply it by i and label it γ^0 for the time-like direction. This matrix is anti-hermitian and satisfies $(\gamma^0)^2 = -\mathbb{1}$. We then relabel the remaining $D-1$ matrices to obtain the Lorentzian set $\gamma^\mu, 0 \leq \mu \leq D-1$. The hermiticity properties of the Lorentzian γ are summarized by

$$\gamma^{\mu\dagger} = \gamma^0 \gamma^\mu \gamma^0 . \tag{3.3}$$

It is fundamental to the Dirac theory that the physics of a spinor field is the same in all equivalent representations of the Clifford algebra. Thus we are really concerned with classes of representations related by conjugacy, i.e.

$$\gamma'^\mu = S \gamma^\mu S^{-1} . \tag{3.4}$$

Since we consider only hermitian representations, in which (3.3) holds, the matrix S must be unitary. Given any two equivalent representations, the transformation matrix S is unique up to a phase factor. Up to this conjugation, there is a unique irreducible representation (irrep) of the Clifford algebra by $2^m \times 2^m$ matrices for even dimension $D = 2m$. Any other representation is reducible and equivalent to a direct sum of copies of the irrep above. One can always choose a hermitian irrep, defined as one which satisfies (3.3). In odd dimensions there are two mathematically inequivalent irreps, which differ only in the sign of the 'final' $\pm\gamma^{2m+1}$. In this book we will always use a hermitian irrep of the γ-matrices. Physical consequences are independent of the particular representation chosen.

3.1.2 The complete Clifford algebra

The full Clifford algebra consists of the identity $\mathbb{1}$, the D generating elements γ^μ, plus all independent matrices formed from products of the generators. Since symmetric products reduce to a product containing fewer γ-matrices by (3.1), the new elements must be antisymmetric products. We thus define

$$\gamma^{\mu_1\ldots\mu_r} = \gamma^{[\mu_1}\ldots\gamma^{\mu_r]}, \qquad \text{e.g.} \qquad \gamma^{\mu\nu} = \tfrac{1}{2}\left(\gamma^\mu\gamma^\nu - \gamma^\nu\gamma^\mu\right), \tag{3.5}$$

where the antisymmetrization indicated with [...] is always with total weight 1. Thus the right-hand side of (3.5) contains the overall factor $1/r!$ times a sum of $r!$ signed permutations of the indices. Non-vanishing tensor components can be written as the products

$$\gamma^{\mu_1\mu_2\ldots\mu_r} = \gamma^{\mu_1}\gamma^{\mu_2}\ldots\gamma^{\mu_r} \qquad \text{where } \mu_1 \neq \mu_2 \neq \cdots \neq \mu_r\,. \tag{3.6}$$

All these matrices are traceless (a proof can be found in Appendix 3A), except for the lowest rank $r = 0$, which is the unit matrix, and the highest rank matrix with $r = D$, which is traceless only for even D as we will see below.

There are C_r^D (binomial coefficients) independent index choices at rank r. For even spacetime dimension the matrices are linearly independent, so that the Clifford algebra is an algebra of dimension 2^D.

Exercise 3.1 *Show that the higher rank γ-matrices can be defined as the alternate commutators or anti-commutators*

$$\begin{aligned}
\gamma^{\mu\nu} &= \tfrac{1}{2}[\gamma^\mu, \gamma^\nu]\,,\\
\gamma^{\mu_1\mu_2\mu_3} &= \tfrac{1}{2}\{\gamma^{\mu_1}, \gamma^{\mu_2\mu_3}\}\,,\\
\gamma^{\mu_1\mu_2\mu_3\mu_4} &= \tfrac{1}{2}[\gamma^{\mu_1}, \gamma^{\mu_2\mu_3\mu_4}]\,,\\
&\text{etc.}
\end{aligned} \tag{3.7}$$

Exercise 3.2 *Show that* $\gamma^{\mu_1\ldots\mu_D} = \frac{1}{2}\left(\gamma^{\mu_1}\gamma^{\mu_2\ldots\mu_D} - (-)^D\gamma^{\mu_2\ldots\mu_D}\gamma^{\mu_1}\right)$. *Thus,* $\operatorname{Tr}\gamma^{\mu_1\ldots\mu_D}$ *vanishes for even D.*

3.1.3 Levi-Civita symbol

We need a short technical digression to introduce the Levi-Civita symbol and derive some of its properties. In every dimension D this is defined as the totally antisymmetric rank D tensor $\varepsilon_{\mu_1\mu_2\ldots\mu_D}$ or $\varepsilon^{\mu_1\mu_2\ldots\mu_D}$ with

$$\varepsilon_{012(D-1)} = 1\,, \qquad \varepsilon^{012(D-1)} = -1\,. \tag{3.8}$$

Indices are raised using the Minkowski metric which leads to the difference in sign above (due to the single time-like direction).

Exercise 3.3 *Prove the contraction identity for these tensors:*

$$\varepsilon_{\mu_1\ldots\mu_n\nu_1\ldots\nu_p}\varepsilon^{\mu_1\ldots\mu_n\rho_1\ldots\rho_p} = -\,p!\,n!\,\delta_{\nu_1\cdots\nu_p}^{\rho_1\cdots\rho_p}\,, \qquad p = D - n\,. \tag{3.9}$$

The antisymmetric p-index Kronecker δ is in turn defined by

$$\delta_{\nu_1\ldots\nu_q}^{\rho_1\cdots\rho_p} \equiv \delta_{\nu_1}^{[\rho_1}\delta_{\nu_2}^{\rho_2}\cdots\delta_{\nu_q}^{\rho_p]}\,, \tag{3.10}$$

which includes a signed sum over $p!$ permutations of the lower indices, each with a coefficient $1/p!$, such that the 'total weight' is 1 (as in (A.8)).

In four dimensions the totally antisymmetric Levi-Civita tensor symbol is written as $\varepsilon^{\mu\nu\rho\sigma}$. Because an antisymmetric tensor of rank 5 necessarily vanishes when $D = 4$, this satisfies the Schouten identity

$$0 = 5\delta_\mu{}^{[\nu}\varepsilon^{\rho\sigma\tau\lambda]} \equiv \delta_\mu{}^\nu\varepsilon^{\rho\sigma\tau\lambda} + \delta_\mu{}^\rho\varepsilon^{\sigma\tau\lambda\nu} + \delta_\mu{}^\sigma\varepsilon^{\tau\lambda\nu\rho} + \delta_\mu{}^\tau\varepsilon^{\lambda\nu\rho\sigma} + \delta_\mu{}^\lambda\varepsilon^{\nu\rho\sigma\tau}. \tag{3.11}$$

3.1.4 Practical γ-matrix manipulation

Supersymmetry and supergravity theories emerge from the concept of fermion spin. It should be no surprise that intricate features of the Clifford algebra are needed to establish and explore the physical properties of these field theories. In this section we explain some useful tricks to multiply γ-matrices. The results are valid for both even and odd D.

Consider first products with index contractions such as

$$\gamma^{\mu\nu}\gamma_\nu = (D-1)\gamma^\mu. \tag{3.12}$$

You can memorize this rule, but it is easier to recall the simple logic behind it: ν runs over all values except μ, so there are $(D-1)$ terms in the sum. Similar logic explains the result

$$\gamma^{\mu\nu\rho}\gamma_\rho = (D-2)\gamma^{\mu\nu}, \tag{3.13}$$

or even more generally

$$\gamma^{\mu_1\ldots\mu_r\nu_1\ldots\nu_s}\gamma_{\nu_s\ldots\nu_1} = \frac{(D-r)!}{(D-r-s)!}\gamma^{\mu_1\ldots\mu_r}. \tag{3.14}$$

Indeed, first we can write $\gamma_{\nu_s\ldots\nu_1}$ as the product $\gamma_{\nu_s}\ldots\gamma_{\nu_1}$ since the antisymmetry is guaranteed by the first factor. Then the index ν_s has $(D-(r+s-1))$ values, while ν_{s-1} has $(D-(r+s-2))$ values, and this pattern continues to $(D-r)$ values for the last one. Note that the second γ on the left-hand side has its indices in opposite order, so that no signs appear when contracting the indices. It is useful to remember the general order reversal symmetry, which is

$$\gamma^{\nu_1\ldots\nu_r} = (-)^{r(r-1)/2}\gamma^{\nu_r\ldots\nu_1}. \tag{3.15}$$

The sign factor $(-)^{r(r-1)/2}$ is negative for $r = 2, 3 \bmod 4$.

Even if one does not sum over indices, similar combinatorial tricks can be used. For example, when calculating

$$\gamma^{\mu_1\mu_2}\gamma_{\nu_1\ldots\nu_D}, \tag{3.16}$$

one knows that the index values μ_1 and μ_2 appear in the set of $\{\nu_i\}$. There are D possibilities for μ_2, and since μ_1 should be different, there remain $D-1$ possibilities for μ_1. Hence the result is

$$\gamma^{\mu_1\mu_2}\gamma_{\nu_1\ldots\nu_D} = D(D-1)\delta^{\mu_2\mu_1}_{[\nu_1\nu_2}\gamma_{\nu_3\ldots\nu_D]}. \tag{3.17}$$

Note that such generalized δ-functions are always normalized with 'weight 1', i.e.

$$\delta^{\mu_2\mu_1}_{\nu_1\nu_2} = \tfrac{1}{2}\left(\delta^{\mu_2}_{\nu_1}\delta^{\mu_1}_{\nu_2} - \delta^{\mu_1}_{\nu_1}\delta^{\mu_2}_{\nu_2}\right). \tag{3.18}$$

This makes contractions easy; e.g. we obtain from (3.17)

$$\gamma^{\mu_1\mu_2}\gamma_{\nu_1\ldots\nu_D}\varepsilon^{\nu_1\ldots\nu_D} = D(D-1)\varepsilon^{\mu_2\mu_1\nu_3\ldots\nu_D}\gamma_{\nu_3\ldots\nu_D}\,. \tag{3.19}$$

We now consider products of γ-matrices without index contractions. The very simplest case is

$$\gamma^{\mu}\gamma^{\nu} = \gamma^{\mu\nu} + \eta^{\mu\nu}\,. \tag{3.20}$$

This follows directly from the definitions: the antisymmetric part of the product is defined in (3.5) to be $\gamma^{\mu\nu}$, while the symmetric part of the product is $\eta^{\mu\nu}$, by virtue of (3.1). This already illustrates the general approach: one first writes the totally antisymmetric Clifford matrix that contains all the indices and then adds terms for all possible index pairings.

Here is another example:

$$\gamma^{\mu\nu\rho}\gamma_{\sigma\tau} = \gamma^{\mu\nu\rho}{}_{\sigma\tau} + 6\gamma^{[\mu\nu}{}_{[\tau}\delta^{\rho]}{}_{\sigma]} + 6\gamma^{[\mu}\delta^{\nu}{}_{[\tau}\delta^{\rho]}{}_{\sigma]}\,. \tag{3.21}$$

This follows the same pattern. We write the indices $\sigma\tau$ in down position to make it easier to indicate the antisymmetry. The second term contains one contraction. One can choose three indices from the first factor and two indices from the second one, which gives the factor 6. For the third term there are also six ways to make two contractions. The δ-functions contract indices that were adjacent, or separated by already contracted indices, so that no minus signs appear.

Exercise 3.4 *As a similar exercise, derive*

$$\gamma^{\mu_1\ldots\mu_4}\gamma_{\nu_1\nu_2} = \gamma^{\mu_1\ldots\mu_4}{}_{\nu_1\nu_2} + 8\gamma^{[\mu_1\ldots\mu_3}{}_{[\nu_2}\delta^{\mu_4]}{}_{\nu_1]} + 12\gamma^{[\mu_1\mu_2}\delta^{\mu_3}{}_{[\nu_2}\delta^{\mu_4]}{}_{\nu_1]}\,. \tag{3.22}$$

Finally, we consider products with both contracted and uncontracted indices. Consider $\gamma^{\mu_1\ldots\mu_4\rho}\gamma_{\rho\nu_1\nu_2}$. The result should contain terms similar to (3.22), but each term has an extra numerical factor reflecting the number of values that ρ can take in this sum. For example, in the second term above there is now one contraction between an upper and lower index, and therefore ρ can run over all D values except the four values $\mu_1,\ldots\mu_4$, and the two values ν_1, ν_2. This counting gives

$$\begin{aligned}\gamma^{\mu_1\ldots\mu_4\rho}\gamma_{\rho\nu_1\nu_2} = {}&(D-6)\gamma^{\mu_1\ldots\mu_4}{}_{\nu_1\nu_2} + 8(D-5)\gamma^{[\mu_1\ldots\mu_3}{}_{[\nu_2}\delta^{\mu_4]}{}_{\nu_1]}\\ &+12(D-4)\gamma^{[\mu_1\mu_2}\delta^{\mu_3}{}_{[\nu_2}\delta^{\mu_4]}{}_{\nu_1]}\,.\end{aligned} \tag{3.23}$$

Exercise 3.5 *Show that*

$$\begin{aligned}\gamma_\nu\gamma^\mu\gamma^\nu &= (2-D)\gamma^\mu\,,\\ \gamma_\rho\gamma^{\mu\nu}\gamma^\rho &= (D-4)\gamma^{\mu\nu}\,.\end{aligned} \tag{3.24}$$

Derive the general form $\gamma_\rho\gamma^{\mu_1\mu_2\ldots\mu_r}\gamma^\rho = (-)^r(D-2r)\gamma^{\mu_1\mu_2\ldots\mu_r}$.

3.1.5 Basis of the algebra for even dimension $D = 2m$

To continue our study we restrict to even-dimensional spacetime and construct an orthogonal basis of the Clifford algebra. It will be easy to extend the results to odd D later.

The basis is denoted by the following list $\{\Gamma^A\}$ of matrices chosen from those defined in Sec. 3.1.2:

$$\{\Gamma^A = \mathbb{1}, \gamma^\mu, \gamma^{\mu_1\mu_2}, \gamma^{\mu_1\mu_2\mu_3}, \ldots, \gamma^{\mu_1\cdots\mu_D}\}. \tag{3.25}$$

Index values satisfy the conditions $\mu_1 < \mu_2 < \ldots < \mu_r$. There are C_r^D distinct index choices at each rank r and a total of 2^D matrices. To see that this is a basis, it is convenient to define the reverse order list

$$\{\Gamma_A = \mathbb{1}, \gamma_\mu, \gamma_{\mu_2\mu_1}, \gamma_{\mu_3\mu_2\mu_1}, \ldots, \gamma_{\mu_D\cdots\mu_1}\}. \tag{3.26}$$

By (3.15) the matrices of this list differ from those of (3.25) by sign factors only.

Exercise 3.6 *Show that $\Gamma^A\Gamma^B = \pm\Gamma^C$, where Γ^C is the basis element whose indices are those of A and B with common indices excluded. Derive the trace orthogonality property*

$$\mathrm{Tr}(\Gamma^A\Gamma_B) = 2^m\, \delta^A_B\,. \tag{3.27}$$

The list (3.25) contains 2^D trace orthogonal matrices in an algebra of total dimension 2^D. Therefore it is a basis of the space of matrices M of dimension $2^m \times 2^m$.

Exercise 3.7 *Show that any matrix M can be expanded in the basis $\{\Gamma^A\}$ as*

$$M = \sum_A m_A\Gamma^A\,, \qquad m_A = \frac{1}{2^m}\,\mathrm{Tr}(M\Gamma_A)\,. \tag{3.28}$$

Readers may already have noted that the signature of spacetime has played little role in the discussion above. The basic conclusion that there is a unique representation of the Clifford algebra of dimension 2^m is true for pseudo-Euclidean metrics of any signature (p, q). Another general fact is that the second rank Clifford elements $\gamma^{\mu\nu}$ are the generators of a representation of the Lie algebra $\mathfrak{so}(p, q)$, with $p+q = D = 2m$; see (1.34) with the metric signature (p, q). Only the hermiticity properties depend on the signature in an obvious fashion.

Exercise 3.8 *Show that*

$$\mathrm{Tr}\,\gamma^\mu\gamma^\nu\gamma^\rho\gamma^\sigma = 2^m[\eta^{\mu\nu}\eta^{\rho\sigma} - \eta^{\mu\rho}\eta^{\nu\sigma} + \eta^{\mu\sigma}\eta^{\nu\rho}]. \tag{3.29}$$

Exercise 3.9 *Count the number of elements in the basis (3.25) for odd dimensions $D = 2m + 1$, and see that it contains twice the number of independent $2^m \times 2^m$ matrices. Check that we already have enough matrices if we consider the matrices up to $\gamma_{\mu_1\ldots\mu_{(D-1)/2}}$. Therefore the results of this section hold only for even dimensions and will have to be modified for odd dimensions; see Sec. 3.1.7.*

3.1.6 The highest rank Clifford algebra element

For several reasons it is useful to study the highest rank tensor element of the Clifford algebra. It provides the link between even and odd dimensions and it is closely related to the chirality of fermions, an important physical property. We define

$$\gamma_* \equiv (-\mathrm{i})^{m+1}\gamma_0\gamma_1\ldots\gamma_{D-1}\,, \tag{3.30}$$

which satisfies $\gamma_*^2 = \mathbb{1}$ in every even dimension and is hermitian. For spacetime dimension $D = 2m$, the matrix γ_* is frequently called γ_{D+1} in the physics literature, as in four dimensions where it is called γ_5.

This matrix occurs as the unique highest rank element in (3.25). For any order of components μ_i, one can write

$$\gamma_{\mu_1\mu_2\cdots\mu_D} = \mathrm{i}^{m+1}\varepsilon_{\mu_1\mu_2\cdots\mu_D}\,\gamma_*\,, \tag{3.31}$$

where the Levi-Civita tensor introduced in Sec. 3.1.3 is used.

Exercise 3.10 *Show that γ_* commutes with all even rank elements of the Clifford algebra and anti-commutes with all odd rank elements. Thus, for example,*

$$\{\gamma_*, \gamma^\mu\} = 0\,, \tag{3.32}$$

$$[\gamma_*, \gamma^{\mu\nu}] = 0\,. \tag{3.33}$$

Since $\gamma_*^2 = \mathbb{1}$ and $\operatorname{Tr}\gamma_* = 0$, it follows that one can choose a representation in which

$$\gamma_* = \begin{pmatrix} \mathbb{1} & 0 \\ 0 & -\mathbb{1} \end{pmatrix}. \tag{3.34}$$

Some exercises follow, which illustrate the properties of a representation of the full Clifford algebra in which γ_* takes the form in (3.34).

Exercise 3.11 *Assume a general block form,*

$$\gamma^\mu = \begin{pmatrix} A & B \\ C & D \end{pmatrix}, \tag{3.35}$$

for the generating elements in a basis where (3.34) holds. Show that (3.32) implies the block off-diagonal form

$$\gamma^\mu = \begin{pmatrix} 0 & \sigma^\mu \\ \bar\sigma^\mu & 0 \end{pmatrix}, \tag{3.36}$$

in which the matrices σ^μ and $\bar\sigma^\mu$ are $2^{m-1}\times 2^{m-1}$ generalizations of the explicit Weyl matrices of (2.2).

Exercise 3.12 *Show that the matrices σ^μ and $\bar\sigma^\mu$ satisfy (2.4) and that* $\operatorname{Tr}(\sigma^\mu\bar\sigma_\nu) = 2^{(m-1)}\delta^\mu_\nu$.

Exercise 3.13 *Show similarly that (3.33) implies that the second rank matrices take the block diagonal form*

$$\Sigma^{\mu\nu} = \frac{1}{2}\gamma^{\mu\nu} = \frac{1}{4}\begin{pmatrix} \sigma^\mu\bar\sigma^\nu - \sigma^\nu\bar\sigma^\mu & 0 \\ 0 & \bar\sigma^\mu\sigma^\nu - \bar\sigma^\nu\sigma^\mu \end{pmatrix}. \tag{3.37}$$

This exercise shows explicitly that the Dirac representation of $\mathfrak{so}(D-1, 1)$, which is generated by $\Sigma^{\mu\nu}$, is reducible (for even D). The matrices of the upper and lower blocks in (3.37) are generators of two subrepresentations, which are inequivalent and irreducible.

(Indeed they are related to the two fundamental spinor representations of D_m denoted by Dynkin integers $(0, 0, \dots, 1, 0)$ and $(0, 0, \dots, 0, 1)$.)

Exercise 3.14 *Show that all requirements are satisfied by generalized Weyl matrices in which the spatial matrices are $\sigma^i = \bar{\sigma}^i$, where the σ^i are hermitian generators of the Clifford algebra in odd dimension $2m - 1$ Euclidean space, and the time matrices are $\sigma^0 = -\bar{\sigma}^0 = \mathbb{1}$. Thus the form of the Weyl matrices in $D = 2m$ dimensions is the same as in $D = 4$.*

It is frequently useful to note that the Weyl fields ψ and χ can be obtained from a Dirac Ψ field by applying the chiral projectors

$$P_L = \tfrac{1}{2}(\mathbb{1} + \gamma_*) , \qquad P_R = \tfrac{1}{2}(\mathbb{1} - \gamma_*) . \tag{3.38}$$

Thus

$$\begin{pmatrix} \psi \\ 0 \end{pmatrix} \equiv P_L \Psi , \qquad \begin{pmatrix} 0 \\ \bar{\chi} \end{pmatrix} \equiv P_R \Psi . \tag{3.39}$$

The specific Weyl representation (3.36) will rarely be used in the rest of this book. However, we will use the projectors P_L and P_R to define the chiral parts of Dirac (and Majorana) spinors in a general representation of the γ-matrices.

Exercise 3.15 *Show that the matrices (3.38) project to orthogonal subspaces, i.e. $P_L P_L = P_L$, $P_R P_R = P_R$ and $P_L P_R = 0$. No specific choice of the Clifford algebra representation is needed.*

3.1.7 Odd spacetime dimension $D = 2m + 1$

The basic idea that we need is that the Clifford algebra for dimension $D = 2m + 1$ can be obtained by reorganizing the matrices in the Clifford algebra for dimension $D = 2m$. In particular we can define *two sets* of $2m + 1$ generating elements by adjoining the highest rank γ_* as follows:

$$\gamma_\pm{}^\mu = (\gamma^0, \gamma^1, \dots, \gamma^{(2m-1)}, \gamma^{2m} = \pm\gamma_*) . \tag{3.40}$$

This gives us two sets of matrices, which each satisfy (2.18) for dimension $D = 2m + 1$. The two sets $\{\gamma_\pm^\mu\}$ are not equivalent, but they lead to equivalent representations of the Lorentz group; see Appendix 3A.3.

The main difference with the case of even dimensions is that the matrices in the list (3.25) are not all independent and are thus an over-complete set. Indeed, the highest element of that list, which is the product of all γ-matrices, is, due to (3.40), a phase factor times the unit matrix. More generally, the rank r and rank $D - r$ sectors are related by the duality relations

$$\gamma_\pm^{\mu_1 \dots \mu_r} = \pm \mathrm{i}^{m+1} \frac{1}{(D-r)!} \varepsilon^{\mu_1 \dots \mu_D} \gamma_{\pm\, \mu_D \dots \mu_{r+1}} . \tag{3.41}$$

Note that the order of the indices in the γ-matrix on the right-hand side is reversed. Otherwise there would be different sign factors.

Exercise 3.16 *Prove the relation (3.41) and the analogous but different relation for* even *dimension:*

$$\gamma^{\mu_1\mu_2\dots\mu_r}\gamma_* = -(-\mathrm{i})^{m+1}\frac{1}{(D-r)!}\varepsilon^{\mu_r\mu_{r-1}\dots\mu_1\nu_1\nu_2\dots\nu_{D-r}}\gamma_{\nu_1\nu_2\dots\nu_{D-r}}\,. \tag{3.42}$$

You can use the tricks explained in Sec. 3.1.4. Show that in four dimensions

$$\gamma_{\mu\nu\rho} = \mathrm{i}\varepsilon_{\mu\nu\rho\sigma}\gamma^\sigma\gamma_*\,. \tag{3.43}$$

Thus, a basis of the Clifford algebra in $D = 2m+1$ dimensions contains the matrices in (3.25) only up to rank m. This agrees with the counting argument in Ex. 3.9. For example, the set $\{\mathbb{1}, \gamma^\mu, \gamma^{\mu\nu}\}$ of $1+5+10 = 16$ matrices is a basis of the Clifford algebra for $D = 5$. Ex. 3.16 shows that it is a rearrangement of the basis $\{\Gamma^A\}$ for $D = 4$.

3.1.8 Symmetries of γ-matrices

In the Clifford algebra of the $2^m \times 2^m$ matrices, for both $D = 2m$ and $D = 2m+1$, one can distinguish between the symmetric and the antisymmetric matrices where the symmetry property is defined in the following way. There exists a unitary matrix, C, called the charge conjugation matrix, such that each matrix $C\Gamma^A$ is either symmetric or antisymmetric. Symmetry depends only on the rank r of the matrix Γ^A, so we can write:

$$(C\Gamma^{(r)})^T = -t_r C\Gamma^{(r)}\,, \qquad t_r = \pm 1\,, \tag{3.44}$$

where $\Gamma^{(r)}$ is a matrix in the set (3.25) of rank r. (The $-$ sign in (3.44) is convenient for later manipulations.) For rank $r = 0$ and 1, one obtains from (3.44)

$$C^T = -t_0 C\,, \qquad \gamma^{\mu T} = t_0 t_1\, C\gamma^\mu C^{-1}\,. \tag{3.45}$$

These relations suffice to determine the symmetries of all $C\gamma^{\mu_1\dots\mu_r}$ and thus all coefficients t_r: e.g. $t_2 = -t_0$ and $t_3 = -t_1$. Further, $t_{r+4} = t_r$.

Exercise 3.17 *A formal proof of the existence of C can be found in [10, 11], but you can check that the following two matrices satisfy (3.45) for even D. They are given in the product representation of (3.2):*[1]

$$\begin{aligned} C_+ &= \sigma_1\otimes\sigma_2\otimes\sigma_1\otimes\sigma_2\otimes\dots, & t_0t_1 &= 1\,,\\ C_- &= \sigma_2\otimes\sigma_1\otimes\sigma_2\otimes\sigma_1\otimes\dots, & t_0t_1 &= -1\,. \end{aligned} \tag{3.46}$$

The values of t_0 and t_1 (thus all t_r) depend on the spacetime dimension D modulo 8 and on the rank r modulo 4, and are given in Table (3.1). The entries in the table are determined by counting the number of independent symmetric and antisymmetric matrices in every dimension; see Appendix 3A.4. An exercise for the simple case $D = 5$ follows below. For even dimension both C_+ and C_- are possible choices. One can go from one to

[1] We consider here only the Minkowski signature of spacetime. A full treatment is given in [12], for which you should set $\epsilon = t_0$ and $\eta = -t_0t_1$.

Table 3.1 Symmetries of γ-matrices. The entries contain the numbers r mod 4 for which $t_r = \pm 1$. For even dimensions, in bold face are the choices that are most convenient for supersymmetry.

D (mod 8)	$t_r = -1$	$t_r = +1$
0	0, 3	2, 1
	0, 1	**2, 3**
1	0, 1	2, 3
2	0, 1	2, 3
	1, 2	**0, 3**
3	1, 2	0, 3
4	**1, 2**	**0, 3**
	2, 3	0, 1
5	2, 3	0, 1
6	2, 3	0, 1
	0, 3	**1, 2**
7	0, 3	1, 2

the other by replacing the charge conjugation matrix C by $C\gamma_*$ (up to a normalizing phase factor). For applications in supersymmetry we need the choice indicated in bold face. For odd dimension, C is unique (again up to a phase factor).

Exercise 3.18 *Check that in five dimensions, where the Clifford algebra basis contains only matrices of rank 0, 1 and 2, the numbers in the table are fixed by counting the number of matrices of each rank. The count must conform to the requirement that there are 10 symmetric and 6 antisymmetric matrices in a basis of* 4×4 *matrices.*

Since we use hermitian representations, which satisfy (3.3), the symmetry property of a γ-matrix determines also its complex conjugation property. To see this, we define the unitary matrix

$$B = \mathrm{i}t_0 C\gamma^0 \,. \tag{3.47}$$

Exercise 3.19 *Derive*

$$\gamma^{\mu *} = -t_0 t_1 \, B\gamma^\mu B^{-1} \,. \tag{3.48}$$

Exercise 3.20 *Prove that* $B^* B = -t_1 \mathbb{1}$.

Exercise 3.21 *Show that, in the Weyl representation (2.19), one can choose* $B = \gamma^0\gamma^1\gamma^3$, *which is real, symmetric, and satisfies* $B^2 = \mathbb{1}$. *Then* $C = \mathrm{i}\gamma^3\gamma^1$.

The properties (3.45) and (3.48) hold for the representation (3.2) using the matrices (3.46) and (3.47). In another representation, related by (3.4), the C and B matrices are given by

$$C' = S^{-1T} C\, S^{-1} \,, \qquad B' = S^{-1T} B\, S^{-1} \,. \tag{3.49}$$

Majorana conjugation Box 3.1

Since symmetries of spinor bilinears are important for supersymmetry, we use the Majorana conjugate to define $\bar{\lambda}$.

3.2 Spinors in general dimensions

In Ch. 2 we used complex spinors. We defined the Dirac adjoint (2.30), which involves the complex conjugate spinor, and used it to obtain a Lorentz invariant bilinear form. In this section we start rather differently. We define the 'Majorana conjugate' of any spinor λ using its transpose and the charge conjugation matrix,

$$\bar{\lambda} \equiv \lambda^T C \,. \tag{3.50}$$

The bilinear form $\bar{\lambda}\chi$ is Lorentz invariant as readers will show in Ex. 3.23 below. It is appropriate to use (3.50) in supersymmetry and supergravity in which the symmetry properties of γ-matrices and of spinor bilinears are very important and these properties are determined by C. For Majorana spinors, to be defined in Sec. 3.3, the definitions (3.50) and (2.30) are equivalent.

Unless otherwise stated, we assume in this book that spinor components are anti-commuting Grassmann numbers. This reflects the important physical relation between spin and statistics.

3.2.1 Spinors and spinor bilinears

Using the definition (3.50) and the property (3.44), we obtain

$$\bar{\lambda}\gamma_{\mu_1\ldots\mu_r}\chi = t_r\bar{\chi}\gamma_{\mu_1\ldots\mu_r}\lambda \,. \tag{3.51}$$

The minus sign obtained by changing the order of Grassmann valued spinor components has been incorporated. The symmetry property (3.51) is valid for Dirac spinors, but its main application for us will be to Majorana spinors. For this reason we use the term 'Majorana flip relations' to refer to (3.51).

We now give some further relations that are useful for spinor manipulations. In fact, the same sign factors can be used for a longer chain of Clifford matrices:

$$\bar{\lambda}\Gamma^{(r_1)}\Gamma^{(r_2)}\cdots\Gamma^{(r_p)}\chi = t_0^{p-1}t_{r_1}t_{r_2}\cdots t_{r_p}\,\bar{\chi}\Gamma^{(r_p)}\cdots\Gamma^{(r_2)}\Gamma^{(r_1)}\lambda \,, \tag{3.52}$$

where $\Gamma^{(r)}$ stands for any rank r matrix $\gamma_{\mu_1\ldots\mu_r}$. Note that the prefactor t_0^{p-1} is not relevant in four dimensions, where $t_0 = 1$.

Exercise 3.22 *One often encounters the special case that the bilinear contains the product of individual γ^μ-matrices. Prove that for the Majorana dimensions $D = 2, 3, 4 \bmod 8$,*

$$\bar{\lambda}\gamma^{\mu_1}\gamma^{\mu_2}\cdots\gamma^{\mu_p}\chi = (-)^p\bar{\chi}\gamma^{\mu_p}\cdots\gamma^{\mu_2}\gamma^{\mu_1}\lambda\,. \tag{3.53}$$

The previous relations imply also the following rule. For any relation between spinors that includes γ-matrices, there is a corresponding relation between the barred spinors,

$$\chi_{\mu_1\ldots\mu_r} = \gamma_{\mu_1\ldots\mu_r}\lambda \implies \bar{\chi}_{\mu_1\ldots\mu_r} = t_0 t_r \bar{\lambda}\gamma_{\mu_1\ldots\mu_r}\,, \tag{3.54}$$

and similar for longer chains,

$$\chi = \Gamma^{(r_1)}\Gamma^{(r_2)}\cdots\Gamma^{(r_p)}\lambda \implies \bar{\chi} = t_0^p t_{r_1} t_{r_2}\cdots t_{r_p}\,\bar{\lambda}\Gamma^{(r_p)}\cdots\Gamma^{(r_2)}\Gamma^{(r_1)}\,. \tag{3.55}$$

In even dimensions we define left-handed and right-handed parts of spinors using the projection matrices (3.38). The definition (3.50) implies that the chirality of the conjugate spinor depends on $t_0 t_D$, and we obtain[2]

$$\chi = P_L\lambda \;\to\; \bar{\chi} = \begin{cases} \bar{\lambda}P_L, & \text{for } D = 0, 4, 8, \ldots, \\ \bar{\lambda}P_R, & \text{for } D = 2, 6, 10, \ldots. \end{cases} \tag{3.56}$$

Exercise 3.23 *Using the 'spin part' of the infinitesimal Lorentz transformation (2.25),*

$$\delta\chi = -\tfrac{1}{4}\lambda^{\mu\nu}\gamma_{\mu\nu}\chi\,, \tag{3.57}$$

prove that the spinor bilinear $\bar{\lambda}\chi$ is a Lorentz scalar.

3.2.2 Spinor indices

For most of this book we do not need spinor indices because they appear contracted within Lorentz covariant expressions. However, in some cases indices are necessary, for example, to write (anti-)commutation relations of supersymmetry generators. The components of the basic spinor λ are indicated as λ_α. The components of the barred spinor defined in (3.50) are indicated with upper indices: λ^α. Sometimes we write $\bar{\lambda}^\alpha$ to stress that these are the components of the barred spinor, but in fact the bar can be omitted. We introduce the raising matrix $\mathcal{C}^{\alpha\beta}$ such that

$$\lambda^\alpha = \mathcal{C}^{\alpha\beta}\lambda_\beta\,. \tag{3.58}$$

Comparing with (3.50) we see that $\mathcal{C}^{\alpha\beta}$ are the components of the matrix C^T. Note that the summation index β in (3.58) appears in a northwest–southeast (NW–SE) line in the

[2] The definition (2.30) would always lead to $\bar{\chi} = \bar{\lambda}P_R$.

equation when adjacent indices are contracted. Therefore, this convention is frequently called the NW–SE spinor convention. This is relevant when the raising matrix is antisymmetric ($t_0 = 1$ in the terminology of Table 3.1). Most applications in the book are for dimensions in which this is the case, e.g. $D = 2, 3, 4, 5, 10, 11$.

We also introduce a lowering matrix such that (again NW–SE contraction)

$$\lambda_\alpha = \lambda^\beta \mathcal{C}_{\beta\alpha} \,. \tag{3.59}$$

In order for these two equations to be consistent, we must require

$$\mathcal{C}^{\alpha\beta}\mathcal{C}_{\gamma\beta} = \delta_\gamma{}^\alpha \,, \qquad \mathcal{C}_{\beta\alpha}\mathcal{C}^{\beta\gamma} = \delta_\alpha{}^\gamma \,. \tag{3.60}$$

Hence $\mathcal{C}_{\alpha\beta}$ are the components of C^{-1}, and the unitarity of C implies then $(\mathcal{C}_{\alpha\beta})^* = \mathcal{C}^{\alpha\beta}$.

When we write a covariant spinor bilinear with components explicitly indicated, the γ-matrices are written as $(\gamma_\mu)_\alpha{}^\beta$. For example, for the simplest case,

$$\bar{\chi}\gamma_\mu\lambda = \chi^\alpha(\gamma_\mu)_\alpha{}^\beta\lambda_\beta \,, \tag{3.61}$$

where again all contractions are NW–SE.

One can now raise or lower indices consistently. For example, one can define

$$(\gamma_\mu)_{\alpha\beta} = (\gamma_\mu)_\alpha{}^\gamma \mathcal{C}_{\gamma\beta} \,. \tag{3.62}$$

These γ-matrices with indices at the 'same level' have a definite symmetry or antisymmetry property, which follows from (3.44):

$$(\gamma_{\mu_1\ldots\mu_r})_{\alpha\beta} = -t_r(\gamma_{\mu_1\ldots\mu_r})_{\beta\alpha} \,. \tag{3.63}$$

An interesting property is that

$$\lambda^\alpha\chi_\alpha = -t_0\lambda_\alpha\chi^\alpha \,. \tag{3.64}$$

Thus, in four dimensions, raising and lowering a contracted index produces a minus sign. The same property can be used when the contracted indices involved are on γ-matrices, e.g. $\gamma_{\mu\alpha}{}^\beta\gamma_{\nu\beta}{}^\gamma = -t_0\gamma_{\mu\alpha\beta}\gamma_\nu{}^{\beta\gamma}$.

Exercise 3.24 *Using this property and (3.63) prove the relation (3.52). Do not forget the sign due to interchange of two (anti-commuting) spinors.*

Exercise 3.25 *Show that, using the index raising and lowering conventions, $\mathcal{C}_\alpha{}^\beta = \delta_\alpha^\beta$, and for $D = 4$ that $\mathcal{C}^\alpha{}_\beta = -\delta_\beta^\alpha$.*

3.2.3 Fierz rearrangement

In this subsection we study an important consequence of the completeness of the Clifford algebra basis $\{\Gamma^A\}$ in (3.25). As we saw in Ex. 3.7 completeness means that any matrix M has a unique expansion in the basis with coefficients obtained using trace orthogonality. The expansion was derived for even $D = 2m$ in Ex. 3.7, but it is also valid for odd $D = 2m + 1$ provided that the sum is restricted to rank $r \leq m$. We saw at the end of Sec. 3.1.7 that the list of (3.25) is complete for odd D when so restricted. The rearrangement properties we derive using completeness are frequently needed in supergravity. These involve changing the pairing of spinors in products of spinor bilinears, which is called a 'Fierz rearrangement'.

Let's proceed to derive the basic Fierz identity. Using spinor indices, we can regard the quantity $\delta_\alpha{}^\beta \delta_\gamma{}^\delta$ as a matrix in the indices $\gamma\ \beta$ with the indices $\alpha\ \delta$ as inert 'spectators'. We apply (3.28) in the detailed form $\delta_\alpha^\beta \delta_\gamma^\delta = \sum_A (m_A)_\alpha^\delta (\Gamma_A)_\gamma{}^\beta$. The coefficients are $(m_A)_\alpha^\delta = 2^{-m} \delta_\alpha{}^\beta \delta_\gamma{}^\delta (\Gamma_A)_\beta{}^\gamma = 2^{-m} (\Gamma_A)_\alpha{}^\delta$. Therefore, we obtain the basic rearrangement lemma

$$\delta_\alpha{}^\beta \delta_\gamma{}^\delta = \frac{1}{2^m} \sum_A (\Gamma_A)_\alpha{}^\delta (\Gamma^A)_\gamma{}^\beta \,. \tag{3.65}$$

Note that the 'column indices' on the left- and right-hand sides have been exchanged.

Exercise 3.26 *Derive the following result:*

$$(\gamma^\mu)_\alpha{}^\beta (\gamma_\mu)_\gamma{}^\delta = \frac{1}{2^m} \sum_A v_A (\Gamma_A)_\alpha{}^\delta (\Gamma^A)_\gamma{}^\beta \,, \tag{3.66}$$

and prove that the explicit values of the expansion coefficients are given by $v_A = (-)^{r_A}(D - 2r_A)$*, where* r_A *is the tensor rank of the Clifford basis element* Γ_A.

Exercise 3.27 *Lower the* β *and* δ *index in the result of the previous exercise and consider the completely symmetric part in* $(\beta\gamma\delta)$*. The left-hand side is only non-vanishing for dimensions in which* $t_1 = -1$*. Consider the right-hand side and use Table 3.1 and the result for* v_A *to prove that for* $D = 3$ *and* $D = 4$ *only rank 1* γ*-matrices contribute to the right-hand side. For* $D = 4$ *you have to use the bold face row in the table to arrive at this result. You can also check that there are no other dimensions where this occurs.*

The previous exercise implies that, for $D = 3$ and $D = 4$,

$$(\gamma_\mu)_{\alpha(\beta} (\gamma^\mu)_{\gamma\delta)} = 0 \,. \tag{3.67}$$

This is called the cyclic identity and is important in the context of string and brane actions. It can be extended to some other dimensions under further restrictions.[3] Multiplying the equations with three spinors λ_1^β, λ_2^γ and λ_3^γ, equation (3.67) can be written as

$$\gamma_\mu \lambda_{[1} \bar{\lambda}_2 \gamma^\mu \lambda_{3]} = 0 \,, \tag{3.68}$$

[3] For $D = 2$ and $D = 10$ this equation holds when contracted with chiral spinors. Owing to (3.56) only odd rank γ-matrices then occur in the sum over A. This is sufficient to extend the result (3.67) to these cases. With the same restrictions of chirality there is for $D = 6$ an analogous identity for the completely antisymmetric part in $[\beta\gamma\delta]$.

where the symmetry of the indices in (3.67) is transformed to an antisymmetry between the three spinors due to the anti-commutating nature of spinors. This result is important to construct supersymmetric Yang–Mills theories; see Sec. 6.3.

The following application of Fierz rearrangement is valid for any set of four anti-commuting spinor fields. The basic Fierz identity (3.65) immediately gives

$$\bar{\lambda}_1\lambda_2\bar{\lambda}_3\lambda_4 = -\frac{1}{2^m}\sum_A \bar{\lambda}_1\Gamma^A\lambda_4\bar{\lambda}_3\Gamma_A\lambda_2 \,. \tag{3.69}$$

This can be generalized to include general matrices M, M' of the Clifford algebra.

Exercise 3.28 *Show that*

$$\begin{aligned}\bar{\lambda}_1 M\lambda_2\bar{\lambda}_3 M'\lambda_4 &= -\frac{1}{2^m}\sum_A \bar{\lambda}_1 M\Gamma_A M'\lambda_4\bar{\lambda}_3\Gamma^A\lambda_2 \\ &= -\frac{1}{2^m}\sum_A \bar{\lambda}_1\Gamma_A M'\lambda_4\bar{\lambda}_3\Gamma^A M\lambda_2 \,. \end{aligned} \tag{3.70}$$

When $\lambda_{1,2,3,4}$ are not all independent, it is frequently the case that some terms in the rearranged sum vanish due to symmetry relations such as (3.51).

One can write the Fierz relation (3.65) in the alternative form:

$$M = 2^{-m}\sum_{k=0}^{[D]} \tfrac{1}{k!}\Gamma_{\mu_1\ldots\mu_k}\,\mathrm{Tr}\left(\Gamma^{\mu_k\ldots\mu_1}M\right) \tag{3.71}$$

where

$$\begin{cases}[D] = D, & \text{for even } D\,,\\ [D] = (D-1)/2, & \text{for odd } D\,.\end{cases}$$

The factor $1/k!$ compensates for the fact that in the sum over $\mu_1\ldots\mu_k$ each matrix of the basis appears $k!$ times.

Exercise 3.29 *Prove the following chiral Fierz identities for $D=4$:*

$$\begin{aligned}P_L\chi\bar{\lambda}P_L &= -\tfrac{1}{2}P_L\left(\bar{\lambda}P_L\chi\right) + \tfrac{1}{8}P_L\gamma^{\mu\nu}\left(\bar{\lambda}\gamma_{\mu\nu}P_L\chi\right)\,,\\ P_L\chi\bar{\lambda}P_R &= -\tfrac{1}{2}P_L\gamma^{\mu}\left(\bar{\lambda}\gamma_{\mu}P_L\chi\right)\,.\end{aligned} \tag{3.72}$$

You will need (3.42) to combine terms in (3.71).

Exercise 3.30 *Prove that for $D=5$ the matrix $\chi\bar{\lambda}-\lambda\bar{\chi}$ can be written as*

$$\chi\bar{\lambda}-\lambda\bar{\chi} = \gamma_{\mu\nu}(\bar{\lambda}\gamma^{\mu\nu}\chi)\,. \tag{3.73}$$

Readers who understand the Majorana flip properties (3.51) and Fierz rearrangement are well equipped for supersymmetry and supergravity!

Box 3.2 **Charge conjugation**

Complex conjugation can be replaced by charge conjugation, an operation that acts as complex conjugation on scalars, and has a simple action on fermion bilinears. For example, it preserves the order of spinor factors.

3.2.4 Reality

In this chapter, we have not yet discussed the complex conjugation of the spinor fields we are working with. Complex conjugation is necessary for such purposes as the verification that a term in the Lagrangian involving spinor bilinears is hermitian. But the complex conjugation of a bilinear[4] is an awkward operation since the hermiticity of both the C matrix in (3.50) and γ-matrices is involved. Therefore we present a related operation called charge conjugation which is much simpler in practice. For any scalar, defined here as a quantity whose spinor indices are fully contracted, charge conjugation and complex conjugation are the same. Since the Lagrangian is a scalar, charge conjugation can be used to manipulate the terms it contains.

First we define the charge conjugate of any spinor as

$$\lambda^C \equiv B^{-1}\lambda^*. \tag{3.74}$$

The barred charge conjugate spinor is then, using (3.50) and (3.47),

$$\overline{\lambda^C} = (-t_0 t_1)\mathrm{i}\lambda^\dagger\gamma^0\,. \tag{3.75}$$

Note that this is the Dirac conjugate as defined in (2.30) except for the numerical factor $(-t_0t_1)$. The meaning of this will become clear below when we discuss Majorana spinors. Note that $(-t_0t_1) = +1$ in 2, 3, 4, 10 or 11 dimensions.[5]

The charge conjugate of any $2^m \times 2^m$ matrix M is defined as

$$M^C \equiv B^{-1}M^*B\,. \tag{3.76}$$

Charge conjugation does not change the order of matrices: $(MN)^C = M^C N^C$. In practice the matrices M we deal with are products of γ-matrices. Hence, we need only the charge conjugation property of the generating γ-matrices, which is

$$(\gamma_\mu)^C \equiv B^{-1}\gamma_\mu^* B = (-t_0t_1)\gamma_\mu\,. \tag{3.77}$$

Exercise 3.31 *Start from (3.77) (and note that charge conjugation on any number is just complex conjugation). Prove that*

$$(\gamma_*)^C = (-)^{D/2+1}\gamma_*\,. \tag{3.78}$$

[4] We use the convention that the order of fermion fields is reversed in the process of complex conjugation. See (A.16) with $\beta = 1$.

[5] For these dimensions the spinor bilinears of Chs. 2 and 3 are related by $(\bar\lambda\chi)_{\text{Ch.2}} = (\overline{\lambda^C}\chi)_{\text{Ch.3}}$.

With these ingredients, we can derive the following rule for complex conjugation of a spinor bilinear involving an arbitrary matrix M:

$$(\bar{\chi} M \lambda)^* \equiv (\bar{\chi} M \lambda)^C = (-t_0 t_1) \overline{\chi^C} M^C \lambda^C . \tag{3.79}$$

A 'hidden' interchange of the order of the fermion fields is needed in the derivation, but there is no change of order in the final result for the charge conjugate of any bilinear. One may think of this relation as the appropriate conjugation property when the conjugate of a spinor is defined as in (3.50).

Exercise 3.32 *It is important that any spinor λ and its conjugate λ^C transform in the same way under a Lorentz transformation. Prove this using (3.57) and the rules above. If the matrix M is a Clifford element of rank r, i.e. $M = \gamma_{\mu_1 \ldots \mu_r}$, then both sides of (3.79) transform as tensors of rank r.*

Exercise 3.33 *Show that for any spinor $(\lambda^C)^C = -t_1 \lambda$, and for any matrix $(M^C)^C = M$.*

Exercise 3.34 *Suppose that $\psi(x)$ is a fermion field which satisfies the free massive Dirac equation $\partial\!\!\!/ \psi = m\psi$ for $D = 4$. Show that the charge conjugate field ψ^C satisfies the same equation. This exercise gives some physical motivation for the definition of Majorana spinors in the next section.*

3.3 Majorana spinors

The concept of supersymmetry is closely tied to the relativistic treatment of particle spin. Indeed the transformation parameters are spinors ϵ_α. It is reasonable to suppose that the simplest supersymmetric field theories in each spacetime dimension D are based on the simplest spinors that are compatible with invariance under the Lorentz group SO$(D-1, 1)$. In even dimension $D = 2m$ we already know that Weyl fields, rather than Dirac fields, transform irreducibly under Lorentz transformations. Weyl fields were first discussed in Sec. 2.6. They have 2^{m-1} complex components while a Dirac field has 2^m complex components. Weyl fields can be obtained by applying the chiral projector P_L or P_R to a Dirac field.

In this section we introduce Majorana fields, which are Dirac fields that satisfy an additional 'reality condition'. This condition reduces the number of independent components by a factor of 2. Thus, like Weyl fields, a Majorana spinor field has half the degrees of freedom and can be viewed as more fundamental than a complex Dirac field. Physically the properties of particles described by a Majorana field are similar to Dirac particles, *except that particles and anti-particles are identical.* The spin states of massive and massless Majorana spinors transform in representations of SO$(D-1)$ and SO$(D-2)$, respectively.

3.3.1 Definition and properties

The results of Sec. 3.2.4 suggest that it might be possible to impose the reality constraint

$$\psi = \psi^C = B^{-1}\psi^* , \qquad \text{i.e.} \qquad \psi^* = B\psi , \tag{3.80}$$

on a spinor field. Ex. 3.32 shows that both sides transform in the same way under Lorentz transformations in any dimension D, so the constraint is compatible with Lorentz symmetry. In fact (3.80) is the defining condition for Majorana spinors. However, there is a subtle and important consistency condition that we now derive, which restricts the spacetime dimension in which Majorana spinors can exist. It is easy to see that the reality condition (3.80) is not automatically consistent. Take the complex conjugate of the second form of the condition and use it again to obtain $\psi = B^*B\,\psi$. Thus the reality condition is mathematically consistent only if $B^*B = \mathbb{1}$. Using Ex. 3.20, we see that this requires[6] $t_1 = -1$.

The two possible values $t_0 = \pm 1$ must be considered, and we begin with the case $t_0 = +1$. Consulting Table 3.1, we see that $t_0 = +1$ holds for spacetime dimension $D = 2, 3, 4$, mod 8. In this case we call the spinors that satisfy (3.80) *Majorana spinors*. It is clear from (3.75) that if $t_0 = 1$ and $t_1 = -1$, the barred (3.50) and Dirac adjoint spinors (2.30) agree for Majorana spinors. In fact, this gives an alternative definition of a Majorana spinor.

Another fact about the Majorana case is that there are representations of the γ-matrices that are explicitly real and may be called really real representations. Here is a really real representation for $D = 4$:

$$\begin{aligned}
\gamma^0 &= \begin{pmatrix} 0 & \mathbb{1} \\ -\mathbb{1} & 0 \end{pmatrix} = \mathrm{i}\sigma_2 \otimes \mathbb{1} , \\
\gamma^1 &= \begin{pmatrix} \mathbb{1} & 0 \\ 0 & -\mathbb{1} \end{pmatrix} = \sigma_3 \otimes \mathbb{1} , \\
\gamma^2 &= \begin{pmatrix} 0 & \sigma_1 \\ \sigma_1 & 0 \end{pmatrix} = \sigma_1 \otimes \sigma_1 , \\
\gamma^3 &= \begin{pmatrix} 0 & \sigma_3 \\ \sigma_3 & 0 \end{pmatrix} = \sigma_1 \otimes \sigma_3 .
\end{aligned} \tag{3.81}$$

Note that the γ^i are symmetric, while γ^0 is antisymmetric. This is required by hermiticity in any real representation. We construct really real representations in all allowed dimensions $D = 2, 3, 4$ mod 8 in Appendix 3A.5.

In such representations (3.48) implies that $B = \mathbb{1}$ (up to a phase). The relation (3.47) then gives $C = \mathrm{i}\gamma^0$. Further, a Majorana spinor field is really real since (3.80) reduces to $\Psi^* = \Psi$.

Really real representations are sometimes convenient, but we emphasize that the physics of Majorana spinors is the same in, and can be explored in, any representation of the Clifford algebra, replacing complex conjugation with charge conjugation. For convenience we

[6] This manipulation is the same as working out Ex. 3.33, and this thus leads to the same result.

often write 'complex conjugation' when in fact we use 'charge conjugation'. For example, the complex conjugate of $\bar{\chi}\gamma_{\mu_1\dots\mu_r}\psi$, where χ and ψ are Majorana, is computed as follows. We follow Sec. 3.2.4 and write

$$(\bar{\chi}\gamma_{\mu_1\dots\mu_r}\psi)^* = (\bar{\chi}\gamma_{\mu_1\dots\mu_r}\psi)^C = \bar{\chi}(\gamma_{\mu_1\dots\mu_r})^C\psi = \bar{\chi}\gamma_{\mu_1\dots\mu_r}\psi\,. \tag{3.82}$$

We used the reality conditions $\psi^C = \psi$ and $\bar{\chi}^C = \bar{\chi}^C$ as well as (3.77) to deduce this result.[7] Hence, bilinears such as $\bar{\chi}\psi$ and $\bar{\chi}\gamma_{\mu_1\dots\mu_r}\psi$ are real.

When $t_0 = -1$ (and still $t_1 = -1$) spinors that satisfy (3.80) are called *pseudo-Majorana* spinors. They are mostly relevant for $D = 8$ or 9. There are no really real representations in these dimensions; instead there are representations of the Clifford algebra in which the generating γ-matrices are imaginary, $(\gamma^\mu)^* = -\gamma^\mu$. In any representation (3.79) and (3.77) hold with $t_0 = t_1 = -1$. This implies that the reality properties of bilinears are different from those of Majorana spinors. Although these differences are significant, the essential property that a complex spinor can be reduced to a real one still holds, and it is common in the literature not to distinguish between Majorana and pseudo-Majorana spinors. However, note the following.

Exercise 3.35 *Show that the mass term $m\bar{\chi}\chi = 0$ for a single pseudo-Majorana field. Pseudo-Majorana spinors must be massless (unless paired).*

We now consider (pseudo-)Majorana spinors in even dimensions $D = 0, 2, 4$ mod 8. We can quickly show using (3.78) that these cases are somewhat different. For $D = 2$ mod 8 we have $(\gamma_*\psi)^C = \gamma_*\psi^C$. Thus the two constraints

$$\text{Majorana:}\quad \psi^C = \psi\,, \qquad \text{Weyl:}\quad P_{L,R}\psi = \psi\,, \tag{3.83}$$

are compatible. It is equivalent to observe that the chiral projections of a Majorana spinor ψ satisfy

$$(P_L\psi)^C = P_L\psi\,, \qquad (P_R\psi)^C = P_R\psi\,. \tag{3.84}$$

Thus the chiral projections of a Majorana spinor are also Majorana spinors. Each chiral projection satisfies both constraints in (3.83) and is called a *Majorana–Weyl spinor*. Such spinors have 2^{m-1} independent 'real' components in dimension $D = 2m = 2$ mod 8 and are the 'most fundamental' spinors available in these dimensions. It is not surprising that supergravity and superstring theories in $D = 10$ dimensions are based on Majorana–Weyl spinors.

For $D = 4$ mod 4 dimensions we have $(\gamma_*\psi)^C = -\gamma_*\psi^C$, so that the equations of (3.84) are replaced by

$$(P_L\psi)^C = P_R\psi\,, \qquad (P_R\psi)^C = P_L\psi\,. \tag{3.85}$$

[7] Notice that Majorana spinors, which are real in the sense of C-conjugation, are not real for the original complex conjugation, not even in the really real representation. In fact, $\bar{\chi}$ is purely imaginary in the really real representation. However, under complex conjugation we should interchange the order of the spinors, which leads to another − sign, compensating the − sign of complex conjugation of $\bar{\chi}$. Neither sign appears explicitly when one uses charge conjugation, independent of the γ-matrix representation. This illustrates how the use of C simplifies the reality considerations.

These equations state that the 'left' and 'right' components of a Majorana spinor are related by charge conjugation. A direct consequence is that, for any expression involving the left-handed projection $P_L\psi$ of a Majorana spinor ψ, the corresponding expression for $P_R\psi$ follows by complex conjugation. Of course there are Weyl spinors that are chiral projections $P_{L,R}\psi$ of a Dirac spinor ψ, but these cannot satisfy the Majorana condition since for Majorana fermions $(P_{L,R}\psi)^C = P_{R,L}\psi$.

3.3.2 Symplectic Majorana spinors

When $t_1 = 1$ we cannot define Majorana spinors, but we can define 'symplectic Majorana spinors'. These consist of an even number of spinors χ^i, with $i = 1, \ldots, 2k$, which satisfy a 'reality condition' containing a non-singular antisymmetric matrix ε^{ij}. The inverse matrix ε_{ij} satisfies $\varepsilon^{ij}\varepsilon_{kj} = \delta^i_k$. Symplectic Majorana spinors satisfy the condition

$$\chi^i = \varepsilon^{ij}(\chi^j)^C = \varepsilon^{ij}B^{-1}(\chi^j)^* . \tag{3.86}$$

The consistency check discussed after (3.80) now works for $t_1 = 1$ because of the antisymmetric ε^{ij}.

Exercise 3.36 *Check that, in five dimensions with symplectic Majorana spinors,* $\bar{\psi}^i\chi_i \equiv \bar{\psi}^i\chi^j\varepsilon_{ji}$ *is pure imaginary while* $\bar{\psi}^i\gamma_\mu\chi_i$ *is real.*

For dimensions $D = 6 \bmod 8$, one can use (3.78) to show that the symplectic Majorana constraint is compatible with chirality. We can therefore define the *symplectic Majorana–Weyl spinors* $P_L\chi^i$ or $P_R\chi^i$.

3.3.3 Dimensions of minimal spinors

The various types of spinors we have discussed are linked to the signs of t_0 and t_1 as follows:

$$\begin{array}{lll} t_1 = -1, & t_0 = 1: & \text{Majorana,} \\ & t_0 = -1: & \text{pseudo-Majorana,} \\ t_1 = 1: & & \text{symplectic Majorana}. \end{array} \tag{3.87}$$

As explained above we no longer distinguish between Majorana and pseudo-Majorana spinors. In any even dimension one can define Weyl spinors, while in dimensions $D = 2 \bmod 4$, one can combine the (symplectic) Majorana condition and Weyl conditions. These facts are summarized in Table 3.2.[8] For each spacetime dimension it is indicated whether Majorana (M), Majorana–Weyl (MW), symplectic (S) or symplectic Weyl (SW) spinors can be defined as the 'minimal spinor'. The number of components of this minimal spinor is given. The table is for Minkowski signature and has a periodicity of 8 in dimension. When D is changed to $D + 8$, the number of spinor components is multiplied by 16. The

[8] For $D = 4 \bmod 4$ we can also define Weyl spinors, but we omit this in the table.

Table 3.2 Irreducible spinors, number of components and symmetry properties.

dim	spinor	min # components	antisymmetric
2	MW	1	1
3	M	2	1,2
4	M	4	1,2
5	S	8	2,3
6	SW	8	3
7	S	16	0,3
8	M	16	0,1
9	M	16	0,1
10	MW	16	1
11	M	32	1,2

final column indicates the ranks of the antisymmetric spinor bilinears, e.g. a 0 indicates that $\bar{\epsilon}_2\epsilon_1 = -\bar{\epsilon}_1\epsilon_2$, and a 2 indicates that $\bar{\epsilon}_2\gamma_{\mu\nu}\epsilon_1 = -\bar{\epsilon}_1\gamma_{\mu\nu}\epsilon_2$. This entry is modulo 4, i.e. if rank 0 is antisymmetric, then so are rank 4 and 8 bilinears. Minimal spinors in dimension $D = 2$ mod 4 must have the same chirality to define a symmetry for their bilinears. The property (3.56) then implies that non-vanishing bilinears contain an odd number of γ-matrices. For $D = 4$ mod 4, there are two possibilities for reality conditions and we have chosen the one that includes rank 1 in the column 'antisymmetric'. This property is needed for the supersymmetry algebra.

3.4 Majorana spinors in physical theories

3.4.1 Variation of a Majorana Lagrangian

In this section we consider a prototype action for a Majorana spinor field in dimension $D = 2, 3, 4$ mod 8. Majorana and Dirac fields transform the same way under Lorentz transformations, but Majorana spinors have half as many degrees of freedom, so we write

$$S[\Psi] = -\tfrac{1}{2}\int \mathrm{d}^D x\, \bar{\Psi}[\gamma^\mu\partial_\mu - m]\Psi(x)\,. \tag{3.88}$$

There is an immediate and curious subtlety due to the symmetries of the matrices C and $C\gamma^\mu$. Using (3.50), we see that the mass and kinetic terms are proportional to $\Psi^T C\Psi$ and $\Psi^T C\gamma^\mu\partial_\mu\Psi$. Suppose that the field components Ψ are conventional commuting numbers. Since C is antisymmetric, the mass term vanishes. Since $C\gamma^\mu$ is symmetric, the kinetic term is a total derivative and thus vanishes when integrated in the action. For commuting field components, there is no dynamics! To restore the dynamics we must assume that Majorana fields are anti-commuting Grassmann variables, which we always assume unless stated otherwise.

Let's derive the Euler–Lagrange equation for Ψ. Field variations must satisfy the Majorana condition (3.80), so that $\delta\Psi$ and $\delta\bar{\Psi}$ are related following Sec. 3.2.1. Initially $\delta S[\Psi]$ contains two terms. However, after a Majorana flip and partial integration, one can see that the two terms are equal, so that $\delta S[\Psi]$ can be written as the single expression

$$\delta S[\Psi] = -\int \mathrm{d}^D x\, \delta\bar{\Psi}[\gamma^\mu \partial_\mu - m]\Psi(x)\,. \tag{3.89}$$

Thus a Majorana field satisfies the conventional Dirac equation.

This fact is no surprise, but it is an example of a more general and simplifying rule for the variation of Majorana spinor actions. If integration by parts is valid, it is sufficient to vary $\bar{\Psi}$ and multiply by 2 to account for the variation of Ψ.

Exercise 3.37 *Derive (3.89) in full detail.*

Exercise 3.38 *A Majorana field is simply a Dirac field subject to the reality condition (3.80). Let's impose that constraint on the plane wave expansion (2.24) for $D = 4$ using the relation $v = u^C = Bu^*$, which holds for the u and v spinors defined in (2.37) and (2.38). In this way one derives $d(\vec{p}, s) = c(\vec{p}, s)$ which proves that a Majorana particle is its own anti-particle. Readers should derive this fact!*

Exercise 3.39 *Show that*

$$v(\vec{p}, s) = u(\vec{p}, s)^C \tag{3.90}$$

holds for the u and v spinors defined for the Weyl representation in Sec. 2.5. This was the motivation for the choice (2.41).

3.4.2 Relation of Majorana and Weyl spinor theories

In even dimensions $D = 0, 2, 4 \bmod 8$, both Majorana and Weyl fields exist and both have legitimate claims to be more fundamental than a Dirac fermion. In fact both fields describe equivalent physics. Let's show this for $D = 4$. We can rewrite the action (3.88) as

$$\begin{aligned} S[\psi] &= -\tfrac{1}{2}\int \mathrm{d}^4 x\, \left[\bar{\Psi}\gamma^\mu \partial_\mu - m\right](P_L + P_R)\Psi \\ &= -\int \mathrm{d}^4 x\, \left[\bar{\Psi}\gamma^\mu \partial_\mu P_L \Psi - \tfrac{1}{2} m \bar{\Psi} P_L \Psi - \tfrac{1}{2} m \bar{\Psi} P_R \Psi\right]. \end{aligned} \tag{3.91}$$

We obtained the second line by a Majorana flip and partial integration. In the second form of the action, the Majorana field is replaced by its chiral projections. In our treatment of chiral multiplets in supersymmetry, we will exercise the option to write Majorana fermion actions in this way.

Exercise 3.40 *Show that the Euler–Lagrange equations that follow from the variation of the second form of the action in (3.91) are*

$$\partial\!\!\!/ P_L \Psi = m P_R \Psi\,, \qquad \partial\!\!\!/ P_R \Psi = m P_L \Psi\,. \tag{3.92}$$

Derive $\Box P_{L,R}\Psi = m^2 P_{L,R}\Psi$ from the equations above.

Majorana and Weyl fields in $D=4$ **Box 3.3**

Any field theory of a Majorana spinor field Ψ can be rewritten in terms of a Weyl field $P_L\Psi$ and its complex conjugate. Conversely, any theory involving the chiral field $\chi = P_L\chi$ and its conjugate $\chi^C = P_R\chi^C$ can be rephrased as a Majorana equation if one defines the Majorana field $\Psi = P_L\chi + P_R\chi^C$. Supersymmetry theories in $D=4$ are formulated in both descriptions in the physics literature.

Let's return to the Weyl representation (2.19) for the final step in the argument to show that the equation of motion for a Majorana field can be reexpressed in terms of a Weyl field and its adjoint. The Majorana condition $\Psi = B^{-1}\Psi^* = \gamma^0\gamma^1\gamma^3\Psi^*$ requires that Ψ take the form

$$\Psi = \begin{pmatrix} \psi_1 \\ \psi_2 \\ \psi_2^* \\ -\psi_1^* \end{pmatrix} . \tag{3.93}$$

With (3.93) and (2.55) in view we define the two-component Weyl fields

$$\psi = \begin{pmatrix} \psi_1 \\ \psi_2 \end{pmatrix} , \qquad \tilde{\psi} = \begin{pmatrix} \psi_2^* \\ -\psi_1^* \end{pmatrix} . \tag{3.94}$$

Using the form of γ^μ (2.19) and γ_* (3.34) in the Weyl representation, we see that we can identify

$$\begin{pmatrix} \psi \\ 0 \end{pmatrix} = P_L\Psi , \qquad \begin{pmatrix} 0 \\ \tilde{\psi} \end{pmatrix} = (P_L\Psi)^C = P_R\Psi . \tag{3.95}$$

The equations of motion (3.92) can then be rewritten as

$$\bar{\sigma}^\mu\partial_\mu\psi = m\tilde{\psi} , \qquad \sigma^\mu\partial_\mu\tilde{\psi} = m\psi . \tag{3.96}$$

These are equivalent to the pair of Weyl equations in (2.56) with the restriction $\tilde{\psi} = \tilde{\chi}$ which comes because we started in this section with a Majorana rather than a Dirac field.

3.4.3 U(1) symmetries of a Majorana field

In Sec. 2.7.1 we considered the U(1) symmetry operation $\Psi \to \Psi' = \mathrm{e}^{\mathrm{i}\theta}\Psi$. This symmetry is obviously *incompatible* with the Majorana condition (3.80). Thus the simplest internal symmetry of a Dirac fermion cannot be defined in a field theory of a (single) Majorana field. However, it is easy to see that $(\mathrm{i}\gamma_*)^C = \mathrm{i}\gamma_*$, so the chiral transformation $\Psi \to \Psi' = \mathrm{e}^{\mathrm{i}\gamma_*\theta}\Psi$ preserves the Majorana condition. Let's ask whether the infinitesimal limit of this transformation is a symmetry of the free massive Majorana action (3.88).

Exercise 3.41 *Use $\delta\bar{\Psi} = \mathrm{i}\theta\bar{\Psi}\gamma_*$ and partial integration to derive the variation*

$$\delta S[\Psi] = \mathrm{i}\theta m \int \mathrm{d}^4x\, \bar{\Psi}\gamma_*\Psi , \tag{3.97}$$

which vanishes only for a massless *Majorana field.*

Thus we have learned the following.

- The conventional vector U(1) symmetry is incompatible with the Majorana condition.
- The axial transformation above is compatible and is a symmetry of the action for a massless Majorana field only.

Exercise 3.42 *Show that the axial current*

$$J_*^\mu = \tfrac{1}{2}\mathrm{i}\bar{\Psi}\gamma^\mu\gamma_*\Psi \tag{3.98}$$

is the Noether current for the chiral symmetry defined above. Use the equations of motion to show that

$$\partial_\mu J_*^\mu = -\mathrm{i}m\,\bar{\Psi}\gamma_*\Psi\,. \tag{3.99}$$

The current is conserved only for massless Majorana fermions.

The dynamics of a Majorana field Ψ can be expressed in terms of its chiral projections $P_{L,R}\Psi$. So can the chiral transformation, which becomes $P_{L,R}\Psi \to P_{L,R}\Psi' = \mathrm{e}^{\pm\mathrm{i}\theta}\Psi$.

Throughout this section we used the simple dynamics of a free massive fermion to illustrate the relation between Majorana and Weyl fields and to explore their U(1) symmetries. It is straightforward to extend these ideas to interacting field theories with nonlinear equations of motion.

Appendix 3A Details of the Clifford algebras for $D = 2m$

3A.1 Traces and the basis of the Clifford algebra

Let us start with the following facts discussed in Sec. 3.1. The Clifford algebra in even dimension $D = 2m$ has a basis of 2^m linearly independent, trace orthogonal matrices, given in (3.25). Any representation by matrices of dimension 2^m is irreducible.

The trace properties of the matrices are important for proofs of these properties which are independent of the explicit construction in (3.2). The matrices Γ^A for tensor rank $1 \leq r \leq D-1$ are traceless. One simple way to see this is to use the Lorentz transformations (2.22) and its extension to general rank

$$L(\lambda)\gamma^{\mu_1\mu_2\ldots\mu_r}L(\lambda)^{-1} = \gamma^{\nu_1\nu_2\ldots\nu_r}\Lambda_{\nu_1}{}^{\mu_1}\ldots\Lambda_{\nu_r}{}^{\mu_r}\,. \tag{3.100}$$

Traces then satisfy the Lorentz transformation law as suggested by their free indices:

$$\operatorname{Tr}\gamma^{\mu_1\mu_2\ldots\mu_r} = \operatorname{Tr}\gamma^{\nu_1\nu_2\ldots\nu_r}\Lambda_{\nu_1}{}^{\mu_1}\ldots\Lambda_{\nu_r}{}^{\mu_r}\,. \tag{3.101}$$

This means that the traces must be *totally antisymmetric Lorentz invariant* tensors. However the only invariant tensors available are the Minkowski metric $\eta^{\mu\nu}$ and the Levi-Civita tensor $\varepsilon^{\mu_1\mu_2\ldots\mu_D}$ introduced in Sec. 3.1.3. No totally antisymmetric tensor can be formed from products of $\eta^{\mu\nu}$. This proves that $\operatorname{Tr}\Gamma^A = 0$ for all elements of rank $1 \leq r \leq D-1$.

The argument does not apply to the highest rank element. However, one can see from the pattern of alternation in (3.7) that this is given by a commutator for even $D = 2m$ and by an anti-commutator for odd $D = 2m + 1$. Thus the trace of the highest rank element vanishes for $D = 2m$ but need not (and does not) vanish for $D = 2m + 1$. This is actually a fundamental distinction between the Clifford algebras for even and odd dimensions. It might have been expected since the second rank elements (see Ex. 2.8) give a representation of the Lorentz algebras $\mathfrak{so}(D-1, 1)$ which are real forms of different Lie algebras in the Cartan classification, namely D_m for even $D = 2m$ and B_m for $D = 2m+1$.

There is another way to prove the traceless property, which does not require information concerning invariant tensors. For rank 1, we simply take the trace of the formula derived in Ex. 2.9. Contraction with $\eta_{\nu\rho}$ immediately gives $\operatorname{Tr}\gamma^\mu = 0$. As an exercise, the reader can extend this argument to higher rank.

The trace property leads also to the proof of independence of the elements of the basis (3.25) for even spacetime dimensions. One uses the 'reverse order' basis of (3.26) and the trace orthogonality property (3.27). We suppose that there is a set of coefficients x_A such that

$$\sum_A x_A \Gamma^A = 0\,. \tag{3.102}$$

Multiply by Γ_B from the right. Take the trace and use the trace orthogonality to obtain

$$\sum_A x_A \operatorname{Tr}\Gamma^A\Gamma^B = \pm x_B \operatorname{Tr}\mathbb{1} = 0\,. \tag{3.103}$$

Hence all $x_A = 0$ and linear independence is proven.

Furthermore, since we have a linearly independent, indeed trace orthogonal, basis of the algebra, the Γ^A are a complete set in the space of $2^m \times 2^m$ matrices.

It now follows that, in any representation of the Clifford algebra for $D = 2m$ spacetime dimensions, the dimension of the $N \times N$ matrices satisfies $N \geq 2^m$. The reason is that no linearly independent set of matrices of any smaller dimension exists. It also follows that any representation of dimension 2^m is irreducible. It can have no non-trivial invariant subspace, since a set of linearly independent matrices of smaller dimension would be realized by projection to this subspace.

3A.2 Uniqueness of the γ-matrix representation

We must now show that there is exactly one irreducible representation up to equivalence. We use the basic properties of representations of finite groups. However, the Clifford algebra is not quite a group because the minus signs that necessarily occur in the set of products $\Gamma^A\Gamma^B = \pm\Gamma^C$ are not allowed by the definition of a group. This problem is solved by doubling the basis in (3.25) to the larger set $\{\Gamma^A, -\Gamma^A\}$. This set is a group of order 2^{2m+1} since all products are contained within the larger set. For $m = 1$, the group obtained is isomorphic to the quaternions, so the groups defined by doubling the Clifford algebras are called generalized quaternionic groups.

Every representation of the Clifford algebra by a set of matrices $D(\Gamma^A)$ extends to a representation of the group if we define $D(-\Gamma^A) = -D(\Gamma^A)$. It is not true that every group representation gives a representation of the algebra. For example, in a one-dimensional group representation, the matrices $D(\gamma^\mu)$ of the Clifford generators cannot satisfy $\{D(\gamma^\mu), D(\gamma^\nu)\} = 2\eta^{\mu\nu}$.

The three basic facts that we need are discussed in many mathematical texts such as [13, 14]. Consider the set of all finite-dimensional irreducible representations and choose one representative within each class of equivalent representations. The set so formed, which may be called the set of all inequivalent irreducible representations, has the following properties:

1. The sum of the squares of the dimensions of these representations is equal to the order of the group.
2. The number of inequivalent irreducible representations is equal to the number of conjugacy classes in the group.
3. The number of inequivalent one-dimensional representations is equal to the index of the commutator subgroup G_c. The index of a subgroup is the ratio of the order of the group divided by the order of the subgroup.

The conjugacy classes of the group are sets of products $\pm\Gamma^B\Gamma^A(\Gamma^B)^{-1}$ (with no sum on B).

Exercise 3.43 *Show that for rank $r \geq 1$ there is a conjugacy class containing the pair $(\Gamma^A, -\Gamma^A)$ for each distinct Γ^A, and that $\mathbb{1}$ and $-\mathbb{1}$ belong to different conjugacy classes.*

Thus there are a total of $2^{2m}+1$ conjugacy classes.

The commutator subgroup is generated by all products of the form $\pm\Gamma^B\Gamma^A(\Gamma^B)^{-1}(\Gamma^A)^{-1}$. But in our case this subgroup contains only $\pm\mathbb{1}$, so the order of the subgroup is 2 and its index is 2^{2m}.

These facts establish that the group has exactly one irreducible representation of dimension 2^m plus 2^{2m} inequivalent one-dimensional representations. We must now show that the 2^m-dimensional representation of the group is also a representation of the algebra. We use the fact that any finite-dimensional algebra has a (reducible) representation called the *regular representation* for which the algebra itself is the carrier space. The dimension is thus the dimension of the algebra, 2^{2m} in our case. The regular representation $\Gamma^A \to T(\Gamma^A)$ is defined by $T(\Gamma^A)\Gamma^B \equiv \Gamma^A\Gamma^B$. This algebra representation, in which $T(-\Gamma^A) = -T(\Gamma^A)$ is necessarily satisfied, is also a group representation. Its decomposition into irreducible components thus cannot contain any one-dimensional group representations in which $D(-\Gamma^A) = +D(\Gamma^A)$. Thus the only possibility is that the regular representation decomposes into 2^m copies of the 2^m-dimensional irreducible representation. This proves the essential fact that there is exactly one irreducible representation of the Clifford algebra for even spacetime dimension. For dimension $D = 2m$, the dimension of the Clifford representation is 2^m.

Another fact from finite group theory is helpful at this point. Any representation of a finite group is equivalent to a representation by unitary matrices. We can and therefore will choose a representation in which the spatial γ-matrices γ^i, $i = 1, \ldots, D-1$, which satisfy $(\gamma^i)^2 = \mathbb{1}$, are hermitian, and γ^0, which satisfies $(\gamma^0)^2 = -\mathbb{1}$, is anti-hermitian.

3A.3 The Clifford algebra for odd spacetime dimensions

We gave in (3.40) two different sets of γ-matrices for odd dimensions. They are inequivalent as representations of the generating elements. Indeed it is easily seen that $S\gamma_+{}^\mu S^{-1} = \gamma_-{}^\mu$ cannot be satisfied. This requires $S\gamma^\mu S^{-1} = \gamma^\mu$ for the first $2m$ components. But then, from the product form in (3.6) and (3.30), we obtain $S\gamma^{2m} S^{-1} = +\gamma^{2m}$, rather than the opposite sign needed.

It follows from Ex. 2.8 that the two sets of second rank elements constructed from the generating elements above, namely

$$\begin{aligned} \Sigma_\pm{}^{\mu\nu} &= \tfrac{1}{4}[\gamma^\mu, \gamma^\nu], \qquad \mu, \nu = 0, \ldots, 2m-1, \\ &= \tfrac{1}{4}[\gamma^\mu, \pm\gamma_*], \qquad \mu = 0, \ldots, 2m-1, \quad \nu = 2m, \end{aligned} \tag{3.104}$$

are each representations of the Lie algebra $\mathfrak{so}(2m, 1)$. The two representations are equivalent, however, since $\gamma_* \Sigma_+{}^{\mu\nu} \gamma_* = \Sigma_-{}^{\mu\nu}$. This representation is irreducible; indeed it is a copy of the unique 2^{2m}-dimensional fundamental irreducible representation with Dynkin designation $(0, 0, \ldots, 0, 1)$. It is associated with the short simple root of the Dynkin diagram for B_m.

We refer readers to[9] [10, 11, 12, 15, 16] for alternative discussions of γ-matrices and Majorana spinors.

3A.4 Determination of symmetries of γ-matrices

We will determine the possible symmetries of γ-matrices for each spacetime dimension $D = 2m$ by showing that each matrix $C\Gamma_A$ formed from the basis (3.25) has a definite symmetry that depends only on the tensor rank r. Then we will count the number of symmetric and antisymmetric matrices in the list $\{C\Gamma_A\}$, which must be equal to $2^{m-1}(2^m \pm 1)$ for $D = 2m$. For given values of t_0 and t_1, the number of antisymmetric matrices in the list $\{C\Gamma_A\}$ is given, using (3.44), by

$$\begin{aligned} N_- &= \sum_{r=0}^{2m} \tfrac{1}{2}[1 + t_r] C_r^{2m} \\ &= 2^{2m-1} + \tfrac{1}{2}t_0 \sum_{s=0}^{m} (-)^s C_{2s}^{2m} + \tfrac{1}{2}t_1 \sum_{s=0}^{m-1} (-)^s C_{2s+1}^{2m} \\ &= 2^{2m-1} + t_0\, 2^{m-1} \cos\frac{m\pi}{2} + t_1\, 2^{m-1} \sin\frac{m\pi}{2} \\ &= 2^{m-1}(2^m - 1). \end{aligned} \tag{3.105}$$

We thus find

$$t_0 \cos\frac{m\pi}{2} + t_1\, 2^{m-1} \sin\frac{m\pi}{2} = -1, \tag{3.106}$$

[9] In [15], the discussion of Majorana spinors is in Sec. 4, pp. 843–851.

which leads to the solutions that are in Table 3.1 for even dimensions.

To understand the situation in odd $D = 2m + 1$ we note that the highest rank Clifford element γ_* in (3.30) has the symmetry determined by t_{2m}. Since we attach $\pm\gamma_* = \gamma^{2m}$ as the last generating element in (3.40) we must require it to have the same symmetry as the other generating γ^μ, and thus t_{2m} should be equal to t_1. This determines which of the two possibilities for even dimensions in Table 3.1 is valid in the next odd dimension.

3A.5 Friendly representations

General construction

In this section we present an explicit recursive construction of the generating γ^μ for any even dimension $D = 2m$. In this representation each generating matrix will be either pure real or pure imaginary. A representation of this type will be called a friendly representation.[10] Using this representation it is also possible to prove the existence of Majorana (and pseudo-Majorana) spinors in a quite simple way [17, 12] (see Appendix B in [17]).

We already know that the γ-matrices in dimension $D = 2m$ are $2^m \times 2^m$ matrices. In the recursive construction the generating matrices γ^μ for dimension $D = 2m$ will be written as direct products of the $\tilde{\gamma}^\mu$ and $\tilde{\gamma}_*$ for dimension $D = 2m - 2$ with the Pauli matrices σ_i.

We start in $D = 2$ and write

$$\gamma^0 = \begin{pmatrix} 0 & 1 \\ -1 & 0 \end{pmatrix} = \mathrm{i}\sigma_2\,, \qquad \gamma^1 = \begin{pmatrix} 0 & 1 \\ 1 & 0 \end{pmatrix} = \sigma_1\,, \tag{3.107}$$

which is a really real, hermitian, and friendly representation. The matrix γ_* is also real:

$$\gamma_* = -\gamma_0\gamma_1 = \sigma_3\,. \tag{3.108}$$

Adding it to (3.107) as γ^2 gives a real representation in $D = 3$.

The recursion relation for moving from a $D = 2m - 2$ representation with $\tilde{\gamma}$ to $D = 2m$ is

$$\begin{aligned} &\gamma^\mu = \tilde{\gamma}^\mu \otimes \mathbb{1}\,, \qquad \mu = 0, \ldots, 2m-3\,, \\ &\gamma^{2m-2} = \tilde{\gamma}_* \otimes \sigma_1\,, \qquad \gamma^{2m-1} = \tilde{\gamma}_* \otimes \sigma_3\,. \end{aligned} \tag{3.109}$$

This gives

$$\gamma_* = -\tilde{\gamma}_* \otimes \sigma_2\,. \tag{3.110}$$

This matrix γ_* can be used as γ^{2m} to define a representation in $D = 2m + 1$ dimensions.

This construction gives a real representation in four dimensions, which is explicitly given in (3.81). This one has an imaginary γ_* and hence this construction will not give real representations for higher dimensions. The matrix B is obtained as the product of all the imaginary γ-matrices.

[10] All of our friends use friendly representations.

We thus obtained representations for all dimensions, and really real for $D = 2, 3, 4$. The latter can be extended to any $D = 10, 11, 12$ or any other dimension that differs from it modulo 8. To see this, consider the following 16×16 matrices:

$$\begin{aligned}
E_1 &= \sigma_1 \otimes \mathbb{1} \otimes \mathbb{1} \otimes \mathbb{1}\,, \\
E_2 &= \sigma_3 \otimes \mathbb{1} \otimes \mathbb{1} \otimes \mathbb{1}\,, \\
E_3 &= \sigma_2 \otimes \sigma_2 \otimes \sigma_1 \otimes \mathbb{1}\,, \\
E_4 &= \sigma_2 \otimes \sigma_2 \otimes \sigma_3 \otimes \mathbb{1}\,, \\
E_5 &= \sigma_2 \otimes \sigma_1 \otimes \mathbb{1} \otimes \sigma_2\,, \\
E_6 &= \sigma_2 \otimes \sigma_3 \otimes \mathbb{1} \otimes \sigma_2\,, \\
E_7 &= \sigma_2 \otimes \mathbb{1} \otimes \sigma_2 \otimes \sigma_1\,, \\
E_8 &= \sigma_2 \otimes \mathbb{1} \otimes \sigma_2 \otimes \sigma_3\,, \\
E_* &= E_1 \ldots E_8 = \sigma_2 \otimes \sigma_2 \otimes \sigma_2 \otimes \sigma_2\,.
\end{aligned} \tag{3.111}$$

This is a real representation for Euclidean γ-matrices in $D = 8$ (or $D = 9$ if one includes E_*). Using this and a representation $\tilde{\gamma}^\mu$ in any D, one can construct a representation γ^μ in $D + 8$ dimensions by

$$\begin{aligned}
\gamma^\mu &= \tilde{\gamma}^\mu \otimes E_*\,, \qquad \mu = 0, \ldots, D-1\,, \\
\gamma^{D-1+i} &= \mathbb{1} \otimes E_i\,, \qquad i = 1, \ldots, 8\,.
\end{aligned} \tag{3.112}$$

When the $\tilde{\gamma}^\mu$ are real, the γ-matrices in $D + 8$ are also real. Hence this gives explicitly real representations in $D = 2, 3, 4$ mod 8. For even dimensions, one obtains

$$\gamma_* = \tilde{\gamma}_* \otimes E_*\,. \tag{3.113}$$

Hence this is real if $\tilde{\gamma}_*$ is real. For the real representations we saw that it is real for $D = 2$ and not in $D = 4$. This shows explicitly that we can define real projections P_L and P_R on real spinors if and only if $D = 2$ mod 8. These are called *Majorana–Weyl* representations.

Exercise 3.44 *We denote a Clifford algebra in s space-like and t time-like directions as $\mathcal{C}(s, t)$ (the ones discussed above are thus of the form $\mathcal{C}(D-1, 1)$, apart from the E_i that correspond to $\mathcal{C}(8, 0)$). See that the above construction proves that the reality properties of $\mathcal{C}(s+8, t)$ are the same as $\mathcal{C}(s, t)$. Further, show that the analogous construction starting with (3.107) shows that also $\mathcal{C}(s+1, t+1)$ has the same properties as $\mathcal{C}(s, t)$.*

4 The Maxwell and Yang–Mills gauge fields

In this chapter we discuss the classical abelian and non-abelian gauge fields. Although our treatment is self-contained, it is best taken as a review for readers who have previously studied the role of the vector potential as the gauge field in Maxwell's electromagnetism and also have some acquaintance with Yang–Mills theory.

We will again take a general dimensional viewpoint, but let's begin the discussion in four dimensions with some remarks about the particle representations of the Poincaré group and the fields usually used to describe elementary particles. A particle is classified by its mass m and spin s, and a massive particle of spin s has $2s + 1$ helicity states. Massless particles of spin $s = 0$ or $s = 1/2$ have one or two helicity states, respectively, in agreement with the counting for massive particles. However, massless particles of spin $s \geq 1/2$ have two helicity states, for all values of s.

Helicity is defined as the eigenvalue of the component of angular momentum in the direction of motion. For a massless particle of spin s, the two helicity states have eigenvalues $\pm s$. For a massive particle of spin s the helicity eigenvalues, called λ, range in integer steps from $\lambda = s$ to $\lambda = -s$.

Let us compare the count of the helicity states with the number of independent functions that must be specified as initial data for the Cauchy initial value problem of the associated field. The first number can be considered to be the *number of on-shell degrees of freedom*, or number of quantum degrees of freedom, while the second is the number of classical degrees of freedom.

Let's do the counting for massless particles that are identified with their anti-particles. The associated fields are real for bosons and satisfy the Majorana condition for fermions. The counting is similar for complex fields. We assume that the equations of motion are second order in time for bosons and first order for fermions. A unique solution of the Cauchy problem for the scalar $\phi(x)$ requires the initial data $\phi(\vec{x}, 0)$ and $\dot{\phi}(\vec{x}, 0)$, the time derivative. For $\Psi_\alpha(x)$, we must specify the initial values $\Psi_\alpha(\vec{x}, 0)$ of all four components, and the first order Dirac equation then determines the future evolution of $\Psi_\alpha(\vec{x}, t)$ and thus the time derivatives $\dot{\Psi}_\alpha(\vec{x}, 0)$. The number of helicity states (number of on-shell degrees of freedom) is 1 for $\phi(x)$ and 2 for $\Psi_\alpha(x)$. The number of classical degrees of freedom is twice the number of helicity states.

We continue this counting, in a naive fashion, for vector $A_\mu(x)$, vector–spinor $\psi_{\mu\alpha}(x)$, and symmetric tensor $h_{\mu\nu}(x)$ fields, the latter describing gravitons in Minkowski space. Following the earlier pattern we would expect to need 8, 16, and 20 functions, respectively, as initial data. These numbers greatly exceed the two helicity states for spin-1, spin-3/2 and spin-2 particles. Something new is required to resolve this mismatch.

The lessons from quantum electrodynamics, Yang–Mills theory, general relativity and supergravity teach us that the only way to proceed is to use very special field equations with gauge invariance. Gauge invariance accomplishes the following goals:

(a) Relativistic covariance is maintained.
(b) The field equations do not determine certain 'longitudinal' field components (such as $\partial^\mu A_\mu$ for vector fields).
(c) A subset of the field equations are constraints on the initial data rather than time evolution equations. The independent initial data are contained in four real functions, thus again two for each helicity state.
(d) The field describes a pure spin-s particle with no lower spin admixtures. Otherwise there would be some negative metric ghosts.
(e) Most important, for $s = 1, 3/2, 2$, gauge invariant interactions can be introduced.[1] Classical dynamics is consistent at the nonlinear level and the theories can be quantized (although power-counting renormalizability is expected to fail except for spin 1).

The dynamics of the gauge fields A_μ, $\psi_{\mu\alpha}$ and $h_{\mu\nu}$ is analyzed in this and subsequent chapters. In every case the purpose is to separate the Euler–Lagrange equations into time evolution equations and constraints and determine the initial data required for a unique solution of the former. In the process we will find that certain field components are harmonic functions in Minkowski space; they satisfy the Laplace equation $\nabla^2\phi(\vec{x}) = 0$, which is time independent. Any combination of gauge field components that satisfies this equation is simply eliminated because the Laplace equation has no normalizable solutions in flat space $\mathbb{R}^{D-1}$. The relevance of the normalizability criterion can be seen by transforming the Laplace equation to momentum space where it becomes $\vec{k}^2\hat{\phi}(\vec{k}) = 0$. The only smooth solution vanishes identically. A smooth solution is one that contains no δ-function-type terms.

4.1 The abelian gauge field $A_\mu(x)$

We now review the elementary features of gauge invariance for spin 1. One purpose is to set the stage for spin 3/2 in the next chapter.

4.1.1 Gauge invariance and fields with electric charge

In Chs. 1 and 2 we discussed the global U(1) symmetry of complex scalar and spinor fields. The abelian gauge symmetry of quantum electrodynamics is an extension in which the phase parameter $\theta(x)$ becomes an arbitrary function in Minkowski spacetime. We generalize the previous discussion slightly and assign an electric charge q, an arbitrary real

[1] There are gauge invariant *free* fields for massless particles of any spin (see [18], for example). It appears to be impossible to introduce consistent interactions for any finite subset of these, but remarkably one can make progress for certain infinite sets of fields and for background spacetimes different from Minkowski space [19, 20, 21].

number at this stage, to each complex field in the system. For a Dirac spinor field of charge q, the gauge transformation, a local change of the phase of the complex field, is[2]

$$\Psi(x) \to \Psi'(x) \equiv e^{iq\theta(x)}\Psi(x). \tag{4.1}$$

The goal is to formulate field equations that transform covariantly under the gauge transformation. This requires the introduction of a new field, the vector gauge field or vector potential $A_\mu(x)$, which is defined to transform as

$$A_\mu(x) \to A'_\mu(x) \equiv A_\mu(x) + \partial_\mu\theta(x). \tag{4.2}$$

One then defines the covariant derivative

$$D_\mu\Psi(x) \equiv (\partial_\mu - iqA_\mu(x))\Psi(x), \tag{4.3}$$

which transforms with the same phase factor as $\Psi(x)$, namely $D_\mu\Psi(x) \to e^{iq\theta(x)}D_\mu\Psi(x)$. The desired field equation is obtained by replacing $\partial_\mu\Psi \to D_\mu\Psi$ in the free Dirac equation (2.16), viz.

$$[\gamma^\mu D_\mu - m]\Psi \equiv [\gamma^\mu(\partial_\mu - iqA_\mu) - m]\Psi = 0. \tag{4.4}$$

This equation is gauge covariant; if $\Psi(x)$ satisfies (4.4) with gauge potential $A_\mu(x)$, then $\Psi'(x)$ satisfies the same equation with gauge potential $A'_\mu(x)$.

The same procedure can be applied to a complex scalar field $\phi(x)$, to which we assign an electric charge q (which may differ from the charge of Ψ). We extend the global U(1) symmetry discussed in Ch. 1 to the local gauge symmetry $\phi(x) \to \phi'(x) = e^{iq\theta(x)}\phi(x)$ by defining the covariant derivative $D_\mu\phi = (\partial_\mu - iqA_\mu)\phi$ and modifying the Klein–Gordon equation to the form

$$[D^\mu D_\mu - m^2]\phi = 0. \tag{4.5}$$

The procedure of promoting the global U(1) symmetry of the Dirac or Klein–Gordon equation to a local or gauge symmetry through the introduction of the vector potential in the covariant derivative is called the principle of minimal coupling. Another part of standard vocabulary is to say that fields with electric charge, such as ϕ or Ψ, are charged 'matter fields', which are minimally coupled to the gauge field A_μ.

Box 4.1 Degrees of freedom

On-shell degrees of freedom = number of helicity states.
Off-shell degrees of freedom = number of field components − gauge transformations.

[2] In the notation of Ch. 1, the 'matrix' generator is $t = -iq$. The U(1) transformation in Sec. 2.7.1 corresponds to the choice $q = 1$.

4.1.2 The free gauge field

It is quite remarkable that the promotion of global to gauge symmetry requires a new field $A_\mu(x)$. In some cases one may wish to consider (4.4) or (4.5) in a fixed external background gauge potential, but it is far more interesting to think of $A_\mu(x)$ as a field that is itself determined dynamically by its coupling to charged matter in a gauge invariant fashion. The resulting theory is quantum electrodynamics, the quantum and Lorentz covariant version of Maxwell's theory of electromagnetism. The predictions of this theory, both classical and quantum, are well confirmed by experiment. There can be no doubt that Nature knows about gauge principles.

Although we expect that readers are familiar with classical electromagnetism, we review the construction because there are similar patterns in Yang–Mills theory, gravity, and supergravity. The first step is the observation that the antisymmetric derivative of the gauge potential, called the field strength

$$F_{\mu\nu}(x) = \partial_\mu A_\nu(x) - \partial_\nu A_\mu(x)\,, \tag{4.6}$$

is invariant under the gauge transformation, a fact that is trivial to verify. In four dimensions $F_{\mu\nu}$ has six components, which split into the electric $E_i = F_{i0}$ and magnetic $B_i = \frac{1}{2}\varepsilon_{ijk}F_{jk}$ fields.

Since A_μ is a bosonic field, we expect it to satisfy a second order wave equation. The only Lorentz covariant and gauge invariant quantity available is $\partial^\mu F_{\mu\nu}$, so the free electromagnetic field satisfies

$$\partial^\mu F_{\mu\nu} = 0\,. \tag{4.7}$$

Since $\partial^\nu\partial^\mu F_{\mu\nu}$ vanishes *identically*,[3] (4.7) comprises $D-1$ independent components in D-dimensional Minkowski spacetime. This is not enough to determine the D components of A_μ, which is not surprising because of the gauge symmetry. One can change $A_\mu \to A_\mu + \partial_\mu\theta$ without affecting $F_{\mu\nu}$. So far, we did not yet use the field equations. Therefore, we will call this number, i.e. $(D-1)$ for the gauge vectors, the *number of off-shell degrees of freedom*.

One deals with this situation by 'fixing the gauge'. This means that one imposes one condition on the D components of A_μ, which eliminates the freedom to change gauge. Different gauge conditions illuminate different physical features of the theory. We will look first at the condition $\partial^i A_i(\vec{x},t) = 0$, which is called the Coulomb gauge condition. Although non-covariant it is a useful gauge to explore the structure of the initial value problem and determine the true degrees of freedom of the system. Note that the time–space split implicit in the initial value problem is also non-covariant.

First let's show that this condition does eliminate the gauge freedom. We check whether there are gauge functions $\theta(x)$ with the property that $\partial^i A'_i = \partial^i(A_i + \partial_i\theta) = 0$ when $\partial^i A_i = 0$. This requires that $\nabla^2\theta = 0$. As explained above, the only smooth solution is $\theta(x) \equiv 0$, so the gauge freedom has been completely fixed.

[3] This is the 'Noether identity', a relation between the field equations that is a consequence of the gauge symmetry.

Let's write out the time ($\mu \to 0$) and space ($\mu \to i$) components of the Maxwell equation (4.7). Using (4.6) and lowering all indices with the Minkowski metric, one finds

$$\begin{aligned} \nabla^2 A_0 - \partial_0(\partial^i A_i) &= 0\,, \\ \Box A_i - \partial_i \partial^0 A_0 - \partial_i(\partial^j A_j) &= 0\,. \end{aligned} \tag{4.8}$$

In the Coulomb gauge, the first equation simplifies to $\nabla^2 A_0 = 0$, and we see that[4] A_0 vanishes. The second equation in (4.8) then becomes $\Box A_i = 0$, so the spatial components A_i satisfy the massless scalar wave equation.

We can now count the classical degrees of freedom, which are the initial data $A_i(\vec{x}, 0)$ and $\dot{A}_i(\vec{x}, 0)$ required for a unique solution of $\Box A_i = 0$. There is a total of $2(D-2)$ independent degrees of freedom, because the initial data must be constrained to obey the Coulomb gauge condition.

This number thus agrees for $D = 4$ with the rule that the classical degrees of freedom are twice the number of on-shell degrees of freedom counted as helicity states. In general, we find for the gauge vectors $(D-1)$ off-shell degrees of freedom and $(D-2)$ on-shell degrees of freedom. These numbers are the dimension of the vector representation of $\mathrm{SO}(D-1)$ off-shell and $\mathrm{SO}(D-2)$ on-shell.

It is instructive to write the solution of $\Box A_i = 0$ as the Fourier transform

$$A_i(x) = \int \frac{\mathrm{d}^{(D-1)}k}{(2\pi)^{(D-1)} 2k^0} \sum_\lambda [\mathrm{e}^{\mathrm{i}k\cdot x} \epsilon_i(\vec{k}, \lambda) a(\vec{k}, \lambda) + \mathrm{e}^{-\mathrm{i}k\cdot x} \epsilon_i^*(\vec{k}, \lambda) a^*(\vec{k}, \lambda)], \tag{4.9}$$

where $\vec{k}$, $k^0 = |\vec{k}|$, is the on-shell energy–momentum vector. The $\epsilon_i(\vec{k}, \lambda)$ are called polarization vectors, which are constrained by the Coulomb gauge condition to satisfy $k^i \epsilon_i(\vec{k}, \lambda) = 0$. So there are $(D-2)$ independent polarization vectors, indexed by λ, and there are $2(D-2)$ independent real degrees of freedom contained in the complex quantities $a(\vec{k}, \lambda)$. As in the case of the plane wave expansions of the free Klein–Gordon and Dirac fields, $a(\vec{k}, \lambda)$ and $a^*(\vec{k}, \lambda)$ are interpreted as Fourier amplitudes in the classical theory and as annihilation and creation operators for particle states after quantization. There are $D-2$ particle states.

To understand these particle states better, we discuss the case $D = 4$ and assume that the spatial momentum is in the 3-direction, i.e. $\vec{k} = (0, 0, k)$ with $k > 0$. The two polarization vectors may be taken to be $\epsilon_i((0, 0, k), \pm) = (1/\sqrt{2})(1, \pm \mathrm{i}, 0)$. We formally add the 0-component $\epsilon_0 = 0$ to form 4-vectors $\epsilon_\mu((0, 0, k), \pm)$, which are eigenvectors of the rotation generator $J_3 = -m_{[12]}$, about the 3-axis (see text above (1.93)), with angular momentum $\lambda = \pm 1$. For general spatial momentum $\vec{k} = k(\sin\beta\cos\alpha, \sin\beta\sin\alpha, \cos\beta)$, we define polarization vectors $\epsilon_\mu(\vec{k}, \pm)$ by applying the spatial rotation with Euler angles α, β, which rotates the 3-axis to the direction of $\vec{k}$. The associated particle states are photons with helicity ± 1.

The same ideas determine the properties of particle states of the gauge field for $D \geq 5$. For spatial momentum in the direction $D-1$, i.e. $\vec{k} = (0, 0, \ldots, k)$, there are $D-2$ independent polarization vectors. We need not specify these in detail; the important point to

[4] When a source current is added to the Maxwell equation (4.7), A_0 no longer vanishes, but it is determined by the source. Thus it is not a degree of freedom of the gauge field system.

note is that these vectors are a basis of the vector representation of the Lie group SO($D-2$), which is the group that 'fixes' the vector $\vec{k}$. The associated particle states also transform in this representation. On the other hand it is clear from (4.9) that the Coulomb gauge vector potential transforms in the vector representation of SO($D-1$).

It should be noted that the equations of the *free* electromagnetic field can be formulated as conditions involving only the field strength components $F_{\mu\nu}$, with the gauge potential A_μ appearing as a derived quantity. In this form of the theory one has the pair of equations

$$\partial^\mu F_{\mu\nu} = 0\,, \tag{4.10}$$

$$\partial_\mu F_{\nu\rho} + \partial_\nu F_{\rho\mu} + \partial_\rho F_{\mu\nu} = 0\,. \tag{4.11}$$

The second equation is called the Bianchi identity. It is easy to see that it is automatically satisfied if $F_{\mu\nu}$ is expressed in terms of A_μ as in (4.6). In a topologically trivial spacetime such as Minkowski space, this is the general solution. This is a consequence of the Poincaré lemma in the theory of differential forms discussed in Ch. 7. Although the manifestly gauge invariant formalism (4.10) and (4.11) is available for the free gauge field, the vector potential is required *ab initio* to describe the minimal coupling to charged matter fields.

This chapter has progressed too far without exercises for readers, so we must now try to remedy this deficiency.

Exercise 4.1 *Derive from (4.3) that*

$$[D_\mu, D_\nu]\Psi \equiv (D_\mu D_\nu - D_\nu D_\mu)\Psi = -\mathrm{i}qF_{\mu\nu}\Psi\,. \tag{4.12}$$

Derive from (4.4) that the charged Dirac field also satisfies the second order equation

$$\left[D^\mu D_\mu - \tfrac{1}{2}\mathrm{i}q\gamma^{\mu\nu}F_{\mu\nu} - m^2\right]\Psi = 0\,. \tag{4.13}$$

Exercise 4.2 *Using only (4.10) and (4.11), show that the field strength tensor satisfies the equation* $\Box F_{\mu\nu} = 0$*. This is a gauge invariant derivation of the fact that the free electromagnetic field describes massless particles.*

4.1.3 Sources and Green's function

Let us now discuss sources for the electromagnetic field. Conventionally one takes an electric source that appears only in (4.10), which is modified to read

$$\partial^\mu F_{\mu\nu} = -J_\nu\,. \tag{4.14}$$

The Bianchi identity (4.11) is unchanged, so that $F_{\mu\nu} = \partial_\mu A_\nu - \partial_\nu A_\mu$. Lorentz covariance requires that the source is a vector, which is called the electric current vector. Since $\partial^\nu\partial^\mu F_{\mu\nu}$ vanishes *identically*, the current must be conserved. The theory is inconsistent unless the current satisfies $\partial^\nu J_\nu = 0$. This condition simply reflects the conventional idea that electric charge cannot be created or destroyed. It is also possible to include sources that carry magnetic charge and appear on the right-hand side of (4.11). However, this requires more sophisticated considerations, which we postpone to Sec. 4.2.3, so we will confine our attention to electric sources.

Exercise 4.3 *Repeat Ex. 4.2 when there is an electric source. Show that*

$$\Box F_{\nu\rho} = -(\partial_\nu J_\rho - \partial_\rho J_\nu)\,. \tag{4.15}$$

Consider first the analogous problem of the scalar field coupled to a source $J(x)$:

$$(\Box - m^2)\phi(x) \;=\; -J(x)\,. \tag{4.16}$$

The response to the source is determined by the Green's function $G(x-y)$, which satisfies the equation

$$(\Box - m^2)G(x-y) = -\delta(x-y)\,. \tag{4.17}$$

The translation symmetry of Minkowski spacetime implies that the Green's function depends only on the coordinate difference $x^\mu - y^\mu$ between observation point x^μ and source point y^μ. Lorentz symmetry implies that it depends only on the invariant quantities $(x-y)^2 = \eta^{\mu\nu}(x-y)^\mu(x-y)^\nu$ and $\mathrm{sgn}(x^0 - y^0)$. In Euclidean space $\mathcal{R}^D$, there is a unique solution of the equation analogous to (4.17), which is damped in the limit of large separation of observation and source points. In Lorentzian signature Minkowski space, there are several choices, which differ in their causal structure, that is in the dependence on $\mathrm{sgn}(x^0 - y^0)$. Many texts on quantum field theory, such as [22, 23, 9], discuss these choices.

The Euclidean Green's function is simplest and sufficient for the purposes of this book. The solution of (4.17) can be written as the Fourier transform

$$G(x-y) = \int \frac{\mathrm{d}^D k}{(2\pi)^D} \frac{\mathrm{e}^{ik\cdot(x-y)}}{k^2 + m^2}\,. \tag{4.18}$$

The integral can be expressed in terms of modified Bessel functions. In the massless case the result simplifies to the power law

$$G(x-y) = \frac{\Gamma(\frac{1}{2}(D-2))}{4\pi^{\frac{1}{2}D}(x-y)^{(D-2)}}\,. \tag{4.19}$$

Here $(x-y)^2 = \delta_{\mu\nu}(x-y)^\mu(x-y)^\nu$ is the Euclidean distance between source point y and observation point x. Given $G(x-y)$, the solution of (4.16) can be expressed as the integral

$$\phi(x) = \int \mathrm{d}^D y\, G(x-y)\, J(y). \tag{4.20}$$

One may note that the Green's function is formally the inverse of the wave operator, i.e. $G = -(\Box - m^2)^{-1}$. In Euclidean space $\Box = \nabla^2$ which is the D-dimensional Laplacian.

Let's continue in Euclidean space and find the Green's function for the gauge field. We might expect to solve (4.14) using a Green's function $G_{\mu\nu}(x-y)$, which is a tensor. However, we run into the immediate difficulty that there is no solution to the equation

$$(\delta^{\mu\rho}\Box - \partial^\mu\partial^\rho)G_{\rho\nu}(x,y) = -\delta^\mu_\nu\,\delta(x-y). \tag{4.21}$$

The Maxwell wave operator $\delta^{\mu\nu}\Box - \partial^\mu\partial^\nu$ is not invertible since any pure gradient $\partial_\nu f(x)$ is a zero mode. This problem is easily resolved. Since the source J_ν is conserved, we can replace (4.21) by the weaker condition

$$(\delta^{\mu\rho}\Box - \partial^\mu\partial^\rho)G_{\rho\nu}(x,y) = -\delta^\mu_\nu\delta(x-y) + \frac{\partial}{\partial y^\nu}\Omega^\mu(x,y)\,, \tag{4.22}$$

where $\Omega^\mu(x,y)$ is an arbitrary vector function. If $\Omega^\mu(x,y)$ and $J_\nu(y)$ are suitably damped at large distance, the effect of the second term in (4.22) cancels (after partial integration) in the formula

$$A_\mu(x) = \int \mathrm{d}^D y\, G_{\mu\nu}(x,y)J^\nu(y)\,, \tag{4.23}$$

which is the analogue of (4.20).

We now derive the precise form of $G_{\mu\nu}(x,y)$. By Euclidean symmetry, we can assume the tensor form

$$G_{\mu\nu}(x,y) = \delta_{\mu\nu}F(\sigma) + (x-y)_\mu(x-y)_\nu\hat{S}(\sigma)\,, \tag{4.24}$$

where $\sigma = \frac{1}{2}(x-y)^2$. It is more useful, but equivalent, to take advantage of gauge invariance and rewrite this ansatz as

$$G_{\mu\nu}(x,y) = \delta_{\mu\nu}F(\sigma) + \partial_\mu\partial_\nu S(\sigma)\,, \tag{4.25}$$

because the pure gauge term involving $S(\sigma)$ has no effect in (4.23) and cancels in (4.22). We may also assume that the gauge term in (4.22) has the Euclidean invariant form $\partial^\mu\partial_\nu\Omega(\sigma)$. Substituting (4.25) in (4.22) we find the two independent tensors δ^μ_ν and $(x-y)^\mu(x-y)_\nu$ and thus two independent differential equations involving F and Ω, namely

$$\begin{aligned} 2\sigma F''(\sigma) + (D-1)F'(\sigma) &= \Omega'(\sigma),\\ F''(\sigma) &= -\Omega''(\sigma)\,. \end{aligned} \tag{4.26}$$

Note that $F'(\sigma) = \mathrm{d}F(\sigma)/\mathrm{d}\sigma$, etc. We have dropped the δ-function term in (4.22), because we will first solve these equations for $\sigma \neq 0$. The second equation in (4.26) may be integrated immediately, giving $F'(\sigma) = -\Omega'(\sigma)$; a possible integration constant is chosen to vanish, so that $F'(\sigma)$ vanishes at large distance. The first equation then becomes $2\sigma F''(\sigma) + DF'(\sigma) = 0$, which has the power-law solution $F(\sigma) \sim \sigma^{1-\frac{1}{2}D}$. However, on any function of σ, the Laplacian acts as $\Box F(\sigma) = 2\sigma F''(\sigma) + DF'(\sigma)$. In our case there is a hidden δ-function in $\Box F(\sigma)$ because the power law is singular. The effect of the δ-function in (4.22) is automatically incorporated if we take $F(\sigma) = G(x-y)$ where G is the massless scalar Green's function in (4.19). The result of this analysis is the gauge field Green's function

$$G_{\mu\nu}(x,y) = \delta_{\mu\nu}G(x-y) + \partial_\mu\partial_\nu S(\sigma)\,. \tag{4.27}$$

The gauge function $S(\sigma)$ is arbitrary and may be taken to vanish. Then (4.27) becomes the usual Feynman gauge propagator.

The gauge field propagators found above are usually derived after gauge fixing in the path integral formalism in quantum field theory texts. The derivation here is purely classical, as appropriate since the response of the gauge field to a conserved current source is a purely classical phenomenon.

It may not be obvious why this method works. To see why, apply $\partial/\partial x^\mu$ to both sides of (4.22), obtaining

$$0 = -\partial_\nu \delta(x-y) - \partial_\nu \Box \Omega(\sigma)\,, \tag{4.28}$$

in which $\partial_\nu = \partial/\partial x^\nu$. This consistency condition is satisfied because the analysis above led us the result $\Omega(\sigma) = -F(\sigma) = -G((x-y)^2)$.

Exercise 4.4 *In $D = 4$ dimensions, consider a point charge at rest, i.e. $J^\mu(x) = \delta^\mu_0 q\, \delta(\vec{x})$. Obtain, using (4.23), that the resulting value of A^0, and therefore of the electric field, is*

$$A^0(x) = \frac{q}{4\pi}\frac{1}{|\vec{x}|}\,, \qquad \vec{E} = \frac{q}{4\pi}\frac{\vec{x}}{|x|^3}\,. \tag{4.29}$$

4.1.4 Quantum electrodynamics

The current vector J_ν in (4.14) may describe a piece of laboratory apparatus, such as a magnetic solenoid. However, we are more interested in the case where the source is the field of a charged elementary particle, such as the Dirac spinor Ψ. This is the theory of quantum electrodynamics, which contains equations that determine both the electromagnetic field A_μ, with $F_{\mu\nu} = \partial_\mu A_\nu - \partial_\nu A_\mu$, and Ψ. In dealing with coupled fields it is generally best to package the dynamics in a Lorentz invariant action. The equations of motion then emerge as the condition for a critical point of the action functional and are guaranteed to be mutually consistent.

It is also advantageous to change notation from that of Sec. 4.1.1 by scaling the vector potential, $A_\mu \to eA_\mu$, where e is the conventional coupling constant of the electromagnetic field to charged fields; $e^2/4\pi \approx 1/137$ is called the fine structure constant. In this notation the relevant equations of Sec. 4.1 read:

$$\begin{aligned} F_{\mu\nu} &\equiv \partial_\mu A_\nu - \partial_\nu A_\mu\,, \\ A_\mu \to A'_\mu &\equiv A_\mu + \frac{1}{e}\partial_\mu \theta\,, \\ D_\mu \Psi &\equiv (\partial_\mu - ieqA_\mu)\Psi\,, \\ [D_\mu, D_\nu]\Psi &= -ieqF_{\mu\nu}\Psi\,. \end{aligned} \tag{4.30}$$

The electric charges q of the various charged fields are then simple rational numbers, for example $q = 1$ for the electron.[5]

The action functional for the electromagnetic field interacting with a field of charge q, which we take to be a massive Dirac field, is the sum of two terms, each gauge invariant,

[5] It is an interesting question why the electric charges of elementary particles in Nature are quantized; that they appear to be integer multiples of a lowest fundamental charge. Two reasons have been found. The first is that quantum theory requires quantization of electric charge if a magnetic monopole exists. Second, electric charge can emerge as an unbroken U(1) generator of a larger non-abelian gauge theory with spontaneous gauge symmetry breaking. These reasons are not independent since monopoles solutions exist when gauge symmetry is broken with residual U(1) symmetry. See Sec. 17A.1 and [24] for discussion of these ideas.

$$S[A_\mu, \bar{\Psi}, \Psi] = \int \mathrm{d}^D x \left[-\tfrac{1}{4} F^{\mu\nu} F_{\mu\nu} - \bar{\Psi}(\gamma^\mu D_\mu - m)\Psi \right] . \tag{4.31}$$

The Euler variation of (4.31) with respect to the gauge potential A_ν is

$$\frac{\delta \mathcal{L}}{\delta A^\nu} = \partial^\mu F_{\mu\nu} + ieq\bar{\Psi}\gamma_\nu \Psi = 0 . \tag{4.32}$$

This is equivalent to (4.14) with the electric current a multiple of the Noether current of the global U(1) phase symmetry discussed in Sec. 2.7.1. It is a typical feature of the various fundamental gauge symmetries in physics that the Noether current of a system with global symmetry becomes the source for the gauge field introduced when the symmetry is gauged. The Euler variation with respect to $\bar{\Psi}$ gives the gauge covariant Dirac equation (4.4), with $D_\mu \Psi$ given in (4.30).

4.1.5 The stress tensor and gauge covariant translations

The situation of the stress tensor of this system is quite curious. The canonical stress tensor, calculated from the Noether formula (1.67) with $\Delta_A \phi^i \to \partial_\nu A_\rho,\ \partial_\nu \bar{\Psi},\ \partial_\nu \Psi$ for the three independent fields, is

$$T^\mu{}_\nu = F^{\mu\rho}\partial_\nu A_\rho + \bar{\Psi}\gamma^\mu \partial_\nu \Psi + \delta^\mu_\nu \mathcal{L} . \tag{4.33}$$

It is conserved on the index μ, but not on ν, not symmetric and not gauge invariant. The situation can be improved by treating fermion terms as in Sec. 2.7.2 and then adding $\Delta T^\mu{}_\nu = -\partial_\rho (F^{\mu\rho} A_\nu)$ in accord with the discussion in Sec. 1.3. The final result is the gauge invariant symmetric stress tensor

$$\Theta_{\mu\nu} = F_{\mu\rho} F_\nu{}^\rho + \tfrac{1}{4}\bar{\Psi}(\gamma_\mu \overleftrightarrow{D}_\nu + \gamma_\nu \overleftrightarrow{D}_\mu)\Psi + \eta_{\mu\nu}\mathcal{L} . \tag{4.34}$$

Exercise 4.5 *Consider the gauge covariant translation, defined by $\delta A_\mu = a^\nu F_{\nu\mu}$ and $\delta \Psi = a^\nu D_\nu \Psi$. Show that they differ from a conventional translation by a gauge transformation with gauge dependent parameter $\theta = -e\, a^\nu A_\nu$. Gauge covariant translations are a symmetry of the action (4.31). What is the Noether current for this symmetry? How is it related to the stress tensor (4.34)?*

4.2 Electromagnetic duality

The subject of electromagnetic duality has several interesting applications in supergravity theories. For example, the symmetry group of black hole solutions of matter-coupled supergravity theories generally contains duality transformations. We recommend that all readers study Secs. 4.2.1 and 4.2.2. However, because the applications of duality are somewhat advanced, the rest of the section can be omitted in the first reading of the book.

4.2.1 Dual tensors

We begin by discussing the duality property of second rank antisymmetric tensors $H_{\mu\nu}$ in four-dimensional Minkowski spacetime. We use the Levi-Civita tensor introduced in Sec. 3.1.3 to define the dual tensor

$$\tilde{H}^{\mu\nu} \equiv -\tfrac{1}{2}\mathrm{i}\varepsilon^{\mu\nu\rho\sigma} H_{\rho\sigma}\,. \tag{4.35}$$

In our conventions the dual tensor is imaginary. The indices of $\tilde{H}$ can be raised and lowered with the Minkowski[6] metric $\eta_{\mu\nu}$. It is also useful to define the linear combinations

$$H^{\pm}_{\mu\nu} = \tfrac{1}{2}(H_{\mu\nu} \pm \tilde{H}_{\mu\nu})\,, \qquad H^{\pm}_{\mu\nu} = (H^{\mp}_{\mu\nu})^*\,. \tag{4.36}$$

Exercise 4.6 *Prove that the dual of the dual is the identity, specifically that*

$$-\tfrac{1}{2}\mathrm{i}\varepsilon^{\mu\nu\rho\sigma}\tilde{H}_{\rho\sigma} = H^{\mu\nu}\,. \tag{4.37}$$

You will need (3.9). The validity of this property is the reason for the i *in the definition (4.35).*

Show that $H^{+}_{\mu\nu}$ and $H^{-}_{\mu\nu}$ are, respectively, self-dual and anti-self-dual, i.e.

$$-\tfrac{1}{2}\mathrm{i}\varepsilon_{\mu\nu}{}^{\rho\sigma}H^{\pm}_{\rho\sigma} = \pm H^{\pm}_{\mu\nu}\,. \tag{4.38}$$

Let $G_{\mu\nu}$ be another antisymmetric tensor with $G^{\pm}_{\mu\nu}$ defined as in (4.36). Prove the following relations (where $(\mu\nu)$ means symmetrization between the indices):

$$G^{+\mu\nu}H^{-}{}_{\mu\nu} = 0\,, \qquad G^{\pm\rho(\mu}H^{\pm\nu)}{}_{\rho} = -\tfrac{1}{4}\eta^{\mu\nu}G^{\pm\rho\sigma}H^{\pm}{}_{\rho\sigma}\,, \qquad G^{+}{}_{\rho[\mu}H^{-}{}_{\nu]}{}^{\rho} = 0\,. \tag{4.39}$$

Hint: you could first prove

$$\tilde{G}^{\rho\mu}\tilde{H}^{\nu}{}_{\rho} = -\tfrac{1}{2}\eta^{\mu\nu}G^{\rho\sigma}H_{\rho\sigma} - G^{\rho\nu}H^{\mu}{}_{\rho}\,. \tag{4.40}$$

Exercise 4.7 *The duality operation can also be applied to matrices of the Clifford algebra. Define the quantity $L_{\mu\nu} = \gamma_{\mu\nu}P_L$. Show that this is anti-self-dual. Hint: check first that $\gamma_{\mu\nu}\gamma_* = \tfrac{1}{2}\mathrm{i}\varepsilon_{\mu\nu\rho\sigma}\gamma^{\rho\sigma}$.*

4.2.2 Duality for one free electromagnetic field

Duality operates as an interesting symmetry of field theories containing one or more abelian gauge fields which may interact with other fields, principally scalars. In this section we discuss the simplest case, namely a single free gauge field. First note that, after contraction with the ε-tensor, the Bianchi identity (4.11) can be expressed as $\partial_\mu \tilde{F}^{\mu\nu} = 0$.

[6] The definition (4.35) is valid in Minkowski space, but must be modified in curved spacetimes as we will discuss in Ch. 7.

So we can temporarily ignore the vector potential and regard $F_{\mu\nu}$ as the basic field variable which must satisfy both the Maxwell and Bianchi equations:

$$\partial_\mu F^{\mu\nu} = 0\,, \qquad \partial_\mu \tilde{F}^{\mu\nu} = 0\,. \tag{4.41}$$

We can now consider the change of variables (the i is included to make the transformation real):

$$F^{\mu\nu} \to F'^{\mu\nu} = \mathrm{i}\tilde{F}^{\mu\nu}\,. \tag{4.42}$$

Since $F'^{\mu\nu}$ also obeys both equations of (4.41) we have defined a symmetry of the free electromagnetic field.

Exercise 4.8 *Show that the symmetry (4.42) exchanges the electric and magnetic fields: $E_i \to E'_i = -B_i$ and $B_i \to B'_i = E_i$.*

It is not possible to extend the symmetry to the vector potentials $F_{\mu\nu} = \partial_\mu A_\nu - \partial_\nu A_\mu$ and $F'_{\mu\nu} = \partial_\mu A'_\nu - \partial_\nu A'_\mu$ because A_μ and A'_μ are not related by any *local* transformation.

Here are some basic exercises involving the duality transform of the field strength tensor $F_{\mu\nu}$.

Exercise 4.9 *Show that the self-dual combinations $F^{\pm}_{\mu\nu}$ contain only photons of one polarization in their plane wave expansions:*

$$F^{\pm}_{\mu\nu} = 2\mathrm{i}\int \frac{\mathrm{d}^3k}{(2\pi)^3 2k^0}\left[\mathrm{e}^{\mathrm{i}k\cdot x}k_{[\mu}\epsilon_{\nu]}(\vec{k},\pm)a(\vec{k},\pm) - \mathrm{e}^{-\mathrm{i}k\cdot x}k_{[\mu}\epsilon^*_{\nu]}(\vec{k},\mp)a^*(\vec{k},\mp)\right]\,. \tag{4.43}$$

To perform this exercise, check first that with the polarization vectors given in Sec. 4.1.2, one has

$$-\tfrac{1}{2}\mathrm{i}\varepsilon^{\mu\nu\rho\sigma}k_\rho\epsilon_\sigma(\vec{k},\pm) = \pm k^{[\mu}\epsilon^{\nu]}(\vec{k},\pm)\,. \tag{4.44}$$

Exercise 4.10 *Show that the quantity $F_{\mu\nu}\tilde{F}^{\mu\nu}$ is a total derivative, i.e.*

$$F_{\mu\nu}\tilde{F}^{\mu\nu} = -\mathrm{i}\partial_\mu(\varepsilon^{\mu\nu\rho\sigma}A_\nu F_{\rho\sigma})\,. \tag{4.45}$$

Show, using (1.45), that under a Lorentz transformation

$$\left(F_{\mu\nu}\tilde{F}^{\mu\nu}\right)(x) \to \det\Lambda^{-1}\left(F_{\mu\nu}\tilde{F}^{\mu\nu}\right)(\Lambda x)\,. \tag{4.46}$$

Thus $F_{\mu\nu}\tilde{F}^{\mu\nu}$ transforms as a scalar under proper Lorentz transformations but changes sign under space or time reflections. Use the Schouten identity (3.11) to prove that

$$F_{\mu\rho}\tilde{F}_\nu{}^\rho = \tfrac{1}{4}\eta_{\mu\nu}F_{\rho\sigma}\tilde{F}^{\rho\sigma}\,. \tag{4.47}$$

4.2.3 Duality for gauge field and complex scalar

The simplest case of electromagnetic duality in an interacting field theory occurs with one abelian gauge field $A_\mu(x)$ and a complex scalar field $Z(x)$. The electromagnetic part of the Lagrangian is

$$\mathcal{L} = -\tfrac{1}{4}(\operatorname{Im} Z)F_{\mu\nu}F^{\mu\nu} - \tfrac{1}{8}(\operatorname{Re} Z)\varepsilon^{\mu\nu\rho\sigma}F_{\mu\nu}F_{\rho\sigma}\,. \tag{4.48}$$

Actions in which the gauge field kinetic term is multiplied by a function of complex scalar fields are quite common in supersymmetry and supergravity. We now define an extension of the duality transformation (4.42) which gives a non-abelian global $\mathrm{SL}(2,\mathbb{R})$ symmetry of the gauge field equations of this theory. In Sec. 7.12.2 we will discuss a generalized scalar kinetic term that is invariant under $\mathrm{SL}(2,\mathbb{R})$. The field $Z(x)$ carries dynamics, and the equations of motion of the combined vector and scalar theory are also invariant.

The gauge Bianchi identity and equation of motion of our theory are

$$\partial_\mu \tilde{F}^{\mu\nu} = 0\,, \qquad \partial_\mu\left[(\operatorname{Im} Z)\,F^{\mu\nu} + \mathrm{i}(\operatorname{Re} Z)\,\tilde{F}^{\mu\nu}\right] = 0\,. \tag{4.49}$$

It is convenient to define the real tensor

$$G^{\mu\nu} \equiv \varepsilon^{\mu\nu\rho\sigma}\frac{\delta S}{\delta F^{\rho\sigma}} = -\mathrm{i}(\operatorname{Im} Z)\tilde{F}^{\mu\nu} + (\operatorname{Re} Z)F^{\mu\nu}\,, \tag{4.50}$$

and to consider the self-dual combinations $F^{\mu\nu\pm}$ and $G^{\mu\nu\pm}$. Note that these are related by

$$G^{\mu\nu-} = ZF^{\mu\nu-}\,, \qquad G^{\mu\nu+} = \bar{Z}F^{\mu\nu+}\,. \tag{4.51}$$

The information in (4.49) can then be reexpressed as

$$\partial_\mu \operatorname{Im} F^{\mu\nu-} = 0\,, \qquad \partial_\mu \operatorname{Im} G^{\mu\nu-} = 0\,. \tag{4.52}$$

We define a matrix of the group $\mathrm{SL}(2,\mathbb{R})$ by

$$\mathcal{S} \equiv \begin{pmatrix} d & c \\ b & a \end{pmatrix}, \qquad ad - bc = 1\,. \tag{4.53}$$

The group $\mathrm{SL}(2,\mathbb{R})$ acts on the tensors F^- and G^- as follows:

$$\begin{pmatrix} F'^- \\ G'^- \end{pmatrix} = \mathcal{S}\begin{pmatrix} F^- \\ G^- \end{pmatrix}. \tag{4.54}$$

Since $\mathcal{S}$ is real, the conjugate tensors F^+ and G^+ also transform in the same way.

Exercise 4.11 *Assume that* $\operatorname{Im} F^-$ *and* $\operatorname{Im} G^-$ *satisfy (4.52), and show that* $\operatorname{Im} F'^-$ *and* $\operatorname{Im} G'^-$ *also obey the same equations. Show that* G'^- *and a transformed scalar* Z' *satisfy* $G'^{\mu\nu-} = Z'F'^{\mu\nu-}$, *if* Z' *is defined as the following nonlinear transform of* Z:

$$Z' = \frac{aZ + b}{cZ + d}\,. \tag{4.55}$$

The two equations (4.54) and (4.55) specify the $\mathrm{SL}(2,\mathbb{R})$ duality transformation on the field strength and complex scalar of our system. The exercise shows that the Bianchi identity and generalized Maxwell equations are duality invariant. In general the duality transform is *not* a symmetry of the Lagrangian or the action integral. The following exercise illustrates this.

Exercise 4.12 *Show that the Lagrangian (4.48) can be rewritten as*

$$\mathcal{L}(F, Z) = -\tfrac{1}{2}\operatorname{Im}(Z F^{-}_{\mu\nu} F^{\mu\nu -}) . \tag{4.56}$$

Consider the SL(2, $\mathbb{R}$) *transformation with parameters $a = d = 1$ and $b = 0$. Show that*

$$\mathcal{L}(F', Z') = -\tfrac{1}{2}\operatorname{Im}\left(Z(1 + c\, Z) F^{-}_{\mu\nu} F^{\mu\nu -}\right) \neq \mathcal{L}(F, Z) . \tag{4.57}$$

The symmetric gauge invariant stress tensor of this theory is

$$\Theta^{\mu\nu} = (\operatorname{Im} Z)\left(F^{\mu\rho} F^{\nu}{}_{\rho} - \frac{1}{4}\eta^{\mu\nu} F_{\rho\sigma} F^{\rho\sigma}\right) . \tag{4.58}$$

As we will see in Ch. 8, when the theory is coupled to gravity, it is this stress tensor that is the source of the gravitational field; see (8.4). It is then important that Im Z is positive, which restricts the domain of Z to the upper half-plane. It is also important that the stress tensor is *invariant* under the duality transformations (4.54) and (4.55). This is the reason for the duality symmetry of many black hole solutions of supergravity,

Exercise 4.13 *Prove that the energy–momentum tensor (4.58) is invariant under duality. Here are some helpful relations which you will need:*

$$\operatorname{Im} Z' = \frac{\operatorname{Im} Z}{(cZ + d)(c\bar{Z} + d)} . \tag{4.59}$$

Further you need again (4.47) and a similar identity (proven by contracting ε-tensors)

$$\tilde{F}_{\mu\rho} \tilde{F}_{\nu}{}^{\rho} = -F_{\mu\rho} F_{\nu}{}^{\rho} + \tfrac{1}{2}\eta_{\mu\nu} F_{\rho\sigma} F^{\rho\sigma} . \tag{4.60}$$

This leads to

$$F'_{\mu\rho} F'_{\nu}{}^{\rho} - \tfrac{1}{4}\eta_{\mu\nu} F'_{\rho\sigma} F'^{\rho\sigma} = |cZ + d|^2 \left[F_{\mu\rho} F_{\nu}{}^{\rho} - \tfrac{1}{4}\eta_{\mu\nu} F_{\rho\sigma} F^{\rho\sigma}\right] . \tag{4.61}$$

When the SL(2, $\mathbb{R}$) duality transformation appears in supergravity, there is also a scalar kinetic term in the Lagrangian which is invariant under the symmetry, specifically under the transformation (4.55). The prototype Lagrangian with this symmetry is the nonlinear σ-model whose target space is the Poincaré plane. This model and its SL(2, $\mathbb{R}$) symmetry group will be discussed in Sec. 7.12; see (7.151) and (7.152). The Poincaré plane is the upper half-plane Im $Z > 0$. The relation (4.59) shows that duality transformations map the upper half-plane into itself. The positive sign is preserved by SL(2, $\mathbb{R}$) transformations and the energy density obtained from the stress tensor Θ^{00} above will be *positive*!

Exercise 4.14 *The free Maxwell theory is the special case of (4.48) with* fixed $Z = \mathrm{i}$. *Suppose that the gauge field is coupled to a conserved current as in (4.14). Check that the electric charge can be expressed in terms of F or G by*

$$q \equiv \int \mathrm{d}^3\vec{x}\, J^0 = \int \mathrm{d}^3\vec{x}\, \partial_i F^{0i} = -\tfrac{1}{2}\int \mathrm{d}^3\vec{x}\, \varepsilon^{ijk}\partial_i G_{jk} . \tag{4.62}$$

A magnetic charge can be introduced in Maxwell theory as the divergence of $\vec{B}$ (recall $E^i = F^{0i}$ and $B^i = \frac{1}{2}\varepsilon^{ijk}F_{jk}$). This leads to a definition[7]

$$p \equiv -\tfrac{1}{2}\int \mathrm{d}^3\vec{x}\,\varepsilon^{ijk}\partial_i F_{jk}\,. \tag{4.63}$$

Show that $\begin{pmatrix} p \\ q \end{pmatrix}$ *is a vector that transforms under* SL(2, $\mathbb{R}$) *in the same way as the tensors* F^- *and* G^- *in (4.54).*

In many applications of electromagnetic duality, magnetic and electric charges appear as sources for the Bianchi 'identity' and generalized Maxwell equations of (4.49). As exemplified in Ex. 4.14 this leads to an SL(2, $\mathbb{R}$) vector of charges. Particles that carry both electric and magnetic charge are called dyons. In quantum mechanics, dyon charges must obey the Schwinger–Zwanziger quantization condition. If a theory contains two dyons with charges (p_1, q_1) and (p_2, q_2), these charges must satisfy $p_1q_2 - p_2q_1 = 2\pi n$, where n is an integer.[8] This condition is invariant under SL(2, $\mathbb{R}$) transformations of the charges. However, one can show [25] that there is a lowest non-zero value of the electric charge and that all allowed charges are restricted to an infinite discrete set of points called the charge lattice. The allowed SL(2) transformations must take one lattice point to another, and this restricts the group parameters in (4.53) to be integers. This restriction defines the subgroup SL(2, $\mathbb{Z}$), often called the modular group.[9] One can show that this subgroup is generated by the following choices of $\mathcal{S}$:

$$\begin{pmatrix} 1 & 0 \\ 1 & 1 \end{pmatrix}, \quad \begin{pmatrix} 0 & 1 \\ -1 & 0 \end{pmatrix},$$
$$Z' = Z + 1, \qquad Z' = -\frac{1}{Z}\,. \tag{4.64}$$

This means that one can express any element of SL(2, $\mathbb{Z}$) as the product of (finitely many) factors of the two generators above and their inverses.

Exercise 4.15 *In (4.48), the kinetic terms of the electromagnetic fields are determined by a variable Z that was treated as a scalar field. Z can also be replaced by a coupling constant, and typically one takes Z to be the imaginary number*[10] i/g^2, *where g is a coupling constant. Observe that the first transformation of (4.64) does not preserve the restriction that Z is imaginary. However, the second one does. Prove that this transformation is of the type (4.42), interchanging the electric and magnetic fields. It transforms g to its inverse, and thus relates the strong and weak coupling descriptions of the theory. In*

[7] In order to obtain a symplectic vector (p, q) and not $(-p, q)$, we changed the sign of the magnetic charge with respect to some classical works. This implies that we have $\vec{\nabla}\cdot\vec{B} = -j^0_{\mathrm{m}}$, where j^0_{m} is the magnetic charge density.

[8] For the case $(p_1, q_1) = (p, 0)$ and $(p_2, q_2) = (0, q)$, this reduces to condition $pq = 2\pi n$ found by Dirac in 1933.

[9] The modular group generated by the matrices (4.64) is in fact PSL(2, $\mathbb{Z}$). In PSL(2, $\mathbb{Z}$), the elements M and $-M$ of SL(2, $\mathbb{Z}$) are identified. Both these elements give in fact the same transformation $Z'(Z)$.

[10] One often adds an extra term that is a real so-called θ-parameter, but we will omit this here.

Secs. 4.1 and 4.2.2 we considered $Z = \mathrm{i}g = \mathrm{i}$. *Check that general duality transformations in this case are of the form*

$$F'^{-}_{\mu\nu} = (d + \mathrm{i}c)F^{-}_{\mu\nu}\,, \qquad \text{i.e.} \qquad F'_{\mu\nu} = dF_{\mu\nu} - \mathrm{i}c\tilde{F}_{\mu\nu}\,. \tag{4.65}$$

4.2.4 Electromagnetic duality for coupled Maxwell fields

In this section we explore how the duality symmetry is extended to systems containing a set of abelian gauge fields $A^A_\mu(x)$, indexed by $A = 1, 2, \ldots, m$ together with scalar fields ϕ^i. Scalars enter the theory through complex functions $f_{AB}(\phi) = f_{BA}(\phi)$. We consider the action

$$S = \int \mathrm{d}^4x\,\mathcal{L}\,, \qquad \mathcal{L} = -\tfrac{1}{4}(\mathrm{Re}\, f_{AB})F^A_{\mu\nu}F^{\mu\nu\,B} + \tfrac{1}{4}\mathrm{i}(\mathrm{Im}\, f_{AB})F^A_{\mu\nu}\tilde{F}^{\mu\nu\,B}\,, \tag{4.66}$$

which is real since $\tilde{F}^{\mu\nu}$ is pure imaginary, as defined in (4.35). The first term is a generalized kinetic Lagrangian for the gauge fields, so we usually require that $\mathrm{Re}\, f_{AB}$ is a positive definite matrix. This ensures that gauge field kinetic energies are positive. Although $F_{\mu\nu}\tilde{F}^{\mu\nu}$ is a total derivative, the second term does contribute to the equations of motion when $\mathrm{Im}\, f_{AB}$ is a function of the scalars ϕ^i. Our discussion will not involve the scalars directly. However, as in Sec. 4.2.3, additional terms to specify the scalar dynamics will appear when theories of this type are encountered in extended $D = 4$ supergravity. The treatment that follows is modeled on Sec. 4.2.3 (where f_{AB} was taken to be $-\mathrm{i}Z$).

Using the self-dual tensors of (4.36), we then rewrite the Lagrangian (4.66) as

$$\begin{aligned}
\mathcal{L}(F^+, F^-) &= -\tfrac{1}{2}\,\mathrm{Re}\left(f_{AB}F^{-A}_{\mu\nu}F^{\mu\nu-B}\right) \\
&= -\tfrac{1}{4}\left(f_{AB}F^{-A}_{\mu\nu}F^{\mu\nu-B} + f^*_{AB}F^{+A}_{\mu\nu}F^{\mu\nu+B}\right),
\end{aligned} \tag{4.67}$$

and define the new tensors

$$\begin{aligned}
G^{\mu\nu}_A &= \varepsilon^{\mu\nu\rho\sigma}\frac{\delta S}{\delta F^{\rho\sigma\,A}} = -(\mathrm{Im}\, f_{AB})F^{\mu\nu\,B} - \mathrm{i}(\mathrm{Re}\, f_{AB})\tilde{F}^{\mu\nu\,B} = G^{\mu\nu+}_A + G^{\mu\nu-}_A\,, \\
G^{\mu\nu-}_A &= -2\mathrm{i}\frac{\delta S(F^+, F^-)}{\delta F^{-A}_{\mu\nu}} = \mathrm{i}f_{AB}F^{\mu\nu-B}\,, \\
G^{\mu\nu+}_A &= 2\mathrm{i}\frac{\delta S(F^+, F^-)}{\delta F^{+A}_{\mu\nu}} = -\mathrm{i}f^*_{AB}F^{\mu\nu+B}\,.
\end{aligned} \tag{4.68}$$

Since the field equation for the action containing (4.67) is

$$0 = \frac{\delta S}{\delta A^A_\nu} = -2\partial_\mu\frac{\delta S}{\delta F^A_{\mu\nu}}\,, \tag{4.69}$$

the Bianchi identity and the equation of motion can be expressed in the concise form

$$\begin{aligned}
\partial^\mu\,\mathrm{Im}\, F^{A-}_{\mu\nu} &= 0 \quad \text{Bianchi identities,} \\
\partial_\mu\,\mathrm{Im}\, G^{\mu\nu-}_A &= 0 \quad \text{equations of motion.}
\end{aligned} \tag{4.70}$$

(The same equations hold for $\mathrm{Im}\, F^{A+}$ and $\mathrm{Im}\, G_A{}^+$.)

Duality transformations are linear transformations of the $2m$ tensors $F^{A\,\mu\nu}$ and $G_A^{\mu\nu}$ (accompanied by transformations of the f_{AB}) which mix Bianchi identities and equations of motion, but preserve the structure that led to (4.70). Since the equations (4.70) are real, we can mix them by a real $2m \times 2m$ matrix. We extend these transformations to the (anti-)self-dual tensors, and consider

$$\begin{pmatrix} F'^- \\ G'^- \end{pmatrix} = \mathcal{S} \begin{pmatrix} F^- \\ G^- \end{pmatrix} \equiv \begin{pmatrix} A & B \\ C & D \end{pmatrix} \begin{pmatrix} F^- \\ G^- \end{pmatrix}, \tag{4.71}$$

with real $m \times m$ submatrices A, B, C, D. Owing to the reality of these matrices, the same relations hold for the self-dual tensors F^+ and G^+. In Sec. 4.2.3, these matrices were just numbers:

$$A = d\,, \qquad B = c\,, \qquad C = b\,, \qquad D = a\,. \tag{4.72}$$

We require that the transformed field tensors F'^A and G'_A are also related by the definitions (4.68), with appropriately transformed f_{AB}. We work out this requirement in the following steps:

$$G'^- = (C + \mathrm{i}Df)F^- = (C + \mathrm{i}Df)(A + \mathrm{i}Bf)^{-1}F'^-\,, \tag{4.73}$$

such that we conclude that

$$\mathrm{i}f' = (C + \mathrm{i}Df)(A + \mathrm{i}Bf)^{-1}\,. \tag{4.74}$$

The last equation gives the symmetry transformation relating f'_{AB} to f_{AB}. If $G'^-_{\mu\nu}$ is to be the variational derivative of a transformed action, as (4.68) requires, then the matrix f' must be symmetric. For a generic[11] symmetric f, this requires that the matrices A, B, C, D satisfy

$$A^T C - C^T A = 0\,, \qquad B^T D - D^T B = 0\,, \qquad A^T D - C^T B = \mathbb{1}\,. \tag{4.75}$$

These relations among A, B, C, D are the defining conditions of a matrix of the symplectic group in dimension $2m$ so we reach the conclusion that

$$\mathcal{S} = \begin{pmatrix} A & B \\ C & D \end{pmatrix} \in \mathrm{Sp}(2m, \mathbb{R})\,. \tag{4.76}$$

The conditions (4.75) may be summarized as

$$\mathcal{S}^T \Omega \mathcal{S} = \Omega \quad \text{where} \quad \Omega = \begin{pmatrix} 0 & \mathbb{1} \\ -\mathbb{1} & 0 \end{pmatrix}. \tag{4.77}$$

[11] If the initial f_{AB} is non-generic, then the matrix $\mathbb{1}$ in the last equation can be replaced by any matrix which commutes with f_{AB}. For generic f_{AB}, this must be a constant multiple of the unit matrix. The constant, which should be positive to preserve the sign of the kinetic energy of the vectors, can be absorbed by rescaling the matrices A, B, C, D.

Duality transformations in $D = 4$ Box 4.2

The duality transformations in four dimensions are transformations in the symplectic group $\mathrm{Sp}(2m, \mathbb{R})$.

The matrix Ω is often called the symplectic metric, and the transformations (4.71) are then called symplectic transformations. This is the main result originally derived in [26]. Duality transformations in four spacetime dimensions are transformations of the group $\mathrm{Sp}(2m, \mathbb{R})$, which is a non-compact group.

Exercise 4.16 *The dimension of the group* $\mathrm{Sp}(2m, \mathbb{R})$ *is the number of elements of the matrix* S*, namely* $4m^2$ *minus the number of independent conditions contained in (4.77). Show that the dimension is* $m(2m+1)$.

Duality transformations have two types of applications: they can describe symmetries of one theory and they can describe transformations from one theory to another. In the first case, the symmetries concerned form a subgroup of the 'maximal' duality group $\mathrm{Sp}(2m, \mathbb{R})$ discussed above. The subgroup consists of transformations (4.74) of $f_{AB}(\phi^i)$ induced by the symmetry transformations of the elementary scalars ϕ^i. These scalar transformations must be symmetries of the scalar kinetic term and other parts of the Lagrangian. The model of Sec. 4.2.3 is one example. The transformation of Z defined in (4.55) is the standard $\mathrm{SL}(2, \mathbb{R})$ symmetry of the Poincaré plane. This could be part of the full symmetry group of all the scalar fields of the theory. In extended supergravities it turns out that all the symmetry transformations that act on the scalars appear also as transformations of the vector kinetic matrix. Hence, the symmetry group is then a subgroup of the 'maximal' group $\mathrm{Sp}(2m, \mathbb{R})$ discussed above.

However, another application is of the type that we encountered in Ex. 4.15. In that case constants that specify the theory under consideration change under the duality transformations. The constants that transform are sometimes called 'spurionic quantities'. The transformations thus relate two different theories. Solutions of one theory are mapped into solutions of the other one. This is the basic idea of dualities in M-theory.

Symplectic transformations always transform solutions of (4.70) into other solutions. However, they are not always invariances of the action. Indeed, writing

$$\mathcal{L} = -\tfrac{1}{2}\,\mathrm{Re}\left(f_{AB}F_{\mu\nu}^{-A}F^{\mu\nu-B}\right) = -\tfrac{1}{2}\,\mathrm{Im}\left(F_{\mu\nu}^{-A}G_A^{\mu\nu-}\right), \tag{4.78}$$

we obtain

$$\mathrm{Im}\,F'^{-}G'^{-} = \mathrm{Im}\left(F^{-}G^{-}\right) + \mathrm{Im}\left[2F^{-}(C^TB)G^{-} + F^{-}(C^TA)F^{-} + G^{-}(D^TB)G^{-}\right]. \tag{4.79}$$

If $C \neq 0$, $B = 0$ the Lagrangian is invariant up to a 4-divergence, since $\mathrm{Im}\,F^{-}F^{-} = -\tfrac{1}{4}\varepsilon^{\mu\nu\rho\sigma}F_{\mu\nu}F_{\rho\sigma}$ and the matrices A and C are real constants. For $B \neq 0$ neither the Lagrangian nor the action is invariant.

Electromagnetic duality has important applications to black hole solutions of extended supergravity theories. Supergravity is also very relevant to the analysis of black hole solutions of string theory. Many black holes are dyons; they carry both magnetic and electric charges for the gauge fields of the system. The general situation is a generalization of what was discussed at the end of Sec. 4.2.3. The charges form a symplectic vector $\begin{pmatrix} q_{\rm m}^{A} \\ q_{{\rm e}\,A} \end{pmatrix}$ which must transform as in (4.71). The Dirac–Schwinger–Zwanziger quantization condition restricts these charges to a lattice. Invariance of this lattice restricts the symplectic transformations of (4.71) to a discrete subgroup $\mathrm{Sp}(2m, \mathbb{Z})$, which is analogous to the $\mathrm{SL}(2, \mathbb{Z})$ group discussed previously.

Finally, we comment that symplectic transformations with $B \neq 0$ should be considered as non-perturbative for the following reasons. A system with no magnetic charges as in classical electromagnetism is transformed to a system with magnetic charges. The elements of f_{AB} may be regarded as coupling constants (see Ex. 4.15), and a system with weak coupling is transformed to one with strong coupling. A duality transformation which mixes electric and magnetic fields cannot be realized by transformation of the vector potential A_μ. One would need a 'magnetic' partner of A_μ to reexpress the $F'_{\mu\nu}$ and $G'_{\mu\nu}$ in terms of potentials.

The important properties of the matrix f_{AB} are that it is symmetric and that $\mathrm{Re}\, f_{AB}$ define a positive definite quadratic form in order to have positive gauge field energy. These properties are preserved under symplectic transformations defined by (4.74).

4.3 Non-abelian gauge symmetry

Yang–Mills theory is based on a non-abelian generalization of the U(1) gauge symmetry. It is the fundamental idea underlying the standard model of elementary particle interactions. We follow the pattern of Sec. 4.1.1, starting with the global symmetry and then gauging it. The focus of our discussion is the derivation of the basic formulas of the classical gauge theory. Readers may need more information on the underlying geometric ideas and the structure and stunning applications of the quantized theory. They are referred to a modern text in quantum field theory.[12]

4.3.1 Global internal symmetry

Suppose that G is a compact simple Lie group of dimension dim_G. Closely associated with the group is its Lie algebra, denoted by $\mathfrak{g}$, which is a real algebra of dimension dim_G. The theory of Lie algebras and Lie groups is an important subject of mathematics with many applications to physics. With some oversimplification we review only the most essential features required by Yang–Mills theory for compact simple groups.

Each compact simple Lie algebra has an infinite number of inequivalent finite-dimensional irreducible representations R of dimension dim_R. In each representation, there is a basis of matrix generators t_A, $A = 1, \ldots, \mathrm{dim}_G$, which are anti-hermitian in the case of a compact gauge group. The commutator of the generators determines the local geometrical structure of the group:

[12] See, for example, Ch. 15 of [9]. This text also reviews aspects of group theory needed in physical applications.

$$[t_A, t_B] = f_{AB}{}^C t_C\,. \tag{4.80}$$

The array of real numbers $f_{AB}{}^C$ are structure constants of the algebra (the same in all representations). They obey the Jacobi identity

$$f_{AD}{}^E f_{BC}{}^D + f_{BD}{}^E f_{CA}{}^D + f_{CD}{}^E f_{AB}{}^D = 0\,. \tag{4.81}$$

The indices can be lowered by the Cartan–Killing metric defined in Appendix B (see (B.6)), and then the f_{ABC} are totally antisymmetric. For simple algebras, the generators can be chosen to be trace orthogonal, $\mathrm{Tr}(t_A t_B) = -c\delta_{AB}$, with c positive for compact groups, and the Cartan–Killing metric is then proportional to this expression.

One important representation is the adjoint representation of dimension $\dim_R = \dim_G$, in which the representation matrices are closely related to the structure constants by $(t_A)^D{}_E = f_{AE}{}^D$. Note that the labels DE denote row and column indices of the matrix t_A. The adjoint representation is a real representation; the representation matrices are real and antisymmetric for compact algebras. For complex representations we will use the notation $(t_A)^\alpha{}_\beta$. Anti-hermiticity then requires $(t_A^*)^\alpha{}_\beta = -(t_A)^\beta{}_\alpha$. The row and column indices will often be suppressed when no ambiguity arises.

Exercise 4.17 *Use (4.81) to show that the matrices $(t_A)^D{}_E = f_{AE}{}^D$ satisfy (4.80) and therefore give a representation.*

The general element of $\mathfrak{g}$ is represented by a superposition of generators $\theta^A t_A$ where the θ^A are $\dim_G$ real parameters. The relation between G and $\mathfrak{g}$ is given by exponentiation, namely $\mathrm{e}^{-\theta^A t_A}$ is an element of G in the representation R.

A theory with global non-abelian internal symmetry contains scalar and spinor fields, each of which transforms in an irreducible representation R. For example, there may be a Dirac spinor[13] field $\Psi^\alpha(x)$, $\alpha = 1, \ldots, \dim_R$, that transforms in the complex representation R as

$$\Psi^\alpha(x) \to (\mathrm{e}^{-\theta^A t_A})^\alpha{}_\beta \Psi^\beta(x)\,. \tag{4.82}$$

The conjugate spinor[14] is denoted by $\bar\Psi_\alpha$ and transforms as

$$\bar\Psi_\alpha \to \bar\Psi_\beta (\mathrm{e}^{\theta^A t_A})^\beta{}_\alpha\,. \tag{4.83}$$

For most of our discussion it is sufficient to restrict attention to the infinitesimal transformations,

$$\begin{aligned}\delta\Psi &= -\theta^A t_A \Psi\,,\\ \delta\bar\Psi &= \bar\Psi\theta^A t_A\,,\\ \delta\phi^A &= \theta^C f_{BC}{}^A \phi^B\,.\end{aligned} \tag{4.84}$$

[13] Note that we use here indices $\alpha, \ldots$ for the representation of the gauge group. They should not be confused with spinor indices, which we usually omit.

[14] The Dirac conjugate (2.30) is used here rather than the Majorana conjugate (3.50).

The first two relations are just the terms of (4.82) and (4.83) that are first order in θ^A. The last relation is the infinitesimal transformation of a field in the adjoint representation, taken here as the set of $\dim_G$ real scalars ϕ^A. Of course, scalars could be assigned to any representation R.

Actions, such as the kinetic action for massive fermion fields,

$$S[\bar{\Psi}, \Psi] = -\int \mathrm{d}^D x \bar{\Psi}[\gamma^\mu \partial_\mu - m]\Psi \,, \tag{4.85}$$

are required to be invariant under (4.82).

Exercise 4.18 *Show that (4.85) is invariant under the transformation (4.82) and (4.83). Consider an infinitesimal transformation and derive the conserved current*

$$J_{A\mu} = -\bar{\Psi} t_A \gamma_\mu \Psi \,, \qquad A = 1, \ldots, \dim_G \,. \tag{4.86}$$

Show that the current transforms as a field in the adjoint representation, i.e.

$$\delta J_{A\mu} = \theta^C f_{CA}{}^B J_{B\mu} \,. \tag{4.87}$$

Show that $\delta(\phi^A J_{A\mu}) = 0$.

4.3.2 Gauging the symmetry

In gauged non-abelian internal symmetry, the group parameter $\theta^A(x)$ is promoted to an arbitrary function of x^μ. The first step in the systematic formulation of gauge invariant field equations is to introduce the gauge potentials, namely a set of vectors $A_\mu^A(x)$ whose infinitesimal transformation rule is

$$\delta A_\mu^A(x) = \frac{1}{g}\partial_\mu \theta^A + \theta^C(x) A_\mu^B(x) f_{BC}{}^A \,. \tag{4.88}$$

The first term is the gradient term similar to that for the abelian gauge field in (4.2), and the second is exactly the transformation of a field in the adjoint representation, as one can see from the third equation in (4.84). The constant g is the Yang–Mills coupling, which replaces the electromagnetic coupling e of Sec. 4.1.4.

Following the pattern of Sec. 4.1.1, we next define the covariant derivative of a field in the representation R with matrix generators t_A. For the fields Ψ^α, $\bar{\Psi}_\alpha$, and ϕ^A of (4.84) we write

$$\begin{aligned}
D_\mu \Psi &= (\partial_\mu + g t_A A_\mu^A)\Psi \,, \\
D_\mu \bar{\Psi} &= \partial_\mu \bar{\Psi} - g \bar{\Psi} t_A A_\mu^A \,, \\
D_\mu \phi^A &= \partial_\mu \phi^A + g f_{BC}{}^A A_\mu^B \phi^C \,.
\end{aligned} \tag{4.89}$$

Note that the gauge transformation (4.88) can be written as $\delta A_\mu^A(x) = (1/g) D_\mu \theta^A$ using the covariant derivative for the adjoint representation.

Exercise 4.19 *Show that the covariant derivatives of the three fields in (4.89) transform in the same way as the fields themselves, and with no derivatives of the gauge parameters. For example* $\delta D_\mu \Psi = -\theta^A t_A D_\mu \Psi$.

Given this result it is easy to see that any globally symmetric action for scalar and spinor matter fields becomes gauge invariant if one replaces $\partial_\mu \to D_\mu$ for all fields. If this is done in (4.85), one obtains the equation of motion

$$\frac{\delta S}{\delta \bar{\Psi}_\alpha} = -[\gamma^\mu D_\mu - m]\Psi^\alpha = 0\,. \tag{4.90}$$

4.3.3 Yang–Mills field strength and action

The next step in the development is to define the quantities that determine the dynamics of the gauge field itself. The simplest way to proceed is to compute the commutator of two covariant derivatives acting on a field in the representation R. We would get the same information, no matter which representation, so we will study just the case $[D_\mu, D_\nu]\Psi \equiv (D_\mu D_\nu - D_\nu D_\mu)\Psi$. A careful computation gives

$$[D_\mu, D_\nu]\Psi = g F_{\mu\nu}^A t_A \Psi\,, \tag{4.91}$$

where

$$F_{\mu\nu}^A = \partial_\mu A_\nu^A - \partial_\nu A_\mu^A + g f_{BC}{}^A A_\mu^B A_\nu^C\,. \tag{4.92}$$

The properties of the covariant derivative guarantee that the right-hand side of (4.91) transforms as a field in the same representation as Ψ. Thus $F_{\mu\nu}^A$ should have simple transformation properties. Indeed, one can derive

$$\delta F_{\mu\nu}^A = \theta^C F_{\mu\nu}^B f_{BC}{}^A\,. \tag{4.93}$$

We see that $F_{\mu\nu}^A$ is an antisymmetric tensor in spacetime, which transforms as a field in the adjoint representation of $\mathfrak{g}$; $F_{\mu\nu}^A$ is the non-abelian generalization of the electromagnetic field strength (4.6). The principal differences between abelian and non-abelian gauge symmetry are that the non-abelian field strength is not gauge invariant, but transforms in the adjoint representation, and that it is *nonlinear* in the gauge potential A_μ^A.

Exercise 4.20 *Derive (4.93).*

Despite these significant differences, it is quite straightforward to formulate the Yang–Mills equations by following the ideas of the electromagnetic case. Since both the current and field strength transform in the adjoint representation, and the covariant derivative does not change the transformation properties, the equation

$$D^\mu F_{\mu\nu}^A = -J_\nu^A \tag{4.94}$$

is both gauge and Lorentz covariant. It is the basic dynamical equation of classical Yang–Mills theory and the analogue of (4.14) for electromagnetism. One important difference, however, is that in the absence of matter sources, when the right-hand side of (4.94) vanishes, that equation is still a (much studied!) nonlinear equation for A^A_μ.

There is also a non-abelian analogue of the Bianchi identity (4.11), which takes the form

$$D_\mu F^A_{\nu\rho} + D_\nu F^A_{\rho\mu} + D_\rho F^A_{\mu\nu} = 0\,, \tag{4.95}$$

where $D_\mu F^A_{\nu\rho} = \partial_\mu F^A_{\nu\rho} + g f_{BC}{}^A A^B_\mu F^C_{\nu\rho}$.

Exercise 4.21 *Show that (4.95) is satisfied identically if $F^A_{\nu\rho}$ is written in the form (4.92).*

Exercise 4.22 *Show that $D^\nu D^\mu F^A_{\mu\nu}$ vanishes* identically *(despite the nonlinearity). This is again a Noether identity: a relation between field equations that follows from the gauge symmetry.*

As in the electromagnetic case, this means that the equation of motion (4.94) is consistent only if the current is covariantly conserved, i.e. only if $D^\nu J^A_\nu = 0$. It also means that (4.94) contains $(D-1)\dim G$ independent equations, which is enough to determine the $D\dim G$ components of A^A_μ up to a gauge transformation. It is usually convenient to 'fix the gauge' by specifying $\dim G$ conditions on the components of A^A_μ.

Note that, in the limit $g \to 0$, equations (4.92), (4.94), and (4.95) reduce to linear equations, which are $\dim G$ copies of the corresponding equations for the free electromagnetic field. The count of degrees of freedom of Sec. 4.1.2 can be repeated in the Coulomb gauge $\partial^i A^A_i(\vec{x},t) = 0$. For each component $A = 1, \ldots, \dim G$, $2(D-2)$ functions must be specified as initial data, and each $A^A_i(x)$ has a Fourier transform identical to (4.9). In this free limit, the gauge field thus describes a particle with $D-2$ polarization states transforming in the adjoint representation of $\mathfrak{g}$.

The equations of motion of the Yang–Mills field A^A coupled to the Dirac field Ψ^α can be obtained from an action functional that is a natural generalization of (4.31):

$$S[A^A_\mu, \bar{\Psi}_\alpha, \Psi^\alpha] = \int \mathrm{d}^D x \left[-\tfrac{1}{4} F^{A\mu\nu} F^A_{\mu\nu} - \bar{\Psi}_\alpha (\gamma^\mu D_\mu - m)\Psi^\alpha \right]. \tag{4.96}$$

The action is gauge invariant. The Euler variation with respect to A^A_ν gives (4.94) with current source (4.86), and the variation with respect to $\bar{\Psi}_\alpha$ gives (4.90).

4.3.4 Yang–Mills theory for $G = \mathrm{SU}(N)$

The most commonly studied gauge group for Yang–Mills theory is SU(N). The generators of the fundamental representation of its Lie algebra are a set of $N^2 - 1$ traceless anti-hermitian $N \times N$ matrices t_A, which are normalized by the bilinear trace relation

$$\mathrm{Tr}(t_A t_B) = -\tfrac{1}{2}\delta_{AB}\,. \tag{4.97}$$

In this section we discuss the special notation that has been developed for this case and is frequently used in the literature. In this notation gauge transformations are *explicitly* realized at the level of the group SU(N) rather than just at the level of its Lie algebra $\mathfrak{su}(N)$ as in the previous sections.

We will use the notation $U(x) = \mathrm{e}^{-\Theta(x)}$, with $\Theta(x) = \theta^A(x)t_A$, to denote an element of the gauge group in the fundamental representation. This may be viewed as a map $x^\mu \to U(x^\mu)$ from Minkowski spacetime into the group SU(N). In this notation the gauge transformation of a spinor field Ψ in the fundamental representation can be written (see (4.82))

$$\Psi(x) \to U(x)\Psi(x). \tag{4.98}$$

Row and column indices of the fundamental representation are consistently omitted in this notation. Usually we will omit the spacetime argument x^μ also, unless useful for special emphasis.

Given any matrix generator t_A, the unitary transformation $U(x)t_AU(x)^{-1}$ gives another traceless anti-hermitian matrix, which must then be a linear combination of the t_B. Therefore we can write

$$U(x)t_AU(x)^{-1} = t_BR(x)^B{}_A\,, \tag{4.99}$$

where $R(x)^B{}_A$ is a real $(N^2-1)\times(N^2-1)$ matrix.

Exercise 4.23 *Consider the product of two gauge group elements U_1 and U_2, which gives a third via $U_1U_2 = U_3$. For each element U_i, there is an associated matrix $(R_i)^B{}_A$, defined by $U_it_AU_i^{-1} = t_B(R_i)^B{}_A$. Prove that $(R_3)^B{}_A = (R_1)^B{}_C(R_2)^C{}_A$, which shows that the matrices $R^B{}_A$ defined by (4.99) are the matrices of an (N^2-1)-dimensional representation of* SU(N). *Use (4.99) to show that, to first order in the gauge parameters θ^C, $R^B{}_A = \delta^B_A + \theta^C f_{AC}{}^B + \ldots$. This shows that the matrices $R^B{}_A$ are exactly those of the adjoint representation.*[15]

Given any set of N^2-1 real quantities X^A, that is any element of the vector space $\mathbb{R}^{N^2-1}$, we can form the matrix $\mathbf{X} = t_AX^A$. For any group element U, we have $U\mathbf{X}U^{-1} = t_BR^B{}_AX^A$. Thus the unitary transformation of the matrix $\mathbf{X}$ contains the information that the quantities $X^A = -2\delta^{AB}\,\mathrm{Tr}(t_B\mathbf{X})$ transform in the adjoint representation, that is as $X^A \to R^A{}_BX^B$. Thus, given any field in the adjoint representation, such as $\phi^A(x)$, we can form the matrix $\Phi(x) = t_A\phi^A(x)$. Gauge transformations can then be implemented as

$$\Phi(x) \to U(x)\Phi(x)U(x)^{-1}. \tag{4.100}$$

One can also form the matrix $\mathbf{A}_\mu(x) = t_AA^A_\mu(x)$ for the gauge potential. Quite remarkably, the gauge transformation of the potential can be expressed in matrix form if we define the transformation by

$$\mathbf{A}_\mu(x) \to \mathbf{A}'_\mu(x) \equiv \frac{1}{g}U(x)\partial_\mu U(x)^{-1} + U(x)\mathbf{A}_\mu(x)U(x)^{-1}. \tag{4.101}$$

[15] The equation (4.99) is true for the generators t_A of *any* representation of *any* Lie algebra $\mathfrak{g}$ and the associated group element $U = \mathrm{e}^{-\theta^Ct_C}$. It follows that the matrices $R^B{}_A$ are those of the adjoint representation of G. A matrix description of Yang–Mills theory for a general gauge group can then be constructed by following the procedure discussed below for the fundamental representation of SU(N).

For infinitesimal transformations this becomes

$$\delta A_\mu(x) = \frac{1}{g}\partial_\mu \Theta(x) + [A_\mu(x), \Theta(x)], \tag{4.102}$$

which agrees with (4.88).

Exercise 4.24 *Suppose that* $\mathbf{A}_\mu \to \mathbf{A}'_\mu$ *by the gauge transformation* $U_2(x)$ *followed by* $\mathbf{A}'_\mu \to \mathbf{A}''_\mu$ *by the gauge transformation* $U_1(x)$. *Show that the combined transformation* $\mathbf{A}_\mu \to \mathbf{A}''_\mu$ *is correctly described by the definition (4.101) for the product matrix* $U_2(x)U_1(x)$. *This result is compatible with (1.23) and with the implementation of gauge transformations by unitary operators in the quantum theory.*

It is easy to define covariant derivatives in which the gauge potential appears in matrix form. For fields Ψ in the fundamental and $\bar\Psi$ in the anti-fundamental representation (and transforming as $\bar\Psi \to \bar\Psi U^{-1}$), the previous definitions in (4.89) can simply be rewritten as

$$\begin{aligned} D_\mu \Psi &\equiv (\partial_\mu + g\mathbf{A}_\mu)\Psi\,, \\ D_\mu \bar\Psi &\equiv \partial_\mu \bar\Psi - g\bar\Psi \mathbf{A}_\mu\,. \end{aligned} \tag{4.103}$$

For a field in the adjoint representation, such as Φ, we define

$$D_\mu \Phi = \partial_\mu \Phi + g[\mathbf{A}_\mu, \Phi]\,, \tag{4.104}$$

which involves the matrix commutator.

Exercise 4.25 *Demonstrate that these covariant derivatives transform correctly, specifically that*

$$D_\mu \Psi \to U(x) D_\mu \Psi\,, \quad D_\mu \bar\Psi \to D_\mu \bar\Psi U(x)^{-1}\,, \quad D_\mu \Phi \to U(x) D_\mu \Phi U(x)^{-1}\,. \tag{4.105}$$

The non-abelian field strength can also be converted to matrix form as

$$\mathbf{F}_{\mu\nu} = t_A F^A_{\mu\nu} = \partial_\mu \mathbf{A}_\nu - \partial_\nu \mathbf{A}_\mu + g[\mathbf{A}_\mu, \mathbf{A}_\nu]\,. \tag{4.106}$$

Exercise 4.26 *Prove this.*

The matrix formalism is a convenient way to express quantities of interest in the theory. For example the Yang–Mills action (4.96) can be written as

$$S[\mathbf{A}_\mu, \bar\Psi, \Psi] = \int \mathrm{d}^D x \left[\frac{1}{2}\mathrm{Tr}(\mathbf{F}^{\mu\nu}\mathbf{F}_{\mu\nu}) - \bar\Psi(\gamma^\mu \mathrm{D}_\mu - \mathrm{m})\Psi \right]. \tag{4.107}$$

The $N^2 - 1$ matrix generators $(t_A)^\alpha{}_\beta$ of the fundamental representation, normalized as in (4.97), together with the matrix $\mathrm{i}\delta^\alpha_\beta$ form a complete set of $N \times N$ anti-hermitian matrices, which are orthogonal in the trace norm. Therefore one can expand any $N \times N$ anti-hermitian matrix $H^\alpha{}_\beta$ in this set as

$$\begin{aligned} H^\alpha{}_\beta &= \mathrm{i}h_0 \delta^\alpha{}_\beta + h^A (t_A)^\alpha{}_\beta\,, \\ h_0 = -\frac{\mathrm{i}}{N}\,\mathrm{Tr}\, H, &\qquad h^A = -2\delta^{AB}\,\mathrm{Tr}(H t_B)\,. \end{aligned} \tag{4.108}$$

Note that there is a sum over the $N^2 - 1$ values of the repeated indices A, B in (4.108) and in the exercise below.

Exercise 4.27 *Use the completeness property (perhaps with Sec. 3.2.3 as a guide) to derive the rearrangement relation*

$$\delta^\alpha_\beta \, \delta^\gamma_\delta = \frac{1}{N} \delta^\alpha_\delta \, \delta^\gamma_\beta - 2(t_A)^\alpha{}_\delta \delta^{AB} (t_B)^\gamma{}_\beta \, . \tag{4.109}$$

4.4 Internal symmetry for Majorana spinors

Majorana spinors play a central role in supersymmetric field theories. In many applications they transform in a representation of a non-abelian internal symmetry group. For example, the spinor fields of super-Yang–Mills theory are denoted as λ^A and transform in the adjoint representation of the gauge group. In the notation of Sec. 4.3.4, we have $\lambda^A \to \lambda'^A = R^A{}_B \lambda^B$. Since the matrix $R^A{}_B$ is real, this transformation rule is consistent with the fact that Majorana spinors obey a reality constraint. Indeed, as shown in Sec. 3A.5, there are really real representations of the Clifford algebra in which the spinors are explicitly real. One can consider the more general situation of a set of Majorana spinors Ψ^α transforming as $\Psi^\alpha \to \Psi'^\alpha = (\mathrm{e}^{-\theta^A t_A})^\alpha{}_\beta \Psi^\beta$. The transformed Ψ'^α must also satisfy the Majorana condition, and this requires that the matrices $\mathrm{e}^{-\theta^A t_A}$ are those of a really real representation of the group G. (Obviously there is a similar requirement on the symmetry transformation of a set of real scalars, such as the ϕ^A of Sec. 4.3.1.)

In $D = 4$ dimensions, the requirement that Majorana spinors transform in a real representation of the gauge group can be bypassed because internal symmetries can include chiral transformations, which involve the highest rank element $\gamma_* = \mathrm{i}\gamma_0\gamma_1\gamma_2\gamma_3$ of the Clifford algebra discussed in Sec. 3.1.6. This matrix is imaginary in a Majorana representation, or in general under the C-operation; see (3.78). We use the chiral projectors P_L and P_R as in (3.38). Suppose that the matrices t_A are generators of a complex representation of the Lie algebra. Then the complex conjugate matrices t^*_A are generators of the conjugate representation. Let χ^α denote a set of Majorana spinors to which we assign the group transformation rule

$$\chi^\alpha \to \chi'^\alpha \equiv (\mathrm{e}^{-\theta^A (t_A P_L + t^*_A P_R)})^\alpha{}_\beta \chi^\beta \, . \tag{4.110}$$

The matrices $t_A P_L + t^*_A P_R$ are generators of a representation of an explicitly real representation of the Lie algebra, so the transformed spinors χ'^α also satisfy the Majorana condition. This is the transformation rule used for Majorana spinors in supersymmetric gauge theories in Ch. 6.

By applying the projectors to (4.110), one can see that the chiral and anti-chiral projections of χ transform as

$$\begin{aligned} P_L\chi &\to P_L\chi' \equiv (\mathrm{e}^{-\theta^A t_A}) P_L \chi \, , \\ P_R\chi &\to P_R\chi' \equiv (\mathrm{e}^{-\theta^A t^*_A}) P_R \chi \, . \end{aligned} \tag{4.111}$$

Exercise 4.28 *What is the covariant derivative $D_\mu \chi$? We now use $\bar{\chi}$ for a Majorana conjugate (3.50), where the transpose includes a transpose in the representation space. When representation indices are needed, $\bar{\chi}$ carries a lower index. Show that then the kinetic Lagrangian density $\bar{\chi}\gamma^\mu D_\mu \chi$ is invariant under the infinitesimal limit of the transformation (4.110) for anti-hermitian t_A, and that the variation of the mass term is*

$$\delta(\bar{\chi}\chi) = -\theta^A \bar{\chi}(t_A + t_A^T)\gamma_* \chi \,. \tag{4.112}$$

The mass term is invariant only for the subset of generators that are antisymmetric, and thus real. This condition defines a subalgebra of the original Lie algebra $\mathfrak{g}$ of the theory, specifically the subalgebra that contains only parity conserving vector-like gauge transformations. For the case $\mathfrak{g} = \mathfrak{su}(N)$, the subalgebra is isomorphic to $\mathfrak{so}(N)$. Non-invariance of the Majorana mass term is a special case of the general idea that chiral symmetry requires massless fermions.

Exercise 4.29 *Show that*

$$\tfrac{1}{2}\int \mathrm{d}^4x\, \bar{\chi}\gamma^\mu D_\mu \chi = \int \mathrm{d}^4x\, \bar{\chi}\gamma^\mu P_L D_\mu \chi = \int \mathrm{d}^4x\, \bar{\chi}\gamma^\mu P_R D_\mu \chi \,. \tag{4.113}$$

Note that $P_{L,R} D_\mu \chi = D_\mu P_{L,R}\chi$.

The free Rarita–Schwinger field 5

In this chapter we begin to assemble the ingredients of supergravity by studying the free spin-3/2 field. Supergravity is the gauge theory of global supersymmetry, which we will usually abbreviate as SUSY. The key feature is that the symmetry parameter of global SUSY transformations is a constant spinor ϵ_α. In supergravity it becomes a general function in spacetime, $\epsilon_\alpha(x)$. The associated gauge field is a vector–spinor $\Psi_{\mu\alpha}(x)$. This field and the corresponding particle have acquired the name 'gravitino'.

Supergravity theories necessarily contain the gauge multiplet, the set of fields required to gauge the symmetry in a consistent interacting theory, and may contain matter multiplets, sets of fields on which global SUSY is realized. The gauge multiplet contains the gravitational field, one or more vector–spinors, and sometimes other fields. This structure is derived from representations of the SUSY algebras in Sec. 6.4.2. In this chapter we are concerned with the free limit, in which the various fields do not interact, and we can consider them separately. In particular we consider $\Psi_\mu(x)$ (omitting the spinor index α) as a free field, subject to the gauge transformation

$$\Psi_\mu(x) \to \Psi_\mu(x) + \partial_\mu \epsilon(x) . \tag{5.1}$$

Furthermore we will assume that Ψ_μ and ϵ are complex spinors with $2^{[D/2]}$ spinor components for spacetime dimension D. This is fine for the *free* theory in any dimension D, but interacting supergravity theories are more restrictive as to the spinor type permitted in a given spacetime dimension (and such theories exist only for $D \leq 11$). We will need to use the required Majorana and/or Weyl spinors when we study these theories in later chapters (and the number $2^{[D/2]}$ must be adjusted to agree with the number of components of each type of spinor).

It is consistent with the pattern set in the previous chapter that the gauge field $\Psi_\mu(x)$ is a field with one more vector index than the gauge parameter $\epsilon(x)$. Furthermore, as in the case of electromagnetism, the antisymmetric derivative $\partial_\mu \Psi_\nu - \partial_\nu \Psi_\mu$ is gauge invariant. An important difference arises because we now seek a gauge invariant *first* order wave equation for the fermion field. It is advantageous to start with the action, which must be (a) Lorentz invariant, (b) first order in spacetime derivatives, (c) invariant under the gauge transformation (5.1) and the simultaneous conjugate transformation of $\bar{\Psi}_\mu$, and (d) hermitian, so that the Euler–Lagrange equation for $\bar{\Psi}_\mu$ is the Dirac conjugate of that for Ψ_μ. It is easy to see that the expression

$$S = -\int \mathrm{d}^D x\, \bar{\Psi}_\mu \gamma^{\mu\nu\rho} \partial_\nu \Psi_\rho\,, \tag{5.2}$$

which contains the third rank Clifford algebra element $\gamma^{\mu\nu\rho}$, has all these properties. Note that the action is gauge invariant but the Lagrangian density is not. Instead its variation is the total derivative $\delta\mathcal{L} = -\partial_\mu(\bar{\epsilon}\gamma^{\mu\nu\rho}\partial_\nu\Psi_\rho)$. The reason is that the fermionic gauge symmetry is the remnant of supersymmetry, and the anti-commutator of two SUSY transformations is a spacetime symmetry.

It should be noted that a physically equivalent theory can be obtained by rewriting (5.2) in terms of the new field variable $\Psi'_\mu \equiv \Psi_\mu + a\gamma_\mu\gamma\cdot\Psi$ where a is an arbitrary parameter.[1] The gauge transformation is modified accordingly. The presentation in (5.1) and (5.2) is universally used in the modern literature, because the gauge transformation is simplest and closely resembles that of electromagnetism. Historically, Rarita and Schwinger invented a wave equation for a *massive* spin-3/2 particle in 1941. The massless limit of the action is a transformed version of (5.2), and Rarita and Schwinger simply noted that it possesses a fermionic gauge symmetry.[2]

The equation of motion obtained from (5.2) reads

$$\gamma^{\mu\nu\rho}\partial_\nu\Psi_\rho = 0\,. \tag{5.3}$$

One can immediately see that it shares some of the properties of the analogous electromagnetic equation (4.7), which is $\partial^\mu F_{\mu\nu} = 0$. Gauge invariance is manifest, and the left-hand side vanishes *identically* when the derivative ∂_μ is applied. Thus (5.3) comprises $(D-1)2^{[D/2]}$ independent equations, which is enough to determine the $2^{[D/2]}D$ components of Ψ_ρ up to the freedom of a gauge transformation. The difference between the number of components of the gauge field and those of the gauge parameter, in this case $(D-1)2^{[D/2]}$, is called the number of off-shell degrees of freedom.

Exercise 5.1 *Show directly that for $D = 3$, the field equation (5.3) implies that $\partial_\nu\Psi_\rho - \partial_\rho\Psi_\nu = 0$. This means that the field has no gauge invariant degrees of freedom and thus no propagating particle modes. This is the supersymmetric counterpart of the situation in gravity for $D = 3$, where the field equation $R_{\mu\nu} = 0$ implies that the full curvature tensor $R_{\mu\nu\rho\sigma} = 0$. Hence no degrees of freedom.*

We notice that (5.3) can be rewritten in an equivalent but simpler form. For this purpose, we use the γ-matrix relation $\gamma_\mu\gamma^{\mu\nu\rho} = (D-2)\gamma^{\nu\rho}$, which implies that $\gamma^{\nu\rho}\partial_\nu\Psi_\rho = 0$ in spacetime dimension $D > 2$. We also note that $\gamma^{\mu\nu\rho} = \gamma^\mu\gamma^{\nu\rho} - 2\eta^{\mu[\nu}\gamma^{\rho]}$. Using this information, it is easy to see that (5.3) implies that

$$\gamma^\mu(\partial_\mu\Psi_\nu - \partial_\nu\Psi_\mu) = 0\,. \tag{5.4}$$

[1] The case $a = -1/D$ requires special treatment since $\gamma\cdot\Psi' = 0$.

[2] One of the present authors met Prof. Schwinger at a cocktail party in the early 1980s. Supergravity came up in the conversation, and Schwinger remarked lightheartedly 'I should have discovered supergravity.'

This is an alternative form of the equation of motion, equivalent to (5.3), but which cannot be obtained *directly* from an action. To see that (5.4) is equivalent, note that one can apply γ^ν and obtain $\gamma^{\nu\rho}\partial_\nu\Psi_\rho = 0$. The previous steps can then be reversed to obtain (5.3) from (5.4). One can also show that the left-hand side of (5.4) vanishes *identically* if $\gamma^\nu \not\partial$ is applied. Finally, let's apply ∂_ρ to (5.4) and antisymmetrize in $\rho\nu$ to obtain

$$\not\partial(\partial_\rho\Psi_\nu - \partial_\nu\Psi_\rho) = 0\,. \tag{5.5}$$

This is a gauge invariant derivation of the fact that the wave equations, either (5.3) or (5.4), describe *massless* particles.

Exercise 5.2 *Do all the manipulations in the preceding paragraph. Do them backwards and forwards.*

5.1 The initial value problem

Let's now study the initial value problem for (5.3) and thus count the number of on-shell degrees of freedom. We must untangle constraints on the initial data from time evolution equations. For this purpose we need to fix the gauge, so we impose the non-covariant condition

$$\gamma^i\Psi_i = 0\,, \tag{5.6}$$

which will play the same role as the Coulomb gauge condition we used in Sec. 4.1.2.

Exercise 5.3 *Show by an argument analogous to that in Sec. 4.1.2 that this condition does fix the gauge uniquely.*

We use the equivalent form (5.4) of the field equations. The $\nu = 0$ and $\nu \to i$ components are

$$\begin{aligned}\gamma^i\partial_i\Psi_0 - \partial_0\gamma^i\Psi_i &= 0\,,\\ \gamma\cdot\partial\Psi_i - \partial_i\gamma\cdot\Psi &= 0\,.\end{aligned} \tag{5.7}$$

Using the gauge condition one can see that $\nabla^2\Psi_0 = 0$, so $\Psi_0 = 0$ according to the discussion on p. 69. The spatial components Ψ_i then satisfy the Dirac equation

$$\gamma\cdot\partial\Psi_i = 0\,, \tag{5.8}$$

which is a time evolution equation. However, there is an additional constraint, $\partial^i\Psi_i = 0$, obtained by contracting (5.8) with γ^i. Thus from the gauge condition and the equation of motion, we find $3 \times 2^{[D/2]}$ independent constraints on the initial data, namely

$$\gamma^i\Psi_i(\vec{x},0) = 0\,, \tag{5.9}$$

$$\Psi_0(\vec{x},0) = 0\,, \tag{5.10}$$

$$\partial^i\Psi_i(\vec{x},0) = 0\,. \tag{5.11}$$

Box 5.1 **Degrees of freedom of the massless Rarita–Schwinger field**

On-shell degrees of freedom $= \frac{1}{2}(D-3)2^{[D/2]}$.
Off-shell degrees of freedom $= (D-1)2^{[D/2]}$.

These constraints imply that there are only $2^{[D/2]}(D-3)$ initial components of Ψ_i to be specified. The time derivatives are already determined by the Dirac equation (5.8). Hence there are $2^{[D/2]}(D-3)$ classical degrees of freedom for the Rarita–Schwinger gauge field in D-dimensional Minkowski space. The on-shell degrees of freedom are half this number. In dimension $D = 4$, with Majorana conditions, we find the two states expected for a massless particle for any spin $s > 0$. We will show below that these states carry helicity $\pm 3/2$. In general dimension, it should be a representation of $\mathrm{SO}(D-2)$ as discussed in Sec. 4.1.2. Indeed, the vector–spinor representation is an irreducible representation after subtraction of the γ-trace. It then contains $\frac{1}{2}(D-3)2^{[D/2]}$ components.

Exercise 5.4 *Analyze the degrees of freedom using the original equation of motion (5.3).*

According to the discussion for $D = 4$ at the beginning of Ch. 4, we would expect the Fourier expansion of the field to contain annihilation and creation operators for states of helicity $\lambda = \pm 3/2$. Let's derive this fact starting from the plane wave

$$\Psi_i(x) = \mathrm{e}^{\mathrm{i}p\cdot x} v_i(\vec{p}) u(\vec{p}) , \tag{5.12}$$

for a positive null energy–momentum vector $p^\mu = (|\vec{p}|, \vec{p})$. Since $\Psi_i(x)$ satisfies the Dirac equation (5.8), the four-component spinor $u(\vec{p})$ must be a superposition of the massless helicity spinors $u(\vec{p}, \pm)$ given in (2.44). Thus we use the Weyl representation (2.19) of the γ-matrices. The vector $v_i(\vec{p})$ may be expanded in the complete set

$$v_i(\vec{p}) = a p_i + b\epsilon_i(\vec{p}, +) + c\epsilon_i(\vec{p}, -) , \tag{5.13}$$

where $\epsilon_i(\vec{p}, \pm)$ are the transverse polarization vectors of Sec. 4.1.2, i.e. they satisfy $p^i \epsilon_i(\vec{p}, \pm) = 0$. The constraint (5.11) requires that $a = 0$. Thus (5.12) is reduced to the form

$$\begin{aligned}\Psi_i(x) = \mathrm{e}^{\mathrm{i}p\cdot x} \big[& b_+ \epsilon_i(\vec{p}, +) u(\vec{p}, +) + c_+ \epsilon_i(\vec{p}, -) u(\vec{p}, +) \\ & + b_- \epsilon_i(\vec{p}, +) u(\vec{p}, -) + c_- \epsilon_i(\vec{p}, -) u(\vec{p}, -) \big] .\end{aligned} \tag{5.14}$$

We must still enforce the constraint $\gamma^i \Psi_i = 0$. Some detailed algebra is needed, which we leave to the reader; the result is that $c_+ = b_- = 0$, while b_+ and c_- are arbitrary. Thus there are two independent physical wave functions $\epsilon_i(\vec{p}, \pm) u(\vec{p}, \pm)$ for each p^μ.

Exercise 5.5 *Do the algebra that was just left for the reader. Show that the resulting vector–spinor wave functions $\epsilon_i(\vec{p}, \pm) u(\vec{p}, \pm)$ carry helicity $\pm 3/2$. Show that the spinor wave functions for the conjugate plane wave are $\epsilon_i^*(\vec{p}, \pm) v(\vec{p}, \pm)$, where $v(\vec{p}, \pm) = B^{-1} u(\vec{p}, \pm)^*$ are the massless v spinors of (2.45).*

The net result of this analysis is that the Rarita–Schwinger field that satisfies the equation of motion and constraints above has the Fourier expansion (we add the trivial 0-components $\epsilon_0 = 0$ to polarization vectors)

$$\Psi_\mu(x) = \int \frac{\mathrm{d}^3\vec{p}}{(2\pi)^3 2p^0} \sum_\lambda [\mathrm{e}^{\mathrm{i}p\cdot x}\epsilon_\mu(\vec{p},\lambda)u(\vec{p},\lambda)c(\vec{p},\lambda) + \mathrm{e}^{-\mathrm{i}p\cdot x}\epsilon^*_\mu(\vec{p},\lambda)v(\vec{p},\lambda)d^*(\vec{p},\lambda)]\,. \tag{5.15}$$

The sum extends over the two physical wave functions of helicity $\pm 3/2$. In the quantum theory the Fourier amplitude $c(\vec{p},\lambda)$ becomes the annihilation operator for helicity $\pm 3/2$ particles, and $d^*(\vec{p},\lambda)$ becomes the creation operator for anti-particles. The situation is similar to that for the Dirac field in (2.24). A Majorana gravitino has the same expansion, with $d^*(\vec{p},\lambda) = c^*(\vec{p},\lambda)$, since there is no distinction between particles and anti-particles.

In dimension $D > 4$ the allowed gravitino modes are obtained by starting with products of the $D-2$ transverse polarization vectors $\epsilon_i(\vec{p}, j)$ and the $\frac{1}{2}2^{[D/2]}$ massless Dirac spinors $u(\vec{p}, s)$. The gauge fixing constraint $\gamma^i\Psi_i = 0$ must then be enforced on linear combinations of these products as was done in (5.14). This leads to $\frac{1}{2}2^{[D/2]}(D-3)$ independent wave functions, which describe the on-shell states of the gravitino.

The canonical stress tensor obtained from (5.2) is

$$T_{\mu\nu} = \bar{\Psi}_\rho\gamma^{\rho\sigma}{}_\mu\partial_\nu\Psi_\sigma - \eta_{\mu\nu}\mathcal{L}\,. \tag{5.16}$$

It is neither symmetric nor gauge invariant under (5.1) (and its Dirac conjugate). It can be made symmetric (see [27]), but gauge non-invariance is intrinsic and cannot be restored by adding terms of the form $\partial_\sigma S^{\sigma\mu\nu}$. The reason is that the gravitino must be joined with gravity in the gauge multiplet of SUSY. In a gravitational theory there is no well-defined energy *density*.

Exercise 5.6 *Show that the total energy–momentum* $P^\nu = \int \mathrm{d}^3\vec{x}\, T^{0\nu}(\vec{x},t)$ *is gauge invariant and given (for* $D = 4$*) by*

$$P^\nu = \int \frac{\mathrm{d}^3\vec{p}}{(2\pi)^3 2p^0} p^\nu \sum_\lambda [c^*(\vec{p},\lambda)c(\vec{p},\lambda) - d(\vec{p},\lambda)d^*(\vec{p},\lambda)]\,. \tag{5.17}$$

5.2 Sources and Green's function

Let's follow the pattern of Sec. 4.1.3 and couple the Rarita–Schwinger field to a vector–spinor source via

$$\gamma^{\mu\nu\rho}\partial_\nu\Psi_\rho = J^\mu\,. \tag{5.18}$$

The contraction of ∂_μ with the left-hand side vanishes identically, which indicates that (5.18) is a consistent equation only if the source current is conserved, i.e. $\partial_\mu J^\mu = 0$. This is the exact analogue of what happens in electromagnetism and Yang–Mills theory. In those theories, the gauge field was later coupled to matter systems, and the source was the Noether current of the global symmetry. Supergravity theories are more complicated.

The same phenomenon occurs, but only as an approximation valid to lowest order in the gravitational coupling. The current J^μ is the Noether supercurrent of the matter multiplets in the theory.

Let's now apply the method of Sec. 4.1.3 to find the Green's function that determines the response of the field to the source. We first solve the simpler problem for the Dirac field,

$$(\not{\partial} - m)\Psi(x) = J(x) . \tag{5.19}$$

Given a Green's function $S(x - y)$ that satisfies

$$(\not{\partial}_x - m)S(x - y) = -\delta(x - y) , \tag{5.20}$$

the solution of (5.19) is given by

$$\Psi(x) = - \int \mathrm{d}^D y \, S(x - y)J(y) . \tag{5.21}$$

Let's solve this problem using the Fourier transform. The symmetries of Minkowski spacetime allow us to assume the Fourier representation

$$S(x - y) = \int \frac{\mathrm{d}^D p}{(2\pi)^D} \mathrm{e}^{\mathrm{i}p\cdot(x-y)} S(p) . \tag{5.22}$$

In momentum space, (5.20) reads

$$(\mathrm{i}\not{p} - m)S(p) = -1, \tag{5.23}$$

and the solution (with Feynman's causal structure) is

$$S(p) = -\frac{1}{\mathrm{i}\not{p} - m} = \frac{\mathrm{i}\not{p} + m}{p^2 + m^2 - \mathrm{i}\epsilon} . \tag{5.24}$$

Comparing with (4.18), we see that we can express $S(x - y)$ in terms of the scalar Green's function as

$$S(x - y) = (\not{\partial}_x + m)G(x - y) . \tag{5.25}$$

This result satisfies (5.20) by inspection and could have been guessed at the start. However, the Fourier transform method is useful as a warmup for the more complicated case of the Rarita–Schwinger field.

We expect the Green's function solution of (5.18) to take the form

$$\Psi_\mu(x) = - \int \mathrm{d}^D y \, S_{\mu\nu}(x - y)J^\nu(y) , \tag{5.26}$$

where $S_{\mu\nu}(x - y)$ is a tensor bispinor. A bispinor has two spinor indices, which are suppressed in our notation, and it can be regarded as a matrix of the Clifford algebra. As in the electromagnetic case, the Rarita–Schwinger operator is not invertible, but we can assume that the Green's function satisfies

$$\gamma^{\mu\sigma\rho} \frac{\partial}{\partial x^\sigma} S_{\rho\nu}(x - y) = -\delta^\mu_\nu \delta(x - y) + \frac{\partial}{\partial y^\nu} \Omega^\mu(x - y) . \tag{5.27}$$

The last term on the right is a 'pure gauge' in the source point index. In momentum space (5.27) reads

$$\mathrm{i}\gamma^{\mu\sigma\rho} p_\sigma S_{\rho\nu}(p) = -\delta^\mu_\nu - \mathrm{i} p_\nu \Omega^\mu(p) \,. \tag{5.28}$$

We will solve (5.28) by writing an appropriate ansatz for $S_{\rho\nu}(p)$ and then find the unknown functions in the ansatz. The matrix $\gamma^{\mu\sigma\rho} p_\sigma$ in (5.28) contains an odd rank element of the Clifford algebra and it is odd under the reflection $p_\sigma \to -p_\sigma$. It is reasonable to guess that the ansatz we need should also involve odd rank Clifford elements and be odd under the reflection. We would also expect that terms that contain the momentum vectors p_ρ or p_ν are 'pure gauges' and thus arbitrary additions to the propagator, which would not be determined by the equation (5.28). So we omit such terms and postulate the ansatz

$$\mathrm{i}\, S_{\rho\nu}(p) = A(p^2)\, \eta_{\rho\nu}\, \not{p} + B(p^2)\, \gamma_\rho\, \not{p}\, \gamma_\nu \,. \tag{5.29}$$

The next step is to substitute the ansatz in (5.28) and simplify the products of γ-matrices that appear. This process yields

$$\begin{aligned} \mathrm{i}\gamma^{\mu\sigma\rho} p_\sigma S_{\rho\nu}(p) &= A\gamma^{\mu\sigma}{}_\nu \not{p} p_\sigma + (D-2)B\gamma^{\mu\sigma} \not{p}\gamma_\nu p_\sigma \\ &= A\left(p^\mu \gamma^\sigma{}_\nu - p^\sigma \gamma^\mu{}_\nu\right) p_\sigma + (D-2)B\left(-p^\mu\gamma^\sigma + p^\sigma\gamma^\mu\right)\gamma_\nu p_\sigma \\ &\quad + \ldots \\ &= [A-(D-2)B]\left(p^\mu \gamma^\sigma{}_\nu - p^\sigma \gamma^\mu{}_\nu\right) p_\sigma + (D-2)Bp^2\delta^\mu_\nu \\ &\quad + \ldots \,. \end{aligned} \tag{5.30}$$

We have omitted terms that are proportional to the vector p_ν, because such terms will be 'matched' in (5.28) by $\Omega^\mu(p)$ rather than by δ^μ_ν. It is now easy to see that the δ^μ_ν term in (5.28) determines the values $A = -1/p^2$ and $B = -1/((D-2)p^2)$. Thus we have found the gravitino propagator

$$S_{\mu\nu}(p) = \mathrm{i}\frac{1}{p^2}\left[\eta_{\mu\nu}\not{p} + \frac{1}{D-2}\gamma_\mu\not{p}\gamma_\nu + C\, p_\mu\gamma_\nu + E\,\gamma_\mu p_\nu + F\, p_\mu \not{p} p_\nu\right], \tag{5.31}$$

in which we have added possible gauge terms that are not determined by this procedure. In position space the propagator is

$$S_{\mu\nu}(x-y) = \left[\eta_{\mu\nu}\not{\partial} + \frac{1}{D-2}\gamma_\mu\not{\partial}\gamma_\nu + C\,\partial_\mu\gamma_\nu + E\,\gamma_\mu\partial_\nu - F\partial_\mu\not{\partial}\partial_\nu\right] G(x-y), \tag{5.32}$$

where $G(x-y)$ is the massless scalar propagator (4.19), and all derivatives are with respect to x.

Exercise 5.7 *Include the omitted p_ν terms in (5.30) and $\Omega(p)$ in the analysis and verify that the gauge terms in the propagator are arbitrary. Show that, for the choice $E = -1/(D-2)$, and arbitrary C and F, the propagator satisfies*

$$\mathrm{i}\gamma^{\mu\sigma\rho} p_\sigma S_{\rho\nu}(p) = -\left(\delta^\mu_\nu - \frac{p^\mu p_\nu}{p^2}\right). \tag{5.33}$$

Show that, for $D = 4$, the propagator, with $C = -1$, takes the 'reverse index' form $S_{\mu\nu}(p) = -\mathrm{i}\frac{1}{2}\gamma_\nu\not{p}\gamma_\mu$, which is the form used in most of the literature on perturbative studies in supergravity [28].

5.3 Massive gravitinos from dimensional reduction

Our aim in this section is quite narrow, but the approach will be broad. The narrow goal is to extend the Rarita–Schwinger equation to describe *massive* gravitinos, but we wish to do it by introducing the important technique of dimensional reduction, which is also called Kaluza–Klein theory. The main idea is that a fundamental theory, perhaps supergravity or string theory, that is formulated in D' spacetime dimensions can lead to an observable spacetime of dimension $D < D'$. In the most common variant of this scenario, there is a stable solution of the equations of the fundamental theory that describes a manifold of the structure $M_{D'} = M_D \times X_d$ with $d = D' - D$. The factor M_D is the spacetime in which we might live, thus non-compact with small curvature, while X_d is a tiny compact manifold of spatial extent L. The compact space X_d can be thought of as hidden dimensions of spacetime that are not accessible to direct observation because of basic properties of wave physics that are coded in quantum mechanics as the uncertainty principle. This principle asserts that it would take wave excitations of energy $E \approx 1/L$ to explore structures of spatial scale L. If L is sufficiently small, this energy scale cannot be achieved by available apparatus. Nevertheless, the dimensional reduction might be confirmed since the presence of X_d has important indirect effects on physics in M_D.

In this section we study an elementary version of dimensional reduction, which still has interesting physics to teach. Instead of obtaining the structure $M_D \times X_d$ from a fundamental theory including gravity, we will simply explore the physics of the various *free* fields we have studied, *assuming* that the $(D+1)$-dimensional spacetime is Minkowski$_D \otimes S^1$. The main feature is that Fourier modes of fields on S^1 are observed as infinite 'towers' of *massive* particles by an observer in Minkowski$_D$. The reduction of the free massless gravitino equation in $D+1$ dimensions will then tell us the correct description of massive gravitinos. Massive gravitinos appear in the physical spectrum of $D = 4$ supergravity when SUSY is spontaneously broken.

5.3.1 Dimensional reduction for scalar fields

Let's change to a more convenient notation and rename the coordinates of the $(D+1)$-dimensional product spacetime $x^0 = t, x^1, \ldots, x^{D-1}, y$, where y is the coordinate of S^1 with range $0 \leq y \leq 2\pi L$. We consider a massive complex scalar field $\phi(x^\mu, y)$ that obeys the Klein–Gordon equation

$$[\Box_{D+1} - m^2]\phi = \left[\Box_D + \left(\frac{\partial}{\partial y}\right)^2 - m^2\right]\phi = 0\,. \tag{5.34}$$

Acceptable solutions must be single-valued on S^1 and thus have a Fourier series expansion

$$\phi(x^\mu, y) = \sum_{k=-\infty}^{\infty} \mathrm{e}^{iky/L}\phi_k(x^\mu)\,. \tag{5.35}$$

It is immediate that the spacetime function associated with the kth Fourier mode, namely $\phi_k(x^\mu)$, satisfies

$$\left[\Box_D - \left(\frac{k}{L}\right)^2 - m^2\right]\phi_k = 0\,. \tag{5.36}$$

Thus it describes a particle of mass $m_k{}^2 = (k/L)^2 + m^2$. So the spectrum of the theory, as viewed in Minkowski$_D$, contains an infinite tower of massive scalars!

There is an even simpler way to find the mass spectrum. Just substitute the plane wave $e^{ip^\mu x_\mu}\, e^{iky/L}$ directly in the $(D+1)$-dimensional equation (5.34). The D-component energy–momentum vector p^μ must satisfy $p^\mu p_\mu = (k/L)^2 + m^2$. The mass shift due to the Fourier wave on S^1 is immediately visible.

5.3.2 Dimensional reduction for spinor fields

We will consider the dimensional reduction process for a complex spinor $\Psi(x^\mu, y)$ for even $D = 2m$ (so that the spinors in $D+1$ dimensions have the same number of components). Two new ideas enter the game. The first just involves the Dirac equation in D dimensions. We remark that if $\Psi(x)$ satisfies

$$[\partial\!\!\!/_D - m]\Psi(x) = 0\,, \tag{5.37}$$

then the new field $\tilde{\Psi} \equiv e^{-i\gamma_*\beta}\Psi$, obtained by applying a chiral phase factor, satisfies

$$[\partial\!\!\!/_D - m(\cos 2\beta + i\gamma_* \sin 2\beta)]\tilde{\Psi} = 0\,. \tag{5.38}$$

Physical quantities are unchanged by the field redefinition, so both equations describe particles of mass m. One simple implication is that the sign of m in (5.37) has no physical significance, since it can be changed by field redefinition with $\beta = \pi/2$.

The second new idea is that a fermion field can be either periodic or anti-periodic $\Psi(x^\mu, y) = \pm\Psi(x^\mu, y + 2\pi L))$. Anti-periodic behavior is permitted because a fermion field is not observable. Rather, bilinear quantities such as the energy density $T^{00} = -\bar{\Psi}\gamma^0\partial^0\Psi$ are observables and they are periodic even when Ψ is anti-periodic. Thus we consider the Fourier series

$$\Psi(x^\mu, y) = \sum_k e^{iky/L}\,\Psi_k(x^\mu), \tag{5.39}$$

where the mode number k is integer or half-integer for periodic or anti-periodic fields, respectively. In either case when we substitute (5.39) in the $(D+1)$-dimensional Dirac equation $[\partial\!\!\!/_{D+1} - m]\Psi((x^\mu, y) = 0$, we find that $\Psi_k(x^\mu)$ satisfies[3]

$$\left[\partial\!\!\!/_D - \left(m - i\gamma_*\frac{k}{L}\right)\right]\Psi_k(x^\mu) = 0\,. \tag{5.40}$$

By applying a chiral transformation with phase $\tan 2\beta = k/(mL)$, we see that $\Psi_k(x^\mu)$ describes particles of mass $m_k{}^2 = (k/L)^2 + m^2$. Again we would observe an infinite tower of massive spinor particles with distinct spectra for the periodic and anti-periodic cases.

[3] Recall from Ch. 3 that for odd spacetime dimension $D = 2m+1$, $\gamma^D = \pm\gamma_*$, where γ_* is the highest rank Clifford element in $D = 2m$ dimensions.

5.3.3 Dimensional reduction for the vector gauge field

We now apply circular dimensional reduction to Maxwell's equation

$$\partial^\nu F_{\nu\mu} = \Box_{D+1} A_\mu - \partial_\mu(\partial^\nu A_\nu) = 0 \tag{5.41}$$

in $D + 1$ dimensions, and we assume a periodic Fourier series representation

$$A_\mu(x, y) = \sum_k e^{iky/L} A_{\mu k}(x) , \qquad A_D(x, y) = \sum_k e^{iky/L} A_{Dk}(x) , \tag{5.42}$$

with k an integer. The analysis simplifies greatly if we assume the gauge conditions $A_{Dk}(x) = 0$ for $k \neq 0$ and vector component D tangent to S^1. It is easy to see that this gauge can be achieved and uniquely fixes the Fourier modes $\theta_k(x)$, $k \neq 0$, of the gauge function. The gauge invariant Fourier mode $A_{D0}(x)$ remains a physical field in the dimensionally reduced theory. A quick examination of the $\mu \to D$ component of (5.41) shows that it reduces to

$$\begin{aligned} k = 0&: \ \Box_{D+1} A_{D0} = \Box_D A_{D0} = 0, \\ k \neq 0&: \ \partial^\mu A_{\mu k} = 0, \end{aligned} \tag{5.43}$$

so the mode $A_{D0}(x)$ simply describes a massless scalar in D dimensions. For $\mu \leq D - 1$, the wave equation (5.41) implies that the vector modes $A_{\mu k}(x)$ satisfy

$$\left[\Box_D - \frac{k^2}{L^2}\right] A_{\mu k} - \partial_\mu(\partial^\nu A_{\nu k}) = 0 . \tag{5.44}$$

For mode number $k = 0$ this is just the Maxwell equation in D dimensions with its gauge symmetry under $A_{\mu 0} \to A_{\mu 0} + \partial_\mu \theta_0$ intact, since the Fourier mode $\theta_0(x)$ remained unfixed in the process above. For mode number $k \neq 0$, (5.44) is the standard equation[4] for a massive vector field with mass $m_k^2 = k^2/L^2$, namely the equation of motion of the action

$$S = \int d^D x \left[-\tfrac{1}{4} F_{\mu\nu} F^{\mu\nu} - \tfrac{1}{2} m^2 A_\mu A^\mu\right] . \tag{5.45}$$

A counting argument similar to that for the massless case in Ch. 4 shows that we have the D-component field $A_{\mu k}$ subject to the single constraint (5.43) and thus giving $D - 1$ quantum degrees of freedom for each Fourier mode $k \neq 0$. The $D - 1$ particle states for each fixed energy–momentum p^μ transform in the vector representation of SO$(D - 1)$ as appropriate for a massive particle. Note that there are three states for $D = 4$, which agrees with $2s + 1$ for spin $s = 1$. The count of states is the same in the massless $k = 0$ sector also, where we have the gauge vector $A_{\mu 0}$ plus the scalar A_{D0} with $(D - 2) + 1$ on-shell degrees of freedom.

5.3.4 Finally $\Psi_\mu(x, y)$

Let's apply dimensional reduction to the massless Rarita–Schwinger field in $D + 1$ dimensions with $D = 2m$. We will assume that the field $\Psi_\mu(x, y)$ is anti-periodic in y so that

[4] Note that the result (5.43) can be obtained by applying ∂^μ to (5.44) and is thus consistent with that equation.

its Fourier series involves modes $\exp(iky/L)\Psi_{\mu k}(x)$ with half-integral k. This assumption simplifies the analysis, since only $k \neq 0$ occurs, and all modes will be massive.

We would like to start with (5.3) in dimension $D+1$ and derive the wave equation of a massive gravitino in Minkowski$_D$. A gauge choice makes this task much easier. All Fourier modes have $k \neq 0$, so we can impose the gauge condition $\Psi_{Dk}(x) = 0$ on all modes and completely eliminate the field component $\Psi_D(x, y)$.

Let's write out the $\mu = D$ and $\mu \leq D-1$ components of (5.3) with $\Psi_D = 0$ (using $\gamma^D = \gamma_*$):

$$\begin{aligned} \gamma^{\nu\rho}\partial_\nu\Psi_{\rho k} &= 0\,, \\ \left[\gamma^{\mu\nu\rho}\partial_\nu - \mathrm{i}\frac{k}{L}\gamma_*\gamma^{\mu\rho}\right]\Psi_{\rho k} &= 0\,. \end{aligned} \tag{5.46}$$

Note that the first equation of (5.46) follows by application of ∂^μ to the second one.

Exercise 5.8 *Show that the chiral transformation $\Psi_{\rho k} = \mathrm{e}^{(-\mathrm{i}\pi\gamma_*/4)}\Psi'_{\rho k}$ leads, after replacing $\Psi' \to \Psi$, to the equation of motion*

$$\left(\gamma^{\mu\nu\rho}\partial_\nu - m\gamma^{\mu\rho}\right)\Psi_\rho = 0\,. \tag{5.47}$$

The last equation is the Euler–Lagrange equation of the action

$$S = -\int \mathrm{d}^D x\, \bar{\Psi}_\mu \left[\gamma^{\mu\nu\rho}\partial_\nu - m\gamma^{\mu\rho}\right]\Psi_\rho\,. \tag{5.48}$$

Exercise 5.9 *The equation of motion (5.47) also contains constraints on the initial data. Obtain $\gamma^{\mu\nu}\partial_\mu\Psi_\nu = 0$, which is not a constraint, by contracting the equation with ∂_μ. Then find the constraint $\gamma^\mu\Psi_\mu = 0$ by contracting with γ_μ. Show that the $\mu = 0$ component of the equation of motion gives the constraint $(\gamma^{ij}\partial_i - m\gamma^j)\Psi_j = 0$.*

Exercise 5.10 *By analysis similar to that which led from (5.3) to (5.4) in the massless case, derive $(\partial\!\!\!/ + m)\Psi_\mu = 0$, which closely resembles the Dirac equation. The constraints of Ex. 5.9 must still be applied to the initial data, but the new equation clearly shows that the field has definite mass m.*

It is useful to recapitulate the equations that we obtained during the analysis (or directly from (5.47)) that determine the counting of the number of initial data and thus the number of degrees of freedom:

$$\begin{aligned} &\gamma^\mu\Psi_\mu = 0\,, \\ &(\gamma^{ij}\partial_i - m\gamma^j)\Psi_j = 0\,, \\ &\left[\partial\!\!\!/ + m\right]\Psi_\mu = 0\,. \end{aligned} \tag{5.49}$$

As in the massless case, the time derivatives are determined by the Dirac equation (last equation of (5.49)). The initial data are thus the values at $t = 0$ of the Ψ_μ restricted by the first two equations of (5.49). Hence, the complex field $\Psi_\mu(x)$ with $D \times 2^{[D/2]}$

degrees of freedom contains $(D-2) \times 2^{[D/2]}$ independent classical degrees of freedom and thus $\frac{1}{2}(D-2) \times 2^{[D/2]}$ on-shell physical states. For $D = 4$ these are the four helicity states required for a massive particle of spin $s = 3/2$. In the situation of dimensional reduction, there is a massive gravitino with $m = |k|/L$ for every Fourier mode k, each with $\frac{1}{2}(D-2)2^{[D/2]}$ states. Note that this is the same as the number of states of a *massless* gravitino in $D+1$ dimensions.

Exercise 5.11 *Study the Kaluza–Klein reduction for the Rarita–Schwinger field assuming periodicity* $\Psi_\mu(x, y + 2\pi) = \Psi_\mu(x, y)$ *in* y*. Show that the spectrum seen in* Minkowski$_D$ *consists of a massive gravitino for each Fourier mode* $k \neq 0$ *plus a massless gravitino and massless Dirac particle for the zero mode.*

The dimensional reduction process has thus taught us the correct action for a massive gravitino. In particular the mass term is $m\bar{\Psi}_\mu \gamma^{\mu\nu}\Psi_\nu$. There is a more general action, namely

$$S = -\int \mathrm{d}^D x\, \bar{\Psi}_\mu \left[\gamma^{\mu\nu\rho}\partial_\nu - m\gamma^{\mu\rho} - m'\eta^{\mu\rho}\right]\Psi_\rho\,, \tag{5.50}$$

which contains an additional Lorentz invariant term with a coefficient m' with the dimension of mass. It is curious that this does not give the correct description of a massive gravitino, because it contains too many degrees of freedom. In the following exercise we ask readers to verify this.

Exercise 5.12 *Derive the equation of motion for the action (5.50). Analyze this equation as in Ex. 5.9, and show that the previous constraint* $\gamma \cdot \Psi = 0$ does not *hold if* $m' \neq 0$*. The field components* $\gamma \cdot \Psi$ *then describe additional degrees of freedom (which propagate as negative Hilbert space metric 'ghosts'). See [28] for an analysis in terms of projection operators.*

$\mathcal{N} = 1$ global supersymmetry in $D = 4$ 6

In global SUSY the scope of symmetries included in quantum field theory is extended from Poincaré and internal symmetry transformations, with charges $M_{[\mu\nu]}$, P_μ, and T_A, to include spinor supercharges Q^i_α, where α is a spacetime spinor index, and $i = 1, \ldots, \mathcal{N}$ is an index labeling the distinct supercharges. We will assume that the Q^i_α are four-component Majorana spinors, although an equivalent formulation using two-component Weyl spinors is also commonly used. In this chapter we will mostly study the simplest case $\mathcal{N} = 1$ where there is a single spinor charge Q_α. This case is called $\mathcal{N} = 1$ SUSY or simple SUSY. Some features of theories with $\mathcal{N} > 1$ spinor charges, called extended SUSY theories, are discussed in Sec. 6.4 and Appendix 6A.

In $\mathcal{N} = 1$ global SUSY the Poincaré generators and Q_α join in a new algebraic structure, that of a superalgebra. A superalgebra contains two classes of elements, even and odd. From the physics viewpoint, they can be called bosonic (B) and fermionic (F). The structure relations include both commutators and anti-commutators in the pattern $[B, B] = B$, $[B, F] = F, \{F, F\} = B$. The bosonic charges span a Lie algebra. In SUSY the subalgebra of the bosonic charges $M_{[\mu\nu]}$ and P_μ is the Lie algebra of the Poincaré group discussed in Ch. 1, while the new structure relations involving Q_α are

$$\begin{aligned} \{Q_\alpha, \bar{Q}^\beta\} &= -\tfrac{1}{2}(\gamma_\mu)_\alpha{}^\beta P^\mu , \\ [M_{[\mu\nu]}, Q_\alpha] &= -\tfrac{1}{2}(\gamma_{\mu\nu})_\alpha{}^\beta Q_\beta , \\ [P_\mu, Q_\alpha] &= 0 . \end{aligned} \tag{6.1}$$

Note that these are the classical (anti-)commutator relations; see Secs. 1.4 and 1.5. We will discuss this further in Ch. 11.

Exercise 6.1 *Use (2.30) to reexpress the supercharge anti-commutator in terms of Q and $Q^\dagger$. Then use the correspondence principle, that is multiply by the imaginary* i, *to obtain the quantum anti-commutator from the classical relation. This procedure gives the operator relation*

$$\left\{Q_\alpha, (Q^\dagger)^\beta\right\}_{\text{qu}} = \tfrac{1}{2}\left(\gamma_\mu\gamma^0\right)_\alpha{}^\beta P^\mu . \tag{6.2}$$

Trace on the spinor indices to obtain the positivity condition

$$\operatorname{Tr}(Q Q^\dagger + Q^\dagger Q) = 2P^0 . \tag{6.3}$$

The energy $E = P^0$ of any state in the Hilbert space of a global supersymmetric field theory must be positive.

Many SUSY theories, but not all, are invariant under a chiral U(1) symmetry called R-symmetry. We denote the generator by T_R. This acts on Q_α via

$$[T_R, Q_\alpha] = -\mathrm{i}(\gamma_*)_\alpha{}^\beta Q_\beta \tag{6.4}$$

but this generator T_R is not required. Other internal symmetries, which commute with Q_α and are frequently called outside charges, can also be included.

There are two important theorems that severely limit the type of charges and algebras that can be realized in an interacting relativistic quantum field theory in $D = 4$ (strictly speaking in a theory with a non-trivial S-matrix in flat space). According to the Coleman–Mandula (CM) theorem [29, 30], in the presence of massive particles, bosonic charges are limited to $M_{[\mu\nu]}$ and P_μ plus (optional) scalar internal symmetry charges, and the Lie algebra is the direct sum of the Poincaré algebra and a (finite-dimensional) compact Lie algebra for internal symmetry.

If superalgebras are admitted, the situation is governed by the Haag–Łopuszański–Sohnius (HLS) theorem [31, 30], and the algebra of symmetries admits spinor charges Q^i_α. If there is only one Q_α, then the superalgebra must agree with the $\mathcal{N} = 1$ Poincaré SUSY algebra in (6.1). When $\mathcal{N} > 1$, the possibilities are restricted to the extended SUSY algebras discussed in Appendix 6A.[1] The main thought that we wish to convey is that SUSY theories realize the most general symmetry possible within the framework of the few assumptions made in the hypotheses of the CM and HLS theorems.[2] They also unify bosons and fermions, the two broad classes of particles found in Nature.

The parameters of global SUSY transformations are *constant* anti-commuting Majorana spinors ϵ_α. In supergravity SUSY is gauged, necessarily with the Poincaré generators, since they are joined in the superalgebra (6.1). This means that gravity is included, so the spinor parameters become arbitrary functions $\epsilon_\alpha(x)$ on a curved spacetime manifold. It is logically possible to skip ahead to Ch. 9 where $\mathcal{N} = 1$, $D = 4$ supergravity is presented. But much important background will be missed, and we encourage only readers quite familiar with global SUSY to do this. We endeavor to give a succinct, pedagogical treatment of *classical* aspects of SUSY field theories. This material is certainly elegant, and part of the reason that SUSY is so appealing. However, there is much more in the deep results that have been discovered in perturbative and non-perturbative *quantum* supersymmetry that we cannot include.

The purpose of this chapter is to move as quickly as possible to an understanding of the structure of the major *interacting* SUSY field theories. At the classical level an interacting field theory is simply one in which the equations of motion are nonlinear. In Sec. 6.4, we give a short survey of the massless particle representations of extended Poincaré SUSY algebras.

[1] They also found the extension with central charges, which will be discussed in Sec. 12.6.2.

[2] In theories that contain only massless fields and are scale invariant at the quantum level, there are the additional possibilities of conformal and superconformal symmetries. The superconformal algebras contain the Poincaré SUSY algebras as subalgebras. They will be discussed later.

6.1 Basic SUSY field theory

SUSY theories contain both bosons and fermions, which are the basis states of a particle representation of the SUSY algebra (6.1)–(6.4). We give a systematic treatment of these representations in Sec. 6.4, but start with an informal discussion here. The states of particles with momentum $\vec{p}$ and energy $E(\vec{p}) = \sqrt{\vec{p}^2 + m^2_{B,F}}$ are denoted by $|\vec{p}, B\rangle$ and $|\vec{p}, F\rangle$, where the labels B and F include particle helicity. SUSY transformations connect these states. Since the spinor Q_α carries angular momentum $1/2$, it transforms bosons into fermions and fermions into bosons. Hence $Q_\alpha|\vec{p}, B\rangle = |\vec{p}, F\rangle$ and $Q_\alpha|\vec{p}, F\rangle \propto |\vec{p}, B\rangle$. Since $[P^\mu, Q_\alpha] = 0$, the transformed states have the same momentum and energy, hence the same mass, so $m^2_B = m^2_F$. We show in Sec. 6.4.1 that a representation of the algebra contains the same number of boson and fermion states.

The simplest representations of the algebra that lead to the most basic SUSY field theories are:

(i) the chiral multiplet, which contains a self-conjugate spin-$1/2$ fermion described by the Majorana field $\chi(x)$ plus a complex spin-0 boson described by the scalar field $Z(x)$. Alternatively, $\chi(x)$ may be replaced by the Weyl spinor $P_L\chi$ and/or $Z(x)$ by the combination $Z(x) = (A(x) + \mathrm{i}B(x))/\sqrt{2}$ where A and B are a real scalar and pseudo-scalar, respectively. A chiral multiplet can be either massless or massive.

(ii) the gauge multiplet consisting of a massless spin-1 particle, described by a vector gauge field $A_\mu(x)$, plus its spin-$1/2$ fermionic partner, the gaugino, described by a Majorana spinor $\lambda(x)$ (or the corresponding Weyl field $P_L\lambda$).

6.1.1 Conserved supercurrents

It follows from our discussion of the Noether formalism for symmetries that the spinor charge should be the integral of a conserved vector–spinor current, the supercurrent $\mathcal{J}^\mu_\alpha$, hence

$$Q_\alpha = \int \mathrm{d}^3x\, \mathcal{J}^0_\alpha(\vec{x}, t)\,. \tag{6.5}$$

If the current is conserved for all solutions of the equations of motion of a theory, then the theory has a fermionic symmetry. By the HLS theorem this symmetry must be supersymmetry!

Therefore we begin the technical discussion of SUSY in quantum field theory by displaying such conserved currents,[3] first for *free* fields and then for one non-trivial *interacting* system. Consider a free scalar field $\phi(x)$ satisfying the Klein–Gordon equation $(\Box - m^2)\phi = 0$ and a spinor field $\Psi(x)$ satisfying the Dirac equation $(\partial\!\!\!/ - m)\Psi = 0$.

Exercise 6.2 *Show that the current $\mathcal{J}^\mu = (\partial\!\!\!/\phi - m\phi)\gamma^\mu\Psi$ is conserved for all field configurations satisfying the Klein–Gordon and Dirac equations.*

[3] The spinor index α on the current and on most spinorial quantities will normally be suppressed.

It is no surprise to find unusual conserved currents in a *free* theory. In fact the current $\mathcal{J}^\mu{}_\nu = (\not\partial\phi - m\phi)\gamma^\mu\partial_\nu\Psi$, which gives a charge that violates the HLS theorem, is conserved. Such currents cannot be conserved in an interacting theory. For similar reasons conservation of $\mathcal{J}^\mu$ at the free level holds whether ϕ and Ψ are real or complex. To extend to interactions we will have to take $\phi \to Z$, a complex scalar, and Ψ a Majorana spinor. Note also that the current in Ex. 6.2 is conserved for any spacetime dimension D; this is another property that fails with interactions.

As the second example let's look at the free gauge multiplet with vector potential A_μ and field strength $F_{\mu\nu} = \partial_\mu A_\nu - \partial_\nu A_\mu$ satisfying the Maxwell equation $\partial^\mu F_{\mu\nu} = 0$ and a spinor λ satisfying $\not\partial\lambda = 0$. Let's show that the current $\mathcal{J}^\mu = \gamma^{\nu\rho}F_{\nu\rho}\gamma^\mu\lambda$ is conserved. We have

$$\partial_\mu\mathcal{J}^\mu = \partial_\mu F_{\nu\rho}\gamma^{\nu\rho}\gamma^\mu\lambda + \gamma^{\nu\rho}F_{\nu\rho}\not\partial\lambda\,. \tag{6.6}$$

The last term vanishes. To treat the first term we manipulate the γ-matrices as discussed in Sec. 3.1.4:

$$\gamma^{\nu\rho}\gamma^\mu = \gamma^{\nu\rho\mu} + 2\gamma^{[\nu}\eta^{\rho]\mu}\,. \tag{6.7}$$

When inserted in the first term of (6.6) we see that the first term vanishes by the gauge field Bianchi identity (4.11), and the second one by the Maxwell equation.

6.1.2 SUSY Yang–Mills theory

With little more work we can now exhibit an important interacting theory, $\mathcal{N}=1$ SUSY Yang–Mills theory and its conserved supercurrent. The theory contains the gauge boson $A^A_\mu(x)$ and its SUSY partner, the Majorana spinor gaugino $\lambda^A(x)$ in the adjoint representation of a simple, compact, non-abelian gauge group G. The action is[4]

$$S = \int \mathrm{d}^4x \left[-\tfrac{1}{4}F^{\mu\nu A}F^A_{\mu\nu} - \tfrac{1}{2}\bar\lambda^A\gamma^\mu D_\mu\lambda^A\right]. \tag{6.8}$$

For details of the notation see Secs. 3.4.1 and 4.3. Note that the gaugino action vanishes unless $\lambda^A(x)$ is anti-commuting! The Euler–Lagrange equations (and gauge field Bianchi identity) are

$$\begin{aligned} D^\mu F^A_{\mu\nu} &= -\tfrac{1}{2}g f_{BC}{}^A\bar\lambda^B\gamma_\nu\lambda^C\,,\\ D_\mu F^A_{\nu\rho} + D_\nu F^A_{\rho\mu} + D_\rho F^A_{\mu\nu} &= 0\,,\\ \gamma^\mu D_\mu\lambda^A &= 0\,. \end{aligned} \tag{6.9}$$

The supercurrent is

$$\mathcal{J}^\mu = \gamma^{\nu\rho}F^A_{\nu\rho}\gamma^\mu\lambda^A\,. \tag{6.10}$$

[4] We assume in this chapter that the Lie algebra has an invariant metric δ_{AB}, so that two 'upper' indices can be contracted.

The proof that it is conserved begins as in the free (abelian) case:

$$\begin{aligned}\partial_\mu \mathcal{J}^\mu &= D_\mu F^A_{\nu\rho}\gamma^{\nu\rho}\gamma^\mu\lambda^A + \gamma^{\nu\rho}F^A_{\nu\rho}\gamma^\mu D_\mu\lambda^A \\ &= -2D^\mu F^A_{\mu\nu}\gamma^\nu\lambda^A \\ &= g f_{ABC}\gamma^\nu\lambda^A\bar\lambda^B\gamma_\nu\lambda^C\,. \end{aligned} \tag{6.11}$$

The right-hand side vanishes due to (3.68) and the supercurrent (6.10) is conserved!

Thus we have established the existence of our first interacting SUSY field theory. Notice how the basic relations of non-abelian gauge symmetry such as the Bianchi identity, the relativistic description of spin by the Dirac–Clifford algebra, and the anti-commutativity of fermion fields are all blended in the proof. Readers whose intellectual curiosity is not excited by this are advised to put this book aside permanently and watch television instead of reading it.

The two main approaches to SUSY field theories are the approach of this chapter, in which we deal with the separate field components describing each physical particle in the theory, and the superspace approach, in which the separate fields are grouped in superfields. The latter approach is not used in this book, but is briefly discussed in Appendix 14A (see references there). A Fierz relation is always required to establish supersymmetry in the 'component' approach to any interacting SUSY theory. This is one reason why the existence and field content of SUSY field theories depend so markedly on the spacetime dimension. The Fierz relation also restricts the type of fermion required in the theory.

Here is an exercise in which readers are asked to show that super-Yang–Mills (SYM) theories with gauge field A_μ plus a specific type of spinor ψ^A and the supercurrent $\mathcal{J}^\mu = \gamma^{\nu\rho}F^A_{\nu\rho}\gamma^\mu\psi^A$ do exist in certain spacetime dimensions [32].

Exercise 6.3 *Study the appropriate Fierz rearrangement and, using the results of Ex. 3.27, show that the supercurrent is conserved in the following cases:*

(i) Majorana spinors in $D = 3$,
(ii) Majorana (or Weyl) spinors in $D = 4$, which is the case analyzed above,
(iii) symplectic Weyl spinors in $D = 6$, and
(iv) Majorana–Weyl spinors in $D = 10$.

Notice that, in every case, the number of on-shell degrees of freedom of the gauge field, namely $D-2$, matches those of the fermion, which are $2\times 2^{[(D-2)/2]}/k$ (real), where $k = 1$ for a complex Dirac fermion, $k = 2$ for a Majorana, a Weyl or a symplectic Weyl fermion, and $k = 4$ for a Majorana–Weyl fermion. Equality of the total number of boson and fermion states is a necessary condition for SUSY. This basic fact will be proved for massive and massless physical states in Sec. 6.4.1 and in general (on- or off-shell) in Appendix 6B.

6.1.3 SUSY transformation rules

Although global SUSY can be formulated using conserved supercurrents as the primary vehicle, as was done above, it is usually more convenient to emphasize the idea of SUSY field variations involving spinor parameters ϵ_α under which actions must be invariant. The field variations are also called SUSY transformation rules. Given the conserved current one can form the supercharge using (6.5) and use the canonical formalism to compute the field variations, i.e.

$$\delta\Phi(x) = \{\bar{\epsilon}^\alpha Q_\alpha, \Phi(x)\}_{\rm PB} = -{\rm i}[\bar{\epsilon}^\alpha Q_\alpha, \Phi(x)]_{\rm qu}\,, \tag{6.12}$$

where Φ denotes any field of the system under study. A brief description of Poisson brackets (PB) and commutation relations in the canonical formalism is given in Secs. 1.4 and 1.5. A link in the opposite direction is provided by the Noether formalism, which produces a conserved supercurrent given field variations under which the action is invariant. One reason to emphasize the field variations, *ab initio*, is that this avoids some subtleties in the canonical formalism for gauge theories and for Majorana spinors.

The next exercise illustrates the link between the supercurrent and field variations. It involves the free scalar–spinor ϕ–Ψ system of Ex. 6.2. The spinors Ψ, the supersymmetry parameters ϵ and the supersymmetry generator Q are Majorana spinors. They all mutually anti-commute. For the canonical formalism, one can either treat Ψ and $\bar{\Psi}$ as independent variables, or use Dirac brackets to obtain

$$\begin{aligned}\{\phi(x), \partial_0\phi(y)\}_{\rm PB} &= -\{\partial_0\phi(x), \phi(y)\}_{\rm PB} = \delta^3(\vec{x}-\vec{y})\,, \\ \left\{\Psi_\alpha(x), \bar{\Psi}^\beta(y)\right\}_{\rm PB} &= \left\{\bar{\Psi}^\beta(x), \Psi_\alpha(y)\right\}_{\rm PB} = \left(\gamma^0\right)_\alpha{}^\beta\delta^3(\vec{x}-\vec{y})\,.\end{aligned} \tag{6.13}$$

Exercise 6.4 *Use* $\bar{Q} = (1/\sqrt{2})\int {\rm d}^3\vec{x}\,\bar{\Psi}\gamma^0(\partial\!\!\!/ + m)\phi$ *or* $Q = (1/\sqrt{2})\int {\rm d}^3\vec{x}\,(\partial\!\!\!/\phi - m\phi)\gamma^0\Psi$ *to obtain*

$$\begin{aligned}\delta\phi(x) &= \{\bar{\epsilon}Q, \phi(x)\}_{\rm PB} = \frac{1}{\sqrt{2}}\bar{\epsilon}\Psi(x)\,, \\ \delta\Psi(x) &= \{\bar{\epsilon}Q, \Psi(x)\}_{\rm PB} = \frac{1}{\sqrt{2}}(\partial\!\!\!/ + m)\phi\epsilon.\end{aligned} \tag{6.14}$$

Note that $[\bar{Q}\epsilon, \Psi_\alpha(x)]_{\rm PB} = -\{\bar{Q}^\beta, \Psi_\alpha(x)\}_{\rm PB}\,\epsilon_\beta$.

6.2 SUSY field theories of the chiral multiplet

The physical fields of the chiral multiplet are a complex scalar $Z(x)$ and the Majorana spinor $\chi(x)$. It simplifies the structure in several ways to bring in a complex scalar auxiliary field $F(x)$. The field equations of F are algebraic, so F can be eliminated from the system at a later stage. The set of fields Z, $P_L\chi$, F constitute a chiral multiplet, and their conjugates[5] $\bar{Z}$, $P_R\chi$, $\bar{F}$ are an anti-chiral multiplet. The treatment is streamlined because we use the chiral projections $P_L\chi$ and $P_R\chi$, but can still regard χ as a Majorana spinor; see Sec. 3.4.2.

Our program is to present the SUSY transformation rules of these multiplets, discuss invariant actions, and then study the SUSY algebra (6.1)–(6.4). The spinor parameter ϵ

[5] Here we use $\bar{Z}$ to denote the complex conjugate rather than *, which was used earlier. Both notations will be used later in the book.

is a Majorana spinor, whose spinor components anti-commute with each other and with components of χ and $\bar{\chi}$.

The transformation rules of the chiral multiplet are

$$\begin{aligned} \delta Z &= \frac{1}{\sqrt{2}}\bar{\epsilon} P_L \chi , \\ \delta P_L \chi &= \frac{1}{\sqrt{2}} P_L (\partial\!\!\!/ Z + F)\epsilon , \\ \delta F &= \frac{1}{\sqrt{2}}\bar{\epsilon}\, \partial\!\!\!/ P_L \chi . \end{aligned} \tag{6.15}$$

The anti-chiral multiplet transformation rules are

$$\begin{aligned} \delta \bar{Z} &= \frac{1}{\sqrt{2}}\bar{\epsilon} P_R \chi , \\ \delta P_R \chi &= \frac{1}{\sqrt{2}} P_R (\partial\!\!\!/ \bar{Z} + \bar{F})\epsilon , \\ \delta \bar{F} &= \frac{1}{\sqrt{2}}\bar{\epsilon}\, \partial\!\!\!/ P_R \chi . \end{aligned} \tag{6.16}$$

Note that the form of the transformation rules for the physical components is similar to those of the 'toy model' in Ex. 6.4.

Exercise 6.5 *Show that the variations $\delta\bar{Z}$, $\delta P_R\chi$, $\delta\bar{F}$ are the complex conjugates of δZ, $\delta P_L\chi$, δF.*

There are two basic actions, which are separately invariant under the transformation rules above. The first is the free kinetic action

$$S_{\rm kin} = \int {\rm d}^4x \left[-\partial^\mu \bar{Z} \partial_\mu Z - \bar{\chi} \partial\!\!\!/ P_L \chi + \bar{F} F \right] , \tag{6.17}$$

in which we have presented the spinor term in chiral form. The interaction is determined by an arbitrary holomorphic function, the superpotential $W(Z)$. Given this we define the action

$$S_F = \int {\rm d}^4x \left[F W'(Z) - \tfrac{1}{2} \bar{\chi} P_L W''(Z) \chi \right] . \tag{6.18}$$

(The reason for the apparent extra derivative will be explained shortly.) Note that S_F involves only the fields of the chiral multiplet and no anti-chiral components. Thus the action S_F is not hermitian, so we must also consider the conjugate action $S_{\bar{F}} = (S_F)^\dagger$. The complete action of the chiral multiplet is the sum

$$S = S_{\rm kin} + S_F + S_{\bar{F}} . \tag{6.19}$$

Exercise 6.6 *Consider the superpotential $W = \frac{1}{2}mZ^2 + \frac{1}{3}gZ^3$, which gives the SUSY theory first considered by Wess and Zumino in 1973 [3]. Obtain the equations of motion for all fields, then eliminate F and $\bar{F}$ and show that the correct equations of motion for the physical fields are obtained if you first eliminate F and $\bar{F}$ by solving their algebraic equations of motion and substituting the result in the action. Substitute $Z = (A + \mathrm{i}B)/\sqrt{2}$ and show that the action (after elimination of auxiliary fields) takes the form*

$$S_{WZ} = \int \mathrm{d}^4x \left[-\frac{1}{2}(\partial_\mu A \partial^\mu A + \partial_\mu B \partial^\mu B) - \frac{1}{2}m^2(A^2 + B^2) - \frac{1}{2}\bar{\chi}(\not{\partial} + m)\chi) \right.$$
$$\left. - \frac{g}{\sqrt{2}}\bar{\chi}(A + \mathrm{i}\gamma_* B)\chi - \frac{mg}{\sqrt{2}}(A^3 + AB^2) - \frac{g^2}{4}(A^2 + B^2)^2 \right]. \quad (6.20)$$

From the viewpoint of a particle theorist this is a parity conserving, renormalizable theory with equal-mass fields and Yukawa plus quartic interactions.

Let's outline the proof that the actions S_{kin} and S_F are invariant under the SUSY transformation (6.15) and (6.16). It is rather intricate, so trusting readers may wish to move ahead. For S_{kin} the work is simplified by an observation that is correct in any representation of the Clifford algebra, but clearest in the Weyl representation (2.19). The projections $P_L\epsilon$ and $\bar{\epsilon}P_L$ involve the same half of the components of the Majorana ϵ, while $P_R\epsilon$ and $\bar{\epsilon}P_R$ involve the conjugate components. We write the total variation $\delta S = \delta_{P_L\epsilon}S + \delta_{P_R\epsilon}S$, temporarily separating the two chiral projections of ϵ in the transformation rules. Since S_{kin} is hermitian, it is sufficient to compute $\delta_{P_L\epsilon}S_{\text{kin}}$; then $\delta_{P_R\epsilon}S$ is its adjoint. In the calculation we temporarily allow $\epsilon(x)$ to be an arbitrary function in Minkowski spacetime since that provides a simple way [33] to obtain the Noether current for SUSY. We also need $\delta\bar{\chi}P_R = -(1/\sqrt{2})\bar{\epsilon}\left(\not{\partial}\bar{Z} - \bar{F}\right)P_R$. Either the Dirac conjugate (2.30) or the Majorana conjugate relations (3.56) and (3.54) can be used. We suggest practice with the latter. (Note that $t_0 = -t_1 = 1$ in four dimensions.)

Now that we have prepared the way, let's calculate

$$\delta_{P_L\epsilon}S_{\text{kin}} = -\frac{1}{\sqrt{2}}\int \mathrm{d}^4x \left[\partial^\mu\bar{Z}\partial_\mu(\bar{\epsilon}P_L\chi) - \bar{\epsilon}(\not{\partial}\bar{Z})\not{\partial}P_L\chi \right.$$
$$\left. + \bar{\chi}\not{\partial}(P_L F\epsilon) - (\bar{\epsilon}\not{\partial}P_R\chi)F\right]. \quad (6.21)$$

We have included all $P_L\epsilon$ and $\bar{\epsilon}P_L$ terms and dropped others. The $\bar{Z}\chi$ and $F\chi$ terms are independent and must vanish separately if we are to have a symmetry (when ϵ is constant). After a Majorana flip in the last term, we find that the $F\chi$ terms combine to (even for $\epsilon(x)$)

$$-\frac{1}{\sqrt{2}}\int \mathrm{d}^4x\, \partial_\mu(\bar{\chi}\gamma^\mu P_L F\epsilon), \quad (6.22)$$

which vanishes. The $\bar{Z}\chi$ terms can then be processed by substituting

$$\partial_\mu(\bar{\epsilon}P_L\chi) = (\partial_\mu\bar{\epsilon})P_L\chi + \bar{\epsilon}P_L\partial_\mu\chi,$$
$$\bar{\epsilon}P_L\gamma^\mu\gamma^\nu(\partial_\mu\bar{Z})\partial_\nu\chi = \bar{\epsilon}P_L[(\partial^\mu\bar{Z})\partial_\mu\chi + \gamma^{\mu\nu}\partial_\nu(\partial_\mu\bar{Z}\chi)] \quad (6.23)$$

in (6.21). Two of the four terms cancel. After partial integration and use of $\eta^{\mu\nu} - \gamma^{\mu\nu} = \gamma^\nu\gamma^\mu$, we find the net result

$$\delta_{P_L\epsilon} S_{\text{kin}} = -\frac{1}{\sqrt{2}} \int \mathrm{d}^4x\, \partial_\mu \bar{\epsilon}\, P_L(\partial\!\!\!/ \bar{Z})\gamma^\mu \chi \,. \tag{6.24}$$

This shows that δS_{kin} vanishes for constant ϵ, which is enough to prove SUSY. The remaining term is a contribution to the supercurrent of the complete theory in (6.19), and we will include it below.

Since SUSY for the *free* action S_{kin} is not worth celebrating, we move on to discuss the interaction term S_F. The variation δS_F under the transformations (6.15) has the structure

$$\delta S_F = \int \mathrm{d}^4x \Big[\delta F W'(Z) + \delta Z F W''(Z) \\ - \delta\bar{\chi} P_L \chi W''(Z) - \tfrac{1}{2}\delta Z \bar{\chi} P_L \chi W'''(Z)\Big] \,, \tag{6.25}$$

where we have taken the derivatives of $W(Z)$ required to include all sources of the δZ variation. After use of (6.15) we combine terms. Two $FP_L\chi$ terms cancel and we are left with the net result

$$\delta S_F = \frac{1}{\sqrt{2}} \int \mathrm{d}^4x \left[\bar{\epsilon}\partial\!\!\!/ \left(W' P_L \chi\right) - \frac{1}{2} W''' \bar{\epsilon} P_L \chi \bar{\chi} P_L \chi\right] . \tag{6.26}$$

The last term vanishes, since $P_L\chi$ has two independent components and any cubic expression vanishes by anti-commutativity! Thus δS_F vanishes for constant ϵ and is supersymmetric. It is clear that $\delta S_{\bar{F}}$ is just the conjugate of (6.26). At last SUSY is established at the interacting level!

The remaining $\partial_\mu \bar{\epsilon}$ terms in (6.24) and (6.26) plus their conjugates combine to give the Noether supercurrent of the interacting theory. This can be written as

$$\mathcal{J}^\mu = \frac{1}{\sqrt{2}} [P_L(\partial\!\!\!/ \bar{Z} - F) + P_R(\partial\!\!\!/ Z - \bar{F})]\gamma^\mu \chi \,, \tag{6.27}$$

in which one must use the auxiliary field equations of motion $F = -\overline{W}'(\bar{Z})$ and $\bar{F} = -W'(Z)$.

Exercise 6.7 *Show that the current is conserved for all solutions of the equations of motion of the theory (6.19).*

Exercise 6.8 *Given the component fields $\bar{Z}$, $P_R\chi$, $\bar{F}$ of an anti-chiral multiplet, show that $\bar{F}$, $P_L\partial\!\!\!/\chi$, $\Box\bar{Z}$ transform in the same way as the Z, $P_L\chi$, F components of a chiral multiplet; see (6.15) and (6.16).*

6.2.1 $\mathrm{U}(1)_R$ symmetry

The R-symmetry is a phase transformation of the fields of the chiral and anti-chiral multiplets. The fields Z, $P_L\chi$, F carry R-charges r, $r_\chi = r - 1$, $r_F = r - 2$ respectively. $\mathrm{U}(1)_R$ is a chiral symmetry, so the charges of component fields of the conjugate anti-chiral multiplet are the opposite of those above. We will discuss below how r is determined.

Infinitesimal transformations with parameter ρ are written as

$$\begin{aligned} \delta_R Z &= \mathrm{i}\rho r Z\,, \\ \delta_R P_L\chi &= \mathrm{i}\rho(r-1)P_L\chi\,, \\ \delta_R F &= \mathrm{i}\rho(r-2)F\,. \end{aligned} \tag{6.28}$$

There are similar variations, with opposite charges, for $\bar{Z}$, $P_R\chi$, $\bar{F}$. The relations $r_\chi = r-1$ and $r_F = r-2$ are required by the commutator (6.4),

$$[\delta_R(\rho), \delta(\epsilon)] = \rho\bar{\epsilon}^\alpha\,[T_R, Q_\alpha] = -\mathrm{i}\rho\bar{\epsilon}^\alpha(\gamma_*)_\alpha{}^\beta Q_\beta\,, \tag{6.29}$$

where $\delta(\epsilon)$ are the supersymmetry transformations (6.15).

Exercise 6.9 *Calculate the variation $\delta_R\mathcal{J}^\mu$ of the supercurrent (6.27) using $\delta_R Z = \mathrm{i}\rho r Z$, $\delta_R P_L\chi = \mathrm{i}\rho r_\chi P_L\chi$, $\delta_R F = \mathrm{i}\rho r_F F$. Show that the result agrees with (6.4) if and only if $r_\chi = r-1$ and $r_F = r-2$.*

The kinetic action $S_{\rm kin}$ is invariant under (6.28), so it is the interaction S_F that controls the situation. We now study the conditions for invariance of S_F. Clearly FW' and $\bar{\chi}W''\chi$ must be separately invariant. The condition for the vanishing of $\delta(FW')$ is

$$\delta_R(FW') = \mathrm{i}\rho F[(r-2)W' + rZW''] = 0\,. \tag{6.30}$$

This must hold for all field configurations, which means that the superpotential must satisfy

$$r(W' + ZW'') = 2W' \;\Rightarrow\; ZW' = \frac{2}{r}W \tag{6.31}$$

(since a constant can be absorbed in the definition of W without changing the physics). Similarly the condition for the vanishing of $\delta_R(\bar{\chi}W''\chi)$ is

$$\delta_R(\bar{\chi}W''\chi) = \mathrm{i}\rho\bar{\chi}[2(r-1)W'' + rZW''']\chi = 0\,. \tag{6.32}$$

However, the quantity in square brackets is the derivative of (6.31) and thus vanishes if $W(Z)$ satisfies (6.31).

The conclusion is that $W(Z)$ must be a homogeneous function of order $2/r$. This means that a theory with monomial superpotential $W(Z) = Z^k$ is $\mathrm{U}(1)_R$ invariant provided we assign $r = 2/k$ as the R-charge of the elementary scalar field Z. In the Wess–Zumino model, $W(Z) = mZ^2/2 + gZ^3/3$. If $g = 0$, then we have $\mathrm{U}(1)_R$ symmetry with $r = 1$. If $m = 0$ we have $\mathrm{U}(1)_R$ symmetry with $r = 2/3$. For general values of m and g, the symmetry is absent.

R-symmetry plays an important role in phenomenological applications of global supersymmetry. For example (see Sec. 28.1 of [30] or Sec. 5.2 of [34]), the related discrete R-parity is used to rule out undesired terms in the minimally supersymmetric standard model. $\mathrm{U}(1)_R$ plays an important role in models with supersymmetry breaking.

6.2.2 The SUSY algebra

In this section we will study the realization of the SUSY algebra on the components of a chiral multiplet. It is convenient and interesting that the $\{Q, \bar{Q}\}$ anti-commutator in (6.1) is realized in classical manipulations as the commutator of two successive variations of the fields with distinct (anti-commuting) parameters ϵ_1, ϵ_2.

The variation of a generic field $\Phi(x)$ is given in (6.12). The commutator of successive variations δ_1, δ_2 of $\Phi(x)$, with parameters ϵ_1, ϵ_2, respectively, is (recall that $\bar{\epsilon}Q = \bar{Q}\epsilon$ for Majorana spinors)

$$
\begin{aligned}
[\delta_1, \delta_2]\Phi(x) &= \left[\bar{\epsilon}_1 Q, \left[\bar{Q}\epsilon_2, \Phi(x)\right]\right] - (\epsilon_1 \leftrightarrow \epsilon_2) \\
&= \bar{\epsilon}_1^\alpha [\{Q_\alpha, \bar{Q}^\beta\}, \Phi(x)]\epsilon_{2\beta} \\
&= -\tfrac{1}{2}\bar{\epsilon}_1 \gamma^\mu \epsilon_2\, \partial_\mu \Phi(x)\,. \qquad (6.33)
\end{aligned}
$$

The standard Jacobi identity has been used to reach the second line and the first relation in (6.1) to obtain the last line. The key result is that the commutator of two SUSY variations is an infinitesimal spacetime translation with parameter $-\frac{1}{2}\,\bar{\epsilon}_1\gamma^\mu\epsilon_2$.

Let's carry out the computation of $[\delta_1, \delta_2]Z(x)$ on the scalar field of a chiral multiplet. Using (6.15) we write

$$
\begin{aligned}
[\delta_1, \delta_2]Z &= \frac{1}{\sqrt{2}}\delta_1(\bar{\epsilon}_2 P_L \chi) - [1 \leftrightarrow 2] \\
&= \tfrac{1}{2}\bar{\epsilon}_2 P_L(\partial\!\!\!/ Z + F)\epsilon_1 - [1 \leftrightarrow 2] \\
&= -\tfrac{1}{2}\bar{\epsilon}_1\gamma^\mu\epsilon_2\partial_\mu Z\,. \qquad (6.34)
\end{aligned}
$$

The symmetry properties of Majorana spinor bilinears (see (3.51)) have been used to reach the final result, which clearly shows the promised infinitesimal translation.

The analogous computation of $[\delta_1, \delta_2]P_L\chi(x)$ is more complex because a Fierz rearrangement is required. We outline it here:

$$
\begin{aligned}
[\delta_1, \delta_2]P_L\chi &= \frac{1}{\sqrt{2}}P_L(\partial\!\!\!/ \delta_1 Z + \delta_1 F)\epsilon_2 - [1 \leftrightarrow 2] \qquad (6.35) \\
&= \tfrac{1}{2}P_L\gamma^\mu\epsilon_2\bar{\epsilon}_1 P_L\partial_\mu\chi + \tfrac{1}{2}P_L\epsilon_2\bar{\epsilon}_1\partial\!\!\!/ P_L\chi - [1 \leftrightarrow 2] \\
&= -\tfrac{1}{8}(\bar{\epsilon}_1\Gamma_A\epsilon_2)P_L(\gamma^\mu\Gamma^A + \Gamma^A\gamma^\mu)P_L\partial_\mu\chi - [1 \leftrightarrow 2]\,.
\end{aligned}
$$

Each term in the second line was reordered as in Ex. 3.28 (with $\bar{\lambda}_1$ of (3.70) removed). We now find a great deal of simplification. Because of the antisymmetrization in $\epsilon_1 \leftrightarrow \epsilon_2$ the only non-vanishing bilinears are $\Gamma_A \to \gamma_\nu$ or $\gamma_{\nu\rho}$. However, only γ^ν survives the chiral projection in the last factor. Thus we find the expected result

$$
[\delta_1, \delta_2]P_L\chi = -\tfrac{1}{2}\bar{\epsilon}_1\gamma^\mu\epsilon_2 P_L\partial_\mu\chi \qquad (6.36)
$$

as the just reward for our labor.

Exercise 6.10 *It is quite simple to demonstrate that*

$$
[\delta_1, \delta_2]F = -\tfrac{1}{2}\bar{\epsilon}_1\gamma^\mu\epsilon_2\partial_\mu F. \qquad (6.37)
$$

Zealous readers should do it.

The auxiliary field F can be eliminated from the action by substituting the value $F = -\overline{W}'(\bar{Z})$, which is the solution of its equation of motion from (6.19), without affecting the classical (or quantum) dynamics. Here is an exercise to show that SUSY is also maintained after elimination.

Exercise 6.11 *Consider the theory after elimination of F and $\bar{F}$. Show that the action*

$$S = \int \mathrm{d}^4x \left[-\partial^\mu \bar{Z} \partial_\mu Z - \bar{\chi} \not{\partial} P_L \chi - \overline{W}' W' - \tfrac{1}{2} \bar{\chi} (P_L W'' + P_R \overline{W}'') \chi \right] \tag{6.38}$$

is invariant under the transformation rules (6.15) and their conjugates (6.16). Show that $[\delta_1, \delta_2] Z$ is exactly the same as in (6.34), but $[\delta_1, \delta_2] P_L \chi$ is modified as follows:

$$[\delta_1, \delta_2] P_L \chi = \bar{\epsilon}_1 \gamma^\mu \epsilon_2 P_L \left[-\tfrac{1}{2} \partial_\mu \chi + \tfrac{1}{4} \gamma_\mu (\not{\partial} + \overline{W}'') \chi \right] . \tag{6.39}$$

We find the spacetime translation plus an extra term that vanishes for any solution of the equations of motion.

Since the commutator of symmetries must give a symmetry of the action[6] and translations are a known symmetry, the remaining transformation, namely

$$\begin{aligned} \delta Z &= 0 , \\ \delta \chi &= v^\mu \gamma_\mu (\not{\partial} + P_L W'' + P_R \overline{W}'') \chi , \end{aligned} \tag{6.40}$$

is itself a symmetry for any constant vector v^μ. However, its Noether charge vanishes when the fermion equation of motion is satisfied, so it has no physical effect. Such a symmetry is sometimes called a 'zilch symmetry'.

Although nothing physically essential is changed by eliminating auxiliary fields, we can nevertheless see that they play a useful role:

(i) It is only with F and $\bar{F}$ included that the form of the SUSY transformation rules (6.15) and (6.16) is independent of the superpotential $W(Z)$.

(ii) The SUSY algebra is also universal on all components of the chiral multiplet when F is included. The phrase used in the literature is that the SUSY algebra is 'closed off-shell' when auxiliary fields are included and 'closed only on-shell' when they are eliminated.

(iii) Auxiliary fields are very useful in determining the terms in a SUSY Lagrangian describing couplings between different multiplets. An example is the general SUSY gauge theory described in the next section.

(iv) In local supersymmetry auxiliary fields simplify the couplings of Faddeev–Popov ghost fields.

[6] The argument is easy: a symmetry is a transformation such that $S_{,i}\delta(\epsilon)\phi^i = 0$, where $S_{,i}$ is the functional derivative with respect to the field ϕ^i. Applying a second transformation gives $S_{,ij}\delta(\epsilon_1)\phi^i\delta(\epsilon_2)\phi^j + S_{,i}\delta(\epsilon_2)\delta(\epsilon_1)\phi^i = 0$. Taking the commutator, the first term vanishes by symmetry, and the second term says that the commutator defines a symmetry.

It is also the case that auxiliary fields are known only for a few extended SUSY theories in four-dimensional spacetime and unavailable for many theories in dimension $D > 4$. Indeed many of the most interesting SUSY theories have no known auxiliary fields.

Although we hope that some readers enjoy the detailed manipulations needed to study SUSY theories, we suspect that many are fed up with Fierz rearrangement and would like a more systematic approach. To a large extent the superspace formalism does exactly that and has many advantages. Unfortunately, complete superspace methods are also unavailable when auxiliary fields are not known.

6.2.3 More chiral multiplets

We conclude this section with a discussion to establish a more general SUSY theory containing several chiral multiplets and their conjugates. We present this theory in the same 'blended' notation used above in which fermions always appear as the chiral projections $P_L\chi$ and $P_R\chi$ of Majorana spinors and the symmetry properties of Majorana bilinears can be used in all manipulations. To make the notation compatible with gauge symmetry in the next section, we denote chiral multiplets[7] by Z^α, $P_L\chi^\alpha$, F^α and their anti-chiral adjoints by $\bar{Z}_\alpha$, $P_R\chi_\alpha$, $\bar{F}_\alpha$. Note that we use lower indices α for the fields of the anti-chiral multiplets.

The interactions of the general theory are determined by an arbitrary *holomorphic* superpotential $W(Z^\alpha)$. We denote derivatives of W by $W_\alpha = \partial W/\partial Z^\alpha$, $W_{\alpha\beta} = \partial^2 W/\partial Z^\alpha \partial Z^\beta$, etc. The kinetic action $S_{\rm kin}$ is the obvious generalization of (6.17) to include a sum over the index α, while the chiral interaction term becomes

$$S_F = \int \mathrm{d}^4x [F^\alpha W_\alpha - \tfrac{1}{2}\bar{\chi}^\alpha P_L W_{\alpha\beta}\chi^\beta], \tag{6.41}$$

and one must add the conjugate action $S_{\bar{F}}$. The transformation rules of each multiplet are unmodified, but the index α is required, e.g. $\delta Z^\alpha = (1/\sqrt{2})\bar{\epsilon}P_L\chi^\alpha$. This general form of S_F explains why W' and W'' appear in (6.18).

Exercise 6.12 *Show that the new actions $S_{\rm kin}$, S_F, $S_{\bar{F}}$ are each invariant. The only essential new feature is that a Fierz rearrangement argument is required to show that the cubic term $W_{\alpha\beta\gamma}\bar{\epsilon}P_L\chi^\alpha\bar{\chi}^\beta P_L\chi^\gamma$, which is the analogue of the last term in (6.26), vanishes.*

After elimination of the auxiliary field using $F^\alpha = \partial\bar{W}/\partial\bar{Z}_\alpha \equiv \overline{W}^\alpha$ one finds the physically equivalent action

$$S = \int \mathrm{d}^4x \Big[-\partial^\mu\bar{Z}_\alpha\partial_\mu Z^\alpha - \bar{\chi}_\alpha \not{\partial} P_L\chi^\alpha - \overline{W}^\alpha W_\alpha$$
$$-\tfrac{1}{2}\bar{\chi}^\alpha P_L W_{\alpha\beta}\chi^\beta - \tfrac{1}{2}\bar{\chi}_\alpha P_R \overline{W}^{\alpha\beta}\chi_\beta\Big]. \tag{6.42}$$

[7] We do not use spinor indices any more, such that the use of $\alpha, \ldots$ to indicate the multiplets should not create confusion.

The $\mathrm{U}(1)_R$ symmetry discussed in Sec. 6.2.1 may be extended to the general chiral multiplet theory provided that the superpotential $W(Z^\alpha)$ satisfies an appropriate condition. To investigate this we assign charges $r_\alpha, r_\alpha - 1, r_\alpha - 2$ to the fields Z^α, $P_L\chi^\alpha$, F^α, so that infinitesimal transformations with parameter ρ are written as

$$\begin{aligned}
\delta_R Z^\alpha &= \mathrm{i}\rho r_\alpha Z^\alpha ,\\
\delta_R P_L\chi^\alpha &= \mathrm{i}\rho(r_\alpha - 1)P_L\chi^\alpha ,\\
\delta_R F^\alpha &= \mathrm{i}\rho(r_\alpha - 2)F^\alpha .
\end{aligned} \tag{6.43}$$

(No sum on α.)

Exercise 6.13 *Show that the general S_F of (6.41) is $\mathrm{U}(1)_R$ invariant for any set of charges r_α such that the superpotential satisfies the homogeneity condition*

$$\sum_\alpha r_\alpha Z^\alpha W_\alpha = 2W . \tag{6.44}$$

To prove this, you must generalize the argument of Sec. 6.2.1. The condition (6.44) is equivalent to the statement that W has definite R-charge $r_W = 2$.

For each specific theory with superpotential $W(Z^\alpha)$ there are several possibilities. There may not be any choice of the r_α for which (6.44) holds. This is the case in the Wess–Zumino model with $m \neq 0$ and $g \neq 0$ discussed at the end of Sec. 6.2.1. In some theories there is a unique set of R-charges, and in others many choices.

6.3 SUSY gauge theories

The basic SUSY gauge theory is the $\mathcal{N}=1$ SYM theory containing the gauge multiplet A_μ^A, λ^A, where A is the index of the adjoint representation of a compact, non-abelian gauge group G. This theory was described in Sec. 6.1.2. The discussion there focused on the conserved supercurrent and will now be extended to include field variations, auxiliary fields, and the SUSY algebra.

We assume that the group has an invariant metric that can be chosen as δ_{AB}. This is the case for 'reductive groups', i.e. products of compact simple groups and abelian factors, i.e. $G = G_1 \otimes G_2 \otimes \ldots$, where each factor is a simple group or U(1). The normalization of the generators is fixed in each factor, which can lead to different gauge coupling constants $g_1, g_2, \ldots$ for each of these factors. We have taken here the normalizations of the generators where these coupling constants do not appear explicitly. One can replace everywhere t_A with $g_i t_A$ and $f_{AB}{}^C$ with $g_i f_{AB}{}^C$, where g_i can be chosen independently in each simple factor, to re-install these coupling constants. Usually one also redefines then the parameters θ^A to $(1/g_i)\theta^A$. This leads to the formulas with coupling constant g in Sec. 4.3. Further note that for these algebras, the structure constants can be written as f_{ABC}, which are completely antisymmetric.

6.3.1 SUSY Yang–Mills vector multiplet

Our first objective is to obtain the SUSY variations δA_μ^A and $\delta\lambda^A$ under which the action (6.8) is invariant. This will give 'on-shell' supersymmetry; then we will add the auxiliary field. We will organize the presentation to make use of previous work in Secs. 6.1.1 and 6.1.2, which established that the supercurrent (6.10) is conserved.

The variation of (6.8) is

$$\delta S = \int \mathrm{d}^4x \left[\delta A_\nu^A D^\mu F_{\mu\nu}^A - \delta\bar\lambda^A \gamma^\mu D_\mu \lambda^A + \tfrac{1}{2} f_{ABC}\delta A_\mu^A \bar\lambda^B \gamma^\mu \lambda^C\right]. \tag{6.45}$$

We first note that the forms

$$\delta A_\mu^A = -\tfrac{1}{2}\bar\epsilon\gamma_\mu\lambda^A, \qquad \delta\lambda^A = \tfrac{1}{4}\gamma^{\rho\sigma}F_{\rho\sigma}^A\epsilon \tag{6.46}$$

are determined, up to constant factors, by Lorentz and parity symmetry and by the dimensions (in units of l^{-1}) of the quantities involved. Denoting the dimension of any quantity x by $[x]$, we have $[\epsilon] = -1/2$, $[A_\mu] = 1$, $[\lambda] = 3/2$. Note that if we use the assumed form for δA_μ, then the last term in (6.45) vanishes by the Fierz rearrangement identity (3.68). We substitute both assumed variations, assuming that $\epsilon(x)$ is a general function, and integrate by parts in the second term of (6.45) to obtain

$$\begin{aligned}\delta S &= -\tfrac{1}{2}\int \mathrm{d}^4x \left[\bar\epsilon\gamma^\nu\lambda^A D^\mu F_{\mu\nu}^A + \tfrac{1}{2}\bar\epsilon\gamma^{\rho\sigma}\gamma^\mu\lambda^A D_\mu F_{\rho\sigma}^A + \tfrac{1}{2}\partial_\mu\bar\epsilon\gamma^{\rho\sigma}\gamma^\mu F_{\rho\sigma}^A\lambda^A\right]\\ &= -\tfrac{1}{2}\int \mathrm{d}^4x\,[\bar\epsilon\gamma^\nu\lambda^A D^\mu F_{\mu\nu}^A - \bar\epsilon\gamma^\nu\lambda^A D^\mu F_{\mu\nu}^A + \tfrac{1}{2}\partial_\mu\bar\epsilon\gamma^{\rho\sigma}\gamma^\mu F_{\rho\sigma}^A\lambda^A],\end{aligned} \tag{6.47}$$

where (6.7) and the gauge field Bianchi identity were used to reach the final line. Thus δS vanishes for constant ϵ, establishing supersymmetry, while the supercurrent $\mathcal{J}^\mu$ of (6.10) appears in the last term![8]

The auxiliary field required for the gauge multiplet is a real pseudo-scalar field D^A in the adjoint representation of G. This fact follows from the superspace formulation. The auxiliary field enters the action and transformation rules in the quite simple fashion

$$S = \int \mathrm{d}^4x \left[-\tfrac{1}{4}F^{\mu\nu A}F_{\mu\nu}^A - \tfrac{1}{2}\bar\lambda^A\gamma^\mu D_\mu\lambda^A + \tfrac{1}{2}D^A D^A\right], \tag{6.48}$$

$$\begin{aligned}\delta A_\mu^A &= -\tfrac{1}{2}\bar\epsilon\gamma_\mu\lambda^A,\\ \delta\lambda^A &= \left[\tfrac{1}{4}\gamma^{\mu\nu}F_{\mu\nu}^A + \tfrac{1}{2}\mathrm{i}\gamma_* D^A\right]\epsilon,\\ \delta D^A &= \tfrac{1}{2}\mathrm{i}\,\bar\epsilon\gamma_*\gamma^\mu D_\mu\lambda^A, \qquad D_\mu\lambda^A \equiv \partial_\mu\lambda^A + \lambda^C A_\mu{}^B f_{BC}{}^A.\end{aligned} \tag{6.49}$$

Exercise 6.14 *Show that $\delta S = 0$. Only terms involving D^A need to be examined.*

[8] This term indicates that the true supercurrent should have been written in (6.10) with a factor $\frac{1}{4}$ included. $\frac{1}{4}\mathcal{J}^\mu$ generates correctly normalized SUSY variations.

The fields of the gauge multiplet transform also under an internal gauge symmetry:

$$\begin{aligned}\delta(\theta)A_\mu^A &= \partial_\mu\theta^A + \theta^C A_\mu{}^B f_{BC}{}^A\,,\\ \delta(\theta)\lambda^A &= \theta^C\lambda^B f_{BC}{}^A\,,\\ \delta(\theta)D^A &= \theta^C D^B f_{BC}{}^A\,.\end{aligned}\tag{6.50}$$

Let us first remark that the commutator of these internal gauge transformations and supersymmetry vanishes.

Exercise 6.15 *Use the transformation rules above to derive the SUSY commutator algebra of the gauge multiplet*

$$\begin{aligned}[\delta_1,\delta_2]\,A_\mu^A &= -\tfrac{1}{2}\bar\epsilon_1\gamma^\nu\epsilon_2 F_{\nu\mu}^A\,,\\ [\delta_1,\delta_2]\,\lambda^A &= -\tfrac{1}{2}\bar\epsilon_1\gamma^\nu\epsilon_2 D_\nu\lambda^A\,,\\ [\delta_1,\delta_2]\,D^A &= -\tfrac{1}{2}\bar\epsilon_1\gamma^\nu\epsilon_2 D_\nu D^A\,.\end{aligned}\tag{6.51}$$

It is no surprise that the commutator of two gauge covariant variations from (6.49) is gauge covariant, but at first glance the result seems to disagree with the spacetime translation required by (6.1) and (6.33). Note that in all three cases in Ex. 6.15 the difference between the covariant result in (6.51) and a translation is a gauge transformation by the field dependent gauge parameter $\theta^A = \frac{1}{2}\bar\epsilon_1\gamma^\nu\epsilon_2 A_\nu^A$. The covariant forms that occur in (6.51) are called gauge covariant translations. The conclusion is that on the fields of a gauge theory the SUSY algebra closes on gauge covariant translations. See [35] and Sec. 4.1.5 for further information on this issue.

6.3.2 Chiral multiplets in SUSY gauge theories

We now present and briefly discuss the class of SUSY theories in which the gauge multiplet A_μ^A, λ^A, D^A is coupled to a chiral matter multiplet Z^α, $P_L\chi^\alpha$, F^α in an arbitrary finite-dimensional representation $\mathbf{R}$ of G with matrix generators $(t_A)^\alpha{}_\beta$. The representation may be either reducible or irreducible. A reducible representation may be decomposed into a direct sum of irreducible components $\mathbf{R}_i$. The matrix generators in each $\mathbf{R}_i$ are denoted by t_{Ai}. Formally the decomposition is expressed by

$$\begin{aligned}\mathbf{R} &= \bigoplus\nolimits_i \mathbf{R}_i\,,\\ t_A &= \bigoplus\nolimits_i t_{Ai}\,.\end{aligned}\tag{6.52}$$

For most purposes and in most formulas below, the decomposition into irreducible representations need not be indicated explicitly, and we use it only where a more detailed notation is required.

The theory necessarily contains the conjugate anti-chiral multiplet $\bar Z_\alpha$, $P_R\chi_\alpha$, $\bar F_\alpha$, and we use lower indices to indicate that these fields transform in the conjugate representation $\bar{\mathbf{R}}$. Under an infinitesimal gauge transformation with parameters $\theta^A(x)$ the fermions transform as

$$\begin{aligned}\delta P_L\chi^\alpha &= -\theta^A(t_A)^\alpha{}_\beta P_L\chi^\beta\,,\\ \delta P_R\chi_\alpha &= -\theta^A(t_A)_\alpha^{*\beta} P_R\chi_\beta\,,\end{aligned}\tag{6.53}$$

with similar rules for the other fields. Representation indices are suppressed in most formulas. Thus we can write covariant derivatives of the various fields as

$$\begin{aligned}
D_\mu \lambda^A &= \partial_\mu \lambda^A + f_{BC}{}^A A_\mu^B \lambda^C \,, \\
D_\mu Z &= \partial_\mu Z + t_A A_\mu^A Z \,, \\
D_\mu P_L \chi &= \partial_\mu P_L \chi + t_A A_\mu^A P_L \chi \,, \\
D_\mu P_R \chi &= \partial_\mu P_R \chi + t_A^* A_\mu^A P_R \chi \,.
\end{aligned} \tag{6.54}$$

The system need not contain a superpotential, but superpotentials $W(Z^\alpha)$, which must be both holomorphic and gauge invariant, are optional. It is useful to express the condition of gauge invariance of $W(Z^\alpha)$ as

$$\delta_{\rm gauge} W = W_\alpha \delta_{\rm gauge} Z^\alpha = -W_\alpha \theta^A (t_A)^\alpha{}_\beta Z^\beta = 0 \,. \tag{6.55}$$

The action of the general theory is the sum of several terms

$$S = S_{\rm gauge} + S_{\rm matter} + S_{\rm coupling} + S_W + S_{\overline{W}} \,. \tag{6.56}$$

The form of some terms agrees with expressions given earlier in this chapter. Since convenience is a virtue and repetition is no sin, we shall write everything here:

$$\begin{aligned}
S_{\rm gauge} &= \int {\rm d}^4 x \left[-\tfrac14 F^{\mu\nu A} F_{\mu\nu}^A - \tfrac12 \bar\lambda^A \gamma^\mu D_\mu \lambda^A + \tfrac12 D^A D^A \right] , &(6.57)\\
S_{\rm matter} &= \int {\rm d}^4 x \left[-D^\mu \bar Z D_\mu Z - \bar\chi \gamma^\mu P_L D_\mu \chi + \bar F F \right] , &(6.58)\\
S_{\rm coupling} &= \int {\rm d}^4 x \left[-\sqrt2 \left(\bar\lambda^A \bar Z t_A P_L \chi - \bar\chi P_R t_A Z \lambda^A \right) + {\rm i}\, D^A \bar Z t_A Z \right] , &(6.59)\\
S_F &= \int {\rm d}^4 x \left[F^\alpha W_\alpha + \tfrac12 \bar\chi^\alpha P_L W_{\alpha\beta} \chi^\beta \right] , &(6.60)\\
S_{\bar F} &= \int {\rm d}^4 x \left[\bar F_\alpha \overline{W}^\alpha + \tfrac12 \bar\chi_\alpha P_R \overline{W}^{\alpha\beta} \chi_\beta \right] . &(6.61)
\end{aligned}$$

The full action is invariant under the SUSY transformation rules given in (6.49) for the gauge multiplet and the following modified gauge covariant transformation rules for the chiral and anti-chiral multiplets:

$$\begin{aligned}
\delta Z &= \frac{1}{\sqrt2} \bar\epsilon P_L \chi \,, \\
\delta P_L \chi &= \frac{1}{\sqrt2} P_L (\gamma^\mu D_\mu Z + F) \epsilon \,, \\
\delta F &= \frac{1}{\sqrt2} \bar\epsilon P_R \gamma^\mu D_\mu \chi - \bar\epsilon P_R \lambda^A t_A Z \,,
\end{aligned} \tag{6.62}$$

and

$$
\begin{aligned}
\delta\bar{Z} &= \frac{1}{\sqrt{2}}\bar{\epsilon}P_R\chi\,,\\
\delta P_R\chi &= \frac{1}{\sqrt{2}}P_R(\gamma^\mu D_\mu\bar{Z}+\bar{F})\epsilon\,,\\
\delta\bar{F} &= \frac{1}{\sqrt{2}}\bar{\epsilon}P_L\gamma^\mu D_\mu\chi-\bar{\epsilon}P_L\lambda^A(t_A)^*\bar{Z}\,.
\end{aligned}
\tag{6.63}
$$

Some modifications in the action and transformation rules above, notably the introduction of gauge covariant derivatives, are clearly required in a gauge theory, but other additions such as the form of the action S_{coupling} are surely not obvious. The best explanation is that they are dictated by the superspace formalism. However, all features can be explained from the component viewpoint. For example, in the SUSY variation δS_{matter} many terms cancel by the same manipulations required to show that δS_{kin} of (6.17) vanishes by simply replacing ∂_μ by D_μ. But there are extra terms due to the variation δA_μ^A,

$$
\delta S_{\text{matter}} = \int \mathrm{d}^4x\,\tfrac{1}{2}\bar{\epsilon}\gamma_\mu\lambda^A\,\left(\bar{\chi}t_A\gamma^\mu P_L\chi-\bar{Z}t_A D^\mu Z+D^\mu\bar{Z}t_A Z\right)\,,
\tag{6.64}
$$

which involves the gauge current of the matter fields, and there is a correction to (6.23) due to the gauge Ricci identity (4.91) applied to $\bar{Z}$. The correction is proportional to $F_{\mu\nu}^A\bar{\epsilon}P_L\gamma^{\mu\nu}\bar{Z}t_A\chi$. These terms are canceled by the variations of Z and χ in $\delta S_{\text{coupling}}$. A complete demonstration that the total action (6.56) is invariant under the variations (6.49) and (6.62) requires quite delicate calculations, which we recommend only for sufficiently diligent readers. The reader will also be invited to explain the extra terms in (6.62) from an algebraic viewpoint below in Ex. 14.2

Exercise 6.16 *Show that the action (6.56) is supersymmetric. How does the variation $\delta F^\alpha W_\alpha(Z)$ induced by the last term in δF cancel?*

The $\mathrm{U}(1)_R$ symmetry extends to SUSY gauge theories as a global symmetry which commutes with gauge transformations. Therefore the R-charges r_α that appear in the transformation (6.43) of chiral fields must be the same for all fields in a given irreducible component $\mathbf{R}_i$ of the full group representation $\mathbf{R}$. Invariance of the Yukawa terms in (6.59) determines the gaugino transformation

$$
\delta_R\lambda^A = \mathrm{i}\rho\gamma_*\lambda^A\,.
\tag{6.65}
$$

If the superpotential satisfies (6.44), then the full gauge theory is also $\mathrm{U}(1)_R$ invariant at the classical level. However, conservation of the Noether current R^μ is typically violated by the quantum level axial anomaly, and this has important consequences. We refer the reader to Sec. 29.3 of [30] and Sec. 2.C. of [36].

Exercise 6.17 *Show that the Noether current R^μ and the gauge currents of the general theory are given by*

$$
\begin{aligned}
R^\mu &= -\tfrac{1}{2}\mathrm{i}\bar{\lambda}^A\gamma^\mu\gamma_*\lambda^A+\mathrm{i}\sum_\alpha\left(r_\alpha(\bar{Z}_\alpha D_\mu Z^\alpha-D_\mu\bar{Z}_\alpha Z^\alpha)-\mathrm{i}(r_\alpha-1)\bar{\chi}_\alpha\gamma^\mu P_L\chi\right),\\
J_A^\mu &= -f_{ABC}\bar{\lambda}^B\gamma^\mu\lambda^C-(\bar{Z}t_A D_\mu Z-D_\mu\bar{Z}t_A Z)+\mathrm{i}\bar{\chi}\gamma^\mu P_L t_A\chi\,.
\end{aligned}
\tag{6.66}
$$

6.4 Massless representations of $\mathcal{N}$-extended supersymmetry

Up till now we have considered the supersymmetry generated by one Majorana spinor Q_α, subject to the structure relations of the superalgebra (6.1). The field theories that realize this algebra are said to possess *simple supersymmetry*. We now consider superalgebras containing $\mathcal{N} > 1$ Majorana spinor charges $Q_{i\alpha}$, $i = 1, \ldots, \mathcal{N}$. These algebras are called *$\mathcal{N}$-extended supersymmetry* algebras.

In the minimal extension, the different supersymmetry generators anti-commute, and each of them separately satisfies (6.1). We will rewrite (6.1) using the chiral components of the Majorana spinors (as justified in Box 3.3). We define $Q_{i\alpha}$ as the left-handed chiral components. They are simplest in the Weyl representation, in which the spinors have the form of (3.93). Then $Q_{i\alpha} = (Q_{i1}, Q_{i2}, 0, 0)$. We write the index i up for their hermitian conjugates, i.e. in Weyl representation: $Q^{\dagger i\alpha} = ((Q_{i1})^*, (Q_{i2})^*, 0, 0)$. Therefore, we can effectively only use the index range $\alpha = 1, 2$. We will use the *quantum* expression, which according to Sec. 1.5 implies a multiplication with i; see (1.84). This gives the algebra

$$
\begin{aligned}
\left\{Q_{i\alpha}, Q^{\dagger j\beta}\right\}_{\text{qu}} &= \tfrac{1}{2}\delta_i^j \left(\gamma_\mu \gamma^0\right)_\alpha{}^\beta p^\mu, \qquad \alpha = 1, 2, \\
\left[M_{[\mu\nu]}, Q_{i\alpha}\right]_{\text{qu}} &= -\tfrac{1}{2}\mathrm{i}(\gamma_{\mu\nu})_\alpha{}^\beta Q_{i\beta}, \\
\left[P_\mu, Q_{i\alpha}\right]_{\text{qu}} &= 0. \qquad (6.67)
\end{aligned}
$$

We refer to Appendix 6A for a more detailed, and representation independent, definition of $Q_{i\alpha}$ and $Q^{\dagger j\beta}$ and the explanation of the anti-commutator in (6.87).

In this chapter, we will restrict attention to the extended supersymmetry algebra in (6.67). More general algebras, e.g. including the concept of 'central charges', will be deferred to Ch. 12.

6.4.1 Particle representations of $\mathcal{N}$-extended supersymmetry

We now discuss the particle representations of the superalgebras (6.67), that is, representations whose carrier space is the Hilbert space of a relativistic quantum field theory. Therefore we use a basis in which particles of energy–momentum $p^\mu = (p^0 = E = \sqrt{m^2 + \vec{p}^2},\ \vec{p})$ and spin s are described by states $|p^\mu, s, h\rangle$. For massive particles the helicity h takes $2s + 1$ equally spaced values in the range $-s \le h \le s$. For a massless particle, there are two helicity values $h = \pm s$ if $s > 0$, and the unique value $h = 0$ if $s = 0$.

The carrier space of a particle representation of supersymmetry consists of the states $|p^\mu, s, h\rangle$ of a set of bosons, $s = 0, 1, 2, \ldots$, and fermions, $s = 1/2, 3/2, 5/2, \ldots$. The basic observation needed to study these representations is that $[Q, P] = 0$. Supersymmetry preserves the energy–momentum p^μ and thus the mass m of any particle state. Thus all that is needed to find the representations of the SUSY algebra is to consider finite sets of Bose and Fermi particles with fixed 4-momentum and various spins s and determine those sets on which the basic anti-commutator of supercharges can be realized irreducibly.

Let's first prove a very general result, namely that any irreducible representation of the SUSY algebra, whether it involves massive or massless particles, contains equal numbers of boson and fermion states. Taking a sum on spinor indices in (6.67) gives

$$Q_{i\alpha} Q^{\dagger j\alpha} + Q^{\dagger j\alpha} Q_{i\alpha} = \delta_i^j P^0 . \tag{6.68}$$

Consider the operator $\mathrm{e}^{-2\pi \mathrm{i} J_3}$ which implements rotations by angle 2π. Clearly its effect on boson and fermion states and on the supercharges is

$$\begin{aligned} \mathrm{e}^{-2\pi \mathrm{i} J_3} |p^\mu, s, h\rangle &= (-)^{2s} |p^\mu, s, h\rangle , \\ \left\{ Q_{i\alpha} , \mathrm{e}^{-2\pi \mathrm{i} J_3} \right\} &= 0 . \end{aligned} \tag{6.69}$$

Multiply (6.68) on the right by $\mathrm{e}^{-2\pi \mathrm{i} J_3}$ and form the Hilbert space expectation value in a particle state $|p^\mu, s, h\rangle$. Finally sum over the spins and helicities of the particles in the carrier space of a representation. One finds

$$\begin{aligned} \sum_{s,h} \langle p^\mu, s, h| \left(Q_{i\alpha} Q^{\dagger j\alpha} + Q^{\dagger j\alpha} Q_{i\alpha} \right) \mathrm{e}^{-2\pi \mathrm{i} J_3} |p^\mu, s, h\rangle \\ = \delta_i^j \sum_{s,h} \langle p^\mu, s, h| P^0 \mathrm{e}^{-2\pi \mathrm{i} J_3} |p^\mu, s, h\rangle . \end{aligned} \tag{6.70}$$

The sum over spins and helicities with fixed p^μ is equivalent to a matrix trace in a finite-dimensional subspace of the Hilbert space and can be manipulated as a conventional matrix trace. Hence we can rewrite (6.70) as

$$\begin{aligned} \mathrm{Tr} \left(Q_{i\alpha} Q^{\dagger j\alpha} \mathrm{e}^{-2\pi \mathrm{i} J_3} + Q^{\dagger j\alpha} Q_{i\alpha} \mathrm{e}^{-2\pi \mathrm{i} J_3} \right) &= \delta_i^j E \, \mathrm{Tr} \, \mathrm{e}^{-2\pi \mathrm{i} J_3} , \\ \mathrm{Tr} \left(Q_{i\alpha} Q^{\dagger j\alpha} \mathrm{e}^{-2\pi \mathrm{i} J_3} - Q^{\dagger j\alpha} \mathrm{e}^{-2\pi \mathrm{i} J_3} Q_{i\alpha} \right) &= \delta_i^j E \, \mathrm{Tr} \, \mathrm{e}^{-2\pi \mathrm{i} J_3} . \end{aligned} \tag{6.71}$$

The left-hand side vanishes by cyclicity of the trace! The trace on the right-hand side can be rewritten as a sum over spins $s = 0, 1/2, 1, \ldots$ weighted by the number of particles n_s of spin s in the representation and the number of helicity states for each s. We thus obtain separate sum rules for massive and massless representations:

$$m > 0, \qquad \sum_{s \geq 0} (-)^{2s} n_s (2s + 1) = 0, \tag{6.72}$$

$$m = 0, \qquad 2 \sum_{s > 0} (-)^{2s} n_s + n_0 = 0 . \tag{6.73}$$

This is the desired result since $(-)^{2s}$ is equal to $+1$ for bosons and -1 for fermions.

There is a small subtlety in the interpretation of (6.73). Lorentz transformations do not change the helicity of a massless particle, so an irreducible representation of the Poincaré group involves the momentum states for a single value of the helicity h. However, the CPT reflection symmetry requires that both helicity states $h = \pm s$ are present in the quantum field theory of a massless particle with spin $s > 0$. This doubling is incorporated in (6.73).

6.4.2 Structure of massless representations

In this section we derive the particle content of unitary irreducible representations involving massless particles. Similar techniques apply to both massless and massive representations, but we focus on the massless representations because they are simpler. The review of Sohnius [37] contains a more complete treatment.

We now use the Weyl representation (2.19) of the γ-matrices explicitly and (6.67) then gives

$$\left\{Q_{i\alpha}, Q^{\dagger j\beta}\right\}_{\text{qu}} = \tfrac{1}{2}\delta_i^j \left(\mathbb{1}P^0 - \vec{\sigma}\cdot\vec{P}\right)_\alpha{}^\beta , \tag{6.74}$$

$$\left\{Q_{i\alpha}, Q_{j\beta}\right\}_{\text{qu}} = 0 , \qquad \left\{Q^{\dagger i\alpha}, Q^{\dagger j\beta}\right\}_{\text{qu}} = 0 , \tag{6.75}$$

$$\left[\vec{J}, Q_{i\alpha}\right]_{\text{qu}} = -\tfrac{1}{2}(\vec{\sigma})_\alpha{}^\beta Q_{i\beta} . \tag{6.76}$$

$\vec{J}$ stands for the space components $J^i = -\frac{1}{2}\varepsilon^{ijk}M_{jk}$.

Since SUSY transformations do not change the 4-momentum, it is sufficient to consider the action of the supercharges on a set of particle states $|\bar{p}, h\rangle$ of fixed energy–momentum $\bar{p}^\mu = (E, 0, 0, E)$. On states of 4-momentum $\bar{p}^\mu$, we find from (6.74) that

$$\begin{aligned} \left\{Q_{i1}, Q^{\dagger j1}\right\}_{\text{qu}} &= 0 , \\ \left\{Q_{i2}, Q^{\dagger j2}\right\}_{\text{qu}} &= E\delta_i^j . \end{aligned} \tag{6.77}$$

We want a *unitary* representation, one in which the Hilbert space norm of all states is positive. The positivity properties of the anti-commutator then require that Q_{i1} and its adjoint must be represented trivially. They have vanishing action on all states of 4-momentum $\bar{p}^\mu$. The remaining non-trivial anti-commutator in our basis involves the $\mathcal{N}$ supercharge components Q_{i2}. Physicists can immediately recognize that this anti-commutator describes the creation and annihilation operators for $\mathcal{N}$ independent fermions. Equivalently the Q_{i2} anti-commutator defines a Clifford algebra with $\mathcal{N}$ complex generators. This is equivalent to the real $2\mathcal{N}$-dimensional Clifford algebra we discussed in Sec. 3.1. The unique irreducible representation of this algebra has dimension $2^{\mathcal{N}}$, so the massless SUSY representation must also have $2^{\mathcal{N}}$ particle states.

The standard Fock space techniques of physics tells us that the unique unitary representation has the following structure. We choose the $Q^{\dagger i2}$ as the creation operators and the Q_{i2} as the annihilators. Note that (6.76) implies that $[J^3, Q^{\dagger i2}] = -\frac{1}{2}Q^{\dagger i2}$. Thus $Q^{\dagger i2}$ lowers the helicity of a state by 1/2. To specify the representation we define its 'Fock vacuum' $|\bar{p}, h_0\rangle$, with h_0 any positive or negative integer or half-integer, as the state that satisfies

$$Q_{i2}|\bar{p}, h_0\rangle = 0 , \qquad J^3|\bar{p}, h_0\rangle = h_0|\bar{p}, h_0\rangle , \qquad \forall i = 1, \dots, \mathcal{N} . \tag{6.78}$$

The basis[9] of the representation consists of the vacuum state together with all states obtained by applying products of creators. Such products are automatically antisymmetric due to (6.75). In more detail the basis is

[9] The particle spin $s = |h|$ is a redundant label, so it is omitted on all states of the Fock basis.

$$
\begin{aligned}
&|\bar{p}, h_0\rangle\,, \\
&|\bar{p}, h_0-\tfrac{1}{2}, i\rangle = Q^{\dagger i2}|\bar{p}, h_0\rangle\,, \\
&|\bar{p}, h_0-1, [ij]\rangle = Q^{\dagger i2}Q^{\dagger j2}|\bar{p}, h_0\rangle\,, \\
&\text{etc.}
\end{aligned}
\tag{6.79}
$$

States of helicity $h_0-\frac{1}{2}m$ have multiplicity $\binom{\mathcal{N}}{m}=\mathcal{N}!/[m!(\mathcal{N}-m)!]$, and the sequence stops at the multiplicity 1 state of lowest helicity $h_0-\frac{1}{2}\mathcal{N}$. The total number of states is $2^{\mathcal{N}}$ as is required by the representation theory of Clifford algebras. One can check that half the states are bosons and half are fermions.

Thus a massless irreducible representation of supersymmetry contains a 'tower' of helicity states of maximum helicity h_0 and minimum helicity $h_0-\frac{1}{2}\mathcal{N}$. A local field theory always contains particles in CPT conjugate pairs with helicities $h=\pm s$. Therefore a supersymmetric field theory typically describes a reducible representation of the algebra in which the CPT conjugate states are added to the states of the basis (6.79). These states are obtained by starting from the CPT conjugate Clifford vacuum $|\bar{p},-h_0\rangle$ and applying products of the operator Q_{i2}, which raises helicity. When $\mathcal{N}=4|h_0|$, the initial sequence (6.79) is already self-conjugate so nothing needs to be added.

Because helicities are always paired, it is simplest to describe the field theory representations in terms of the number of particle states of a given spin. In Table 6.1 [38] we list the spin content[10] of all representations whose maximum spin satisfies $s_{\rm max}\leq 2$. It is this set of field theories that can incorporate nonlinear interactions.

Exercise 6.18 *Show that the spin content of representations with $\mathcal{N}=4s_{\rm max}$ and $\mathcal{N}=4s_{\rm max}-1$ is the same (see footnote 10).*

The $\mathcal{N}=1$ multiplets with maximum spin $s_{\rm max}=1/2$ and $s_{\rm max}=1$ are the chiral and gauge multiplets whose interactions are discussed earlier in this chapter. In principle the next $\mathcal{N}=1$ multiplet has spins $(1,3/2)$. There is a corresponding *free* field theory, but no *interacting* field theory is known for this multiplet without supergravity. The reason, discussed in Ch. 5, is that field theories for spin-3/2 fields involve a local supersymmetry. The supersymmetry algebra would contain local translations, and hence general relativity. Therefore, we find the spin-3/2 particle only in the multiplet $(3/2,2)$. This is the supergravity multiplet that we will consider in Ch. 9.

The table is limited to $\mathcal{N}\leq 8$ because of the great difficulty of consistent higher spin interactions. For $\mathcal{N}\geq 9$ massless representations necessarily contain some particles of higher spin $s\geq 5/2$. Despite much effort, no interacting field theories in Minkowski spacetime exist. The content of the table may be summarized by the statements:

1. For $\mathcal{N}\leq 4$ there are interacting theories with global SUSY and $s_{\rm max}\leq 1$.
2. For $\mathcal{N}\leq 8$, there are theories with local SUSY, which involve one spin-2 graviton, $\mathcal{N}$ spin-3/2 gravitinos, and, for $\mathcal{N}\geq 2$, lower spin particles. These are supergravity theories.
3. For $\mathcal{N}\geq 9$ there is the higher spin desert.

[10] There is a subtle hermiticity requirement for $\mathcal{N}=2$, which requires that the multiplet $(-1/2,0,0,1/2)$ must be doubled although it is self-conjugate.

Table 6.1 Spin content of representations of supersymmetry with maximal spin $s_{\max} \leq 2$.

		$s=2$	$s=3/2$	$s=1$	$s=1/2$	$s=0$
$\mathcal{N}=1$	$s_{\max}=2$	1	1			
	$s_{\max}=3/2$		1	1		
	$s_{\max}=1$			1	1	
	$s_{\max}=1/2$				1	1+1
$\mathcal{N}=2$	$s_{\max}=2$	1	2	1		
	$s_{\max}=3/2$		1	2	1	
	$s_{\max}=1$			1	2	1+1
	$s_{\max}=1/2$				2	2+2
$\mathcal{N}=3$	$s_{\max}=2$	1	3	3	1	
	$s_{\max}=3/2$		1	3	3	1+1
	$s_{\max}=1$			1	3+1	3+3
$\mathcal{N}=4$	$s_{\max}=2$	1	4	6	4	1+1
	$s_{\max}=3/2$		1	4	6+1	4+4
	$s_{\max}=1$			1	4	6
$\mathcal{N}=5$	$s_{\max}=2$	1	5	10	10+1	5+5
	$s_{\max}=3/2$		1	5+1	10+5	10+10
$\mathcal{N}=6$	$s_{\max}=2$	1	6	15+1	20+6	15+15
	$s_{\max}=3/2$		1	6	15	20
$\mathcal{N}=7$	$s_{\max}=2$	1	7+1	21+7	35+21	35+35
$\mathcal{N}=8$	$s_{\max}=2$	1	8	28	56	70

Exercise 6.19 *The reader should check Table 6.1.*

Appendix 6A Extended supersymmetry and Weyl spinors

It is often useful to discuss extended supersymmetry using Weyl spinors and their conjugates rather than Majorana spinors. The equivalence is discussed in Box 3.3, where we showed that one can represent Majorana spinors in terms of Weyl spinors and their conjugates, i.e. $Q = P_L Q + P_R Q$. The chirality of the (Majorana) conjugate spinors can be obtained from (3.56), which implies that

$$\overline{(P_R Q)} \equiv (P_R Q)^T C = \overline{Q} P_R. \qquad (6.80)$$

Hence applying a chiral projector to (6.1) teaches us that the two supercharges should have opposite chirality in order to have a non-vanishing anti-commutator:

$$\left\{(P_L Q)_\alpha, \overline{(P_R Q)}^\beta\right\} = -\tfrac{1}{2}(P_L\gamma^\mu)_\alpha{}^\beta P_\mu\,. \tag{6.81}$$

It is convenient to use the up or down position of the index $i=1,\ldots,\mathcal{N}$ to indicate at once the chiral projections of the supersymmetry generators for extended supersymmetry:

$$Q_i = P_L Q_i\,, \qquad Q^i = P_R Q^i\,. \tag{6.82}$$

The Majorana spinors are thus $Q^i + Q_i$, and Q_i is the charge conjugate of Q^i. From (6.80) we obtain

$$\overline{Q}_i = \overline{(P_L Q_i)} = \overline{Q}_i P_L\,, \qquad \overline{Q}^i = \overline{(P_R Q^i)} = \overline{Q}^i P_R\,. \tag{6.83}$$

The minimal extended algebra is then

$$\begin{aligned}
\{Q_{i\alpha}, \bar Q^{j\beta}\} &= -\tfrac12\delta_i^j(P_L\gamma_\mu)_\alpha{}^\beta P^\mu\,, & \left\{Q^i_\alpha, \bar Q_j^\beta\right\} &= -\tfrac12\delta^i{}_j(P_R\gamma_\mu)_\alpha{}^\beta P^\mu\,,\\
\left\{Q_{i\alpha}, \bar Q_j^\beta\right\} &= 0\,, & \{Q^i_\alpha, \bar Q^{j\beta}\} &= 0\,,\\
\left[M_{[\mu\nu]}, Q_{i\alpha}\right] &= -\tfrac12(\gamma_{\mu\nu})_\alpha{}^\beta Q_{i\beta}\,, & \left[M_{[\mu\nu]}, Q^i_\alpha\right] &= -\tfrac12(\gamma_{\mu\nu})_\alpha{}^\beta Q^i_\beta\,,\\
\left[P_\mu, Q_{i\alpha}\right] &= 0, & \left[P_\mu, Q^i_\alpha\right] &= 0\,.
\end{aligned} \tag{6.84}$$

Exercise 6.20 *Lower the spinor indices in the first anti-commutator of (6.84) and check that the equation is consistent with the one obtained by taking the charge conjugate.*

In Sec. 6.4 we used the complex conjugate spinors. First, remark that Q is Majorana, and we can thus also use $\overline{Q} = \mathrm{i}Q^\dagger\gamma^0$ in (6.80) to write

$$\overline{(P_R Q)} = \mathrm{i}Q^\dagger\gamma^0 P_R = \mathrm{i}Q^\dagger P_L\gamma^0 = \mathrm{i}(P_L Q)^\dagger\gamma^0\,, \tag{6.85}$$

which illustrates again that $(P_R Q)$ is not Majorana. Using the notation (6.82), this gives

$$\overline{Q}^i = \mathrm{i}(Q_i)^\dagger\gamma^0 = \mathrm{i}Q^{\dagger i}\gamma^0\,, \qquad Q^{\dagger i\alpha} \equiv (Q_{i\alpha})^\dagger\,. \tag{6.86}$$

In the last equation we define $Q^{\dagger i}$, with upper i index, as the hermitian conjugate of Q_i, which implies (omitting again spinor indices) $Q^{\dagger i} = Q^{\dagger i}P_L$. This leads to

$$\left\{Q_{i\alpha}, Q^{\dagger j\beta}\right\} = -\tfrac12\mathrm{i}\delta_i^j\left(P_L\gamma_\mu\gamma^0\right)_\alpha{}^\beta P^\mu\,, \qquad \alpha = 1,2\,. \tag{6.87}$$

In the Weyl representation, $Q_{i\alpha} = (Q_{i1}, Q_{i2}, 0, 0)$, and $\bar Q^i$ are their right-handed conjugates, i.e. $\bar Q^{i\alpha} = (0,0,(\mathrm{i}Q_{i1})^*, \mathrm{i}(Q_{i2})^*)$. The extra factor i to go to the quantum bracket thus leads to (6.67).

Appendix 6B On- and off-shell multiplets and degrees of freedom

It has been shown in Sec. 6.4.1 that on-shell multiplets have equal numbers of bosonic and fermionic degrees of freedom. But the reader may have noticed in the explicit examples in

Box 6.1

Equal number of bosonic and fermionic degrees of freedom

There are equal numbers of bosonic and fermionic degrees of freedom in any realization of a supersymmetry algebra of the form $\{Q, Q\} = P$.

this chapter that also off-shell the multiplets have equal numbers of bosonic and fermionic degrees of freedom. More strictly stated, the theorem of Box 6.1 holds. The theorem is proven in [37]. Off-shell equality of bosonic and fermionic degrees of freedom holds for some extended supersymmetry and higher dimensional theories, but it is not always true. It is valid when the algebra (6.1) holds. In practice this holds when the theory has auxiliary fields which 'close the algebra off-shell'.

Consider first the example of the chiral multiplet. We have discussed the chiral multiplet first with the fields $\{Z, \chi, F\}$, and satisfying this algebra. Z and F are complex fields, and thus there are four real off-shell (since we did not use field equations) bosonic degrees of freedom. These are balanced by the four components of the Majorana spinor χ. We say that the chiral multiplet is a $4+4$ off-shell multiplet.

On the other hand, we have seen in Sec. 6.2.2 that the algebra is also valid when the equations of motion are used. Then F is no longer an independent field, we count two bosonic degrees of freedom for the complex Z, and also the fermions have two on-shell degrees of freedom. So the chiral multiplet is also a $2+2$ on-shell multiplet. These two ways of counting are called *on-shell counting* and *off-shell counting*.

To interpret this theorem we remind the reader of the terminology of on-shell and off-shell degrees of freedom that we introduced in Box 4.1. To illustrate that the relevant definition of off-shell degrees of freedom indeed should contain the subtraction of gauge transformations, we consider the gauge multiplet. The off-shell counting is easily established: the gauge field A_μ and the gaugino λ both describe two on-shell degrees of freedom. To apply the theorem in the off-shell case, we have to remember that at the end of Sec. 6.3.1 we saw that the anti-commutator of two supersymmetries involves also a gauge transformation. Therefore, we can only apply the theorem on gauge invariant states (or identify states that differ by a gauge transformation), i.e. we have to subtract the gauge transformations

Table 6.2 The number of off-shell and on-shell degrees of freedom of the basic fields.

field	off-shell		on-shell	
		$D=4$		$D=4$
ϕ	1	1	1	1
λ	$2^{[D/2]}$	4	$\frac{1}{2}2^{[D/2]}$	2
A_μ	$D-1$	3	$D-2$	2
ψ_μ	$(D-1)2^{[D/2]}$	12	$\frac{1}{2}(D-3)2^{[D/2]}$	2
$g_{\mu\nu}$	$\frac{1}{2}D(D-1)$	6	$\frac{1}{2}D(D-3)$	2

in the counting. Thus, the gauge vector A_μ counts off-shell for three degrees of freedom, which together with the one real degree of freedom of the auxiliary field D balance the four off-shell ones of the gaugino.

Since this counting can be used to understand the structure of many realizations of supersymmetry, we end with Table 6.2 that summarizes the results for the degrees of freedom for the scalar field ϕ, a Majorana fermion λ, the gauge field A_μ, the Majorana Rarita–Schwinger field ψ_μ and the graviton field $g_{\mu\nu}$.

Exercise 6.21 *Check that the entries of the table correspond with the results on degrees of freedom obtained in previous chapters. The results for the graviton will be obtained in Sec. 8.2.*

PART II

DIFFERENTIAL GEOMETRY AND GRAVITY

7 Differential geometry

In this chapter we collect the ideas of differential geometry that are required to formulate general relativity and supergravity. There are several books, written for physicists, which explore this subject at greater length and greater depth [39, 40, 41, 42, 43, 44].

In general relativity spacetime is viewed as a differentiable manifold of dimension $D \geq 4$ with a metric of Lorentzian signature $(-, +, +, \ldots, +)$ indicating one time dimension and $D - 1$ space dimensions. We assume that readers of this book are not intimidated by the idea of $D - 4$ hidden dimensions that are not directly observed. We will also consider manifolds of purely Euclidean signature $(+, +, \ldots, +)$, which may appear in the hidden extra dimensions and as the target space of nonlinear σ-models.

We will give a reasonably rigorous definition of a manifold and then introduce the various quantities that 'live on it' in a less formal manner, emphasizing the way that the quantities transform under changes of coordinates. Invariance under coordinate transformations is one of the key principles that underlie general relativity. The most important structures we need are the metric, connection, and curvature. But other quantities such as vector and tensor fields and differential forms are also very useful. We will discuss them first since they require only the manifold structure.

It would be good if readers have already encountered some of the more elementary ideas before, perhaps in a course on general relativity. Our primary purpose is to collect the necessary ideas and explain them, hopefully clearly albeit non-rigorously, and thus to prepare readers for later chapters where the ideas are applied. Readers who do the suggested exercises will achieve the most thorough preparation.

7.1 Manifolds

A D-dimensional manifold is a topological space M together with a family of open sets M_i that cover it, i.e. $M = \cup_i M_i$. The M_i are called coordinate patches. On each patch there is a $1:1$ map ϕ_i, called a chart, from $M_i \to \mathbb{R}^D$. In more concrete language a point $p \in M_i \subset M$ is mapped to $\phi_i(p) = (x^1, x^2, \ldots, x^D)$. We say that the set $(x^1, x^2, \ldots, x^D)$ are the local coordinates of the point p in the patch M_i. If $p \in M_i \cap M_j$, then the map $\phi_j(p) = (x'^1, x'^2, \ldots, x'^D)$ specifies a second set of coordinates for the point p. The compound map $\phi_j \circ \phi_i^{-1}$ from $\mathbb{R}^D \to \mathbb{R}^D$ is then specified by the set of functions $x'^\mu(x^\nu)$. These functions, and their inverses $x^\nu(x'^\mu)$, are required to be smooth, usually C^∞. See Fig. 7.1 for an illustration of the ideas just discussed.

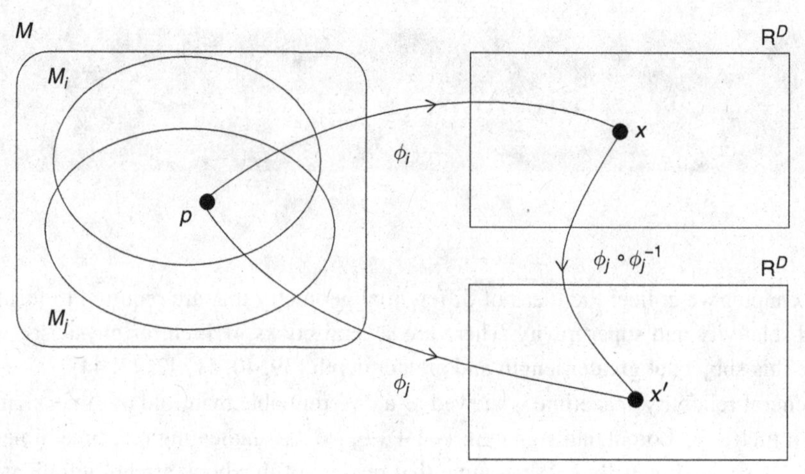

Fig. 7.1 Two charts in $\mathbb{R}^D$ for subsets M_i and M_j of the space M, and the compound map.

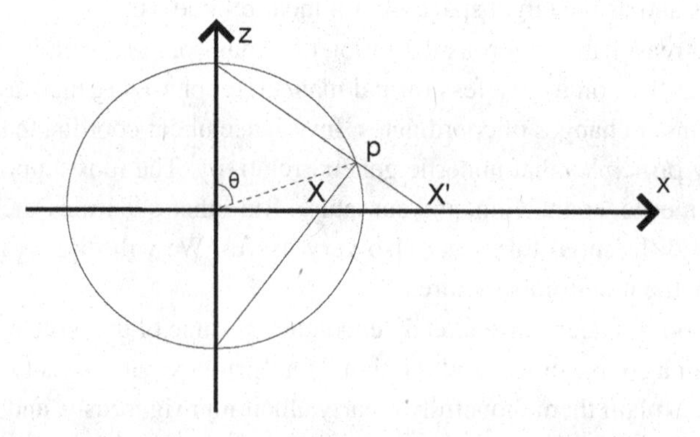

Fig. 7.2 Stereographic projection of the x–z plane of the 2-sphere. The coordinates of the point p are $(x, y, z) = (\sin\theta, 0, \cos\theta)$.

We now describe the unit 2-sphere S^2 as an interesting and useful example of a manifold. Initially it may be defined as the surface $x^2 + y^2 + z^2 = 1$ embedded in $\mathbb{R}^3$. It is common to use the usual spherical polar coordinates θ, ϕ with $z = \cos\theta$, $x = \sin\theta\cos\phi$, $y = \sin\theta\sin\phi$. This is fine for some purposes, but it does not define a good coordinate chart at the poles $\theta = 0, \pi$, since these points have no unique values of ϕ.

There are many ways to introduce coordinate charts to define a manifold structure. One useful way is to use the stereographic projection illustrated in Fig. 7.2. There are two patches whose union covers the sphere, namely M_1, consisting of the sphere with south pole deleted, and M_2, which is the sphere with north pole deleted. From the plane geometry of the triangles in Fig. 7.2, one defines the maps ϕ_1 and ϕ_2 to the central plane in the figure. These maps take the point with polar coordinates θ, ϕ to points X, Y and X', Y' respectively. The maps are given by

$$\phi_1:\; X + \mathrm{i}Y = \mathrm{e}^{\mathrm{i}\phi}\tan(\theta/2)\,,$$
$$\phi_2:\; X' + \mathrm{i}Y' = \mathrm{e}^{\mathrm{i}\phi}\cot(\theta/2)\,. \tag{7.1}$$

On the overlap, we see that

$$\phi_2 \circ \phi_1^{-1}(X, Y) = X' + \mathrm{i}Y' = 1/(X - \mathrm{i}Y)\,. \tag{7.2}$$

Exercise 7.1 *Derive (7.1) and (7.2).*

7.2 Scalars, vectors, tensors, etc.

The simplest objects to define on a manifold M are scalar functions f that map $M \to \mathbb{R}$. We say that the point p maps to $f(p) = z \in \mathbb{R}$. On each coordinate patch M_i we can define the compound map $f \circ \phi_i^{-1}$ from $\mathbb{R}^D \to \mathbb{R}$ as $f_i(x) \equiv f \circ \phi_i^{-1}(x) = z$, where x stands for $\{x^\mu\}$. On the overlap $M_i \cap M_j$ of two patches with local coordinates x^μ and x'^ν of the point p, the two descriptions of f must agree. Thus $f_i(x) = f_j(x')$.

We now define the properties of scalar functions in the less formal way we will use for most of the objects that live on M. We no longer refer to a covering by coordinate patches. Instead we conceive of the manifold as a set whose points may be described by many different coordinate systems, say $(x^0, x^1, \ldots, x^{D-1})$ and $(x'^0, x'^1, \ldots, x'^{D-1})$. Any two sets of coordinates are related by a set of C^∞ functions, e.g. $x'^\mu(x^\nu)$ with non-singular Jacobian $\partial x'^\mu/\partial x^\nu$. We refer to such a change of coordinates as a general coordinate transformation. A scalar function, also called a scalar field, is described by $f(x)$ in one set of coordinates and $f'(x')$ in the second set. The two functions must be pointwise equal, i.e.

$$f'(x') = f(x)\,. \tag{7.3}$$

Locally, at least, the informal definition agrees with the more formal one above.

In the same fashion, a contravariant vector field is described by D functions $V^\mu(x)$ in one coordinate system and D functions $V'^\mu(x')$ in the second. They are related by

$$V'^\mu(x') = \frac{\partial x'^\mu}{\partial x^\nu} V^\nu(x)\,, \tag{7.4}$$

with a summation convention on the repeated index ν. We go on to define covariant vector fields $\omega_\mu(x)$ and (mixed) tensors $T_\nu^\mu(x)$ by their behavior under coordinate transformations, namely

$$\omega'_\mu(x') = \frac{\partial x^\nu}{\partial x'^\mu}\omega_\nu(x)\,,$$
$$T'^\mu_\nu(x') = \frac{\partial x'^\mu}{\partial x^\sigma}\frac{\partial x^\rho}{\partial x'^\nu} T^\sigma_\rho(x)\,. \tag{7.5}$$

We leave it to the reader to devise the analogous definitions of higher rank tensors such as $T^{\mu\nu}(x)$, $S_{\mu\nu\rho}$, etc. A tensor field with p contravariant and q covariant indices is called a tensor of type (p, q) and rank $p + q$.

At this point in the development, contravariant and covariant quantities are unrelated objects, which transform differently. However a contravariant and covariant index can be contracted (i.e. summed) to define tensorial quantities of lower rank.

Exercise 7.2 *Given $V^\mu(x)$, $\omega_\mu(x)$, $T_\nu^\mu(x)$, show that $V^\mu(x)\omega_\mu(x)$ transforms as a scalar field and that $T_\nu^\mu(x)V^\nu(x)$ transforms as a contravariant vector.*

One can proceed with concrete local definitions of this type to obtain a physically satisfactory formulation of general relativity. However, there is much richness to be gained, and considerable practical advantage, if we develop the ideas further and incorporate some of the concepts of a more mathematical treatment of differential geometry.

Given a contravariant vector field $V^\mu(x)$, one can consider the system of differential equations

$$\frac{\mathrm{d}x^\mu}{\mathrm{d}\lambda} = V^\mu(x)\,. \tag{7.6}$$

A solution $x^\mu(\lambda)$ is a map from $\mathbb{R} \to M$, which is a curve on M, called an integral curve of the vector field. There is an integral curve through every point of any open subset of M in which the vector field does not vanish. If the manifold is $\mathbb{R}^D$, then we know that the vector $\mathrm{d}x^\mu/\mathrm{d}\lambda$ is tangent to the curve $x^\mu(\lambda)$, and we make the same interpretation for a general manifold.

Let $x^\mu(\lambda)$ be the integral curve through the point p of M with coordinates $x^\mu(\lambda_0)$. Then $\mathrm{d}x^\mu/\mathrm{d}\lambda|_{\lambda_0} = V^\mu(x(\lambda_0))$ is the tangent vector to the curve $x^\mu(\lambda)$ at p. We can now consider $D-1$ other vector fields $\tilde{V}^\mu(x)$ whose values $\tilde{V}^\mu(x(\lambda_0))$, together with the first $V^\mu(x(\lambda_0))$, fill out a basis of $\mathbb{R}^D$. Each $\tilde{V}^\mu(x(\lambda_0))$ is the tangent vector of an integral curve $\tilde{x}(\lambda)$ through p. Thus the vector fields evaluated at p determine the D-dimensional vector space $T_p(M)$, the tangent space to the manifold at point p. A vector field $V^\mu(x)$ may then be thought of as a smooth assignment of a tangent vector in each $T_p(M)$ as p varies over M. We shall use the notation $T(M)$ to denote the space of contravariant vector fields on M.

One important structure that one can form using the components $V^\mu(x)$ of a contravariant vector field is the differential operator $V = V^\mu(x)\partial/\partial x^\mu$. It follows from the transformation property (7.4) and the chain rule that V is constructed in the same way in all coordinate systems, e.g. $V = V'^\mu(x')\partial/\partial x'^\mu$. In this sense it is invariant under coordinate transformations. The differential operator V acts naturally on a scalar field $f(x)$, yielding another scalar field

$$\mathcal{L}_V f(x) \equiv V^\mu(x)\frac{\partial f}{\partial x^\mu}\,. \tag{7.7}$$

On the manifold $\mathbb{R}^D$, this operation is just the directional derivative $V \cdot \nabla f$, and it has the same interpretation on a general manifold M. At each point p with coordinates x^m, $\mathcal{L}_V f(x)$ is the derivative of $f(x)$ in the direction of the tangent of the integral curve of $V^\mu(x)$ through p.

Locally, there is a $1:1$ correspondence between contravariant vector fields $V^\mu(x)$ and differential operators. In mathematical treatments a vector field is viewed as a smooth assignment of a differential operator at each point p. The set of elementary operators $\{\partial/\partial x^\mu,\ \mu = 1, \ldots, D\}$ are a basis in this view of the tangent space $T_p(M)$. This is consistent with our discussion since $\partial f/\partial x^\mu$ for a given value of μ is the derivative in the direction of the tangent to the curve on which the single coordinate x^μ changes, but the other coordinates x^ν for $\nu \neq \mu$ are constant. The basis $\{\partial/\partial x^\mu, \mu = 1, \dots, D\}$ is called a coordinate basis because these operators differentiate along such coordinate curves at each p.

The derivative $\mathcal{L}_V f(x)$ defined in (7.7) may be extended to vector and tensor fields of any type (p, q), always yielding another tensor of the same type. For the vectors and tensors in (7.4) and (7.5), the precise definition is

$$\begin{aligned}
\mathcal{L}_V U^\mu &= V^\rho \partial_\rho U^\mu - (\partial_\rho V^\mu) U^\rho , \\
\mathcal{L}_V \omega_\mu &= V^\rho \partial_\rho \omega_\mu + (\partial_\mu V^\rho) \omega_\rho , \\
\mathcal{L}_V T_\nu^\mu &= V^\rho \partial_\rho T_\nu^\mu - (\partial_\rho V^\mu) T_\nu^\rho + (\partial_\nu V^\rho) T_\rho^\mu .
\end{aligned} \tag{7.8}$$

The derivative defined in this way is called the Lie derivative. Its definition requires a vector field, but not a connection; yet it preserves the tensor transformation property.

Exercise 7.3 *Show explicitly that $\mathcal{L}_V U^\mu$, $\mathcal{L}_V \omega_\mu$, and $\mathcal{L}_V T_\nu^\mu$ defined in (7.8) do transform under coordinate transformations as required by (7.4) and (7.5).*

The Lie derivative of a contravariant vector field has special significance because it occurs in the commutator of the corresponding differential operators $U = U^\mu(x)\partial/\partial x^\mu$ and $V = V^\mu(x)\partial/\partial x^\mu$. An elementary calculation gives

$$[U, V] = W = W^\mu(x)\frac{\partial}{\partial x^\mu} , \tag{7.9}$$

with $W^\mu = \mathcal{L}_U V^\mu = -\mathcal{L}_V U^\mu$. The new vector field W^μ is called the Lie bracket of U^μ and V^μ. This discussion also shows that the contravariant tensor fields on M naturally form a Lie algebra.

Let us consider the transformation properties of (7.3)–(7.5) for infinitesimal coordinate transformations, namely those for which $x'^\mu = x^\mu - \xi^\mu(x)$. To first order in $\xi^\mu(x)$, the previous transformation rules can be expressed in terms of Lie derivatives as

$$\begin{aligned}
\delta\phi(x) &\equiv \phi'(x) - \phi(x) = \mathcal{L}_\xi \phi , \\
\delta U^\mu(x) &\equiv U'^\mu(x) - U^\mu(x) = \mathcal{L}_\xi U^\mu , \\
\delta\omega_\mu(x) &\equiv \omega'_\mu(x) - \omega_\mu(x) = \mathcal{L}_\xi \omega_\mu , \\
\delta T_\nu^\mu(x) &\equiv T_\nu'^\mu(x) - T_\nu^\mu(x) = \mathcal{L}_\xi T_\nu^\mu .
\end{aligned} \tag{7.10}$$

Thus one of the useful roles of Lie derivatives is in the description of infinitesimal coordinate transformations.

Exercise 7.4 *Show that the transformations (7.10) follow from (7.3)–(7.5).*

Next we focus attention on covariant vector fields, such as $\omega_\mu(x)$. We already noted in Ex. 7.2 that the contraction $\omega_\mu(x)V^\mu(x)$ with any contravariant vector field gives a scalar field. Thus at any point p with coordinates x^ν, $\omega_\mu(x)$ can be regarded as an element of the dual space $T_p^*(M)$, a linear functional that maps $T_p(M) \to \mathbb{R}$. The space $T_p^*(M)$ is usually called the cotangent space at p.

In parallel to the way in which we associated contravariant vector fields $V^\mu(x)$ with differential operators $V = V^\mu(x)\partial/\partial x^\mu$, we use the coordinate differentials $\mathrm{d}x^\mu$ to write $\Omega = \omega_\mu(x)\mathrm{d}x^\mu$. Note that both $\omega_\mu(x)$ and $\mathrm{d}x^\mu$ transform under coordinate transformations, but $\Omega = \omega'_\mu(x')\mathrm{d}x'^\mu$ is constructed in the same way in any coordinate system. Ω is called a differential 1-form on M. Note that the gradient $\partial_\mu\phi(x)$ of any scalar transforms as a covariant vector and that the associated differential 1-form $\mathrm{d}\phi = \partial_\mu\phi\mathrm{d}x^\mu$ is just the differential of calculus. We can think of the set of coordinate differentials $\{\mathrm{d}x^\mu,\ \mu = 1,\ldots,D\}$ as a basis of the space of 1-forms.

The notion of the cotangent space $T_p^*(M)$ of linear functionals on $T_p(M)$ is naturally extended to the level of 1-forms and differential operators. We define the pairing of basis elements as $\langle \mathrm{d}x^\mu|\partial/\partial x^\nu\rangle \equiv \delta^\mu_\nu$. This is extended using linearity to any general 1-form Ω and differential operator V, so that we then have $\langle\Omega|V\rangle = \omega_\mu(x)V^\mu(x)$. This agrees with the initial definition as the contraction of component indices.

7.3 The algebra and calculus of differential forms

Among the various fields defined on M, the scalars ϕ, covariant vectors ω_μ, and totally antisymmetric tensors such as $\omega_{\mu\nu} = -\omega_{\nu\mu}$ have a particularly useful structure when considered together. Note that antisymmetry is preserved under coordinate transformations so it is a tensorial property. Using the coordinate differentials $\mathrm{d}x^\mu$, we can construct differential p-forms for $p = 1, 2, \ldots, D$ as

$$\begin{aligned}
\omega^{(1)} &= \omega_\mu(x)\mathrm{d}x^\mu\,,\\
\omega^{(2)} &= \frac{1}{2}\omega_{\mu\nu}(x)\mathrm{d}x^\mu \wedge \mathrm{d}x^\nu\,,\\
&\vdots\\
\omega^{(p)} &= \frac{1}{p!}\omega_{\mu_1\mu_2\cdots\mu_p}\mathrm{d}x^{\mu_1} \wedge \mathrm{d}x^{\mu_2} \wedge \cdots \wedge \mathrm{d}x^{\mu_p}\,.
\end{aligned} \tag{7.11}$$

The wedge product is defined as antisymmetric; that is, $\mathrm{d}x^\mu \wedge \mathrm{d}x^\nu = -\mathrm{d}x^\nu \wedge \mathrm{d}x^\mu$, $\mathrm{d}x^\mu \wedge \mathrm{d}x^\nu \wedge \mathrm{d}x^\rho = -\mathrm{d}x^\rho \wedge \mathrm{d}x^\nu \wedge \mathrm{d}x^\mu$, etc. At each point we have an element of the p-fold antisymmetric tensor product of the cotangent space, so the differential form $\omega^{(p)}$ is a smooth assignment of an element of this tensor product as the point varies over M. The space of p-forms is denoted by $\Lambda^p(M)$. By convention the scalars are considered to be 0-forms.

There is an exterior algebra and calculus of p-forms, which we will not develop in detail. See [40, 41, 42, 44, 45] for more complete discussions. Rather we will state some key properties without proof and write the specific examples needed later to discuss frames, connections, and curvature. In the exterior algebra, a p-form $\omega^{(p)}$ and a q-form $\omega^{(q)}$ can be multiplied to give a $(p+q)$-form if $p+q \leq D$. The product vanishes if $p+q > D$. The product satisfies $\omega^{(p)} \wedge \omega^{(q)} = (-)^{pq}\omega^{(q)} \wedge \omega^{(p)}$ and it is associative. Some examples are

$$\begin{aligned}
\omega^{(1)} \wedge \tilde{\omega}^{(1)} &= \omega_\mu \mathrm{d}x^\mu \wedge \tilde{\omega}_\nu \mathrm{d}x^\nu \\
&= \tfrac{1}{2}(\omega_\mu \tilde{\omega}_\nu - \omega_\nu \tilde{\omega}_\mu)\mathrm{d}x^\mu \wedge \mathrm{d}x^\nu , \\
\omega^{(1)} \wedge \omega^{(2)} &= \omega_\mu \mathrm{d}x^\mu \wedge \tfrac{1}{2}\omega_{\nu\rho}(x)\mathrm{d}x^\nu \wedge \mathrm{d}x^\rho \\
&= \tfrac{1}{6}(\omega_\mu \omega_{\nu\rho} + \omega_\nu \omega_{\rho\mu} + \omega_\rho \omega_{\mu\nu})\mathrm{d}x^\mu \wedge \mathrm{d}x^\nu \wedge \mathrm{d}x^\rho . \qquad (7.12)
\end{aligned}$$

The explicit antisymmetrization in the second line of each example is not necessary, since it is implicit in the wedge products of the $\mathrm{d}x^\mu$. But it is convenient to indicate that the covariant tensor field associated with each form is antisymmetric.

The exterior calculus is based on the exterior derivative, which maps p-forms into $(p+1)$-forms as follows:

$$\mathrm{d}\omega^{(p)} = \frac{1}{p!}\partial_\mu \omega_{\mu_1\mu_2\cdots\mu_p}\mathrm{d}x^\mu \wedge \mathrm{d}x^{\mu_1} \wedge \mathrm{d}x^{\mu_2} \wedge \cdots \wedge \mathrm{d}x^{\mu_p} . \qquad (7.13)$$

Exercise 7.5 *Show that the operation* d *is nilpotent, i.e.* $\mathrm{d}(\mathrm{d}\omega^{(p)}) = 0$ *on any* p*-form, and that it satisfies the distributive property*

$$\mathrm{d}(\omega^{(p)} \wedge \omega^{(q)}) = \mathrm{d}\omega^{(p)} \wedge \omega^{(q)} + (-)^p \omega^{(p)} \wedge \mathrm{d}\omega^{(q)} . \qquad (7.14)$$

On forms of degree 0, 1, 2

$$\begin{aligned}
\mathrm{d}\phi &= \partial_\mu \phi \mathrm{d}x^\mu , \\
\mathrm{d}\omega^{(1)} &= \tfrac{1}{2}(\partial_\mu \omega_\nu - \partial_\nu \omega_\mu)\mathrm{d}x^\mu \wedge \mathrm{d}x^\nu , \\
\mathrm{d}\omega^{(2)} &= \tfrac{1}{6}(\partial_\mu \omega_{\nu\rho} + \partial_\nu \omega_{\rho\mu} + \partial_\rho \omega_{\mu\nu})\mathrm{d}x^\mu \wedge \mathrm{d}x^\nu \wedge \mathrm{d}x^\rho . \qquad (7.15)
\end{aligned}$$

A p-form that satisfies $\mathrm{d}\omega^{(p)} = 0$ is called closed. A p-form $\omega^{(p)}$ that can be expressed as $\omega^{(p)} = \mathrm{d}\omega^{(p-1)}$ is called exact. The Poincaré lemma implies that locally any closed p-form can be expressed as $\mathrm{d}\omega^{(p-1)}$, but $\omega^{(p-1)}$ may not be well defined globally on M.

We saw that the exterior derivative is a map from p-forms into $(p+1)$-forms. There is also an interior derivative, which maps p-forms into $(p-1)$-forms. The latter depends on a vector V and is denoted as i_V. It is defined as follows:

$$\begin{aligned}
(i_V \omega^{(p)}) &= \frac{1}{(p-1)!}V^\mu \omega_{\mu\mu_1\ldots\mu_{p-1}}\mathrm{d}x^{\mu_1} \wedge \mathrm{d}x^{\mu_2} \wedge \cdots \wedge \mathrm{d}x^{\mu_{p-1}} , \\
i_V(\mathrm{d}x^{\mu_1} &\wedge \mathrm{d}x^{\mu_2} \wedge \cdots \wedge \mathrm{d}x^{\mu_p}) \\
&= V^{\mu_1}\mathrm{d}x^{\mu_2} \wedge \cdots \wedge \mathrm{d}x^{\mu_p} - V^{\mu_2}\mathrm{d}x^{\mu_1} \wedge \mathrm{d}x^{\mu_3} \cdots \wedge \mathrm{d}x^{\mu_p} + \ldots . \qquad (7.16)
\end{aligned}$$

Exercise 7.6 *Prove that the interior derivative is also nilpotent, i.e.* $i_V i_V = 0$.

Like the internal and external derivatives, the Lie derivative, introduced in Sec. 7.2 as a derivative on tensor fields, has a simple action on p-forms. It maps p-forms to p-forms via the formula

$$\mathcal{L}_V = \mathrm{d}i_V + i_V\mathrm{d}\,. \tag{7.17}$$

It is instructive to work out the example $\mathcal{L}_V\omega^{(1)} = (\mathrm{d}i_V + i_V\mathrm{d})\omega^{(1)}$:

$$\begin{aligned}(\mathrm{d}i_V + i_V\,\mathrm{d})\omega^{(1)} &= \mathrm{d}(V^\mu\omega_\mu) + i_V\tfrac{1}{2}(\partial_\mu\omega_\nu - \partial_\nu\omega_\mu)\mathrm{d}x^\mu \wedge \mathrm{d}x^\nu \\ &= (\partial_\nu V^\mu\omega_\mu + V^\mu\partial_\nu\omega_\mu)\mathrm{d}x^\nu + V^\mu(\partial_\mu\omega_\nu - \partial_\nu\omega_\mu)\mathrm{d}x^\nu \\ &= (V^\mu\partial_\mu\omega_\nu + \partial_\nu V^\mu\omega_\mu)\mathrm{d}x^\nu = (\mathcal{L}_V\omega)_\nu\mathrm{d}x^\nu\,. \end{aligned} \tag{7.18}$$

The Lie derivative of the covariant vector field ω_μ, which contains the components of the 1-form $\omega^{(1)}$, was defined in (7.8) and appears in the final result.[1]

Exercise 7.7 *Use the formula (7.17) to calculate the Lie derivative of a 0-form (where the first term vanishes by definition) and a 2-form. The final result should again contain the components of the Lie derivative as defined in Sec. 7.2.*

Differential forms have a natural application to the theories of electromagnetism, to Yang–Mills theory, and to the antisymmetric tensor gauge theories that appear in higher dimensional supergravity. However, we need to bring in some other ideas in the next section before discussing these physical applications.

7.4 The metric and frame field on a manifold

We now introduce the additional structure of a metric on a manifold M. In general relativity the metric is of primary importance in describing the geometry of spacetime and the dynamics of gravity. In theories such as supergravity where there are fermions coupled to gravity, one must use an auxiliary quantity, the frame field (more commonly called the vierbein or vielbein), which we discuss in detail. The metric tensor is quadratically related to the frame field.

7.4.1 The metric

A metric or inner product on a real vector space V is a non-degenerate bilinear map from $V \otimes V \to \mathbb{R}$. The inner product of two vectors $u,\ v \in V$ is a real number denoted by (u, v). The inner product must satisfy the following properties:

(i) *bilinearity*, $(u, c_1v_1 + c_2v_2) = c_1(u, v_1) + c_2(u, v_2)$ and $(c_1v_1 + c_2v_2, u) = c_1(v_1, u) + c_2(v_2, u)$;
(ii) *non-degeneracy*, if $(u, v) = 0$ for all $v \in V$, then $u = 0$;
(iii) *symmetry*, $(u, v) = (v, u)$.

[1] The distributive formula $\mathcal{L}_V(\omega_\mu\mathrm{d}x^\mu) = (\mathcal{L}_V\omega_\mu)\mathrm{d}x^\mu + \omega_\mu\mathcal{L}_V\mathrm{d}x^\mu$ can be used if it is interpreted carefully. Both terms are non-vanishing and can be calculated using (7.17) and (7.16). The latter equation requires that each component of $\mathcal{L}_V\omega_\mu = i_V\mathrm{d}\omega_\mu$ is calculated as the Lie derivative of a 0-form.

The metric on a manifold is a smooth assignment of an inner product map on each $T_p(M) \otimes T_p(M) \to \mathbb{R}$. In local coordinates the metric is specified by a covariant second rank symmetric tensor field $g_{\mu\nu}(x)$, and the inner product of two contravariant vectors $U^\mu(x)$ and $V^\mu(x)$ is $g_{\mu\nu}(x)U^\mu(x)V^\nu(x)$, which is a scalar field. In particular the metric gives a formula for the length s of a curve $x^\mu(\lambda)$ with tangent vector $\mathrm{d}x^\mu/\mathrm{d}\lambda$:

$$s_{12} = \int_{\lambda_1}^{\lambda_2} \mathrm{d}\lambda \sqrt{g_{\mu\nu}(x(\lambda))(\mathrm{d}x^\mu/\mathrm{d}\lambda)(\mathrm{d}x^\nu/\mathrm{d}\lambda)}\,. \tag{7.19}$$

Thus it is most convenient to summarize the properties of a given metric by the line element

$$\mathrm{d}s^2 = g_{\mu\nu}(x)\mathrm{d}x^\mu \mathrm{d}x^\nu\,. \tag{7.20}$$

Non-degeneracy means that $\det g_{\mu\nu} \neq 0$, so the inverse metric $g^{\mu\nu}(x)$ exists as a rank 2 symmetric contravariant tensor, which satisfies

$$g^{\mu\rho}g_{\rho\nu} = g_{\nu\rho}g^{\rho\mu} = \delta^\mu_\nu\,. \tag{7.21}$$

The metric tensor and its inverse may be used to lower and raise indices, e.g. $V_\mu(x) = g_{\mu\nu}V^\nu(x)$ and $\omega^\mu(x) = g^{\mu\nu}(x)\omega_\nu(x)$, thus providing a natural isomorphism between the spaces of contravariant and covariant vectors and tensors.

In a gravity theory in spacetime, the metric has signature $- + + \cdots +$. Concretely this means that the metric tensor $g_{\mu\nu}$ may be diagonalized by an orthogonal transformation, i.e. $(O^{-1})_\mu{}^a = O^a{}_\mu$ and

$$g_{\mu\nu} = O^a{}_\mu D_{ab} O^b{}_\nu\,, \tag{7.22}$$

with positive eigenvalues λ^a in $D_{ab} = \mathrm{diag}(-\lambda^0, \lambda^1, \ldots, \lambda^{D-1})$.

Exercise 7.8 *Show that $\lambda^a(x) > 0$ holds throughout M if the metric is non-degenerate. In another coordinate system the transformed metric $g'_{\rho\sigma} = (\mathrm{d}x^\mu/\mathrm{d}x'^\rho)(\mathrm{d}x^\nu/\mathrm{d}x'^\sigma)g_{\mu\nu}$ may be diagonalized giving another set of eigenvalues λ'^a, in general different from the λ^a. Show that the $\lambda'^a > 0$. Thus the signature of a metric is a global invariant.*

7.4.2 The frame field

The construction above, which involved only matrix linear algebra, allows us to define an important auxiliary quantity in a theory of gravity, namely

$$e^a_\mu(x) \equiv \sqrt{\lambda^a(x)}O^a{}_\mu(x)\,. \tag{7.23}$$

In four dimensions this quantity is commonly called the tetrad or vierbein. In general dimension the term vielbein is frequently used, but we prefer the term frame field for reasons that will become clear as we discuss its properties.

Note that

$$g_{\mu\nu}(x) = e^a_\mu(x)\eta_{ab}e^b_\nu(x)\,, \tag{7.24}$$

where $\eta_{ab} = \text{diag}(-1, 1, \ldots, 1)$ is the metric of flat D-dimensional Minkowski spacetime. It is (7.24) that states the general relation between metric and frame field. For a given metric tensor $g_{\mu\nu}(x)$, the frame field $e^a_\mu(x)$ of (7.23), obtained by diagonalization, is not the only solution. Given any x-dependent matrix $\Lambda^a{}_b(x)$ which leaves η_{ab} invariant, in other words, given a local Lorentz transformation, we can construct another solution of (7.24), namely

$$e'^a_\mu(x) = \Lambda^{-1\,a}{}_b(x) e^b_\mu(x) . \tag{7.25}$$

All choices of frame fields related by local Lorentz transformations are viewed as equivalent. So we require that the frame field and geometrical quantities derived from it must be used in a way that is covariant with respect to the transformation (7.25).

Local Lorentz transformations in curved spacetime differ from the global Lorentz transformations of Minkowski space discussed in Ch. 1. Only frame indices $a, b, \ldots$ of a quantity transform, coordinate indices $\mu, \nu, \ldots$ are inert, and the spacetime coordinate does not change. Instead, (7.24) requires that the frame field e^a_μ transforms as a covariant vector under diffeomorphisms (coordinate transformations), viz.

$$e'^a_\mu(x') = \frac{\partial x^\rho}{\partial x'^\mu} e^a_\rho(x) , \tag{7.26}$$

while the frame index is inert.[2]

Since e^a_μ is a non-singular $D \times D$ matrix, with $\det e^a_\mu = \sqrt{-\det g} \neq 0$, there is an inverse frame field $e^\mu_a(x)$, which satisfies $e^a_\mu e^\mu_b = \delta^a_b$ and $e^\mu_a e^a_\nu = \delta^\mu_\nu$.

Exercise 7.9 *Show that*

$$e^\mu_a = g^{\mu\nu} \eta_{ab} e^b_\nu , \qquad e^\mu_a g_{\mu\nu} e^\nu_b = \eta_{ab} . \tag{7.27}$$

The last relation shows that the (inverse) frame field can be used to relate a general metric of signature $- + + \cdots +$ to the Minkowski metric. Show that, under local Lorentz and coordinate transformations,

$$e'^\mu_a(x) = \Lambda^{-1}{}_a{}^b e^\mu_b(x) , \qquad e'^\mu_a(x') = \frac{\partial x'^\mu}{\partial x^\rho} e^\rho_a(x) . \tag{7.28}$$

Frame indices are raised and lowered using the Minkowski metric.

The second relation of (7.27) indicates that the e^μ_a form an orthonormal set of vectors in the tangent space of M at each point. Since $\det e^\mu_a \neq 0$, we have a basis of each tangent space. Any contravariant vector field has a unique expansion in the new basis, i.e. $V^\mu(x) = V^a(x) e^\mu_a(x)$ with $V^a(x) = V^\mu(x) e^a_\mu(x)$. The $V^a(x)$ are the frame components of the original vector field $V^\mu(x)$. They transform as a set of D scalar fields under coordinate transformations, and as a vector under Lorentz transformations, i.e. $V'^a(x) = \Lambda^{-1\,a}{}_b(x) V^b(x)$. The same may be done for covariant vectors, i.e. $\omega_\mu(x) = \omega_a(x) e^a_\mu(x)$ with $\omega_a(x) = \omega_\mu(x) e^\mu_a(x)$. These constructions may be extended to tensor fields of any rank in a straightforward way.

Thus we may use e^a_μ and e^μ_a to transform vector and tensor fields back and forth between a coordinate basis with indices $\mu, \nu, \ldots$ and a local Lorentz basis with indices $a, b, \ldots$ in

[2] The relation between local and global Lorentz transformations is discussed further in Sec. 11.3.1.

which the metric is η_{ab}. Invariants such as the inner product may be calculated in either basis.

Exercise 7.10 *Show that*

$$U^\mu(x)V_\mu(x) = g_{\mu\nu}(x)U^\mu(x)V^\nu(x) = \eta_{ab}U^a(x)V^b(x) = U^a(x)V_a(x)\,. \qquad (7.29)$$

At the level of differential operators the change of basis in the tangent space is expressed as

$$E_a \equiv e_a^\mu(x)\frac{\partial}{\partial x^\mu}\,. \qquad (7.30)$$

This makes it clear that the local Lorentz basis is a non-coordinate basis. If there were local coordinates y^a such that $E_a = \partial/\partial y^a$, these differential operators would commute. However, the commutator

$$[E_a, E_b] = -\Omega_{ab}{}^c E_c\,, \qquad (7.31)$$

where $\Omega_{ab}{}^c = -e^c_\mu \mathcal{L}_{e_a} e^\mu_b = e^c_\mu \mathcal{L}_{e_b} e^\mu_a$ are the frame components of the Lie bracket, which do not vanish in a general manifold, and are called 'anholonomy coefficients'.

Exercise 7.11 *Show that* $\Omega_{ab}{}^c = e_a^\mu e_b^\nu(\partial_\mu e^c_\nu - \partial_\nu e^c_\mu)$.

We can also use the frame field e^a_μ to define a new basis in the spaces $\Lambda^p(M)$ of differential forms. The local Lorentz basis of 1-forms is

$$e^a \equiv e^a_\mu(x)\mathrm{d}x^\mu\,. \qquad (7.32)$$

This is the dual basis to (7.30), since the pairing is given by $\langle e^a|E_b\rangle = \delta^a_b$. For 2-forms the basis consists of the wedge products $e^a \wedge e^b$, and so on.

In a field theory containing only bosonic fields, which are always vectors or tensors, the use of local frames is unnecessary, although it is an option that is convenient for some purposes. Local frames are a necessity to treat the coupling of fermion fields to gravity, because spinors are defined by their special transformation properties under Lorentz transformations.

7.4.3 Induced metrics

In many applications of differential geometry one encounters a manifold of dimension D which can be viewed as a surface embedded in flat Minkowski or Euclidean space of dimension $D+1$. We discuss the Euclidean case for $D = 2$. Suppose that our surface is described by the equation

$$f(x, y, z) = 0\,. \qquad (7.33)$$

On the surface the differential vanishes, viz.

$$\mathrm{d}f = \frac{\partial f}{\partial x}\mathrm{d}x + \frac{\partial f}{\partial y}\mathrm{d}y + \frac{\partial f}{\partial z}\mathrm{d}z = 0\,. \qquad (7.34)$$

The intrinsic geometry of the surface is determined by the Euclidean metric

$$ds^2 = dx^2 + dy^2 + dz^2 . \tag{7.35}$$

To find it one can, in principle, solve (7.33) to eliminate one variable and then use (7.34) to find a relation among the coordinate differentials. When this information is inserted in (7.35), one has the induced metric. Voila!

This is often easier said than done, so we confine our discussion to the solvable and instructive example of the unit 2-sphere for which the embedding equation (7.33) is

$$x^2 + y^2 + z^2 = 1 . \tag{7.36}$$

Let's proceed using spherical coordinates:

$$z = r\cos\theta , \qquad x = r\sin\theta\cos\varphi , \qquad y = r\sin\theta\sin\varphi . \tag{7.37}$$

The embedding equation becomes simply $r^2 = 1$, so we can eliminate the coordinate r and write the differentials:

$$\begin{aligned} dz &= -\sin\theta\, d\theta , \\ dx &= \cos\theta\cos\varphi\, d\theta - \sin\theta\sin\varphi\, d\varphi , \\ dy &= \cos\theta\sin\varphi\, d\theta + \sin\theta\cos\varphi\, d\varphi . \end{aligned} \tag{7.38}$$

Upon substitution in (7.35) one finds the induced metric

$$ds^2 = d\theta^2 + \sin^2\theta\, d\varphi^2 . \tag{7.39}$$

This is a commonly used and quite useful metric on S_2, but it is evidently singular at the north and south poles where the metric tensor is not invertible. One can do somewhat better using one of the two sets of coordinates defined by the stereographic projection in Sec. 7.1, and this is the subject of the following exercise.

Exercise 7.12 *Reexpress the metric (7.39) in the coordinates* $X = \cos\varphi\tan(\theta/2)$, $Y = \sin\varphi\tan(\theta/2)$. *Show that the new metric is*

$$ds^2 = \frac{4(dX^2 + dY^2)}{(1 + X^2 + Y^2)^2} . \tag{7.40}$$

7.5 Volume forms and integration

The equations of motion in any field theory are most conveniently packaged in the action integral. In a gravitational theory this requires integration over the curved spacetime manifold. We thus need a procedure for integration that is invariant under coordinate transformations. The volume form is the key to this procedure.

On a D-dimensional manifold, one may choose *any* top degree D-form $\omega^{(D)}$ as a volume form and define the integral

$$\begin{aligned} I &= \int \omega^{(D)} \\ &= \frac{1}{D!}\int \omega_{\mu_1\cdots\mu_D}(x)\mathrm{d}x^{\mu_1}\wedge\cdots\wedge\mathrm{d}x^{\mu_D} \\ &= \int \omega_{01\cdots D-1}\mathrm{d}x^0\mathrm{d}x^1\ldots\mathrm{d}x^{D-1}\,. \end{aligned} \tag{7.41}$$

The antisymmetric tensor $\omega_{\mu_1\cdots\mu_D}(x)$ has only one independent component, and we have used this fact in the last line above to write the integral so that it may be performed by the rules of multi-variable calculus. For the same reason any two D-forms $\tilde{\omega}(D)$ and $\omega^{(D)}$ must be related by $\tilde{\omega}(D) = f\,\omega^{(D)}$, where $f(x)$ is a scalar field. Thus the definition (7.41) includes $\int f\,\omega^{(D)}$.

Exercise 7.13 *Show that in a new coordinate system with coordinates $x'^{\mu}(x^{\nu})$ the integral I in (7.41) takes the form*

$$I = \frac{1}{D!}\int \omega'_{\mu_1\cdots\mu_D}(x')\mathrm{d}x'^{\mu_1}\wedge\cdots\wedge\mathrm{d}x'^{\mu_D} \tag{7.42}$$

and is thus coordinate invariant.

Although there are many possible volume forms, there are two types that usually appear in the context of physics. The first, which is the more specialized, occurs when the physical theory contains form fields. As an example, on a 3-manifold the wedge product $\omega^{(1)}\wedge\omega^{(2)}$ can be chosen as a volume form. Using (7.12) we see that

$$\begin{aligned} I &= \int \omega^{(1)}\wedge\omega^{(2)} \\ &= \tfrac{1}{6}\int (\omega_\mu\omega_{\nu\rho} + \omega_\nu\omega_{\rho\mu} + \omega_\rho\omega_{\mu\nu})\mathrm{d}x^\mu\wedge\mathrm{d}x^\nu\wedge\mathrm{d}x^\rho \\ &= \int (\omega_0\omega_{12} + \omega_1\omega_{20} + \omega_2\omega_{01})\mathrm{d}x^0\mathrm{d}x^1\mathrm{d}x^2\,. \end{aligned} \tag{7.43}$$

The integral is coordinate invariant, and it does not involve the metric on M. The action integral of the simplest Chern–Simons field theory, in which $\omega^{(2)} = \mathrm{d}\omega^{(1)}$, takes this form.

The second type of volume form is far more common in physics and we call it the canonical volume form. There are several ways to introduce it, and we will use the frame field $e^a_\mu(x)$ and the basis of frame 1-forms e^a for this purpose. As a preliminary we define the Levi-Civita alternating symbol in local frame components:

$$\varepsilon_{a_1a_2\cdots a_D} = \begin{cases} +1, & a_1a_2\cdots a_D \text{ an even permutation of } 01\ldots(D-1), \\ -1, & a_1a_2\cdots a_D \text{ an odd permutation of } 01\ldots(D-1), \\ 0, & \text{otherwise.} \end{cases} \tag{7.44}$$

Under (proper) Lorentz transformations, i.e. $\det\Lambda^a{}_b = 1$, this is an invariant tensor that takes the same form in any Lorentz frame. As usual Lorentz indices are raised with η^{ab}. Note that $\varepsilon^{01\cdots(D-1)} = -1$.

Note that the Levi-Civita symbol provides a useful formula for the determinant of any $D\times D$ matrix $A^a{}_b$, namely

$$\det A\, \varepsilon_{b_1 b_2 \cdots b_D} = \varepsilon_{a_1 a_2 \cdots a_D} A^{a_1}{}_{b_1} A^{a_2}{}_{b_2} \cdots A^{a_D}{}_{b_D}\,, \tag{7.45}$$

and that there are systematic identities for the contraction of p of the $D = p + q$ indices, as we saw in (3.9).

The Levi-Civita form in the coordinate basis is defined by contracting with frame fields and inserting factors of $e = \det e_\mu^a$ or e^{-1}:

$$\begin{aligned} \varepsilon_{\mu_1 \mu_2 \cdots \mu_D} &\equiv e^{-1} \varepsilon_{a_1 a_2 \cdots a_D} e_{\mu_1}^{a_1} e_{\mu_2}^{a_2} \cdots e_{\mu_D}^{a_D}\,, \\ \varepsilon^{\mu_1 \mu_2 \cdots \mu_D} &\equiv e\, \varepsilon^{a_1 a_2 \cdots a_D} e_{a_1}^{\mu_1} e_{a_2}^{\mu_2} \cdots e_{a_D}^{\mu_D}\,. \end{aligned} \tag{7.46}$$

Note that these definitions ensure that $\varepsilon^{\mu_1 \cdots \mu_D}$ and $\varepsilon_{\mu_1 \cdots \mu_D}$ take the constant values given on the right-hand side of (7.44). This can be seen using (7.45). The quantities defined in (7.46) are called *tensor densities*. It is important to recognize that $\varepsilon^{\mu_1 \mu_2 \cdots \mu_D}$ cannot be obtained by raising the indices of $\varepsilon_{\mu_1 \mu_2 \cdots \mu_D}$ in the usual way using the inverse of the metric. Therefore expressions like $\varepsilon^{\mu_1 \cdots \mu_p}{}_{\mu_{p+1} \cdots \mu_D}$ are not well defined. There is no such problem for $\varepsilon^{a_1 \cdots a_p}{}_{b_{p+1} \cdots b_D}$.

Exercise 7.14 *Prove, using (7.45), that both $\varepsilon^{\mu_1 \mu_2 \cdots \mu_D}$ and $\varepsilon_{\mu_1 \mu_2 \cdots \mu_D}$ take values ± 1 for any choice of frame field e_μ^a. This guarantees that they are invariant under infinitesimal changes of the frame field. Show also directly that $\delta \varepsilon^{\mu_1 \mu_2 \cdots \mu_D} = 0$ for any δe_a^μ using the general matrix formula*

$$\delta \det M = (\det M)\, \mathrm{Tr}(M^{-1} \delta M)\,, \tag{7.47}$$

and the Schouten identity; see (3.11).

With these preliminaries, the canonical volume form is defined as

$$\begin{aligned} \mathrm{d}V &\equiv e^0 \wedge e^1 \wedge \cdots \wedge e^{D-1} \\ &= \frac{1}{D!} \varepsilon_{a_1 \cdots a_D} e^{a_1} \wedge \cdots \wedge e^{a_D} \\ &= \frac{1}{D!} e\, \varepsilon_{\mu_1 \cdots \mu_D} \mathrm{d}x^{\mu_1} \wedge \cdots \wedge \mathrm{d}x^{\mu_D} \\ &= e\, \mathrm{d}x^0 \ldots \mathrm{d}x^{D-1} \\ &= \mathrm{d}^D x \sqrt{-\det g}\,. \end{aligned} \tag{7.48}$$

Note that the determinant of the frame field e_μ^a appears in a natural fashion. In the last line we give the abbreviated notation we will use in most applications. For example, given the Lagrangian of a system of fields, such as the kinetic Lagrangian $L = \frac{1}{2} g^{\mu\nu} \partial_\mu \phi \partial_\nu \phi$ of a scalar field, the action integral is written as

$$S = \int \mathrm{d}V\, L = \int \mathrm{d}^D x \sqrt{-\det g}\, L\,. \tag{7.49}$$

7.6 Hodge duality of forms

Since p- and q-forms have the same number of components when $p+q = D$, it is possible to define a $1:1$ map between them. This map is the Hodge duality map from $\Lambda^p(M) \to \Lambda^q(M)$, and it is quite useful in the physics of supergravity. The map is denoted by $\Omega^{(q)} = {}^*\omega^{(p)}$.

Since the map is linear we can define it on a basis of p-forms and then extend to a general form. It is convenient to use the local frame basis initially and define

$$^*e^{a_1} \wedge \cdots \wedge e^{a_p} = \frac{1}{q!} e^{b_1} \wedge \cdots \wedge e^{b_q} \varepsilon_{b_1 \cdots b_q}{}^{a_1 \cdots a_p} \,. \tag{7.50}$$

A general p-form can be expressed in this basis, and we can proceed to define its dual via

$$\begin{aligned} \Omega^{(q)} = {}^*\omega^{(p)} &= {}^*\left(\frac{1}{p!}\omega_{a_1 \cdots a_p} e^{a_1} \wedge \cdots \wedge e^{a_p}\right) \\ &= \frac{1}{p!}\omega_{a_1 \cdots a_p} {}^*e^{a_1} \wedge \cdots \wedge e^{a_p} \,. \end{aligned} \tag{7.51}$$

Exercise 7.15 *Show that the frame components of $\Omega^{(q)}$ are given by*

$$\Omega_{b_1 \cdots b_q} = \left({}^*\omega\right)_{b_1 \cdots b_q} = \frac{1}{p!}\varepsilon_{b_1 \cdots b_q}{}^{a_1 \cdots a_p} \omega_{a_1 \cdots a_p} \,. \tag{7.52}$$

These formulas are far less complicated than they look since there is only one independent term in each sum. For example, for $D = 4$ the dual of a 3-form is a 1-form. For basis elements we have $^*e^1 \wedge e^2 \wedge e^3 = e^0$ and $^*e^0 \wedge e^1 \wedge e^2 = e^3$. For components, $({}^*\omega)_0 = \omega_{123}$ and $({}^*\omega)_3 = \omega_{012}$.

The duality has an important involutive property, which can be inferred from the following sequence of operations on basis elements:

$$\begin{aligned} {}^*({}^*e^{a_1} \wedge \cdots \wedge e^{a_p}) &= \frac{1}{q!}{}^*e^{b_1} \wedge \cdots \wedge e^{b_q} \varepsilon_{b_1 \cdots b_q}{}^{a_1 \cdots a_p} \\ &= \frac{1}{p!q!} e^{c_1} \wedge \cdots \wedge e^{c_p} \varepsilon_{c_1 \cdots c_p}{}^{b_1 \cdots b_q} \varepsilon_{b_1 \cdots b_q}{}^{a_1 \cdots a_p} \\ &= -(-)^{pq} e^{c_1} \wedge \cdots \wedge e^{c_p} \delta_{c_1 \cdots c_p}^{a_1 \cdots a_p} \\ &= -(-)^{pq} e^{a_1} \wedge \cdots \wedge e^{a_p} \,. \end{aligned} \tag{7.53}$$

This leads to the general relation $^*({}^*\omega^{(p)}) = -(-)^{pq}\omega^{(p)}$. This is the correct relation for a Lorentzian signature manifold. For Euclidean signature the involution property is $^*({}^*\omega^{(p)}) = (-)^{pq}\omega^{(p)}$.

For even dimension $D = 2m$, it is possible to impose the constraint of self-duality (or anti-self-duality) on forms of degree m, i.e. $\Omega^{(m)} = \pm{}^*\Omega^{(m)}$. In a given dimension this condition is consistent only if duality is a strict involution, i.e. $-(-)^{m^2} = -(-)^m = +1$

for Lorentzian signature and $(-)^m = +1$ for Euclidean signature. Thus it is possible to have self-dual Yang–Mills instantons in four Euclidean dimensions. A self-dual $F^{(5)}$ is possible in $D = 10$ Lorentzian signature, and it indeed appears in Type IIB supergravity.

The duality relations defined above in a frame basis are easily transformed to a coordinate basis using the relations $e^a = e^a_\mu(x)\mathrm{d}x^\mu$ and $\mathrm{d}x^\mu = e^\mu_a(x)e^a$. For coordinate basis elements the duality map is

$$ {}^*(\mathrm{d}x^{\mu_1} \wedge \cdots \wedge \mathrm{d}x^{\mu_p}) = \frac{1}{q!} e\, g^{\mu_1\rho_1} \ldots g^{\mu_p\rho_p} \mathrm{d}x^{\nu_1} \wedge \cdots \wedge \mathrm{d}x^{\nu_p} \varepsilon_{\nu_1\cdots\nu_q\rho_1\cdots\rho_p} . \tag{7.54} $$

For antisymmetric tensor components, we have

$$ ({}^*\omega)_{\mu_1\cdots\mu_q} = \frac{1}{p!} e\, \varepsilon_{\mu_1\cdots\mu_q\rho_1\cdots\rho_p} g^{\nu_1\rho_1} \ldots g^{\nu_p\rho_p} \omega_{\nu_1\cdots\nu_p} . \tag{7.55} $$

Following the discussion in Sec. 7.5, we may take as a volume form on M the wedge product ${}^*\omega^{(p)} \wedge \omega^{(p)}$ of any p-form and its Hodge dual. The integral of this volume form is simply the standard invariant norm of the tensor components of $\omega^{(p)}$, i.e.

$$ \int {}^*\omega^{(p)} \wedge \omega^{(p)} = \frac{1}{p!} \int \mathrm{d}^D x \sqrt{-g}\, \omega^{\mu_1\cdots\mu_p} \omega_{\mu_1\cdots\mu_p} . \tag{7.56} $$

Exercise 7.16 *Prove (7.56). Use the definitions above and those in Sec. 7.5 and the fact that*

$$ e^{a_1} \wedge \cdots \wedge e^{a_q} \wedge e^{b_1} \wedge \cdots \wedge e^{b_p} = -\varepsilon^{a_1\cdots a_q b_1\cdots b_p} \mathrm{d}V , \tag{7.57} $$

where $\mathrm{d}V$ *is the canonical volume element of (7.48).*

Exercise 7.17 *Show that the volume form* $\mathrm{d}V$ *can also be written as* ^{*1}.

Exercise 7.18 *Compare these definitions with Sec. 4.2.1, to obtain*

$$ \tilde{F}_{\mu\nu} = -\mathrm{i}\left({}^*F\right)_{\mu\nu} . \tag{7.58} $$

Show that the factor i *ensures that the tilde operation squares to the identity. Self-duality is then possible for complex 2-forms.*

Exercise 7.19 *For applications to gauge field theories it is useful to record the relation between the components of the field strength 2-form and its dual:*

$$ {}^*F_{\mu\nu} \equiv \tfrac{1}{2}\sqrt{-g}\, \varepsilon_{\mu\nu\rho\sigma} F^{\rho\sigma} , \qquad {}^*F^{\mu\nu} = \frac{1}{2\sqrt{-g}} \varepsilon^{\mu\nu\rho\sigma} F_{\rho\sigma} . \tag{7.59} $$

Verify the second relation. Since both $F_{\mu\nu}$ *and* ${}^*F_{\mu\nu}$ *are tensors, their indices are raised by* $g^{\mu\nu}$.

7.7 Stokes' theorem and electromagnetic charges

Suppose that M is a manifold of dimension D, and that Σ_p with boundary $\Sigma_{p-1} = \partial\Sigma_p$ is a submanifold of dimension $p \leq D$. Suppose further that ω^{p-1} is a $(p-1)$-form that satisfies certain smoothness properties which we omit here.[3] Stokes' theorem asserts that

$$\int_{\Sigma_p} \mathrm{d}\omega^{p-1} = \int_{\Sigma_{p-1}} \omega^{p-1}. \tag{7.60}$$

The integrals can be evaluated using any choice of coordinates.

In Ex. 4.14 electric and magnetic charges in three-dimensional flat spacetime were expressed as volume integrals. We will use Stokes' theorem to convert these to surface integrals. However, we generalize the discussion and consider a spacetime metric $g_{\mu\nu}(x)$ on M and a conserved current J^ν to which the gauge field is coupled. The coordinate invariant action that describes this coupling is

$$S = \int \mathrm{d}^4x\sqrt{-g}\left[-\tfrac{1}{4}g^{\mu\rho}g^{\nu\sigma}F_{\mu\nu}F_{\rho\sigma} + A_\nu J^\nu\right]. \tag{7.61}$$

The gauge field equation of motion (4.49) and the tensor $G^{\mu\nu}$ of (4.50) generalize to

$$\frac{\delta S}{\delta A_\nu} = \partial_\mu(\sqrt{-g}F^{\mu\nu}) + \sqrt{-g}J^\nu = 0\,, \tag{7.62}$$

$$G_{\mu\nu} \equiv \varepsilon_{\mu\nu\rho\sigma}\frac{\delta S}{\delta F_{\rho\sigma}} = -\tfrac{1}{2}\sqrt{-g}\varepsilon_{\mu\nu\rho\sigma}F^{\rho\sigma} = -{}^*F_{\mu\nu}\,. \tag{7.63}$$

The volume integrals (4.62) and (4.63) for the electric and magnetic charge contained in a region Σ_3 with boundary Σ_2 now become

$$\begin{aligned} q &= \int_{\Sigma_3} \mathrm{d}^3x\,\sqrt{-g}J^0 = \int_{\Sigma_3} \mathrm{d}^3x\,\partial_i\sqrt{-g}F^{0i} = -\tfrac{1}{2}\int_{\Sigma_3} \mathrm{d}^3x\,\varepsilon^{ijk}\partial_i G_{jk}, \\ p &= -\tfrac{1}{2}\int_{\Sigma_3} \mathrm{d}^3x\varepsilon^{ijk}\partial_i F_{jk}\,. \end{aligned} \tag{7.64}$$

The integrands in the final expressions are each the exterior derivatives of 2-forms on Σ_3, so we can apply Stokes' theorem and rewrite them as

$$\begin{pmatrix} p \\ q \end{pmatrix} = -\frac{1}{2}\int_{\Sigma_3} \mathrm{d}x^i \wedge \mathrm{d}x^j \wedge \mathrm{d}x^k \partial_i \begin{pmatrix} F_{jk} \\ G_{jk} \end{pmatrix} = -\frac{1}{2}\int_{\Sigma^2} \mathrm{d}x^\mu \wedge \mathrm{d}x^\nu \begin{pmatrix} F_{\mu\nu} \\ G_{\mu\nu} \end{pmatrix}. \tag{7.65}$$

The detailed form of $G_{\mu\nu} = -{}^*F_{\mu\nu}$ in terms of the components of $F_{\mu\nu}$ is given in (7.59).

Exercise 7.20 *The components of $F_{\mu\nu}$ which describe point charges are quite basic quantities. Derive them for flat spacetime using polar coordinates with metric*

$$\mathrm{d}s^2 = -\mathrm{d}t^2 + \mathrm{d}r^2 + r^2\left(\mathrm{d}\theta^2 + \sin^2\theta\,\mathrm{d}\phi^2\right). \tag{7.66}$$

[3] See [43] for details and proof of the theorem. Stokes' theorem dates from 1850 and 1854.

Show that a field configuration whose only non-vanishing component is

$$F_{\theta\phi} = -\frac{p}{4\pi}\sin\theta \tag{7.67}$$

is a solution of Maxwell's equation (7.62) and has magnetic charge p. Show that a field configuration whose only non-vanishing component is

$$E_r = F_{rt} = \frac{q}{4\pi r^2} \tag{7.68}$$

is a solution of Maxwell's equation (7.62) which describes an electric point charge q (use $\varepsilon_{tr\theta\phi} = 1$*).*

7.8 p-form gauge fields

Using the ideas of Sec. 7.6, we can rewrite the simplest kinetic actions of scalars and gauge vectors as integrals of differential forms:

$$\begin{aligned} S_0 &= -\tfrac{1}{2}\int {}^*F^{(1)} \wedge F^{(1)}\,, \qquad F^{(1)} \equiv \mathrm{d}\phi\,, \\ S_1 &= -\tfrac{1}{2}\int {}^*F^{(2)} \wedge F^{(2)}\,, \qquad F^{(2)} \equiv \mathrm{d}A^{(1)}\,. \end{aligned} \tag{7.69}$$

In each case there is a Bianchi identity, $\mathrm{d}F^{(1)} = 0$ and $\mathrm{d}F^{(2)} = 0$, which implies that the field strengths can be written as differentials of a lower form. For the form $A^{(1)}$, which describes the photon, there is a gauge transformation that can be written as $\delta A^{(1)} = \mathrm{d}\Lambda^{(0)}$. We can interpret the actions of (7.69) as the definition of field theories for 0-form and 1-form 'potentials'.

This suggests a generalization. We can describe a p-form 'potential' in terms of a $(p+1)$-form 'field strength' and write the action

$$S_p = -\tfrac{1}{2}\int {}^*F^{(p+1)} \wedge F^{(p+1)}\,, \qquad F^{(p+1)} \equiv \mathrm{d}A^{(p)}\,. \tag{7.70}$$

Again there is a gauge transformation $\delta A^{(p)} = \mathrm{d}\Lambda^{(p-1)}$, and these transformations of the p-form gauge potential leave $F^{(p+1)}$ and the action invariant.

Exercise 7.21 *Show that the action (7.70) can be expressed in form components as*

$$\begin{aligned} S_p &= -\frac{1}{2\,(p+1)!}\int \mathrm{d}^D x\,\sqrt{-g}\,F^{\mu_1\cdots\mu_{p+1}}F_{\mu_1\cdots\mu_{p+1}}\,, \\ F_{\mu_1\cdots\mu_{p+1}} &= (p+1)\partial_{[\mu_1}A_{\mu_2\ldots\mu_{p+1}]}\,. \end{aligned} \tag{7.71}$$

We now determine the number of degrees of freedom of a p-form gauge field. The number of independent components in $\Lambda^{(p-1)}$ is $\binom{D}{p-1}$. However, not all the components of

$\Lambda^{(p-1)}$ are independent symmetries, since transformations where $\Lambda^{(p-1)} = \mathrm{d}\Lambda'^{(p-2)}$ have no effect on the gauge field. However, not all components of the latter are to be subtracted from the gauge symmetries. Indeed, gauge parameters of the form $\Lambda'^{(p-2)} = \mathrm{d}\Lambda''^{(p-3)}$ are annihilated by the d operation. Thus gauge symmetries reduce the number of independent components of a p-form to

$$\binom{D}{p} - \binom{D}{p-1} + \binom{D}{p-2} - \ldots = \binom{D-1}{p}, \tag{7.72}$$

the number of components of a p-form in $D-1$ dimensions. We thus find that the off-shell degrees of freedom (field variables minus symmetries) correspond to a (rank p antisymmetric tensor) representation of $\mathrm{SO}(D-1)$, as we saw before for other fields. The field equations for the action (7.70) impose a further reduction, and the independent components that remain transform in the rank p antisymmetric tensor representation of $\mathrm{SO}(D-2)$. The dimension of this representation is $\binom{D-2}{p}$, which is the number of on-shell degrees of freedom or, equivalently, the number of physical degrees of freedom of the p-form gauge field.

The fact that $\binom{D-2}{p} = \binom{D-2}{D-p-2}$ tells us that the numbers of on-shell degrees of freedom of p- and $(D-p-2)$-form gauge fields are the same. In fact the theories of these forms are physically equivalent. We will show that there is a 'duality transformation' that relates them.

The following argument gives one indication of the equivalence. The field equation and Bianchi identity for $F^{(p+1)}$ read

$$\mathrm{d}^* F^{(p+1)} = 0\,, \qquad \mathrm{d}F^{(p+1)} = 0\,. \tag{7.73}$$

We define the dual form $G^{(D-p-1)} = {}^* F^{(p+1)}$, and observe that the two equations can be rewritten as

$$\mathrm{d}G^{(D-p-1)} = 0\,, \qquad \mathrm{d}^* G^{(D-p-1)} = 0\,. \tag{7.74}$$

The first equation implies that $G^{(D-p-1)}$ is the field strength of a $(D-p-2)$-form gauge potential, while the second equation is exactly the equation of motion which follows from the action

$$S_{(D-p-2)} = -\tfrac{1}{2}\int {}^*G^{(D-p-1)} \wedge G^{(D-p-1)}\,, \qquad G^{(D-p-1)} \equiv \mathrm{d}B^{(D-p-2)}\,. \tag{7.75}$$

The equations of the $F^{(p+1)}$ and $G^{(D-p-1)}$ are thus very similar. The equation of motion of one transforms into the Bianchi identity for the other under the duality map $G^{(D-p-1)} = {}^* F^{(p+1)}$.

We can prove the duality by the showing that the action (7.70) can be rewritten in a way that can be transformed into (7.75). We rewrite (7.70) as

$$S_p = -\int \left[\tfrac{1}{2}{}^* F^{(p+1)} \wedge F^{(p+1)} + b^{(D-p-2)} \wedge \mathrm{d}F^{(p+1)}\right], \tag{7.76}$$

in which $F^{(p+1)}$ and $b^{(D-p-2)}$ are viewed as independent fields. The latter is a Lagrange multiplier, and its field equation implies that $F^{(p+1)}$ is the field strength of a p-form potential. Therefore the information in the action (7.76) is equivalent to that in the original form

Box 7.1 **Duality between p-forms**

The p-form gauge field with action (7.70) is equivalent to the similar $(D-p-2)$-form gauge field.

(7.70). To show that (7.76) is also equivalent to (7.75), we consider the Euler variation for $F^{(p+1)}$, which states that

$$ {}^*F^{(p+1)} = (-)^{D-p}\, \mathrm{d}b^{(D-p-2)}\,. \tag{7.77}$$

This is effectively an algebraic field equation, so it can be substituted in the action to obtain (7.75). This proves the equivalence of p-form and $(D-2-p)$-form gauge potentials.

Consider the simple case of a rank 2 antisymmetric tensor gauge field (a 2-form) in four dimensions. Its dual is a scalar field (a 0-form). Thus the simplest kinetic actions for the antisymmetric tensor and scalar fields are equivalent. However, duality arguments are valid only for the simplest actions, typically for abelian gauge fields. It is not generally true that gauge theories for p- and $(D-p-2)$-form gauge fields are equivalent.

The duality transformation for gauge vectors in four dimensions is a *self-duality*. A 1-form is dual to another 1-form. In fact, it transforms electric into magnetic components, and this is the duality that we discussed in Sec. 4.2.

7.9 Connections and covariant derivatives

A covariant derivative on a manifold is a rule to differentiate a tensor of type (p,q) producing a tensor of type $(p,q+1)$. It is well known that one needs to introduce the affine connection $\Gamma^\rho_{\mu\nu}(x)$ to accomplish this. On vector fields the covariant derivative is defined by

$$\begin{aligned} \nabla_\mu V^\rho &= \partial_\mu V^\rho + \Gamma^\rho_{\mu\nu} V^\nu ,\\ \nabla_\mu V_\nu &= \partial_\mu V_\nu - \Gamma^\rho_{\mu\nu} V_\rho , \end{aligned} \tag{7.78}$$

and it is straightforward to extend this definition to tensors. Most discussions of the affine connection in the physics literature are based on the idea of parallel transport and use a coordinate basis; see [46, 47].

Since supergravity requires the frame field $e^a_\mu(x)$, and we have used the frame 1-forms e^a extensively, it is natural to introduce the affine connection in this framework and then make contact with the more common treatment. In frames the affine connection is specified by the 1-forms $\omega^{ab} = \omega_\mu{}^{ab}(x)\mathrm{d}x^\mu$. Although more general connections can be considered, the connection required for gravitational theories with fermions is antisymmetric in Lorentz indices, $\omega^{ab} = -\omega^{ba}$, and we impose this condition *ab initio*. As we will see, antisymmetry means that ω^{ab} is a connection for the Lorentz group $\mathrm{O}(D-1,1)$. Indeed covariance under local Lorentz transformations (7.25) of the frame is the guiding principle of the discussion. See [39] for a roughly similar treatment. The components $\omega_\mu{}^{ab}(x)$ of ω^{ab} are usually called the spin connection because they are essential in the description of spinors on manifolds.

7.9.1 The first structure equation and the spin connection $\omega_{\mu ab}$

Given the frame 1-forms e^a, we examine the 2-forms

$$de^a = \tfrac{1}{2}(\partial_\mu e^a_\nu - \partial_\nu e^a_\mu)\, dx^\mu \wedge dx^\nu\,. \tag{7.79}$$

The antisymmetric components transform as a (0, 2) tensor under coordinate transformations, but not as a local Lorentz vector. This is most quickly seen at the 2-form level, where, using (7.25),

$$de'^a = d(\Lambda^{-1a}{}_b e^b) = \Lambda^{-1a}{}_b de^b + d\Lambda^{-1a}{}_b \wedge e^b\,. \tag{7.80}$$

The second term spoils the vector transformation property. To cancel it we add the contribution from a 2-form involving the spin connection and consider

$$de^a + \omega^a{}_b \wedge e^b \equiv T^a\,. \tag{7.81}$$

If $\omega^a{}_b$ is defined to transform under local Lorentz transformations as

$$\omega'^a{}_b = \Lambda^{-1a}{}_c d\Lambda^c{}_b + \Lambda^{-1a}{}_c\, \omega^c{}_d\, \Lambda^d{}_b\,, \tag{7.82}$$

then T^a does indeed transform as a vector, i.e. $T'^a = \Lambda^{-1a}{}_b T^b$. The 2-form T^a is called the torsion 2-form of the connection, and (7.81) is called the first Cartan structure equation.

Exercise 7.22 *Confirm that T^a transforms as a Lorentz vector if the connection transforms as in (7.82).*

There is a lot to be said about the connection 1-form ω^{ab}, and we begin with some properties that should be familiar from the study of non-abelian gauge theories. The components $\omega_\mu{}^{ab}(x)$ transform as a covariant vector under coordinate transformations, while (7.82) implies that

$$\omega'_\mu{}^a{}_b = \Lambda^{-1a}{}_c\, \partial_\mu \Lambda^c{}_b + \Lambda^{-1a}{}_c\, \omega_\mu{}^c{}_d\, \Lambda^d{}_b\,. \tag{7.83}$$

These are exactly the gauge transformation properties of a Yang–Mills potential for the group $O(D-1,1)$. The situation is even more familiar for Euclidean signature manifolds in which frame fields transform under local $SO(D)$ rotations, and an antisymmetric $\omega_\mu{}^{ab}$ transforming as in (7.83) is a gauge potential for the compact group $SO(D)$.

Thus local Lorentz covariance is implemented like Yang–Mills gauge covariance for the gauge group $O(d-1,1)$.

In Sec. 7.5 we showed that any (vector or) tensor field on M can be transformed from a coordinate basis to a local Lorentz basis, where the tensor components of a type (p, q) tensor take the form $T^{a_1\cdots a_p}_{b_1\cdots b_q}(x)$. Let us consider the simplest cases of vectors V^a, U_a and type (0, 2) tensors T_{ab}, which transform as

$$\begin{aligned}
V'^a(x) &= \Lambda^{-1a}{}_b(x) V^b(x)\,,\\
U'_a(x) &= U_b(x) \Lambda^b{}_a(x)\,,\\
T'_{ab}(x) &= T_{cd}(x) \Lambda^c{}_a(x) \Lambda^d{}_b(x)\,,
\end{aligned} \tag{7.84}$$

respectively. The extension of these local Lorentz transformation rules to Lorentz tensors of type (p, q) is straightforward.

We use the spin connection to define local Lorentz covariant derivatives as

$$\begin{aligned} D_\mu V^a &= \partial_\mu V^a + \omega_\mu{}^a{}_b V^b\,, \\ D_\mu U_a &= \partial_\mu U_a - U_b \omega_\mu{}^b{}_a = \partial_\mu U_a + \omega_{\mu a}{}^b U_b\,, \\ D_\mu T_{ab} &= \partial_\mu T_{ab} - T_{cb}\omega_\mu{}^c{}_a - T_{ac}\omega_\mu{}^c{}_b\,. \end{aligned} \tag{7.85}$$

The extension to type (p, q) Lorentz tensors involves p connection terms with $\omega_\mu{}^{a_i}{}_c$ contracted from the left and q terms with $-\omega_\mu{}^c{}_{b_i}$ contracted on the right. Recall that $\omega_\mu{}^{ab} = \omega_\mu{}^a{}_c\eta^{cb}$ and $\omega_{\mu a}{}^b = \eta_{ac}\omega_\mu{}^c{}_d\eta^{db}$, etc.

Exercise 7.23 *Use (7.82) to show that $D_\mu V^a$, $D_\mu U_a$ and $D_\mu T_{ab}$ do transform as Lorentz tensors, i.e. as in (7.84). They also transform as covariant vectors under coordinate transformations.*

Exercise 7.24 *Show that the Lorentz covariant derivative obeys the Leibnitz product rule, e.g. $D_\mu(V^a U_b) = (D_\mu V^a)U_b + V^a D_\mu U_b$. Show also that it commutes with index contractions, e.g. $\delta^c_a D_\mu(V^a T_{bc}) = D_\mu(V^c T_{bc})$.*

Spinor fields $\Psi(x)$ are absolutely vital for supergravity. As discussed in Ch. 2, a spinor field has $2^{[D/2]}$ components in D-dimensional spacetime, and the generators of Lorentz transformations are the second rank Clifford matrices $\frac{1}{2}\gamma^{ab}$. In a gravitational theory, spinors must be described through their local frame components. As in the case of vector fields above, the local Lorentz transformation rule

$$\Psi'(x) = \exp\left(-\tfrac{1}{4}\lambda^{ab}(x)\gamma_{ab}\right)\Psi(x) \tag{7.86}$$

determines the covariant derivative

$$D_\mu\Psi(x) = \left(\partial_\mu + \tfrac{1}{4}\omega_{\mu ab}(x)\gamma^{ab}\right)\Psi(x)\,. \tag{7.87}$$

Exercise 7.25 *Show that $D_\mu\Psi$ also transforms as a spinor, namely into $\exp\left(-\frac{1}{4}\lambda^{ab}(x)\gamma_{ab}\right)D_\mu\Psi(x)$. You will need (7.82) and (2.22) for this purpose. Note that indices ρ, σ in (2.22) are to be interpreted as local Lorentz indices a, b.*

Let's apply (7.85) to the Lorentz metric tensor η_{ab}. We have

$$D_\mu\eta_{ab} = -\eta_{cb}\omega_\mu{}^c{}_a - \eta_{ac}\omega_\mu{}^c{}_b = -\omega_{\mu ba} - \omega_{\mu ab} = 0\,. \tag{7.88}$$

The metric has vanishing covariant derivative because η_{ab} is an invariant tensor of the Lorentz group. A direct consequence is that scalar products $V^a U_a$ are preserved under parallel transport[4] with the spin connection [44]. Thus our connection is metric-preserving.

The Cartan structure equation (7.81) is important, both conceptually and practically. The least familiar element may be the torsion 2-form T^a or its components, the torsion

[4] The infinitesimal parallel transport of a vector from the point with coordinates x^μ to the nearby point $x^\mu + \Delta x^\mu$ is defined by $\tilde{V}^a(x) = V^a(x) + \omega_\mu{}^a{}_b V^b(x)\Delta x^\mu$.

tensor $T_{\mu\nu}{}^a = -T_{\nu\mu}{}^a$. Indeed in most applications of differential geometry to gravity, the torsion vanishes, and one deals with a torsion-free, metric-preserving connection. This is also called the Levi-Civita connection, for which the structure equation reads $de^a + \omega^a{}_b \wedge e^b = 0$. However, non-vanishing torsion can arise from coupling of gravity to certain matter fields, and this does occur in supergravity. The geometrical effect of torsion is seen in the properties of an infinitesimal 'parallelogram' constructed by the parallel transport of two vector fields. The parallelogram closes for a torsion-free connection, but not if there is torsion; see [44].

Let's examine the tensor components of the structure equation (7.81). It is convenient to refer all quantities to a coordinate basis. The goal is to express $\omega_{\mu[\rho\nu]} = \omega_{\mu ab} e^a_\rho e^b_\nu$ in terms of the anholonomy coefficients

$$\Omega_{[\mu\nu]\rho} = (\partial_\mu e^a_\nu - \partial_\nu e^a_\mu)\, e_{a\rho} \qquad (7.89)$$

and the torsion tensor $T_{[\mu\nu]\rho} = T_{\mu\nu}{}^a e_{a\rho}$. We use [..] to indicate the antisymmetric pair of indices. The structure equation then reads

$$T_{[\mu\nu]\rho} = \Omega_{[\mu\nu]\rho} + \omega_{\mu[\rho\nu]} - \omega_{\nu[\rho\mu]}\,. \qquad (7.90)$$

We simply have $\frac{1}{2}D^2(D-1)$ equations for the $\frac{1}{2}D^2(D-1)$ independent 'unknowns' $\omega_{\mu[\rho\nu]}$. It is a standard exercise, outlined below, to find the unique solution

$$\omega_{\mu[\nu\rho]} = \omega_{\mu[\nu\rho]}(e) + K_{\mu[\nu\rho]}\,, \qquad (7.91)$$

$$\omega_{\mu[\nu\rho]}(e) = \tfrac{1}{2}(\Omega_{[\mu\nu]\rho} - \Omega_{[\nu\rho]\mu} + \Omega_{[\rho\mu]\nu}) = \omega_{\mu ab}(e) e^a_\nu e^b_\rho\,, \qquad (7.92)$$

$$\omega_\mu{}^{ab}(e) = 2e^{\nu[a}\partial_{[\mu}e_{\nu]}{}^{b]} - e^{\nu[a}e^{b]\sigma}e_{\mu c}\partial_\nu e_\sigma{}^c\,, \qquad (7.93)$$

$$K_{\mu[\nu\rho]} = -\tfrac{1}{2}(T_{[\mu\nu]\rho} - T_{[\nu\rho]\mu} + T_{[\rho\mu]\nu})\,. \qquad (7.94)$$

The unique torsion-free spin connection appears in (7.93), and what is conventionally called the contortion tensor is defined in (7.94).

Exercise 7.26 *Obtain the results in (7.91)–(7.94) from the following combination of (7.90) and two permutations:* $T_{[\mu\nu]\rho} - T_{[\nu\rho]\mu} + T_{[\rho\mu]\nu}$.

Exercise 7.27 *The fact that an infinitesimal variation $\delta\omega_{\mu ab}$ of the spin connection transforms covariantly under local Lorentz transformations follows easily from (7.83). In the next chapter we will need the variation $\delta\omega_{\mu ab}$ of the torsion-free spin connection due to a small change δe^a_μ of the frame field. To calculate this, consider the variation of the Cartan structure equation (7.81) (without torsion):* $d\delta e^a + \omega^a{}_b \wedge \delta e^b + \delta\omega^a{}_b \wedge e^b = 0$. *This 2-form equation is equivalent to the component relation*

$$D_{[\mu}\delta e^a_{\nu]} + (\delta\omega_{[\mu}{}^{ab})e_{\nu]b} = 0 \quad \textit{with} \quad D_\mu \delta e^a_\nu \equiv \partial_\mu \delta e^a_\nu + \omega_\mu{}^{ab}\delta e_{b\nu}\,. \qquad (7.95)$$

With the structure of (7.92) in mind, you should be able to derive

$$e^a_\nu e^b_\rho \delta\omega_{\mu ab} = (D_{[\mu}\delta e^a_{\nu]})e_{\rho a} - (D_{[\nu}\delta e^a_{\rho]})e_{\mu a} + (D_{[\rho}\delta e^a_{\mu]})e_{\nu a}\,. \qquad (7.96)$$

7.9.2 The affine connection $\Gamma^{\rho}_{\mu\nu}$

Our next task is to transform Lorentz covariant derivatives of vector and tensor frame fields to the coordinate basis where they become covariant derivatives with respect to general coordinate transformations and take the familiar form in (7.78), but include torsion. It should be emphasized that no new structure on the manifold is required. We simply reexpress the information in the spin connection $\omega_{\mu}{}^{a}{}_{b}$ in a coordinate basis where it is contained in the affine connection $\Gamma^{\rho}_{\mu\nu}$.

We note that the quantity $\nabla_{\mu}V^{\nu} \equiv e^{\nu}_{a} D_{\mu} V^{a}$ is the transform to coordinate basis of a frame vector field and coordinate covariant vector field. It is necessarily a type $(1, 1)$ tensor under coordinate transformations. The following manipulations will bring it to the form in (7.78):

$$\begin{aligned}\nabla_{\mu}V^{\rho} &\equiv e^{\rho}_{a} D_{\mu} V^{a} \\ &= e^{\rho}_{a} D_{\mu}(e^{a}_{\nu} V^{\nu}) \\ &= \partial_{\mu}V^{\rho} + e^{\rho}_{a}(\partial_{\mu}e^{a}_{\nu} + \omega_{\mu}{}^{a}{}_{b}e^{b}_{\nu})V^{\nu}\,. \end{aligned} \tag{7.97}$$

Exercise 7.28 *Show by similar manipulation that* $\nabla_{\mu}V_{\nu} \equiv e^{a}_{\nu} D_{\mu} V_{a}$ *also takes the form in (7.78), namely*

$$\nabla_{\mu}V_{\nu} = \partial_{\mu}V_{\nu} - e^{\rho}_{a}(\partial_{\mu}e^{a}_{\nu} + \omega_{\mu}{}^{a}{}_{b}e^{b}_{\nu})V_{\rho}\,. \tag{7.98}$$

These results can be extended to Lorentz tensors of type (p, q). The conclusion is that the transformation to coordinate basis given by

$$\nabla_{\mu}T^{\rho_1\cdots\rho_p}_{\nu_1\cdots\nu_q} \equiv e^{\rho_1}_{a_1}\ldots e^{\rho_p}_{a_p}\, e^{b_1}_{\nu_1}\ldots e^{b_q}_{\nu_q}\, D_{\mu}T^{a_1\cdots a_p}_{b_1\cdots b_q} \tag{7.99}$$

defines a tensor of type $(p, q+1)$ with the properties of the conventional covariant derivative. The affine connection is related to the spin connection by

$$\Gamma^{\rho}_{\mu\nu} = e^{\rho}_{a}(\partial_{\mu}e^{a}_{\nu} + \omega_{\mu}{}^{a}{}_{b}e^{b}_{\nu})\,. \tag{7.100}$$

We now show that this definition of $\Gamma^{\rho}_{\mu\nu}$ does satisfy the expected properties, noting first that it can be rewritten as

$$\partial_{\mu}e^{a}_{\nu} + \omega_{\mu}{}^{a}{}_{b}e^{b}_{\nu} - \Gamma^{\sigma}_{\mu\nu}e^{a}_{\sigma} = 0\,. \tag{7.101}$$

This property is called the 'vielbein postulate' in some discussions of the spin connection. The vielbein postulate leads to the more familiar metric postulate or metricity property,

$$\nabla_{\mu}g_{\nu\rho} \equiv \partial_{\mu}g_{\nu\rho} - \Gamma^{\sigma}_{\mu\nu}g_{\sigma\rho} - \Gamma^{\sigma}_{\mu\rho}g_{\nu\sigma} = 0. \tag{7.102}$$

Note that (7.101) and (7.102) are valid whether or not the connection carries torsion.

Exercise 7.29 *Obtain (7.102) by contracting (7.101) with* $e_{a\rho}$ *and adding the same expression with* ν *and* ρ *interchanged.*

The metric postulate means that the metric tensor is covariantly constant. Hence lengths and scalar products of vectors are preserved under parallel transport, and covariant differentiation commutes with index raising, e.g. $\nabla_\mu V^\rho = g^{\rho\nu}\nabla_\mu V_\nu$.

The affine connection $\Gamma^\rho_{\mu\nu}$ does not transform like a tensor. It has a special transformation law, which can be obtained from the definition (7.100), that ensures that the operation ∇_μ transforms any (p, q) tensor into a type $(p+1, q)$ tensor. It is worthwhile for readers to find this transformation law.

Exercise 7.30 *Consider the definition (7.100) in two different coordinate systems and show that $\Gamma'^\rho_{\mu\nu}(x')$ and $\Gamma^\rho_{\mu\nu}(x)$ are related by the conventional transformation property*

$$\Gamma'^\rho_{\mu\nu}(x') = \frac{\partial x'^\rho}{\partial x^\sigma}\left(\frac{\partial^2 x^\sigma}{\partial x'^\mu \partial x'^\nu} + \frac{\partial x^\alpha}{\partial x'^\mu}\frac{\partial x^\beta}{\partial x'^\nu}\Gamma^\sigma_{\alpha\beta}\right). \tag{7.103}$$

Note that the first term in (7.103) cancels in the difference between any two connections, such as the infinitesimal variation $\delta\Gamma^\rho_{\mu\nu}(x)$, which occurs in the variational principles needed for gravitational field theories. Thus $\delta\Gamma^\rho_{\mu\nu}(x)$ transforms as a tensor. It follows from (7.83) that the same property holds for the variation $\delta\omega_{\mu ab}$ of the spin connection, which transforms as a type (0, 2) local Lorentz tensor. This tensor was computed in Ex. 7.27.

Last, but hardly least, we substitute the explicit form (7.91) of the spin connection in the definition (7.100). After some calculation we obtain the explicit formula for $\Gamma^\rho_{\mu\nu}$:

$$\Gamma^\rho_{\mu\nu} = \Gamma^\rho_{\mu\nu}(g) - K_{\mu\nu}{}^\rho\,, \tag{7.104}$$

$$\Gamma^\rho_{\mu\nu}(g) = \tfrac{1}{2}g^{\rho\sigma}(\partial_\mu g_{\sigma\nu} + \partial_\nu g_{\mu\sigma} - \partial_\sigma g_{\mu\nu})\,. \tag{7.105}$$

(Remember that K is antisymmetric in the last two indices, i.e. $K_{\mu\nu}{}^\rho = -K_\mu{}^\rho{}_\nu$.) The first term, written in detail in (7.105), is the torsion-free connection, frequently called the Christoffel symbol and denoted by $\{{}^{\rho}_{\mu\nu}\}$ instead of our $\Gamma^\rho_{\mu\nu}(g)$. It is well known that it is the unique *symmetric* affine connection that satisfies (7.102). When torsion is present the affine connection is not symmetric, rather

$$\Gamma^\rho_{\mu\nu} - \Gamma^\rho_{\nu\mu} = -K_{\mu\nu}{}^\rho + K_{\nu\mu}{}^\rho = T_{\mu\nu}{}^\rho\,. \tag{7.106}$$

Exercise 7.31 *Below (7.103), we argued that $\delta\Gamma^\rho_{\mu\nu}$ is a tensor. Find a simple expression for $\delta\Gamma^\rho_{\mu\nu}$ by varying (7.102) or (7.105).*

The Lie derivative operation, defined for some examples in (7.8), maps any type (p, q) tensor into another type (p, q) tensor, yet does not require a metric or connection. In most physical applications of differential geometry there is a natural metric and connection, and it is convenient to rewrite Lie derivatives in terms of covariant derivatives. *If the connection is symmetric*, and thus given by (7.105), this is easily done since the connection cancels pairwise among the various terms in (7.8), leading to

$$
\begin{aligned}
\mathcal{L}_V U^\mu &= V^\rho \nabla_\rho U^\mu - (\nabla_\rho V^\mu) U^\rho , \\
\mathcal{L}_V \omega_\mu &= V^\rho \nabla_\rho \omega_\mu + (\nabla_\mu V^\rho)\omega_\rho , \\
\mathcal{L}_V T_\nu^\mu &= V^\rho \nabla_\rho T_\nu^\mu - (\nabla_\rho V^\mu) T_\nu^\rho + (\nabla_\nu V^\rho) T_\rho^\mu .
\end{aligned} \tag{7.107}
$$

If there is torsion then there are additional terms.

Exercise 7.32 *Derive (7.107) from (7.8). Show that, if the T and K tensors vanish, the Lie derivative of the metric tensor is*

$$
\mathcal{L}_V g_{\mu\nu} = \nabla_\mu V_\nu + \nabla_\nu V_\mu . \tag{7.108}
$$

For 'mixed' quantities with both coordinate and frame indices, it is useful to distinguish between local Lorentz and coordinate covariant derivatives. Thus, for a vector–spinor field $\Psi_\mu(x)$, we define both:

$$
\begin{aligned}
D_\mu \Psi_\nu &\equiv \left(\partial_\mu + \tfrac{1}{4}\omega_{\mu ab}\gamma^{ab}\right)\Psi_\nu , \\
\nabla_\mu \Psi_\nu &= D_\mu \Psi_\nu - \Gamma^\rho_{\mu\nu}\Psi_\rho .
\end{aligned} \tag{7.109}
$$

The first transforms as a local Lorentz spinor, but not a tensor, and the second transforms as a spinor and type (0, 2) tensor. After antisymmetrization in $\mu\nu$, both derivatives yield (0, 2) tensors, but they differ by a torsion term, viz.

$$
\nabla_\mu \Psi_\nu - \nabla_\nu \Psi_\mu = D_\mu \Psi_\nu - D_\nu \Psi_\mu - T_{\mu\nu}{}^\rho \Psi_\rho . \tag{7.110}
$$

Similarly, the vielbein transforms as a Lorentz vector and a coordinate vector, such that ∇ contains both connections, and the vielbein postulate (7.101) can be written as

$$
\nabla_\mu e_\nu^a = \partial_\mu e_\nu^a + \omega_\mu{}^a{}_b e_\nu^b - \Gamma^\sigma_{\mu\nu} e_\sigma^a = 0 . \tag{7.111}
$$

7.9.3 Partial integration

In Minkowski spacetime the integral $\int \mathrm{d}^D x\, \partial_\mu V^\mu = 0$ if the vector field $V^\mu(x)$ vanishes at large distance. This property validates the partial integration operations which are frequently needed in quantum field theory. In curved spacetime the integral is replaced by the coordinate invariant form $\int \mathrm{d}^D x \sqrt{-g}\, \nabla_\mu V^\mu$. Using (7.78) and (7.104) and the key property

$$
\partial_\mu \sqrt{-g} = \sqrt{-g}\, \Gamma^\rho_{\rho\mu}(g) , \tag{7.112}
$$

of the torsion-free (Christoffel) connection, we find

$$
\int \mathrm{d}^D x \sqrt{-g}\, \nabla_\mu V^\mu = \int \mathrm{d}^D x\, \partial_\mu \left(\sqrt{-g}\, V^\mu\right) - \int \mathrm{d}^D x \sqrt{-g}\, K_{\nu\mu}{}^\nu V^\mu . \tag{7.113}
$$

The first term vanishes just as in flat space quantum field theory, but the second does not. It is proportional to

$$
K_{\nu\mu}{}^\nu = -T_{\nu\mu}{}^\nu . \tag{7.114}
$$

This leads to the conclusion mentioned in Box 7.2.

Integration by parts Box 7.2

Conventional integration by parts is valid if the connection is torsion-free, but there is a correction term, which involves $K_{\nu\mu}{}^{\nu} = -T_{\nu\mu}{}^{\nu}$, when there is torsion.

Exercise 7.33 *Observe how the connection components $\Gamma^{\rho}_{\mu\nu}(g)$ required in the covariant derivative on the left-hand side of (7.113) are 'automatically constructed' from derivatives of the metric on the right-hand side.*

7.10 The second structure equation and the curvature tensor

In Sec. 7.9.1 we discussed the fact that the spin connection $\omega_{\mu ab}$ transforms as a Yang–Mills gauge potential for the group $O(D-1, 1)$; see (7.83). It then follows that the quantity

$$R_{\mu\nu ab} \equiv \partial_\mu \omega_{\nu ab} - \partial_\nu \omega_{\mu ab} + \omega_{\mu ac}\omega_\nu{}^c{}_b - \omega_{\nu ac}\omega_\mu{}^c{}_b \tag{7.115}$$

has the properties of a Yang–Mills field strength, transforming as a type (0, 2) Lorentz tensor under local Lorentz transformations; see (7.85). Because of antisymmetry in $\mu\nu$, it is also a (0, 2) tensor under general coordinate transformation, called the curvature tensor. Thus we can define the curvature 2-form

$$\rho^{ab} = \frac{1}{2} R_{\mu\nu}{}^{ab}(x)\mathrm{d}x^\mu \wedge \mathrm{d}x^\nu \,. \tag{7.116}$$

Using (7.115) it is easy to see that the curvature 2-form is related to the connection 1-form by

$$\mathrm{d}\omega^{ab} + \omega^a{}_c \wedge \omega^{cb} = \rho^{ab} \,. \tag{7.117}$$

This equation is known as the second Cartan structure equation. Needless to say, it is the metric, curvature, and (if present) the torsion tensor which carry the basic local information about the spacetime geometry that is needed for gravitational physics.

Exercise 7.34 *Working with the forms and the Cartan structure relations is often easier than using component relations as in (7.93). Consider the metric*

$$\mathrm{d}s^2 = \mathrm{d}r^2 + C^2(r)\mathrm{d}\theta^2 \tag{7.118}$$

for an arbitrary function $C(r)$. The metric can be brought to standard form using the frame

1-forms

$$e^1 = \mathrm{d}r\,, \qquad e^2 = C(r)\mathrm{d}\theta\,. \tag{7.119}$$

Using (7.81) (with no torsion) check that the spin connection 1-form is $\omega^{12} = C'(r)\mathrm{d}\theta$. *Then use (7.117) to find* $\rho^{12} = C''(r)\mathrm{d}r \wedge \mathrm{d}\theta$, *which leads to* $R_{r\theta}{}^{12} = C''(r)$.

Exercise 7.35 *Consider the effect on* $R_{\mu\nu ab}$ *of an infinitesimal variation of the spin connection. Show that*

$$\delta R_{\mu\nu ab} = D_\mu \delta\omega_{\nu ab} - D_\nu \delta\omega_{\mu ab}\,. \tag{7.120}$$

This fact is related to the discussion below Ex. 7.30, which argued that $\delta\omega_{\mu ab}$ *transforms as a tensor.*

One immediate application of the Cartan structure equations (7.81) and (7.117) is to derive the Bianchi identities for the curvature tensor.

Exercise 7.36 *Apply the exterior derivative to (7.81) and (7.117) to obtain the 3-form relations*

$$\begin{aligned}
\rho^{ab} \wedge e_b &= \mathrm{d}T^a + \omega^{ab} \wedge T_b\,, \\
\mathrm{d}\rho^{ab} + \omega^a{}_c \wedge \rho^{cb} - \rho^{ac} \wedge \omega_c{}^b &= 0\,.
\end{aligned} \tag{7.121}$$

Show that the components of these 3-form equations read, using $R_{\mu\nu\rho}{}^a = R_{\mu\nu b}{}^a e^b_\rho$,

$$\begin{aligned}
R_{\mu\nu\rho}{}^a + R_{\nu\rho\mu}{}^a + R_{\rho\mu\nu}{}^a &= -D_\mu T_{\nu\rho}{}^a - D_\nu T_{\rho\mu}{}^a - D_\rho T_{\mu\nu}{}^a\,, \\
D_\mu R_{\nu\rho}{}^{ab} + D_\nu R_{\rho\mu}{}^{ab} + D_\rho R_{\mu\nu}{}^{ab} &= 0\,.
\end{aligned} \tag{7.122}$$

The derivatives D_μ etc. in these equations are Lorentz covariant derivatives and contain only the spin connection. However, each relation is a type (0, 3) coordinate tensor because of antisymmetry. The first relation is the conventional first Bianchi identity for the curvature tensor, but corrected by torsion terms on the right-hand side. This has no analogue in Yang–Mills theory. The second relation is the usual Bianchi identity of non-abelian gauge theory, and is called the second Bianchi identity for the curvature.

The commutator of Lorentz covariant derivatives leads to important relations, in both gauge theory and gravity, which we call Ricci identities. We write the identity for a field $\Phi(x)$ transforming in a representation of the proper Lorentz group with generators M^{ab} together with special cases of interest, namely frame vector fields $V^a(x)$ and spinor fields $\Psi(x)$,

$$\begin{aligned}
[D_\mu, D_\nu]\Phi &= \tfrac{1}{2} R_{\mu\nu ab} M^{ab} \Phi\,, \\
[D_\mu, D_\nu]V^a &= R_{\mu\nu}{}^a{}_b V^b\,, \\
[D_\mu, D_\nu]\Psi &= \tfrac{1}{4} R_{\mu\nu ab} \gamma^{ab} \Psi\,.
\end{aligned} \tag{7.123}$$

There are also generalized Ricci identities for the commutator of coordinate covariant derivatives, but they contain torsion terms. For example, for vector fields $V^\rho(x)$, one can derive by direct computation

$$[\nabla_\mu, \nabla_\nu]V^\rho = R_{\mu\nu}{}^\rho{}_\sigma V^\sigma - T_{\mu\nu}{}^\sigma \nabla_\sigma V^\rho\,, \tag{7.124}$$

in which the curvature tensor appears as

$$R_{\mu\nu}{}^{\rho}{}_{\sigma} = \partial_\mu \Gamma^{\rho}_{\nu\sigma} - \partial_\nu \Gamma^{\rho}_{\mu\sigma} + \Gamma^{\rho}_{\mu\tau}\Gamma^{\tau}_{\nu\sigma} - \Gamma^{\rho}_{\nu\tau}\Gamma^{\tau}_{\mu\sigma} . \tag{7.125}$$

Exercise 7.37 *Verify (7.124). Show that $R_{\mu\nu}{}^{\rho}{}_{\sigma} = R_{\mu\nu ab} e^{a\rho} e^{b}_{\sigma}$ by evaluating $[\nabla_\mu, \nabla_\nu] e^{\rho}_{a} = 0$. Prove the generalized second Bianchi identity*

$$\nabla_\mu R_{\nu\rho}{}^{\sigma\tau} + \nabla_\nu R_{\rho\mu}{}^{\sigma\tau} + \nabla_\rho R_{\mu\nu}{}^{\sigma\tau} = T_{\mu\nu}{}^{\xi} R_{\xi\rho}{}^{\sigma\tau} + T_{\nu\rho}{}^{\xi} R_{\xi\mu}{}^{\sigma\tau} + T_{\rho\mu}{}^{\xi} R_{\xi\nu}{}^{\sigma\tau} . \tag{7.126}$$

Exercise 7.38 *Derive*

$$\delta R_{\mu\nu}{}^{\rho}{}_{\sigma} = \nabla_\mu \delta\Gamma^{\rho}_{\nu\sigma} - \nabla_\nu \delta\Gamma^{\rho}_{\mu\sigma} , \tag{7.127}$$

which is closely related to (7.120).

The Ricci tensor is defined as $R_{\mu\nu} = R_{\mu}{}^{\sigma}{}_{\nu\sigma}$ and the curvature scalar is $R = g^{\mu\nu} R_{\mu\nu}$.

Exercise 7.39 *Show that $R_{\mu\nu} = R_{\nu\mu}$ if and only if there is no torsion. Show that the same holds for the symmetry condition $R_{\mu\nu\rho\sigma} = R_{\rho\sigma\mu\nu}$.*

Below is an exercise designed to show that the Hilbert action for general relativity, $S \sim \int \mathrm{d}^D x \sqrt{-g} R$, can be written as the integral of a volume form involving the frame and curvature forms. It is this form of the action that reveals how the mathematical framework of connections with torsion is realized in the physical setting of gravity coupled to fermions. We will explore this in the next chapter, but the exercise brings together some of the ideas discussed in this chapter.

Exercise 7.40 *Show that*

$$\frac{1}{(D-2)!} \int \varepsilon_{abc_1 \ldots c_{(D-2)}} e^{c_1} \wedge \cdots \wedge e^{c_{(D-2)}} \wedge \rho^{ab} = \int \mathrm{d}^D x \sqrt{-g} R . \tag{7.128}$$

Hint: express $\rho^{ab} = \frac{1}{2} R_{cd}{}^{ab} e^{c} \wedge e^{d}$ and use (7.57) and (7.48).

7.11 The nonlinear σ-model

The principal application of differential geometry in this book is to theories of gravity and supergravity in spacetime, viewed as a differentiable manifold. The matter couplings in these theories, including fermions and p-form gauge fields, also require the ideas of differential geometry reviewed in this chapter. There is still another physical application of differential geometry that is important in supergravity. This application involves the dynamics of scalar fields in flat spacetime. Physicists usually call it the nonlinear σ-model because it first appeared as a description of the low-energy interactions of the triplet of $\vec{\pi}$ mesons which involve the composite field $\sigma = (f_\pi^2 + \vec{\pi}^2)^{1/2}$. The notation $\vec{\pi}$ indicates that

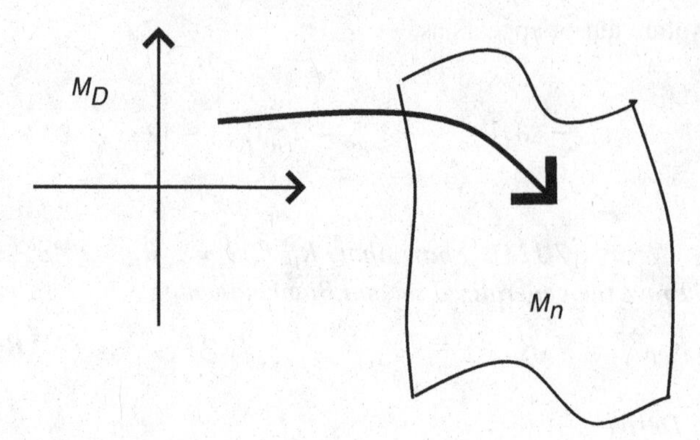

Fig. 7.3 Scalar fields as maps from spacetime to the target manifold.

the three pion fields transform in the fundamental representation of an SO(3) internal symmetry. In mathematics the same type of field theory may be called the theory of harmonic maps.

To begin the discussion let us consider an n-dimensional Riemannian manifold M_n. The name Riemannian means that M_n has a smooth Euclidean signature metric. In local coordinates called ϕ^i, $i = 1, 2, \ldots, n$, the metric tensor is $g_{ij}(\phi)$. In the nonlinear σ-model the coordinates are fields $\phi(x)$ in which the x^μ, $\mu = 1, 2, \ldots, D$, are the Cartesian coordinates of a flat spacetime M_D. We assume that M_D is of Minkowski signature, but our discussion requires little or no change for Euclidean signature. Thus we deal with maps from spacetime to the internal space or target space M_n; see Fig. 7.3.

We postulate that the dynamics of these maps is governed by the action

$$S[\phi] = -\frac{1}{2}\int \mathrm{d}^D x\, g_{ij}(\phi)\eta^{\mu\nu}\partial_\mu\phi^i\partial_\nu\phi^j\,. \tag{7.129}$$

Spacetime indices are raised and lowered with the Minkowski metric $\eta_{\mu\nu}$.

Exercise 7.41 *Compute the Euler variation of this action. Observe that the Christoffel connection (7.105) is obtained in the process and express the equation of motion in the form*

$$\Box\phi^i + \Gamma^i_{jk}(\phi)\partial^\mu\phi^j\partial_\mu\phi^k = 0\,. \tag{7.130}$$

For Euclidean signature the d'Alembertian $\Box$ is replaced by the Laplacian ∇^2. This is why the solutions are called harmonic maps; they are a nonlinear generalization of harmonic functions. Indeed the conventional Laplace equation is obtained in the special case when the target space is also flat.

Suppose that we use a different set of coordinates $\phi'^i(\phi)$ on the target space M_n. In these coordinates the metric tensor becomes

$$g'_{ij}(\phi') = \frac{\partial\phi^k}{\partial\phi'^i} g_{k\ell}(\phi) \frac{\partial\phi^\ell}{\partial\phi'^j}\,, \tag{7.131}$$

and the action changes to

$$S'[\phi'] = -\frac{1}{2}\int \mathrm{d}^D x\, g'_{ij}(\phi')\eta^{\mu\nu}\partial_\mu\phi'^i\partial_\nu\phi'^j\,. \tag{7.132}$$

This means that the equation of motion for $\phi'^i(x)$ has the same structure form as in (7.130), but with metric and connection expressed in the new coordinates. Our viewpoint is that significant physical information may be expressed in either set of coordinates.

It is interesting to consider the case when the 'spacetime' degenerates to the one-dimensional line with a time coordinate t. The action (7.129) is still perfectly sensible and the equation of motion (7.130) reduces to a set of n coupled nonlinear differential equations

$$\frac{\mathrm{d}^2\phi^i}{\mathrm{d}t^2} + \Gamma^i_{jk}\frac{\mathrm{d}\phi^j}{\mathrm{d}t}\frac{\mathrm{d}\phi^k}{\mathrm{d}t} = 0\,. \tag{7.133}$$

Readers should recognize this as the equation for geodesic[5] curves on M_n.

Physically, the nonlinear σ-model on the line describes the motion of a particle of mass m in a curved target space. Indeed, any solution $\phi^i(t)$ of (7.133) describes a possible trajectory of a particle. For the special case when the ϕ^i are Cartesian coordinates of a flat target space, equation (7.133) reduces to the statement that $\mathrm{d}^2\phi^i/\mathrm{d}t^2 = 0$. The particle moves freely with no acceleration. In this case the kinetic energy $E = m\delta_{ij}(\mathrm{d}\phi^i/\mathrm{d}t)(\mathrm{d}\phi^j/\mathrm{d}t)/2$ is conserved. There is also a conserved energy for motion on a general target space.

Exercise 7.42 *Show that $E \equiv \frac{1}{2}m\, g_{ij}(\phi)(\mathrm{d}\phi^i/\mathrm{d}t)(\mathrm{d}\phi^j/\mathrm{d}t)$ is conserved, i.e. $\mathrm{d}E/\mathrm{d}t = 0$, for any solution of the equation of motion (7.133).*

Many different target spaces appear in specific applications of the nonlinear σ-model. They may be compact, for example, the spheres S^n, or non-compact. As an example that will be relevant later in the book, we discuss the Poincaré plane. This is the non-compact two-dimensional manifold with coordinates X, Y with $Y > 0$. The Riemannian metric is defined by the line element

$$\mathrm{d}s^2 = \frac{\mathrm{d}X^2 + \mathrm{d}Y^2}{Y^2}\,. \tag{7.134}$$

The action that describes particle motion on the Poincaré plane is

$$S = \frac{1}{2}\int \mathrm{d}t \frac{1}{Y^2}(\dot{X}^2 + \dot{Y}^2)\,, \tag{7.135}$$

in which we have introduced the notation $\dot{X} = \mathrm{d}X/\mathrm{d}t$.

[5] Geodesics are an important subject in differential geometry for both true spacetimes and Euclidean signature manifolds. We do not discuss geodesics since they are peripheral to the main thrust of this book. Readers are referred to a general relativity textbook such as [46, 47].

The purpose of the following exercise is to illustrate a method to compute the connection components of a specific metric by what might be called the 'geodesic' method. It is frequently faster to use this method rather than the general definition (7.105).

Exercise 7.43 *Compute the variation of the action (7.135) quite directly to obtain*

$$\delta S = -\int \mathrm{d}t \Big(\delta X \Big[\frac{1}{Y^2}\ddot{X} - \frac{2}{Y^3}\dot{Y}\dot{X} \Big] + \delta Y \Big[\frac{1}{Y^2}\ddot{Y} + \frac{1}{Y^3}(\dot{X}^2 - \dot{Y}^2) \Big] \Big) . \tag{7.136}$$

The equations of motion then read:

$$\ddot{X} - \frac{2}{Y}\dot{Y}\dot{X} = 0 , \qquad \ddot{Y} + \frac{1}{Y}(\dot{X}^2 - \dot{Y}^2) = 0 . \tag{7.137}$$

Compare with the general form (7.133) and identify the connection coefficients for the Poincaré plane metric in an obvious notation as

$$\begin{aligned} \Gamma^X_{XY} = \Gamma^X_{YX} = -\frac{1}{Y} , \qquad & \Gamma^X_{XX} = \Gamma^X_{YY} = 0 , \\ \Gamma^Y_{XX} = \frac{1}{Y} = -\Gamma^Y_{YY} , \qquad & \Gamma^Y_{XY} = \Gamma^Y_{YX} = 0 . \end{aligned} \tag{7.138}$$

Exercise 7.44 *Here is an exercise on geodesics of the Poincaré plane metric. Show that there are geodesics that are the straight vertical lines* $X = x_0$, $Y = y_0 e^{kt}$. *Can you show that the general geodesic curve is a semicircle of any radius* r_0 *whose center is at any point* $(X_0, 0)$ *on the* X*-axis of the plane? If not, then wait until the next section.*

7.12 Symmetries and Killing vectors

It is frequently and correctly said that the equations of a gravitational theory are invariant under general coordinate transformations, and the same is true for the equations of the nonlinear σ-model. What this really means, however, is that the equations are constructed from the same elements, vectors, tensors, covariant derivatives, etc. in any coordinate system. This has some content, but it is only a special class of coordinate transformations that lead to symmetries of a system. As discussed in Ch. 1, a symmetry is a transformation that takes one solution of the equations of motion into another. The equations themselves must take the same form in two coordinate transformations related by a symmetry. For each global symmetry that leaves the action invariant, there is a conserved Noether current.

7.12.1 σ-model symmetries

Although the situation is similar for both gravitation and the nonlinear σ-model, we prefer to discuss symmetries in the latter setting. The equations of motion (7.130) (and also (7.133)) take the same form in the two sets of coordinates ϕ^i and ϕ'^i if the metric g'_{ij} in (7.131) has the same functional form as the original metric, i.e. if

$$g'_{ij}(\phi) = g_{ij}(\phi). \tag{7.139}$$

A symmetry transformation of the metric is frequently called an isometry.

We now assume that the coordinate transformation is continuous. This means that the functional relation $\phi'^i(\phi)$ depends continuously on parameters θ^A. In the limit of small θ^A it is assumed that the transformation is close to the identity and thus approximated by $\phi'^i = \phi^i + \theta^A k_A{}^i(\phi)$. The $k_A{}^i(\phi)$ are vector fields, which are the generators of the coordinate transformation. Thus we have an infinitesimal coordinate transformation, whose effect on the various tensor fields on the target space is given by Lie derivatives as defined in (7.10). In particular, the condition (7.139) for a symmetry holds to first order in θ^A if

$$\delta g_{ij} \equiv g'_{ij}(\phi) - g_{ij}(\phi) = -\theta^A \mathcal{L}_{k_A} g_{ij} = 0. \tag{7.140}$$

This means that each $k_A{}^i$ is a vector field that satisfies

$$\nabla_i k_{jA} + \nabla_j k_{iA} = 0, \qquad k_{iA} = g_{ij} k^j_A, \qquad \nabla_i k_{jA} = \partial_i k_{jA} - \Gamma^k_{ij}(g) k_{kA}, \tag{7.141}$$

where ∂_i is the ordinary derivative with respect to ϕ^i and $\Gamma^k_{ij}(g)$ is the Christoffel connection on M_n (we are not dealing with torsion here). A vector field with this property is called a Killing vector, and each Killing vector is associated with a symmetry. To summarize, for each continuous isometry of the target space metric there is a Killing vector k^i_A.

Symmetries of the nonlinear σ-model are Noether symmetries[6] as discussed in Sec. 1.3. The Killing symmetries are a special case of (1.63) in which the infinitesimal internal symmetry variations of the fields $\phi^i(x)$ take the form

$$\delta(\theta)\phi^i = \theta^A k_A{}^i(\phi). \tag{7.142}$$

As opposed to the symmetries discussed in Ch. 1, the Killing symmetries are frequently nonlinear in the fields. It is easy to see that this transformation leaves the σ-model action invariant. To first order in θ^A, the transformation of (7.129) is[7]

$$\begin{aligned}
\delta S[\phi] &= S[\phi'] - S[\phi] \\
&= -\tfrac{1}{2}\int \mathrm{d}^D x\, \theta^A \left[\frac{\partial g_{ij}}{\partial \phi_k} k_A{}^k\, \partial_\mu \phi^i\, \partial^\mu \phi^j \right. \\
&\quad \left. + g_{ij}(\phi)\frac{\partial k_A{}^i}{\partial \phi^k}\partial_\mu \phi^k\, \partial^\mu \phi^j + g_{ij}(\phi)\,\partial_\mu \phi^i \frac{\partial k_A{}^j}{\partial \phi^k}\partial_\mu \phi^k\right] \\
&= -\tfrac{1}{2}\int \mathrm{d}^D x\, \theta^A \left[(\nabla_i k_{jA} + \nabla_j k_{iA})\partial^\mu \phi^i \partial_\mu \phi^j\right] = 0.
\end{aligned} \tag{7.143}$$

[6] Killing vectors are not related to the quantity K^μ_A of Sec. 1.3. That quantity denoted the total derivative that appears for spacetime transformations of a Lagrangian system. Killing symmetries are internal symmetries, so K^μ_A vanishes.

[7] The symmetry acts on the dynamical variables ϕ^i of the nonlinear σ-model, and the transformation of $g_{ij}(\phi)$ is induced by the variations $\delta\phi^i$.

Thus the formula (1.67) applies immediately and tells us that

$$J^\mu_A = g_{ij}(\partial^\mu\phi^i)k_A{}^j(\phi) \tag{7.144}$$

is a conserved Noether current for the symmetry.

Exercise 7.45 *It is an instructive exercise to show explicitly that the current satisfies $\partial_\mu J^\mu_A = 0$, if k^i_A is a Killing vector and $\phi^i(x)$ is any solution of the σ-model equations of motions (7.130).*

As described in Sec. 7.2 it is frequently useful to encode a Killing vector in a differential operator

$$k_A \equiv k^j_A \frac{\partial}{\partial\phi^j}\,. \tag{7.145}$$

One can then define the action of the symmetry transformation of any scalar function $f(\phi)$ on the target space as $\delta f \equiv \theta^A k_A f$. The elementary variation (7.142) is then the special case when $f(\phi) = \phi^i$.

The isometry group of a given metric on M_n is frequently non-abelian. Then there are several linearly independent Killing vectors. As in Sec. 7.2, the Lie algebra is determined from the commutators of the differential operators, i.e.

$$[k_A\,,\, k_B] = f_{AB}{}^C k_C = f_{AB}{}^C k^i_C \frac{\partial}{\partial\phi^i}\,. \tag{7.146}$$

The vector field k^i_C is also a Killing vector. With the transformation parameters included, one finds that

$$\left[\theta^A_1 k_A\,,\, \theta^B_2 k_B\right] = \theta^C_3 k_C\,, \qquad \theta^C_3 = \theta^B_2\theta^A_1 f_{AB}{}^C\,. \tag{7.147}$$

Thus nonlinear Killing symmetries compose in the same way as the linear internal symmetries in (1.22). Although Killing vectors $k_A{}^i(\phi)$ are generically nonlinear functions of the ϕ^k, the isometry group of a manifold M_n can include linearly realized matrix symmetries, which act in the same way as the internal symmetries of Sec. 1.2.2. In this case the Killing vectors $k_A{}^i$ are related to the matrix generators t_A by $k_A{}^i(\phi) = -(t_A)^i{}_j\phi^j$ and to the differential operators by

$$k_A = -(t_A)^i{}_j\phi^j \frac{\partial}{\partial\phi^i}\,. \tag{7.148}$$

Exercise 7.46 *Check that for products of Killing vectors of the form (7.148) we have*

$$k_A k_B = (t_B t_A)^i{}_j\phi^j \frac{\partial}{\partial\phi^i} + [\text{a term symmetric in } A \leftrightarrow B]. \tag{7.149}$$

This interchange of the matrices is the underlying reason for introducing − signs in the definitions of transformations defined by matrices already started in (1.15).

Finally we discuss how to recognize when a given metric has symmetries and how to obtain the Killing vectors. One simple situation occurs when the metric tensor g_{ij} does not depend on a particular coordinate, say ϕ^k. Then the 'translation' $\phi^k \to \phi^k + c$ is an isometry, and $k_k = \partial/\partial\phi^k$ is the differential operator that generates the symmetry. In general, it may not be easy to examine a metric tensor and determine its isometries. Indeed there are target space metrics that have important applications in supergravity and string theory and have no continuous isometries.[8]

7.12.2 Symmetries of the Poincaré plane

The isometry group of the Poincaré plane metric (7.134) is the Lie group SL(2, $\mathbb{R}$). Its applications in supergravity and string theory are fundamental, so it is worthwhile to study it. The group may be defined as the group of 2×2 real matrices of the form

$$\begin{pmatrix} a & b \\ c & d \end{pmatrix}, \qquad ad - bc = 1\,. \tag{7.150}$$

First we will study the action of finite group transformations and then identify three Killing vectors as infinitesimal symmetries. For this purpose it is very convenient to use a complex coordinate $Z = X + \mathrm{i}Y$ on the upper half-plane. The line element (7.134) becomes

$$\mathrm{d}s^2 = \frac{\mathrm{d}Z\mathrm{d}\bar{Z}}{Y^2}\,. \tag{7.151}$$

SL(2, $\mathbb{R}$) transformations act as nonlinear maps

$$Z \to Z' = \frac{aZ + b}{cZ + d} = X' + \mathrm{i}Y' \tag{7.152}$$

of the upper half-plane onto itself. These are the same transformations used in (4.55) in a field theory with electromagnetic duality. It is a matter of straightforward algebra to show that

$$\begin{aligned} X' &= \frac{ac(X^2 + Y^2) + (ad + bc)X + bd}{|cZ + d|^2}\,, \\ Y' &= \frac{Y}{|cZ + d|^2}\,, \\ \mathrm{d}Z' &= \frac{\mathrm{d}Z}{(cZ + d)^2}\,. \end{aligned} \tag{7.153}$$

Then, by direct substitution we find that the line element is *invariant*:

$$\mathrm{d}s^2 = \frac{\mathrm{d}Z\mathrm{d}\bar{Z}}{Y^2} = \frac{\mathrm{d}Z'\mathrm{d}\bar{Z}'}{Y'^2}\,. \tag{7.154}$$

Exercise 7.47 *The straightforward algebra is highly recommended. Note that Y' is positive whenever Y is positive. This shows that the transformation maps the upper half-plane into itself. It would fail for complex a, b, c, d.*

We can use the finite isometry (7.152) in a simple and elegant way to obtain the general geodesic of the Poincaré metric. In Ex. 7.44 readers found that vertical lines with

[8] For interested readers we note that compact Calabi–Yau metrics have no isometries.

exponential t dependence are particularly simple geodesics. For the present purpose, it is sufficient to consider the special case $Z_0(t) = \mathrm{i}e^t$. Under the isometry (7.152) this curve is mapped to

$$Z(t) = \frac{a\mathrm{i}e^t + b}{c\mathrm{i}e^t + d}. \tag{7.155}$$

Since a symmetry maps any solution of the equations of motion into another solution, the curves $Z(t) = X(t) + \mathrm{i}Y(t)$ are also solutions of (7.133) for every choice of a, b, c, d. Thus we obtain a large family of geodesics! In the theory of complex variables, the maps are well-known conformal transformations that map straight lines into circles.

One way to obtain the Killing vectors of the Poincaré plane metric is to expand (7.152) around the identity transformation. However, as an illustrative example, the reader is invited to obtain the Killing vectors directly from the metric in the following exercise.

Exercise 7.48 *Consider Z and $\bar{Z}$ as the independent fields, rather than X and Y, and use the line element (7.154). The metric components are*

$$g_{ZZ} = g_{\bar{Z}\bar{Z}} = 0, \qquad g_{Z\bar{Z}} = g_{\bar{Z}Z} = -\frac{2}{(Z-\bar{Z})^2}. \tag{7.156}$$

Show that the only non-vanishing components of the Christoffel connection are Γ^Z_{ZZ} and its complex conjugate $\Gamma^{\bar{Z}}_{\bar{Z}\bar{Z}}$. Calculate them and then show that there are three Killing vectors,

$$k_1^Z = 1, \qquad k_2^Z = Z, \qquad k_3^Z = Z^2, \tag{7.157}$$

each with conjugate components $k_A^{\bar{Z}}$. Show that their Lie brackets give a Lie algebra whose non-vanishing structure constants are

$$f_{12}{}^1 = 1, \qquad f_{13}{}^2 = 2, \qquad f_{23}{}^3 = 1. \tag{7.158}$$

This is a standard presentation of the Lie algebra of $\mathfrak{su}(1,1) = \mathfrak{so}(2,1) = \mathfrak{sl}(2)$.

The σ-model Lagrangian that corresponds to the metric (7.151) is

$$\mathcal{L} = 2\frac{\partial_\mu Z \partial^\mu \bar{Z}}{(Z-\bar{Z})^2}. \tag{7.159}$$

It has the same form in complex coordinates related by (7.152), so finite SL(2, $\mathbb{R}$) transformations are isometries and there is a conserved current for each Killing vector of (7.157).

8 The first and second order formulations of general relativity

The successes of the theory of general relativity give more than enough reasons for physicists to learn and master differential geometry at least at the level of discussion in Ch. 7. It is most important to realize that, above all, general relativity is a physical theory that incorporates Newtonian gravity in the limit of low velocity $|\vec{v}| \ll c$ and makes specific predictions about relativistic effects. The theory has been confirmed by experiments over the huge range of scales from 1 mm to cosmological distances. Unfortunately, the focus of this book allows us to discuss only some formal features of general relativity at the expense of the rich physics. It is a great injustice that general relativity is viewed here merely as $\mathcal{N} = 0$ supergravity. More balanced treatments can be found in many texts, such as [46, 47, 48].

In this chapter we discuss and compare the first and second order formulations of gravitational dynamics. The most familiar setup is the second order formalism in which the metric tensor or the frame field is the dynamical variable describing gravity. If fermions are present one must use the frame field. The curvature tensor and covariant derivatives are constructed from the torsion-free connections $\Gamma^\rho_{\mu\nu}(g)$ and $\omega_{\mu ab}(e)$; see (7.93) and (7.104). The name 'second order' refers to the fact that the gravitational field equation is second order in derivatives of $g_{\mu\nu}$ or e^a_μ.

In the first order (or Palatini) formalism one starts with an action in which e^a_μ and $\omega_{\mu ab}$ are independent variables, and the Euler–Lagrange equations are first order in derivatives. Without matter, the solution of the $\omega_{\mu ab}$ field equation simply sets $\omega_{\mu ab} = \omega_{\mu ab}(e)$. When this result is substituted in the field equation for e^a_μ, one finds the conventional Einstein equations, exactly as they emerge in the second order formulation for frame fields. When gravity is coupled to spinor fields, the $\omega_{\mu ab}$ field equation contains terms bilinear in the spinors, and its solution is $\omega_{\mu ab} = \omega_{\mu ab}(e) + K_{\mu ab}$, as in (7.91), with contortion tensor determined as a bilinear expression in the spinor fields. It is in this way that the mathematical formalism of connections with torsion is realized in physics.

When the result $\omega_{\mu ab} = \omega_{\mu ab}(e) + K_{\mu ab}$ is substituted in the other field equations, one finds that the full effects of the torsion could have been obtained in the second order formalism with an added set of quartic fermion terms in the action. One may then ask, why bother? Why introduce the complication of torsion when its physical effects can be described more conventionally? The answer is that the proof that a supergravity theory is invariant under local supersymmetry transformations is greatly simplified by the fact that quartic terms in the gravitino field can be organized within the connection $\omega_{\mu ab} = \omega_{\mu ab}(e) + K_{\mu ab}$.

8.1 Second order formalism for gravity and bosonic matter

In this section we review the conventional treatment of gravity coupled to bosonic matter. To be definite we consider the simplest and most common matter fields, a real scalar field $\phi(x)$ and an abelian gauge field with gauge potential $A_\mu(x)$ and field strength $F_{\mu\nu} = \partial_\mu A_\nu - \partial_\nu A_\mu$. The action functional of the coupled system is the sum of the Hilbert action for gravity plus the action for matter fields:

$$S = \int \mathrm{d}^D x \sqrt{-\det g}\left(\frac{1}{2\kappa^2} g^{\mu\nu} R_{\mu\nu}(g) + L\right),$$
$$L = -\tfrac{1}{2} g^{\mu\nu} \partial_\mu \phi \partial_\nu \phi - \tfrac{1}{4} g^{\mu\rho} g^{\nu\sigma} F_{\mu\nu} F_{\rho\sigma}\,. \tag{8.1}$$

The constant $\kappa^2 = 8\pi G$ is the gravitational coupling constant; see (A.11). The Ricci tensor $R_{\mu\nu}(g)$ contains the torsion-free connection.

Exercise 8.1 *Show that κ^2 must have dimension of a length to the power $(D-2)$ to make the action dimensionless.*

The matter field Lagrangian was obtained from the Minkowski space kinetic Lagrangians discussed in Chs. 1 and 4 by the minimal coupling prescription. This consists of the following three rules:

(i) Replace the Minkowski metric $\eta_{\mu\nu}$ by the spacetime metric tensor $g_{\mu\nu}(x)$.
(ii) Replace each derivative ∂_μ by the appropriate covariant derivative ∇_μ with connection $\Gamma^\rho_{\mu\nu}(g)$.
(iii) Use the canonical volume form (7.48).

These rules incorporate the equivalence principle of general relativity and the principle of general covariance. The Lagrangian transforms as a scalar under coordinate transformations, so the matter action is invariant. The second rule is not really needed for the scalar and vector gauge fields of our matter system. For scalars, no connection is needed since $\partial_\mu \phi = \nabla_\mu \phi$. The same is true for the gauge field, since $F_{\mu\nu}$ can be written as $F_{\mu\nu} = \nabla_\mu A_\nu - \nabla_\nu A_\mu$. The Christoffel connection cancels by symmetry.

The equations of motion are obtained by requiring that the action is stationary with respect to variations of the three independent fields.

Exercise 8.2 *Show that the scalar and gauge field equations are*

$$\partial_\mu(\sqrt{-\det g}\, g^{\mu\nu} \partial_\nu \phi) = \sqrt{-\det g}\, g^{\mu\nu} \nabla_\mu \partial_\nu \phi = 0\,,$$
$$\partial_\mu(\sqrt{-\det g}\, F^{\mu\nu}) = \sqrt{-\det g}\, \nabla_\mu F^{\mu\nu} = 0\,. \tag{8.2}$$

Use (7.113) without the torsion term.

Exercise 8.3 *Obtain the Einstein field equations*

$$G_{\mu\nu} \equiv R_{\mu\nu} - \frac{1}{2} g_{\mu\nu} R = \kappa^2 T_{\mu\nu}\,, \tag{8.3}$$

$$T_{\mu\nu} \equiv -2\frac{1}{\sqrt{-\det g}}\frac{\delta(\sqrt{-\det g}\,L)}{\delta g^{\mu\nu}} = \partial_\mu\phi\partial_\nu\phi + F_\mu{}^\rho F_{\nu\rho} + g_{\mu\nu}L \tag{8.4}$$

from

$$\delta S = \frac{1}{2\kappa^2}\int \mathrm{d}^D x[\delta(\sqrt{-\det g}g^{\mu\nu})R_{\mu\nu} + \sqrt{-\det g}g^{\mu\nu}\delta R_{\mu\nu} + \ldots], \tag{8.5}$$

where ... indicates the metric variation of the matter field terms that give the stress tensor[1] *$T_{\mu\nu}$ in (8.4). Note that (7.127) implies that $\delta R_{\mu\nu} = \nabla_\rho \delta\Gamma^\rho_{\mu\nu} - \nabla_\mu \delta\Gamma^\rho_{\nu\rho}$. Thus the term in (8.5) containing $\delta R_{\mu\nu}$ is the integral of a total derivative, which vanishes.*

Two relations that are helpful in variational calculations like those in Ex. 8.3 are

$$\begin{aligned}\delta g^{\mu\nu} &= -g^{\mu\rho}\delta g_{\rho\sigma}g^{\sigma\nu}, \\ \delta\sqrt{-\det g} &= \frac{1}{2}\sqrt{-\det g}\, g^{\mu\nu}\delta g_{\mu\nu} = -\frac{1}{2}\sqrt{-\det g}g_{\mu\nu}\,\delta g^{\mu\nu}.\end{aligned} \tag{8.6}$$

The first relation follows from (7.21), and the second one uses (7.47).

Since the covariant divergence of the Einstein tensor vanishes by the contracted second Bianchi identity (7.126), the Einstein equations are consistent only if the matter stress tensor is covariantly conserved, i.e. $\nabla^\mu T_{\mu\nu} = 0$. In Minkowski spacetime this conservation law follows from Noether's theorem for global spacetime translations. The covariant form follows because the matter action is invariant under coordinate transformations, specifically under infinitesimal transformations for which the field variations are given by Lie derivatives; see (7.10), (7.107), and (7.108). The variation of the matter action is

$$\begin{aligned}&\delta\int \mathrm{d}^D x\sqrt{-\det g}\,L \\ &= \int \mathrm{d}^D x\,\sqrt{-\det g}\left[\frac{1}{2}T^{\mu\nu}(\nabla_\mu\xi_\nu + \nabla_\nu\xi_\mu) + \frac{\delta L}{\delta\phi}\mathcal{L}_\xi\phi + \frac{\delta L}{\delta A_\mu}\mathcal{L}_\xi A_\mu\right] \\ &= 0.\end{aligned} \tag{8.7}$$

This vanishes *identically*, whether or not equations of motion are satisfied, if the explicit forms of the Lie derivatives $\mathcal{L}_\xi\phi$ and $\mathcal{L}_\xi A_\mu$ are inserted. This is a correct result but one that is not immediately useful. However, if ϕ and A_μ do satisfy (8.2), then (8.7) reduces to $\int \mathrm{d}^D x\sqrt{-\det g}\,T^{\mu\nu}\nabla_\mu\xi_\nu = 0$. Upon partial integration, we deduce, since ξ^μ is an arbitrary vector field, $\nabla^\mu T_{\mu\nu} = 0$, which is the covariant conservation law.

Exercise 8.4 *Show explicitly that $T_{\mu\nu}$ in (8.4) is conserved for any solution of the matter equations of motion (8.2).*

Exercise 8.5 *Derive the 'Ricci form' of the Einstein field equation (8.3), namely*

$$R_{\mu\nu} = \kappa^2\left[T_{\mu\nu} - \frac{1}{D-2}g_{\mu\nu}T^\rho_\rho\right]. \tag{8.8}$$

[1] The stress tensor that we obtain in this way is the 'improved' stress tensor, which was denoted by $\Theta_{\mu\nu}$ in Chs. 2 and 4.

This form is frequently simpler in applications. For example, when only the scalar field is present, it reads $R_{\mu\nu} = \kappa^2 \partial_\mu \phi \partial_\nu \phi$.

Exercise 8.6 *Consider a theory in which gravity is coupled to a real scalar field whose own dynamics is specified by a potential* $V(\phi)$. *The matter Lagrangian is* $L = -\frac{1}{2}\partial_\mu \phi \partial^\mu \phi - V(\phi)$. *Derive the Ricci form of the gravitational field equation*

$$R_{\mu\nu} = \kappa^2 \left[\partial_\mu \phi \partial_\nu \phi + g_{\mu\nu} \frac{2}{D-2} V \right]. \tag{8.9}$$

In Ch. 15 we consider a more general form of the coupling of a scalar field to gravity that is invariant under the local scale transformation (or Weyl transformation)

$$g_{\mu\nu} \to \mathrm{e}^{-2\sigma(x)} g_{\mu\nu}, \qquad \phi \to \mathrm{e}^{[(D-2)/2]\sigma(x)} \phi, \tag{8.10}$$

where $\sigma(x)$ is an arbitrary function. The modified action is

$$S = -\frac{1}{2} \int \mathrm{d}^D x \sqrt{-\det g}\, [\partial_\mu \phi \partial^\mu \phi + B \phi^2 R], \qquad B = \frac{D-2}{4(D-1)}. \tag{8.11}$$

Exercise 8.7 *Show that the action (8.11) is invariant under the transformations of (8.10). Obtain the equations of motion*

$$\Box \phi = B R \phi, \tag{8.12}$$
$$-B\phi^2 (R_{\mu\nu} - \frac{1}{2} g_{\mu\nu} R) = \partial_\mu \phi \partial_\nu \phi - \frac{1}{2} g_{\mu\nu} \partial_\rho \phi \partial^\rho \phi - B(\nabla_\mu \partial_\nu - g_{\mu\nu} \Box)\phi^2 .$$

Show that the tensor equation is traceless. Thus the number of independent equations of motion is one less than for the conventional scalar–gravity coupling in (8.1). The reduction occurs because of the extra gauge symmetry (8.10).

The field equations of a gravitational theory are frequently complicated. The second order formulation outlined in this section is the most convenient for many applications, such as classical solutions of supergravity theories when fermions can be neglected. However, fermions are essential in the formulation of supergravity, and in Sec. 8.3 we discuss how to include them.

8.2 Gravitational fluctuations of flat spacetime

In the absence of matter, the gravitational field satisfies the equation

$$R_{\mu\nu} = 0. \tag{8.13}$$

One solution is flat Minkowski spacetime with metric tensor $g_{\mu\nu} = \eta_{\mu\nu}$. In this section we study weak gravitational perturbations of Minkowski space, perturbations that are described by metrics of the form

$$g_{\mu\nu}(x) = \eta_{\mu\nu} + \kappa h_{\mu\nu}(x). \tag{8.14}$$

We will work systematically to first order in the gravitational coupling constant κ. The symmetric tensor fluctuation $h_{\mu\nu}(x)$ is a gauge field. We will count the physical degrees of freedom and determine its propagator, as we have done for the free vector A_μ and vector–spinor ψ_μ fields in Chs. 4 and 5, respectively. The free equation of motion for $h_{\mu\nu}$ is the linearization of the exact Ricci tensor, obtained by substituting the metric ansatz (8.14) and retaining terms of order κ. We ask readers to do this as an exercise.

Exercise 8.8 *Obtain the linearized Christoffel connection*

$$\Gamma^{\rho\,\mathrm{Lin}}_{\mu\nu} = \tfrac{1}{2}\kappa\,\eta^{\rho\sigma}(\partial_\mu h_{\sigma\nu} + \partial_\nu h_{\mu\sigma} - \partial_\sigma h_{\mu\nu})\,. \tag{8.15}$$

Indices are raised using $\eta^{\mu\nu}$. Then use the variational formula $\delta R_{\mu\nu} = \nabla_\rho \delta\Gamma^\rho_{\mu\nu} - \nabla_\mu \delta\Gamma^\rho_{\nu\rho}$ to determine the linearized Ricci tensor

$$R^{\mathrm{Lin}}_{\mu\nu} = -\tfrac{1}{2}\kappa\,[\Box h_{\mu\nu} - \partial^\rho(\partial_\mu h_{\rho\nu} + \partial_\nu h_{\mu\rho}) + \partial_\mu\partial_\nu h^\rho_\rho]\,. \tag{8.16}$$

To obtain the gauge properties of the free field system we first write the exact transformation rules under infinitesimal diffeomorphisms

$$\begin{aligned}\delta g_{\mu\nu} &= \kappa\,(\nabla_\mu \xi_\nu + \nabla_\nu \xi_\mu)\,,\\ \delta R_{\mu\nu} &= \kappa\,(\xi^\rho \nabla_\rho R_{\mu\nu} + \nabla_\mu \xi^\rho R_{\rho\nu} + \nabla_\nu \xi^\rho R_{\mu\rho})\,,\end{aligned} \tag{8.17}$$

in which the arbitrary vector $\xi(x)$ of (7.10) was scaled by the factor κ. The $\eta_{\mu\nu}$ term in (8.14) is then invariant. The lowest order terms in κ determine the linearization

$$\delta h_{\mu\nu} = \partial_\mu \xi_\nu + \partial_\nu \xi_\mu\,, \qquad \delta R^{\mathrm{Lin}}_{\mu\nu} = 0\,. \tag{8.18}$$

Thus we learn that the gravitational fluctuation field is a gauge field with gauge transformation (8.18), which obeys the gauge invariant wave equation

$$R^{\mathrm{Lin}}_{\mu\nu} = 0\,. \tag{8.19}$$

Exercise 8.9 *Show explicitly that $R^{\mathrm{Lin}}_{\mu\nu}$ in (8.16) is invariant under the gauge transformation of $\delta h_{\mu\nu}$ in (8.18).*

For most applications of the linearized equation (8.19), it is desirable to fix the gauge. The Lorentz covariant de Donder gauge condition $\partial^\rho h_{\rho\nu} - \partial_\nu h^\rho_\rho/2 = 0$ is commonly used. It leads to the simple wave equation $\Box h_{\mu\nu} = 0$, which shows convincingly that the field describes massless particles. However, the de Donder gauge condition does not fix the gauge completely since $\delta(\partial^\rho h_{\rho\nu} - \partial_\nu h^\rho_\rho/2) = \Box\xi_\nu$. This makes the argument for counting physical degrees of freedom somewhat indirect; see [46].

Instead, we will choose a non-covariant gauge condition, namely the D conditions

$$\partial^i h_{i\mu} = 0\,, \tag{8.20}$$

where the sum includes only the space coordinates $i = 1, \ldots, D-1$. We will show that this fixes the gauge completely and then proceed to count the physical modes of the field.

First note that the variation of the gauge condition is

$$\delta(\partial^i h_{i\mu}) = \nabla^2 \xi_\mu + \partial^i \partial_\mu \xi_i = 0\,. \tag{8.21}$$

Let $\mu \to j$ and contract with ∂^j to learn that $\nabla^2 \partial^i \xi_i = 0$. The argument on p. 69 concerning harmonic functions in flat space tells us that $\partial^i \xi_i$ vanishes. With this information incorporated, (8.21) informs us that $\xi_\mu(x) \equiv 0$.

We now turn to the wave equation specified by (8.16) and (8.19). To count the degrees of freedom, we use the gauge condition (8.20) to write it as

$$\Box h_{\mu\nu} - \partial^0(\partial_\mu h_{0\nu} + \partial_\nu h_{\mu 0}) + \partial_\mu \partial_\nu (h_{ii} - h_{00}) = 0\,. \tag{8.22}$$

We now distinguish the specific components

$$\begin{aligned} \mu = \nu = 0:&\ \ \nabla^2 h_{00} + \partial_0^2 h_{ii} = 0\,, \\ \mu = \nu = i:&\ \ 2\nabla^2 h_{ii} - \partial_0^2 h_{ii} - \nabla^2 h_{00} = 0\,, \end{aligned} \tag{8.23}$$

in which the indices i are summed. The sum of the two equations tells us that $\nabla^2 h_{ii} = 0$, so that $h_{ii} = 0$. Going back to the 00 equation, we learn that $h_{00} \equiv 0$. With the information just learned incorporated, the $\mu = 0,\ \nu = i$ component of (8.22) becomes $\nabla^2 h_{0i} = 0$, so that $h_{0i} \equiv 0$.

Only components h_{ij} remain, and they satisfy the simple wave equation

$$\Box h_{ij} = 0\,. \tag{8.24}$$

There are $D(D-1)/2$ distinct components h_{ij}, but they satisfy the constraints $\partial^i h_{ij} = 0$ from the gauge condition (8.20) and the trace condition $h_{ii} = 0$, which was found above. Thus the number of independent functions that must be supplied as initial data is $D(D-1)/2 - (D-1) - 1 = D(D-3)/2$ together with their time derivatives. A graviton in D spacetime dimensions thus has $D(D-3)/2$ degrees of freedom. As discussed in Ch. 4, the states of a massless particle in D dimensions carry an irreducible representation of the orthogonal group SO$(D-2)$. For the graviton, this is the traceless symmetric tensor representation, which indeed has dimension $D(D-3)/2$.

In Ex. 4.2, readers showed using only the Maxwell equation (4.10) and the Bianchi identity (4.11) that the gauge field strength satisfies the d'Alembertian equation $\Box F_{\mu\nu} = 0$. Thus one learns, entirely by gauge invariant manipulation, that the field describes massless particles. We will obtain the analogous result for the free gravitational field, but we must first ascertain where the gauge invariant degrees of freedom of the linearized field reside.

For this purpose, we note that the exact curvature tensor $R_{\mu\nu\rho\sigma}$ may be decomposed as follows:

$$\begin{aligned} R_{\mu\nu\rho\sigma} = &\ C_{\mu\nu\rho\sigma} + \frac{1}{D-2}\left(g_{\mu\rho} R_{\nu\sigma} - g_{\nu\rho} R_{\mu\sigma} - g_{\mu\sigma} R_{\nu\rho} + g_{\nu\sigma} R_{\mu\rho}\right) \\ &- \frac{1}{(D-1)(D-2)}\left(g_{\mu\rho} g_{\nu\sigma} - g_{\mu\sigma} g_{\nu\rho}\right) R\,. \end{aligned} \tag{8.25}$$

The tensor $C_{\mu\nu\rho\sigma}$ has the same symmetries as $R_{\mu\nu\rho\sigma}$, but is traceless, viz. $g^{\nu\sigma}C_{\mu\nu\rho\sigma} = 0$. It is called the Weyl tensor because $C_{\mu\nu}{}^{\rho}{}_{\sigma}$ is invariant under the Weyl rescaling of the metric, $g_{\mu\nu} \to g'_{\mu\nu} = \exp{(2f(x))}g_{\mu\nu}$. The remaining terms in (8.25) are trace parts of the curvature tensor, which are entirely determined by the Ricci tensor and scalar curvature.

In the limit of a weak gravitational field the curvature tensor and Weyl tensor are linear in the fluctuation $h_{\mu\nu}$, and they are gauge invariant, e.g. $\delta R^{\text{Lin}}_{\mu\nu\rho\sigma} = 0$ under (8.18). If the fluctuation satisfies its equation of motion $R^{\text{Lin}}_{\mu\nu} = 0$, then clearly $R^{\text{Lin}}_{\mu\nu\rho\sigma} = C^{\text{Lin}}_{\mu\nu\rho\sigma}$. Thus we learn that the gauge invariant content of a weak gravitational field lies in the Weyl components of the linearized curvature tensor. These remarks set the stage for the following exercise.

Exercise 8.10 *Use the equation of motion $R^{\text{Lin}}_{\mu\nu} = 0$, the Bianchi identities and symmetry properties of the curvature tensor to show that $\Box R^{\text{Lin}}_{\mu\nu\rho\sigma} = 0$. Only gauge invariant manipulations are allowed in this exercise, which shows that the linearized graviton is massless.*

8.2.1 The graviton Green's function

Let us consider the linearization of the Ricci form of the Einstein field equation (8.8), which determines the gravitational fluctuation $h_{\mu\nu}$ of Minkowski spacetime due to a conserved tensor source $T_{\mu\nu}$. Inserting (8.16), we find that the equation to be solved is

$$-\Box h_{\mu\nu} + \partial^{\rho}(\partial_{\mu}h_{\rho\nu} + \partial_{\nu}h_{\mu\rho}) - \partial_{\mu}\partial_{\nu}h^{\rho}_{\rho} = 2\kappa\left(T_{\mu\nu} - \frac{1}{D-2}\eta_{\mu\nu}T^{\rho}_{\rho}\right). \qquad (8.26)$$

We will represent the solution for a general source as the spacetime integral

$$h_{\mu\nu}(x) = \kappa\int \mathrm{d}^D y\, G_{\mu\nu;\mu'\nu'}(u)T^{\mu'\nu'}(y), \qquad (8.27)$$

in which $G_{\mu\nu;\mu'\nu'}$ is the Green's function. The Green's function is a bitensor that depends on the coordinates of both the observation point x^{μ} and source point $y^{\mu'}$. In a general spacetime background, we would write $G_{\mu\nu;\mu'\nu'}(x, y)$, and note that under changes of coordinates $\hat{x}^{\rho}(x)$ of the background, the bitensor transforms with products of distinct Jacobian factors $(\partial\hat{x}^{\rho}/\partial x^{\mu})\,(\partial\hat{x}^{\sigma}/\partial x^{\nu})$ and $(\partial\hat{y}^{\rho'}/\partial y^{\mu'})\,(\partial\hat{y}^{\sigma'}/\partial y^{\nu})$ in the observation and source coordinates. Because of the high symmetry of Minkowski space, the Green's function in (8.27) depends on the single Poincaré invariant variable $u = (x-y)^2$. If we restrict coordinate changes to global Lorentz transformations, we can ignore the bitensor complication.

We proceed to find $G_{\mu\nu;\mu'\nu'}(u)$ by a method similar to those used for the vector and vector–spinor gauge fields in Chs. 4 and 5. We note that (8.26) is solved by the integral in (8.27) if the Green's function satisfies the partial differential equation

$$\begin{aligned}&-\Box G_{\mu\nu;\mu'\nu'} + \partial^{\rho}(\partial_{\mu}G_{\rho\nu;\mu'\nu'} + \partial_{\nu}G_{\mu\rho;\mu'\nu'}) - \partial_{\mu}\partial_{\nu}G^{\rho}{}_{\rho;\mu'\nu'} \qquad (8.28)\\ &= \left(\eta_{\mu\mu'}\eta_{\nu\nu'} + \eta_{\mu\nu'}\eta_{\nu\mu'} - \frac{2}{D-2}\eta_{\mu\nu}\eta_{\mu'\nu'}\right)\delta(x-y) + \partial_{\mu'}\Lambda_{\mu\nu;\nu'} + \partial_{\nu'}\Lambda_{\mu\nu;\mu'},\end{aligned}$$

in which all derivatives are with respect to x. Although the differential operator on the left-hand side is non-invertible because of the gauge symmetry (8.18), the added terms involving Λ on the right-hand side allow a solution for $G_{\mu\nu;\mu'\nu'}$. These terms are 'pure gauge' in the source coordinates and have no effect in the integral (8.27) because the source is conserved.

The symmetries of Minkowski space and gauge properties of the equations permit the representation

$$G_{\mu\nu;\mu'\nu'}(u) = (\eta_{\mu\mu'}\eta_{\nu\nu'} + \eta_{\mu\nu'}\eta_{\nu\mu'})G(u) + \eta_{\mu\nu}\eta_{\mu'\nu'}H(u) + \ldots . \tag{8.29}$$

Here ... indicates three independent tensors, which are 'pure gauge' in x and thus change $h_{\mu\nu}$ in a physically inessential way or 'pure gauge' in y and thus make no contribution to the integral in (8.27).

We need only determine the two invariant scalar functions $G(u)$ and $H(u)$. We proceed to substitute the representation (8.29) in (8.27) and apply the differential operator of (8.28) inside the integral. Then after straightforward tensor algebra, we see that (8.26) is satisfied if

$$\begin{aligned}&\int \mathrm{d}^D y \left\{2\Box G(u)T_{\mu\nu}(y) + \eta_{\mu\nu}\Box H(u)T^\rho_\rho(y) + \partial_\mu\partial_\nu\left[(D-2)H(u) + 2G(u)\right]T^\rho_\rho(y)\right\}\\ &= -2\left(T_{\mu\nu} - \frac{1}{D-2}\eta_{\mu\nu}T^\rho_\rho\right).\end{aligned} \tag{8.30}$$

This equation is satisfied if we require that

$$\begin{aligned}\Box G(u) &= -\delta(x-y),\\ H(u) &= -\frac{2}{D-2}G(u).\end{aligned} \tag{8.31}$$

We choose the solution with Feynman causal structure. Thus $G(u)$ is the propagator of a massless scalar field, given explicitly in (4.19).

There are faster ways to find the graviton propagator in Minkowski space than the method used above. For example, in the de Donder gauge, all terms on the left-hand side of (8.26) vanish except for $-\Box h_{\mu\nu}$. It is then immediate to write the solution (8.29) and (8.31) which we have found more laboriously. However, our method applies to the Green's function for vector and tensor fields in anti-de Sitter space, where it is much simpler than other methods; see [49].

8.3 Second order formalism for gravity and fermions

In this section we explore the most commonly used framework for the coupling of spinor fields to gravity. The frame field $e^a_\mu(x)$ and covariant derivatives constructed with the torsion-free connection $\omega_\mu{}^{ab}(e)$ play an essential role. It is instructive to consider the simplest case of a massless Dirac field $\Psi(x)$. The action functional is

$$S = S_2 + S_{1/2} = \int \mathrm{d}^D x\, e\left[\frac{1}{2\kappa^2}e^\mu_a e^\nu_b R_{\mu\nu}{}^{ab}(\omega) - \frac{1}{2}\bar\Psi\gamma^\mu\nabla_\mu\Psi + \frac{1}{2}\bar\Psi\overleftarrow{\nabla}_\mu\gamma^\mu\Psi\right]. \tag{8.32}$$

Box 8.1 **γ-matrices**

The γ^a are numerical matrices. The $\gamma^\mu(x) = e^\mu_a(x)\gamma^a$ are matrices that contain the frame field.

The first term contains the curvature tensor (7.115),

$$R_{\mu\nu ab} = \partial_\mu \omega_{\nu ab} - \partial_\nu \omega_{\mu ab} + \omega_{\mu ac}\omega_\nu{}^c{}_b - \omega_{\nu ac}\omega_\mu{}^c{}_b\,, \tag{8.33}$$

with $\omega \to \omega(e)$; see (7.93).

Exercise 8.11 *Calculate $\delta S_2(e,\omega)/\delta\omega$ using (7.120), and obtain that it is proportional to $D_{[\mu}e^a_{\nu]}$. Remember that $\omega(e)$ is defined in (7.81) as the solution of the requirement that this vanishes. Conclude from this that*

$$\left.\frac{\delta S_2(e,\omega)}{\delta\omega_{\mu ab}}\right|_{\omega=\omega(e)} = 0\,. \tag{8.34}$$

The fermion action contains the curved space γ^μ matrix, whose construction and properties we discuss below, and the covariant derivatives

$$\begin{aligned} \nabla_\mu \Psi &= D_\mu \Psi = (\partial_\mu + \tfrac{1}{4}\omega_\mu{}^{ab}\gamma_{ab})\Psi\,, \\ \bar\Psi \overleftarrow{\nabla}_\mu &= \bar\Psi \overleftarrow{D}_\mu = \bar\Psi(\overleftarrow{\partial}_\mu - \tfrac{1}{4}\omega_\mu{}^{ab}\gamma_{ab})\,. \end{aligned} \tag{8.35}$$

Note that the total covariant derivative ∇_μ and the local Lorentz derivative coincide for a spinor field Ψ, but not for a field such as the gravitino Ψ_μ with additional coordinate indices; see (7.109).

The fermion action is a covariant version of the antisymmetric derivative form introduced in Sec. 2.7.2. The necessary calculations are simpler using this form. In the present second order formalism it differs by a total derivative from the covariant version of (2.32).

The procedure to covariantize Dirac–Clifford matrices begins with the observation that the Clifford algebra is closely linked to the properties of spinors on the spacetime manifold and is therefore defined in local frames. The generators are the *constant* matrices γ^a, which satisfy $\{\gamma^a, \gamma^b\} = 2\eta^{ab}$, and higher rank elements γ^{ab}, $\gamma^{abc}, \ldots$ are defined as antisymmetric products of the generators as in Ch. 3.

Frame fields are used to transform frame *vector* indices to a coordinate basis. For example, $\gamma_\mu = e_{a\mu}\gamma^a$ or $\gamma^\mu = e^\mu_a\gamma^a = g^{\mu\nu}\gamma_\nu$. Thus γ_μ transforms as a covariant vector under coordinate transformations. But it also has (suppressed) row and column *spinor* indices and is therefore a Lorentz bispinor.[2] The covariant derivative of γ_μ is therefore

$$\nabla_\mu \gamma_\nu = \partial_\mu \gamma_\nu + \tfrac{1}{4}\omega_\mu{}^{ab}[\gamma_{ab}, \gamma_\nu] - \Gamma^\rho_{\mu\nu}\gamma_\rho\,. \tag{8.36}$$

The spin connection appears with the commutator as required for a bispinor.

[2] The archetypal example of a bispinor is the product $\Psi_\alpha\bar\Psi_\beta$ of a Dirac spinor and its adjoint, where α, β are spinor indices. Under infinitesimal local Lorentz transformations, this bispinor transforms as $\delta(\Psi\,\bar\Psi) = -\frac{1}{4}\lambda_{ab}(x)[\gamma^{ab}, \Psi\,\bar\Psi]$.

Exercise 8.12 *Derive the very useful result that the covariant derivative of γ_ν vanishes. Specifically show that*

$$\nabla_\mu \gamma_\nu = \gamma^a (\partial_\mu e_{a\nu} + \omega_{\mu ab} e^b_\nu - \Gamma^\rho_{\mu\nu} e_{a\rho}) = 0 . \tag{8.37}$$

In the last step use (7.101). Note that $\nabla_\mu \gamma_\nu = 0$ holds for any affine connection, with or without torsion, provided that one uses the total covariant derivative.

The result (8.37) implies that covariant derivatives commute with multiplication by γ-matrices. For example, if $\Psi(x)$ is a Dirac spinor field, then $\nabla_\mu(\gamma_\nu \Psi) = \gamma_\nu \nabla_\mu \Psi$.

Exercise 8.13 *Show that $\bar\Psi \gamma^\mu \nabla_\mu \Psi$ is invariant under infinitesimal local Lorentz transformations $\delta\Psi = -\frac{1}{4}\lambda_{ab}\gamma^{ab}\Psi$, $\delta\bar\Psi = \frac{1}{4}\lambda_{ab}\bar\Psi\gamma^{ab}$, $\delta e^\mu_a = -\lambda_a{}^b e^\mu_b$. Show that it transforms as a scalar under infinitesimal coordinate transformations $\delta\Psi = \xi^\rho \partial_\rho \Psi$, $\delta e^\mu_a = \xi^\rho \partial_\rho e^\mu_a - \partial_\rho \xi^\mu e^\rho_a$, etc. These transformations are Lie derivatives, as in (7.8), under which local frame indices are inert.*

The fermion equation of motion obtained from (8.32) is the (massless) covariant Dirac equation

$$\gamma^\mu \nabla_\mu \Psi = 0 . \tag{8.38}$$

Exercise 8.14 *Use (8.37), the Ricci identity (7.123), properties of $R_{\mu\nu ab}$, and γ-matrix manipulation to show that the square of the covariant Dirac operator is*

$$\gamma^\mu \nabla_\mu \gamma^\nu \nabla_\nu \Psi = \left(g^{\mu\nu} \nabla_\mu \nabla_\nu - \tfrac{1}{4} R \right) \Psi . \tag{8.39}$$

We now outline the derivation of the Einstein equation starting from the variation of (8.32) with respect to $e^{a\mu}$. The goal is to bring this equation to the form (8.3) with conserved symmetric fermion stress tensor $T_{\mu\nu}$, which will turn out to be the covariant extension of the stress tensor (2.64) of Ch. 2. The variation of the action is

$$\begin{aligned} \delta S = \int \mathrm{d}^D x\, e \bigg[& \frac{1}{\kappa^2} \left(e^{b\nu} R_{\mu\nu ab} - \frac{1}{2} e_{a\mu} R \right) \delta e^{a\mu} \\ & - \frac{1}{2} \bar\Psi \gamma^a \overleftrightarrow{\nabla}_\mu \Psi \delta e^{a\mu} - \frac{1}{8} \bar\Psi \{\gamma^\mu, \gamma^{ab}\} \Psi \delta\omega_{\mu ab} \bigg] . \end{aligned} \tag{8.40}$$

We have used (8.34) and dropped a term proportional to the fermion Lagrangian L because we assume that $\Psi(x)$ satisfies (8.38). We continue to omit terms that vanish by the fermion equation of motion.

We now study the last term, involving the variation $\delta\omega_{\mu ab}$ of the spin connection. When integrated by parts this term will contribute to the stress tensor. The anti-commutator in this term is equal to twice the third rank Clifford matrix $e^\mu_c \gamma^{cab}$. In Ex. 7.27 we obtained the expression (7.95), which we can now use since $\gamma^{\mu\nu\rho}$ is totally antisymmetric. We can thus write the last term of (8.40) as

$$-\tfrac{1}{4} \bar\Psi \gamma^{\mu\nu\rho} \Psi e^a_\nu e^b_\rho \delta\omega_{\mu ab} = -\tfrac{1}{4} \bar\Psi \gamma^{\mu\nu\rho} \Psi e^b_\rho \nabla_\mu \delta e_{\nu b} . \tag{8.41}$$

The next step is to integrate (8.41) by parts, including the sign change that this brings in (8.40). We then use the distributive property of ∇_μ, the Dirac equation (8.38), and γ-matrix manipulation to rewrite (8.41) as

$$-\tfrac{1}{4}e_{a\rho}\delta e^a_\nu \bar{\Psi}(\gamma^\nu \overleftrightarrow{\nabla}^\rho - \gamma^\rho \overleftrightarrow{\nabla}^\nu)\Psi = \tfrac{1}{4}\bar{\Psi}(\gamma_a \overleftrightarrow{\nabla}_\mu - \gamma_\mu \overleftrightarrow{\nabla}_\rho e^\rho_a)\Psi\,\delta e^{a\mu}\,. \tag{8.42}$$

To reach the last form we use $e_{a\rho}\delta e^a_\nu = -e_{b\rho}e^b_\mu \delta e^\mu_a e^a_\nu = -g_{\rho\mu}e^a_\nu \delta e^\mu_a$. Finally we insert this result in (8.40) and combine terms to obtain

$$\delta S = \int \mathrm{d}^D x \left[\frac{1}{\kappa^2}\left(e^{b\nu}R_{\mu\nu ab} - \frac{1}{2}e_{a\mu}R\right)\delta e^{a\mu} - \frac{1}{4}\bar{\Psi}\left[\gamma_a \overleftrightarrow{\nabla}_\mu + \gamma_\mu e^\rho_a \overleftrightarrow{\nabla}_\rho\right]\Psi \delta e^{a\mu}\right]. \tag{8.43}$$

The variational condition $e^a_\nu \delta S/\delta e^{a\mu} = 0$ then gives the Einstein equation in the form

$$R_{\mu\nu} - \tfrac{1}{2}g_{\mu\nu}R = \kappa^2 T_{\mu\nu} \equiv \kappa^2 \tfrac{1}{4}\bar{\Psi}\left[\gamma_\mu \overleftrightarrow{\nabla}_\nu + \gamma_\nu \overleftrightarrow{\nabla}_\mu\right]\Psi\,. \tag{8.44}$$

The stress tensor is indeed the covariant version of the symmetric $T_{\mu\nu}$ derived in Ch. 2.

The manipulations needed to obtain the detailed form (8.44) of the Einstein equation of motion were quite arduous for the Dirac spinor and are far more complex for the gravitino. It is worth pointing out that the key properties of the stress tensor, namely conservation, $\nabla^\mu T_{\mu\nu} = 0$, and symmetry, $T_{\mu\nu} = T_{\nu\mu}$, follow from the general requirements that the matter action be invariant under coordinate transformations and local Lorentz transformations. The argument in Sec. 8.1 that $\nabla^\mu T_{\mu\nu} = 0$ is easily extended to include fermion fields, and it will not be repeated.

We now proceed in the same spirit and show that $T_{\mu\nu} = T_{\nu\mu}$ if the matter action is local Lorentz invariant. Suppose that the action has the schematic form

$$S = \int \mathrm{d}^D x\, \mathcal{L}(e, \omega(e), \psi)\,, \tag{8.45}$$

where ψ is a generic spinor field, perhaps the Dirac field Ψ treated above or the gravitino ψ_μ that appears in supergravity. The independent fields transform as

$$\begin{aligned}
\delta e_{a\mu} &= -\lambda_a{}^b e_{b\mu}\,,\\
\delta\Psi &= -\tfrac{1}{4}\lambda_{ab}\gamma^{ab}\Psi\,,\\
\delta\psi_\mu &= -\tfrac{1}{4}\lambda_{ab}\gamma^{ab}\psi_\mu\,,
\end{aligned} \tag{8.46}$$

under local Lorentz transformations specified by infinitesimal parameters $\lambda_{ab}(x)$. We assume that S is invariant so that its variation

$$\delta S = \int \mathrm{d}^D x \left[\frac{\delta S}{\delta e_{a\mu}}\delta e_{a\mu} + \frac{\delta S}{\delta \psi}\delta\psi\right] = 0 \tag{8.47}$$

vanishes *identically*. The first term gives the stress tensor of the frame field formalism,

$$T^{a\mu} \equiv \frac{1}{e}\frac{\delta S}{\delta e_{a\mu}}\,. \tag{8.48}$$

It is essential that the variation of the connection, integrated by parts, is included in this definition. If the fermion equations of motion are satisfied then (8.48) becomes

$$\begin{aligned} \delta S &= \int \mathrm{d}^D x\, e\, T^{\nu\mu} e^a_\nu \delta e_{a\mu} \\ &= \int \mathrm{d}^D x\, e\, T^{\nu\mu} e^a_\nu e^b_\mu \lambda_{ab}(x) = 0\,. \end{aligned} \tag{8.49}$$

Since $\lambda_{ab}(x)$ is antisymmetric in ab, but otherwise arbitrary, the stress tensor must be symmetric! Symmetry is guaranteed only when the fermion equations of motion are satisfied.

8.4 First order formalism for gravity and fermions

Let us now describe the first order formalism for the coupling of fermions to the gravitational field. As discussed at the beginning of this chapter, the key points are that the frame field e^a_μ and spin connection $\omega_{\mu ab}$ appear as independent variables in the first order action. The field equation for $\omega_{\mu ab}$ may then be solved, giving a connection with torsion, $\omega_{\mu ab} = \omega_{\mu ab}(g) + K_{\mu ab}$. The contortion tensor is bilinear in the fermions. It is interesting to see how connections with torsion arise in a physical setting.

We can still use (8.32) and (8.35) for the first order action and covariant derivatives with the understanding that the connection ω_μ^{ab} is an independent quantity, and that the curvature tensor $R_{\mu\nu}{}^{ab}$ is constructed from this connection; see (8.33). Note that the first order spinor actions with antisymmetric and right-acting D_μ are not equivalent, since a torsion term appears upon partial integration; see Sec. 7.9.3.

Our goal is to solve the field equations for the connection. We will obtain $\delta S/\delta\omega_\mu{}^{ab}$, which is the change in the action due to a small variation $\delta\omega_\mu{}^{ab}$ of the connection. Note that $\delta\omega_\mu{}^{ab}$ is a local Lorentz tensor and covariant coordinate vector.

The variation of the gravitational action, using (7.120), is

$$\delta S_2 = \frac{1}{2\kappa^2} \int \mathrm{d}^D x\, e\, e^\mu_a e^\nu_b \left(D_\mu \delta\omega_\nu{}^{ab} - D_\nu \delta\omega_\mu{}^{ab} \right). \tag{8.50}$$

Owing to the antisymmetry in $\mu\nu$, we can replace the Lorentz covariant derivatives with fully covariant derivatives if the torsion correction similar to that in (7.110) is included. Since there is antisymmetry in ab, the result is

Box 8.2 First and second order formalism

First order formalism: e^a_μ and $\omega_{\mu ab}$ are independent variables.
The field equation for $\omega_{\mu ab}$ gives a connection with torsion: $\omega_{\mu ab} = \omega_{\mu ab}(e) + K_{\mu ab}(\phi)$.

Second order formalism: by definition $\omega_{\mu ab}$ is a function of the other fields. This can be $\omega \equiv \omega_{\mu ab}(e)$ or the solution found in first order formalism: $\omega_{\mu ab} \equiv \omega_{\mu ab}(e) + K_{\mu ab}(\phi)$. The latter leads to new terms in the action that are linear and quadratic in K, and is equivalent to the first order action.

$$\delta S_2 = \frac{1}{2\kappa^2}\int \mathrm{d}^D x\, e\, e_a^\mu e_b^\nu \left(2\nabla_\mu \delta\omega_\nu{}^{ab} + T_{\mu\nu}{}^\rho \delta\omega_\rho{}^{ab}\right). \tag{8.51}$$

Using (7.111), the covariant derivatives commute with the vielbeins, and the first term is a total covariant derivative. Comparing with (7.113), and omitting the boundary terms, we obtain the connection variation

$$\begin{aligned}\delta S_2 &= \frac{1}{2\kappa^2}\int \mathrm{d}^D x\, e\, \left(-2K_{\rho\mu}{}^\rho e_a^\mu e_b^\nu \delta\omega_\nu{}^{ab} + T_{ab}{}^\rho \delta\omega_\rho{}^{ab}\right)\\ &= \frac{1}{2\kappa^2}\int \mathrm{d}^D x\, e\, \left(T_{\rho a}{}^\rho e_b^\nu - T_{\rho b}{}^\rho e_a^\nu + T_{ab}{}^\nu\right)\delta\omega_\nu{}^{ab},\end{aligned} \tag{8.52}$$

where we used (7.114) and made the $[ab]$ antisymmetry explicit.

Exercise 8.15 *It is instructive to obtain the formula (8.52) from the connection variation of the wedge product form of the gravitational action in (7.128). Please do so.*

The connection variation of the spinor action is simpler. Using (8.35) we find

$$\begin{aligned}\delta S_{1/2} &= -\tfrac{1}{8}\int \mathrm{d}^D x\, e\, \bar\Psi\{\gamma^\nu, \gamma_{ab}\}\Psi\, \delta\omega_\nu^{ab}\\ &= -\tfrac{1}{4}\int \mathrm{d}^D x\, e\, \bar\Psi\, \gamma^\nu{}_{ab}\,\Psi\, \delta\omega_\nu^{ab}.\end{aligned} \tag{8.53}$$

The connection field equation can now be identified as the coefficient of $\delta\omega_\nu{}^{ab}$ in $\delta S_2 + \delta S_{1/2} = 0$. We find directly an equation for the torsion tensor

$$T_{ab}{}^\nu - T_{a\rho}{}^\rho\, e_b^\nu + T_{b\rho}{}^\rho\, e_a^\nu = \tfrac{1}{2}\kappa^2 \bar\Psi\, \gamma_{ab}{}^\nu\, \Psi. \tag{8.54}$$

Since the trace of the right-hand side vanishes, the trace of the torsion vanishes too, and the torsion of the spinor field with $\overleftrightarrow{D}_\mu$ kinetic term is simply given by the totally antisymmetric tensor

$$T_{ab}{}^\nu = \tfrac{1}{2}\kappa^2 \bar\Psi\, \gamma_{ab}{}^\nu\, \Psi = -2K^\nu{}_{ab}. \tag{8.55}$$

The physical effects of torsion are rather unexciting, but they are worth discussing for pedagogical reasons. We substitute $\omega = \omega(e) + K$ in the first order action (8.32):

$$\begin{aligned}S = \frac{1}{2\kappa^2}\int \mathrm{d}^D x\, e\, \Big[&R(g) - \kappa^2 \bar\Psi \gamma^\mu \overleftrightarrow{\nabla}_\mu \Psi\\ &- 2\nabla_\mu K_\nu{}^{\nu\mu} + K_{\mu\nu\rho}K^{\nu\mu\rho} - K_\rho{}^\rho{}_\mu K_\sigma{}^{\sigma\mu} - \tfrac{1}{2}\bar\Psi \gamma_{\mu\nu\rho}\Psi\, K^{\mu\nu\rho}\Big],\end{aligned} \tag{8.56}$$

The connection and curvature that appear in (8.56) are torsion-free, so the ∇K term is a total derivative, which can be dropped. We substitute the specific form of the torsion tensor from (8.55) and obtain the physically equivalent second order action

$$S = \frac{1}{2}\int \mathrm{d}^D x\, e\, \left[\frac{1}{\kappa^2}R(g) - \bar\Psi\gamma^\mu \overleftrightarrow{\nabla}_\mu \Psi + \frac{1}{16}\kappa^2 (\bar\Psi \gamma_{\mu\nu\rho}\Psi)(\bar\Psi \gamma^{\mu\nu\rho}\Psi)\right]. \tag{8.57}$$

Note that the $(K_\rho{}^\rho{}_\mu)^2$ term vanishes in this theory because the torsion is totally antisymmetric. Physical effects in the fermion theories with and without torsion differ only by the presence of the quartic Ψ^4 term, which is a dimension 6 operator suppressed by the Planck scale. This term generates four-point contact Feynman diagrams in fermion–fermion scattering amplitudes. The contact diagrams modify the conventional contribution of graviton exchange diagrams for large angle scattering. The contact terms could thus, in principle, be detected in experiments, so the fermion theories with and without torsion are physically inequivalent. We advise against wagers on this question.

PART III

BASIC SUPERGRAVITY

$\mathcal{N} = 1$ pure supergravity in four dimensions 9

In earlier chapters we have reviewed the ideas and implementation of Lorentz invariance, relativistic spin, gauge principles, global supersymmetry, and spacetime geometry. We will now begin to study how these elements combine in supergravity. The key idea is that supersymmetry holds *locally* in a supergravity theory. The action is invariant under SUSY transformations in which the spinor parameters $\epsilon(x)$ are arbitrary functions of the spacetime coordinates. The SUSY algebra (see (6.1) and (6.33)) will then involve local translation parameters $\bar{\epsilon}_1\gamma^\mu\epsilon_2$ which must be viewed as diffeomorphisms. Thus local supersymmetry requires gravity. The 'converse' is also true. In any supersymmetric theory which includes gravity, SUSY must be realized locally. The reason is that a constant spinor ϵ is not compatible with the symmetries required in a theory of gravity with fermions. One must extend $\epsilon \to \epsilon(x)$.

A supergravity theory is a nonlinear, and thus interacting, field theory that necessarily contains the gauge or gravity multiplet plus, optionally, other matter multiplets of the underlying global supersymmetry algebra. The gauge multiplet consists of the frame field $e_\mu^a(x)$ describing the graviton, plus a specific number $\mathcal{N}$ of vector–spinor fields $\Psi_\mu^i(x)$, $i = 1, \ldots, \mathcal{N}$, whose quanta are the gravitinos, the supersymmetric partners of the graviton. In the basic case of $\mathcal{N} = 1$ supergravity in $D = 4$ spacetime dimensions, the gauge multiplet consists entirely of the graviton and one Majorana spinor gravitino. In all other cases, both $\mathcal{N} \geq 2$ in $D = 4$ dimensions and $\mathcal{N} \geq 1$ for $D \geq 5$, additional fields are required in the gauge multiplet.

Supergravity theories exist for spacetime dimensions $D \leq 11$. For each dimension D, a specific type of spinor is required, e.g. Majorana or Weyl. For $D = 4$, theories exist for $\mathcal{N} = 1, 2, \ldots, 8$. Beyond these limits local supersymmetry fails and the classical equations of motion are inconsistent.

In this chapter we will concentrate on the basic $\mathcal{N} = 1$, $D = 4$ supergravity theory. We will discuss the form of the action and transformation rules and prove local supersymmetry [4]. The principal terms of the $\mathcal{N} = 1$, $D = 4$ action are an important part of the structure of all supergravity theories, and the initial steps in the proof of local SUSY are universal, that is, applicable in any dimension. We discuss these universal steps in the next section, and then refocus and complete the proof for $\mathcal{N} = 1$, $D = 4$ supergravity.

The approach in this chapter is one of the two approaches to supergravity in this book. An alternative treatment of $\mathcal{N} = 1$, $D = 4$ supergravity based on the superconformal tensor calculus is developed in Chs. 11, 15 and 16 and further developed for $\mathcal{N} = 2$, $D = 4$ in Ch. 20. Both approaches are valuable. The approach below is most easily extended to $\mathcal{N} > 2$, $D = 4$ and to higher dimension, while the superconformal approach is better suited to the derivation and understanding of matter couplings.

9.1 The universal part of supergravity

The part of the supergravity action we will study in this section consists of the sum of the Hilbert action for gravity in second order formalism plus a local Lorentz and diffeomorphism invariant extension of the free gravitino action of (5.2). This action is multiplied by $1/2$ because we deal with a Majorana gravitino and rescaled by the factor $1/\kappa^2$ for convenience.[1]

The action is

$$\begin{aligned} S &= S_2 + S_{3/2}\,, \\ S_2 &= \frac{1}{2\kappa^2}\int \mathrm{d}^D x\, e\, e^{a\mu} e^{b\nu} R_{\mu\nu ab}(\omega)\,, \\ S_{3/2} &= -\frac{1}{2\kappa^2}\int \mathrm{d}^D x\, e\, \bar{\psi}_\mu \gamma^{\mu\nu\rho} D_\nu \psi_\rho\,, \end{aligned} \tag{9.1}$$

where e stands for the determinant of e_μ^a. The gravitino covariant derivative is given by

$$D_\nu \psi_\rho \equiv \partial_\nu \psi_\rho + \tfrac{1}{4}\omega_{\nu ab}\gamma^{ab}\psi_\rho\,. \tag{9.2}$$

It is the torsion-free spin connection $\omega_{\nu ab}(e)$, given in (7.93), that is used exclusively in this section. We need not include the Christoffel connection term $\Gamma^\sigma_{\nu\rho}(g)\psi_\sigma$ in (9.2) because the connection (7.105) is symmetric, and the term vanishes in the action $S_{3/2}$ since $\gamma^{\mu\nu\rho}$ is antisymmetric.

We also need transformation rules, and we postulate the rules

$$\delta e_\mu^a = \tfrac{1}{2}\bar{\epsilon}\gamma^a\psi_\mu\,, \tag{9.3}$$

$$\delta\psi_\mu = D_\mu\epsilon(x) \equiv \partial_\mu\epsilon + \tfrac{1}{4}\omega_{\mu ab}\gamma^{ab}\epsilon\,. \tag{9.4}$$

The gravitino is the gauge field of local supersymmetry, so it is natural to postulate (9.4) as the curved space generalization of (5.1). For the frame field, (9.3) is the simplest form consistent with the tensor structure required and the Bose–Fermi character of supersymmetry.[2]

[1] The derivation we present in this section uses transposition properties of γ-matrices (in particular the case $t_3 = 1$ of Table 3.1) and reality properties that are strictly valid only for Majorana spinors in $D = 4, 10, 11$ mod 8 dimensions. However, the method can be easily modified to apply to any desired type of spinor.

[2] The possible form $\delta e_\mu^a \sim \bar{\epsilon}\gamma_\mu\psi^a$ may be considered explicitly and shown to be incompatible with local supersymmetry.

(In Ch. 11 these transformation rules will be derived from the viewpoint of gauging the SUSY algebra; see also [50].)

Exercise 9.1 *Deduce from (9.3) that*

$$\delta e_a^\mu = -\tfrac{1}{2}\bar{\epsilon}\gamma^\mu\psi_a\,, \qquad \delta e = \tfrac{1}{2}e\,(\bar{\epsilon}\gamma^\rho\psi_\rho). \tag{9.5}$$

The action and transformation rules above are not quite the whole story even for the $\mathcal{N}=1, D=4$ theory, but they can be completed by incorporating the first order formalism with torsion. This is done in the next section. For all other theories one must also add terms describing the other fields in the gauge multiplet. The variation of the action (9.1) consists of linear terms in ψ_μ from the frame field variation of S_2 and the gravitino variation of $S_{3/2}$ and cubic terms from the frame field variation of $S_{3/2}$. The linear terms are universal, and we now proceed to calculate them and show that they cancel.

The elements of the proof are the algebra of Dirac γ-matrices which underlies the relativistic treatment of particle spin and some of the key identities of differential geometry. At a later stage we will need a Fierz rearrangement, which brings in fermion anticommutativity. It is because of the interplay of these principles that supergravity exists! The stage has been set for this proof earlier in the book where some of the ideas and manipulations were discussed. We refer readers back when necessary.

The variation of the gravitational action under (9.3) is[3]

$$\begin{aligned}\delta S_2 &= \frac{1}{2\kappa^2}\int \mathrm{d}^D x\,\Big[(2e\,(\delta e^{a\mu})e^{b\nu} + (\delta e)e^{a\mu}e^{b\nu})R_{\mu\nu ab} + e\,e^{a\mu}e^{b\nu}\delta R_{\mu\nu ab}\Big]\\ &= \frac{1}{2\kappa^2}\int \mathrm{d}^D x\,e\,\Big(R_{\mu\nu} - \tfrac{1}{2}g_{\mu\nu}R\Big)\left(-\bar{\epsilon}\gamma^\mu\psi^\nu\right).\end{aligned} \tag{9.6}$$

The symmetry $R_{\mu\nu ab} = R_{\nu\mu ba}$ of the Riemann tensor gives the factor of 2 in the first line. In the second line we insert the variations of (9.5) and observe that the $\delta R_{\mu\nu ab}$ term is a total covariant derivative (see (7.120)), whose integral vanishes. It is no surprise that the Einstein tensor appears in the variation of the gravitational action.

We now study the gravitino variation. In the second order formalism, partial integration is valid, so it is sufficient to vary $\delta\bar{\psi}_\mu$ and multiply by 2, obtaining

$$\begin{aligned}\delta S_{3/2} &= -\frac{1}{\kappa^2}\int \mathrm{d}^D x\,e\,\bar{\epsilon}\overleftarrow{D}_\mu\gamma^{\mu\nu\rho}D_\nu\psi_\rho\\ &= \frac{1}{\kappa^2}\int \mathrm{d}^D x\,e\,\bar{\epsilon}\gamma^{\mu\nu\rho}D_\mu D_\nu\psi_\rho = \frac{1}{8\kappa^2}\int \mathrm{d}^D x\,e\,\bar{\epsilon}\gamma^{\mu\nu\rho}R_{\mu\nu ab}\gamma^{ab}\psi_\rho\,.\end{aligned} \tag{9.7}$$

We integrated by parts and used (8.37) to move to the second line[4] and then used the Ricci identity (7.123) to obtain the last expression.

We now need some Dirac algebra to evaluate the product $\gamma^{\mu\nu\rho}\gamma^{ab}$. This product was explained as an example in (3.21). It is a sum of fifth, third, and first order elements of the Clifford algebra. After contraction with the Riemann tensor, we find

[3] The same calculation determined the gravitational terms in (8.40).
[4] We can replace ∇_μ by D_μ in the second line because of the antisymmetry and absence of torsion.

$$
\begin{aligned}
\gamma^{\mu\nu\rho}\gamma^{ab}R_{\mu\nu ab} &= \gamma^{\mu\nu\rho ab}R_{\mu\nu ab} + 6R_{\mu\nu}{}^{[\rho}{}_{b}\gamma^{\mu\nu]b} + 6\gamma^{[\mu}R_{\mu\nu}{}^{\rho\nu]} \\
&= \gamma^{\mu\nu\rho ab}R_{\mu\nu ab} + 2R_{\mu\nu}{}^{\rho}{}_{b}\gamma^{\mu\nu b} + 4R_{\mu\nu}{}^{\mu}{}_{b}\gamma^{\nu\rho b} \\
&\quad + 4\gamma^{\mu}R_{\mu\nu}{}^{\rho\nu} + 2\gamma^{\rho}R_{\mu\nu}{}^{\nu\mu} \,.
\end{aligned} \tag{9.8}
$$

In the second line we 'unpack' the cyclic terms of $[\rho\mu\nu]$. The first and second terms vanish because of the first Bianchi identity (7.122) (without torsion). The third term vanishes because of the symmetry clash between the symmetric Ricci tensor $R_{\nu b} = R_{\mu\nu}{}^{\mu}{}_{b}$ and the antisymmetric $\gamma^{\nu\rho b}$. We are thus left with

$$
\delta S_{3/2} = \frac{1}{2\kappa^2}\int \mathrm{d}^D x\, e\,(R_{\mu\nu} - \tfrac{1}{2}g_{\mu\nu}R)(\bar{\epsilon}\gamma^{\mu}\psi^{\nu}) \,. \tag{9.9}
$$

We now observe the exact cancelation between (9.6) and (9.9) which shows that local supersymmetry holds to linear order in ψ_μ for any spacetime dimension[5] D that allows Majorana spinors (see footnote 1).

Although most symmetries are presented as transformations that leave the action invariant, a symmetry can also be described at the level of the equations of motion. A symmetry of a dynamical system is a transformation of its variables which takes any solution of its equations of motion into another solution. This means that an infinitesimal variation of any equation of motion must give a linear combination of equations of motion.

To see how this idea operates in global supersymmetry, consider the simple system of the free gauge multiplet discussed in Sec. 6.3. The gauge field A_μ satisfies the Maxwell equation $\partial^\mu F_{\mu\nu} = 0$ and the Bianchi identity $\partial_\mu F_{\nu\rho} +$ cyclic perms $= 0$, and has the SUSY variation as in (6.49), $\delta A_\mu = -\tfrac{1}{2}\bar{\epsilon}\gamma_\mu\lambda$. The gaugino λ satisfies the Dirac equation $\gamma^\mu\partial_\mu\lambda = 0$ and has SUSY variation $\delta\lambda = \tfrac{1}{4}\gamma^{\nu\rho}F_{\nu\rho}\epsilon$. In global SUSY ϵ is a constant spinor. The SUSY transform of the Dirac equation is

$$
0 = \gamma^\mu\partial_\mu\delta\lambda = \tfrac{1}{4}\gamma^\mu\gamma^{\nu\rho}\partial_\mu F_{\nu\rho}\epsilon \,. \tag{9.10}
$$

Indeed, with straightforward Dirac algebra (see (6.7)), we see that the right-hand side is a sum of terms that vanishes by the Maxwell equation and Bianchi identity.

The same principle works in supergravity, both for exact local SUSY transformations and in the linear approximation used in this section. In fact, manipulations very similar to those used to calculate $\delta S_{3/2}$ above can be used to show that the transform of the gravitino equation of motion is

$$
0 = \gamma^{\mu\nu\rho}D_\nu\delta\psi_\rho = -\tfrac{1}{2}(R^{\mu\nu} - \tfrac{1}{2}g^{\mu\nu}R)\gamma_\nu\epsilon \,. \tag{9.11}
$$

The right-hand side certainly vanishes if the Einstein equation is satisfied.

Exercise 9.2 *Derive (9.11). Show that the supergravity extension of the alternative form (5.4) of the equation of motion is*

$$
\gamma^\mu(D_\mu\psi_\nu - D_\nu\psi_\mu) = 0 \,, \tag{9.12}
$$

[5] Restrictions on D appear if local SUSY is imposed beyond linear order.

where $D_\mu \psi_\nu$ *is given by (9.2). Show that the (linear) SUSY variation of this equation of motion is*

$$0 = \gamma^\mu (D_\mu \delta\psi_\nu - D_\nu \delta\psi_\mu) = \tfrac{1}{2} R_{\nu\rho} \gamma^\rho \epsilon \,, \tag{9.13}$$

so the right-hand side again vanishes by the Ricci form of the gravitational field equation.

It must also be true that the supersymmetry transform of the Einstein equation vanishes if the gravitino satisfies its equation of motion. For linear fluctuations of the gravitational field about Minkowski spacetime, this will be true if the SUSY transformation of the metric

$$\delta g_{\mu\nu} = \tfrac{1}{2} \bar{\epsilon} (\gamma_\mu \psi_\nu + \gamma_\nu \psi_\mu) \tag{9.14}$$

is used in the global limit in which ϵ and γ_μ are constant.

Exercise 9.3 *Show that the SUSY transform of the linearized Ricci tensor of (8.16) vanishes if the metric is transformed by (9.14) in the global limit and* ψ_μ *satisfies the free gravitino equation of motion.*

9.2 Supergravity in the first order formalism

Beyond linear order it becomes complicated to establish local supersymmetry for any supergravity theory. It is therefore very useful to recognize simplifications which lead to convenient organization of terms. One important simplification [5] is to express the action and transformation rules in the first order formalism. We continue to work with the D-dimensional action (9.1) and transformation rules (9.3) and (9.4), but we regard the spin connection $\omega_{\mu ab}$ as an independent variable. There is no Christoffel connection term in $S_{3/2}$. It would be inconsistent with local supersymmetry to include it.[6]

As discussed in Ch. 8, in the first order formalism the equation of motion for the spin connection can be solved to obtain a connection with torsion. This result can then be substituted in the action to obtain the physically equivalent second order form of the theory with torsion-free connection and explicit four-fermion contact terms. We now carry out this process for the supergravity action (9.1). We will use the connection variation (8.52) of S_2 derived in Ch. 8.

For $S_{3/2}$ we easily obtain

$$\delta S_{3/2} = -\frac{1}{8\kappa^2} \int d^D x \, e \, (\bar{\psi}_\mu \gamma^{\mu\nu\rho} \gamma_{ab} \psi_\rho) \delta\omega_\nu{}^{ab} \,. \tag{9.15}$$

The Clifford algebra relation needed to simplify the spinor bilinear in (9.15) is again the one given in (3.21). It contains γ-matrices of rank 1, 3, and 5. We work with Majorana spinors in dimensions $D = 2, 3, 4, 10$, and 11. Table 3.1 tells us that spinor bilinears of

[6] Note that a Christoffel term is permitted by diffeomorphism invariance since the antisymmetric part of $\Gamma^\sigma_{\nu\rho}$ is proportional to the torsion tensor; see (7.106).

rank 3 are then symmetric. So the rank 3 terms from (3.21) vanish because of antisymmetry in the gravitino indices μ, ρ. Therefore, we obtain

$$\bar{\psi}_\mu \gamma^{\mu\nu\rho} \gamma_{ab} \psi_\rho = \bar{\psi}_\mu \left(\gamma^{\mu\nu\rho}{}_{ab} + 6\gamma^{[\mu} e^\nu{}_{[b} e^{\rho]}{}_{a]} \right) \psi_\rho \,. \tag{9.16}$$

It is now very easy to solve the connection field equation

$$\delta S_2 + \delta S_{3/2} = 0 \,, \tag{9.17}$$

using (8.52). The trace structure in the two terms matches exactly, so the unique solution for the torsion is

$$T_{ab}{}^\nu = \tfrac{1}{2} \bar{\psi}_a \gamma^\nu \psi_b + \tfrac{1}{4} \bar{\psi}_\mu \gamma^{\mu\nu\rho}{}_{ab} \psi_\rho \,. \tag{9.18}$$

The fifth rank tensor term is one of the complications of supergravity for $D \geq 5$, but it simply vanishes when $D = 4$.

For gravity coupled to a spin-1/2 Dirac field, we showed in (8.56) and (8.57) how to obtain the physically equivalent second order form of the theory by substituting the value of the torsion tensor in the first order action. Here is an exercise to do the same for supergravity.

Exercise 9.4 *For $D = 4$ substitute the torsion tensor (9.18) in the action (9.1) to obtain the second order action of supergravity,*

$$S = \frac{1}{2\kappa^2} \int \mathrm{d}^4 x \, e \left[R(e) - \bar{\psi}_\mu \gamma^{\mu\nu\rho} D_\nu \psi_\rho + \mathcal{L}_{\mathrm{SG,torsion}} \right] , \tag{9.19}$$

$$\mathcal{L}_{\mathrm{SG,torsion}} = -\frac{1}{16} \left[(\bar{\psi}^\rho \gamma^\mu \psi^\nu)(\bar{\psi}_\rho \gamma_\mu \psi_\nu + 2\bar{\psi}_\rho \gamma_\nu \psi_\mu) - 4(\bar{\psi}_\mu \gamma \cdot \psi)(\bar{\psi}^\mu \gamma \cdot \psi) \right] ,$$

in which the curvature $R(e)$ and the gravitino covariant derivative now contain the torsion-free connection

$$D_\nu \psi_\rho \equiv \partial_\nu \psi_\rho + \tfrac{1}{4} \omega_{\nu ab}(e) \gamma^{ab} \psi_\rho \,. \tag{9.20}$$

The theory expressed by (9.19) contains four-point contact diagrams for gravitino scattering. Their physical effects are certainly not dramatic, but they are necessary for the consistency of the theory, which does not otherwise obey the requirement of local supersymmetry. Local supersymmetry means that the action (9.19) is invariant under the transformation rules (9.3) and (9.4), with $\omega = \omega(e) + K$; see (7.91).

Box 9.1 **Torsion in supergravity**

The first order formalism determines a connection with torsion for supergravity. Rewriting the result in terms of a torsion-free connection leads to four-fermion terms in the supergravity action.

1.5 order formalism Box 9.2

$\delta\omega$ can be neglected in the variation of the action if ω takes the value $\omega(e, \psi)$ determined by its field equation.

9.3 The 1.5 order formalism

The second order action (9.19) for $\mathcal{N} = 1$, $D = 4$ supergravity is complete, and it possesses complete local supersymmetry under the transformation rules (9.3) and (9.4) with the connection

$$\begin{aligned} \omega_{\mu ab} &= \omega_{\mu ab}(e) + K_{\mu ab}\,, \\ K_{\mu\nu\rho} &= -\tfrac{1}{4}(\bar{\psi}_\mu\gamma_\rho\psi_\nu - \bar{\psi}_\nu\gamma_\mu\psi_\rho + \bar{\psi}_\rho\gamma_\nu\psi_\mu)\,, \end{aligned} \tag{9.21}$$

that includes the gravitino torsion. We will prove this invariance property in the next section.

Let us think schematically about the structure of the proof. The variation of the action contains terms that are first, third, and fifth order in the gravitino field. They are independent and must cancel separately. In the first construction of the theory [4], which used the second order formalism, the first and third order terms were treated analytically, but it required a computer calculation to show that the complicated order $(\psi_\mu)^5$ term vanishes.[7]

In the first order form of the theory [5], the fifth order variation is avoided, but one must specify a new transformation rule $\delta\omega_{\mu ab}$, since the connection is an independent variable. This approach becomes quite complicated when matter multiplets are coupled to supergravity.

For the reasons above, the simplest treatment of supergravity uses a formalism intermediate between the first and second order versions we have discussed. It is therefore called the 1.5 order formalism [51, 52], and it is applicable to supergravity theories in any dimension.

One is really working in the second order formalism since there are only two independent fields, e^a_μ and ψ_μ. However, since the second order Lagrangian is obtained by substituting ω from (9.21) in the first order action, one can simplify the proof of invariance by retaining the original grouping of terms. To see more concretely how this works, let's consider an action that is a functional of the three variables $S[e, \omega, \psi]$. We use the chain rule to calculate its variation in the second order formalism, viz.

[7] Such terms come from the variation of the order $(\psi_\mu)^4$ contact terms in (9.19) with respect to the frame field and the torsion part of $\delta\psi_\mu$.

$$\delta S[e,\omega(e)+K,\psi]=\int \mathrm{d}^D x\left[\frac{\delta S}{\delta e}\delta e+\frac{\delta S}{\delta\omega}\delta(\omega(e)+K)+\frac{\delta S}{\delta\psi}\delta\psi\right]. \tag{9.22}$$

The variation $\delta(\omega(e)+K)$ should be calculated using (9.3) and (9.21). However, no calculation is needed since the second term vanishes because the expression for ω in (9.21) is obtained by solving the algebraic field equation $\delta S/\delta\omega=0$. Thus, in the 1.5 order formalism, we can neglect *all* $\delta\omega$ variations as we proceed to establish local supersymmetry.

We can now summarize the prescription for δS in the 1.5 order formalism:

1. Use the first order form of the action $S[e,\omega,\psi]$ and the transformation rules δe and $\delta\psi$ with connection ω unspecified.
2. Ignore the connection variation and calculate

$$\delta S=\int \mathrm{d}^D x\left[\frac{\delta S}{\delta e}\delta e+\frac{\delta S}{\delta\psi}\delta\psi\right]. \tag{9.23}$$

3. Substitute ω from (9.21) in the result, which must vanish for a consistent supergravity theory.

9.4 Local supersymmetry of $\mathcal{N}=1, D=4$ supergravity

In this section[8] we will use the 1.5 order formalism to prove that $\mathcal{N}=1$, $D=4$ supergravity is a consistent gauge theory, invariant under the transformation rules (9.3) and (9.21) with arbitrary $\epsilon(x)$. The proof is specific to $D=4$ spacetime dimensions, so we first simplify the gravitino action by introducing the highest rank Clifford element $\gamma_*=\mathrm{i}\gamma_0\gamma_1\gamma_2\gamma_3$. One can then reexpress the third rank Clifford matrices as

$$\gamma^{abc}=-\mathrm{i}\varepsilon^{abcd}\gamma_*\gamma_d\,,\qquad \gamma^{\mu\nu\rho}=-\mathrm{i}e^{-1}\varepsilon^{\mu\nu\rho\sigma}\gamma_*\gamma_\sigma\,. \tag{9.24}$$

The first relation holds in local frames and the second, which we need, in the coordinate basis of spacetime. The Levi-Civita tensor density, with $\varepsilon^{0123}=-1$, appears in both relations. Using (9.24), we can rewrite the gravitino action (9.1) as

$$S_{3/2}=\frac{\mathrm{i}}{2\kappa^2}\int \mathrm{d}^4x\,\varepsilon^{\mu\nu\rho\sigma}\,\bar\psi_\mu\gamma_*\gamma_\sigma D_\nu\psi_\rho\,. \tag{9.25}$$

This is the way that supergravity was written before it was 'discovered' that the universe has six or seven hidden dimensions! (Forgive this joke.) The advantage of this form is that the frame field variation δe is needed only in γ_σ rather than in four positions in $e\,\gamma^{\mu\nu\rho}$.

We can ignore the $\delta\omega$ variation of $R_{\mu\nu ab}$ in the 1.5 order formalism. So we must consider the four terms

$$\delta S=\delta S_2+\delta S_{3/2,e}+\delta S_{3/2,\psi}+\delta S_{3/2,\bar\psi}\,, \tag{9.26}$$

[8] An alternative proof is developed in Chs. 11, 15, and 16.

where the first term is the variation of the gravity action, the second one is due to the frame variation of $S_{3/2}$, while the third and fourth are the variations of ψ and $\bar{\psi}$, respectively. We must obtain the ψ and $\bar{\psi}$ variations separately because partial integration of the local Lorentz derivative D_μ must be done carefully. Indeed, we will encounter a number of subtleties because of the connection with torsion.

The Ricci tensor with torsion is not symmetric, and we call it $R_{\mu\nu}(\omega)$ as a reminder. But it is easy to check that (9.6) is still valid. It can be rewritten as

$$\delta S_2 = \frac{1}{2\kappa^2}\int d^4x\, e\left(R_{\mu\nu}(\omega) - \frac{1}{2}g_{\mu\nu}R(\omega)\right)(-\bar{\epsilon}\gamma^\mu\psi^\nu). \tag{9.27}$$

The second term in (9.26) contains the variation of γ_σ, i.e.

$$\delta S_{3/2,e} = \frac{\mathrm{i}}{4\kappa^2}\int d^4x\, \varepsilon^{\mu\nu\rho\sigma}(\bar{\epsilon}\gamma^a\psi_\sigma)(\bar{\psi}_\mu\gamma_*\gamma_a D_\nu\psi_\rho). \tag{9.28}$$

The ψ variation of $S_{3/2}$ is quite simple:

$$\begin{aligned}\delta S_{3/2,\psi} &= \frac{\mathrm{i}}{2\kappa^2}\int d^4x\, \varepsilon^{\mu\nu\rho\sigma}\bar{\psi}_\mu\gamma_*\gamma_\sigma D_\nu D_\rho\epsilon \\ &= \frac{\mathrm{i}}{16\kappa^2}\int d^4x\, \bar{\psi}_\mu\varepsilon^{\mu\nu\rho\sigma}\gamma_*\gamma_\sigma\gamma^{ab}R_{\nu\rho ab}(\omega)\epsilon.\end{aligned} \tag{9.29}$$

The curvature tensor appears through the commutator of covariant derivatives.

Next we write the $\bar{\psi}_\mu$ variation and exchange the spinors $D_\mu\epsilon$ and ψ_ρ (see (3.51) with $t_3=1$) to obtain

$$\delta S_{3/2,\bar{\psi}} = \frac{\mathrm{i}}{2\kappa^2}\int d^4x\, \varepsilon^{\mu\nu\rho\sigma}\bar{\psi}_\rho\overleftarrow{D}_\nu\gamma_*\gamma_\sigma D_\mu\epsilon. \tag{9.30}$$

With some thought one can see that the left-acting derivative $\bar{\psi}_\rho\overleftarrow{D}_\nu = \partial_\nu\bar{\psi}_\rho - \frac{1}{4}\bar{\psi}_\rho\omega_{\nu ab}\gamma^{ab}$ can be partially integrated and acts distributively to give

$$\begin{aligned}\delta S_{3/2,\bar{\psi}} &= \frac{-\mathrm{i}}{2\kappa^2}\int d^4x\, \varepsilon^{\mu\nu\rho\sigma}\bar{\psi}_\rho\gamma_*[(D_\nu\gamma_\sigma)D_\mu\epsilon + \gamma_\sigma D_\nu D_\mu\epsilon] \\ &= \frac{-\mathrm{i}}{2\kappa^2}\int d^4x\, \varepsilon^{\mu\nu\rho\sigma}\bar{\psi}_\rho\gamma_*\left[\frac{1}{2}T_{\nu\sigma}{}^a\gamma_a D_\mu\epsilon - \frac{1}{8}\gamma_\sigma\gamma^{ab}R_{\mu\nu ab}(\omega)\epsilon\right].\end{aligned} \tag{9.31}$$

As indicated in (8.37) the full covariant $\nabla_\nu\gamma_\sigma = 0$. But supergravity employs only the local Lorentz covariant derivative D_ν. When we add back the Christoffel connection and use antisymmetry in $\nu\sigma$, we obtain using (7.106) the torsion tensor in (9.31).

Note that the $R(\omega)$ terms in (9.29) and (9.31) are equal, so that we obtain

$$\delta S_{3/2,\psi} + \delta S_{3/2,\bar{\psi}} = \frac{-\mathrm{i}}{2\kappa^2}\int d^4x\, \varepsilon^{\mu\nu\rho\sigma}\bar{\psi}_\rho\gamma_*\left[\frac{1}{2}T_{\nu\sigma}{}^a\gamma_a D_\mu\epsilon - \frac{1}{4}\gamma_\sigma\gamma_{ab}R_{\mu\nu}{}^{ab}(\omega)\epsilon\right]. \tag{9.32}$$

The last term can be treated using the methods for γ-matrix manipulations of Sec. 3.1.4:

$$\gamma_\sigma\gamma_{ab} = \gamma_{\sigma ab} + 2e_{\sigma[a}\gamma_{b]} = \mathrm{i}e_\sigma^d\varepsilon_{abcd}\gamma_*\gamma^c + 2e_{\sigma[a}\gamma_{b]}\,. \tag{9.33}$$

We now consider these two terms separately, beginning with the first. We thus encounter the contraction of two Levi-Civita symbols and Riemann tensor. Using (3.9), we write

$$\begin{aligned}\varepsilon^{\mu\nu\rho\sigma}\varepsilon_{abcd}e_\sigma^d R_{\nu\rho}{}^{ab}(\omega) &= -2e\left[e_a^\mu e_b^\nu e_c^\rho + e_b^\mu e_c^\nu e_a^\rho + e_c^\mu e_a^\nu e_b^\rho\right]R_{\nu\rho}{}^{ab}(\omega)\\ &= 4e\left[R_c{}^\mu(\omega) - \tfrac{1}{2}e_c^\mu R(\omega)\right].\end{aligned} \tag{9.34}$$

When this relation is inserted in (9.32) and we use $\bar\psi_\mu\gamma^c\epsilon = -\bar\epsilon\gamma^c\psi_\mu$, the result exactly cancels δS_2 in (9.27). The work so far has brought us to the level of linear local supersymmetry proven in Sec. 9.1, although the present cancelation includes cubic terms from the torsion contribution to (9.27).

The contribution of the last term in (9.33) to the integrand of (9.32) involves the factor

$$\varepsilon^{\mu\nu\rho\sigma}R_{\nu\rho\sigma b}(\omega) = -\varepsilon^{\mu\nu\rho\sigma}D_\nu T_{\rho\sigma b}\,, \tag{9.35}$$

in which we have used the modified first Bianchi identity, derived in (7.122), where it is shown that the derivative D_ν contains only the spin connection acting on the index b. This leaves us with

$$\delta S_2 + \delta S_{3/2,\psi} + \delta S_{3/2,\bar\psi} = \frac{-\mathrm{i}}{4\kappa^2}\int \mathrm{d}^4x\,\varepsilon^{\mu\nu\rho\sigma}\bar\psi_\mu\gamma_*\gamma_a\left[T_{\rho\sigma}{}^a D_\nu\epsilon + \left(D_\nu T_{\rho\sigma}{}^a\right)\epsilon\right]. \tag{9.36}$$

The other remaining variations are in (9.28), which we will now rewrite using the torsion tensor. Therefore, we first reorder the spinors in the integrand using Fierz rearrangement technique of Sec. 3.2.3. Using (3.68) we can write

$$\begin{aligned}(\bar\epsilon\gamma^a\psi_{[\sigma})(\bar\psi_{\mu]}\gamma_a\gamma_*D_\nu\psi_\rho) &= \tfrac{1}{2}(\bar\epsilon\gamma_a\gamma_*D_\nu\psi_\rho)(\bar\psi_\mu\gamma^a\psi_\sigma)\\ &= (\bar\epsilon\gamma_a\gamma_*D_\nu\psi_\rho)T_{\mu\sigma}{}^a\,.\end{aligned} \tag{9.37}$$

On the left-hand side, we indicated the antisymmetrization in $[\mu\sigma]$, due to the contraction with $\varepsilon^{\mu\nu\rho\sigma}$ in (9.28). We have rewritten the last line using the $D=4$ torsion tensor (9.18). We insert this in (9.28), reorder the $(\bar\epsilon\ldots D\psi)$ bilinear, and exchange the indices $\mu\rho$ to obtain

$$\delta S_{3/2,e} = \frac{-\mathrm{i}}{4\kappa^2}\int \mathrm{d}^4x\,\varepsilon^{\mu\nu\rho\sigma}T_{\rho\sigma}{}^a\bar\psi_\mu\overleftarrow{D}_\nu\gamma_*\gamma_a\epsilon\,. \tag{9.38}$$

We have now reached the final step of the proof in which we combine these with the uncanceled torsion terms in (9.36) to obtain the sum:

$$\delta S = \frac{-\mathrm{i}}{4\kappa^2}\int \mathrm{d}^4x\,\varepsilon^{\mu\nu\rho\sigma}\partial_\nu[T_{\rho\sigma}{}^b\bar\psi_\mu\gamma_*\gamma_b\epsilon] \equiv 0\,. \tag{9.39}$$

The integrand is a total derivative because the local Lorentz derivative works distributively. Indeed, the spin connection cancels among the three terms from (9.36) and (9.38). This proves that $\mathcal{N} = 1,\ D = 4$ supergravity is locally supersymmetric and thus consistent as a classical theory of the graviton and gravitino! The basic relations of differential geometry with torsion and the Clifford algebra and spinor anti-commutativity combine in the proof in a very striking way!

It is actually true that the integrand in (9.39) vanishes as follows from (3.68). This is the same Fierz rearrangement identity required in super-Yang–Mills theory; see Sec. 6.1.2. In supergravity it is not required for invariance of the action, since it appears as a total derivative in (9.39). Note that the Lagrangian density (9.19) is *not invariant* under local SUSY since there is another total derivative in the partial integration that led to (9.31).

9.5 The algebra of local supersymmetry

The commutator of two local SUSY transformations should realize an algebra which is compatible with that of global SUSY. On any component field Φ of a chiral multiplet we found in Ch. 6 (see (6.33))

$$[\delta_1, \delta_2]\,\Phi = -\tfrac{1}{2}\bar{\epsilon}_1\gamma^\mu\epsilon_2\partial_\mu\Phi\,. \tag{9.40}$$

It is natural to expect that the local extension of the global result (9.40) should be a general coordinate transformation with parameter $\xi^\mu(x) = -\frac{1}{2}\bar{\epsilon}_1(x)\gamma^\mu\epsilon_2(x)$. However, the general formalism for symmetries requires only that the commutator closes on a sum of the gauge symmetries of the theory, in this case a sum of general coordinate, local Lorentz, and local SUSY transformations. Furthermore, the gauge parameters that appear in the commutator can be field dependent, a phenomenon already encountered in SUSY gauge theories in Sec. 6.3.1. Thus it is not immediately clear what we will find in supergravity, and it is well advised to do the computation.

The computation is quite simple for the frame field. Here it is:

$$\begin{aligned}
[\delta_1, \delta_2]\, e_\mu^a &= \tfrac{1}{2}\delta_1\bar{\epsilon}_2\gamma^a\psi_\mu - (1 \leftrightarrow 2) = \tfrac{1}{2}\bar{\epsilon}_2\gamma^a\delta_1\psi_\mu - (1 \leftrightarrow 2)\\
&= \tfrac{1}{2}\bar{\epsilon}_2\gamma^a D_\mu\epsilon_1 - (1 \leftrightarrow 2)\\
&= \tfrac{1}{2}(\bar{\epsilon}_2\gamma^a D_\mu\epsilon_1 + D_\mu\bar{\epsilon}_2\gamma^a\epsilon_1)\\
&= D_\mu\xi^a\,,\\
\xi^a &= \tfrac{1}{2}\bar{\epsilon}_2\gamma^a\epsilon_1 = -\tfrac{1}{2}\bar{\epsilon}_1\gamma^a\epsilon_2\,.
\end{aligned} \tag{9.41}$$

Under the expected general coordinate transformation $x'^\mu = x^\mu - \xi^\mu(x)$, the frame field transforms as a covariant vector; see (7.26). The infinitesimal transformation is

$$\delta_\xi e_\mu^a = \xi^\rho\partial_\rho e_\mu^a + \partial_\mu\xi^\rho e_\rho^a\,. \tag{9.42}$$

Let us 'covariantize' the derivatives by adding and subtracting the ω and Γ connection terms. This must be done with some care since there is torsion. The symbol ∇_ρ includes all appropriate connections:

$$
\begin{aligned}
\delta_\xi e_\mu^a &= \xi^\rho \nabla_\rho e_\mu^a - \xi^\rho \omega_\rho{}^a{}_b e_\mu^b + \xi^\rho \Gamma^\sigma_{\rho\mu} e_\sigma^a + \nabla_\mu \xi^\rho \, e_\rho^a - \Gamma^\rho_{\mu\sigma} \xi^\sigma e_\rho^a \\
&= \nabla_\mu \xi^\rho e_\rho^a - \xi^\rho \omega_\rho{}^a{}_b e_\mu^b + \xi^\rho T_{\rho\mu}{}^a \,.
\end{aligned} \tag{9.43}
$$

Since $\nabla_\rho e_\mu^a = 0$ (see (7.101) which is valid with torsion), we have dropped it in moving to the second line. Since $e_\rho^a \nabla_\mu \xi^\rho = D_\mu \xi^a$, the first term in (9.43) matches the supergravity result (9.41). The last two terms seem mysterious, but, in the light of the remarks above, let's try to interpret them as field dependent symmetry transformations. The second term in (9.43) is simply a local Lorentz transformation of the frame field with field dependent parameter $\hat{\lambda}_{ab} = \xi^\rho \omega_{\rho ab}$. To interpret the third term, we use the explicit form (9.18) of the torsion tensor to write

$$
\xi^\rho T_{\rho\mu}{}^a = \tfrac{1}{2}(\xi^\rho \bar{\psi}_\rho)\gamma^a \psi_\mu \,. \tag{9.44}
$$

This is just a local SUSY transformation of e_μ^a with field dependent $\hat{\epsilon}$. Thus we have derived

$$
[\delta_1, \delta_2]\, e_\mu^a = \left(\delta_\xi - \delta_{\hat{\lambda}} - \delta_{\hat{\epsilon}}\right) e_\mu^a \,, \qquad \hat{\lambda}_{ab} = \xi^\rho \omega_{\rho ab} \,, \qquad \hat{\epsilon} = \xi^\rho \psi_\rho \,. \tag{9.45}
$$

The combination of symmetries that is on the right-hand side is reminiscent of the transformations that we discussed in Ex. 4.5. This combination is a *covariant general coordinate transformation*, which we will discuss more systematically in Sec. 11.3.2.

It is more difficult to calculate the SUSY commutator on the gravitino field largely because Fierz rearrangement is required. The result is

$$
[\delta_1, \delta_2]\, \psi_\mu = \xi^\rho (D_\rho \psi_\mu - D_\mu \psi_\rho) + \ldots \,. \tag{9.46}
$$

We will discuss the omitted terms ... momentarily. First we note that, as done above, one can manipulate the formula for a general coordinate transformation of a vector–spinor field to bring (9.46) to the same form as (9.45), namely

$$
[\delta_1, \delta_2]\, \psi_\mu = \left(\delta_\xi - \delta_{\hat{\lambda}} - \delta_{\hat{\epsilon}}\right) \psi_\mu + \ldots \,. \tag{9.47}
$$

Exercise 9.5 *Derive (9.46) including the terms ... and verify (9.47). If you need details, see [53].*

The omitted terms ... vanish when ψ_μ satisfies its equation of motion. Such terms do not affect the commutator algebra on physical states, so they can be dropped for most purposes. Similar equation of motion terms also appear in the commutator algebra of global SUSY, after elimination of auxiliary fields (see Ex. 6.11 and the discussion that follows it). Their presence in (9.47) means that the local SUSY algebra also 'closes only on-shell'. As in global SUSY, the physical fields e_μ^a and ψ_μ are only an 'on-shell multiplet' which can be completed to an 'off-shell multiplet' by adding auxiliary fields (for $\mathcal{N}=1$, $D=4$ only!). The auxiliary fields are needed to formulate systematic methods for the coupling of chiral and gauge multiplets to supergravity, as we discussed already in Sec. 6.2.2. We will obtain a set of auxiliary fields for $\mathcal{N}=1$, $D=4$ supergravity in Ch. 16.

Supergravity algebra Box 9.3

The commutator of supersymmetry transformations on the frame field leads to covariant general coordinate transformations. On the gravitino this is true modulo equations of motion, or one has to add auxiliary fields.

Finally we note that the field dependent gauge transformations of this section can be reinterpreted as modifications of the algebra of local supersymmetry leading to the 'soft algebras' discussed in Ch. 11.

9.6 Anti-de Sitter supergravity

In this section we describe a very simple extension of supergravity. No new fields are required. Just as the simplest classical solution of the theory of (9.1) is D-dimensional Minkowski spacetime, the simplest solution of the extended theory we now describe is D-dimensional anti-de Sitter space AdS$_D$. (The gravitino field vanishes in both solutions.) The geometry of AdS$_D$ is described in detail in Ch. 22. We follow [54]. We first work at the 'universal level' and modify the action and transformation rules of Sec. 9.1.

Let's begin by defining a modified covariant derivative $\hat{D}_\mu$ which acts on spinors. On $\epsilon(x)$ it acts as

$$\hat{D}_\mu\epsilon \equiv \left(D_\mu - \frac{1}{2L}\gamma_\mu\right)\epsilon = \left(\partial_\mu + \frac{1}{4}\omega_{\mu ab}\gamma^{ab} - \frac{1}{2L}\gamma_\mu\right)\epsilon\,. \tag{9.48}$$

Note that the commutator of two such derivatives is

$$\begin{aligned}\left[\hat{D}_\mu, \hat{D}_\nu\right]\epsilon &= \frac{1}{4}\left(R_{\mu\nu ab}(\omega) + \frac{1}{L^2}(e_{a\mu}e_{b\nu} - e_{b\mu}e_{a\nu})\right)\gamma^{ab}\epsilon \\ &\equiv \frac{1}{4}\hat{R}_{\mu\nu ab}(\omega)\gamma^{ab}\epsilon\,.\end{aligned} \tag{9.49}$$

The last line contains the implicit definition of the modified curvature tensor $\hat{R}_{\mu\nu ab}$, and we can define its Ricci and scalar contractions by

$$\begin{aligned}\hat{R}_{\mu a} &\equiv \hat{R}_{\mu\nu ab}e^{b\nu} = R_{\mu a} + \frac{D-1}{L^2}e_{a\mu}\,, \\ \hat{R} &\equiv \hat{R}_{\mu a}e^{a\mu} = R + \frac{D(D-1)}{L^2}\,.\end{aligned} \tag{9.50}$$

Readers may wish to peek ahead at (22.11) and observe that AdS$_D$ is a solution of the equation $\hat{R}_{\mu\nu} = 0$.

There are two steps necessary to obtain the AdS supergravity action from (9.1). The first step is to replace D_ν in (9.1) by $\hat{D}_\nu$. The second step is to add a cosmological term to (9.1) with value chosen so that the Ricci form of the graviton equation of motion is $\hat{R}_{\mu\nu} = 0$. We write the result as the single expression

$$
\begin{aligned}
S &\equiv \frac{1}{2\kappa^2}\int \mathrm{d}^D x\, e\left(R - \bar\psi_\mu\gamma^{\mu\nu\rho}\hat{D}_\nu\psi_\rho + \frac{(D-1)(D-2)}{L^2}\right) \\
&= \frac{1}{2\kappa^2}\int \mathrm{d}^D x\, e\left(R - \bar\psi_\mu\gamma^{\mu\nu\rho}D_\nu\psi_\rho - \frac{D-2}{2L}\bar\psi_\mu\gamma^{\mu\nu}\psi_\nu + \frac{(D-1)(D-2)}{L^2}\right).
\end{aligned} \tag{9.51}
$$

We thus find a constant negative contribution to the potential, i.e. a negative cosmological constant. This negative value is the origin of the name 'anti'-de Sitter gravity; the opposite sign appears in de Sitter gravity used in realistic cosmological models. Furthermore, we see that, in a supersymmetric action, anti-de Sitter gravity is accompanied by a mass-like term for the gravitino (see (5.48)), with $m_{3/2} = (D-2)(2L)^{-1}$. However, the correct interpretation [55] is that (9.51) describes a *massless* gravitino in an AdS_4 background geometry.

The derivative $\hat{D}_\mu$ also appears in the new transformation rules:

$$
\delta e_\mu^a = \tfrac{1}{2}\bar\epsilon\gamma^a\psi_\mu, \qquad \delta\psi_\mu = \hat{D}_\mu\epsilon(x). \tag{9.52}
$$

There are new terms in the variation δS due to the modifications we have made. It is very easy to show that the new 'universal' terms, those linear in $\epsilon\psi$, vanish for every spacetime dimension D. There are also new order $\epsilon\psi^3$ terms in δS, and these vanish when $D=4$ provided that the spin connection in (9.49) includes the torsion (9.18). Life would be easy if all modifications of the $\mathcal{N}=1$, $D=4$ supergravity theory were as simple as this one, but it would also be less interesting. Please read on (after doing the following exercise).

Exercise 9.6 *Check that the 'universal' $1/L$ and $1/L^2$ terms in δS vanish.*

$D = 11$ supergravity

10

The basic $\mathcal{N} = 1$, $D = 4$ supergravity theory discussed in Ch. 9 has been generalized in several ways. In four dimensions, one can couple the gravity multiplet (e^a_μ, ψ_μ) to gauge (A^A_μ, λ^A) and chiral $(z^\alpha, P_L\chi^\alpha)$ multiplets and promote the global SUSY theories of Ch. 6 to local SUSY. A systematic approach to these matter-coupled supergravity theories is presented in Ch. 17 of this book.

Another important generalization of supergravity is to spacetime dimension $5 \leq D \leq 11$. The two different classical $D = 10$ supergravities, called Type IIA and Type IIB respectively, are the low energy limits of the superstring theories of the same name. Type IIB and gauged $D = 5$ supergravities have important applications to the AdS/CFT correspondence. Most higher dimensional supergravities are quite complicated, but the maximum dimension $D = 11$ theory has a relatively simple structure. It is an important theory for at least two reasons. First, many interesting lower dimensional cases, such as the maximal $\mathcal{N} = 8$, $D = 4$ theory, can be obtained through dimensional reduction. Second, $D = 11$ supergravity, together with its $M2$- and $M5$-brane solutions, is the basis of the extended object theory called M-theory, which is widely considered to be the master theory that contains the various string theories. Thus we devote this chapter to $D = 11$ supergravity.

10.1 $D \leq 11$ from dimensional reduction

Before we embark on a technical discussion of the theory, let's review the argument why $D = 11$ is the largest spacetime dimension allowed for supergravity. The argument is based on a generalization of the dimensional reduction technique we discussed in Sec. 5.3. There we studied Kaluza–Klein compactifications of $(D+1)$-dimensional fields on Minkowski$_D \times S^1$. That discussion did not include symmetric tensor fields, but it is clear that the Fourier modes of a symmetric tensor h_{MN} in $D+1$ dimensions give rise to symmetric tensor fields $h_{\mu\nu k}$, vector fields $h_{\mu D k}$, and scalar fields h_{DDk} in D dimensions. Here k is the Fourier mode number.

More generally we can consider the compactification of a D'-dimensional theory on a product spacetime $M_{D'} = M_D \times X_d$, with $D' = D + d$ and X_d a compact d-dimensional internal space. Fields of the lower dimensional theory arise from harmonic expansion on X_d of the various higher dimensional fields. In a Kaluza–Klein compactification one keeps the entire infinite set of harmonic modes, which describe both massless and massive fields in D dimensions. In the related process called dimensional reduction one keeps only a finite set of modes which must be a *consistent truncation* of the full set. Usually the modes that are kept are the massless or light modes, and the omitted modes are heavy modes. A consistent truncation is one in which the field equations of the omitted heavy modes are not sourced by the light modes that are kept. Thus setting the heavy modes to zero is consistent with the field equations.[1]

To show that $D = 11$ is the maximal dimension [58], we consider the toroidal compactification of a D'-dimensional theory to $M_4 \times T^{D'-4}$. Fourier modes on the torus are generalized Fourier modes labeled by integers $k_1, \ldots, k_{D'-4}$. Only the lowest modes with $k_i = 0$ are massless in four dimensions, and these are retained in the four-dimensional theory. We don't yet know the full field content of the putative D'-dimensional supergravity, but we can anticipate that it must contain the metric tensor and at least one gravitino of the simplest spinor type (e.g. Majorana) permitted in dimension D'.

We need to see which four-dimensional fields arise from the gravitino in the truncation.[2] Suppose that $D' = 11$, in which the simplest spinor is a 32-component Majorana spinor. The 11 matrices Γ^M that generate the Dirac–Clifford algebra in 11 dimensions can be represented as tensor products of 4×4 γ^μ and 8×8 $\hat{\gamma}^i$ as

$$\begin{aligned} \Gamma^\mu &= \gamma^\mu \times \mathbb{1}, \qquad && \mu = 0, 1, 2, 3, \\ \Gamma^i &= \gamma_* \times \hat{\gamma}^i, \qquad && i = 4, 5, 6, 7, 8, 9, 10. \end{aligned} \tag{10.1}$$

Here $\mathbb{1}$ is the 8×8 unit matrix, and $\gamma_* = \mathrm{i}\gamma_0\gamma_1\gamma_2\gamma_3$, while the $\hat{\gamma}^i$ are the generating elements of the Clifford algebra in seven-dimensional Euclidean space. In this basis the 11-dimensional gravitino field is labelled as $\Psi_{M\,\alpha a}$ in which $\alpha = 1, 2, 3, 4$ is a four-dimensional spinor index and $a = 1, \ldots, 8$ is the index on which the $\hat{\gamma}^i$ (and their products) act.

From the four-dimensional standpoint $\Psi_{\mu\,\alpha a}$ transforms under Lorentz transformations as a set of eight gravitinos, while $\Psi_{i\,\alpha a}$ transforms as a set of $7 \times 8 = 56$ spin-1/2 fields. However, note that eight Majorana gravitinos plus 56 Majorana graviphotinos is the complete fermion content of the $s_{\max} = 2$ (maximal spin 2) particle representation of the $\mathcal{N} = 8$ SUSY algebra; see Sec. 6.4.2 and Table 6.1. If we started with a single gravitino in $D' \geq 12$ dimensions or more gravitinos in 11 dimensions, then the dimensionally reduced theory would contain more than eight gravitinos which can only be accommodated in a representation that involves spins $\geq 5/2$ for which no consistent interactions are known.

Exercise 10.1 *Show that the product representation of the 11 Γ^μ matrices defined above does satisfy $\{\Gamma^M, \Gamma^N\} = 2\eta^{MN}$. Compute commutators of the set of 28 matrices consisting of the 21 independent $\hat{\gamma}^{ij} = \hat{\gamma}^{[i}\hat{\gamma}^{j]}$, plus the 7 matrices $\mathrm{i}\hat{\gamma}^k$ and show that they are a basis for an eight-dimensional representation of the Lie algebra of the* SO(8) *group that acts on the eight gravitinos of the reduced theory.* SO(8) *has three inequivalent eight-dimensional representations.*

[1] The truncation to massless modes on a torus is consistent [56, 57].

[2] In this section and the next, we use upper case $M, N, \ldots$ to denote a vector index in D' dimensions, $i, j, \ldots$ for a direction on the torus, and reserve $\mu, \nu, \ldots$ for four-dimensional vector indices.

Maximal dimension for supergravity Box 10.1

$D = 11$ is the maximal dimension for supergravity since any higher dimension would lead after dimensional reduction on a torus to $\mathcal{N} > 8$ in four dimensions.

10.2 The field content of $D = 11$ supergravity

The previous argument has taught us quite a bit about the theory. We know that it contains the gravitational field whose quantum excitations (see Sec. 8.3) transform in the traceless symmetric tensor representation of SO$(D-2)$ of dimension $D(D-3)/2$. In 11 dimensions this contains 44 bosonic states. There is also one Majorana spinor gravitino, whose excitations (see Sec. 5.1) transform in a vector–spinor representation of SO$(D-2)$ of dimension $(D-3)\,2^{[(D-2)/2]}$. For $D = 11$ this contains 128 real fermion states.

The theory must contain an equal number of boson and fermion states, so we are missing 84 bosons. Where are they? Recall the discussion in Sec. 7.8 that bosons in spacetime dimension D can be described by p-form gauge fields or equivalently antisymmetric tensor $A_{M_1,M_2,...,M_p}$ potentials of rank $p < D$. The excitations of such a field transform in the rank p antisymmetric tensor representation of $SO(D-2)$, which contains $\binom{D-2}{p}$ states. For rank 3 in 11 dimensions, this contains exactly 84 quantum degrees of freedom. Thus, Cremmer, Julia, and Scherk [59], who first formulated supergravity in 11 dimensions, made the elegant hypothesis that the theory should contain the metric tensor g_{MN}, the 3-form potential A_{MNP}, and the Majorana vector–spinor Ψ_M.

We already know that upon dimensional reduction on T^7, the vector–spinor produces the eight gravitinos and 56 spin-1/2 fermions of $\mathcal{N} = 8,\ D = 4$ supergravity. The metric tensor components $g_{\mu\nu}$ give the four-dimensional spacetime metric, while $g_{\mu i}$ produces seven spin-1 particles, and g_{ij} contains 28 scalars. The $\mathcal{N} = 8$ theory contains 28 vectors, and the missing 21 are supplied by the 3-form components $A_{\mu ij}$. The form components A_{ijk} contain 35 scalars, and the components $A_{\mu\nu i}$ give an additional seven scalars. In this way the field assignment of [59] accounts for the 70 scalars of the dimensionally reduced theory;[3] see Table 10.1.

10.3 Construction of the action and transformation rules

We now know the field content of the theory, and we need to be more precise and determine the Lagrangian and transformation rules. To find them we start with an initial ansatz for the action compatible with the expected symmetries and use some of the ideas of Chs. 6 and 9 to finalize the construction. Several additional calculations are needed to demonstrate local

[3] The field components $A_{\mu\nu\rho}$ contain no degrees of freedom in $D = 4$.

Table 10.1 The reduction of the fields of $D=11$ supergravity to $\mathcal{N}=8, D=4$.

$D=11$	spin 2		spin 1		spin 0	
g_{MN}	$g_{\mu\nu}$	1	$g_{\mu i}$	7	g_{ij}	28
A_{MNP}			$A_{\mu ij}$	21	$A_{\mu\nu i}$	7
					A_{ijk}	35
		1		28		70
	spin 3/2		spin 1/2			
$\Psi_{M\alpha a}$	$\Psi_{\mu\alpha a}$	8	$\Psi_{i\alpha a}$	56		

SUSY completely. We present some of these and refer readers to the literature [59, 60] for the rest.

To start we note that a theory containing the 3-form potential $A_{\mu\nu\rho}$ must be invariant under a gauge transformation involving a gauge parameter $\theta_{\nu\rho}$ which is a 2-form. The theory thus involves a gauge invariant 4-form field strength $F_{\mu\nu\rho\sigma}$. The basic equations are

$$\begin{aligned}\delta A_{\mu\nu\rho} &= 3\partial_{[\mu}\theta_{\nu\rho]} \equiv \partial_\mu\theta_{\nu\rho} + \partial_\nu\theta_{\rho\mu} + \partial_\rho\theta_{\mu\nu},\\ F_{\mu\nu\rho\sigma} &= 4\partial_{[\mu}A_{\nu\rho\sigma]} \equiv \partial_\mu A_{\nu\rho\sigma} - \partial_\nu A_{\rho\sigma\mu} + \partial_\rho A_{\sigma\mu\nu} - \partial_\sigma A_{\mu\nu\rho},\\ \partial_{[\tau}F_{\mu\nu\rho\sigma]} &\equiv 0.\end{aligned} \tag{10.2}$$

The last equation contains five terms when written in full. It is the Bianchi identity which follows from the fact that $F = \mathrm{d}A$ when expressed as a differential form.

Exercise 10.2 *Readers should prove this identity!*

Another important ingredient we need is the exchange property of (here Majorana) spinor bilinears $\bar{\chi}\Gamma^A\lambda$ where Γ^A is a general element of rank r of the Clifford algebra. As we saw in (3.51), we have the same properties as in four dimensions:

$$\bar{\chi}\gamma^{\mu_1\mu_2\ldots\mu_r}\lambda = t_r\bar{\lambda}\gamma^{\mu_1\mu_2\ldots\mu_r}\chi, \qquad t_0 = t_3 = 1, \quad t_1 = t_2 = -1, \quad t_{r+4} = t_r. \tag{10.3}$$

We postulate that the action contains the universal graviton and gravitino terms of Sec. 9.1 plus the covariant kinetic action for the 3-form plus additional terms (denoted by ...) which we must find. Thus we write

$$S = \frac{1}{2\kappa^2}\int \mathrm{d}^{11}x\, e\left[e^{a\mu}e^{b\nu}R_{\mu\nu ab} - \bar{\psi}_\mu\gamma^{\mu\nu\rho}D_\nu\psi_\rho - \frac{1}{24}F^{\mu\nu\rho\sigma}F_{\mu\nu\rho\sigma} + \ldots\right]. \tag{10.4}$$

Initially we use the second order formalism with torsion-free spin connection $\omega_{\mu ab}(e)$. We also need transformation rules and make the ansatz

$$\begin{aligned}\delta e^a_\mu &= \tfrac{1}{2}\bar{\epsilon}\gamma^a\psi_\mu,\\ \delta\psi_\mu &= D_\mu\epsilon + \left(a\,\gamma^{\alpha\beta\gamma\delta}{}_\mu + b\,\gamma^{\beta\gamma\delta}\delta^\alpha_\mu\right)F_{\alpha\beta\gamma\delta}\epsilon,\\ \delta A_{\mu\nu\rho} &= -c\,\bar{\epsilon}\gamma_{[\mu\nu}\psi_{\rho]} = -\tfrac{1}{3}c\,\bar{\epsilon}(\gamma_{\mu\nu}\psi_\rho + \gamma_{\nu\rho}\psi_\mu + \gamma_{\rho\mu}\psi_\nu).\end{aligned} \tag{10.5}$$

For $\delta A_{\mu\nu\rho}$ and the new terms of $\delta\psi_\mu$ we have postulated general expressions consistent with coordinate and gauge symmetries which contain the numerical constants a, b, c.

We will determine these constants and other useful information by temporarily treating the fields ψ_μ and $A_{\mu\nu\rho}$ as a *free* system with global SUSY in D-dimensional Minkowski space. The free action is (dropping here an irrelevant factor κ^2)

$$S_0 = \frac{1}{2}\int \mathrm{d}^{11}x \left[-\bar{\psi}_\mu\gamma^{\mu\nu\rho}\partial_\nu\psi_\rho - \frac{1}{24}F^{\mu\nu\rho\sigma}F_{\mu\nu\rho\sigma}\right]. \tag{10.6}$$

Of course there is no true supermultiplet containing only ψ_μ and $A_{\mu\nu\rho}$, but for a *free* theory, this need not be an obstacle. We now follow the same steps as in the discussion of super-Yang–Mills theory in Sec. 6.3.1. Using (3.54) we have

$$\delta\bar{\psi}_\mu = \bar{\epsilon}\left(-a\gamma^{\alpha\beta\gamma\delta}{}_\mu + b\gamma^{\beta\gamma\delta}\delta^\alpha_\mu\right)F_{\alpha\beta\gamma\delta}. \tag{10.7}$$

We compute the variation

$$\begin{aligned}\delta S_0 &= \int \mathrm{d}^{11}x\,\bar{\epsilon}\left[\left(a\gamma^{\alpha\beta\gamma\delta}{}_\mu - b\gamma^{\beta\gamma\delta}\delta^\alpha_\mu\right)F_{\alpha\beta\gamma\delta}\gamma^{\mu\nu\rho}\partial_\nu\psi_\rho - \tfrac{1}{6}c\gamma_{\nu\rho}\psi_\sigma\partial_\mu F^{\mu\nu\rho\sigma}\right]\\ &= \int \mathrm{d}^{11}x\bar{\epsilon}\left[\left(-a\gamma^{\alpha\beta\gamma\delta}{}_\mu + b\gamma^{\beta\gamma\delta}\delta^\alpha_\mu\right)\partial_\nu F_{\alpha\beta\gamma\delta}\gamma^{\mu\nu\rho}\psi_\rho - \tfrac{1}{6}c\gamma_{\nu\rho}\psi_\sigma\partial_\mu F^{\mu\nu\rho\sigma}\right].\end{aligned} \tag{10.8}$$

We used partial integration to obtain the last line, assuming that ϵ is constant as in global SUSY.

We need matrix identities to reduce the products of γ-matrices in (10.8) to sums over rank 6, rank 4, and rank 2 elements Γ^A of the Clifford algebra. In the spirit of the discussion at the end of Sec. 3.1.4 (the first identity is (3.23)), we write

$$\begin{aligned}\gamma^{\alpha\beta\gamma\delta}{}_\mu\gamma^{\mu\nu\rho}F_{\alpha\beta\gamma\delta} &= (D-6)\gamma^{\alpha\beta\gamma\delta\nu\rho}F_{\alpha\beta\gamma\delta} + 8(D-5)\gamma^{\alpha\beta\gamma[\nu}F^{\rho]}{}_{\alpha\beta\gamma}\\ &\quad -12(D-4)\gamma^{\alpha\beta}F_{\alpha\beta}{}^{\nu\rho},\\ \gamma^{\beta\gamma\delta}\gamma^{\mu\nu\rho}F_{\mu\beta\gamma\delta} &= -\gamma^{\nu\rho\alpha\beta\gamma\delta}F_{\alpha\beta\gamma\delta} - 6\gamma^{\alpha\beta\gamma[\nu}F^{\rho]}{}_{\alpha\beta\gamma} + 6\gamma^{\alpha\beta}F_{\alpha\beta}{}^{\nu\rho}.\end{aligned} \tag{10.9}$$

Exercise 10.3 *Conscientious readers should verify these identities.*

When inserted in the first term of (10.8) both rank 6 terms vanish due to the Bianchi identity. The sum of the two rank 4 contributions must vanish while the rank 2 terms must cancel with the second term of (10.8). These conditions lead to the following two numerical relations among a, b, c:

$$\begin{aligned}8(D-5)a + 6b &= 0,\\ 12(D-4)a + 6b &= \tfrac{1}{6}c.\end{aligned} \tag{10.10}$$

It is only the case $D = 11$ that is relevant here, and the solution of (10.10) in this case is $a = c/216$, $b = -8a$. Thus for the free theory we have found the transformation rules

$$\delta\psi_\mu = \partial_\mu\epsilon + \frac{c}{216}\left(\gamma^{\alpha\beta\gamma\delta}{}_\mu - 8\gamma^{\beta\gamma\delta}\delta^\alpha_\mu\right)F_{\alpha\beta\gamma\delta}\epsilon,$$
$$\delta A_{\mu\nu\rho} = -c\bar{\epsilon}\gamma_{[\mu\nu}\psi_{\rho]}. \tag{10.11}$$

To fix the remaining parameter c, we examine the commutator of two SUSY transformations and require that this agree with the local supergravity algebra discussed in Sec. 9.5. It is simplest to work with the gauge potential, so we write (for constant ϵ)

$$[\delta_1, \delta_2]A_{\mu\nu\rho} = -\frac{1}{216}c^2\bar{\epsilon}_2\gamma_{[\mu\nu}\left(\gamma^{\alpha\beta\gamma\delta}{}_{\rho]} - 8\gamma^{\beta\gamma\delta}\delta^\alpha_{\rho]}\right)\epsilon_1 F_{\alpha\beta\gamma\delta} - (1 \leftrightarrow 2). \tag{10.12}$$

It would be good practice to work out the detailed identities for the products of γ-matrices in (10.12), which involve contributions from Clifford elements of rank 1, 3, 5, and 7, but this task can be simplified by the following observations:

(i) Since the spinor parameters ϵ_1 and ϵ_2 are antisymmetrized, only the rank 1 and rank 5 terms can contribute.

(ii) The first product has no rank 1 part. It would be necessary to contract three pairs of indices to obtain a non-vanishing rank 1 term, and the antisymmetrizations do not allow this.

(iii) Except for index changes, the second product is already given in (3.21).

The SUSY algebra must contain a spacetime translation involving the rank 1 bilinear $\bar{\epsilon}_1\gamma^\sigma\epsilon_2$ from the product $\gamma_{\mu\nu}\gamma^{\beta\gamma\delta}$. Using (3.21) we obtain the rank 1 contribution to the commutator,

$$[\delta_1, \delta_2]A_{\mu\nu\rho} = -\tfrac{4}{9}c^2\bar{\epsilon}_1\gamma^\sigma\epsilon_2 F_{\sigma\mu\nu\rho}. \tag{10.13}$$

SUSY requires that the rank 5 contribution actually cancels between the two terms of (10.12), and it does, as the following exercise shows.

Exercise 10.4 *Show that the rank 5 terms cancel in the expression* $\gamma_{[\mu\nu}(\gamma_{\rho]}{}^{\alpha\beta\gamma\delta} - 8\gamma^{\beta\gamma\delta}\delta^\alpha_{\rho]})F_{\alpha\beta\gamma\delta}$.

The interpretation of the result (10.13) is straightforward if we refer to the detailed form of the field strength in (10.2). The first term $\partial_\sigma A_{\mu\nu\rho}$ is the spacetime translation we are looking for, while the remaining terms just add up to a gauge transformation of the 3-form potential with field dependent gauge parameter proportional to $\theta_{\mu\nu} = -\bar{\epsilon}_1\gamma^\sigma\epsilon_2 A_{\sigma\mu\nu}$. Such field dependent gauge transformations were already found in SUSY gauge theories (see Ex. 6.15), and were found also in the local algebra in Sec. 9.5. We must normalize the coefficient of the translation term to agree with these results, which were (and should be) uniform for all fields. Thus we fix the parameter $c^2 = 9/8$, and we choose the positive root $c = 3/2\sqrt{2}$.

Recall that we have been studying the global supersymmetry of the free system of ψ_μ and $A_{\mu\nu\rho}$ in flat spacetime. There is another important piece of information from that study, namely the effective supercurrent of the system, obtained by allowing the spinor parameter ϵ in (10.8) to depend on x^μ. After partial integration we find that δS_0 contains a term proportional to $D_\nu\epsilon$ whose coefficient is the supercurrent

$$\mathcal{J}^\nu = \frac{\sqrt{2}}{96}\left(\gamma^{\alpha\beta\gamma\delta\nu\rho}F_{\alpha\beta\gamma\delta} + 12\gamma^{\alpha\beta}F_{\alpha\beta}{}^{\nu\rho}\right)\psi_\rho. \tag{10.14}$$

As we will see below this provides a new term in the $D = 11$ supergravity action.

Exercise 10.5 *Show that $\partial_\nu \mathcal{J}^\nu = 0$ if $F_{\alpha\beta\gamma\delta}$ and ψ_ρ satisfy their free equations of motion (and Bianchi identity).*

To extend results on the free ψ_μ, $A_{\mu\nu\rho}$ system to the interacting supergravity theory, we introduce a general frame field $e^a_\mu(x)$ and consider general $\epsilon(x)$. We then have the transformation rules

$$\begin{aligned}
\delta e^a_\mu &= \frac{1}{2}\bar{\epsilon}\gamma^a\psi_\mu, \\
\delta\psi_\mu &= D_\mu\epsilon + \frac{\sqrt{2}}{288}\left(\gamma^{\alpha\beta\gamma\delta}{}_\mu - 8\gamma^{\beta\gamma\delta}\delta^\alpha_\mu\right)F_{\alpha\beta\gamma\delta}\epsilon, \\
\delta A_{\mu\nu\rho} &= -\frac{3\sqrt{2}}{4}\bar{\epsilon}\gamma_{[\mu\nu}\psi_{\rho]},
\end{aligned} \tag{10.15}$$

and the action

$$\begin{aligned}
S = \frac{1}{2\kappa^2}\int \mathrm{d}^{11}x\, e \Bigg[& e^{a\mu}e^{b\nu}R_{\mu\nu ab} - \bar{\psi}_\mu\gamma^{\mu\nu\rho}D_\nu\psi_\rho - \frac{1}{24}F^{\mu\nu\rho\sigma}F_{\mu\nu\rho\sigma} \\
& - \frac{\sqrt{2}}{96}\bar{\psi}_\nu\left(\gamma^{\alpha\beta\gamma\delta\nu\rho}F_{\alpha\beta\gamma\delta} + 12\gamma^{\alpha\beta}F_{\alpha\beta}{}^{\nu\rho}\right)\psi_\rho + \ldots\Bigg].
\end{aligned} \tag{10.16}$$

The previous discussion ensures that all terms in δS of the form $\bar{\epsilon}R_{\mu\nu ab}\psi_\rho$ and $\bar{\epsilon}F_{\alpha\beta\gamma\delta}\psi_\rho$ cancel. The curvature terms vanish by the universal manipulations of Sec. 9.1. For terms linear in $F_{\alpha\beta\gamma\delta}$, the calculations done above in the free limit are essentially the same in a general background geometry. The only new feature is the term $(D_\nu\bar{\epsilon})\mathcal{J}^\nu$ which is canceled by the $\delta\bar{\psi}_\nu = D_\nu\bar{\epsilon}$ variation of the last term written in (10.16). We still leave ... in (10.16) because the action is not yet complete.

For the next step it is simpler to rewrite (10.16) as the integral of a Lagrangian, namely

$$S = \frac{1}{\kappa^2}\int \mathrm{d}^{11}x\, e\, L, \tag{10.17}$$

and to study variations of the Lagrangian δL. We consider terms in δL of order $\bar{\epsilon}F^2\psi$ which come from the frame field variation of the order F^2 term in L and the $\delta\psi \sim F\epsilon$ variation of the order $\bar{\psi}F\psi$ term. The two contributions are

$$\delta L_{FF} = \frac{1}{48}\left(4\bar{\epsilon}\gamma^\mu\psi^\nu - \frac{1}{2}g^{\mu\nu}\bar{\epsilon}\gamma\cdot\psi\right)F_\mu{}^{\rho\sigma\tau}F_{\nu\rho\sigma\tau}, \tag{10.18}$$

$$\begin{aligned}
\delta L_{\bar{\psi}F\psi} = \frac{1}{96\times 144}&\bar{\epsilon}\left(\gamma^{\alpha'\beta'\gamma'\delta'}{}_\nu + 8\gamma^{\beta'\gamma'\delta'}\delta^{\alpha'}_\nu\right)F_{\alpha'\beta'\gamma'\delta'} \\
&\times\left(\gamma^{\alpha\beta\gamma\delta\nu\rho}F_{\alpha\beta\gamma\delta} + 12\gamma^{\alpha\beta}F_{\alpha\beta}{}^{\nu\rho}\right)\psi_\rho.
\end{aligned} \tag{10.19}$$

The products of γ-matrices in (10.19) contain sums of rank 9, 7, 5, 3, and 1 terms. A detailed treatment requires rather complicated identities, so we will be content here to summarize the results and refer readers who need more information to the literature [59, 60].

It turns out that the rank 1 terms cancel between (10.18) and (10.19), and the several rank 3, 5 and 7 terms cancel within (10.19). However, there are rank 9 terms in (10.19) which can be obtained from the products

$$\begin{aligned} \gamma^{\alpha'\beta'\gamma'\delta'}{}_{\nu}\gamma^{\alpha\beta\gamma\delta\nu\rho} &= (D-9)\gamma^{\alpha'\beta'\gamma'\delta'\alpha\beta\gamma\delta\rho} + \ldots = 2\gamma^{\alpha'\beta'\gamma'\delta'\alpha\beta\gamma\delta\rho} + \ldots, \\ \gamma^{\beta'\gamma'\delta'}\gamma^{\alpha\beta\gamma\delta\alpha'\rho} &= -\gamma^{\alpha'\beta'\gamma'\delta'\alpha\beta\gamma\delta\rho}, \end{aligned} \tag{10.20}$$

where ... indicate lower rank contributions which we omit. Thus the rank 9 term

$$\delta L_{FF} + \delta L_{\bar\psi F\psi} = -\frac{1}{16\times 144}\bar\epsilon\gamma^{\alpha'\beta'\gamma'\delta'\alpha\beta\gamma\delta\rho}\psi_\rho\, F_{\alpha'\beta'\gamma'\delta'}F_{\alpha\beta\gamma\delta} \tag{10.21}$$

survives in the sum of (10.18) and (10.19), and we must find a way to cancel it.

To cancel the high-rank γ-matrix, recall the discussion of Sec. 3.1.7 of the Dirac–Clifford algebra for spacetimes of odd dimension $D = 2m+1$. For $D = 11$, the generating matrices are the 32×32 matrices $\gamma^0, \gamma^1, \ldots, \gamma^9, \gamma^{10} = \gamma_*$, with $\gamma_* = \gamma^0\gamma^1\ldots\gamma^9$ (i.e. we take the + sign in (3.40)). The rank 9 Clifford element is related to rank 2 by (3.41):

$$\gamma^{\alpha'\beta'\gamma'\delta'\alpha\beta\gamma\delta\rho} = -\frac{1}{2e}\,\varepsilon^{\alpha'\beta'\gamma'\delta'\alpha\beta\gamma\delta\rho\mu\nu}\gamma_{\nu\mu}. \tag{10.22}$$

This allows us to rewrite (10.21) as

$$\begin{aligned} e\left(\delta L_{FF} + \delta L_{\bar\psi F\psi}\right) &= \frac{1}{32\times 144}\varepsilon^{\alpha'\beta'\gamma'\delta'\alpha\beta\gamma\delta\rho\mu\nu}\bar\epsilon\gamma_{\nu\mu}\psi_\rho\, F_{\alpha'\beta'\gamma'\delta'}F_{\alpha\beta\gamma\delta}, \\ &= \frac{4}{3\sqrt{2}\times 32\times 144}\varepsilon^{\alpha'\beta'\gamma'\delta'\alpha\beta\gamma\delta\rho\mu\nu}(\delta A_{\mu\nu\rho})\, F_{\alpha'\beta'\gamma'\delta'}F_{\alpha\beta\gamma\delta}. \end{aligned} \tag{10.23}$$

This suggests that one can add a term in the Lagrangian of the following form to cancel this variation:

$$\begin{aligned} S_{\text{C-S}} &= -\frac{\sqrt{2}}{(144\kappa)^2}\int \mathrm{d}^{11}x\,\varepsilon^{\alpha'\beta'\gamma'\delta'\alpha\beta\gamma\delta\mu\nu\rho}F_{\alpha'\beta'\gamma'\delta'}F_{\alpha\beta\gamma\delta}A_{\mu\nu\rho} \\ &= -\frac{\sqrt{2}}{6\kappa^2}\int F^{(4)}\wedge F^{(4)}\wedge A^{(3)}. \end{aligned} \tag{10.24}$$

We used form notation in the last line, with $F^{(4)} = \mathrm{d}A^{(3)}$, to simplify the formula. The result is called a Chern–Simons term. It has very special properties. First, using integration by parts, one sees that a variation of $A^{(3)}$ gives the three similar terms

$$\begin{aligned} \delta\int F^{(4)}\wedge F^{(4)}\wedge A^{(3)} &= \int\left[2\mathrm{d}\delta A^{(3)}\wedge F^{(4)}\wedge A^{(3)} + F^{(4)}\wedge F^{(4)}\wedge\delta A^{(3)}\right] \\ &= 3\int F^{(4)}\wedge F^{(4)}\wedge\delta A^{(3)}, \end{aligned} \tag{10.25}$$

where we used the Bianchi identity $\mathrm{d}F^{(4)} = 0$. This shows how such a variation of (10.24) cancels (10.23). Further, this term does not produce other variations, since there are no frame fields in the expression.

One might also wonder whether such a term is invariant under gauge transformations of the form (10.2) in view of the explicit presence of the gauge field. To check this it is simplest to write the gauge transformation as $\delta A^{(3)} = \mathrm{d}\theta^{(2)}$. Plugging this into (10.25), integrating by parts and using the Bianchi identity as above, the gauge invariance is guaranteed. These are typical properties of Chern–Simons actions.

We have now reached the point where the major terms in the Lagrangian and transformation rules have been determined. Although there is more work to be done to complete the theory and establish local supersymmetry, the further modifications are quite simple. We prefer to write the full action and transformation rules and then interpret the changes below.

The full action is

$$
\begin{aligned}
S = \frac{1}{2\kappa^2}\int \mathrm{d}^{11}x\, e\Big[& e^{a\mu}e^{b\nu}R_{\mu\nu ab}(\omega) - \bar\psi_\mu\gamma^{\mu\nu\rho}D_\nu\left(\tfrac12(\omega+\hat\omega)\right)\psi_\rho - \frac{1}{24}F^{\mu\nu\rho\sigma}F_{\mu\nu\rho\sigma}\\
&-\frac{\sqrt2}{192}\bar\psi_\nu\left(\gamma^{\alpha\beta\gamma\delta\nu\rho}+12\gamma^{\alpha\beta}g^{\gamma\nu}g^{\delta\rho}\right)\psi_\rho(F_{\alpha\beta\gamma\delta}+\hat F_{\alpha\beta\gamma\delta})\\
&-\frac{2\sqrt2}{(144)^2}\varepsilon^{\alpha'\beta'\gamma'\delta'\alpha\beta\gamma\delta\mu\nu\rho}F_{\alpha'\beta'\gamma'\delta'}F_{\alpha\beta\gamma\delta}A_{\mu\nu\rho}\Big].
\end{aligned}
\qquad (10.26)
$$

The ‘hatted’ connection and field strength that appear above are given by

$$
\begin{aligned}
\omega_{\mu ab} &= \omega_{\mu ab}(e) + K_{\mu ab},\\
\hat\omega_{\mu ab} &= \omega_{\mu ab}(e) - \tfrac14(\bar\psi_\mu\gamma_b\psi_a - \bar\psi_a\gamma_\mu\psi_b + \bar\psi_b\gamma_a\psi_\mu),\\
K_{\mu ab} &= -\tfrac14(\bar\psi_\mu\gamma_b\psi_a - \bar\psi_a\gamma_\mu\psi_b + \bar\psi_b\gamma_a\psi_\mu) + \tfrac18\bar\psi_\nu\gamma^{\nu\rho}{}_{\mu ab}\psi_\rho,\\
\hat F_{\mu\nu\rho\sigma} &= 4\,\partial_{[\mu}A_{\nu\rho\sigma]} + \tfrac32\sqrt2\,\bar\psi_{[\mu}\gamma_{\nu\rho}\psi_{\sigma]}.
\end{aligned}
\qquad (10.27)
$$

This action is invariant under the transformation rules

$$
\begin{aligned}
\delta e^a_\mu &= \frac12\bar\epsilon\gamma^a\psi_\mu,\\
\delta\psi_\mu &= D_\mu(\hat\omega)\epsilon + \frac{\sqrt2}{288}\left(\gamma^{\alpha\beta\gamma\delta}{}_\mu - 8\gamma^{\beta\gamma\delta}\delta^\alpha_\mu\right)\hat F_{\alpha\beta\gamma\delta}\epsilon,\\
\delta A_{\mu\nu\rho} &= -\frac{3\sqrt2}{4}\bar\epsilon\gamma_{[\mu\nu}\psi_{\rho]}.
\end{aligned}
\qquad (10.28)
$$

The connection ω incorporates the torsion tensor (9.18).

The ‘hatted’ connection and field strength in (10.27) are ‘supercovariant’. In this context a supercovariant quantity[4] is one whose local SUSY transformation does not contain the

[4] We discuss covariant quantities further in Sec. 11.2.

derivative $\partial_\mu \epsilon$ of the SUSY parameter $\epsilon(x)$. The principle that guides the way in which these quantities are introduced in the action is that the equations of motion must transform supercovariantly, i.e. without a derivative on the supersymmetry parameter. Like any symmetry, local supersymmetry transforms the equations of motion of the theory among themselves. The field equation for the gravitino that results after a long calculation (involving three-gravitino Fierz relations; see e.g. [61]) is

$$\gamma^{\mu\nu\rho} D_\nu(\hat{\omega})\psi_\rho - \frac{\sqrt{2}}{288}\gamma^{\mu\nu\rho}\left(\gamma^{\alpha\beta\gamma\delta}{}_\nu - 8\gamma^{\beta\gamma\delta}\delta^\alpha_\nu\right)\psi_\rho \hat{F}_{\alpha\beta\gamma\delta} = 0. \tag{10.29}$$

10.4 The algebra of $D = 11$ supergravity

It is again instructive to evaluate the commutator of two supersymmetry transformations. The result, given by

$$\left[\delta_Q(\epsilon_1), \delta_Q(\epsilon_2)\right] = \delta_{\text{gct}}(\xi^\mu) + \delta_L(\lambda^{ab}) + \delta_Q(\epsilon_3) + \delta_A(\theta_{\mu\nu}), \tag{10.30}$$

is the sum of a general coordinate transformation (gct), plus field dependent local Lorentz, supersymmetry and 3-form gauge transformations (as given in (10.2)). The parameters of these transformations are

$$\begin{aligned}
\xi^\mu &= \frac{1}{2}\bar{\epsilon}_2\gamma^\mu\epsilon_1, \\
\lambda^{ab} &= -\xi^\mu\hat{\omega}_\mu{}^{ab} + \frac{1}{288}\sqrt{2}\bar{\epsilon}_1\left(\gamma^{ab\mu\nu\rho\sigma}\hat{F}_{\mu\nu\rho\sigma} + 24\gamma_{\mu\nu}\hat{F}^{ab\mu\nu}\right)\epsilon_2, \\
\epsilon_3 &= -\xi^\mu\psi_\mu, \\
\theta_{\mu\nu} &= -\xi^\rho A_{\rho\mu\nu} + \frac{1}{4}\sqrt{2}\bar{\epsilon}_1\gamma_{\mu\nu}\epsilon_2.
\end{aligned} \tag{10.31}$$

As was the case in $D = 4$ supergravity, the algebra is realized on the gravitino only if its field equations are satisfied. In previous similar cases this could be remedied by adding auxiliary fields to the theory. Much effort has been devoted to a search for auxiliary fields for $D = 11$ supergravity, but no solution has ever been found.

The terms in (10.31) which contain the spinor bilinears $\bar{\epsilon}_1\Gamma^{(2)}\epsilon_2$ and $\bar{\epsilon}_1\Gamma^{(6)}\epsilon_2$ have special significance. (The $\Gamma^{(6)}$ and $\Gamma^{(5)}$ bilinears are duals in $D = 11$ dimensions and thus equivalent.) These terms are non-vanishing in the classical $M2$- and $M5$- brane solutions of the theory. The BPS property (to be discussed in Sec. 22.2) means that the brane solutions

Box 10.2 **The $D = 11$ superalgebra**

The commutator of supersymmetries in $D = 11$ supergravity contains, apart from the covariant general coordinate transformation, terms with $\bar{\epsilon}_1\Gamma^{(2)}\epsilon_2$ and $\bar{\epsilon}_1\Gamma^{(6)}\epsilon_2$. These are non-vanishing for BPS solutions such as $M2$- and $M5$-branes.

preserve a global supersymmetry algebra in which the non-vanishing $\Gamma^{(2)}$ and $\Gamma^{(5)}$ terms are central charges.

It should be noted that $D = 11$ supergravity is a fixed and unalterable classical theory. For example, there are no 'matter multiplets' which might otherwise be added, and the 'cosmological modification' of Sec. 9.6 cannot be applied here.

11 General gauge theory

In the previous chapters we studied the construction of the simplest supergravity theories, namely $D = 4$ and $D = 11$, $\mathcal{N} = 1$ supergravity, and we proved invariance under local supersymmetry. But $\mathcal{N} = 1$ supergravity in four dimensions contains much more. Chiral multiplets and gauge multiplets of global supersymmetry (see Ch. 6) can be coupled to supergravity, yielding the rich structure of matter-coupled $\mathcal{N} = 1$ supergravities with many applications.

To investigate matter-coupled theories, we will use more advanced methods, which will be developed in the chapters ahead. To prepare for this, we will sharpen our knives in this chapter and discuss a general formulation of gauge theory of the type needed for supergravity. This consists largely of the refinement and extension of the notations and concepts used for symmetries earlier in this book. We will formalize several manipulations and concepts that we encountered in the previous sections. This will allow us to perform calculations with covariant derivatives more effectively. However, we will also see how the supergravity transformation rules postulated in (9.3) and (9.4) are actually determined by the structure constants of the Poincaré supersymmetry algebra. In Sec. 11.3, we will explain how the general rules of gauge theory can be applied in gravity theories. We will see that some formulas must be modified.

11.1 Symmetries

In Sec. 1.2 we defined symmetries as maps of the configuration space such that solutions of the equations of motion are transformed into new solutions. As we did there, we restrict the discussion to continuous symmetries that leave the action invariant. We consider only infinitesimal transformations, and hence are concerned with the Lie algebra rather than the Lie group. However, we need also extensions of these mathematical concepts, e.g. allowing structure constants to depend on fields and thus becoming rather 'structure functions'. We use the same formalism to describe both spacetime and internal symmetries and also supersymmetries.

11.1.1 Global symmetries

The algebra

An infinitesimal symmetry transformation is determined by a parameter, which we denote in general by ϵ^A, and an operation $\delta(\epsilon)$, which depends linearly on the parameter, and acts on the fields of the dynamical system under study. For global symmetries, often called rigid symmetries, the parameters do not depend on the spacetime point where the symmetry operation is applied. Since the symmetry operation is linear in ϵ we can write it in general as

$$\delta(\epsilon) = \epsilon^A T_A, \tag{11.1}$$

in which T_A is an operator on the space of fields.[1] It describes the symmetry transformation with the parameter stripped. We call T_A the field space generator of the transformation. The notation of (11.1) and the formalism we develop below apply to all the types of symmetries encountered in this book. Internal, spacetime, and supersymmetry can be viewed as special cases. However, to be concrete, we begin by rephrasing the discussion of (linearly realized) internal symmetry of Sec. 1.2.2 in the present notation (see also Sec. 4.3.1). Then we will discuss how spacetime and supersymmetries fit the pattern.

Suppose that the matrices $(t_A)^i{}_j$ are the matrix generators of a representation of a Lie algebra with commutator

$$[t_A, t_B] = f_{AB}{}^C t_C, \tag{11.2}$$

and suppose that the system contains a set of fields ϕ^i that transforms in this representation. Then the transformation rule (1.19) can be expressed as the action of the generator T_A as follows:

$$T_A \phi^i = -(t_A)^i{}_j \phi^j. \tag{11.3}$$

The Lie algebra is determined by the commutator of two such transformations, so it is important to define the product of two symmetry operations carefully. By the product $\delta(\epsilon_1)\delta(\epsilon_2)\phi^i$, we mean that we first make a transformation with parameter ϵ_2 followed by another with parameter ϵ_1. In the notation of (11.1), the product operation reads

$$\delta(\epsilon_1)\delta(\epsilon_2)\phi^i = \epsilon_1^A T_A \left(\epsilon_2^B T_B \phi^i\right). \tag{11.4}$$

As discussed in Sec. 1.2.2, a symmetry transformation acts on the fields, which are the dynamical variables of the system, and not on the matrices, which are the result of a prior transformation. Therefore, using (11.3), the explicit form of the product operation becomes

$$\begin{aligned} \delta(\epsilon_1)\delta(\epsilon_2)\phi^i &= \epsilon_1^A T_A \epsilon_2^B \left[-(t_B)^i{}_j \phi^j\right] \\ &= \epsilon_1^A \epsilon_2^B (-t_B)^i{}_j T_A \phi^j \\ &= \epsilon_1^A \epsilon_2^B (-t_B)^i{}_j (-t_A)^j{}_k \phi^k. \end{aligned} \tag{11.5}$$

[1] When the theory has a Hamiltonian description and Poisson brackets, $T_A\phi$ is the operator $\Delta_A\phi$ as defined by Poisson brackets in (1.79).

It is important to realize that, in the second line, the transformation operator acts on the field ϕ^j, and not on the 'numbers' $(t_B)^i{}_j$. This is similar to the way in which we calculate commutators as, for example, in (9.41) where δ_1 is similar to the T_A here and acts on ψ_μ and goes after γ^a. The commutator is then

$$\begin{aligned}[]
[\delta(\epsilon_1), \delta(\epsilon_2)]\phi^i &= \epsilon_2^B \epsilon_1^A [T_A, T_B]\phi^i \\
&= \epsilon_2^B \epsilon_1^A ([t_B, t_A]\phi)^i = -\epsilon_2^B \epsilon_1^A f_{AB}{}^C (t_C\phi)^i \\
&= \epsilon_2^B \epsilon_1^A f_{AB}{}^C T_C \phi^i .
\end{aligned} \tag{11.6}$$

Therefore the commutator of the generators T_A, namely

$$[T_A, T_B] = f_{AB}{}^C T_C, \tag{11.7}$$

conforms to the matrix commutator (11.2). The minus sign in the transformation rule (11.3) is necessary to obtain the same result in (11.7) as in (11.2), and it is a general feature of symmetry transformations defined as left multiplication by a matrix.

Although every case of internal symmetry in which the symmetry transformations act linearly on the fields (linearly realized internal symmetry) can be expressed as matrix transformations, it is sometimes convenient to use the more general notation with T_A. For example the transformation rules for fields Ψ^α in a complex representation of a compact symmetry group, their conjugates $\bar{\Psi}_\alpha$, and fields ϕ^B in the adjoint representation were defined in (4.84). In each case we obtain the generator T_A by stripping the parameter ϵ^A and write

$$\begin{aligned}
T_A \Psi^\alpha &= -(t_A)^\alpha{}_\beta \Psi^\beta , \\
T_A \bar{\Psi}_\alpha &= \bar{\Psi}_\beta (t_A)^\beta{}_\alpha , \\
T_A \phi^B &= -f_{AC}{}^B \phi^C .
\end{aligned} \tag{11.8}$$

Here the matrix generator t_A acts by left multiplication on Ψ^α, but by right multiplication on $\bar{\Psi}_\alpha$.[2] On adjoint fields we again have left action by the matrix generators $(t_A)^B{}_C = f_{AC}{}^B$.

Exercise 11.1 *Show that (11.7) holds for repeated symmetry transformations of $\bar{\Psi}_\alpha$ and ϕ^B. Recall that the second transformation always acts only on the fields.*

Our conventions are easily extended to include spacetime symmetries. The operators $M_{\mu\nu}$ for Lorentz transformations and P_μ for translations have been defined in (1.83) and conform to the Lie algebra (1.57) of the Poincaré group (provided they are calculated with the convention that the second transformation acts only on the fields resulting from the first).

Exercise 11.2 *Readers should verify this.*

Let's return to the equations (11.5) and (11.6). The order in which the symmetry parameters are written is irrelevant for internal symmetries since the parameters commute.

[2] Because the matrices t_A are anti-hermitian for a compact group, this is equivalent to left action by $-(t_A)^\dagger$; see Sec. 4.3.1.

However, the order written in (11.6) will also be valid for fermionic symmetries for which the parameters anti-commute. Indeed we now proceed to formulate global supersymmetry in the present framework.

Here we need the notation for spinor indices that was explained in Sec. 3.2.2. Remember that ordinary spinors carry a lower spinor index, while conjugate (barred) spinors, defined in (3.50), carry an upper spinor index. Gamma-matrix indices are raised and lowered with the charge conjugation matrix. Symmetries of γ-matrices are determined by (3.63), such that in four dimensions $(\gamma^\mu)_{\alpha\beta}$ is symmetric. Flipping indices up–down gives a sign specified by (3.64). In four dimensions this is a minus sign.

For supersymmetry the generators T_A are replaced by the four-component Majorana spinor Q_α and the parameters ϵ^A by the anti-commuting conjugate (Majorana) spinor $\bar{\epsilon}^\alpha$. A supersymmetry transformation is an operation on fields denoted by[3]

$$\delta(\epsilon) = \bar{\epsilon}^\alpha Q_\alpha. \tag{11.9}$$

As an explicit example of supersymmetry we rewrite the transformation rule (6.15) of the scalar field $Z(x)$ of a chiral multiplet:

$$Q_\alpha Z = \frac{1}{\sqrt{2}}(P_L\chi)_\alpha. \tag{11.10}$$

In parallel with (11.4), we write the product of two SUSY transformations as

$$\delta(\epsilon_1)\delta(\epsilon_2) = (\bar{\epsilon}_1)^\alpha Q_\alpha (\bar{\epsilon}_2)^\beta Q_\beta = (\bar{\epsilon}_2)^\beta (\bar{\epsilon}_1)^\alpha Q_\alpha Q_\beta. \tag{11.11}$$

Note that there is no sign change when $(\bar{\epsilon}_2)_\beta$ is moved through the bosonic quantity $(\bar{\epsilon}_1)_\alpha Q^\alpha$. It is then easy to check that the commutator of two transformations is

$$[\delta(\epsilon_1), \delta(\epsilon_2)] = (\bar{\epsilon}_2)^\beta (\bar{\epsilon}_1)^\alpha \left(Q_\alpha Q_\beta + Q_\beta Q_\alpha \right). \tag{11.12}$$

When applied to any field of a supersymmetric theory this result has the same structure as the first line of (11.6), except that the anti-commutator appears. Of course, this is entirely expected for the fermionic elements of a superalgebra, whose structure relation reads

$$\{Q_\alpha, Q_\beta\} = f_{\alpha\beta}{}^C T_C, \tag{11.13}$$

with structure constants $f_{\alpha\beta}{}^C$, which are symmetric in α, β and connect fermionic elements to a sum of bosonic elements T_C.

In Sec. 6.2.2, we computed the commutator of SUSY variations on the fields of a chiral multiplet and found

$$[\delta(\epsilon_1), \delta(\epsilon_2)] = -\tfrac{1}{2}\bar{\epsilon}_1\gamma^\mu\epsilon_2 P_\mu = \tfrac{1}{2}\bar{\epsilon}_2\gamma^\mu\epsilon_1 P_\mu = -\tfrac{1}{2}\epsilon_2^\beta (\gamma^\mu)_{\beta\alpha}\epsilon_1^\alpha P_\mu, \tag{11.14}$$

where P_μ acts on the fields of the chiral multiplet as ∂_μ. Thus the structure constants of supersymmetry are given by

[3] We assume here that $\bar{\epsilon}Q$ is real, which is certainly the case for Majorana spinors. In Sec. 20.2.1, we will see that, for example, for $D = 5$ we need an extra factor i for reality.

$$\{Q_\alpha, Q_\beta\} = -\tfrac{1}{2}(\gamma^\mu)_{\alpha\beta} P_\mu \quad \Rightarrow \quad f_{\alpha\beta}{}^\mu = -\tfrac{1}{2}(\gamma^\mu)_{\alpha\beta} = f_{\beta\alpha}{}^\mu . \tag{11.15}$$

Finally we would like to show that the treatments of supersymmetry and internal symmetry can be united in a common notation. Towards this end we insert the definition (11.13) in (11.12) and obtain

$$[\delta(\epsilon_1), \delta(\epsilon_2)] = (\bar\epsilon_2)^\beta (\bar\epsilon_1)^\alpha f_{\alpha\beta}{}^C T_C . \tag{11.16}$$

This relation has the same structure as the last line of (11.6). Thus the result

$$[\delta(\epsilon_1), \delta(\epsilon_2)] = \delta(\epsilon_3^C = \epsilon_2^B \epsilon_1^A f_{AB}{}^C) \tag{11.17}$$

holds for both bosonic and fermionic symmetries. The presence of a fermionic symmetry is signaled only by the fact that its parameter (called $\bar\epsilon^\alpha$ for supersymmetry) is anti-commuting. The left-hand side of (11.17) then contains the necessary anti-commutator of the generators as in (11.12).

A superalgebra also contains structure relations of bosonic and fermionic elements, and they are realized by commutators of the generators. In supersymmetry the commutator of a Lorentz generator and a supercharge component was defined in (6.1). Here is a relevant exercise for the reader.

Exercise 11.3 *Use (11.10) and (1.83) to show that the commutator of the field space generators $M_{[\mu\nu]}$ and Q_α conforms to (6.1) when acting on $Z(x)$, specifically*

$$\left[M_{[\mu\nu]}, Q_\alpha\right] Z = -\tfrac{1}{2}(\gamma_{\mu\nu})_\alpha{}^\beta Q_\beta Z . \tag{11.18}$$

The nonlinear σ-model and Killing symmetries

Internal symmetries do not always act linearly. We now consider symmetries in the nonlinear σ-model, which was discussed in Sec. 7.11. It contains scalar fields that transform as in (7.142), where the Killing vectors $k_A^i(\phi)$ should satisfy (7.141). Hence, in this case the symmetry generator is[4]

$$T_A \phi^i = k_A^i(\phi). \tag{11.19}$$

The Lie brackets (7.9) of the set of Killing vector fields determine the Lie algebra of the isometry group of M_n. The Lie bracket, which is the commutator of the differential operators k_A, reads as in (7.146)

$$[k_A, k_B] = f_{AB}{}^C k_C , \tag{11.20}$$

or, in components,

$$k_A^j \partial_j k_B^i - k_B^j \partial_j k_A^i = f_{AB}{}^C k_C^i . \tag{11.21}$$

It is then clear that the commutator of the field space generators T_A conforms to (11.7). The following exercise confirms this.

[4] For a linear Killing symmetry, $k_A^i(\phi) = -(t_A)^i{}_j \phi^j$.

Exercise 11.4 *Use (11.19) to obtain*

$$\delta(\epsilon_1)\delta(\epsilon_2)\phi^i = \delta(\epsilon_1)\epsilon_2^A k_A^i = \epsilon_2^A \epsilon_1^B k_B^j \partial_j k_A^i. \tag{11.22}$$

Use (11.21) to show that, for any smooth scalar function $f(\phi^i)$ on M_n,

$$[T_A, T_B] f(\phi^i) = f_{AB}{}^C T_C f(\phi^i). \tag{11.23}$$

11.1.2 Local symmetries and gauge fields

We now consider gauge symmetries and use the notation of the previous section in which symmetries are indexed by $A, B, C, \ldots$. For each symmetry there is a field space generator T_A, but the parameters $\epsilon^A(x)$ are arbitrary functions on spacetime. To realize local symmetry in Lagrangian field theory, one needs a gauge field, which we will generically denote by $B_\mu^A(x)$, for every gauged symmetry. The gauge fields transform as

$$\delta(\epsilon) B_\mu{}^A \equiv \partial_\mu \epsilon^A + \epsilon^C B_\mu{}^B f_{BC}{}^A. \tag{11.24}$$

It is easy to check that these transformations satisfy the algebra (11.17). This definition is modeled on the transformation of Yang–Mills fields in (4.88), but the definition is valid for internal and spacetime symmetries and for supersymmetry. The gauge fields for each case transform in the same way, although each case has its own specific notation in which the generic index A is appropriately changed.

Exercise 11.5 *Show that the commutator of two gauge transformations (11.24) conforms to the structure of (11.17). Specifically, assume that the generators T_A form a bosonic Lie algebra and show that*

$$[\delta(\epsilon_1), \delta(\epsilon_2)]\, B_\mu^A = \partial_\mu \epsilon_3^A + \epsilon_3^C B_\mu{}^B f_{BC}{}^A, \tag{11.25}$$

where $\epsilon_3^C = \epsilon_2^B \epsilon_1^A f_{AB}{}^C$. The proof requires the Jacobi identity (4.81) for Lie algebras. The result (11.25) is also valid for the more general situation of a superalgebra. Here you must take the anti-commutativity of parameters into account and use the graded Jacobi identity

$$\epsilon_2^B \epsilon_1^A \epsilon_3^C f_{AB}{}^D f_{CD}{}^E + (\text{cyclic } 1 \to 2 \to 3) = 0, \qquad \textit{assuming constant } f_{AB}{}^C. \tag{11.26}$$

We will be applying (11.24) in theories of gravity. Therefore, we distinguish between coordinate indices $\mu, \nu, \rho, \ldots$ and local frame indices $a, b, c, \ldots$, and we now use $M_{[ab]}$ and P_a to denote the generators of local Lorentz transformations and translations. There are some subtleties for local translations, which are actually general coordinate transformations, which we will discuss in Sec. 11.3, but the other major ingredients of our treatment

Table 11.1 Gauge transformations, parameters and gauge fields.

generic gauge symmetry T_A	parameter ϵ^A	gauge field B_μ^A
local translations P_a	ξ^a	e_μ^a
Lorentz transformations $M_{[ab]}$	λ^{ab}	$\omega_\mu{}^{ab}$
supersymmetry Q_α	$\bar{\epsilon}^\alpha$	$\bar{\psi}_\mu^\alpha$
internal symmetry T_A	θ^A	A_μ^A

Table 11.2 Useful commutators, structure constants and third parameter in the commutation relation (11.17).

(anti-)commutators	structure constants	third parameter
$\left[M_{[ab]}, M_{[cd]}\right] = 4\eta_{[a[c}M_{d]b]]}$	$f_{[ab][cd]}{}^{[ef]} = 8\eta_{[c[b}\delta_{a]}^{[e}\delta_{d]}^{f]}$	$\lambda_3^{ab} = -2\lambda_1{}^{[a}{}_c\lambda_2{}^{b]c}$
$\left[P_a, M_{[bc]}\right] = 2\eta_{a[b}P_{c]}$	$f_{a,[bc]}{}^d = 2\eta_{a[b}\delta_{c]}^d$	$\xi_3^a = -\lambda_2^{ab}\xi_{1b} + \lambda_1^{ab}\xi_{2b}$
$\left[P_a, P_b\right] = 0$		
$\{Q_\alpha, Q_\beta\} = -\frac{1}{2}(\gamma^a)_{\alpha\beta}P_a$	$f_{\alpha\beta}{}^a = -\frac{1}{2}(\gamma^a)_{\alpha\beta}$	$\xi_3^a = \frac{1}{2}\bar{\epsilon}_2\gamma^a\epsilon_1$
$\left[M_{[ab]}, Q\right] = -\frac{1}{2}\gamma_{ab}Q$	$f_{[ab],\alpha}{}^\beta = -\frac{1}{2}(\gamma_{ab})_\alpha{}^\beta$	$\epsilon_3 = \frac{1}{4}\lambda_1^{ab}\gamma_{ab}\epsilon_2 - \frac{1}{4}\lambda_2^{ab}\gamma_{ab}\epsilon_1$
$[P_a, Q] = 0$		

of gauge symmetries are developed in this section. Table 11.1 indicates both the generic notation and the particular cases we are concerned with.[5]

The structure constants $f_{BC}{}^A$ are the important data in the formula (11.24), and we extract them from the symmetry algebras in each specific case.

Table 11.2 gives examples for the most common (anti-)commutators (remember that the calculation takes into account factors of 2 in summations over pairs of antisymmetric indices as in Ex. 1.4).

Exercise 11.6 *The spin connection field $\omega_\mu{}^{ab}$ transforms under Lorentz transformations[6] as*

$$\delta\omega_\mu{}^{ab} = \partial_\mu\lambda^{ab} + 2\omega_{\mu c}{}^{[a}\lambda^{b]c}. \tag{11.27}$$

[5] The same index A is used both for the general case and for internal symmetry, but the distinction will be clear from the context of each application. For the other symmetries, the index A is replaced by either an index a, or an antisymmetric pair $[ab]$ or a spinor index. For example, $\xi^a = e_\mu^a\xi^\mu$ is the parameter of translations. Our convention for spinors is that barred spinors carry an upper index. Thus we have written barred spinors in the table. But we consider only Majorana spinors in supersymmetry, so they are linearly related to ϵ or ψ_μ.

[6] The orbital part of Lorentz transformations is not included here. Why this can be omitted will be explained in Sec. 11.3.1.

Check that this equation corresponds to (11.24). The gauge field B^A_μ is then replaced by $\omega_\mu{}^{ab}$. You must replace each of the indices A, B, C by antisymmetric pairs $[ab]$, etc., and insert factors $\frac{1}{2}$ as in Ex. 1.4.

When we started with $\mathcal{N} = 1$ supergravity, we simply *assumed* the transformations of the frame field and the gravitino given in (9.3) and (9.4). However, now we can consider this again in the context of the gauged super-Poincaré algebra, and it turns out that this ansatz is actually *determined* by (11.24) as the following exercise shows.

Exercise 11.7 *Use (11.24) to calculate the supersymmetry transform of the vierbein e^a_μ, using information from Table 11.2. Since the exercise asks only for the supersymmetry transformation, the index on the parameter can be restricted to a spinor index α. On the other hand, the A index is a for translations. Therefore, the first term of (11.24) does not contribute. From the second term you should obtain (9.3). Consider now in the same way the transformation of the gravitino and re-obtain (9.4).*

Although the results of this exercise are encouraging, there are also some puzzling features. As Ex. 11.5 shows, the transformations (11.24) do automatically satisfy the commutator algebra. But for spacetime symmetries, these commutator transformations are not always the ones that we expect. For example, the supersymmetry commutator $[\delta_Q(\epsilon_1), \delta_Q(\epsilon_2)]$ vanishes on the gravitino. This result is correct since Table 11.2 and (11.24) tell us that $\delta_P \psi_{\mu\alpha} = 0$, but it disagrees with the results for the local supersymmetry algebra found in Sec. 9.5. Similarly $\delta_Q \omega_\mu{}^{ab} = 0$. This is consistent because the spin connection is an independent field here, but the result differs from Ch. 9 in which the spin connection is related to frame field and gravitino. Therefore we will need to modify the present setup before we can apply it to construct gravitational theories. The new ingredients will be discussed in Sec. 11.3.

First we will discuss some other issues which will concern us in applications to gravity and supergravity. They are situations in which the commutator of symmetry transformations contains the terms corresponding to the conventional structure constants $f_{AB}{}^C$ of the algebra *plus other terms*. Algebras modified in this way frequently occur in supersymmetry and supergravity.

11.1.3 Modified symmetry algebras

An important aspect of supergravity is the fact that the symmetry structure is not that of a Lie algebra in the mathematical sense. Rather it has a modified symmetry structure, which we want to highlight in this section.

Soft algebras

In supersymmetry we often encounter a generalization of standard Lie algebras, which we call 'soft algebras'. This means that the usual structure constants $f_{AB}{}^C$ can depend on fields and are thus called 'structure functions'. We already encountered this phenomenon when we studied the commutator of SUSY transformations in supersymmetric gauge theories. Consider the result (6.51), which we can write as

$$\left[\delta_Q(\epsilon_1), \delta_Q(\epsilon_2)\right] = \delta_P(\xi) - \delta_{\text{gauge}}(\theta^A = A_\mu^A \xi^\mu), \qquad \xi^\mu = \tfrac{1}{2}\bar{\epsilon}_2 \gamma^\mu \epsilon_1. \tag{11.28}$$

The first term is a translation, which acts as $\xi^\mu \partial_\mu$ on the fields. The second term was called a field dependent gauge transformation in Ch. 6. However, we now interpret it as a modification of the SUSY algebra in which the commutator of supercharges contains a gauge transformation giving the new 'structure function'

$$f_{\alpha\beta}{}^A = \tfrac{1}{2} A_\mu^A (\gamma^\mu)_{\alpha\beta}, \tag{11.29}$$

which depends on the gauge field A_μ.

It is not hard to find the reason for the modified algebra in (11.28). The SUSY transformation rules (6.62) and (6.49) of the chiral and gauge multiplets are covariant under non-abelian gauge transformations, so their commutator is also covariant. However, the action of δ_P on any component field is not covariant. Instead its gauge transform contains the derivative of the gauge parameter $\theta^A(x)$. The second term in (11.28) restores gauge covariance.

The extension to soft algebras does not affect most of the formulas we have developed for gauge symmetries. Jacobi identities are modified due to the variation of the structure functions, but we will not need this explicitly. For the interested reader we point out that this type of generalized gauge theory is quite naturally described in the framework of the Batalin–Vilkovisky or field–antifield formalism; see [62, 63, 64].

When one has a solution of the field equations of the full theory, one may plug in the values of the fields into this soft algebra. Then the structure constants of a conventional algebra appear, and these encode the remaining symmetries for that particular solution. The soft algebra thus allows for classical solutions with various supersymmetry algebras such as the anti-de Sitter algebra or algebras with central charges.

Zilch symmetries

A zilch symmetry is one whose transformation rules vanish on solutions of the equations of motion. The corresponding Noether current then vanishes. Any action with at least two fields has zilch symmetries. Indeed consider an action $S(\phi)$ containing fields ϕ^i. Consider the transformation

$$\delta\phi^i = \epsilon h^{ij} \frac{\delta S}{\delta \phi^j} \tag{11.30}$$

for any antisymmetric matrix $h^{ij} = -h^{ji}$. One finds that

$$\delta S = \frac{\delta S}{\delta \phi^i} \epsilon h^{ij} \frac{\delta S}{\delta \phi^j} = 0. \tag{11.31}$$

We assumed bosonic fields here, but the idea extends to fermions if we change the symmetry properties of h. This is in fact the way that we encountered such a zilch symmetry in (6.40). In that case the h^{ij} is represented by $v^\mu (\gamma_\mu)^{\alpha\beta}$, which is symmetric in the indices

α, β that refer to the spinor fields. The transformation (6.40) is a zilch symmetry since it vanishes if the field χ satisfies the equation of motion.

The existence of zilch symmetries implies that symmetries are not uniquely defined. One can add to any symmetry transformation $T_A\phi^i$ a zilch symmetry $h_A^{ij}\,\delta S/\delta\phi^j$, and it is still a symmetry. We view this as a change of basis in the algebra of symmetries.

Open algebras

As we saw already in Ch. 6, zilch symmetries sometimes occur in the commutators of conventional symmetries such as SUSY. We found this in the algebra of supersymmetry transformations of a chiral multiplet after elimination of the auxiliary field; see Ex. 6.11 and the following discussion.

From first principles (see footnote 6 in Sec. 6.2.2) it follows that the commutator of two symmetries of the action is also a symmetry. The result of the commutator can thus be expanded in the set of all symmetries, as indicated in (11.17). However, this expansion may also include the zilch symmetries. As such it makes sense to write the algebra

$$[\delta(\epsilon_1), \delta(\epsilon_2)]\,\phi^i = \text{minimal SUSY algebra} + \eta^{ij}(\epsilon_1, \epsilon_2)\,\frac{\delta S}{\delta\phi^j}. \tag{11.32}$$

Thus, in a general gauge theory, a commutator of symmetries of the form (11.17) will only be valid after equations of motion are used. In this case the algebra of field transformations is said to be closed only 'on-shell'. If one has a basis such that (11.17) is valid without use of the equations of motion, then the commutator closes for any configuration of fields, whether the equations of motion are satisfied or not. In this case we say that the algebra is closed 'off-shell'. One also uses the terminology *closed supersymmetry algebra* when the algebra holds without zilch symmetries. When zilch symmetries enter the algebra, it is called an *open supersymmetry algebra*. But remember that in this case the algebra does close when the field equations are satisfied or when the (infinite set of) zilch symmetries are included.

11.2 Covariant quantities

We now want to consider field theories in which the Lagrangian contains both gauge fields B_μ^A and other fields ϕ^i, sometimes called 'matter fields', whose transformation rules

$$\delta(\epsilon)\phi^i(x) = \epsilon^A(T_A\phi^i)(x) \tag{11.33}$$

do not involve derivatives of the gauge parameters. Readers are already familiar with the important case of gauged non-abelian internal symmetry discussed in Sec. 4.3.2. In that section we saw how to define and use covariant derivatives and field strengths, which also

Box 11.1 **Definition of a covariant quantity**

A *covariant quantity* is a local function that transforms under all local symmetries with no derivatives of a transformation parameter.

transform without derivatives of the ϵ^A, and which are the building blocks of the physical gauge theories. The Yang–Mills action (4.96) is directly constructed from such covariant building blocks.

In this section we want to define such quantities within the more general framework considered in this chapter. We begin with the definition in Box 11.1. Our considerations apply in a straightforward way to gauged internal symmetry, Lorentz transformations and SUSY, but local translations require special care and are mainly discussed in Sec. 11.3.

11.2.1 Covariant derivatives

One very important covariant quantity is the covariant derivative of a field ϕ^i (whether elementary or composite) whose gauge transformation rule is of the form in (11.33). The ordinary derivative $\partial_\mu\phi^i$ is certainly not a covariant quantity since

$$\delta(\epsilon)\partial_\mu\phi^i = \epsilon^A\partial_\mu(T_A\phi^i) + (\partial_\mu\epsilon^A)T_A\phi^i . \tag{11.34}$$

The second term spoils covariance but we can correct for this by adding a term involving the gauge fields and defining

$$\begin{aligned} \mathcal{D}_\mu\phi^i &\equiv \left(\partial_\mu - \delta(B_\mu)\right)\phi^i \\ &= \left(\partial_\mu - B_\mu^A T_A\right)\phi^i . \end{aligned} \tag{11.35}$$

The notation $\delta(B_\mu)$ means that the covariant derivative is constructed by the specific prescription to subtract the gauge transform of the field with the gauge field itself as the symmetry parameter. As is clear from the second line of (11.35) there is a sum over all gauge transformations of the theory. As an example of this construction we rewrite the spinor covariant derivative of (4.89) and transformation rule in the present notation[7] as

$$\delta(\theta)\Psi = -\theta^A t_A\Psi, \qquad \mathcal{D}_\mu\Psi = \left(\partial_\mu + A_\mu^A t_A\right)\Psi. \tag{11.36}$$

It is easy to check using (11.24) that the gauge transformation of the covariant derivative does not contain derivatives of the gauge parameters. Hence, the covariant derivative is a covariant quantity. We now prove the stronger result that gauge transformations commute with covariant differentiation on fields ϕ for which the algebra is closed.[8] Thus we can

[7] The gauge coupling of Ch. 4 is now set to $g = 1$.
[8] Remember that this is not a trivial statement; see the remarks above on open algebras. Thus, for example, in the chiral multiplet without auxiliary fields, (6.40) implies that the statements below do apply on Z, but not for χ.

Transformation of covariant derivatives **Box 11.2**

Gauge transformations and covariant derivatives commute on fields on which the algebra is off-shell closed.

apply (11.17). We first write an auxiliary relation, which is (11.17) applied to ϕ as in (11.33), with ϵ_1 replaced by B_μ and ϵ_2 by ϵ:

$$\epsilon^A \delta(B_\mu)(T_A\phi) - B_\mu^A \delta(\epsilon)(T_A\phi) = \epsilon^B B_\mu^A f_{AB}{}^C(T_C\phi). \tag{11.37}$$

Here, we use that the algebra is satisfied without adding equations of motion. We then write, using (11.24) in the second line and (11.37) in the third,

$$\begin{aligned}\delta(\epsilon)\mathcal{D}_\mu\phi &= \partial_\mu\left(\epsilon^A(T_A\phi)\right) - B_\mu^A\delta(\epsilon)(T_A\phi) - \left(\delta(\epsilon)B_\mu^A\right)(T_A\phi)\\ &= \epsilon^A\partial_\mu(T_A\phi) - B_\mu^A\delta(\epsilon)(T_A\phi) - \epsilon^C B_\mu^B f_{BC}{}^A(T_A\phi)\\ &= \epsilon^A\left[\partial_\mu(T_A\phi) - \delta(B_\mu)(T_A\phi)\right] = \epsilon^A\mathcal{D}_\mu(T_A\phi).\end{aligned} \tag{11.38}$$

Therefore, we obtain the result stated in Box 11.2. In summary, we repeat the definition of the covariant derivative in slightly more general form.

The covariant derivative of any covariant quantity is given by the operator

$$\mathcal{D}_\mu = \partial_\mu - \delta(B_\mu) \tag{11.39}$$

acting on that quantity. The instruction $\delta(B_\mu)$ means compute all gauge transformations of the quantity, with the potential B_μ^A as the gauge parameter.

Exercise 11.8 *Symmetries of the nonlinear σ-model are generated by Killing vectors $k_A^i(\phi)$, which in general determine a Lie algebra as defined in (11.20). The elementary fields ϕ^j are local coordinates of the target space. Suppose that the symmetry is gauged. Show that the covariant derivative $\mathcal{D}_\mu\phi^i = \partial_\mu\phi^i - A_\mu^A k_A^i$ transforms as*

$$\delta\mathcal{D}_\mu\phi^i = \theta^A\mathcal{D}_\mu k_A^i = \theta^A\left(\partial_j k_A^i\right)\mathcal{D}_\mu\phi^j. \tag{11.40}$$

11.2.2 Curvatures

We can use the covariant derivative to define the next important set of quantities in any gauge theory of an algebra. For each generator of the algebra the curvature $R_{\mu\nu}{}^A$ is a second rank antisymmetric tensor which is also a covariant quantity. In Yang–Mills theory the curvature was called a field strength and was defined in (4.91) and (4.92) as a commutator of covariant derivatives acting on a covariant field.

We now proceed in the same way in this more general framework. Using (11.38) and (11.37) (with now ϵ replaced by B_ν), we obtain

$$[\mathcal{D}_\mu, \mathcal{D}_\nu] = -\delta(R_{\mu\nu}),$$
$$R_{\mu\nu}{}^A = 2\partial_{[\mu} B_{\nu]}{}^A + B_\nu{}^C B_\mu{}^B f_{BC}{}^A. \qquad (11.41)$$

Here again $\delta(R_{\mu\nu})$ means that one takes a sum over all gauge symmetries and replaces the parameters ϵ^A with $R_{\mu\nu}{}^A$.

Curvatures are covariant quantities, which transform as

$$\delta(\epsilon) R_{\mu\nu}{}^A = \epsilon^C R_{\mu\nu}{}^B f_{BC}{}^A. \qquad (11.42)$$

They satisfy Bianchi identities, which are, using the definition (11.39),

$$\mathcal{D}_{[\mu} R_{\nu\rho]}{}^A = 0. \qquad (11.43)$$

The curvatures and Bianchi identities discussed in Sec. 7.10 are a special case of (11.41) in which the Lie algebra is the algebra of the Lorentz group $\mathfrak{so}(D-1,1)$. Here is a small exercise to verify this.

Exercise 11.9 *Obtain the curvature tensor (7.115) for Lorentz symmetry using the approach of this chapter. Use the definition (11.41) and the structure constants in Table 11.2.*

We mentioned earlier that gauged local translations, generated by P_a, are more subtle and require special treatment. The following exercise will illustrate this.

Exercise 11.10 *Translations occur on the right-hand side of the commutators $[M_{ab}, P_c]$ and $\{Q_\alpha, Q_\beta\}$; see Table 11.2. Use this information to show that the curvature for the translation generator P^a is*

$$R_{\mu\nu}(P^a) = 2\partial_{[\mu} e_{\nu]}{}^a + 2\omega_{[\mu}{}^{ab} e_{\nu]b} - \tfrac{1}{2}\bar{\psi}_\mu \gamma^a \psi_\nu. \qquad (11.44)$$

The last term is the torsion tensor of $\mathcal{N}=1$, $D=4$ (see (9.18)) which, by (7.106), is the antisymmetric part of the connection. Hence, the equation (11.44) is just the antisymmetric part of (7.101) in $[\mu\nu]$, which is equivalent to the first Cartan structure equation (7.81). Therefore, the curvature for spacetime translations vanishes:

$$R_{\mu\nu}(P^a) = 0. \qquad (11.45)$$

The last exercise indicates already that something is special when local translations are included. In the next section we will discuss the special features of such theories.

11.3 Gauged spacetime translations

The main issue that we have to address here is the reconciliation of the concept of gauge theories developed earlier in this chapter with general coordinate transformations, which we studied in Ch. 7. For this we will have to redefine the spacetime transformations and covariant derivatives and curvatures. This is more difficult than the previous part of this chapter, which will require special attention from the reader. The concepts developed in this section are, however, necessary to understand later the manipulations of covariant derivatives in supergravity theories with matter couplings.

11.3.1 Gauge transformations for the Poincaré group

The Lie algebra of the Poincaré group includes both translations P_a and Lorentz transformations $M_{[ab]}$. In theories of gravity both symmetries are gauged by parameters $\xi^a(x)$ and $\lambda^{ab}(x)$, which are arbitrary functions on the spacetime manifold.

Special issues arise for *local translations*. The first issue is quite simple. Under global transformations of the Poincaré group, it was shown in Ch. 1 that a scalar field in Minkowski spacetime transforms as

$$\delta(a,\lambda)\phi(x) = \left[a^\mu\partial_\mu - \tfrac{1}{2}\lambda^{\mu\nu}L_{[\mu\nu]}\right]\phi(x) = \left[a^\mu + \lambda^{\mu\nu}x_\nu\right]\partial_\mu\phi(x). \tag{11.46}$$

In curved spacetime we replace $a^\mu \to \xi^\mu(x)$, which is an arbitrary spacetime dependent function. The effect of the second term, which is sometimes called the 'orbital part' of a Lorentz transformation, can be included in $\xi^\mu(x)$. In this way we perform a change of basis in the set of gauge transformations. Originally the independent ones were parametrized by $a^\mu(x)$ and $\lambda^{ab}(x)$. From now on we use $\xi^\mu(x) = a^\mu(x) + \lambda^{\mu\nu}(x)x_\nu$ and $\lambda^{ab}(x)$ as a basis of independent transformations. Thus (11.46) is replaced by

$$\delta_{\rm gct}(\xi)\phi(x) = \xi^\mu(x)\partial_\mu\phi(x) \tag{11.47}$$

in curved spacetime.

This defines the local translations, and we have seen that they have absorbed the 'orbital part' of Lorentz transformations. The 'spin part' of Lorentz transformations is implemented as *local Lorentz transformations* in curved spacetime. The rule for these is simple to state and is implicit in the discussion of Ch. 7; see Box 11.3.

Local Poincaré transformations **Box 11.3**

Local translations are replaced by general coordinate transformations.
Local Lorentz transformations: Only fields carrying local frame indices transform under local Lorentz transformations. The transformation rule involves the appropriate matrix generator.

Let us make this explicit for various fields. For scalars, rather than the global transformations in (11.46), we thus write the local Poincaré transformation

$$\delta(\xi,\lambda)\phi(x) = \xi^\mu(x)\partial_\mu\phi(x)\,; \tag{11.48}$$

the local Lorentz transformation has no effect on a scalar field.

For spinors, the global Poincaré transformations have been defined as

$$\delta(a,\lambda)\Psi(x) = \left[a^\mu + \lambda^{\mu\nu}x_\nu\right]\partial_\mu\Psi(x) - \tfrac{1}{4}\lambda^{ab}\gamma_{ab}\Psi(x). \tag{11.49}$$

The local Poincaré transformations are then written as

$$\delta(\xi,\lambda)\Psi(x) = \xi^\mu(x)\partial_\mu\Psi(x) - \tfrac{1}{4}\lambda^{ab}(x)\gamma_{ab}\Psi(x). \tag{11.50}$$

According to the rules of Sec. 1.2.3, the Poincaré transformations for vectors are

$$\delta(a,\lambda)V_\mu = \left[a^\nu + \lambda^{\nu\rho}x_\rho\right]\partial_\nu V_\mu - \lambda_\mu{}^\nu V_\nu. \tag{11.51}$$

Since V_μ has no local Lorentz indices, the complete transformation is given by a general coordinate transformation implemented by the Lie derivative, as explained in (7.10), namely

$$\delta(\xi,\lambda)V_\mu(x) = \mathcal{L}_\xi V_\mu \;=\; \xi^\nu(x)\partial_\nu V_\mu(x) + V_\nu(x)\partial_\mu\xi^\nu(x). \tag{11.52}$$

Thus there are no separate Lorentz transformations for V_μ. The global Lorentz transformations reappear in the global limit using the identification $\xi^\mu = a^\mu + \lambda^{\mu\nu}x_\nu$.

For a vector field we have a choice of using either the coordinate basis or local frame. The components are related by $V_\mu = e_\mu^a V_a$. Frame vectors transform as

$$\begin{aligned}\delta(\xi,\lambda)V^a(x) &= \xi^\mu(x)\partial_\mu V^a(x) - \lambda^a{}_b(x)V^b(x),\\ \delta(\xi,\lambda)V_a(x) &= \xi^\mu(x)\partial_\mu V_a(x) + V_b(x)\lambda^b{}_a(x).\end{aligned} \tag{11.53}$$

Fields referred to local frames effectively transform as scalars under diffeomorphisms. Note that (11.52) and (11.53) are compatible if the transformation rule of the frame field is

$$\delta e_\mu^a = \xi^\nu\partial_\nu e_\mu^a + e_\nu^a\partial_\mu\xi^\nu - \lambda^{ab}e_{\mu b}. \tag{11.54}$$

We have combined Lorentz and coordinate transformations, and we have simply applied the previously stated rules to a field with both coordinate and frame indices. To complete the list of transformation rules we include the rule for the transformation of the spin connection ω_μ^{ab}, namely

$$\delta\omega_\mu{}^{ab} = \xi^\nu\partial_\nu\omega_\mu{}^{ab} + \omega_\nu{}^{ab}\partial_\mu\xi^\nu + \partial_\mu\lambda^{ab} - \lambda^a{}_c\omega_\mu{}^{cb} + \omega_\mu{}^{ac}\lambda_c{}^b. \tag{11.55}$$

In order to apply Table 11.1, we also introduce the frame vector parameter $\xi^a(x)$ related to ξ^μ by $\xi^a = \xi^\mu e_\mu^a$. Then (11.47) can be rewritten as

$$\delta_{\rm gct}\phi(x) \;=\; \xi^a(x)e_a^\mu(x)\partial_\mu\phi(x) \;=\; \xi^a(x)\partial_a\phi(x), \tag{11.56}$$

with $\partial_a\phi \equiv e_a^\mu \partial_\mu\phi$. Thus the main results of this section are to establish ξ^a and λ^{ab} as the 'basis' for gauge parameters of the Poincaré group and to define the associated transformation rules of the fields we will encounter.

11.3.2 Covariant derivatives and general coordinate transformations

In this section, we first show why the definitions of covariant derivatives made in Sec. 11.2 need to be modified when applied to an algebra with spacetime symmetries. Then we will improve the definitions. Though this makes the formalism more involved, a good and clear definition of covariant derivatives is needed for supergravity. When one calculates transformations of covariant quantities, the improved formalism avoids (with the help of lemmas to be introduced in Sec. 11.3.3) needless heavy calculations of terms that later cancel. We will use these lemmas later in the book, especially in Chs. 15, 16, 17 and 20. Indeed, in Ch. 15 we will show how general relativity is derived using these concepts.

It is easy to see that the definition (11.39) needs to be changed when the gauge group includes local translations. Otherwise we would write the covariant derivative of a scalar field (with no internal symmetry present) as

$$\mathcal{D}_\mu\phi = \partial_\mu\phi - e_\mu^a(x)\partial_a\phi(x) = 0. \tag{11.57}$$

So the previous definition is rather useless. It is not hard to repair the definition by removing general coordinate transformations from the sum over gauge transformations. Thus, from now on, we define the covariant derivative of any covariant field ϕ as

$$\mathcal{D}_\mu\phi \equiv \partial_\mu\phi - B_\mu{}^A T_A\phi, \tag{11.58}$$

with general coordinate transformations omitted in the sum over the gauge group indices A. The same restriction applies to the gauge transformation $\delta(\epsilon)\phi$ of (11.1). We discuss below how this changes the rules established in Sec. 11.2. We use the term 'standard gauge transformations' to refer to the set of gauge transformations that remain in (11.58). These include local Lorentz, local supersymmetry, and local internal symmetry transformations. However, in later chapters we will include conformal and superconformal transformations. In any case we now use the general notation $B_\mu{}^A$ for all such standard gauge fields.

A second peculiar feature of translations was encountered in Ex. 11.10. Namely the curvature component for local translations in gravity or supergravity vanishes; see (11.44). *From now on we will always impose (11.45) as a constraint*, which determines the connection in the presence of torsion due to the gravitino field.[9] It was shown in Ex. 7.26 how to solve this constraint, which we found there as the first Cartan structure equation, to express the spin connection in terms of the frame field and the torsion tensor. Thus the spin connection becomes a 'composite gauge field' in the formalism we are now developing. Other examples of composite gauge fields will appear in Ch. 15 in the gauging of the conformal group and in Ch. 16 when we gauge the superconformal group. In both cases they will be determined by curvature constraints.

[9] For application to ordinary gravity, we just set $\psi_\mu = 0$.

Box 11.4 Covariant general coordinate transformations and standard gauge transformations

Always use *covariant* general coordinate transformations. All other transformations are 'standard gauge transformations' and the sum over the gauge group indices A for covariant derivatives and curvatures should be restricted to these.

We now make a further change in the basis of gauge transformations. Namely we replace general coordinate transformations by *covariant* general coordinate transformations (cgct). To motivate this we consider a set of scalar fields ϕ^i transforming under an internal symmetry as $\delta(\theta)\phi^i(x) = -\theta^A(x)t_A{}^i{}_j\phi^j$. Using the previous definition their transformation under a coordinate transformation would be

$$\delta(\xi)\phi^i = \xi^\mu \partial_\mu \phi^i. \tag{11.59}$$

This is correct, but it has the undesirable property that it does not transform covariantly under internal symmetry. We fix this by adding a field dependent gauge transformation and thus define

$$\delta_{\rm cgct}(\xi)\phi^i \equiv \xi^\mu \partial_\mu \phi^i + (\xi^\mu A_\mu{}^A)t_A{}^i{}_j\phi^j. \tag{11.60}$$

A good indication that this is a quite natural modification already came in Ex. 6.15, where readers showed that such field dependent gauge transformations occur in the commutator of global SUSY transformations in a SUSY gauge theory. This led to the concept of soft algebras in (11.28). It also appeared in Sec. 9.5 when we first discussed the algebra of local supersymmetry.

We thus conclude that, for any field that transforms under one or more of the standard gauge transformations, we define its *covariant general coordinate transformation* [35, 65] by

$$\delta_{\rm cgct}(\xi) = \delta_{\rm gct}(\xi) - \delta(\xi^\mu B_\mu). \tag{11.61}$$

For every standard gauge transformation this contains a term in which the parameter of the transformation is replaced by the scalar product of ξ^μ with the standard gauge field $B_\mu{}^A$.

We now discuss the form of the covariant general coordinate transformations on the several types of fields that we use.

Coordinate scalars. Note that the two terms in the example (11.60) can be grouped into the standard covariant derivative, so we can write

$$\delta_{\rm cgct}(\xi)\phi = \xi^\mu \mathcal{D}_\mu \phi = \xi^a \mathcal{D}_a \phi. \tag{11.62}$$

We thus find that cgct transform the scalars to

$$\mathcal{D}_a\phi = e_a{}^\mu \mathcal{D}_\mu \phi, \qquad \mathcal{D}_\mu\phi = \partial_\mu\phi - B_\mu^A(T_A\phi). \tag{11.63}$$

The same situation occurs for the spinor fields of $\mathcal{N}=1$ multiplets, which are also general coordinate scalars when they are coupled to supergravity. Consider the fermion χ of a chiral multiplet and the gaugino λ of an abelian gauge multiplet. For them the gauge covariant translation and subsequent covariant derivative require a field dependent SUSY transformation and thus involve their SUSY partners Z, F, $A_\mu(x)$ and D, as well as the gravitino ψ_μ:

$$\begin{aligned}\delta_{\rm cgct}(\xi)P_L\chi &= \xi^\mu P_L(\partial_\mu\chi + \frac{1}{4}\omega_\mu{}^{ab}\gamma_{ab}\chi - \frac{1}{\sqrt{2}}\gamma^\nu\mathcal{D}_\nu Z\psi_\mu - F\psi_\mu)\\ &= \xi^\mu P_L\mathcal{D}_\mu\chi,\\ \delta_{\rm cgct}(\xi)\lambda &= \xi^\mu\left(\partial_\mu\lambda + \tfrac{1}{4}\omega_\mu{}^{ab}\gamma_{ab}\lambda + \tfrac{1}{4}\gamma^{\rho\sigma}F_{\rho\sigma}\psi_\mu - \tfrac{1}{2}{\rm i}\gamma_*D\right) = \xi^\mu\mathcal{D}_\mu\lambda.\end{aligned} \tag{11.64}$$

Gauge fields. We now compute the cgct for the gauge field $B_\mu{}^A$ of a standard gauge transformation. For simplicity we assume that the structure constant $f_{aB}{}^A=0$, where A is the particular symmetry index of interest, the index a indicates local translations, and B is general. This assumption is satisfied in the $\mathcal{N}=1$ SUSY algebra,[10] as shown in Table 11.2. The cgct of B_μ^A is as the sum of a Lie derivative plus field dependent standard gauge transformation:

$$\delta_{\rm cgct}(\xi)B_\mu^A = \xi^\nu\partial_\nu B_\mu^A + B_\nu^A\partial_\mu\xi^\nu - \partial_\mu\left(\xi^\nu B_\nu^A\right) - \xi^\nu B_\nu^C B_\mu{}^B f_{BC}{}^A = \xi^\nu R_{\nu\mu}{}^A. \tag{11.65}$$

Thus the cgct of a gauge field involves its curvature. An example already appeared in the commutator of global SUSY transformations on the Yang–Mills potential in (6.51).

The frame field. The frame field e_μ^a is the gauge field of local translations. Its standard gauge transformations are defined in (11.24), where the index A still included all the gauge transformations. When we split these into a (with gauge parameter ξ^a) and A for the standard gauge transformations, the transformations of the frame field thus read (assuming $f_{ab}{}^c=0$, since translations do not commute to translations)

$$\begin{aligned}\delta e_\mu{}^a &= \partial_\mu\xi^a + \xi^c B_\mu{}^B f_{Bc}{}^a\\ &\quad + \epsilon^C\left(B_\mu{}^B f_{BC}{}^a + e_\mu{}^b f_{bC}{}^a\right).\end{aligned} \tag{11.66}$$

The computation of the cgct is a formalization of the computation that we did in Sec. 9.5, and leads to a simple result:

$$\begin{aligned}\delta_{\rm cgct}(\xi)e_\mu^a &= \xi^\nu\partial_\nu e_\mu^a + e_\nu^a\partial_\mu\xi^\nu - \xi^\nu B_\nu^C\left(B_\mu{}^B f_{BC}{}^a + e_\mu{}^b f_{bC}{}^a\right)\\ &= \partial_\mu\xi^a + \xi^\nu\left(2\partial_{[\nu}e_{\mu]}^a - B_\nu^C B_\mu{}^B f_{BC}{}^a - B_\nu^C e_\mu^b f_{bC}{}^a - e_\nu^c B_\mu^B f_{Bc}{}^a\right)\\ &\quad + \xi^c B_\mu^B f_{Bc}{}^a\\ &= \partial_\mu\xi^a + \xi^c B_\mu{}^B f_{Bc}{}^a - \xi^\nu R_{\mu\nu}{}^a.\end{aligned} \tag{11.67}$$

[10] In some situations that occur later in this book, this simplification is not generally valid. In Sec. 11.3.3 we will learn how to treat these cases.

As discussed above, the last term vanishes as a constraint on the spin connection. The result is that the cgct of the frame field is given by the first two terms, which agree with the first line of (11.66). This confirms the identification of the cgct as the appropriate modification of local translations. For the super-Poincaré algebra, the only gauge field that contributes in the first line of (11.66) is the Lorentz connection $\omega_{\mu ab}$, so that the covariant gct of the frame field becomes the local Lorentz covariant derivative $\delta e_\mu{}^a = \partial_\mu \xi^a + \omega_\mu{}^a{}_b \xi^b$. For the superconformal algebra there is another contribution also.

11.3.3 Covariant derivatives and curvatures in a gravity theory

In this section we will find the modifications needed in the treatment of covariant derivatives and curvatures in Sec. 11.2. We maintain the same definition of a covariant quantity used on p. 222. In particular, since the general coordinate transformation of a covariant quantity should not involve a derivative of the parameter ξ^μ, a covariant quantity must be a world scalar. Hence, the covariant derivative that we will use is $\mathcal{D}_a\phi$ as defined in (11.63), rather than $\mathcal{D}_\mu\phi$.

The modifications made for covariant derivatives have their counterpart for curvatures. First of all we have to distinguish again translations from standard gauge transformations. Hence, we define now curvatures

$$r_{\mu\nu}{}^A = 2\partial_{[\mu} B_{\nu]}{}^A + B_\nu{}^C B_\mu{}^B f_{BC}{}^A, \tag{11.68}$$

where the sums over B and C involve only standard gauge transformations.

But there is more. There are important cases in supergravity in which (11.24) (with indices restricted to standard gauge transformations) is not the complete transformation law. Indeed, the transformation (11.24) involves only gauge fields, and multiplets are often composed of gauge fields together with non-gauge fields. Therefore, often (11.24) has to be completed with additional terms. To allow for this, we write the more general form

$$\delta(\epsilon) B_\mu^A = \partial_\mu \epsilon^A + \epsilon^C B_\mu{}^B f_{BC}{}^A + \epsilon^B \mathcal{M}_{\mu B}{}^A. \tag{11.69}$$

One case is where gauge transformations do not satisfy $f_{aB}{}^A = 0$, and the latter term thus contains terms of the form $\epsilon^B e_\mu^a f_{aB}{}^A$. Such structure constants do not appear in the algebra that we considered up to here, but will appear later in the book (for conformal algebras). Another example is the Yang–Mills gauge field A_μ^A, whose transformation law under supersymmetry in (6.49) is not related to its gauge properties. This is due to the presence of the non-gauge field λ^A in the gauge multiplet. In this case the modified gauge transformation law reads

$$\delta A_\mu^A = \partial_\mu \theta^A + \theta^C A_\mu{}^B f_{BC}{}^A - \tfrac{1}{2}\bar{\epsilon}\gamma_\mu \lambda^A. \tag{11.70}$$

In this case the new term involves only the local SUSY parameter $\bar{\epsilon}^\alpha$. Formally, we thus have for the latter case (with spinor index α referring to supersymmetry)

$$\mathcal{M}_{\mu\alpha}{}^A = -\tfrac{1}{2}\left(\gamma_\mu \lambda^A\right)_\alpha. \tag{11.71}$$

Principles of transformations of covariant quantities **Box 11.5**

1. The covariant derivative $\mathcal{D}_a$ of a covariant quantity is a covariant quantity, and so is the curvature $\widehat{R}_{ab}$.
2. The gauge transformation of a covariant quantity does not involve a derivative of a parameter.
3. If the algebra closes on the fields, then the transformation of a covariant quantity is a covariant quantity, i.e. gauge fields only appear included either in covariant derivatives or in curvatures.

In practical calculations it turns out that the $\mathcal{M}_{\mu B}{}^A$ are the important terms, while the other parts of the transformations (11.69) automatically appear in the formalism of covariant derivatives.

If gauge fields transform with extra terms as in (11.69), one has to 'covariantize' also for these in the curvatures. Hence the modified curvature is

$$\widehat{R}^A_{\mu\nu} = r_{\mu\nu}{}^A - 2B_{[\mu}{}^B \mathcal{M}_{\nu]B}{}^A. \tag{11.72}$$

These already appear in (11.65). In case the gauge fields transform as in (11.69), we have

$$\delta_{\rm cgct}(\xi) B^A_\mu = \xi^\nu \widehat{R}_{\nu\mu}{}^A. \tag{11.73}$$

Finally, as argued above for the covariant derivative, the covariant curvature should carry Lorentz indices rather than world indices in order to be a world scalar. Thus

$$\widehat{R}_{ab}{}^A = e_a{}^\mu e_b{}^\nu \widehat{R}_{\mu\nu}{}^A. \tag{11.74}$$

11.3.4 Calculating transformations of covariant quantities

Covariant quantities are the building blocks of general supergravity theories and the construction of these theories frequently requires the gauge transform of one or more covariant quantities. An *ab initio* computation in every case involves very tedious details. Fortunately, there is a common structure in such calculations, and the purpose of this section is to exploit it to obtain useful shortcuts which will be applied many times in the next few chapters.

The common structure and the shortcuts derived from it require the application of the principles written in Box 11.5.[11] We postpone the proof of these statements to Appendix 11A. Instead we illustrate how they are applied here. The importance of these facts is that they tell us in advance that explicit gauge fields only occur inside covariant derivatives and curvatures. This simplifies our (lives and our) calculations.

[11] The second one is not a derived result; rather, it is the *definition* of a covariant quantity, but we include it here to have all the important facts together for further reference.

Consider first a set of scalar fields ϕ^i in the setting of a gravity theory with a gauged internal symmetry group. The scalars transform under gauge transformations as

$$\delta\phi^i = \theta^A T_A \phi^i = -\theta^A (t_A)^i{}_j \phi^j . \qquad (11.75)$$

Their covariant derivatives are given in (11.63) with $B_\mu^A = A_\mu^A$, the usual Yang–Mills gauge fields. We want to calculate the standard transformation of the covariant derivative $\mathcal{D}_a\phi$. In this example the standard transformations include only local Lorentz and internal symmetry. Distributing the transformations on the various fields gives

$$\delta\mathcal{D}_a\phi = \left(\delta e_a{}^\mu\right)\mathcal{D}_\mu\phi + e_a{}^\mu\partial_\mu\delta\phi - e_a{}^\mu(\delta A_\mu^A)(T_A\phi) - e_a{}^\mu A_\mu^A\delta(T_A\phi). \qquad (11.76)$$

In the first term we need the local Lorentz part of (11.54), which we can also find using the commutator $[P, M]$ from Table 11.2:

$$\delta e_\mu{}^a = -\lambda^{ab} e_{\mu b} \qquad \Rightarrow \qquad \delta e_a{}^\mu = -\lambda_{ab} e^{b\mu} . \qquad (11.77)$$

The first term then becomes

$$\delta_{\rm Lor}(\lambda)\mathcal{D}_a\phi^i = -\lambda_a{}^b\mathcal{D}_b\phi^i . \qquad (11.78)$$

Rather than work out the remaining terms explicitly we wish to show how the principles above eliminate much of the work. We can forget the $\partial_\mu\theta^A$ which appears when (11.75) is inserted in the second term of (11.76) because the first principle tells us that $\mathcal{D}_a\phi$ is a covariant quantity and the second principle says that its transformations cannot contain derivatives of a gauge parameter. So the derivative term must cancel at the end and we can simply drop it at the beginning. The second term of (11.76) then reduces to $-e_a^\mu\theta^A(t_A)^i{}_j\partial_\mu\phi^j$. The remaining parts of the third and fourth terms of (11.76) contain the gauge field explicitly, and the third principle tells us that gauge fields can only 'covariantize' the derivative $\partial_\mu\phi^i$. Thus we can write the final answer as

$$\delta\mathcal{D}_a\phi^i = -\lambda_a{}^b\mathcal{D}_b\phi^i - \theta^A(t_A)^i{}_j\mathcal{D}_a\phi^j . \qquad (11.79)$$

In this example, it is easy to perform a complete calculation to verify this result. However, for more complicated calculations in supergravity we will be happy to have principles that save us a lot of calculational work. An illustration is provided in Appendix 11A.2.

Exercise 11.11 *One can derive a general formula for the transformation of curvatures, correcting (11.42) for the effects that gauge fields transform with matter-like terms. To apply the methods explained earlier, the decomposition in (11.72) is most useful. Indeed, explicit gauge fields appear in $r_{ab}{}^A$ only quadratically. Show that*

$$\delta(\epsilon)\widehat{R}_{ab}{}^A = \epsilon^B\widehat{R}_{ab}{}^C f_{CB}{}^A + 2\epsilon^B\mathcal{D}_{[a}\mathcal{M}_{b]B}{}^A - 2\epsilon^C\mathcal{M}_{[aC}{}^B\mathcal{M}_{b]B}{}^A . \qquad (11.80)$$

To do this, you should use the principles of this section, i.e. first deleting all terms that have explicit gauge fields, and at the end replacing ordinary derivatives with covariant ones. Similarly, show that the Bianchi identity becomes

$$\mathcal{D}_{[a}\widehat{R}_{bc]}{}^A - 2\widehat{R}_{[ab}{}^B\mathcal{M}_{c]B}{}^A = 0. \qquad (11.81)$$

Exercise 11.12 *Check that the commutator of the covariant derivatives in theories with gravity still gives the curvature, i.e.*

$$[\mathcal{D}_a, \mathcal{D}_b]\phi = -\widehat{R}_{ab}{}^A(T_A\phi). \tag{11.82}$$

You can use the principles mentioned in this section. That means that all terms with undifferentiated gauge fields can be omitted, and $2\partial_{[\mu}B_{\nu]}{}^A$ *can be replaced by the curvature. Furthermore, you will need that the curvature of translations vanishes to eliminate (the covariantization of)* $\partial_{[\mu}e_{\nu]}{}^c$.

Appendix 11A Manipulating covariant derivatives

11A.1 Proof of the main lemma

Lemma on covariant derivatives. *If a covariant quantity* ϕ *transforms into covariant quantities under standard gauge transformations, its covariant derivative* $\mathcal{D}_a\phi$ *given by (11.63) is a covariant quantity. Moreover, if the algebra closes on the field* ϕ *then the gauge transformations of* $\mathcal{D}_a\phi$ *involve only covariant quantities.*

The proof closely resembles (11.38), but the separation of cgct from the other gauge transformations and the modification (11.69) change a few steps. There is one extra term due to the last term in (11.69). Furthermore one should take into account that the sum over C on the right-hand side of (11.37) includes translations. There are no other modifications because we only consider the standard gauge transformations. The result is therefore

$$\delta(\epsilon)\mathcal{D}_\mu\phi = \epsilon^A\mathcal{D}_\mu(T_A\phi) - \epsilon^B\mathcal{M}_{\mu B}{}^A(T_A\phi) + \epsilon^B B_\mu^A f_{AB}{}^c\mathcal{D}_c\phi. \tag{11.83}$$

The last term can be recognized in the transformation of the vielbein,

$$\delta(\epsilon)e_\mu^c = \epsilon^B B_\mu^A f_{AB}{}^c + \epsilon^B e_\mu^a f_{aB}{}^c, \tag{11.84}$$

where we again distinguish standard gauge transformations and translations in the sum over B or b. Therefore, the result can be reexpressed as the transformation of $\mathcal{D}_a\phi$:

$$\delta(\epsilon)\mathcal{D}_a\phi = \epsilon^A\mathcal{D}_a(T_A\phi) - \epsilon^B\mathcal{M}_{aB}{}^A(T_A\phi) - \epsilon^A f_{aA}{}^b\mathcal{D}_b\phi. \tag{11.85}$$

This is exactly what we want and like. Indeed, $\mathcal{D}_\mu\phi$ is not a covariant quantity because it is not a world scalar, and thus transforms under gct with a derivative on the parameter ξ. On the other hand, $\mathcal{D}_a\phi$ is again a world scalar. Moreover, we see that the right-hand side of (11.85) does not contain explicit gauge fields (which do occur on the right-hand side of (11.83)). Hence, the transformation of $\mathcal{D}_a\phi$ is a covariant quantity.

This is a very important result that simplifies many calculations in supergravity. We find that standard gauge transformations commute with a $\mathcal{D}_a$ covariant derivative up to two types of terms:

- The terms generated by the extra transformations of gauge fields (last term in (11.69)) hidden in the covariant derivative, i.e. terms from $-e_a{}^\mu(\delta B_\mu{}^A)(T_A\phi)$.
- The terms generated by transformations $(\delta e_a{}^\mu)\mathcal{D}_\mu\phi$, when these are not proportional to gauge fields of standard gauge transformations. This occurs for gauge symmetries whose commutator with translations leads to another translation (e.g. Lorentz transformation). The transformation of these symmetries of $\mathcal{D}_a$ is similar to the action of that generator on P_a in the algebra.

Hence, the result can be obtained from the rules given in Sec. 11.3.4.

11A.2 Examples in supergravity

We illustrate here the use of the tricks with covariant derivatives and curvatures in a setting that is a bit advanced for the readers of this chapter because it uses supergravity.

The first exercise shows how useful the lemma is.

Exercise 11.13 *Consider the scalars Z of chiral multiplets that we will embed in supergravity in Ch. 17. We assume moreover that they transform under a Yang–Mills gauge group as* $\delta_{\rm YM}Z = -\theta^A t_A Z$. *This means that the full set of gauge transformations contains (covariant) general coordinate transformations and as standard gauge transformations: supersymmetry, Lorentz transformations and Yang–Mills transformations. The definition of the covariant derivative on Z is therefore*

$$\mathcal{D}_a Z = e_a{}^\mu\left(\partial_\mu Z - \frac{1}{\sqrt{2}}\bar\psi_\mu P_L\chi + A_\mu^A t_A Z\right). \tag{11.86}$$

Prove that (assuming that the gravitino has only the gauge transformations that follow from the algebra) this leads to

$$\begin{aligned}\delta(\epsilon)\mathcal{D}_a Z = {}& \frac{1}{\sqrt{2}}\bar\epsilon\mathcal{D}_a P_L\chi - \theta^A t_A\mathcal{D}_a Z\\ &+\tfrac12\bar\epsilon\gamma_a\lambda^A t_A Z\\ &-\lambda_a{}^b\mathcal{D}_b Z.\end{aligned} \tag{11.87}$$

The first line is easy. It follows from commuting transformations and covariant derivatives. The second line is due to the extra term in the supersymmetry transformation of the gauge vector, and the third line is the one due to transformation of the frame field as written above.

Exercise 11.14 *To illustrate (11.74), show that in a supersymmetric theory the covariant field strength of a vector in an abelian vector multiplet is*

$$\widehat{F}_{ab} = e_a{}^\mu e_b{}^\nu\left(2\partial_{[\mu}A_{\nu]} + \bar\psi_{[\mu}\gamma_{\nu]}\lambda\right). \tag{11.88}$$

We now demonstrate the calculation of the transformation of a covariant field strength by calculating the supersymmetry transform of $\widehat{F}_{ab}$. Principle 1 says that $\widehat{F}_{ab}$ is a covariant quantity. Though we have not yet derived the coupling of the gauge multiplet to supergravity, we assume that the full supersymmetry rule of the gauge field A_μ is the one given in

(6.49). Furthermore let us assume that the supersymmetry transformation of the gravitino has, apart from the transformation determined by the algebra, an extra term denoted by Υ_μ. Its form is irrelevant for this exercise:

$$\delta_{\rm susy}(\epsilon)\psi_\mu = \left(\partial_\mu + \tfrac{1}{4}\gamma^{ab}\omega_{\mu ab}\right)\epsilon + \Upsilon_\mu\epsilon. \tag{11.89}$$

When considering the transformation of A_ν in the first term of (11.88), we can ignore terms in which the derivative ∂_μ acts on the gauge parameter $\bar\epsilon^\alpha$. Indeed, principle 2 implies that such terms must cancel. The derivative thus acts only on the gaugino λ. Since the supersymmetry algebra is supposed to be closed on the vector multiplet, principle 3 implies that this derivative should appear as a covariant derivative. This leads to

$$\delta_{\rm susy}(\epsilon)\widehat{F}_{ab} = \bar\epsilon\gamma_{[a}\mathcal{D}_{b]}\lambda + \ldots . \tag{11.90}$$

Now we still have to consider the second term in (11.88). An important simplification is that we do not have to act with $\delta_{\rm susy}$ on λ. Indeed, such terms would leave an explicit ψ_μ in the result, and principle 3 says that such terms anyway cancel. Thus, we need only consider (11.89). With the same principles, one sees that only the last term is relevant as the others contain a derivative on the parameter or an explicit gauge field $\omega_{\mu ab}$. Hence the final result is

$$\delta_{\rm susy}(\epsilon)\widehat{F}_{ab} = \bar\epsilon\gamma_{[a}\mathcal{D}_{b]}\lambda + \bar\lambda\gamma_{[a}\Upsilon_{b]}\epsilon. \tag{11.91}$$

The same principles also determine the modification due to supergravity of the transformation law of the gaugino given in (6.49). The modified transformation law will contain the covariant curvature, i.e.

$$\delta\lambda = \left[\tfrac{1}{4}\gamma^{ab}\widehat{F}_{ab} + \tfrac{1}{2}{\rm i}\gamma_* D\right]\epsilon. \tag{11.92}$$

Since we have already calculated the supersymmetry transformation of $\widehat{F}_{ab}$, it is easy to calculate the commutator of two supersymmetries on the gaugino. In fact, it can be compared to the calculation in the global case. If the reader still has his/her calculation for Ex. 6.15, he/she has only to replace each F by $\widehat{F}$ and D_μ by $\mathcal{D}_\mu$, and nothing changes, except for the contribution of the Υ term to (11.91). This shows how these tricks are used in practice.

12 Survey of supergravities

Supergravity theories are field theories with gauged or local supersymmetry. In Chs. 9 and 10 we introduced the basic 'pure' supergravity theories in spacetime dimensions 4 and 11. There are many other types of supergravity, for example, extended supergravity theories with $2 \leq \mathcal{N} \leq 8$ in four dimensions and higher dimensional theories in dimension $5 \leq D \leq 10$. A D-dimensional supergravity theory embodies the properties of spinors in dimension D, which we discussed in detail in Ch. 3. The primary purpose of this chapter is to develop an overview of supergravity starting from the spinors and supersymmetry algebras on which the various theories are based. Much of the material will not be treated further in later chapters and is presented for the general perspective of readers. In particular, most of the material is not strictly needed to understand the conformal and physical structure of $\mathcal{N} = 1$, $D = 4$ supergravity which is developed in detail in Chs. 13–19.

12.1 The minimal superalgebras

12.1.1 Four dimensions

We started the discussion of supersymmetry with the simplest algebra (see (6.1)) or with indices pulled down as in (11.15) (see also Sec. 3.2.2):

$$\{Q_\alpha, Q_\beta\} = -\tfrac{1}{2}(\gamma^a)_{\alpha\beta} P_a \,. \qquad (12.1)$$

The supersymmetries commute with translations and transform as a spinor under Lorentz transformations. The anti-commutator (12.1) expresses that Q is roughly speaking a square root of translations, and this is the main feature that defines 'supersymmetry'. It is not a general feature of superalgebras.

We have seen that in many cases the algebra of transformations in supergravity is more complicated due to the appearance of structure functions. But let us for now neglect this complication, to which we will return at the end of this chapter.

In Sec. 6A we discussed the minimal extended supersymmetry algebras, written in a convenient notation with chiral spinor supercharges, where the position of the index i, $1 \leq i \leq \mathcal{N}$, indicates the chirality; see (6.82). This leads to (6.84), in which the supercharges (with lowered spinor indices) satisfy

$$\begin{aligned} \{Q_{i\alpha}, Q^j_\beta\} &= -\tfrac{1}{2}\delta_i^j (P_L\gamma^a)_{\alpha\beta} P_a \,, \\ \{Q_{i\alpha}, Q_{j\beta}\} &= 0 \,, \qquad \{Q^i_\alpha, Q^j_\beta\} = 0 \,. \end{aligned} \qquad (12.2)$$

One should note that the Poincaré supersymmetry algebras exist for all $\mathcal{N}$. However, as discussed in Sec. 6.4.2, there are consistent interacting field theories only for $\mathcal{N} \leq 8$. Thus for physical reasons we consider algebras with at most $4 \times 8 = 32$ real components of the supercharges.

12.1.2 Minimal superalgebras in higher dimensions

We now consider superalgebras for spacetime dimension $D > 4$, and we point out first that the physical restriction to at most 32 supercharges also applies. Indeed the maximum $D = 11$ supergravity constructed in Ch. 10 is based on 32-component Majorana spinors and the supercharges satisfy (12.1). As discussed in Sec. 10.1, with more supercharges, dimensional reduction to four dimensions would violate the limit $\mathcal{N} \leq 8$. In particular we must not consider extended SUSY in $D = 11$.

Since the left-hand side of (12.1) is symmetric in $(\alpha\beta)$, we find from (3.63) that this equation is only consistent if $t_1 = -1$. This is exactly the condition to allow (pseudo-) Majorana spinors. On the other hand, the chirality considerations in (6.81) apply only for four or eight dimensions. Indeed, (3.56) implies that for six or 10 dimensions (6.80) is not valid. Thus, we can use (12.2) only for $D = 4$ or $D = 8$. For $D = 8$, where the minimal spinor has 16 components, we have only $\mathcal{N} = 1$ or $\mathcal{N} = 2$ supergravity theories.

In $D = 9$, there are Majorana spinors (i.e. $t_1 = -1$) with 16 real components, but there is no chirality. There are theories with $\mathcal{N} = 2$ supersymmetry. For these it suffices to add an index $i = 1, 2$ to the generators and add a δ-function

$$D = 9,\ \mathcal{N} = 2: \left\{Q^i_\alpha, Q^j_\beta\right\} = -\tfrac{1}{2}\delta^{ij}(\gamma^a)_{\alpha\beta}P_a\,, \qquad i = 1, 2. \tag{12.3}$$

The remaining dimension with Majorana spinors is $D = 10$, where the most elementary spinors are Majorana–Weyl, which have 16 real components. Owing to (3.56) the chirality of both supersymmetry operators should be equal in order to have a non-vanishing anti-commutator, rather than opposite as we saw in $D = 4$.

Exercise 12.1 *Make this argument explicit, using $t_0 = 1$, $t_{10} = t_2 = -1$, which should bring you first to*

$$(P_L Q)_\alpha = (P_L)_\alpha{}^\beta Q_\beta = Q^\beta (P_R)_{\beta\alpha}\,. \tag{12.4}$$

Continue then the argument that if one of the supercharges is left-chiral in (12.1), the other should also be left-chiral. This exercise involves mostly the application of Sec. 3.2.2.

For SUSY algebras in $D = 10$, there are three different cases to consider:

1. one chiral supercharge: $Q = P_L Q$ or $Q = P_R Q$;
2. two chiral supercharges of *opposite* chirality: $Q^1 = P_L Q^1$, $Q^2 = P_R Q^2$;
3. two chiral supercharges of *the same* chirality: $Q^1 = P_L Q^1$, $Q^2 = P_L Q^2$.

The three algebras are called 'Type I', 'Type IIA' and 'Type IIB', since they are related to the symmetry of the superstring theories with the same names. The anti-commutator has the same form as the one in $D = 9$ in (12.3) in both $\mathcal{N} = 2$ cases.

This leaves us with the case $t_1 = 1$, i.e. the dimensions $D = 5, 6$ and 7 where we must use symplectic Majorana spinors. Then (12.1) is not consistent with the symmetry of an anti-commutator. However, the symplectic Majorana condition (3.86) involves an antisymmetric tensor ε^{ij} that solves the problem. Let us start first with $D = 5$ and $D = 7$. We can then write

$$\left\{Q^i_\alpha, Q^j_\beta\right\} = -\tfrac{1}{2}(\gamma^a)_{\alpha\beta}\varepsilon^{ij} P_a \,. \tag{12.5}$$

For $D = 5$ the relevant extensions are $\mathcal{N} = 2, 4, 6$ and 8. For $D = 7$ we can have $\mathcal{N} = 2$ and 4. Note that ε^{ij} is always an invertible antisymmetric matrix, and the choice is a matter of normalization. For the case $\mathcal{N} = 2$ it can be taken to be the usual Levi-Civita tensor.

In the case of six dimensions, we can use again (12.5), but the supersymmetries can be symplectic Majorana–Weyl spinors. The argument as for $D = 10$ implies that the two supercharges in the anti-commutator should have equal chirality for the right-hand side to be non-vanishing. Thus in the simplest case, $\mathcal{N} = 2$, both supercharges have the same chirality. This symmetry algebra is therefore also denoted as $(1, 0)$. For $\mathcal{N} = 4$, the chiralities should occur in pairs and the possible algebras are indicated as $(2, 0)$ or $(1, 1)$ supersymmetry. In general we have algebras of type (p, q) with $2p$ chiral and $2q$ anti-chiral supersymmetries. The limit of at most 32 real components of supersymmetries gives $p + q \leq 4$.

Exercise 12.2 *The reader should consult Table 12.2 (later in this chapter) and check that the above arguments lead to the entries given in that table. The extra entries in the table will be explained as we progress in this chapter, as will the absence of $\mathcal{N} = 7$ in $D = 4$.*

12.2 The R-symmetry group

In Sec. 6.2.1, we discussed the R-symmetry of the $\mathcal{N} = 1$ SUSY algebra in four spacetime dimensions. The main characteristic of this symmetry is (6.4) (or (6.29)): it does not commute with Q_α. This distinguishes it from the gauge symmetries of the SUSY gauge theories in Sec. 6.3, which do commute with supersymmetry. We will now generalize this to extended supersymmetry and to higher dimensions. The general feature is that R-symmetry commutes with Lorentz and translation generators but not with the supercharges. R-symmetry is an automorphism of the SUSY algebra.

We begin with a systematic investigation of R-symmetry in $D = 4$. Suppose that T_A is a real generator of the Lie algebra of the R-symmetry group. Its action on the chiral supercharges $Q_{\alpha i}$ and Q^i_α is determined by matrices $(U_A)_i{}^j$ and $(U_A)^i{}_j$ as follows:

$$[T_A, Q_{\alpha i}] = (U_A)_i{}^j Q_{\alpha j} \,, \qquad \left[T_A, Q^i_\alpha\right] = (U_A)^i{}_j Q^j_\alpha \,. \tag{12.6}$$

The second equation of (12.6) is related to the first one by charge conjugation. Hence, $(U_A)^i{}_j$ is the complex conjugate of $(U_A)_i{}^j$. The structure relations for any superalgebra must be consistent with the super-Jacobi identities; see Appendix B.3. We first consider

$$[[T_A, T_B], Q] = [T_A, [T_B, Q]] - [T_B, [T_A, Q]] \quad \Rightarrow \quad f_{AB}{}^C U_C = [U_B, U_A], \quad (12.7)$$

where $f_{AB}{}^C$ are the structure constants of the R-symmetry algebra we seek. This implies that the matrices U_A form a representation of this algebra. We also need the super-Jacobi identity

$$0 = [T_A, \{Q_{\alpha i}, Q_\beta{}^j\}] = \{[T_A, Q_{\alpha i}], Q_\beta{}^j\} + \{[T_A, Q_\beta{}^j], Q_{\alpha i}\}. \quad (12.8)$$

The left-hand side vanishes because $[T_A, P_a] = 0$. Thus we obtain

$$0 = (U_A)_i{}^k \delta_k^j + (U_A)^j{}_k \delta_i^k, \qquad \text{i.e.} \quad (U_A)_i{}^j = -(U_A)^j{}_i = -\left((U_A)_j{}^i\right)^*. \quad (12.9)$$

The explicit Kronecker δ-function that originates in (12.2) is included since this is the relevant ingredient. We thus conclude that the matrices U are anti-hermitian matrices, i.e. they are the generating matrices of $\mathrm{U}(\mathcal{N})$. This identifies the R-symmetry group in $D = 4$ as $\mathrm{U}(\mathcal{N})$.

We now compare with (6.4) for $\mathcal{N} = 1$ SUSY. Up to normalization, we obtain from (12.9) that $U^1{}_1 = -U_1{}^1 = \mathrm{i}$. Denoting the corresponding generator as T_R, we obtain for the Majorana generator $Q = Q_L + Q_R = Q_1 + Q^1$

$$[T_R, Q] = -\mathrm{i}P_L Q + \mathrm{i}P_R Q = -\mathrm{i}\gamma_* Q. \quad (12.10)$$

This agrees with (6.4).

We can generalize this analysis to the other dimensions. Obviously, it already applies also to $D = 8$ since we saw in the previous section that they also have chiral supersymmetry generators and anti-commutators as in (12.2).

There is no chirality in odd dimensions. Hence we have to use the Majorana form of the supersymmetry generators, and for consistency the matrices that appear instead of the $U^i{}_j$ in (12.6) are real. We still find that they should form a representation of the R-symmetry algebra. Finally, the Jacobi identities that are similar to (12.8) now determine that these matrices should be antisymmetric. The result is thus $\mathrm{SO}(\mathcal{N})$. Hence there is no R-symmetry group for $D = 11$ or $D = 9, \mathcal{N} = 1$, and there is just one SO(2) generator for $D = 9, \mathcal{N} = 2$.

For $D = 10$, there are left- and right-handed supersymmetry generators. They are Majorana spinors, so the matrices that enter in $[T_A, Q]$ must be real for consistency. Furthermore, they cannot mix left- and right-chiral operators. Therefore, in each chiral sector there can be an orthogonal group. But from the three superalgebras that we saw for $D = 10$, this leaves only a non-zero result for the Type IIB algebra, where the R-symmetry is SO(2).

If there are symplectic Majorana (S) conditions, then the same argument of the Jacobi identity $[TQQ]$ leads to the preservation of the symplectic metric, and the automorphism algebra is thus reduced to $\mathrm{USp}(\mathcal{N})$ (with even $\mathcal{N}$).[1] If there are symplectic Weyl (SW) spinors, then there are two such factors, for the left- and for the right-handed sector.

In summary, the R-symmetry automorphism groups obtained from the analysis above are listed in Table 12.1.

[1] See Appendix B for the notations of the groups.

Table 12.1 *R*-symmetry automorphism groups.

dimension	spinor type	*R*-symmetry
$D = 10$	MW	$\mathrm{SO}(\mathcal{N}_L) \times \mathrm{SO}(\mathcal{N}_R)$
$D = 9$	M and D odd	$\mathrm{SO}(\mathcal{N})$
$D = 8$ and $D = 4$	M and D even	$\mathrm{U}(\mathcal{N})$
$D = 7$ and $D = 5$	S	$\mathrm{USp}(\mathcal{N})$
$D = 6$	SW	$\mathrm{USp}(\mathcal{N}_L) \times \mathrm{USp}(\mathcal{N}_R)$

12.3 Multiplets

12.3.1 Multiplets in four dimensions

The massless particle representations of supersymmetry were listed in Table 6.1. In the discussion that followed we commented that the $s_{\max} = 3/2$ representations do not lead to independent interacting field theories. For field theories we include the following multiplets:

The supergravity multiplets: $s_{\max} = 2$. They are the multiplets that contain the graviton plus $\mathcal{N}$ gravitinos together with the other fields needed to represent the SUSY algebras.

Vector or gauge multiplets: $s_{\max} = 1$. They exist for $\mathcal{N} \leq 4$. The gauge fields of these multiplets can gauge an extra Yang–Mills group that commutes with supercharges[2] as we saw in Sec. 6.3.

Chiral and hypermultiplets: $s_{\max} = 1/2$. For $\mathcal{N} = 1$ these are the chiral multiplets that we know from Sec. 6.2. For $\mathcal{N} = 2$ a similar multiplet is called the 'hypermultiplet'. Such multiplets do not exist for $\mathcal{N} > 2$. These multiplets may also transform under the gauge group defined by the vector multiplet, as we have illustrated in Sec. 6.3.2. They form thus a representation of these gauge groups.

The vector and chiral multiplets are called *matter multiplets*. These exist for global supersymmetry, and they can be coupled to supergravity theories (which is then called 'matter-coupled supergravity') as we will see later in the book.

Notice that the field contents of $\mathcal{N} = 7$ and $\mathcal{N} = 8$ supergravity multiplets are the same. In fact, it turns out that when one constructs a supergravity theory with $\mathcal{N} = 7$, it automatically has an eighth local supersymmetry. That is why it is not mentioned in Table 12.2. Similarly, the vector multiplets of $\mathcal{N} = 3$ and $\mathcal{N} = 4$ have the same field content. If one constructs a global supersymmetric theory with $\mathcal{N} = 3$ in four dimensions, it automatically has a fourth supersymmetry. However, in supergravity one can make three supersymmetries local, and not the fourth one. Thus $\mathcal{N} = 3$ is only meaningful in supergravity. This explains the lowest line of the table.

[2] In supergravity, the commutator may not vanish, as we will see in Sec. 17.3.9.

Table 12.2 Supersymmetry and supergravity theories in dimensions 4 to 11. An entry represents the possibility to have supergravity theories in a specific dimension D with the number of (real) supersymmetries indicated in the top row. We first repeat for every dimension the type of spinors that can be used. Theories with up to 16 (real) supersymmetry generators allow 'matter' multiplets. The possibility of vector multiplets is indicated with ♡. Tensor multiplets in $D = 6$ are indicated by ◇. Multiplets with only scalars and spin-1/2 fields are indicated with ♣. At the bottom is indicated whether these theories exist only in supergravity, or also with just global supersymmetry.

D	SUSY	32	24	20	16	12	8	4
11	M	M						
10	MW	IIA | IIB			I ♡			
9	M	$\mathcal{N}=2$			$\mathcal{N}=1$ ♡			
8	M	$\mathcal{N}=2$			$\mathcal{N}=1$ ♡			
7	S	$\mathcal{N}=4$			$\mathcal{N}=2$ ♡			
6	SW	(2, 2) | (3, 1) | (4, 0)	(2, 1) | (3, 0)		(1, 1) ♡ | (2, 0) ◇		(1, 0) ♡, ◇, ♣	
5	S	$\mathcal{N}=8$	$\mathcal{N}=6$		$\mathcal{N}=4$ ♡		$\mathcal{N}=2$ ♡, ♣	
4	M	$\mathcal{N}=8$	$\mathcal{N}=6$	$\mathcal{N}=5$	$\mathcal{N}=4$ ♡	$\mathcal{N}=3$ ♡	$\mathcal{N}=2$ ♡, ♣	$\mathcal{N}=1$ ♡, ♣
		SG	SG	SG	SG/SUSY	SG	SG/SUSY	SG/SUSY

Note that this classification is made according to the physical spin of the fields and is applicable for massless fields. Particles in these multiplets can sometimes be represented by different fields: scalars may be represented by antisymmetric tensors. Massive multiplets can be obtained by combination of massless ones. This is a supersymmetric version of the BEH (Brout–Englert–Higgs) effect. For example, one can combine an $\mathcal{N} = 1$ gauge multiplet with a chiral multiplet, and one of the two scalars of the chiral multiplet can be eaten by the gauge vector which then becomes massive. Hence we have then a massive multiplet with spin content $(1, 1/2, 1/2, 0)$; see e.g. [66, 67].

12.3.2 Multiplets in more than four dimensions

We now discuss the multiplets in more than four dimensions. As we progress, readers can understand more details of Table 12.2. We restrict ourselves to dimensions $D \geq 4$, to Minkowski spacetimes, and to theories with positive definite kinetic terms. The paper of Strathdee [68] that analyzes the representations of supersymmetries is a useful reference.

Supergravity multiplets with 32 supercharges

We first discuss the supergravity multiplets with the maximal supersymmetry. The 11-dimensional theory [59] is the basis of 'M-theory', and is therefore indicated as M in the table. Let us start now from this one, and dimensionally reduce it on tori. In this way we will get supergravity multiplets in any $D < 11$ with 32 real supersymmetries. The three fields, graviton, gravitino and 3-index gauge field, are representations of the Lorentz group in 11 dimensions, and split in different representations in the lower dimension. See Sec. 10.2 where the reduction to four dimensions was discussed in detail.

When we reduce to 10 dimensions, the supersymmetry of 11 dimensions will split into a left- and right-chiral one (each of 16 components), and thus the multiplet becomes the supergravity multiplet for Type IIA supergravity in 10 dimensions. This is the theory of the massless sector of Type IIA string theory, and that is why we have indicated it as IIA [69, 70, 71].

The other theory with 32 supercharges in 10 dimensions (IIB supergravity) is the one with two supercharges of the same chirality [72, 73, 74]. This cannot be obtained by dimensional reduction from 11 dimensions, as indicated by its place in Table 12.2. It is the massless sector of IIB superstring theory. The theory contains a self-dual 5-form field strength. This makes it difficult to define a covariant action, but the equations of motion are locally supersymmetric. Duality between the IIA and IIB theories can be treated in a 'democratic' formulation, where all the fields are introduced together with their magnetic duals [75]; see also [76] for a recent account.

The analysis of superalgebras in Sec. 12.1.2 leads only to one 32-component supersymmetry in $D = 9$: the $\mathcal{N} = 2$ theory. Therefore both the IIA and the IIB theories of $D = 10$ must reduce to this unique theory in $D = 9$, and indeed that can be explicitly verified by using the techniques of dimensional reduction. The fact that both $D = 10$ supergravities are mapped to the same $D = 9$ theory is a basic ingredient in the understanding of dualities between superstring theories.

In $D = 6$ there are superalgebras of chirality type (2, 2), (3, 1) and (4, 0) which all contain 32 real supercharges. In dimensional reduction, the higher dimensional supercharges split into an equal number of left- and right-handed supersymmetries (as in the $D = 11$ to $D = 10$, IIA reduction). Hence, the dimensional reduction always gives a (2, 2) theory. The (3, 1) and (4, 0) theories exist in principle [77]. However, the graviton field is represented not by a metric tensor $g_{\mu\nu}(x)$, but by a more complicated tensor field [78]. Thus, these theories are different in the sense that they are not based on a dynamical metric tensor. They have not been constructed beyond the linear level.

When there are 32 supersymmetries, the supergravity multiplet for any theory is unique and this multiplet is only known on-shell. There is no known way to add auxiliary fields to obtain off-shell closure.

The supergravity multiplet for less than 32 supersymmetries

There are 24 real-component supersymmetries possible for dimensions up to six. There are two options in the six-dimensional case. In the (2, 1) theory the graviton is described by a metric tensor, but for a (3, 0) theory one needs another representation of the Lorentz group, and there is no nonlinear action known. The reduction to $D = 5$ and $D = 4$ gives the unique $\mathcal{N} = 6$ theories in these dimensions.

The highest spacetime with a superalgebra containing 16 real supercharges is $D = 10$. There is one chiral Majorana spinor. The reduction to lower dimensions is unique. For $D = 6$, the reduction gives the (1, 1) theory, but there is also a (2, 0) supergravity. It contains a self-dual antisymmetric tensor, such that, as in the IIB theory in $D = 10$, the construction of an action is not straightforward.

Supergravities with eight real-component supersymmetries will be discussed in Ch. 20. Finally, with four supersymmetries there is the gravitational multiplet that we discussed in Ch. 9. This multiplet and its coupling to matter are the subjects of Chs. 16 to 19.

Vector multiplets

With 16 supersymmetries or less, global supersymmetry is possible. The highest dimension is $D = 10$ which allows a vector multiplet containing a vector and one Majorana–Weyl spinor. This multiplet can be reduced along the vertical line in Table 12.2. The reduction to $D = 4$ is the vector multiplet, $\mathcal{N} = 4$, $s_{\text{max}} = 1$ in Table 6.1.

For eight supersymmetries, the vector multiplet can be constructed for $D = 6$, 5 and 4. In six dimensions it has just a vector and a symplectic–chiral spinor. Reducing it to five dimensions gives a multiplet that has also a real scalar, and in $\mathcal{N} = 2$ in four dimensions it has a complex scalar, the remnant of the two components of the six-dimensional vector in the compactified directions. Vector multiplets for $\mathcal{N} = 1$ in four dimensions were already treated in Ch. 6.

Tensor multiplets

There exist also multiplets with fields that are antisymmetric tensors $T_{\mu\nu}$. We have seen in Sec. 7.8 that in four dimensions an antisymmetric tensor is equivalent to a scalar, and

Box 12.1 Basic theories, kinetic terms and deformations

Basic theories determine the kinetic terms. For a given $D, \mathcal{N}$ and number of matter multiplets, these kinetic terms are uniquely defined if there are more than eight supercharges. These basic theories can be deformed, e.g. by gauging, and this can produce a potential for the scalars.

thus we can omit them and refer to multiplets with scalars. This is at least true when the antisymmetric tensors have the standard kinetic terms as described in (7.71). We will here restrict attention to this case, but one should be aware that more general theories do exist when antisymmetric tensor fields cannot be dualized.

In five dimensions, an antisymmetric tensor is dual to a vector. Therefore, for the trivial kinetic terms, we can again omit them. However, vectors have to be in adjoint representations, while the antisymmetric tensors can be in other representations, and hence they cannot be neglected for a complete treatment of $D = 5$ supersymmetry.

In six dimensions, however, antisymmetric tensors can have self-dual properties. They are therefore physically different fields from the vectors or scalars. They can be real and can have self-dual field strength. Matter multiplets (multiplets that do not contain the graviton) with antisymmetric tensors are called 'antisymmetric tensor multiplets'. They exist, for example, for minimal supersymmetry (1, 0). The non-chiral (1, 1) supersymmetry does not allow such multiplets, but they do occur for (2, 0) supersymmetry. On the other hand, (2, 0) supersymmetry does not allow vector multiplets.

Hypermultiplets

For $D > 4$, the only multiplets that have only scalars and spinors are the hypermultiplets, which exist for $D = 5, 6$.

12.4 Supergravity theories: towards a catalogue

Earlier in this chapter, we discussed the superalgebras and multiplets from which supergravity theories in spacetime dimensions $4 \leq D \leq 11$ can be constructed. Our discussion led to the entries in Table 12.2. In this section we discuss the various types of actions that can appear. We restrict ourselves to Minkowski signature, and to positive metric kinetic terms with two spacetime derivatives for bosons and one derivative for fermions.

12.4.1 The basic theories and kinetic terms

For every entry in Table 12.2, the field content is determined by specifying the number and type of multiplet that the theory contains. One may wonder whether this information determines the Lagrangian completely. For example, the discussion of Ch. 10 shows that the $D = 11$ supergravity is unique. However, at the other end of the table, the action of chiral multiplets is not unique, since one can choose an arbitrary holomorphic superpotential in (6.18). Even for pure supergravity there is a simple way to obtain the alternative Lagrangian of Sec. 9.6. In these two cases, the different actions have the same kinetic terms. Only mass terms and interactions are changed. This type of modification is called a 'deformation'. One can consider non-abelian gauge theories as deformations of the actions of a system of abelian gauge fields. Theories containing matter fields charged under abelian symmetries are also 'deformations'. In fact, many deformations are of this type and are denoted by the term 'gauged supergravities'. We will discuss such deformations in Sec. 12.4.2.

For the theories so far described, there is a 'basic theory' in which there is no potential for the scalar fields, and no cosmological constant. For the $\mathcal{N} = 1$ supersymmetry theories with auxiliary fields that we discussed in Ch. 6, these auxiliary fields vanish in the basic theory.

With the information in earlier parts of the book, we can write down a general structure for the basic $D = 4$ supergravities. They contain the graviton, represented by the vierbein e_μ^a, a number of vectors A_μ^A with field strengths $F_{\mu\nu}^A$, a number of scalars φ^u, $\mathcal{N}$ gravitinos ψ_μ^i, and a number of spin-1/2 fermions. The pure bosonic terms of such an action are[3]

$$e^{-1}\mathcal{L}_{\text{bos}} = \tfrac{1}{2}R + \tfrac{1}{4}(\text{Im}\,\mathcal{N}_{AB})F_{\mu\nu}^A F^{\mu\nu B} - \tfrac{1}{8}(\text{Re}\,\mathcal{N}_{AB})e^{-1}\varepsilon^{\mu\nu\rho\sigma}F_{\mu\nu}^A F_{\rho\sigma}^B$$
$$-\tfrac{1}{2}g_{uv}(\varphi)\,\partial_\mu\varphi^u\,\partial^\mu\varphi^v\,. \qquad (12.11)$$

The first term gives the pure gravity action (we put $\kappa = 1$ here). Then there are the kinetic terms for the spin-1 fields,[4] which we discussed already in Sec. 4.2. The scalar kinetic terms are determined by a symmetric σ-model metric $g_{uv}(\varphi)$ discussed in Sec. 7.11; see (7.129).

In gauged supergravities, the $F_{\mu\nu}^A$ can be deformed to non-abelian field strengths. Scalars then couple 'minimally' to the vectors with a covariant derivative for the gauge symmetries, and there is an extra potential term $-V(\varphi)$, which appears also for other deformations.

The first question that we pose now is whether the kinetic terms are uniquely determined by the field content. In fact this is true for all theories with more than eight real supercharges. For more than 16 supercharges there are no matter multiplets. In these cases, the kinetic action is unique. For the theories with 16 supercharges, we have the choice of the number of vector multiplets coupled to supergravity (or tensor multiplets when we consider (2, 0) in six dimensions), but once this number is fixed, the kinetic terms of the supergravity theory are determined. Also for the global supersymmetric field theories with these multiplets, the kinetic terms are fixed once the field content is given.

The kinetic terms of theories with eight or four supersymmetries are not fixed by the discrete choice of the number of multiplets. Instead the σ-model can depend on one or more *arbitrary functions*. We will study this for $\mathcal{N} = 1$ global supersymmetry in Ch. 14. We will find that the kinetic terms of chiral multiplets can be generalized such that they depend on an arbitrary function $\mathcal{K}(Z, \bar{Z})$. This determines a geometry on the scalar manifold, called

[3] We use here the notation $\mathcal{N}_{AB}$ for the matrix appearing in the kinetic terms of the vectors, as it is mostly used in extended supersymmetry. This matrix thus replaces $f_{AB} = -\mathrm{i}\mathcal{N}_{AB}$ that we used in Sec. 4.2.

[4] Some theories may have more general terms, called Chern–Simons terms, which have some similarities with those that we saw for $D = 11$ in Sec. 10.3.

Kähler geometry, which we will discuss at length in Ch. 13. For theories with eight supersymmetries the kinetic terms are not fixed by giving the field content. There are arbitrary functions related to choices of 'special geometries', discussed in Ch. 20.

12.4.2 Deformations and gauged supergravities

There are many deformations of the basic supergravity theories. Many of them are 'gauged supergravities', i.e. the deformations are obtained by introducing a gauge group. But we saw already other deformations above (the superpotential and the anti-de Sitter supergravity for $\mathcal{N} = 1$ in $D = 4$). Similarly, a massive theory of $D = 10$ supergravity was already found by Romans [79]. It is also possible to gauge the R-symmetry in extended supergravity [80, 81].

In the context of string theory, gauged supergravities appear when branes carry fluxes. The terminology related to 'gauged supersymmetry/supergravity' can be confusing, so we try to define it clearly here.

Ungauged supersymmetry: means that the super-Poincaré group is not gauged and the only gauge groups permitted are abelian, $\delta A_\mu = \partial_\mu \theta$.

Gauged supersymmetry: vector fields in matter multiplets gauge a Yang–Mills group. Transformations of the non-abelian group can act on fields in other multiplets. Gauged supersymmetry does not mean that the supersymmetry is gauged! In fact, that is only the case for supergravity.

Ungauged supergravity: the super-Poincaré group is gauged, but there are no other gauged symmetries (as for ungauged SUSY).

Gauged supergravity: supergravity in which vector fields gauge a Yang–Mills group, or matter fields are charged under the (non-)abelian gauge group.

The number of generators of the gauge group is equal to the number of vector fields. This count includes vectors in the supergravity multiplet as well as those in gauge multiplets. In general in supergravity the kinetic terms of these vectors are mixed. For example, in $\mathcal{N} = 2$, $D = 4$ there is one vector in the supergravity multiplet and one in each gauge multiplet. We will see in Ch. 20 that the matrices $\mathcal{N}_{AB}$ in (12.11) are non-diagonal between both types of vectors. Therefore, we do not make a distinction between the vectors in the supergravity multiplets and those in gauge multiplets. The gauge group is in principle arbitrary, but the requirement of positive kinetic terms gives restrictions. In supersymmetry this restricts us to compact groups; in supergravity some non-compact gauge groups are possible without spoiling the positivity of the kinetic energies. However, the list of possible non-compact groups is restricted for any $\mathcal{N}$.

Such deformations have various consequences:

1. The supersymmetry transformations of the fermions acquire new terms not present in the undeformed theory. In global supersymmetry these terms arise because auxiliary

fields no longer vanish. These terms in the fermion supersymmetry transformations are sometimes called 'fermion shifts'.

2. A scalar potential is generated, which can be expressed as sums of squares of the fermion shifts.
3. Similarly, there are other new terms in the action such as fermion mass terms. For example, see (6.59), which becomes a mass term when Z gets a non-zero vacuum value.

Gauged supergravities can also be approached from the point of view of the isometry group of the scalar sector. The basic theories lead to the kinetic terms for the scalars, which define a metric. The latter have isometries, according to our discussion in Sec. 7.12. These isometries are global symmetries. It turns out that they nearly always extend to global symmetries of the full supergravity action. In the four-dimensional theories they have an embedding in the symplectic group of dualities (see Sec. 4.2). The gauged supergravity is then obtained by gauging a subgroup of these global symmetries.

This is the essential idea of the 'embedding tensor formalism' [82, 83, 84]. The embedding tensor projects a subset of the global symmetries to a basis of the gauge symmetries, and should satisfy a set of constraints for consistency. A complete catalogue of deformations of the supergravity theories is not yet known. However, we believe that all supersymmetric field theories (with a finite number of fields and field equations that are at most quadratic in derivatives) belong to one of the entries in Table 12.2. The embedding tensor formalism is an attempt to obtain a complete catalogue of supergravities. Another approach starts from an infinite-dimensional symmetry group [85, 86, 87, 88, 89, 90].

12.5 Scalars and geometry

As mentioned, the last term of (12.11) determines the geometry of the scalar manifold. It also appears in higher dimensions. This metric is part of the 'data' that define a particular theory. This is to be contrasted with $g_{\mu\nu}(x)$, the metric of spacetime, which is a dynamical field describing the graviton. This scalar geometry characterizes all the basic supergravities. For $\mathcal{N} > 1,\ D = 4$ (and for $D > 4$) it determines the vector kinetic terms. This is not true for $\mathcal{N} = 1$ where the σ-model metric and f_{AB} (or $\mathcal{N}_{AB}$) are independent.

We learned in Sec. 7.12 that the scalar manifold can have isometries, generated by Killing vectors. In many cases, the manifold itself can be characterized by the isometries. That is the case for 'homogeneous spaces', a concept that we will now explain.

A *homogeneous space* is a manifold in which any point can be reached from any other point by a symmetry operation. In order that the symmetries connect all neighboring points, there must be n independent Killing vectors at any point in a homogeneous space of dimension n. The Poincaré plane is of this type. The manifold is two-dimensional. In Ex. 7.48 we found three Killing vectors, but at any point in the domain of the scalar fields ($\text{Im}\, Z > 0$) only two of them are independent, and there is one generator that leaves the point invariant. The subgroup of the isometry group G that leaves a point invariant is called the isotropy group H. The manifold in this case can be identified with the coset G/H. Hence the

Box 12.2 Symmetric spaces in supergravity

The scalar manifolds of all supergravities with more than eight real-component supersymmetries are symmetric spaces G/H, where G is non-compact and H is its maximal compact subgroup. H is the R-symmetry group in the case of pure supergravities, or has the R-symmetry group as a factor when there are matter multiplets.

Poincaré plane is the manifold SU(1, 1)/U(1). We will see in Sec. 22.1.3 that anti-de Sitter space is a homogeneous space $SO(D-1,2)/SO(D-1,1)$.

Exercise 12.3 *Consider an arbitrary point in the manifold of Ex. 7.48, and find the Killing vector $c^A K_A$ that vanishes. Check that the other two Killing vectors in that point are independent.*

Exercise 12.4 *Why do the isotropy generators define a group? How do you associate the manifold to the coset space?*

In the example above, the structure of the algebras $\mathfrak{g}$ of the group G and $\mathfrak{h}$ of the group H has special properties. We can define a complementary space $\mathfrak{k}$ to $\mathfrak{h}$ such that any $g \in \mathfrak{g}$ can be written as

$$g = h + k, \qquad h \in \mathfrak{h}, \qquad k \in \mathfrak{k}, \tag{12.12}$$

and for any $h_1, h_2 \in \mathfrak{h}$ and $k_1, k_2 \in \mathfrak{k}$

$$[h_1, h_2] \in \mathfrak{h}, \qquad [h_1, k_1] \in \mathfrak{k}, \qquad [k_1, k_2] \in \mathfrak{h}. \tag{12.13}$$

When we use the Cartan–Killing metric in the algebra $\mathfrak{g}$, the decomposition is orthogonal. Any simply connected homogeneous space for which the isomorphism and isotropy algebra have this structure is called a *symmetric space*. Such manifolds are also characterized by the fact that their curvature tensor is covariantly constant.

Exercise 12.5 *Check that the Poincaré plane is a symmetric space.*

Exercise 12.6 *Which of the equations (12.13) are specific for a symmetric space, and how are these for a general homogeneous space?*

It turns out that symmetric spaces are very important in supergravity; see Box 12.2. See Table 12.3 where all the geometries for supergravities with more than eight real supercharges are displayed.

The scalar geometries of $\mathcal{N}=1$ theories in four dimensions are all complex manifolds with a (closed) complex structure, i.e. Kähler manifolds. We will study these in Ch. 13, and explain how they occur in $\mathcal{N}=1$ global supersymmetry in Ch. 14. In supergravity, these Kähler manifolds have an extra property that will be explained in Sec. 17A and are then called Kähler–Hodge manifolds. They can be homogeneous or symmetric (with a U(1) factor in the isotropy group), but this is not required.

The scalar manifolds of theories with eight real supersymmetries ($\mathcal{N} = 2$) belong to a class that was baptized *special geometries* [91, 92]. This beautiful class of geometries contains real [93], Kähler [94] and quaternionic manifolds [95]. Those that appear in couplings of $D = 5$ vector multiplets are called 'very special real manifolds'. Those appearing in couplings of $D = 4$ vector manifolds are the 'special Kähler manifolds'. They are a subclass of the Kähler manifolds that have a structure related to the duality transformations of the vectors discussed in Sec. 4.2. The scalar manifolds of $\mathcal{N} = 2$ theories are always the direct product of those of the scalars of hypermultiplets, and those of tensor multiplets ($D = 6$), or tensor/vector multiplets ($D = 5$) or vector multiplets ($D = 4$). This leads to

$$
\begin{aligned}
D = 6&: \quad \frac{\mathrm{O}(1, n)}{\mathrm{O}(n)} \times \text{quaternionic-Kähler manifold}, \\
D = 5&: \quad \text{very special real manifold} \times \text{quaternionic-Kähler manifold}, \\
D = 4&: \quad \text{special Kähler manifold} \times \text{quaternionic-Kähler manifold}.
\end{aligned} \tag{12.14}
$$

There exist also versions of these geometries for global supersymmetry, leading to 'rigid Kähler manifolds' [96, 97] and hyper-Kähler manifolds. We will discuss all these special geometries in Ch. 20.

12.6 Solutions and preserved supersymmetries

Given the action of a supergravity theory, it is generally useful to search for solutions of the classical equations of motion. It is most useful to obtain solutions that can be interpreted as backgrounds or vacua. Fluctuations above the background are then treated quantum mechanically. The backgrounds that are considered have vanishing values of fermions, and are thus determined by a value of the metric, the vector fields (or higher forms) and scalar fields. One common background is Minkowski space, but there are others such as anti-de Sitter space, certain black holes, cosmic strings, branes or pp-waves, which are all supersymmetric, i.e. they 'preserve some supersymmetry'. This means that the background is invariant under a subset of the local supersymmetries of the supergravity theory. For a preserved supersymmetry, the local SUSY variations of all fields must vanish when the background solution is substituted. This leads to conditions of the generic form

$$
\delta(\epsilon)\,\text{boson} = \epsilon\,\text{fermion} = 0\,, \qquad \delta(\epsilon)\,\text{fermion} = \epsilon\,\text{boson} = 0\,. \tag{12.15}
$$

Since the fermions vanish in these backgrounds, the right-hand side of the first condition is always satisfied, and the relevant condition is the second one. For a given background this equation restricts the form of $\epsilon(x)$. Thus if the background is flat Minkowski space, the supersymmetry transformation of the gravitino implies that the $\epsilon(x)$ are restricted to be constant. In general the solution $\epsilon(x)$ can be expressed in terms of a number of constant parameters. If the original theory has Q local real-component supersymmetries (e.g. $Q = 8$ for $\mathcal{N} = 2$ in four dimensions), and the vanishing of (12.15) can be expressed in terms of Q_0 constant parameters, the theory is said to preserve a fraction Q_0/Q supersymmetry.

Table 12.3 Scalar geometries in theories with more than eight supersymmetries (and dimension ≥ 4). The theories are ordered as in Table 12.2. Note that the R-symmetry group, mentioned in Table 12.1, is always a factor in the isotropy group. For more than 16 supersymmetries, there is only a unique supergravity (up to gaugings irrelevant to the geometry), while for 16 and 12 supersymmetries there is a number n indicating the number of vector multiplets in the theory.

D	32	24	20	16	12
10	$O(1,1)$; $\frac{SU(1,1)}{U(1)}$				
9	$\frac{SL(2)}{SO(2)} \otimes O(1,1)$			$\frac{O(1,n)}{O(n)} \otimes O(1,1)$	
8	$\frac{SL(3)}{SU(2)} \otimes \frac{SL(2)}{U(1)}$			$\frac{O(2,n)}{U(1) \times O(n)} \otimes O(1,1)$	
7	$\frac{SL(5)}{USp(4)}$			$\frac{O(3,n)}{USp(2) \times O(n)} \otimes O(1,1)$	
6	$\frac{O(5,5)}{USp(4) \times USp(4)}$; $\frac{F_{4,4}}{USp(6) \times USp(2)}$; $\frac{E_{6,6}}{USp(8)}$	$\frac{SU^*(4)}{USp(4)}$; $\frac{SU^*(6)}{USp(6)}$		$\frac{O(4,n)}{O(n) \times SO(4)} \otimes O(1,1)$; $\frac{O(5,n)}{O(n) \times USp(4)}$	
5	$\frac{E_{6,6}}{USp(8)}$	$\frac{SU^*(6)}{USp(6)}$		$\frac{O(5,n)}{USp(4) \times O(n)} \otimes O(1,1)$	
4	$\frac{E_{7,7}}{SU(8)}$	$\frac{SO^*(12)}{U(6)}$	$\frac{SU(1,5)}{U(5)}$	$\frac{SU(1,1)}{U(1)} \times \frac{SO(6,n)}{SU(4) \times SO(n)}$	$\frac{SU(3,n)}{U(3) \times SU(n)}$

Preserved supersymmetries Box 12.3

Preserved supersymmetries of a solution of the theory are determined by the vanishing of the supersymmetry transformations of fermions. These are global supersymmetries that are a subset of the local supersymmetries of the supergravity action.

This terminology is confusing, since the original symmetries are local and the preserved supersymmetries are global.

The algebra of the preserved supersymmetry is in many cases (except Minkowski flat space solutions) different from (12.1). In fact, for any massive object it must be different. Indeed, this minimal algebra can be written as in (6.68). When this is applied to a solution that has some supersymmetry (vanishes under the action of some Q) then this algebra implies that the solution has zero energy.

Below we give some examples of such different superalgebras that can be obtained as algebras of symmetries preserved by solutions. Several examples of backgrounds with preserved supersymmetries are discussed in Sec. 22.2.

12.6.1 Anti-de Sitter superalgebras

(Anti-)de Sitter algebras are characterized by the fact that the translations commute to Lorentz transformations

$$\left[P_\mu, P_\nu\right] = \mp\frac{1}{4L^2}M_{\mu\nu}, \tag{12.16}$$

where the upper and lower signs are for de Sitter and anti-de Sitter space, respectively, and L is a length scale. The Lie algebras are $SO(D,1)$ for de Sitter and $SO(D-1,2)$ for anti-de Sitter.

Only the anti-de Sitter algebra can be embedded in a supersymmetric algebra. In the simplest case there is one Majorana generator Q_α, and the superalgebra has the structure

$$\begin{aligned}\left[P_\mu, Q_\alpha\right] &= \frac{1}{4L}(\gamma_\mu Q)_\alpha, \\ \left\{Q_\alpha, Q_\beta\right\} &= -\frac{1}{2}(\gamma^\mu)_{\alpha\beta}P_\mu - \frac{1}{8L}(\gamma^{\mu\nu})_{\alpha\beta}M_{\mu\nu}.\end{aligned} \tag{12.17}$$

These (anti-)commutators satisfy the super-Jacobi identities if $D = 4$. The Lie algebras of $SO(3,2)$ and $Sp(4)$ are isomorphic (see (B.7)), and the superalgebra is called $OSp(1|4)$. For $\mathcal{N}$-extended algebras, the generators of the $SO(\mathcal{N})$ R-symmetry group are also present on the right-hand side of the anti-commutator. The anti-de Sitter superalgebras are then $OSp(\mathcal{N}|4)$. For $D = 5, 6$ and 7 the superalgebras are, respectively, $SU(2,2|\mathcal{N})$, $F^2(4)$ and $OSp(8^*|\mathcal{N})$. There are no simple superalgebras whose bosonic subalgebra is the sum of a de Sitter algebra and a compact R-symmetry group.

Exercise 12.7 *Check that the $[P_\mu, P_\nu, Q]$ Jacobi identity fixes the constant in the first line of (12.17) to $1/(4L)$, and that the identity cannot be satisfied if we use the*

Box 12.4 **(Anti-)de Sitter superalgebras**

For dimensions $D \leq 7$ there are simple superalgebras with bosonic subalgebra a direct sum of an anti-de Sitter algebra and a compact R-symmetry group. Such super-de Sitter algebras do not exist.

upper sign in (12.16). This shows already that only the anti-de Sitter algebra can be supersymmetrized.

12.6.2 Central charges in four dimensions

We now consider the superalgebras with *central charges*. Central charges appeared first in the classical work of Haag, Łopuszański, and Sohnius [31]. The simplest example occurs for $\mathcal{N} = 2$, where we can insert a new operator $\mathcal{Z}$ in the anti-commutator of two supercharges of the same chirality:

$$\left\{Q_{\alpha i}, Q_{\beta j}\right\} = -\tfrac{1}{2}\varepsilon_{ij} P_{L\alpha\beta}\mathcal{Z}\,, \qquad \left\{Q_{\alpha}{}^{i}, Q_{\beta}{}^{j}\right\} = -\tfrac{1}{2}\varepsilon^{ij} P_{R\alpha\beta}\bar{\mathcal{Z}}\,. \tag{12.18}$$

The matrices P_L are antisymmetric in the spinor indices, and thus one needs an antisymmetric ε^{ij} for consistency of the superalgebra. The other relations in (12.2) are not changed.

Exercise 12.8 *Check that these anti-commutator relations are consistent with symmetry in $(i\alpha) \leftrightarrow (j\beta)$ and with chirality projections.*

The generator $\mathcal{Z}$ and its complex conjugate $\bar{\mathcal{Z}}$ commute with all other generators in the superalgebra, so they are really 'central' in the mathematical sense. Owing to the presence of ε^{ij}, such an algebra is only possible for extended supersymmetry. In the case of $\mathcal{N} = 4$ one can have six independent complex central charges (the number of independent antisymmetric 4×4 matrices).

Physical states of the theory can be chosen to be eigenstates of the central charges. Indeed, we will see in Ch. 22 that for black hole solutions the central charge can be expressed in terms of electric and magnetic charges. In Ex. 6.1 we saw how to prove that the energy in supersymmetric theories is positive. We will now use the same technique to derive a bound on the value of the central charge. To do so, we first convert (12.18) to quantum anti-commutators, as we did in (6.87) This gives[5]

$$\left\{Q_{i\alpha}, Q_j{}^{\beta}\right\}_{\text{qu}} = -\tfrac{1}{2}\mathrm{i}\varepsilon_{ij} P_{L\alpha}{}^{\beta}\mathcal{Z}\,, \qquad \left\{Q^{\dagger i}{}_{\alpha}, Q^{\dagger j\beta}\right\}_{\text{qu}} = \tfrac{1}{2}\mathrm{i}\varepsilon^{ij}(P_L)_{\alpha}{}^{\beta}\bar{\mathcal{Z}}\,, \tag{12.19}$$

where we raised and lowered spinor indices using the techniques of Sec. 3.2.2. Note that $Q^{\dagger j}{}_{\alpha} = Q^{\dagger j\beta}\mathcal{C}_{\beta\alpha} = -(\mathcal{C}^{\alpha\beta}Q_{j\beta})^{\dagger} = -(Q_j{}^{\alpha})^{\dagger}$. Define the operators (for an arbitrary phase θ)

$$A_{i\alpha} \equiv Q_{i\alpha} + \mathrm{i}e^{\mathrm{i}\theta}\varepsilon_{ij}Q^{\dagger j}{}_{\alpha}\,, \qquad A^{\dagger i\alpha} \equiv (A_{i\alpha})^{\dagger} = Q^{\dagger i\alpha} + \mathrm{i}e^{-\mathrm{i}\theta}\varepsilon^{ij}Q_j{}^{\alpha}\,. \tag{12.20}$$

[5] For $\mathcal{N} > 2$ we assume that ε^{ij} is the complex conjugate of ε_{ij} and that $\varepsilon^{ik}\varepsilon_{jk} = \delta^i_j$.

Using (6.67) and (12.19) we obtain, using a sum over spinor indices,

$$\left\{A_{i\alpha}, A^{\dagger j\alpha}\right\}_{\text{qu}} = \delta_i^j \left(2P^0 + \mathrm{e}^{-\mathrm{i}\theta}\mathcal{Z} + \mathrm{e}^{\mathrm{i}\theta}\bar{\mathcal{Z}}\right). \tag{12.21}$$

Therefore, the quantity in brackets should be positive for all θ, which implies that $P^0 \geq |\mathcal{Z}|$. This implies [98] that

$$M \geq |\mathcal{Z}|. \tag{12.22}$$

This is the Bogomol'nyi–Prasad–Sommerfield (BPS) bound, which is a very important result for solutions of supergravity theories. We will encounter it again in Ch. 22. Solutions that saturate this bound, i.e. $M = |\mathcal{Z}|$, are called BPS solutions. It follows from this argument that supersymmetric solutions are BPS solutions. The bound (12.22) implies that central charges are necessary to allow massive supersymmetric solutions.

When central charges are present, the commutator of two SUSY transformations (see (11.17)) can be written as

$$[\delta(\epsilon_1), \delta(\epsilon_2)] = \tfrac{1}{2}\left(\bar{\epsilon}_{2i}\gamma^a\epsilon_1^i + \bar{\epsilon}_2^i\gamma^a\epsilon_{1i}\right)P_a + \tfrac{1}{2}\bar{\epsilon}_2^i\epsilon_1^j\varepsilon_{ij}\mathcal{Z} + \tfrac{1}{2}\bar{\epsilon}_{2i}\epsilon_{1j}\varepsilon^{ij}\bar{\mathcal{Z}}. \tag{12.23}$$

12.6.3 'Central charges' in higher dimensions

The name 'central charges' has been generalized to include other generators that can appear in the anti-commutator of supersymmetries, much as in (12.18), but they may not be Lorentz scalars. For example, in $D = 11$ the properties of the spinors allow us to extend the anti-commutator as [99]

$$\left\{Q_\alpha, Q_\beta\right\} = \gamma^\mu_{\alpha\beta}P_\mu + \gamma^{\mu\nu}_{\alpha\beta}\mathcal{Z}_{\mu\nu} + \gamma^{\mu_1\cdots\mu_5}_{\alpha\beta}\mathcal{Z}_{\mu_1\cdots\mu_5}. \tag{12.24}$$

The allowed structures on the right-hand side are determined by the last entry in Table 3.2 (remember that the entries there are valid modulo 4, which thus allows the 5-index object). The 'central charges' $\mathcal{Z}$ are no longer Lorentz scalars, and thus do not commute with the Lorentz generators. They are therefore not 'central' in the group theoretical meaning of the word, but their physical interpretation is the same as those in (12.18). They also allow the construction of supersymmetric solutions by the mechanism mentioned in Sec. 12.6.2, so they are still called 'central charges'. The second and third terms in (12.24) are in this way important for the existence of $M2$- and $M5$-branes of $D = 11$ supergravity. In general, branes can be classified by considering the possibilities for central charges.

12C.3 Central charges in higher dimensions

PART IV

COMPLEX GEOMETRY AND GLOBAL SUSY

13 Complex manifolds

The material of this chapter is largely mathematical, although we will include several physically motivated examples of the geometric constructions. The major prerequisite is Ch. 7 on the differential geometry of real manifolds. A complex manifold is a real manifold of even dimension $2n$ on which one can choose n complex coordinates z^α in a smooth fashion. More rigorously there is a cover of M_{2n} by open sets U_I. On each U_I there is a $1:1$ continuous map $\psi_I(p) = (z^1, z^2, \ldots, z^n)$, where $z^\alpha \in \mathbb{C}$. On intersections the compound maps $\psi_J \circ \psi_I^{-1}$ are analytic. It is not always possible to define such a complex structure on a real M_{2n}. There are important topological restrictions which, in the main, are beyond the scope of this book. (For interested readers we suggest the references [100] and [101].) Locally it is always possible to introduce complex coordinates z^α by combining real coordinates ϕ^i, and this is the approach we take in our initial technical discussion below.

Complex manifolds are included in this book for a very simple reason. The scalar fields of supersymmetric theories in four spacetime dimensions are a set of complex fields z^α which can be viewed as coordinates of an important type of complex manifold known as a Kähler manifold. We will discuss Kähler manifolds extensively in this chapter. They appear in a natural way in both global supersymmetry and supergravity.

13.1 The local description of complex and Kähler manifolds

For local considerations,[1] one can view an n-dimensional complex manifold as a $2n$-dimensional real manifold parametrized by n complex coordinates. To obtain complex coordinates one can start with a real coordinate set $\phi^1, \ldots, \phi^n, \phi^{n+1}, \ldots, \phi^{2n}$ and define

$$\begin{aligned} z^\alpha &= \phi^\alpha + \mathrm{i}\phi^{\alpha+n}, \qquad \alpha = 1, 2, \ldots, n, \\ z^{\bar\alpha} &= \phi^\alpha - \mathrm{i}\phi^{\alpha+n} = \bar z^\alpha. \end{aligned} \tag{13.1}$$

We then take z^a to be the set of $2n$ complex coordinates where the index a runs first through the n unbarred or 'holomorphic' coordinates and then through the n barred or 'anti-holomorphic' coordinates. We consider the map $\phi^i \to z^a$ defined in (13.1) as a coordinate transformation of the type normally considered in differential geometry, and we use the standard transformation formulas of Ch. 7. Complexity causes no difficulty.

[1] We follow the pedagogical treatment of [102].

We assume that M possesses a Euclidean signature metric structure, with Christoffel connection, and curvature tensor, all as defined in Ch. 7. For convenience we record the relevant formulas again:

$$ds^2 = g_{ij}\mathrm{d}\phi^i \mathrm{d}\phi^j\,, \tag{13.2}$$

$$\Gamma^i_{jk} = \tfrac{1}{2}g^{im}(\partial_j g_{km} + \partial_k g_{jm} - \partial_m g_{jk})\,, \tag{13.3}$$

$$R_{ij}{}^k{}_l = \partial_i \Gamma^k_{jl} - \partial_j \Gamma^k_{il} + \Gamma^k_{im}\Gamma^m_{jl} - \Gamma^k_{jm}\Gamma^m_{il}\,. \tag{13.4}$$

The transformation rules of vector fields under a real coordinate change $\phi^i \to \phi'^i$ are

$$V'^i(\phi') = \frac{\partial \phi'^i}{\partial \phi^j} V^j(\phi)\,,$$
$$V'_i(\phi') = \frac{\partial \phi^j}{\partial \phi'^i} V_j(\phi)\,. \tag{13.5}$$

Covariant derivatives are defined (see (7.78)) by

$$\nabla_j V^i = \partial_j V^i + \Gamma^i_{jk} V^k\,,$$
$$\nabla_j V_i = \partial_j V_i - \Gamma^k_{ji} V_k\,. \tag{13.6}$$

The transformation rules and covariant derivatives extend to higher rank tensors as discussed in Ch. 7.

All quantities above can be expressed in complex coordinates by applying the transformation formulas to the map $\phi^i \to z^a$. For example,

$$V^i(\phi) \to \tilde{V}^a(z) = \frac{\partial z^a}{\partial \phi_j} V^j(\phi)\,. \tag{13.7}$$

We then separate the unbarred and barred components:

$$\tilde{V}^\alpha = V^\alpha + \mathrm{i}V^{\alpha+n}\,, \qquad \tilde{V}^{\bar{\alpha}} = V^\alpha - \mathrm{i}V^{\alpha+n}\,, \qquad \alpha = 1, \dots, n\,. \tag{13.8}$$

The 'splitting' of an index a into α and $\bar{\alpha}$ is not preserved by general transformation of complex coordinates $z'^\alpha = f^\alpha(z, \bar{z})$, but it is preserved under the special class of 'holomorphic' coordinate transformations $z'^\alpha = f^\alpha(z)$. Under this subgroup of diffeomorphisms the holomorphic indices α of any tensor transform into holomorphic indices α' and anti-holomorphic $\bar{\alpha}$ into $\bar{\alpha}'$.

The Riemannian metric is expressed in complex coordinates as

$$ds^2 = g_{ab}\mathrm{d}z^a \mathrm{d}z^b = g_{ij}\frac{\partial \phi^i}{\partial z^a}\frac{\partial \phi^j}{\partial z^b}\mathrm{d}z^a \mathrm{d}z^b\,. \tag{13.9}$$

The complex metric tensor $g_{ab} = g_{ba}$ is obtained from g_{ij} as indicated. The general form of the line element is

$$ds^2 = 2g_{\alpha\bar{\beta}}\mathrm{d}z^\alpha \mathrm{d}\bar{z}^{\bar{\beta}} + g_{\alpha\beta}\mathrm{d}z^\alpha \mathrm{d}z^\beta + g_{\bar{\alpha}\bar{\beta}}\mathrm{d}\bar{z}^{\bar{\alpha}} \mathrm{d}\bar{z}^{\bar{\beta}}\,. \tag{13.10}$$

This form is real. This is ensured by the transformation in (13.9) which implies that $g_{\alpha\bar{\beta}} = \bar{g}_{\beta\bar{\alpha}}$ and $g_{\alpha\beta} = \bar{g}_{\bar{\alpha}\bar{\beta}}$. Covariant derivatives are written as

$$\nabla_b \tilde{V}^a = \partial_b \tilde{V}^a + \Gamma^a_{bc} \tilde{V}^c \,,$$
$$\nabla_b \tilde{V}_a = \partial_b V_a - \Gamma^c_{ba} \tilde{V}_c \,. \tag{13.11}$$

The connection Γ^a_{bc} is just the Christoffel connection (13.3) expressed in terms of g_{ab}:

$$\Gamma^a_{bc} = \tfrac{1}{2} g^{ad} (\partial_b g_{cd} + \partial_c g_{bd} - \partial_d g_{bc}) \,. \tag{13.12}$$

Note that $g^{ad} g_{bd} = \delta^a_b$.

We now define two conditions on the metric g_{ab} which are preserved by holomorphic coordinate transformations. The metric is said to be hermitian if there are choices of coordinates in which $g_{\alpha\beta} = g_{\bar{\alpha}\bar{\beta}} = 0$. The line element then takes the hermitian form

$$\mathrm{d}s^2 = 2 g_{\alpha\bar{\beta}} \mathrm{d}z^\alpha \mathrm{d}\bar{z}^{\bar{\beta}} \,. \tag{13.13}$$

Coordinate systems in which this form holds are said to be adapted to the hermitian structure. The hermitian form is a restriction on the metric. It is not possible to transform the general complex form (13.9) to hermitian form.

Exercise 13.1 *Formulate the equations for a coordinate transformation $z'^a = f^a(z, \bar{z})$ from coordinates in which the line element takes the general form in (13.9) to coordinates in which $g'_{\alpha\beta} = 0$. Show that in general there are too many equations to possess a solution.*

As is customary, we write the holomorphic index of $g_{\alpha\bar{\beta}}$ or its inverse $g^{\alpha\bar{\beta}}$ on the left. Thus $g^{\alpha\bar{\gamma}} g_{\beta\bar{\gamma}} = \delta^\alpha_\beta$.

Given a hermitian metric, we can define the fundamental 2-form:

$$\Omega = -2\mathrm{i} g_{\alpha\bar{\beta}} \mathrm{d}z^\alpha \wedge \mathrm{d}\bar{z}^{\bar{\beta}} \,. \tag{13.14}$$

Exercise 13.2 *Show that Ω is a real 2-form, i.e. $\Omega = \bar{\Omega}$.*

A manifold with hermitian metric is a Kähler manifold if its fundamental form (then called the Kähler form) is closed, i.e. $\mathrm{d}\Omega = 0$. The exterior derivative is given by

$$\mathrm{d}\Omega = -\mathrm{i}(\partial_\gamma g_{\alpha\bar{\beta}} - \partial_\alpha g_{\gamma\bar{\beta}}) \mathrm{d}z^\gamma \wedge \mathrm{d}z^\alpha \wedge \mathrm{d}\bar{z}^{\bar{\beta}} + \text{c.c.} \,, \tag{13.15}$$

so that the necessary and sufficient condition for a Kähler manifold is

$$\partial_\gamma g_{\alpha\bar{\beta}} - \partial_\alpha g_{\gamma\bar{\beta}} = 0 \,. \tag{13.16}$$

This condition implies that locally (i.e. in each coordinate patch) the metric can be represented as

Box 13.1 **Kähler manifold**

Definition 1: A Kähler manifold is a complex manifold with hermitian metric whose fundamental 2-form is closed. Locally the metric is then determined by a Kähler potential. (See below.)
Definition 2: A Kähler manifold is a manifold with covariantly constant complex structure and hermitian metric. (See Sec. 13.2.)

$$g_{\alpha\bar\beta} = \partial_\alpha \partial_{\bar\beta} K(z,\bar z)\,. \tag{13.17}$$

The real function $K(z,\bar z)$ is called the Kähler potential. It is not uniquely determined since the change

$$K(z,\bar z) \to K'(z,\bar z) = K(z,\bar z) + f(z) + \bar f(\bar z) \tag{13.18}$$

leaves $g_{\alpha\bar\beta}$ invariant. On a topologically non-trivial Kähler manifold, it is frequently the case that there is no globally defined potential. Instead the potentials $K(z,\bar z)$ and $K'(z',\bar z)$ on overlapping coordinate charts are related by a transformation similar to (13.18). These transformations are called Kähler transformations.

The hermitian and Kähler conditions on the metric lead to simplifications of the connection Γ^a_{bc}. Hermiticity implies that the connection components $\Gamma^\alpha_{\bar\beta\bar\gamma}$ (and conjugates $\Gamma^{\bar\alpha}_{\beta\gamma}$) vanish. For a Kähler metric, there are the additional conditions $\Gamma^\alpha_{\beta\bar\gamma} = \Gamma^{\bar\alpha}_{\bar\beta\gamma} = 0$, so the only non-vanishing connection components are those of the form

$$\Gamma^\alpha_{\beta\gamma} = g^{\alpha\bar\delta}\partial_\beta g_{\gamma\bar\delta}\,, \qquad \Gamma^{\bar\alpha}_{\bar\beta\bar\gamma} = g^{\delta\bar\alpha}\partial_{\bar\beta} g_{\delta\bar\gamma}\,. \tag{13.19}$$

These formulas are much simpler than the conventional real (13.3) and general complex (13.12) forms.

Exercise 13.3 *Take the trace of the first relation in (13.19) and derive*

$$\Gamma^\gamma_{\alpha\gamma} = \partial_\alpha \log\det g_{\beta\bar\delta}\,, \tag{13.20}$$

which is very similar to its analogue for Riemannian manifolds.

The curvature tensor on a complex manifold is written as R_{abcd}. It is defined by applying the standard formula (13.4) in complex coordinates z^a with Γ^a_{bc}, and it has the usual index symmetries. On a Kähler manifold, there are important simplifications because many connection components vanish. The non-vanishing curvature components are

$$\begin{aligned} &R_{\gamma\bar\delta}{}^\alpha{}_\beta = -R_{\bar\delta\gamma}{}^\alpha{}_\beta\,, \qquad R_{\gamma\bar\delta}{}^{\bar\alpha}{}_{\bar\beta} = -R_{\bar\delta\gamma}{}^{\bar\alpha}{}_{\bar\beta}\,,\\ &R_{\alpha\bar\beta\gamma\bar\delta} = -R_{\bar\beta\alpha\gamma\bar\delta} = R_{\gamma\bar\delta\alpha\bar\beta}\,, \end{aligned} \tag{13.21}$$

and these are given by the simple formulas

$$R_{\gamma\bar{\delta}}{}^{\alpha}{}_{\beta} = -\partial_{\bar{\delta}}\Gamma^{\alpha}_{\beta\gamma}\,,$$
$$R_{\alpha\bar{\beta}\gamma\bar{\delta}} = \partial_{\gamma}\partial_{\bar{\delta}}g_{\alpha\bar{\beta}} - g^{\eta\bar{\epsilon}}\partial_{\gamma}g_{\alpha\bar{\epsilon}}\partial_{\bar{\delta}}g_{\eta\bar{\beta}}\,. \tag{13.22}$$

Exercise 13.4 *Prove the symmetry properties $R_{\alpha\bar{\beta}\gamma\bar{\delta}} = R_{\gamma\bar{\beta}\alpha\bar{\delta}}$ and $R_{\gamma\bar{\delta}}{}^{\alpha}{}_{\beta} = R_{\beta\bar{\delta}}{}^{\alpha}{}_{\gamma}$. Show that the form $R_{\alpha}{}^{\gamma}{}_{\beta}{}^{\delta}$, with two pairs of holomorphic indices, is symmetric in $(\alpha\beta)$ and in $(\gamma\delta)$.*

There are simplifications in the Bianchi identities (7.122) satisfied by the curvature tensor. The forms given for Riemannian geometry in (7.122) are also valid in complex coordinates, but some terms vanish due to the restrictions on curvature components for a Kähler manifold. Note that there is a form of the curvature tensor with two pairs of symmetrized holomorphic indices. The symmetry properties and Bianchi identities of the curvature tensor are summarized as follows:

$$R_{\alpha\bar{\beta}\gamma\bar{\delta}} = R_{\gamma\bar{\beta}\alpha\bar{\delta}} = g_{\epsilon\bar{\beta}}g_{\zeta\bar{\delta}}R_{\alpha}{}^{\epsilon}{}_{\gamma}{}^{\zeta}\,, \qquad R_{\alpha}{}^{\epsilon}{}_{\gamma}{}^{\zeta} = R_{\gamma}{}^{\epsilon}{}_{\alpha}{}^{\zeta} = R_{\alpha}{}^{\zeta}{}_{\gamma}{}^{\epsilon}\,,$$
$$\nabla_{\epsilon}R_{\alpha\bar{\beta}\gamma\bar{\delta}} = \nabla_{\gamma}R_{\alpha\bar{\beta}\epsilon\bar{\delta}}\,. \tag{13.23}$$

Finally, note the Ricci identities (see (7.124) for vanishing torsion):

$$[\nabla_{\alpha},\nabla_{\beta}]V^{\gamma} = 0\,, \qquad [\nabla_{\alpha},\nabla_{\beta}]V^{\bar{\gamma}} = 0,$$
$$\left[\nabla_{\alpha},\overline{\nabla}_{\bar{\beta}}\right]V^{\gamma} = R_{\alpha\bar{\beta}}{}^{\gamma}{}_{\delta}V^{\delta}\,, \qquad [\nabla_{\alpha},\overline{\nabla}_{\bar{\beta}}]V^{\bar{\gamma}} = R_{\alpha\bar{\beta}}{}^{\bar{\gamma}}{}_{\bar{\delta}}V^{\bar{\delta}}\,,$$
$$\left[\nabla_{\alpha},\overline{\nabla}_{\bar{\beta}}\right]V_{\gamma} = R_{\alpha\bar{\beta}\gamma}{}^{\delta}V_{\delta}\,, \qquad [\nabla_{\alpha},\overline{\nabla}_{\bar{\beta}}]V_{\bar{\gamma}} = R_{\alpha\bar{\beta}\bar{\gamma}}{}^{\bar{\delta}}V_{\bar{\delta}}\,. \tag{13.24}$$

The Ricci tensor of a Kähler metric is defined in the usual way, namely as $R_{ab} = g^{cd}R_{acbd} = R_{ba}$. It can then be shown that the components $R_{\alpha\beta}$ and $R_{\bar{\alpha}\bar{\beta}}$ vanish and that mixed components can be written in the form

$$R_{\alpha\bar{\beta}} = g^{\bar{\gamma}\gamma}R_{\alpha\bar{\gamma}\bar{\beta}\gamma} = -R_{\alpha\bar{\beta}\gamma}{}^{\gamma} = -\partial_{\alpha}\partial_{\bar{\beta}}(\log\det g_{\gamma\bar{\delta}})\,. \tag{13.25}$$

Exercise 13.5 *Derive these properties of the Ricci tensor.*

13.2 Mathematical structure of Kähler manifolds

Most mathematical treatments of complex differential geometry (for example, Ch. IX of [100]) start with the definition of an almost complex structure. This is a (real-valued) tensor $J_i{}^j(\phi)$ on the tangent space of a real manifold M_d with the property $J_i{}^k J_k{}^j = -\delta_i{}^j$. The analogy with $\mathrm{i}\cdot\mathrm{i} = -1$ is obvious.

Exercise 13.6 *Show that this property can be satisfied only if the dimension is even, $d = 2n$, and that the eigenvalues of J are $\pm\mathrm{i}$.*

The almost complex structure maps the tangent space $T(M_{2n})$ onto itself, acting on basis vectors as

$$J\frac{\partial}{\partial\phi^i} \equiv J_i{}^j\frac{\partial}{\partial\phi^j}\,. \tag{13.26}$$

The almost complex structure allows one to express the conditions that define a Kähler metric directly in terms of real coordinates on M_{2n}. This can be useful in physical applications. We summarize the mathematical treatment in this section using both real and complex coordinates.

One can show [100, 103] that a real manifold M_{2n} equipped with an almost complex structure is a complex manifold (of dimension n) if and only if the Nijenhuis tensor vanishes:

$$N_{ij}{}^k \equiv J_i{}^\ell(\partial_\ell J_j{}^k - \partial_j J_\ell{}^k) - J_j{}^\ell(\partial_\ell J_i{}^k - \partial_i J_\ell{}^k) = 0\,. \tag{13.27}$$

Exercise 13.7 *Show that $N_{ij}{}^k$ transforms as a tensor under diffeomorphisms despite the absence of covariant derivatives.*

The almost complex structure is then called the complex structure of M_{2n}. There is then a covering of M_{2n} by coordinate charts U_I with complex coordinates z^α, $\bar{z}^{\bar\alpha} = \overline{z^\alpha}$ as used in Sec. 13.1. In the coordinate basis $\partial/\partial z^\alpha$, $\partial/\partial\bar{z}^{\bar\alpha}$ the almost complex structure takes the form

$$J = \begin{pmatrix} \mathrm{i}\delta_\alpha{}^\beta & 0 \\ 0 & -\mathrm{i}\delta_{\bar\alpha}{}^{\bar\beta} \end{pmatrix}. \tag{13.28}$$

We now suppose that M_{2n} has a torsion-free (i.e. symmetric) connection Γ^k_{ij} and that $J_i{}^j$ is covariantly constant, i.e.

$$\nabla_k J_i{}^j = \partial_k J_i{}^j - \Gamma^\ell_{ki}J_\ell{}^j + \Gamma^j_{k\ell}J_i{}^\ell = 0\,. \tag{13.29}$$

This is sufficient to ensure that $N_{ij}{}^k = 0$.

Exercise 13.8 *Show that in complex coordinates the condition (13.29) implies that $\Gamma^\beta_{\gamma\bar\alpha} = 0$ and $\Gamma^{\bar\beta}_{\gamma\alpha} = 0$. The connection thus has only pure holomorphic and anti-holomorphic components $\Gamma^\alpha_{\beta\gamma}$ and $\Gamma^{\bar\alpha}_{\bar\beta\bar\gamma}$.*

We now assume that M_{2n} has a Riemannian metric $g_{ij}(\phi)$. This metric is hermitian if it is invariant under the action of the almost complex structure, namely if

$$J_i{}^k g_{k\ell} J_j{}^\ell = g_{ij}\,, \qquad \text{i.e.} \qquad J\,g\,J^T = g\,. \tag{13.30}$$

The second condition is equivalent to the first, but stated in matrix notation.

Exercise 13.9 *Express this condition in complex coordinates and show that it is equivalent to $g_{\alpha\beta} = g_{\bar\alpha\bar\beta} = 0$, as used in Sec. 13.1.*

Finally, we demand that the affine connection in (13.29) is the Levi-Civita connection. We then have two covariantly constant tensors:

$$\nabla_k J_i{}^j = \partial_k J_i{}^j - \Gamma^\ell_{ki} J_\ell{}^j + \Gamma^j_{k\ell} J_i{}^\ell = 0\,,$$
$$\nabla_k g_{ij} = \partial_k g_{ij} - \Gamma^\ell_{ki} g_{\ell j} - \Gamma^\ell_{kj} g_{i\ell} = 0\,. \tag{13.31}$$

We then define a Kähler manifold as a real M_{2n} with almost complex structure J and hermitian metric g such that J is covariantly constant with respect to the Levi-Civita connection (see Box 13.1). These conditions imply that a Kähler manifold admits complex coordinate charts and is thus a complex manifold.

In real coordinates the Kähler form is defined as the 2-form

$$\Omega = -J_{ij}\mathrm{d}\phi^i \wedge \mathrm{d}\phi^j\,, \qquad J_{ij} = J_i{}^k g_{kj}\,. \tag{13.32}$$

Exercise 13.10 *Use (13.31) to show that Ω is a closed 2-form. Conversely, show that $\mathrm{d}\Omega = 0$ implies that $J_i{}^j$ is covariantly constant. Show that (13.32) is equivalent to (13.14) in complex coordinates.*

We can now see that the approach to Kähler manifolds in this section has led us to the conditions (13.31) and then to the key result that the Kähler 2-form is closed. This is exactly the condition used to *define* a Kähler manifold in Sec. 13.1. Thus all other results in that section apply. In particular, the properties of the curvature tensor obtained in Sec. 13.1 are valid.

The main role of Kähler manifolds in supersymmetry and supergravity is that they serve as the scalar field target space in supersymmetric versions of the nonlinear σ-model discussed in Sec. 7.11. For global $\mathcal{N} = 1$ SUSY in four spacetime dimensions the σ-model can be defined without any further conditions or structure. The coupling to $\mathcal{N} = 1$ supergravity requires the further condition that the target space is a Kähler–Hodge manifold. This is discussed in Sec. 17A. For $\mathcal{N} = 2$ global supersymmetry the target space must have three covariantly constant complex structures $J_{Ii}{}^j$, $I = 1, 2, 3$. Manifolds with this structure are called hyper-Kähler manifolds and are described in Sec. 20.3.4 together with the quaternionic-Kähler manifolds required for $\mathcal{N} = 2$ supergravity. A subclass of Kähler manifolds called special Kähler manifolds are needed for supersymmetric theories of the $\mathcal{N} = 2$ vector multiplet. We discuss them in Sec. 20.3.3 when these theories are explained.

Finally, we point out that Kähler manifolds are special cases of *symplectic manifolds*, which are defined as real manifolds M_{2n} with a closed invertible 2-form (Ω in this case). Among other applications, symplectic manifolds play an important role in Hamiltonian mechanics. For example, see [104].

13.3 The Kähler manifolds CP^n

The complex projective spaces called CP^n, $n = 1, 2, \ldots$, are an interesting yet simple example of Kähler manifolds. The manifold CP^n is defined as the space of complex lines in flat $\mathbb{C}^{n+1}$. Two points in $\mathbb{C}^{n+1}$, denoted by the complex $(n+1)$-tuples $(X^1, X^2, \ldots, X^{n+1})$

and $(X'^1, X'^2, \ldots, X'^{n+1})$ respectively, are defined to lie on the same complex line if there is a complex number ρ such that

$$(X^1, X^2, \ldots, X^{n+1}) = (\rho X'^1, \rho X'^2, \ldots, \rho X'^{n+1}) . \tag{13.33}$$

The space of lines thus defined has complex dimension n. If $X^{n+1} \neq 0$, we can define a set of n coordinates $z^\alpha = X^\alpha / X^{n+1}$. These can be viewed as intrinsic coordinates which describe all lines with $X^{n+1} \neq 0$, and one can proceed to cover the space with analogously defined coordinates, $z^\alpha_{\{I\}} = X^\alpha / X^I$ with $I = 1, \ldots, n+1$, that describe all lines with $X^I \neq 0$.

We now use the ideas of induced metrics and the nonlinear σ-model, discussed in Secs. 7.4 and 7.11, respectively, to obtain a Kähler metric on CP^n. Let $X^I(x)$, $I = 1, 2, \ldots, n+1$, denote a set of complex scalar fields over flat Minkowski spacetime of real dimension D. The scalars X^I define an $(n+1)$-tuple or point in $\mathbb{C}^{n+1}$. We denote the complex conjugates of X^I as $\bar{X}_I$ and impose the constraint

$$X^I \bar{X}_I = 1 , \qquad I = 1, 2, \ldots, n+1 . \tag{13.34}$$

This can be thought of as part of the process of choosing a representative of the complex line associated with the $(n+1)$-tuple $(X^1, X^2, \ldots, X^{n+1})$. Indeed, from the definition (13.33) it is clear that two points that satisfy (13.34) and lie on the same line must be related by (13.33) with $|\rho| = 1$. We now incorporate the fact that the common phase of the fields $X^I(x)$ is irrelevant by introducing a U(1) gauge potential $\mathcal{A}_\mu(x)$ and covariant derivative in the Lagrangian

$$\mathcal{L} = -(\partial_\mu + \mathrm{i}\mathcal{A}_\mu)\bar{X}_I \, (\partial^\mu - \mathrm{i}\mathcal{A}^\mu) X^I . \tag{13.35}$$

The gauge potential $\mathcal{A}_\mu$ is an auxiliary non-dynamical field in this system. Its Euler–Lagrange equation is algebraic, with solution, after use of the constraint (13.34),

$$\mathcal{A}_\mu = -\tfrac{1}{2}\mathrm{i}\bar{X}_I \overleftrightarrow{\partial}_\mu X^I . \tag{13.36}$$

This result may be inserted in (13.35) to obtain the equivalent Lagrangian (with constrained fields)

$$\mathcal{L} = -\partial_\mu \bar{X}_I \partial^\mu X^I - \tfrac{1}{4}(\bar{X}_I \overleftrightarrow{\partial}_\mu X^I)(\bar{X}_J \overleftrightarrow{\partial}^\mu X^J) , \qquad \bar{X}_I X^I = 1 . \tag{13.37}$$

Exercise 13.11 *Show that the Lagrangian (13.37) is gauge invariant under the* U(1) *transformation* $X^I(x) \to X'^I(x) = \mathrm{e}^{\mathrm{i}\theta(x)} X^I(x)$. *Show that any solution* $X^I(x)$ *of the new system (13.37) is also a solution of the equations of motion of (13.35) and constraint (13.34).*

We can now simultaneously fix the gauge and solve the constraint by expressing the X^I in terms of n independent complex fields $z^\alpha(x)$ and their complex conjugates $\bar{z}_\alpha$ defined by

$$(X^1, \ldots, X^n, X^{n+1}) = \frac{1}{\sqrt{1 + \bar{z}_\alpha z^\alpha}} (z^1, \ldots, z^n, 1) . \tag{13.38}$$

Note that $z^\alpha = X^\alpha/X^{n+1}$, as in the coordinate chart defined in the first paragraph of this section. We substitute (13.38) in the Lagrangian of (13.37) and find the Lagrangian

$$\mathcal{L} = -\frac{1}{1+\bar{z}z}\left(\delta_\alpha^\beta - \frac{\bar{z}_\alpha z^\beta}{1+\bar{z}z}\right)\partial_\mu z^\alpha \partial^\mu \bar{z}_\beta = -\frac{1}{1+\bar{z}z}\left(\delta_{\alpha\bar{\beta}} - \frac{\bar{z}_\alpha z_{\bar{\beta}}}{1+\bar{z}z}\right)\partial_\mu z^\alpha \partial^\mu \bar{z}^{\bar{\beta}},$$
$$\bar{z}z \equiv \bar{z}_\alpha z^\alpha = z^\alpha \delta_{\alpha\bar{\beta}}\bar{z}^{\bar{\beta}}. \tag{13.39}$$

Note that we raise and lower indices here with $\delta_{\alpha\bar{\beta}}$ and not with the full metric $g_{\alpha\bar{\beta}}$. We can now interpret this result in the light of the discussion of the nonlinear σ-model in Sec. 7.11. There we associated Lagrangians of the form

$$\mathcal{L} = -\tfrac{1}{2}g_{ij}(\phi)\partial_\mu\phi^i\partial^\mu\phi^j \tag{13.40}$$

with maps from Minkowski space to a Riemannian target manifold with coordinates ϕ^i and metric

$$\mathrm{d}s^2 = g_{ij}(\phi)\mathrm{d}\phi^i\mathrm{d}\phi^j. \tag{13.41}$$

The form of (13.39) suggests that we interpret z^α and $\bar{z}^{\bar{\beta}}$ as local coordinates of a complex manifold with hermitian metric defined by

$$\mathrm{d}s^2 = 2g_{\alpha\bar{\beta}}\mathrm{d}z^\alpha\mathrm{d}\bar{z}^{\bar{\beta}} = 2\frac{1}{1+\bar{z}z}\left(\delta_{\alpha\bar{\beta}} - \frac{\bar{z}_\alpha z_{\bar{\beta}}}{1+\bar{z}z}\right)\mathrm{d}z^\alpha\mathrm{d}\bar{z}^{\bar{\beta}}. \tag{13.42}$$

Exercise 13.12 *Show that the metric tensor of (13.42) satisfies the Kähler condition (13.16). Show that the metric tensor can be obtained as in (13.17) from the Kähler potential* $K = \ln(1+\bar{z}z)$.

This Kähler metric on CP^n is known as the Fubini–Study metric, named after the Italian and German mathematicians who first found it.

Exercise 13.13 *Use (13.25) to calculate the Ricci tensor of the CP^n metric (13.42) and show that it satisfies*

$$R_{\alpha\bar{\beta}} = (n+1)g_{\alpha\bar{\beta}}. \tag{13.43}$$

A Kähler metric whose Ricci tensor satisfies $R_{\alpha\bar{\beta}} = kg_{\alpha\bar{\beta}}$ is called a Kähler–Einstein metric.

Exercise 13.14 *The metric tensor of the Poincaré plane was defined in real form in (7.134) and in the complex hermitian form*

$$\mathrm{d}s^2 = \mathrm{d}Z\mathrm{d}\bar{Z}/(\mathrm{Im}Z)^2 \tag{13.44}$$

in (7.151). Show that the metric of this manifold of complex dimension 1 is a Kähler metric. What is the Kähler potential?

Exercise 13.15 *A complex coordinate $z = X + \mathrm{i}Y$ for the unit 2-sphere was introduced using stereographic projection in Sec. 7.1. Calculate the induced metric from the embedding of S^2 in Euclidean 3-space using the coordinates z, $\bar{z}$. Compare to the case $n = 1$ of the CP^n metric in (13.42).*

13.4 Symmetries of Kähler metrics

13.4.1 Holomorphic Killing vectors and moment maps

Since a complex manifold is a special case of a real manifold, the formalism of Killing vectors that was introduced in Sec. 7.12 certainly applies. However, Kähler manifolds are defined using two basic structures: the hermitian metric and the covariantly constant complex structure. We consider symmetries that *preserve both structures*. For each continuous symmetry there is a Killing vector, which generates an infinitesimal variation of local coordinates. It can be expressed equivalently in real or complex form as

$$\delta\phi^i = \theta k^i(\phi) \qquad \text{or} \qquad \delta z^a = \theta k^a(z, \bar{z}) . \tag{13.45}$$

We now require that the Lie derivative of the metric tensor *and* the complex structure *both* vanish:

$$\mathcal{L}_k\, g_{ij} = \nabla_i k_j + \nabla_j k_i = 0 , \tag{13.46}$$
$$\mathcal{L}_k\, J_i{}^j = \nabla_i k^\ell J_\ell{}^j - \nabla_\ell k^j J_i{}^\ell = 0 . \tag{13.47}$$

The first condition is the definition of a Killing vector for any manifold, real or complex, and the second is special to Kähler manifolds.

In complex coordinates J takes the form (13.28), so the holomorphic components, i.e. $i \to \alpha$, $j \to \beta$, of the condition (13.47) vanish trivially. However, when we replace i by $\bar{\alpha}$ and j by β, we find the condition $\partial_{\bar{\alpha}} k^\beta = 0$, which says that the components k^β are functions of the z^α and not the $\bar{z}^{\bar{\alpha}}$. For this reason a Killing vector that satisfies (13.47) is called a *holomorphic Killing vector*. In other words, the symmetry map $\delta z^a = \theta k^a$ preserves the split between holomorphic and anti-holomorphic coordinates, and we have

$$k^a = \{k^\alpha(z),\ k^{\bar{\alpha}}(\bar{z})\} . \tag{13.48}$$

The Kähler manifolds are symplectic manifolds, as we mentioned at the end of Sec. 13.2, because $\mathrm{d}\Omega = 0$. The standard example in physics of a symplectic manifold is the Hamiltonian phase space with invertible closed 2-form $\mathrm{d}q \wedge \mathrm{d}p$. The symmetries, which preserve this form, are canonical transformations, which are all characterized by a generating function. The same applies to holomorphic symmetries of Kähler metrics, and the corresponding functions are called the *moment maps* [100].

The existence of moment maps (which are also called Killing potentials) can be simply derived from the fact that (13.46) and (13.47) imply that the Lie derivative of the 2-form Ω vanishes. Indeed this gives, using the definition (7.17), and then the vanishing of $\mathrm{d}\Omega$,

Box 13.2 Symmetries of Kähler metrics; moment maps

Symmetries of Kähler metrics (and their complex structures) are defined by holomorphic Killing vectors, which are determined in turn by real 'moment maps'.

$$0 = \mathcal{L}_k\Omega = (i_k\mathrm{d} + \mathrm{d}i_k)\,\Omega = \mathrm{d}i_k\Omega\,. \tag{13.49}$$

The Poincaré lemma, mentioned in Sec. 7.3, then implies that there exists a function $\mathcal{P}$ with the property

$$i_k\Omega = -2\mathrm{d}\mathcal{P}\,. \tag{13.50}$$

The scalar function $\mathcal{P}(z,\bar{z})$, which is real-valued since (13.50) is a real equation, is the moment map we have been seeking. Note that $\mathcal{P}$ is not determined uniquely; one is free to add a real constant, $\mathcal{P}(z,\bar{z}) \to \mathcal{P}(z,\bar{z}) + \xi$.

The condition (13.50) involves 1-forms, but it can also be expressed as an equality of covariant vectors. In complex coordinates $z^a = \{z^\alpha,\ \bar{z}^{\bar{\alpha}}\}$, this equality reads

$$\begin{aligned} k_\alpha &= g_{\alpha\bar{\beta}}\, k^{\bar{\beta}}(\bar{z}) = \mathrm{i}\partial_\alpha\, \mathcal{P}(z,\bar{z})\,, \\ k_{\bar{\alpha}} &= g_{\beta\bar{\alpha}}\, k^{\beta}(z) = -\mathrm{i}\partial_{\bar{\alpha}}\, \mathcal{P}(z,\bar{z})\,. \end{aligned} \tag{13.51}$$

The Killing equation (13.46) splits in these coordinates into two conditions:

$$\nabla_\alpha k_\beta + \nabla_\beta k_\alpha = 0\,, \tag{13.52}$$

$$\nabla_a k_{\bar{\beta}} + \overline{\nabla}_{\bar{\beta}} k_\alpha = 0\,. \tag{13.53}$$

The first of these is automatic since $\nabla_\alpha g_{\beta\bar{\gamma}} V^{\bar{\gamma}}(\bar{z}) = g_{\beta\bar{\gamma}}\partial_\alpha V^{\bar{\gamma}}(\bar{z}) = 0$ for a holomorphic vector.[2] The second condition is satisfied whenever we have a moment map $\mathcal{P}(z,\bar{z})$ due to (13.51). Hence, the symmetries of a Kähler manifold are characterized by real functions $\mathcal{P}(z,\bar{z})$ such that

$$k^\alpha(z) = -\mathrm{i}g^{\alpha\bar{\beta}}\partial_{\bar{\beta}}\mathcal{P}(z,\bar{z}) \tag{13.54}$$

is holomorphic. Note that $\mathcal{P}$ is always a function of both z^α and $z^{\bar{\alpha}}$ because the lower vector components k_α and $k_{\bar{\alpha}}$ also have this property.

As a conclusion, all symmetries of a Kähler manifold can be found by searching for the real functions $\mathcal{P}(z,\bar{z})$ that satisfy the property (we take for convenience the complex conjugate of (13.54) and drop an invertible factor)

$$\nabla_\alpha\partial_\beta\mathcal{P}(z,\bar{z}) = 0\,. \tag{13.55}$$

Exercise 13.16 *In the special case of the flat Kähler metric* $g_{\alpha\bar{\beta}} = \delta_{\alpha\bar{\beta}}$, *solve the holomorphy condition of (13.54) and determine that the moment map is in general of the form*

[2] The vector V^γ is holomorphic but need not be a Killing vector. Note that $\partial_\alpha V^{\bar{\gamma}}$ is covariant since the connection coefficients $\Gamma^{\alpha}_{\bar{\beta}\gamma}$ vanish for Kähler metrics.

$$\mathcal{P} = z^\alpha p_{\alpha\bar{\beta}} \bar{z}^{\bar{\beta}} + z^\alpha q_\alpha + \bar{z}^{\bar{\alpha}} \bar{q}_{\bar{\alpha}} + \xi\,. \tag{13.56}$$

Determine the reality conditions on the constant coefficients $p_{\alpha\bar{\beta}}$, q_α and ξ. Check explicitly that the related vectors k^α are indeed symmetries of the Lagrangian $\partial_\mu z^\alpha \delta_{\alpha\bar{\beta}} \partial^\mu \bar{z}^{\bar{\beta}}$. Note that the symmetries are holomorphic, but the moment map depends on both z^α and $z^{\bar{\alpha}}$.

The Killing vector relations (13.52) and (13.53) state the fact that the Kähler metric $g_{\alpha\bar{\beta}}$ is invariant under the isometry. However, the Kähler potential need not be invariant. It must satisfy only the weaker condition

$$\delta K = \theta \left(k^\alpha \partial_\alpha + k^{\bar{\alpha}} \partial_{\bar{\alpha}} \right) K(z, \bar{z}) = \theta \left[r(z) + \bar{r}(\bar{z}) \right], \tag{13.57}$$

with an arbitrary holomorphic function $r(z)$. If this function does not vanish, it means that $K(z, \bar{z})$ changes by a Kähler transformation under the action of the symmetry. This is sufficient for an invariant metric. This fact allows us to find a solution to (13.54) when we know the Killing vector:

$$\mathcal{P}(z, \bar{z}) = \mathrm{i}\left[k^\alpha \partial_\alpha K(z, \bar{z}) - r(z) \right] = -\mathrm{i}\left[k^{\bar{\alpha}} \partial_{\bar{\alpha}} K(z, \bar{z}) - \bar{r}(\bar{z}) \right]. \tag{13.58}$$

This solution is general, since we can absorb the real constant ξ in $r(z) \to r(z) + \mathrm{i}\xi$, which does not influence (13.57).

Exercise 13.17 *Prove (13.58) by applying $\partial_{\bar{\beta}}$ to deduce (13.54). The second term is unnecessary for this calculation, but it is needed in order to have a real $\mathcal{P}$. Prove this using (13.57).*

13.4.2 Algebra of holomorphic Killing vectors

In many cases the (holomorphic) symmetry group of a Kähler metric is non-abelian. To discuss this important case, we consider a set of holomorphic Killing vectors $k_A^\alpha(z)$. For each Killing vector there is a Killing potential $\mathcal{P}_A(z, \bar{z})$. For convenience we repeat the main formulas for the Killing vectors and moment maps indicating the extra index:

$$\begin{aligned}
k_A{}^\alpha(z) &= -\mathrm{i} g^{\alpha\bar{\beta}} \partial_{\bar{\beta}} \mathcal{P}_A(z, \bar{z}), \\
\left[k_A^\alpha(z) \partial_\alpha + k_A^{\bar{\alpha}}(\bar{z}) \partial_{\bar{\alpha}} \right] K(z, \bar{z}) &= r_A(z) + \bar{r}_A(\bar{z}), \\
\mathcal{P}_A(z, \bar{z}) &= \mathrm{i}\left[k_A^\alpha \partial_\alpha K(z, \bar{z}) - r_A(z) \right] = -\mathrm{i}\left[k_A^{\bar{\alpha}} \partial_{\bar{\alpha}} K(z, \bar{z}) - \bar{r}_A(\bar{z}) \right].
\end{aligned} \tag{13.59}$$

The holomorphic Killing vectors $k_A^\alpha(z)$, $k_A^{\bar{\alpha}}(\bar{z})$ generate a Lie algebra. The Lie algebra structure is obtained by expressing the Lie bracket relation (7.146) in complex coordinates, assuming holomorphy. This gives

$$k_A^\beta \partial_\beta k_B^\alpha - k_B^\beta \partial_\beta k_A^\alpha = f_{AB}{}^C k_C^\alpha, \tag{13.60}$$

together with the complex conjugate. Note that if k_A and k_B are holomorphic vectors, then their Lie bracket is necessarily holomorphic.

By adjustment of the additive constants, the $\mathcal{P}_A$ can be chosen to transform in the adjoint representation,

$$(k_A^\alpha \partial_\alpha + k_A^{\bar\alpha}\partial_{\bar\alpha})\mathcal{P}_B(z,\bar z) = f_{AB}{}^C\mathcal{P}_C . \tag{13.61}$$

The $\mathcal{P}_A$ are then uniquely fixed for (simple) non-abelian symmetries, but the ambiguity $\mathcal{P} \to \mathcal{P} + \xi$ remains for U(1) factors of the symmetry group.

We have introduced the Killing potentials $\mathcal{P}_A$, the possible constants ξ_A, and the quantities r_A using purely mathematical considerations. It is remarkable that they also occur naturally in supersymmetry, as we will discuss in Ch. 14. We will see later that the moment maps that are used in supersymmetry should satisfy the extra condition (13.61), sometimes denoted as 'equivariance relation', which can, using the last equation of (13.59), also be written as

$$k_A{}^\alpha g_{\alpha\bar\beta} k_B{}^{\bar\beta} - k_B{}^\alpha g_{\alpha\bar\beta} k_A{}^{\bar\beta} = \mathrm{i} f_{AB}{}^C\mathcal{P}_C . \tag{13.62}$$

13.4.3 The Killing vectors of CP^1

It is time for an example, and we will discuss the holomorphic Killing vectors of CP^1 which span the Lie algebra of SU(2). One way to see that this is the expected symmetry group is to observe that the elementary SU(2) transformations of the doublet (X^1, X^2) give a holomorphic global symmetry of the Lagrangian (13.37). In these variables the symmetries are generated by the differential operators

$$k_A = -\frac{1}{2}\mathrm{i}(\sigma_A)^i{}_j X^j \frac{\partial}{\partial X^i} , \tag{13.63}$$

where the σ_A are the Pauli matrices. We use the relation $z = X^1/X^2$ and the chain rule to write

$$\begin{aligned}
k_1 &= -\frac{1}{2}\mathrm{i}\left(X^1\frac{\partial z}{\partial X^2} + X^2\frac{\partial z}{\partial X^1}\right)\frac{\partial}{\partial z} = -\mathrm{i}\frac{1}{2}(1-z^2)\frac{\partial}{\partial z} ,\\
k_2 &= -\frac{1}{2}\left(X^1\frac{\partial z}{\partial X^2} - X^2\frac{\partial z}{\partial X^1}\right)\frac{\partial}{\partial z} = \frac{1}{2}(1+z^2)\frac{\partial}{\partial z} ,\\
k_3 &= -\frac{1}{2}\mathrm{i}\left(X^1\frac{\partial z}{\partial X^1} - X^2\frac{\partial z}{\partial X^2}\right)\frac{\partial}{\partial z} = -\mathrm{i}z\frac{\partial}{\partial z} .
\end{aligned} \tag{13.64}$$

The components[3] $k_A^z(z)$ can be read immediately from the expressions on the right.

Exercise 13.18 *Show that the Lie brackets of the vectors $k_A = k_A{}^z\partial/\partial z$ satisfy the structure relations $[k_A\, ,\, k_B] = \varepsilon_{ABC}k_C$ of* SU(2).

[3] We use the common notation in which vector and tensor components are denoted by the name of the coordinate, in this case z, rather than the number

The next step is to find the Killing potentials. Although not all 'data' are needed, we record the Kähler potential, metric tensor, and connection:

$$K = \ln(1+z\bar{z}), \qquad g_{z\bar{z}} = (1+z\bar{z})^{-2}, \qquad \Gamma^z_{zz} = -2\bar{z}(1+z\bar{z})^{-1}. \tag{13.65}$$

The Killing potentials can be obtained by solving the three differential equations

$$k_A^z = -\mathrm{i}g^{z\bar{z}}\partial_{\bar{z}}\mathcal{P}_A = -\mathrm{i}(1+z\bar{z})^2\,\partial_{\bar{z}}\mathcal{P}_A, \tag{13.66}$$

together with their complex conjugates. The solutions are

$$\mathcal{P}_1 = \frac{1}{2}\frac{z+\bar{z}}{1+z\bar{z}}, \qquad \mathcal{P}_2 = -\mathrm{i}\frac{1}{2}\frac{z-\bar{z}}{1+z\bar{z}}, \qquad \mathcal{P}_3 = -\frac{1}{2}\frac{1-z\bar{z}}{1+z\bar{z}}. \tag{13.67}$$

Exercise 13.19 *Consider $\mathcal{P}_3' = \mathcal{P}_3 + \xi$, where ξ is a constant. Obviously also $\mathcal{P}_3'$ satisfies (13.66). Show that the requirement that the $\mathcal{P}_A$ satisfy (13.61) with the structure constants $f_{AB}{}^C = \varepsilon_{ABC}$ of* SU(2) *fixes $\xi = 0$.*

Exercise 13.20 *Apply (13.59) to obtain the $r_A(z)$:*

$$r_1 = \tfrac{1}{2}\mathrm{i}z, \qquad r_2 = \tfrac{1}{2}z, \qquad r_3 = -\tfrac{1}{2}\mathrm{i}. \tag{13.68}$$

Note that the Kähler potential is invariant under the third isometry k_3, but still $r_3 \neq 0$. Its value is fixed by the requirement (13.61).

We mentioned in Sec. 12.5 that Kähler manifolds are not in general symmetric spaces (see (12.13) for the essential property). In fact, a general Kähler manifold need not possess any Killing vectors. This is, for example, the case for Calabi–Yau manifolds, which we will discuss in Sec. 21.4.3. The non-compact Kähler manifolds that are symmetric spaces are products of the irreducible symmetric Kähler spaces listed below:

$$\frac{\mathrm{SU}(p,q)}{\mathrm{SU}(p)\times\mathrm{SU}(q)\times\mathrm{U}(1)}, \qquad \frac{\mathrm{SO}^*(2n)}{\mathrm{U}(n)}, \qquad \frac{\mathrm{Sp}(2n)}{\mathrm{U}(n)},$$
$$\frac{\mathrm{SO}(n,2)}{\mathrm{SO}(n)\times\mathrm{SO}(2)}, \qquad \frac{\mathrm{E}_{6,-14}}{\mathrm{SO}(10)\times\mathrm{U}(1)}, \qquad \frac{\mathrm{E}_{7,-25}}{\mathrm{E}_6\times\mathrm{U}(1)}. \tag{13.69}$$

General actions with $\mathcal{N}=1$ supersymmetry 14

In this chapter we will introduce the multiplet calculus of global $\mathcal{N}=1$ supersymmetry which provides a procedure to construct more general actions than those discussed in Ch. 6. The theories considered there all have standard quadratic kinetic terms. Here we will find kinetic terms involving non-trivial metrics constructed from the scalar fields of chiral multiplets. These include the Kähler metrics discussed in Ch. 13. Study of the multiplet calculus at the level of global SUSY will make it easier to understand the superconformal multiplet calculus, which we will use to construct supergravity theories later in the book.[1]

The basic multiplet calculus constructions are derived in Secs. 14.1–14.3. They are combined and extended in Sec. 14.4 to obtain the general $\mathcal{N}=1$ supersymmetric gauge theory in which Killing symmetries of the Kähler metric are gauged. In Sec. 14.5 we discuss some physical properties of SUSY gauge theories, notably the important question of spontaneous breakdown of supersymmetry.

Multiplet calculus constructions are rigorous and complete, but they can be technically complicated. The same constructions are simpler and more natural in the well-known superspace formalism of $\mathcal{N}=1$, $D=4$ global supersymmetry. This insightful and elegant method allows an easy construction of supersymmetric actions and has other benefits. We do not use superspace methods in this book because it is very complicated to extend them to supergravity. We do include a short appendix (at the end of this chapter) in which the basic superfields are discussed and related to the multiplet formulas in the main part of the chapter.

14.1 Multiplets

A supermultiplet of fields is a set of boson and fermion fields which transform among themselves under supersymmetry, such that the commutator algebra of two SUSY variations is

$$[\delta(\epsilon_1), \delta(\epsilon_2)] = -\tfrac{1}{2}\bar{\epsilon}_1\gamma^{\mu}\epsilon_2\partial_{\mu}. \tag{14.1}$$

The simplest example is the chiral multiplet consisting of two complex scalars Z and F and the left-chiral projection of a Majorana spinor χ. The SUSY transformation rules were given in (6.15), and the algebra was studied in Sec. 6.2.2. In the superspace formalism the component fields, such as Z, $P_L\chi$, and F, are packaged in a single superfield.

[1] We consider only chiral and gauge multiplets. There are other multiplets, such as the antisymmetric tensor multiplet [105, 106], which has the same physical content as the chiral multiplet.

In our approach the SUSY transformation rules of a supermultiplet are built by starting from the lowest dimension field and introducing the additional fields needed to obtain a closed realization of the algebra of (14.1).

14.1.1 Chiral multiplets

To illustrate our approach to the multiplet calculus, let us reconsider the chiral multiplet. A chiral multiplet is now *defined* as one that contains a complex scalar Z whose SUSY transformation involves only the chiral projection $P_L\epsilon$, or equivalently $\bar{\epsilon}P_L$, of the constant Majorana SUSY parameter ϵ. We therefore write the transformation rule

$$\delta Z = \frac{1}{\sqrt{2}}\bar{\epsilon}P_L\chi. \tag{14.2}$$

The field that multiplies $\bar{\epsilon}P_L$ is *defined* as the spinor component $P_L\chi$ of the multiplet. Lorentz invariance requires that it is a spinor rather than, say, a vector–spinor. Since $P_L\chi$ must also transform, we postulate the most general Lorentz covariant form

$$\delta P_L\chi = \frac{1}{\sqrt{2}}P_L\left(F + \gamma^\mu X_\mu + \gamma^{\mu\nu}T_{\mu\nu}\right)\epsilon, \tag{14.3}$$

in which F, X_μ and $T_{\mu\nu}$ are to be determined by demanding that (14.1) holds for the field Z. This simple calculation requires the symmetry properties of spinor bilinears (see e.g. Table 3.1). The result is (i) that $T_{\mu\nu} = 0$, (ii) that $X_\mu = \partial_\mu Z$, and (iii) that F drops out of the commutator and is thus not constrained. So we add F as a new field of the multiplet. Its transformation rule must take the form

$$\delta F = \frac{1}{\sqrt{2}}\bar{\epsilon}\left(P_L\lambda + P_R\psi\right), \tag{14.4}$$

where λ and ψ are to be determined by imposing the supersymmetry algebra on χ. This time we need a Fierz rearrangement identity; those of (3.72) are most appropriate. We find that λ must vanish in order to remove unwanted terms of the form $\bar{\epsilon}_1 P_R\gamma^{\mu\nu}\epsilon_2$ in the commutator and that $P_R\psi = \not{\partial}P_L\chi$. Next we must check that the SUSY commutator on F is consistent with (14.1), and it is.

To summarize, we constructed the chiral multiplet transformation rules (6.15) as a closed realization of supersymmetry starting with the assumption that its lowest dimension component is a complex scalar Z whose SUSY transformation involves only $P_L\epsilon$.

Exercise 14.1 *Construct the anti-chiral multiplet defined as one that contains a complex scalar called $\bar{Z}$, whose SUSY transformation involves only $P_R\epsilon$. Your results should agree with (6.16).*

Box 14.1 Chiral and real multiplets

Multiplets are sets of fields on which the supersymmetry algebra is realized. A chiral multiplet is a multiplet in which the transformation of the lowest (complex scalar) component involves only $P_L\epsilon$. A real multiplet is a multiplet in which the lowest component is a real scalar.

Exercise 14.2 *In (6.62) we showed that SUSY variations of chiral multiplets are modified when gauge multiplets are present. In this exercise we outline how to rework the construction of chiral multiplets in this more general situation. Simply repeat the process above with the ordinary derivative ∂_μ in (14.1) replaced by the gauge covariant derivative D_μ defined in (6.54), i.e.*

$$[\delta(\epsilon_1), \delta(\epsilon_2)] = -\tfrac{1}{2}\bar{\epsilon}_1\gamma^\mu\epsilon_2 D_\mu. \tag{14.5}$$

Check that, in the first step, the modification leads to what one should expect: $X_\mu = D_\mu Z$ rather than $\partial_\mu Z$. In the second step, when calculating the supersymmetry commutator on $P_L\chi$, one needs to transform A_μ^A, which appears in the covariant derivative. Its transformation was given in (6.49), and involves λ^A. Show that this leads to the modified transformation of F as shown in (6.62).

Although the fields Z, $P_L\chi$ and F of a chiral multiplet can certainly be elementary fields in a Lagrangian, it is important to realize that a chiral multiplet is a more general object whose components can be composites of elementary fields. For example, two elementary chiral multiplets can be multiplied to form a composite multiplet, as the following exercise shows.

Exercise 14.3 *Consider two chiral multiplets with components Z^i, $P_L\chi^i$ and F^i, $i = 1, 2$. Show that the quadratic combinations $Z^3 = Z^1Z^2$, $P_L\chi^3 = P_L(Z^1\chi^2 + Z^2\chi^1)$ and $F^3 = F^1Z^2 + F^2Z^1 - \bar{\chi}^1P_L\chi^2$ also transform under (6.15). Thus we have constructed a new chiral multiplet which can be considered to be the product of the first two.*

In Sec. 6.2 we introduced interactions of an elementary chiral multiplet using the superpotential $W(Z)$, an arbitrary holomorphic function. Let's now show that $W(Z)$ is the lowest component of a composite chiral multiplet. Using the chain rule we can obtain its SUSY variation

$$\delta W(Z) = \frac{1}{\sqrt{2}}W'(z)\bar{\epsilon}P_L\chi. \tag{14.6}$$

Comparing with (14.2), we see that $W(Z)$ transforms with the chiral projection $\bar{\epsilon}P_L$, and we can thus identify the χ-component of the composite multiplet:

$$\chi(W) \equiv W'(Z)\chi. \tag{14.7}$$

Exercise 14.4 *Complete the construction of this multiplet as follows.*

(i) Compute $\delta\chi(W)$ using the chain rule and a Fierz rearrangement. Compare with (6.15) to identify the F-component

$$F(W) = W'F - \tfrac{1}{2}W''\bar{\chi}P_L\chi. \tag{14.8}$$

(ii) As the final check show that $\delta F(W) = \not{\partial}P_L\chi(W)/\sqrt{2}$.

14.1.2 Real multiplets

Another basic multiplet of $\mathcal{N}=1$ SUSY is the real multiplet, which corresponds to a real superfield. By *definition*, the lowest component is a real scalar, $C(x)$, and one adds the additional components necessary to satisfy the SUSY algebra (12.1) on all fields. The SUSY variation of C is the most general real expression: $\delta C = \bar{\epsilon}\chi$, which is real if χ is a Majorana fermion. To agree with common practice we rename $\chi \to \mathrm{i}\gamma_*\zeta$. We now proceed as in the previous section, writing a general ansatz for $\delta\zeta$, and restricting it by requiring that the SUSY algebra is preserved on C. After several such steps we find the closed *real multiplet* containing the component fields

$$(C, \zeta, \mathcal{H}, B_\mu, \lambda, D), \tag{14.9}$$

with the transformation rules

$$\begin{aligned}
\delta C &= \tfrac{1}{2}\mathrm{i}\bar{\epsilon}\gamma_*\zeta, \\
\delta P_L\zeta &= \tfrac{1}{2}P_L\left(\mathrm{i}\mathcal{H} - \not{B} - \mathrm{i}\not{\partial}C\right)\epsilon, \\
\delta\mathcal{H} &= -\mathrm{i}\bar{\epsilon}P_R\left(\lambda + \not{\partial}\zeta\right), \\
\delta B_\mu &= -\tfrac{1}{2}\bar{\epsilon}\left(\gamma_\mu\lambda + \partial_\mu\zeta\right), \\
\delta\lambda &= \tfrac{1}{2}\left[\gamma^{\rho\sigma}\partial_\rho B_\sigma + \mathrm{i}\gamma_* D\right]\epsilon, \\
\delta D &= \tfrac{1}{2}\mathrm{i}\,\bar{\epsilon}\gamma_*\gamma^\mu\partial_\mu\lambda.
\end{aligned} \tag{14.10}$$

The fermions ζ and λ are Majorana fields, the bosons C, B_μ and D are real, while $\mathcal{H}$ is complex. Chiral projectors are used in the second and third entries for convenience, and the conjugate relations are easily obtained.

Exercise 14.5 *Show that* $\delta\bar{\mathcal{H}} = \mathrm{i}\bar{\epsilon}P_L(\lambda + \not{\partial}\zeta)$ *and that* $\delta P_R\zeta = \frac{1}{2}P_R(-\mathrm{i}\bar{\mathcal{H}} - \not{B} + \mathrm{i}\not{\partial}C)\epsilon$.

The real multiplet has two major applications to the SUSY gauge theories discussed in Ch. 6. In the first application the components of the real multiplet are viewed as elementary fields which include the gluon, gluino, and D-auxiliary fields of a physical gauge multiplet. To see how this works, we observe first that the variations of the fields B_μ, λ and D are very similar to the abelian case of (6.49), and they would agree exactly if the term $\partial_\mu\zeta$ in δB_μ were absent. In fact the components C, ζ and $\mathcal{H}$ can be eliminated by a supersymmetric generalization of a gauge transformation. To see how this is done, suppose we are given a chiral multiplet Z, $P_L\chi$, F and its anti-chiral conjugate $\bar{Z}$, $P_R\chi$, $\bar{F}$. Then $\mathrm{Im}\,Z$ is a real scalar, which becomes the lowest component of a real multiplet determined by the chiral transformation laws (and their conjugates). The fields of this multiplet are viewed as parameters of the 'supergauge' transformation, which is defined as the shift

$$\left(C \to C+\mathrm{Im}\,Z,\ \zeta \to \zeta - \frac{1}{\sqrt{2}}\chi,\ \mathcal{H} \to \mathcal{H}+\mathrm{i}F,\ B_\mu \to B_\mu + \partial_\mu\,\mathrm{Re}\,Z,\ \lambda \to \lambda,\ D \to D\right). \tag{14.11}$$

Clearly we can choose parameters to make the transformed C, ζ, $\mathcal{H}$ components vanish, while Re Z acts as an abelian gauge parameter for the vector B_μ. Of course, we then rename $B_\mu \to A_\mu$ and Re $Z \to \theta$, so that A_μ, λ and D are the standard fields of the gauge multiplet with the usual gauge transformation. The 'supergauge' transformation just described is said to take the theory to Wess–Zumino gauge. The corresponding transformation in the superspace formalism is described in Appendix 14A. The non-abelian extension of this procedure gives the result that the real multiplet in Wess–Zumino gauge includes the fields A_μ^A, λ^A and D^A. Their SUSY transformations are exactly those of (6.49) with the gauge covariant algebra given in (6.51).

14.2 Generalized actions by multiplet calculus

In the superspace formalism, the actions of supersymmetric field theories are expressed as integrals over both the x^μ and θ^α coordinates of superspace. To construct the same actions using components, we need only note in (6.15) and (14.10) that the SUSY variations of the highest dimension fields, F for a chiral multiplet and D for a real multiplet, are total spacetime derivatives. Therefore the integrals

$$S_F = \int \mathrm{d}^4x\, F, \qquad S_D = \int \mathrm{d}^4x\, D \tag{14.12}$$

are invariant under SUSY,

$$\begin{aligned} \delta S_F &= \int \mathrm{d}^4x\, \delta F = \frac{1}{\sqrt{2}} \int \mathrm{d}^4x\, \bar{\epsilon} \not{\partial} P_L \chi = 0, \\ \delta S_D &= \int \mathrm{d}^4x\, \delta D = \frac{1}{2}\mathrm{i} \int \mathrm{d}^4x\, \bar{\epsilon}\gamma_* \not{\partial}\lambda = 0. \end{aligned} \tag{14.13}$$

The actions of supersymmetric field theories are easily obtained by choosing F and D to be components of composite multiplets constructed from elementary chiral and gauge multiplets. The SUSY field theories discussed in Ch. 6 and some important extensions can be obtained in this way.

14.2.1 The superpotential

We start with the simplest case. Suppose that we have a set of n elementary chiral multiplets with components $(Z^\alpha, \chi^\alpha, F^\alpha)$, $\alpha = 1, \ldots, n$. We now consider the composite chiral

Lagrangians from highest components — Box 14.2

Actions can be constructed as integrals of the highest components of composite multiplets constructed from elementary ones. There are F-type actions for composite chiral multiplets, and D-type actions for composite real multiplets.

multiplet whose lowest component is a holomorphic function $W(Z^\alpha)$ of the scalars. By the same construction as in (14.6)–(14.8), we find the χ and F components

$$\chi(W) = W_\alpha \chi^\alpha, \qquad W_\alpha \equiv \frac{\partial W}{\partial Z^\alpha},$$
$$F(W) = W_\alpha F^\alpha - \frac{1}{2} W_{\alpha\beta} \bar{\chi}^\alpha P_L \chi^\beta, \qquad W_{\alpha\beta} \equiv \frac{\partial^2 W}{\partial Z^\alpha \partial Z^\beta}. \tag{14.14}$$

The integral of $F(W)$ is a SUSY invariant, indeed the same action considered in (6.41).

14.2.2 Kinetic terms for chiral multiplets

We now describe a similar construction which involves the D-term of a composite real multiplet whose lowest component is $K(Z^\alpha, \bar{Z}^{\bar{\alpha}})$, an arbitrary real function of chiral multiplet scalars Z^α and their anti-chiral complex conjugates which we call $\bar{Z}^{\bar{\alpha}}$. It is convenient to include an overall factor of $1/2$. We list the components of this multiplet which can be obtained using the chain rule, the variations of (6.15) and (6.16), and then comparing with the general transformations of (14.10). Derivatives of K are denoted by subscripts $\alpha, \bar{\beta}, \ldots$ as in (14.14). We find that the components of this composite real multiplet are

$$\begin{aligned}
C(\tfrac{1}{2}K) &= \tfrac{1}{2}K, \\
\zeta(\tfrac{1}{2}K) &= -\frac{1}{\sqrt{2}} \mathrm{i} P_L K_\alpha \chi^\alpha + \frac{1}{\sqrt{2}} \mathrm{i} P_R K_{\bar{\alpha}} \chi^{\bar{\alpha}}, \\
\mathcal{H}(\tfrac{1}{2}K) &= -K_\alpha F^\alpha + \frac{1}{2} K_{\alpha\beta} \bar{\chi}^\alpha P_L \chi^\beta, \\
B_\mu(\tfrac{1}{2}K) &= \frac{1}{2} \mathrm{i} K_\alpha \partial_\mu Z^\alpha - \frac{1}{2} \mathrm{i} K_{\bar{\alpha}} \partial_\mu \bar{Z}^{\bar{\alpha}} + \frac{1}{2} \mathrm{i} K_{\alpha\bar{\beta}} \bar{\chi}^\alpha P_L \gamma_\mu \chi^{\bar{\beta}}, \\
P_R \lambda(\tfrac{1}{2}K) &= \frac{1}{\sqrt{2}} \mathrm{i} K_{\alpha\bar{\beta}} P_R \left[(\not{\partial} \bar{Z}^{\bar{\beta}}) \chi^\alpha - F^\alpha \chi^{\bar{\beta}} \right] + \frac{\mathrm{i}}{2\sqrt{2}} K_{\alpha\beta\bar{\gamma}} \chi^{\bar{\gamma}} \bar{\chi}^\alpha P_L \chi^\beta, \\
D(\tfrac{1}{2}K) &= K_{\alpha\bar{\beta}} \left(-\partial_\mu Z^\alpha \partial^\mu \bar{Z}^{\bar{\beta}} - \tfrac{1}{2} \bar{\chi}^\alpha P_L \not{\partial} \chi^{\bar{\beta}} - \tfrac{1}{2} \bar{\chi}^{\bar{\beta}} P_R \not{\partial} \chi^\alpha + F^\alpha \bar{F}^{\bar{\beta}} \right) \\
&\quad + \tfrac{1}{2} \left[K_{\alpha\beta\bar{\gamma}} \left(-\bar{\chi}^\alpha P_L \chi^\beta \bar{F}^{\bar{\gamma}} + \bar{\chi}^\alpha P_L (\not{\partial} Z^\beta) \chi^{\bar{\gamma}} \right) + \text{h.c.} \right] \\
&\quad + \tfrac{1}{4} K_{\alpha\beta\bar{\gamma}\bar{\delta}} \bar{\chi}^\alpha P_L \chi^\beta \bar{\chi}^{\bar{\gamma}} P_R \chi^{\bar{\delta}}.
\end{aligned} \tag{14.15}$$

The general result (14.13) guarantees that $S = \int \mathrm{d}^4 x\, D(K/2)$ is a SUSY invariant. It is a generalized kinetic term for chiral multiplets. Indeed the first term of $D(K/2)$ suggests that we interpret $K(Z^\alpha, \bar{Z}^{\bar{\alpha}})$ as the Kähler potential for the metric $g_{\alpha\bar{\beta}}$. Then

$$S(K) = -\int \mathrm{d}^4 x\, g_{\alpha\bar{\beta}} \partial_\mu Z^\alpha \partial^\mu \bar{Z}^{\bar{\beta}} + \ldots, \qquad g_{\alpha\bar{\beta}} = K_{\alpha\bar{\beta}} = \partial_\alpha \partial_{\bar{\beta}} K. \tag{14.16}$$

The first term is the kinetic action for the nonlinear σ-model on a Kähler target space, and the remaining terms give the $\mathcal{N}=1$ supersymmetric extension of this σ-model. We will discuss its properties in Sec. 14.3. In the special case $K = \sum_\alpha Z^\alpha \bar{Z}^{\bar{\alpha}}$, the Kähler metric $g_{\alpha\bar{\beta}} = \delta_{\alpha\bar{\beta}}$ is flat, and $S(K)$ reduces to the conventional chiral multiplet kinetic term in (6.17).

14.2.3 Kinetic terms for gauge multiplets

Our final application gives a *generalized kinetic action for gauge multiplets*, which includes interactions with chiral multiplets. The result is a supersymmetric extension of the theories considered in Sec. 4.2. Suppose that we have a set of *abelian* gauge multiplets with components A_μ^A, λ^A, D^A labelled by upper indices $A, B = 1, \ldots, n$. From the gaugino components λ^A, λ^B of any pair of these multiplets, we construct the complex Lorentz scalar $\bar{\lambda}^A P_L \lambda^B$. From (6.49) one can see that its SUSY transformation involves only the chiral projection $P_L\epsilon$. Hence, there is a composite chiral multiplet whose lowest component is $\bar{\lambda}^A P_L \lambda^B$. We also consider another set of composite chiral multiplets whose lowest components $f_{AB}(Z^\alpha) = f_{BA}(Z^\alpha)$ are holomorphic functions of chiral multiplet scalars Z^α. (The trivial case in which f_{AB} is simply a symmetric matrix of complex numbers is also allowed.) By the method of Ex. 14.3, one can show that the product of these two multiplets (multiplied by the convenient factor 1/4) has the following components:

$$
\begin{aligned}
Z(f) &= -\tfrac{1}{4} f_{AB} \bar{\lambda}^A P_L \lambda^B, \\
P_L\chi(f) &= \frac{1}{2\sqrt{2}} f_{AB} \left(\tfrac{1}{2}\gamma^{\mu\nu} F_{\mu\nu}^{-A} - \mathrm{i}D^A\right) P_L\lambda^B - \frac{1}{4} f_{AB\alpha} P_L\chi^\alpha \bar{\lambda}^A P_L \lambda^B, \\
F(f) &= \frac{1}{4} f_{AB}\left(-2\bar{\lambda}^A P_L \not{\partial}\lambda^B - F_{\mu\nu}^{-A} F^{\mu\nu-B} + D^A D^B\right) \\
&\quad + \frac{1}{2\sqrt{2}} f_{AB\alpha}\bar{\chi}^\alpha \left(-\frac{1}{2}\gamma^{\mu\nu} F_{\mu\nu}^{-A} + \mathrm{i}D^A\right) P_L\lambda^B - \frac{1}{4} f_{AB\alpha} F^\alpha \bar{\lambda}^A P_L \lambda^B \\
&\quad + \frac{1}{8} f_{AB\alpha\beta}\bar{\chi}^\alpha P_L \chi^\beta \bar{\lambda}^A P_L \lambda^B .
\end{aligned}
\tag{14.17}
$$

We use the notation $f_{AB\alpha} = \partial_\alpha f_{AB}$, etc. and the (anti-)self-dual tensor $F_{\mu\nu}^-$ introduced in (4.36). The self-dual field strength often appears 'automatically' owing to the following chain of equalities that use the results of Ex. 4.7 and (4.39):

$$
\gamma^{\mu\nu} F_{\mu\nu} P_L = \tfrac{1}{2}\gamma^{\mu\nu} F_{\mu\nu}(1+\gamma_*) = \tfrac{1}{2}\left(\gamma^{\mu\nu} - \tilde{\gamma}^{\mu\nu}\right) F_{\mu\nu} = \gamma^{-\mu\nu} F_{\mu\nu} = \gamma^{\mu\nu} F_{\mu\nu}^- .
\tag{14.18}
$$

It follows from (14.13) that $\int \mathrm{d}^4x\, F(f)$ is a supersymmetric action. In the simplest case, namely $f_{AB} = \delta_{AB}$, this action reduces to the kinetic action of a set of n free gauge multiplets, specifically the action (6.8) with gauge coupling $g = 0$. Note also that imaginary constant terms in f_{AB} give total derivatives in $F(f)$, which give vanishing contribution to the action integral. The F-component of a chiral multiplet is complex, so in the general case $f_{AB}(Z^\alpha)$, we must take the sum of $\int \mathrm{d}^4x\, F(f)$ plus its complex conjugate to obtain a hermitian action.

14.3 Kähler geometry from chiral multiplets

We now discuss the properties of the action $\int d^4x\, D(K/2)$, obtained from (14.15), in more detail. We interpret $K(Z, \bar{Z})$ as a Kähler potential and $g_{\alpha\bar{\beta}} = K_{\alpha\bar{\beta}}$ as the metric.[2] The action then describes the $\mathcal{N} = 1$ extension of the nonlinear σ-model with a Kähler manifold as target space. It is interesting to see how the mathematics of Kähler geometry, discussed in Ch. 13, emerges from the construction of the theory via the multiplet calculus. We will use the formulas (13.19)–(13.22) for the connection and curvature tensor of a Kähler metric $g_{\alpha\bar{\beta}}$.

We will bring the Lagrangian $D(K/2)$ into a form in which its geometrical content is manifest. The first step is to note the solution of the auxiliary field equation in (14.15):

$$F^\alpha = \tfrac{1}{2} g^{\alpha\bar{\beta}} K_{\gamma\beta\bar{\beta}} \bar{\chi}^\gamma P_L \chi^\beta = \tfrac{1}{2} \Gamma^\alpha_{\gamma\beta} \bar{\chi}^\gamma P_L \chi^\beta . \tag{14.19}$$

After substitution of this result in the action and some rearrangement, one finds the equivalent action

$$S(K)|_F = \int d^4x \left[g_{\alpha\bar{\beta}} \left(-\partial_\mu Z^\alpha \partial^\mu \bar{Z}^{\bar{\beta}} - \tfrac{1}{2} \bar{\chi}^\alpha P_L \not{\nabla} \chi^{\bar{\beta}} - \tfrac{1}{2} \bar{\chi}^{\bar{\beta}} P_R \not{\nabla} \chi^\alpha \right) \right.$$
$$\left. + \tfrac{1}{4} R_{\alpha\bar{\gamma}\beta\bar{\delta}}\, \bar{\chi}^\alpha P_L \chi^\beta \bar{\chi}^{\bar{\gamma}} P_R \chi^{\bar{\delta}} \right] . \tag{14.20}$$

Fermion derivatives are defined by

$$P_L \nabla_\mu \chi^\alpha = P_L \left(\partial_\mu \chi^\alpha + \Gamma^\alpha_{\beta\gamma} \chi^\gamma \partial_\mu Z^\beta \right),$$
$$P_R \nabla_\mu \chi^{\bar{\alpha}} = P_R \left(\partial_\mu \chi^{\bar{\alpha}} + \Gamma^{\bar{\alpha}}_{\bar{\beta}\bar{\gamma}} \chi^{\bar{\gamma}} \partial_\mu \bar{Z}^{\bar{\beta}} \right). \tag{14.21}$$

As we will see very soon these derivatives are covariant under reparametrizations of the target space.

Let us compare the result (14.20) with (6.17). We see that the Kähler σ-model contains new nonlinear interactions, including quartic fermion terms. These couplings are described by the connection and curvature tensor of the target space, quantities that have geometrical significance. The situation is similar to string theory in which couplings typically have a geometric interpretation.

In fact we now show that each of the four terms in (14.20) is invariant under transformations of the fields Z^α, $P_L\chi^\alpha$, which are natural extensions of reparametrizations of the target space to include the fermions. Bosons transform as $Z'^\alpha = Z'^\alpha(Z)$. The Z'^α will have fermion partners χ'^α determined by the chiral multiplet transformation rules and related to χ^α by the chain rule. We write

[2] Note that $D(K/2)$ is invariant under the Kähler transformation $K \to K + f(Z) + \bar{f}(\bar{Z})$ discussed in Ch. 13. The Kähler transformation is implemented as the shift of the real multiplet of (14.15) by the real part of a chiral multiplet. This is similar to the transformation of the elementary real multiplet to Wess–Zumino gauge discussed in Sec. 14.1.2.

$$\begin{aligned}\delta Z'^{\alpha} &= \frac{1}{\sqrt{2}}\bar{\epsilon} P_L \chi'^{\alpha} \\ &= \frac{\partial Z'^{\alpha}}{\partial Z^{\beta}}\delta Z^{\beta} = \frac{\partial Z'^{\alpha}}{\partial Z^{\beta}}\frac{1}{\sqrt{2}}\bar{\epsilon} P_L \chi^{\beta}.\end{aligned} \tag{14.22}$$

Thus reparametrization of the Z^α is accompanied by a transformation of the fermions, and we have the formulas

$$Z'^{\alpha} = Z'^{\alpha}(Z), \qquad P_L \chi'^{\alpha} = \frac{\partial Z'^{\alpha}}{\partial Z^{\beta}} P_L \chi^{\beta}. \tag{14.23}$$

We see that the χ^α transform as tangent vectors on the target space.

Exercise 14.6 *Show that the spacetime derivatives $\partial_\mu Z^\alpha$ and $\nabla_\mu \chi^\alpha$ also transform as tangent vectors.*

The transformations of the fields $\bar{Z}^{\bar{\alpha}}$ and $\chi^{\bar{\alpha}}$ are the conjugates of those above. It is then quite obvious that the four terms in (14.20) are each invariant under reparametrization.

Exercise 14.7 *Obtain the transformation of the F^α auxiliary fields under reparametrization from the SUSY transform of $P_L\chi'^\alpha$ as follows:*

$$\delta P_L \chi'^{\alpha} \equiv \frac{1}{\sqrt{2}} P_L \left(\partial\!\!\!/ Z'^{\alpha} + F'^{\alpha}\right) = \delta\left(\frac{\partial Z'^{\alpha}}{\partial Z^{\beta}} P_L \chi^{\beta}\right). \tag{14.24}$$

Work out the SUSY variation of the product in the last term. After a Fierz rearrangement you should find that

$$F'^{\alpha} = \frac{\partial Z'^{\alpha}}{\partial Z'^{\beta}} F^{\beta} - \frac{1}{2}\frac{\partial^2 Z'^{\alpha}}{\partial Z^{\beta}\partial Z^{\gamma}}(\bar{\chi}^{\beta} P_L \chi^{\gamma}). \tag{14.25}$$

The connection-like term in the transformation is compatible with (and required by) the solution (14.19).

The solution (14.19) applies to the case when the complete theory under study is specified by the action integral of the D-term $D(K/2)$ in (14.15). The solution for F^α will change if we modify the theory by adding a superpotential or include interactions with gauge multiplets. To allow for these generalizations it is convenient to define the new auxiliary field

$$\tilde{F}^{\alpha} = F^{\alpha} - \tfrac{1}{2} g^{\alpha\bar{\beta}} K_{\gamma\beta\bar{\beta}} \bar{\chi}^{\gamma} P_L \chi^{\beta} = F^{\alpha} - \tfrac{1}{2}\Gamma^{\alpha}_{\gamma\beta}\bar{\chi}^{\gamma} P_L \chi^{\beta}. \tag{14.26}$$

The full kinetic action integral of $D(K/2)$ can then be written as

$$S(K) = S(K)|_F + \int \mathrm{d}^4 x\, g_{\alpha\bar{\beta}} \tilde{F}^{\alpha} \bar{\tilde{F}}^{\bar{\beta}}, \tag{14.27}$$

where $S(K)|_F$ is the expression in (14.20). This form of the action is manifestly invariant under reparametrization of the target space since the first term $S(K)|_F$, given in (14.20), is constructed from covariant elements and the auxiliary $\tilde{F}^\alpha$ now transforms as a tangent vector.[3]

[3] The supersymmetry transformations of (6.15), which we have used so far, are correct, but contain terms that are not covariant under reparametrization. In Appendix 14B we define modified covariant SUSY transformation rules, which are useful to understand the geometrical structure of supersymmetry.

14.4 General couplings of chiral multiplets and gauge multiplets

In the actions discussed in Sec. 14.2, all gauge multiplets are effectively abelian, i.e. $F_{\mu\nu} = \partial_\mu A_\nu - \partial_\nu A_\mu$, and possible internal symmetries of chiral multiplets are all global symmetries. A theory of this type is said to possess 'ungauged supersymmetry'; see Sec. 12.4.2. The main purpose of this section is to derive more general supersymmetric theories in which internal symmetries are gauged. In these new theories the fields of gauge and chiral multiplets are coupled in a more intimate way, but the actions are still built from the previous three SUSY invariants, determined by a real function $K(Z,\bar{Z})$, a holomorphic function $W(Z)$ and a holomorphic symmetric tensor $f_{AB}(Z)$.

First, it is useful to combine the previous results for ungauged supersymmetry into the single action:

$$\begin{aligned}
S &= S(K) + S(W) + S(f),\\
S(K) &= \int \mathrm{d}^4x\, D(\tfrac{1}{2}K) = \int \mathrm{d}^4x\, \mathcal{L}_{\text{kin,chir}},\\
S(W) &= \int \mathrm{d}^4x\, F(W) + \text{h.c.} = \int \mathrm{d}^4x\, \mathcal{L}_{\text{pot,chir}},\\
S(f) &= \int \mathrm{d}^4x\, F(f) + \text{h.c.} = \int \mathrm{d}^4x\, \mathcal{L}_{\text{kin,gauge}}.
\end{aligned} \tag{14.28}$$

These describe, respectively, kinetic and potential terms for chiral multiplets, and kinetic terms of the gauge multiplets, which can also depend on fields of the chiral multiplets. Note that we now include the hermitian conjugates of $S(W)$ and $S(f)$. Using results of the previous sections, we find the following explicit Lagrangians:

$$\begin{aligned}
\mathcal{L}_{\text{kin,chir}} &= g_{\alpha\bar\beta}\Big[-\partial_\mu Z^\alpha \partial^\mu \bar{Z}^{\bar\beta} - \tfrac{1}{2}\bar\chi^\alpha P_L \nabla\!\!\!\!/\, \chi^{\bar\beta} - \tfrac{1}{2}\bar\chi^{\bar\beta} P_R \nabla\!\!\!\!/\, \chi^\alpha\\
&\qquad + \Big(F^\alpha - \tfrac{1}{2}\Gamma^\alpha_{\gamma\beta}\bar\chi^\gamma P_L\chi^\beta\Big)\Big(\bar{F}^{\bar\beta} - \tfrac{1}{2}\Gamma^{\bar\beta}_{\bar\gamma\bar\alpha}\bar\chi^{\bar\gamma} P_R\chi^{\bar\alpha}\Big)\Big]\\
&\quad + \tfrac{1}{4}R_{\alpha\bar\gamma\beta\bar\delta}\,\bar\chi^\alpha P_L\chi^\beta\,\bar\chi^{\bar\gamma} P_R\chi^{\bar\delta},\\
\mathcal{L}_{\text{pot,chir}} &= W_\alpha F^\alpha - \tfrac{1}{2}W_{\alpha\beta}\bar\chi^\alpha P_L\chi^\beta + \text{h.c.},\\
\mathcal{L}_{\text{kin,gauge}} &= -\frac{1}{4}\,\text{Re}\, f_{AB}\Big(2\bar\lambda^A \not{D}\lambda^B + F^A_{\mu\nu}F^{\mu\nu\,B} - 2D^A D^B\Big)\\
&\quad + \frac{1}{8}(\text{Im}\, f_{AB})\varepsilon^{\mu\nu\rho\sigma}F^A_{\mu\nu}F^B_{\rho\sigma} + \frac{1}{4}\mathrm{i}(\partial_\mu\, \text{Im}\, f_{AB})\bar\lambda^A\gamma_*\gamma^\mu\lambda^B\\
&\quad + \Big[\frac{1}{2\sqrt{2}}f_{AB\alpha}\bar\chi^\alpha\Big(-\frac{1}{2}\gamma^{\mu\nu}F^{-A}_{\mu\nu} + \mathrm{i}D^A\Big)P_L\lambda^B - \frac{1}{4}f_{AB\alpha}F^\alpha\bar\lambda^A P_L\lambda^B\\
&\quad + \frac{1}{8}f_{AB\alpha\beta}\bar\chi^\alpha P_L\chi^\beta\bar\lambda^A P_L\lambda^B + \text{h.c.}\Big].
\end{aligned} \tag{14.29}$$

The Kähler metric $g_{\alpha\bar\beta}$ of the scalar target space appears together with its connection and curvature tensor; $\nabla_\mu\chi$ is defined in (14.21). In the gauge multiplet Lagrangian we have split the contributions from the real parts and the imaginary parts of f_{AB}, and integrated by parts in one term. This shows that imaginary *constant* terms in f_{AB} do not contribute.

It also makes clear that Re f_{AB} can be interpreted as a metric in the gauge multiplet sector. Thus our theories contain two metrics, $g_{\alpha\bar{\beta}}$ for chiral multiplets and Re f_{AB} for gauge multiplets. It is a physical requirement that both metrics are positive definite, so that the energy of any classical field configuration is positive.

The introduction of gauged internal symmetries is called a 'deformation' of the ungauged theory; see Sec. 12.4.2. The process involves several steps, which are discussed in the next two subsections:

1. The target space Kähler metric must have a Lie group of symmetries generated by holomorphic Killing vectors. A subgroup is chosen as the gauge group.
2. The theory must contain gauge multiplets for this group. These are promoted from Maxwell vector fields to Yang–Mills vectors.
3. Gauge and chiral multiplets are then coupled. This requires the 'reconstruction' of the D-term $D(K/2)$ using the SUSY transformation rules of gauged supersymmetry.

14.4.1 Global symmetries of the SUSY σ-model

We now assume that the target space Kähler metric has a continuous isometry group. As discussed in Ch. 13, this means that there are a set of holomorphic Killing vectors $k_A{}^\alpha(Z)$, which are related to real scalar moment maps $\mathcal{P}_A(Z,\bar{Z})$ by

$$k_A{}^\alpha(Z) = -\mathrm{i}g^{\alpha\bar{\beta}}\partial_{\bar{\beta}}\mathcal{P}_A(Z,\bar{Z}). \tag{14.30}$$

As in (13.60) Lie brackets of the Killing vectors close on a Lie algebra with structure constants $f_{AB}{}^C$.

The simplest example of holomorphic Killing vectors occurs for a flat Kähler metric $g_{\alpha\bar{\beta}} = \delta_{\alpha\bar{\beta}}$. In this case the linear transformations of the coordinates, given by

$$\delta(\theta)Z^\alpha = -\theta^A(t_A)^\alpha{}_\beta Z^\beta, \tag{14.31}$$

are symmetries. The $(t_A)^\alpha{}_\beta$ are the matrix generators of a Lie algebra such as SU(n). The Killing vectors and their moment maps are (see Ex. 13.16)

$$k_A{}^\alpha = -(t_A)^\alpha{}_\beta Z^\beta, \qquad \mathcal{P}_A = -\mathrm{i}\bar{Z}^{\bar{\alpha}}\delta_{\bar{\alpha}\alpha}(t_A)^\alpha{}_\beta Z^\beta. \tag{14.32}$$

A more general Killing symmetry is really a special case of a reparametrization of the target space whose effect on chiral multiplet fields Z^α, χ^α, F^α was derived in Sec. 14.3. It is the special case of an infinitesimal reparametrization generated by a holomorphic Killing vector. Thus we write $Z'^\alpha(Z) = Z^\alpha + \theta^A k_A{}^\alpha(Z)$ and work to first order in the constant scalar parameters θ^A. Therefore we define the symmetry transformation of the fields Z^α by

$$\delta(\theta)Z^\alpha = \theta^A k_A{}^\alpha(Z). \tag{14.33}$$

The analogous transformations of χ^α and F^α are just the infinitesimal versions of (14.23) and (14.25), namely

$$\delta(\theta)P_L\chi^\alpha = \theta^A \frac{\partial k_A{}^\alpha(Z)}{\partial Z^\beta} P_L\chi^\beta, \qquad (14.34)$$

$$\delta(\theta)F^\alpha = \theta^A \frac{\partial k_A{}^\alpha(Z)}{\partial Z^\beta} F^\beta - \frac{1}{2}\theta^A \frac{\partial^2 k_A{}^\alpha(Z)}{\partial Z^\beta \partial Z^\gamma} \bar{\chi}^\beta P_L\chi^\gamma. \qquad (14.35)$$

These expressions can also be derived from the requirement that internal $\delta(\theta)$ and SUSY transformations $\delta(\epsilon)$ commute. An alternative version of these transformation rules, which are explicitly covariant under reparametrizations of the target space, is derived in Appendix 14B; see (14.122).

So far we have been working at the level of global internal symmetry. The action $S(K)$ is invariant under the transformations (14.33), (14.34) and (14.35), while $S(W)$ is invariant if $W(Z)$ is invariant, i.e. if

$$W_\alpha k_A{}^\alpha = 0. \qquad (14.36)$$

14.4.2 Gauge and SUSY transformations for chiral multiplets

We now discuss how to promote some or all of the Killing symmetries to gauge symmetries, with arbitrary functions $\theta^A(x)$ as gauge parameters, while maintaining $\mathcal{N}=1$ supersymmetry. Of course, the subset of symmetries that we choose to gauge should form a closed algebra.[4] From now on, the index A refers only to the Killing symmetries that are gauged.

Gauge coupling constants can be introduced according to the discussion in the introduction to Sec. 6.3. To do this for Killing symmetries, one should replace $k_A{}^\alpha$ by $g_i k_A{}^\alpha$ and $\mathcal{P}_A$ by $g_i\mathcal{P}_A$ with the g_i for simple or U(1) factors of the gauge group. Gauge coupling constants are implicitly included in this way in the remainder of the book.

Gauging is an involved process, but the first steps are clear. The theory must contain a gauge multiplet A_μ^A, λ^A, D^A for each Killing vector. In general there are both non-abelian and abelian gauge multiplets whose SUSY and gauge transformation rules and gauge covariant algebra were discussed in Sec. 6.3.1; see (6.49), (6.50) and (6.51).

The next step is to define gauge covariant derivatives for nonlinear Killing symmetries:

$$D_\mu Z^\alpha = \partial_\mu Z^\alpha - A_\mu^A k_A{}^\alpha(Z), \qquad (14.37)$$

$$D_\mu P_L\chi^\alpha = \partial_\mu P_L\chi^\alpha - A_\mu^A \frac{\partial k_A{}^\alpha(Z)}{\partial Z^\beta} P_L\chi^\beta. \qquad (14.38)$$

With these ingredients we can obtain the SUSY transformation rules of chiral multiplet components. We extend the method of Ex. 14.2 to Killing symmetries, always imposing the gauge covariant SUSY algebra (14.5). This procedure leads to transformation rules which include the covariant derivatives (14.37) and (14.38) and which generalize (6.62):

[4] We consider only 'electric gaugings', but we refer readers to [84] for an interesting generalization which includes both 'electric' and 'magnetic' gauge potentials. The new formulation embodies the duality symmetry discussed in Sec. 4.2.

$$\begin{aligned}
\delta Z^\alpha &= \frac{1}{\sqrt{2}}\bar{\epsilon} P_L \chi^\alpha, \\
\delta P_L \chi^\alpha &= \frac{1}{\sqrt{2}} P_L (\not{D} Z^\alpha + F^\alpha)\epsilon, \\
\delta F^\alpha &= \frac{1}{\sqrt{2}}\bar{\epsilon}\, \not{D} P_L \chi^\alpha + \bar{\epsilon} P_R \lambda^A k_A{}^\alpha(Z).
\end{aligned} \tag{14.39}$$

Exercise 14.8 *Show that both $D_\mu Z^\alpha$ and $D_\mu P_L \chi^\alpha$ transform without derivatives of $\theta^A(x)$ under a gauge transformation. For χ^α you should obtain*

$$\delta(\theta) D_\mu P_L \chi^\alpha = \theta^A \frac{\partial k_A{}^\alpha(Z)}{\partial Z^\beta} D_\mu P_L \chi^\beta. \tag{14.40}$$

14.4.3 Actions of chiral multiplets in a gauge theory

We can now generalize our results for supersymmetric actions to the situation in which chiral multiplets transform under gauged Killing symmetries and are therefore coupled to gauge multiplets. The potential terms in $S(W)$ of (14.28) are the easiest to discuss. The superpotential $W(Z)$ must be gauge invariant, i.e. it must satisfy (14.36). The previous construction of the composite chiral multiplet is then unchanged, and the Lagrangian $\mathcal{L}_{\text{pot,chir}}$ is correct in gauged supersymmetry.

Gauged Killing symmetry does require modification of the kinetic action. To find the changes we need to reconstruct the D-term of the composite real multiplet $D(K/2)$ of (14.15) using the new transformation rules (14.39) for chiral multiplets. Since these rules involve gauge multiplet components, we also need their transformation rules (6.49) in the process.

The scalar $K(Z, \bar{Z})$ is the Kähler potential. We saw in Sec. 13.4 that this is not necessarily invariant under a Killing symmetry, but can change by a Kähler transformation as shown in (13.57). However, it simplifies the reconstruction of the real multiplet in gauged supersymmetry to assume that K is gauge invariant. This allows us to use the formulas of (14.10) which were derived by imposing the SUSY algebra (12.1) for gauge invariant quantities. We will show that the final result for the new D-term is valid even for non-invariant K.

We now outline the construction of the modified $D(K/2)$ focusing on the new terms required in gauged supersymmetry. We start with the component $C(K/2) = K(Z, \bar{Z})$, compute the variation $\delta C(K/2)$ using δZ^α in (14.39), use (14.10) to identify the component $\zeta(K/2)$, and then repeat this process for higher dimension components of the composite multiplet. The components C, ζ and $\mathcal{H}$ are the same as in (14.15). In $B_\mu(K/2)$ the covariant derivative (14.37) appears rather than ∂_μ. This was to be expected, and similar 'trivial covariantizations' also occur in $\lambda(K/2)$ and $D(K/2)$. The first essential change occurs in the step when we compute $\delta\mathcal{H}(K/2)$ using δF^α in (14.39). This leads to the new expression for the λ-component:

$$\begin{aligned}
P_R \lambda(\tfrac{1}{2}K) = {}& \frac{1}{\sqrt{2}} \mathrm{i} K_{\alpha\bar{\beta}} P_R \left[(\not{D}\bar{Z}^{\bar{\beta}})\chi^\alpha - F^\alpha \chi^{\bar{\beta}} \right] + \frac{\mathrm{i}}{2\sqrt{2}} K_{\alpha\beta\bar{\gamma}} \chi^{\bar{\gamma}} \bar{\chi}^\alpha P_L \chi^\beta \\
& - \mathrm{i} P_R \lambda^A k_A{}^\alpha K_\alpha .
\end{aligned} \tag{14.41}$$

The next step is to compute $\delta P_R\lambda(K/2)$ and use (14.10) to identify the new form of the D-component. There is no need to repeat previous work, so we concentrate on new terms due to the gauged Killing symmetry. These terms come from three sources: (i) from the $\delta A_\mu{}^A$ variation of the covariant derivative in (14.41), (ii) from the gauge multiplet modification of δF^α, and (iii) from the variation of the last term in (14.41). Fierz rearrangement identities, given in (3.72), are needed at this stage. The result for the new D-component is

$$D(\tfrac{1}{2}K) = \ldots - \mathrm{i}D^A k_A{}^\alpha K_\alpha - \sqrt{2}g_{\alpha\bar\beta}\bar\lambda^A\left(P_L\chi^\alpha k_A{}^{\bar\beta} + P_R\chi^{\bar\beta}k_A{}^\alpha\right), \tag{14.42}$$

where the $\ldots$ stand for the terms already written in (14.15) with derivatives replaced by covariant derivatives.

To complete our discussion we need some properties of moment maps derived in Sec. 13.4.2. We are assuming that the Kähler potential is invariant under the Killing symmetries which are gauged. Then $r(z)$ in (13.57) is restricted to be an imaginary constant, and (13.58) requires that moment maps take the form

$$\mathcal{P}_A = \mathrm{i}k_A^\alpha K_\alpha + p_A, \tag{14.43}$$

where the p_A are arbitrary real constants. The p_A are called Fayet–Iliopoulos (FI) constants [107]. They are related to a simple new type of supersymmetric action available for abelian gauge multiplets. Indeed if $D(x)$ is the auxiliary field of an abelian gauge multiplet and $\lambda(x)$ is the gaugino in this multiplet, then $\delta D = \bar\epsilon\partial\!\!\!/\lambda$ is a total derivative; see (6.49). The integral $\int \mathrm{d}^4x\, D(x)$ is then invariant under SUSY.

To explore the physics of FI terms in a more general way, we consider the conditions under which we can replace the action integral $D(K/2)$ in (14.42) by the integral of

$$D(\tfrac{1}{2}K) = \ldots - D^A\mathcal{P}_A - \sqrt{2}\bar\lambda^A\left(P_L\chi^\alpha k_{A\alpha} + P_R\chi^{\bar\beta}k_{A\bar\beta}\right). \tag{14.44}$$

The difference is the integral

$$S_{\rm FI} = -\int \mathrm{d}^4x\, p_A D^A, \tag{14.45}$$

and the replacement is valid if the integral $S_{\rm FI}$ is invariant under supersymmetry and gauge transformations. Both conditions require

$$p_A f_{BC}{}^A = 0, \tag{14.46}$$

which must hold for all choices of the indices B, C. Thus, the constants p_A vanish for generators that occur at the right-hand side of commutation relations. (In mathematics terminology this is the derived algebra.) This is the case for generators in simple non-abelian factors of the gauge group. Conversely, if A is the index of an abelian factor of the gauge group, we can choose $p_A \neq 0$ as a possible FI coupling. In general the linear equations (14.46) admit $n_{\rm FI}$ non-trivial linearly independent 'vectors' p_A, and the gauged theory contains $n_{\rm FI}$ independent FI couplings.

Exercise 14.9 *Consider a simple non-abelian gauge group and suppose that K is gauge invariant. Then the moment maps are given by $\mathcal{P}_A = \mathrm{i}k_A^\alpha\partial_\alpha K$. Show that the equivariance relation (13.61) is satisfied. Hint: carefully commute derivatives.*

The discussion above shows that we can use $D(K/2)$ of (14.44) in the action integral of chiral multiplets in gauged supersymmetry if the target space Kähler potential $K(Z,\bar{Z})$ is gauge invariant. Fayet–Iliopoulos terms appear as an integral of the form $S_{\rm FI}$ contained within

$$S(K)=\int \mathrm{d}^4x\, D\left(\tfrac{1}{2}K\right). \tag{14.47}$$

It is quite interesting that the mathematics of constant shifts in moment maps $\mathcal{P}_A$ for abelian Killing symmetries corresponds to new terms in the general action integral of gauged supersymmetry.

Let's consider the situation when K is not invariant, and the moment map is corrected by the extra term shown in (13.58). We now outline an argument to show that the spacetime integral of the D-term (14.44) is invariant under SUSY, whether or not K is gauge invariant. We apply the component transformation rules of (14.39) and (6.49) to the first term of (14.44), obtaining

$$-\delta\int \mathrm{d}^4x\, D^A\mathcal{P}_A=-\int \mathrm{d}^4x\left[\frac{1}{2}\mathrm{i}\bar{\epsilon}\gamma_*\gamma^\mu\left(D_\mu\lambda^A\right)\mathcal{P}_A\right.$$
$$\left.+\frac{1}{\sqrt{2}}\mathrm{i}D^A\,\bar{\epsilon}\left(P_L\chi^\alpha k_{A\alpha}-P_R\chi^{\bar{\beta}}k_{A\bar{\beta}}\right)\right]. \tag{14.48}$$

We will show that the first term cancels with part of the $\delta P_L\chi^\alpha\sim P_L\gamma^\mu D_\mu Z$ plus conjugate $\delta P_R\bar{\chi}^{\bar{\beta}}$ variations of the second term in (14.44). These variations are

$$-\int \mathrm{d}^4x\bar{\lambda}^A\left(P_L D_\mu Z^\alpha k_{A\alpha}+P_R D_\mu\bar{Z}^{\bar{\beta}}k_{A\bar{\beta}}\right)\gamma^\mu\epsilon. \tag{14.49}$$

We isolate the axial vector term $\bar{\lambda}\gamma_*\gamma^\mu\epsilon$ of this expression. After a Majorana flip and use of (13.51), it becomes

$$\begin{aligned}
&-\tfrac{1}{2}\mathrm{i}\bar{\epsilon}\gamma_*\gamma^\mu\lambda^A\left(D_\mu Z^\alpha\partial_\alpha+D_\mu\bar{Z}^{\bar{\alpha}}\partial_{\bar{\alpha}}\right)\mathcal{P}_A\\
&=-\tfrac{1}{2}\mathrm{i}\bar{\epsilon}\gamma_*\gamma^\mu\lambda^A\left[\partial_\mu\mathcal{P}_A-A_\mu^B\left(k_B^\alpha\partial_\alpha+k_B^{\bar{\alpha}}\partial_{\bar{\alpha}}\right)\mathcal{P}_A\right]\\
&=-\tfrac{1}{2}\mathrm{i}\bar{\epsilon}\gamma_*\gamma^\mu\lambda^A\left[\partial_\mu\mathcal{P}_A-A_\mu^B f_{BA}{}^C\mathcal{P}_C\right].
\end{aligned} \tag{14.50}$$

In the last step we used the equivariance relation (13.61) to write the covariant derivative $D_\mu\mathcal{P}_A=\partial_\mu\mathcal{P}_A-A_\mu^B f_{BA}{}^C\mathcal{P}_C$. The $D_\mu\lambda^A$ term in (14.48) then combines with the $D_\mu\mathcal{P}_A$ term in (14.50) into a total derivative which vanishes in the integrated $\delta S(K)$. Thus equivariant moment maps are necessary for the consistent gauging of Killing symmetries. The moment maps $\mathcal{P}_A$ used in (14.44) must be equivariant!

There are several other terms in the variation $\delta S(K)$ that depend on the Killing vector, and they must all cancel. For example, the $\bar{\lambda}\gamma^\mu\epsilon$ terms in (14.49) cancel with terms that originate from the variation of A_μ^A in the kinetic term $D_\mu ZD^\mu\bar{Z}$ of $\mathcal{L}_{\rm kin,chiral}$ with derivatives replaced by covariant derivatives. Similarly, the second term in (14.48) cancels against the $\delta\lambda^A\sim\bar{\epsilon}D^A$ variation of the second term in (14.44). The cancelation of these variations and others which we have not mentioned do not depend on the assumption that $K(Z,\bar{Z})$ is gauge invariant. The conclusion of this argument is that the action $S(K)$ (14.47) is invariant under SUSY transformations if we use the D-term of (14.44) with moment maps that satisfy the equivariance condition (13.61) or, equivalently, (13.62). The addition of FI constants p_A satisfying (14.46) is consistent with this equivariance equation.

14.4.4 General kinetic action of the gauge multiplet

The general kinetic terms for abelian gauge multiplets derived in Sec. 14.2.3 include holomorphic functions $f_{AB}(Z)$ of chiral multiplet scalars. The Lagrangian given in (14.29) is also correct for non-abelian gauge multiplets provided that several changes required for non-abelian gauge invariance are made. These changes are as follows:

(i) Use the non-abelian gauge field strength in $F_{\mu\nu}^{-A}$.
(ii) Replace $\partial_\mu \lambda^A$ by the Yang–Mills covariant derivative $D_\mu \lambda^A$.
(iii) The quantity $f_{AB}(Z^\alpha)$ is no longer arbitrary but must transform as the direct product of adjoint representations, specifically as[5]

$$\delta f_{AB}(Z) = f_{AB\alpha}\theta^C k_C^\alpha = 2\theta^C f_{C(A}{}^D f_{B)D}(Z). \tag{14.51}$$

This last requirement is needed both for gauge invariance of the action and for global SUSY. Indeed, an action cannot be supersymmetric invariant unless it is gauge invariant. The latter statement follows from the fact that the commutator of two supersymmetries includes a gauge symmetry; see e.g. (11.28).

It is quite common to assume that the 'tensor' f_{AB} is proportional to the Cartan–Killing metric on each simple factor of the gauge group. For a compact simple Lie algebra, one can choose a basis in which f_{AB} is proportional to δ_{AB}. In string theory applications, there is often still a gauge invariant proportionality factor depending on moduli fields. In the simplest case $f_{AB} = \delta_{AB}$, the Lagrangian $\mathcal{L}_{\text{kin,gauge}}$ in (14.29) reduces to that of (6.8) and is gauge invariant.

As mentioned after (4.66), the real part of f_{AB} must be positive definite so that gauge field kinetic terms of the vectors are positive definite. When f_{AB} is proportional to the Cartan–Killing metric, the gauge group must be compact (and one must use the negative of the Cartan–Killing metric; see Appendix B).

14.4.5 Requirements for an $\mathcal{N}=1$ SUSY gauge theory

Box 14.3 summarizes the results derived earlier in this section. The possible ways to gauge the theory are characterized by the following:

1. The choice of a gauge group which must be a subgroup of the holomorphic isometry group of the Kähler target space. Information on the Lie algebra of this group is encoded in the real structure constants $f_{AB}{}^C = -f_{BA}{}^C$, which should satisfy Jacobi identities as in (4.81).

[5] There is a more general possibility related to the fact that imaginary constant parts in $f_{AB}(Z)$ do not contribute to the action. The gauge transformation may have extra terms $\mathrm{i}\theta^C C_{AB,C}$, where the $C_{AB,C} = C_{BA,C}$ are real constants. If the completely symmetric component vanishes, $C_{(AB,C)} = 0$, additional Chern–Simons terms are needed to restore supersymmetry. Otherwise, gauge anomalies (which are accompanied by supersymmetry anomalies) play a role. See [108, 109].

General $\mathcal{N} = 1, D = 4$ supersymmetry actions **Box 14.3**

General actions of chiral and gauge multiplets are determined by three functions of the chiral multiplet scalars: the Kähler potential $K(Z, \bar{Z})$ (up to Kähler transformations (13.18)), a superpotential $W(Z)$, and a holomorphic symmetric matrix $f_{AB}(Z)$ (up to imaginary constants). Holomorphic Killing symmetries of the Kähler metric $g_{\alpha\bar{\beta}} = \partial_\alpha \partial_{\bar{\beta}} K$ can be gauged. These symmetry transformations of chiral multiplets are determined by moment maps and the holomorphic Killing vectors that they generate.

2. Gauge transformations of the chiral multiplet, determined by real moment maps $\mathcal{P}_A(Z, \bar{Z})$.

We summarize the conditions that these should satisfy. The moment maps should satisfy

$$\nabla_\alpha \partial_\beta \mathcal{P}_A(Z, \bar{Z}) = 0. \tag{14.52}$$

This ensures that the Killing vectors

$$k_A{}^\alpha(Z) = -\mathrm{i} g^{\alpha\bar{\beta}} \partial_{\bar{\beta}} \mathcal{P}_A(Z, \bar{Z}) \tag{14.53}$$

are holomorphic. Furthermore, they should satisfy the 'equivariance relation' (13.62), i.e.

$$k_A{}^\alpha g_{\alpha\bar{\beta}} k_B^{\bar{\beta}} - k_B^\alpha g_{\alpha\bar{\beta}} k_A{}^{\bar{\beta}} = \mathrm{i} f_{AB}{}^C \mathcal{P}_C. \tag{14.54}$$

This condition restricts the constant contributions to the moment maps, and ensures the closure of the algebra of Killing vectors.

A superpotential may be included if it is gauge invariant. Thus we require (14.36), i.e.

$$k_A{}^\alpha \partial_\alpha W(Z) = 0. \tag{14.55}$$

Finally, we require gauge invariant kinetic terms for vector multiplets, so the condition

$$k_C^\alpha(Z) f_{AB\alpha}(Z) = 2 f_{C(A}{}^D f_{B)D}(Z) + \mathrm{i} C_{AB,C} \tag{14.56}$$

must be satisfied. The role of the constants $C_{AB,C}$ is discussed briefly in footnote 5.

In physical applications of global supersymmetry, kinetic terms must be positive, so the matrices $g_{\alpha\bar{\beta}}$ and $\mathrm{Re}\, f_{AB}$ must be positive definite.

When all these requirements are satisfied, we have an $\mathcal{N} = 1$ supersymmetric gauge theory with Lagrangian

$$\begin{aligned}
\mathcal{L} &= \mathcal{L}_{\text{kin,chir}} + \mathcal{L}_{\text{kin,gauge}} + \mathcal{L}_{\text{pot,chir}}, \\
\mathcal{L}_{\text{kin,chir}} &= g_{\alpha\bar{\beta}} \Big[-D_\mu Z^\alpha D^\mu \bar{Z}^{\bar{\beta}} - \tfrac{1}{2} \bar{\chi}^\alpha P_L \not{D} \chi^{\bar{\beta}} - \tfrac{1}{2} \bar{\chi}^{\bar{\beta}} P_R \not{D} \chi^\alpha \\
&\qquad + \left(F^\alpha - \tfrac{1}{2} \Gamma^\alpha_{\gamma\beta} \bar{\chi}^\gamma P_L \chi^\beta \right) \left(\bar{F}^{\bar{\beta}} - \tfrac{1}{2} \Gamma^{\bar{\beta}}_{\bar{\gamma}\bar{\alpha}} \bar{\chi}^{\bar{\gamma}} P_R \chi^{\bar{\alpha}} \right) \Big] \\
&\quad + \tfrac{1}{4} R_{\alpha\bar{\beta}\gamma\bar{\delta}}\, \bar{\chi}^\alpha P_L \chi^\gamma \bar{\chi}^{\bar{\beta}} P_R \chi^{\bar{\delta}} \\
&\quad - D^A \mathcal{P}_A - \sqrt{2} \bar{\lambda}^A \left(P_L \chi^\alpha k_{A\alpha} + P_R \chi^{\bar{\beta}} k_{A\bar{\beta}} \right), \\
\mathcal{L}_{\text{pot,chir}} &= W_\alpha F^\alpha - \tfrac{1}{2} W_{\alpha\beta} \bar{\chi}^\alpha P_L \chi^\beta + \text{h.c.},
\end{aligned}$$

$$
\begin{aligned}
\mathcal{L}_{\text{kin,gauge}} = & -\frac{1}{4}\operatorname{Re} f_{AB}\left(2\bar{\lambda}^A \not{D}\lambda^B + F_{\mu\nu}^A F^{\mu\nu\,B} - 2D^A D^B\right) \\
& + \frac{1}{8}(\operatorname{Im} f_{AB})\varepsilon^{\mu\nu\rho\sigma} F_{\mu\nu}^A F_{\rho\sigma}^B + \frac{1}{4}\mathrm{i}(\operatorname{Im} D_\mu f_{AB})\bar{\lambda}^A \gamma_* \gamma^\mu \lambda^B \\
& + \left[\frac{1}{2\sqrt{2}} f_{AB\alpha}\bar{\chi}^\alpha \left(-\frac{1}{2}\gamma^{\mu\nu} F_{\mu\nu}^{-A} + \mathrm{i}D^A\right) P_L\lambda^B - \frac{1}{4} f_{AB\alpha} F^\alpha \bar{\lambda}^A P_L \lambda^B \right. \\
& \left. + \frac{1}{8} f_{AB\alpha\beta}\bar{\chi}^\alpha P_L \chi^\beta \bar{\lambda}^A P_L \lambda^B + \text{h.c.}\right],
\end{aligned}
\tag{14.57}
$$

where

$$
\begin{aligned}
D_\mu Z^\alpha &= \partial_\mu Z^\alpha - A_\mu{}^A k_A{}^\alpha, \\
P_L \nabla_\mu \chi^\alpha &= \partial_\mu P_L \chi^\alpha - A_\mu^A \frac{\partial k_A{}^\alpha(Z)}{\partial Z^\beta} P_L \chi^\beta + \Gamma^\alpha_{\beta\gamma} P_L \chi^\gamma D_\mu Z^\beta, \\
D_\mu \lambda^A &= \partial_\mu \lambda^A + A_\mu{}^B f_{BC}{}^A \lambda^C, \\
F_{\mu\nu}^A &= \partial_\mu A_\nu^A - \partial_\nu A_\mu^A + f_{BC}{}^A A_\mu^B A_\nu^C, \\
D_\mu f_{AB} &= \partial_\mu f_{AB} - 2A_\mu^C f_{C(A}{}^D f_{B)D}.
\end{aligned}
\tag{14.58}
$$

Exercise 14.10 *Write the full action for one chiral multiplet and one gauge multiplet, with Kähler potential $K = Z\bar{Z}$, kinetic matrix $f_{11} = 1$, and moment map $\mathcal{P}_1 = -gZ\bar{Z}+\xi$. Show that the superpotential can only be a constant.*

14.5 The physical theory

14.5.1 Elimination of auxiliary fields

The Lagrangian of (14.57) contains the complex auxiliary fields F^α of chiral multiplets, their conjugates $\bar{F}^{\bar{\beta}}$, and the real auxiliaries D^A of gauge multiplets. The Euler–Lagrange equations of these fields do not contain spacetime derivatives. They are purely algebraic and can be easily solved to express auxiliary fields in terms of the physical components. The dynamical content of the theory is preserved if we substitute the solutions for F^α and D^A back in the Lagrangian and transformation rules. The SUSY algebra changes in this process, but the physical content is unchanged. We saw examples in Ch. 6. We now discuss the elimination of auxiliary fields in the general Lagrangian (14.57) of gauged supersymmetry.

We first discuss the elimination of the auxiliary fields F^α (and $\bar{F}^{\bar{\beta}}$). The relevant terms in the Lagrangian of (14.57) are

$$
\mathcal{L}_{\text{aux},F} = g_{\alpha\bar{\beta}} F^\alpha \bar{F}^{\bar{\beta}} - F^\alpha f_\alpha - \bar{F}^{\bar{\beta}} \bar{f}_{\bar{\beta}},
\tag{14.59}
$$

with

$$
f_\alpha = -W_\alpha + \tfrac{1}{2} g_{\alpha\bar{\beta}} \Gamma^{\bar{\beta}}_{\bar{\gamma}\bar{\alpha}} \bar{\chi}^{\bar{\gamma}} P_R \chi^{\bar{\alpha}} + \tfrac{1}{4} f_{AB\alpha}\bar{\lambda}^A P_L \lambda^B.
\tag{14.60}
$$

The equations of motion,

$$g_{\alpha\bar{\beta}}\bar{F}^{\bar{\beta}} = f_\alpha, \qquad g_{\alpha\bar{\beta}}F^\alpha = \bar{f}_{\bar{\beta}}, \tag{14.61}$$

can be immediately solved using the inverse Kähler metric $g^{\alpha\bar{\beta}}$,

$$\bar{F}^{\bar{\beta}} = g^{\alpha\bar{\beta}}f_\alpha, \qquad F^\alpha = g^{\alpha\bar{\beta}}\bar{f}_{\bar{\beta}}. \tag{14.62}$$

When these results are substituted in (14.59), one obtains

$$\mathcal{L}_{\mathrm{aux},F} = -g^{\alpha\bar{\beta}}f_\alpha\bar{f}_{\bar{\beta}}. \tag{14.63}$$

The negative sign and the appearance of an inverse metric are general features of the elimination process. We will study the physical content of (14.63), but we first consider the elimination of the D^A auxiliaries.

The Lagrangian (14.57) contains the following terms involving the D^A:

$$\begin{aligned}\mathcal{L}_{\mathrm{aux},D} &= \frac{1}{2}(\mathrm{Re}\, f_{AB})D^A D^B \\ &+ D^A\left[-\mathcal{P}_A + \frac{1}{2\sqrt{2}}\mathrm{i} f_{AB\alpha}\bar{\chi}^\alpha P_L\lambda^B - \frac{1}{2\sqrt{2}}\mathrm{i}\bar{f}_{AB\bar{\alpha}}\bar{\chi}^{\bar{\alpha}}P_R\lambda^B\right].\end{aligned} \tag{14.64}$$

The solution of the field equation for D^A thus gives

$$\mathrm{Re}\, f_{AB}D^B = \mathcal{P}_A - \frac{1}{2\sqrt{2}}\mathrm{i} f_{AB\alpha}\bar{\chi}^\alpha P_L\lambda^B + \frac{1}{2\sqrt{2}}\mathrm{i}\bar{f}_{AB\bar{\alpha}}\bar{\chi}^{\bar{\alpha}}P_R\lambda^B. \tag{14.65}$$

Substitution in (14.64) gives the physically equivalent action

$$\begin{aligned}\mathcal{L}_{\mathrm{aux},D} = -\frac{1}{2}(\mathrm{Re}\, f)^{-1\,AB}&\left(\mathcal{P}_A - \frac{1}{2\sqrt{2}}\mathrm{i} f_{AC\alpha}\,\bar{\chi}^\alpha P_L\lambda^C + \frac{1}{2\sqrt{2}}\mathrm{i}\bar{f}_{AC\bar{\alpha}}\,\bar{\chi}^{\bar{\alpha}}P_R\lambda^C\right) \\ \times&\left(\mathcal{P}_B - \frac{1}{2\sqrt{2}}\mathrm{i} f_{BD\beta}\,\bar{\chi}^\beta P_L\lambda^D + \frac{1}{2\sqrt{2}}\mathrm{i}\bar{f}_{BD\bar{\beta}}\,\bar{\chi}^{\bar{\beta}}P_R\lambda^D\right).\end{aligned} \tag{14.66}$$

Again a negative sign appears together with the inverse metric $(\mathrm{Re}\, f)^{-1\,AB}$.

14.5.2 The scalar potential

The scalar potential in a classical field theory is defined as the sum of all terms in the Lagrangian that contain only scalar fields and no spacetime derivatives. The sum is multiplied by -1 because the basic structure of a Lagrangian in classical mechanics is $\mathcal{L} = T - V$, kinetic minus potential. The Hamiltonian is $T + V$. The lowest energy state of the theory is found by minimizing the Hamiltonian. It is usually assumed that this vacuum state is Lorentz invariant and translation invariant. Lorentz invariance requires that gauge fields and fermions vanish in the vacuum, and translation invariance means that scalar kinetic terms, which contain spacetime derivatives, can be ignored. Thus the classical approximation to the vacuum state is obtained by minimizing the scalar potential. The values of scalar fields at the minimum are usually called 'vacuum expectation values'.

In $\mathcal{N}=1$, $D=4$ supersymmetric field theories, the scalar potential arises entirely from the elimination of auxiliary fields. We retain only the scalar terms in (14.63) and (14.66) and write the potential

$$V = g^{\alpha\bar{\beta}} W_\alpha \overline{W}_{\bar{\beta}} + \tfrac{1}{2} (\operatorname{Re} f)^{-1\,AB} \mathcal{P}_A \mathcal{P}_B . \tag{14.67}$$

The first term is called the F-term and the second term the D-term in accordance with their origin from the auxiliary fields. Minimization of this potential determines the vacuum state of the supersymmetric theory.

It is useful to recognize that this form arises from the auxiliary scalar contributions to fermion transformations. The general structure for any supersymmetric theory is[6]

$$V = (\delta_s \text{fermion})\,(\text{metric})\,(\delta_s \text{fermion})\,, \tag{14.68}$$

where $\delta_s$fermion refers to the scalar parts of the supersymmetry transformations of all fermions. These are often called *fermion shifts*. Specifically, in gauged global supersymmetry,

$$\begin{aligned} \delta_s P_L \chi^\alpha &\equiv \frac{1}{\sqrt{2}} F^\alpha = -\frac{1}{\sqrt{2}} g^{\alpha\bar{\beta}} \overline{W}_{\bar{\beta}} , \\ \delta_s P_L \lambda^A &\equiv \frac{1}{2} \mathrm{i} D^A = \frac{1}{2} \mathrm{i} (\operatorname{Re} f)^{-1\,AB} \mathcal{P}_B . \end{aligned} \tag{14.69}$$

Only the scalar parts of the on-shell values of the auxiliary fields are included. Using (14.69) and the complex conjugate expressions for the P_R projections of the fermions, (14.67) can be rewritten as

$$V = 2 \left(\delta_s P_L \chi^\alpha\right) g_{\alpha\bar{\beta}} \left(\delta_s P_R \chi^{\bar{\beta}}\right) + 2 \left(\delta_s P_L \lambda^A\right) \operatorname{Re} f_{AB} \left(\delta_s P_R \lambda^B\right) . \tag{14.70}$$

The structure (14.68) is very important in supersymmetry, and we will see later that it applies also in supergravity. Thus it is generally valid.

Notice that the fermion variations in (14.67) and (14.70) are contracted with the appropriate metric in the chiral and gauge sectors of the theory. These metrics are positive definite. It then follows that the potential is positive semi-definite. Positivity of the potential is universal in global supersymmetry.[7] This is a consequence of the SUSY algebra; see (6.3). So the expectation value of the Hamiltonian in any state of the theory is non-negative!

[6] The general rule is that the metric in all cases is determined by the fermion kinetic terms.
[7] We will see later that there are negative contributions in supergravity because gravitinos are gauge fields.

14.5.3 The vacuum state and SUSY breaking

As discussed in the previous section, the vacuum state is determined in the classical approximation by minimizing the scalar potential $V(Z, \bar{Z})$ given in (14.67) with respect to the scalar fields Z^α of the chiral multiplets in the theory. There may be multiple configurations of the Z^α which realize the *same* minimum value of the potential. In this case the vacuum configuration is not unique, and any one of these configurations determines a possible vacuum state.

One important physical question of interest is whether symmetries of the action are broken in the vacuum. If so we say that the symmetry is *spontaneously* broken. We assume that readers are familiar with the consequences of spontaneously broken *internal* symmetry. If the symmetry in question is a global symmetry, then the Goldstone theorem tells us that the theory contains a massless scalar particle. If the broken symmetry is a gauge symmetry, then, according to the Higgs mechanism, the Goldstone boson disappears but there is a massive spin-1 particle. In the classical approximation this information about the particle mass spectrum is contained in the quadratic terms in the power series expansion of the Lagrangian about the vacuum configuration. There is a more formal way (see [110]), to characterize the possibility of breaking of a global internal symmetry. The symmetry is unbroken if the charge operator T_A annihilates the vacuum state, i.e. if $T_A|0\rangle = 0$, and the symmetry is broken if $T_A|0\rangle \neq 0$.

In this section we are primarily interested in whether supersymmetry is spontaneously broken in the vacuum state. If the elementary particles and interactions observed in Nature come from a supersymmetric theory then SUSY breaking is vital. The reason is that unbroken supersymmetry requires that for every bosonic particle, elementary or composite, there is a fermion of the same mass and conversely. This is certainly not the situation we observe, so if supersymmetry is relevant to experiment, it must appear as a broken symmetry.

At the quantum level the question of SUSY breaking is the question whether the supercharge components Q_α annihilate the vacuum state or not, i.e. whether $Q_\alpha|0\rangle = 0$ or $Q_\alpha|0\rangle \neq 0$. The key to this question is contained in (6.3). The left-hand side is the sum of four non-negative terms. Applying the anti-commutator to the vacuum state we can see the following:

1. If all four supercharges annihilate the vacuum state, then so does the Hamiltonian. Thus *unbroken supersymmetry requires that the vacuum energy vanishes.*
2. If at least one component of the supercharge does not annihilate the vacuum, then the Hamiltonian does not annihilate the vacuum. Thus *broken supersymmetry requires positive vacuum energy.*

In the classical approximation this leads to an astonishingly simple criterion for spontaneous SUSY breaking, stated in Box 14.4. Since the potential $V(Z, \bar{Z})$ in (14.67) is a non-negative quadratic form in the quantities $W_\alpha(Z)$ and $\mathcal{P}_A(Z, \bar{Z})$, we can restate the criterion in the following simple way. The vacuum is supersymmetric if and only if there is a configuration of the scalar fields Z^α, $\bar{Z}^{\bar{\alpha}}$ such that the algebraic equations

Box 14.4 **Spontaneous SUSY breaking**

If the minimum value of the scalar potential $V(Z, \bar{Z})$ is zero, then supersymmetry is unbroken. If the minimum value is positive, then supersymmetry is broken.

$$W_\alpha(Z) = 0 \quad \text{and} \quad \mathcal{P}_A(Z, \bar{Z}) = 0 \tag{14.71}$$

have a common solution. As we can see from (14.62) and (14.65), this is equivalent to saying that supersymmetry is preserved if all the auxiliary fields F^α and D^A vanish in the vacuum and is broken otherwise. (Recall that fermion fields vanish in a Lorentz invariant vacuum.)

Readers may question whether the classical approximation provides the correct answer to the properties of the vacuum of the theory. Fortunately there is a remarkable non-renormalization theorem [111] which ensures that there are no perturbative quantum corrections to the classical superpotential $W(Z)$. The equations (14.71) are then valid to all orders in quantum perturbation theory. In some SUSY gauge theories there are non-perturbative additions to the superpotential which change the situation [112, 113].

Much is known about the structure of the equations (14.71) and whether or not there is a solution that produces exact supersymmetry.[8] It is worthwhile to mention one general theorem (see Sec. 27.4 of [30]), which applies to gauged supersymmetry. In that framework, a common solution of the $W_\alpha = 0$ equations is part of an 'orbit' of solutions related by complexified gauge transformations. There is always a point on the orbit where the moment maps $\mathcal{P}_A(Z, \bar{Z})$ also vanish, unless the moment maps contain Fayet–Iliopoulos terms. Thus the question of SUSY breaking is controlled by the F-term conditions unless there are Fayet–Iliopoulos couplings.

We study the equations first in the simplest context of ungauged supersymmetry with canonical Kähler metric $g_{\alpha\bar{\beta}} = \delta_{\alpha\bar{\beta}}$. Suppose that the theory contains n chiral multiplets. The vacuum is supersymmetric if there is a common root of the n equations $\partial_\alpha W(Z) = 0$. These are n *complex* equations in n *complex* variables, so for a generic superpotential there is a common solution. The superpotential must be special in some way to obtain broken supersymmetry.

Here is an example from [30] and [113] of the O'Raifeartaigh mechanism [114], which shows how this can happen. Suppose that there are three complex scalars which we call X_1, X_2 and Z and that the superpotential is

$$W = X_1 g_1(Z) + X_2 g_2(Z). \tag{14.72}$$

The vacuum would be supersymmetric if the three equations

$$\partial_{X_1} W = g_1(Z) = 0, \qquad \partial_{X_2} W = g_2(Z) = 0, \qquad \partial_Z W = X_1 g_1'(Z) + X_2 g_2'(Z) = 0 \tag{14.73}$$

[8] A useful recent reference is [113].

have a common root. However, the first two equations cannot both be satisfied unless the holomorphic functions $g_1(Z)$ and $g_2(Z)$ are specially related. Thus the situation is that, once the form (14.72) is chosen, supersymmetry breaking is generic.

Once it is established that there is no solution of the $\partial_\alpha W(Z) = 0$ conditions, one proceeds to find the SUSY-breaking vacuum by minimizing the potential. In this example the potential is

$$V = |g_1(Z)|^2 + |g_2(Z)|^2 + |X_1 g_1'(Z) + X_2 g_2'(Z)|^2. \tag{14.74}$$

Generically, the stationary conditions $\partial V/\partial X_i = 0$ are solved by fixing the ratio X_1/X_2 so that the third term of V vanishes. The vacuum expectation value of Z is then fixed at the minimum of the first two terms. Only the ratio of X_1 and X_2 has been fixed, so their common complex scale is not determined. Thus there is a degenerate subspace of SUSY-breaking vacua in this model at least in the classical approximation.[9]

It is useful to discuss the role of $U(1)_R$ symmetry in this model. In Sec. 6.2.1 we saw that $U(1)_R$ is a global symmetry of supersymmetric field theories. The $U(1)_R$ transformation rules of the fields of chiral multiplets are given in (6.43). (In gauged supersymmetry gaugino fields must also transform as shown in (6.65).) A supersymmetric theory with superpotential $W(Z)$ is $U(1)_R$ invariant provided that there is a choice of the charges r_α of the scalar fields Z^α such that $W(Z)$ has overall charge 2:

$$\sum_\alpha r_\alpha Z^\alpha W_\alpha = 2W. \tag{14.75}$$

The superpotential of (14.72) satisfies this condition if we choose $r_{X_i} = 2$ and $r_Z = 0$. Thus $U(1)_R$ is spontaneously broken in a generic vacuum of this model in which the X_i fields are non-vanishing. (The exceptional vacuum in which $X_1 = X_2 = 0$ is $U(1)_R$ invariant.) The properties of the vacua in this model are typical of supersymmetric theories in which SUSY breaking is determined by the superpotential. Spontaneously broken $U(1)_R$ symmetry is important in such models as shown in [115]. (See [116] for a pedagogical discussion.)

14.5.4 Supersymmetry breaking and the Goldstone fermion

We now discuss an important feature of spontaneous breakdown of SUSY, namely the fact that the theory necessarily contains a massless fermion commonly called the Goldstino. The Goldstino field is a linear combination of the elementary fermion fields in the Lagrangian, which we will identify in the classical approximation. We work first in the simple context of ungauged supersymmetry with flat Kähler metric using the notation of (6.42), i.e. with indices raised by the inverse Kähler metric. Then we will extend the result to the general case of gauged supersymmetry.

To be definite we take a model which contains n chiral multiplets Z^α, $P_L\chi^\alpha$, F^α with holomorphic superpotential $W(Z)$. The physical Lagrangian, with auxiliary fields eliminated, is given in (6.42). Since we are interested in SUSY breaking, we assume that the

[9] Classical degeneracies of SUSY-breaking vacua are expected to be 'lifted' by perturbative radiative corrections.

n equations $W_\alpha(Z)=0$ have no common solution. Instead we proceed to minimize the scalar potential

$$V(Z,\bar{Z})=W_\alpha(Z)\overline{W}^\alpha(\bar{Z}) \tag{14.76}$$

by finding the solution $Z^\alpha=Z_0^\alpha$ of the stationary condition

$$\partial_\alpha V=W_{\alpha\beta}(Z_0)\overline{W}^\beta(\bar{Z}_0)=0, \tag{14.77}$$

which gives the global minimum with $V(Z_0,\bar{Z}_0)>0$. The condition (14.77) shows that, at minimum, the complex symmetric matrix $W_{\alpha\beta}$ has a zero mode with eigenvector $\overline{W}^\beta$. This vector is non-vanishing because $V>0$. The matrix $W_{\alpha\beta}$, evaluated at minimum, is the mass matrix of the fermions of the model, so there is a massless fermion in the spectrum associated with the null eigenvector. This massless fermion is the Goldstino.

To identify it precisely, and to prepare for the more complicated case of gauged SUSY, let us write the Lagrangian of a 'toy model' containing a set of n free chiral fermions $P_L\psi^i$ (and their charge conjugates $P_R\psi_i$), namely

$$\mathcal{L}=-\bar{\psi}_i\partial\!\!\!/ P_L\psi^i+\tfrac{1}{2}M_{ij}\bar{\psi}^i P_L\psi^j+\tfrac{1}{2}\bar{M}^{ij}\bar{\psi}_i P_R\psi_j, \tag{14.78}$$

with $\bar{M}^{ij}$ being the complex conjugate of M_{ij}. The equations of motion are

$$\partial\!\!\!/ P_L\psi^i=\bar{M}^{ij}P_R\psi_j,\qquad \partial\!\!\!/ P_R\psi_i=M_{ij}P_L\psi^j. \tag{14.79}$$

Suppose now that the mass matrix has a zero mode eigenvector $\bar{v}^i$. Thus $M_{ij}\bar{v}^j=0$. The linear combination

$$P_L v\equiv v_i P_L\psi^i \tag{14.80}$$

is the corresponding massless fermion. To see this we simply compute

$$\partial\!\!\!/ P_L v=v_i\partial\!\!\!/ P_L\psi^i=v_i\overline{M}^{ij}P_R\psi_j=0. \tag{14.81}$$

In the last step we used the complex conjugate of the zero mode equation.

The toy model result can be applied immediately to the fermions of the model of (6.42). Using (the conjugate of) the null eigenvector $\overline{W}^\alpha$, we identify the Goldstino field

$$P_L v=-\frac{1}{\sqrt{2}}W_\alpha P_L\chi^\alpha=P_L\chi^\alpha\,\delta_s P_R\chi_\alpha. \tag{14.82}$$

We used the fermion shifts (14.69) (and $P_R\chi_\alpha=g_{\alpha\bar{\beta}}P_R\chi^{\bar{\beta}}$) to write the last form which turns out to generalize to gauged supersymmetry. The Goldstone fermion transforms non-trivially under supersymmetry,

$$\delta_s P_L v=\delta_s P_L\chi^\alpha g_{\alpha\bar{\beta}}\delta_s P_R\chi^{\bar{\beta}}. \tag{14.83}$$

We will now discuss the Goldstino in the general gauged supersymmetric theory with Lagrangian given in (14.57). After elimination of auxiliary fields one obtains the scalar potential (14.67). We assume that SUSY is broken in the vacuum, so there is no common solution to the equations (14.71). The vacuum then corresponds to the minimum positive value of V. The stationary condition $\partial_\alpha V = 0$ is very relevant and we will write it explicitly below.

The major complication in the general theory is that the fermions $P_L\chi^\alpha$ and $P_L\lambda^A$ mix. This is seen in the fermion mass term which is obtained from (14.57), (14.60), (14.63) and (14.66),

$$\begin{aligned}
\mathcal{L}_{m\,\text{fermions}} &= -\tfrac{1}{2}m_{\alpha\beta}\bar{\chi}^\alpha P_L\chi^\beta - m_{\alpha A}\bar{\chi}^\alpha P_L\lambda^A - \tfrac{1}{2}m_{AB}\bar{\lambda}^A P_L\lambda^B + \text{h.c.},\\
m_{\alpha\beta} &= \nabla_\alpha\partial_\beta W \equiv \partial_\alpha\partial_\beta W - \Gamma^\gamma_{\alpha\beta}\partial_\gamma W,\\
m_{\alpha A} &= \mathrm{i}\sqrt{2}\left[\partial_\alpha\mathcal{P}_A - \tfrac{1}{4}f_{AB\,\alpha}\,(\mathrm{Re}\,f)^{-1\,BC}\,\mathcal{P}_C\right] = m_{A\alpha},\\
m_{AB} &= -\tfrac{1}{2}f_{AB\,\alpha}g^{\alpha\bar{\beta}}\overline{W}_{\bar{\beta}}.
\end{aligned}\tag{14.84}$$

The quantities $m_{\alpha\beta}$, $m_{\alpha A}$, m_{AB} are evaluated in the vacuum configuration $Z^\alpha = Z^\alpha_0$. Thus we find the fermion mass matrix of the system in the block form

$$\begin{pmatrix} m_{\alpha\beta} & m_{\alpha B} \\ m_{A\beta} & m_{AB} \end{pmatrix}.\tag{14.85}$$

We will show that the vector

$$-\sqrt{2}\begin{pmatrix} \delta_s P_L\chi^\beta \\ \delta_s P_L\lambda^B \end{pmatrix} = \begin{pmatrix} \overline{W}^\beta \\ -\mathrm{i}\mathcal{P}^B/\sqrt{2} \end{pmatrix} = \begin{pmatrix} g^{\beta\bar{\gamma}}\overline{W}_{\bar{\gamma}} \\ -\mathrm{i}(\mathrm{Re}\,f)^{-1\,BC}\mathcal{P}_C/\sqrt{2} \end{pmatrix}\tag{14.86}$$

is a zero eigenvector of the mass matrix and is non-vanishing because $V > 0$ in the SUSY-breaking vacuum. To show this we need two facts. First we note that the stationary condition $\partial_\alpha V = 0$ can be written neatly in terms of the elements of the mass matrix, namely as

$$\partial_\alpha V = m_{\alpha\beta}\overline{W}^\beta + m_{\alpha A}(-\mathrm{i}\mathcal{P}^A/\sqrt{2}) = 0.\tag{14.87}$$

This is just the statement that the 'top row' of the zero mode equation is satisfied. The 'bottom row' condition is also satisfied, since

$$m_{\beta A}\overline{W}^\beta + m_{AB}(-\mathrm{i}\mathcal{P}^B/\sqrt{2}) = \sqrt{2}k^{\bar{\alpha}}_A\overline{W}_{\bar{\alpha}} = 0.\tag{14.88}$$

Here we have used the relation (13.51) between Killing vectors and moment maps. The last expression vanishes simply because the superpotential must be gauge invariant!

The toy model then tells us that the Goldstino field is obtained by contraction of χ^α and λ^A with the appropriate components of the conjugate of the null eigenvector,

$$P_L\upsilon = -\frac{1}{\sqrt{2}}P_L\left[W_\alpha\chi^\alpha + \frac{1}{\sqrt{2}}\mathrm{i}\mathcal{P}_A\lambda^A\right].\tag{14.89}$$

It is useful to incorporate the results in (14.69) so that we recognize the general structure

$$P_L v = P_L\left[\chi^\alpha g_{\alpha\bar\beta}\delta_s P_R\chi^{\bar\beta} + \lambda^A(\mathrm{Re}\, f)_{AB}\delta_s P_R\lambda^B\right]. \tag{14.90}$$

The Goldstone fermion is always the linear combination in which all elementary fermion fields are multiplied by the fermion shifts and summed with the appropriate metric. Note that (14.87) and (14.88) can also be reexpressed as contractions with the fermion shifts.

14.5.5 Mass spectra and the supertrace sum rule

After finding the vacuum state of a supersymmetric theory, the next step is to study the dynamics of fluctuations about the vacuum. In this section we are concerned with the mass spectra of these fluctuations. Masses are determined by quadratic terms in an expansion of the Lagrangian about the vacuum configuration. Since the vacuum values of spinor and vector fields vanish in a Lorentz invariant vacuum, quadratic terms in these fields can be read directly from the Lagrangian (14.57) after elimination of auxiliary fields. For scalars we define the fluctuations as the deviation from the vacuum values Z_0^α, $\bar Z_0^{\bar\alpha}$, i.e. as

$$u^\alpha(x) \equiv Z^\alpha(x) - Z_0^\alpha, \qquad \bar u^{\bar\alpha}(x) \equiv \bar Z^{\bar\alpha}(x) - \bar Z_0^{\bar\alpha}. \tag{14.91}$$

Quadratic terms in u^α, $\bar u^{\bar\alpha}$ in the expansion of the potential V of (14.67) determine the scalar masses. (Linear terms in u^α, $\bar u^{\bar\alpha}$ vanish, since we are expanding about the vacuum.)

From the quadratic terms for scalars, spinors, and vectors we can find the matrix of squared masses for each type of particle. The eigenvalues of these matrices are the physical (squared) masses of the particles of the theory. Our main purpose is to outline the derivation of the mass supertrace sum rule, which involves a weighted sum of the scalar, spinor, and vector masses [117, 118]. This sum rule can be presented in a uniform fashion so as to include the various possibilities for broken or unbroken supersymmetry and broken or unbroken gauge symmetry. To simplify our discussion we will assume that gauge multiplet kinetic terms are 'canonical', which means that we take

$$f_{AB} = \delta_{AB}. \tag{14.92}$$

The scalar mass matrix is actually the most complicated case because quadratic terms in the expansion of the potential (14.67) include both 'hermitian' terms involving products $u^\alpha \bar u^{\bar\beta}$ and 'non-hermitian' terms with products $u^\alpha u^\beta$ (and their conjugates).[10] The trace of the scalar mass matrix $\mathcal{M}_0^2$, which is the sum of the physical squared masses, involves the contraction of the inverse kinetic matrix with the second derivative of the potential. Since the scalar kinetic terms in (14.29) have the form $\partial Z\partial\bar Z$, the trace involves only the hermitian terms of the potential, which give the result

[10] See Sec. 27.5 of [30] for more detail.

$$
\begin{aligned}
\operatorname{tr}\mathcal{M}_0^2 &= g^{ab}D_a\partial_b V = 2g^{\alpha\bar\alpha}\partial_\alpha\partial_{\bar\alpha}V \\
&= 2g^{\alpha\bar\alpha}\Big(\nabla_\alpha W_\beta g^{\beta\bar\beta}\overline{\nabla}_{\bar\alpha}\overline{W}_{\bar\beta} + W_\beta g^{\beta\bar\beta}\nabla_\alpha\overline{\nabla}_{\bar\alpha}\overline{W}_{\bar\beta} \\
&\quad +\partial_\alpha\mathcal{P}_A\,\delta^{AB}\,\partial_{\bar\alpha}\mathcal{P}_B + \mathcal{P}_A\delta^{AB}\partial_\alpha\partial_{\bar\alpha}\mathcal{P}_B\Big).
\end{aligned}
\tag{14.93}
$$

Since $\nabla_\alpha\overline{W}_{\bar\beta} = \partial_\alpha\overline{W}_{\bar\beta} = 0$, the double covariant derivative on $\overline{W}_{\bar\beta}$ reduces to a commutator, which is then expressed in terms of the Ricci tensor of the target space; see (13.24). The second term in (14.93) thus simplifies, viz.

$$
2g^{\alpha\bar\alpha}g^{\beta\bar\beta}W_\beta\nabla_\alpha\overline{\nabla}_{\bar\alpha}\overline{W}_{\bar\beta} = 2g^{\alpha\bar\alpha}g^{\beta\bar\beta}W_\beta R_{\alpha\bar\alpha\bar\beta}{}^{\bar\gamma}\overline{W}_{\bar\gamma} = 2W_\beta R^{\beta\bar\gamma}\overline{W}_{\bar\gamma}. \tag{14.94}
$$

Exercise 14.11 *To check the result that* $\operatorname{tr}\mathcal{M}_0^2$ *involves only the hermitian terms of the mass matrix, consider the free scalar Lagrangian*

$$
\mathcal{L} = -\partial_\mu Z\partial^\mu\bar Z - \left(aZ^2 + \bar a\bar Z^2 + bZ\bar Z\right), \tag{14.95}
$$

in which b *is a positive real parameter and* a *is complex. Show that* $\operatorname{tr}\mathcal{M}_0^2 = 2b$. *Hint: express the theory in real fields using* $Z = (A + \mathrm{i}B)/\sqrt{2}$.

To study spinor masses it is useful to return to the toy model of the previous section and write the second order wave equation, obtained by applying $\not\partial$ to (14.79):

$$
\Box P_L\psi^i = \overline{M}^{ik}M_{kj}P_L\psi^j. \tag{14.96}
$$

Thus we can define the spinor squared mass matrix $(\mathcal{M}_{1/2}^2)^i{}_j \equiv \overline{M}^{ik}M_{kj}$. In the general gauged SUSY model, $(\mathcal{M}_{1/2}^2)^i{}_j$ can be obtained from (14.84). Since $m_{AB} = 0$ under the assumption (14.92), the trace is

$$
\begin{aligned}
\operatorname{tr}\mathcal{M}_{1/2}^2 &= g^{\alpha\bar\alpha}m_{\alpha\beta}g^{\beta\bar\beta}\overline{m}_{\bar\beta\bar\beta} + 2g^{\alpha\bar\alpha}m_{\alpha A}\delta^{AB}m_{\bar\alpha B} \\
&= g^{\alpha\bar\alpha}\nabla_\alpha W_\beta g^{\beta\bar\beta}\overline{\nabla}_{\bar\alpha}\overline{W}_{\bar\beta} + 4g^{\alpha\bar\alpha}\partial_\alpha\mathcal{P}_A\delta^{AB}\partial_{\bar\alpha}\mathcal{P}_B.
\end{aligned}
\tag{14.97}
$$

If gauge symmetries are spontaneously broken in the vacuum state, then vector fields acquire mass by the Higgs mechanism. Their masses come from non-derivative terms in the kinetic Lagrangian of chiral multiplet scalars. From (14.37) and $\mathcal{L}_{\rm kin,chir}$ in (14.57) we find the trace of the vector squared mass matrix,

$$
\begin{aligned}
\operatorname{tr}\mathcal{M}_1^2 &= 2g_{\alpha\bar\alpha}k_A{}^\alpha\delta^{AB}k_B^{\bar\alpha} \\
&= 2g^{\alpha\bar\alpha}\partial_\alpha\mathcal{P}_A\delta^{AB}\partial_{\bar\alpha}\mathcal{P}_B.
\end{aligned}
\tag{14.98}
$$

To obtain the supertrace sum rule we add the three contributions above, each multiplied by the number of spin states of the particle it describes, and with bosons and fermions counted with opposite signs. This leads to

$$
\begin{aligned}
\operatorname{Supertr}\mathcal{M}^2 &\equiv \sum_J(-)^{2J}(2J+1)m_J^2 = \operatorname{tr}\mathcal{M}_0^2 - 2\operatorname{tr}\mathcal{M}_{1/2}^2 + 3\operatorname{tr}\mathcal{M}_1^2 \\
&= 2W_\alpha R^{\alpha\bar\alpha}\overline{W}_{\bar\alpha} + 2g^{\alpha\bar\alpha}\mathcal{P}_A\delta^{AB}\partial_\alpha\partial_{\bar\alpha}\mathcal{P}_B \\
&= 2W_\alpha R^{\alpha\bar\alpha}\overline{W}_{\bar\alpha} + 2\mathrm{i}D^A\nabla_\alpha k_A{}^\alpha.
\end{aligned}
\tag{14.99}
$$

Box 14.5 **Mass supertrace sum rule**

The supertrace of masses vanishes for flat Kähler manifolds unless supersymmetry is spontaneously broken by non-vanishing D-terms.

The conclusion is in Box 14.5. For linearly realized internal symmetries as in (14.32), the result becomes $-2\mathrm{i}D^A \operatorname{tr} t_A$. The trace of any generator t_A of a simple non-abelian factor of the gauge group vanishes, but the trace of a U(1) generator vanishes only if the sum of the charges of the chiral multiplets in the theory vanishes. Vanishing total charge is required in order to cancel the gravitational anomaly in the one-loop three-point function of the U(1) gauge current and two stress tensors. Otherwise the quantum theory would not be consistent.

Supersymmetric gauge theories have been very widely studied as the possible answer to the vital question of the new physical laws which may govern elementary particles at energy scales of 100 GeV and higher, energies beyond the scale at which the standard model is currently tested. It is tempting to think that the simplest type of theory would be applicable, namely a theory with flat Kähler metric and linear gauge symmetries. We have just shown that the supertrace sum rule gives a vanishing result in this setting. As discussed in more detail in Sec. 28.3 of [30], a vanishing sum rule implies that masses of the undiscovered SUSY partners of some known particles are well below experimental limits. Thus there must be other contributions to the supertrace sum rule. Several mechanisms have been studied. For example, new contributions do arise when global supersymmetry theories are coupled to supergravity and supersymmetry breaking occurs due to supergravity effects. These effects will be discussed in Ch. 17.

14.5.6 Coda

In most of this chapter we emphasized scalar target spaces with general Kähler metrics and nonlinear Killing symmetries. In this way we showed how complex differential geometry can arise even in global supersymmetry. This can be viewed as a step towards supergravity in which the differential geometry of spacetime enters the picture. The Kähler metric $g_{\alpha\bar{\beta}}$ for chiral multiplets and the general metric $\operatorname{Re} f_{AB}$ for gauge multiplets also appear naturally in supergravity, especially in supergravity models which are the low energy limits of string theory constructions.

Appendix 14A Superspace

In this appendix we briefly discuss the superspace formalism and its relation to the basic multiplets treated in the main part of the chapter. The superspace method was first formulated in [119, 105] and it is presented in detail in many textbooks (for example, see [120, 121, 122, 123, 30, 124, 125]). In the superspace approach Minkowski spacetime is extended to an eight-dimensional 'supermanifold' whose coordinates are the usual commuting x^μ plus four anti-commuting coordinates θ_α, which transform as a Majorana spinor under Lorentz transformations. It is frequently convenient to use the chiral projections $P_L\theta$ and $P_R\theta$; which transform as two-component Weyl spinors. They are denoted by θ_α and $\bar{\theta}_{\dot{\alpha}}$ in many presentations of four-dimensional superspace. Supersymmetry transformations are 'motions' in the superspace under which coordinates are shifted, viz. $x^\mu \to x'^\mu = x^\mu + \frac{1}{4}\bar{\epsilon}\gamma^\mu\theta, \theta \to \theta' = \theta - \epsilon$.

The multiplets discussed earlier in this chapter are described by functions on superspace, denoted by $\Phi(x,\theta)$, which are called superfields. A superfield can be expanded in a power series in θ_α, and such expansions terminate at order $(\theta)^4$ because the θ_α anti-commute. The various component fields of a multiplet are the coefficients in this series expansion. To illustrate this, we consider a real superfield, which satisfies $\Phi(x,\theta) = \bar{\Phi}(x,\theta)$. Its series expansion is

$$\begin{aligned}\Phi(x,\theta) = C &+ \tfrac{1}{2}\mathrm{i}\bar{\theta}\gamma_*\zeta - \tfrac{1}{8}\bar{\theta}P_L\theta\mathcal{H} - \tfrac{1}{8}\bar{\theta}P_R\theta\bar{\mathcal{H}} - \tfrac{1}{8}\mathrm{i}\bar{\theta}\gamma_*\gamma^\mu\theta B_\mu \\ &- \tfrac{1}{8}\mathrm{i}\bar{\theta}\theta\bar{\theta}\gamma_*\left(\lambda + \tfrac{1}{2}\partial\!\!\!/\zeta\right) + \tfrac{1}{32}\bar{\theta}P_L\theta\bar{\theta}P_R\theta\left(D + \tfrac{1}{2}\Box C\right).\end{aligned} \quad (14.100)$$

The value at $\theta = 0$ defines a real scalar field, i.e. $\Phi(x,0) = C(x)$, which we identify with the lowest component of the real multiplet (14.9). Higher components of (14.9) appear in the θ-dependent terms, and the entire structure encodes the component transformation rules (14.10) in the precise way we will describe below.

Exercise 14.12 *Check that $\bar{\theta}\theta\bar{\theta}\theta = 2\bar{\theta}P_L\theta\bar{\theta}P_R\theta$. Hint: since there are only two components in $P_L\theta$, the bilinear $\bar{\theta}P_L\theta$ exhausts the left-chiral components; hence, for example, $\theta\bar{\theta}P_L\theta = P_R\theta\bar{\theta}P_L\theta$.*

Supersymmetry transformations, defined above as shifts in superspace, are implemented by the differential operators[11]

$$\begin{aligned}\mathbb{Q}_\alpha &= \frac{\vec{\partial}}{\partial\bar{\theta}^\alpha} - \frac{1}{4}(\gamma^\mu\theta)_\alpha\frac{\partial}{\partial x^\mu}, \\ \overline{\mathbb{Q}}^\alpha \equiv C^{\alpha\beta}\mathbb{Q}_\beta &= -\frac{\vec{\partial}}{\partial\theta_\alpha} + \frac{1}{4}(\bar{\theta}\gamma^\mu)^\alpha\frac{\partial}{\partial x^\mu}.\end{aligned} \quad (14.101)$$

The anti-commutator of the operators $\mathbb{Q}$ is

$$\{\mathbb{Q}_\alpha, \overline{\mathbb{Q}}^\beta\} = \tfrac{1}{2}(\gamma^\mu)_\alpha{}^\beta\partial_\mu. \quad (14.102)$$

The variation of a superfield is defined as

$$\delta\Phi \equiv \bar{\epsilon}\mathbb{Q}\,\Phi = \overline{\mathbb{Q}}\epsilon\,\Phi. \quad (14.103)$$

[11] The sign of the derivative of a bosonic quantity with respect to θ, a fermionic quantity, depends on whether one derives from the left or right. Therefore we write $\vec{\partial}$ to indicate that the derivative acts from the left.

When one commutes the supersymmetry operators, one should apply the operators on the fields, as we have seen in Ch. 1 for iterated symmetry transformations; see (1.59). This leads to

$$\delta(\epsilon_1)\delta(\epsilon_2)\Phi = \bar{\epsilon}_2^\alpha \mathbb{Q}_\alpha \overline{\mathbb{Q}}^\beta \epsilon_{1\beta}\,\Phi,$$
$$[\delta(\epsilon_1),\delta(\epsilon_2)]\,\Phi = \bar{\epsilon}_2\{\mathbb{Q}_\alpha, \overline{\mathbb{Q}}^\beta\}\epsilon_{1\beta}\,\Phi = \tfrac{1}{2}\bar{\epsilon}_2\gamma^\mu\epsilon_1\partial_\mu\Phi. \tag{14.104}$$

Note the difference with the calculation in (11.11)–(11.12), which is the origin of the difference of sign between (14.102) and (11.15). The result agrees with the SUSY algebra studied in Sec. 6.2.2 and also with (11.14).

To identify the SUSY variations of components from the superfield variation, we write the θ expansion of $\delta\Phi$:

$$\delta\Phi \equiv \bar{\epsilon}\mathbb{Q}\,\Phi \;=\; \delta C + \tfrac{1}{2}\mathrm{i}\bar{\theta}\gamma_*\delta\zeta + \ldots. \tag{14.105}$$

By computing $\mathbb{Q}\Phi$ explicitly and comparing with the expansion (14.100), it is easy to identify $\delta C = \frac{1}{2}\mathrm{i}\bar{\epsilon}\gamma_*\zeta$ and $\delta\zeta = -\frac{1}{2}\mathrm{i}\gamma_*\partial\!\!\!/ C\epsilon + \ldots$, where the $\ldots$ indicate contributions from the $\mathcal{H}$ and B_μ components of Φ. These results agree with (14.10). The complete set of component transformations can be obtained in this way, but the process is tedious. It is easier for the chiral superfield; see Ex. 14.13 below.

The 'covariant derivative'

$$\mathbb{D} = \frac{\overrightarrow{\partial}}{\partial\bar{\theta}} + \frac{1}{4}\gamma^\mu\theta\frac{\partial}{\partial x^\mu} \tag{14.106}$$

has many applications in the superspace formalism. This operator anti-commutes with the supersymmetry generator $\mathbb{Q}$. Therefore it can be used to impose constraints on a superfield that are compatible with supersymmetry. The most common and most useful constraint defines a chiral superfield. It is defined as a superfield that satisfies

$$P_R\mathbb{D}\Phi = 0. \tag{14.107}$$

At $\theta = 0$ this constraint implies that the linear term in the θ expansion of $\Phi(x,\theta)$ involves only $P_L\theta$. Therefore, the SUSY transform of the lowest component involves only $P_L\epsilon$. This is exactly how we defined the chiral multiplet in components! The remainder of the constraint is a covariantization, i.e. making it consistent with the supersymmetry algebra, as we did in Sec. 14.1.1.

To solve the constraint (14.107), it is useful to redefine the superspace coordinates. Specifically we introduce the shifted bosonic coordinate $x_+^\mu = x^\mu + \frac{1}{8}\bar{\theta}\gamma_*\gamma^\mu\theta$. The chain rule then gives

$$\left.\frac{\overrightarrow{\partial}}{\partial\bar{\theta}}\right|_{x+} = \left.\frac{\overrightarrow{\partial}}{\partial\bar{\theta}}\right|_{x} - \tfrac{1}{4}\gamma_*\gamma^\mu\theta\frac{\partial}{\partial x^\mu}. \tag{14.108}$$

Therefore, when x_+ is used as independent variable, the supersymmetries and covariant derivatives become

$$P_L\mathbb{Q} = P_L\frac{\overrightarrow{\partial}}{\partial\bar{\theta}}, \qquad P_R\mathbb{Q} = P_R\left(\frac{\overrightarrow{\partial}}{\partial\bar{\theta}} - \frac{1}{2}\gamma^\mu\theta\frac{\partial}{\partial x_+^\mu}\right),$$

$$P_L\mathbb{D} = P_L\left(\frac{\overrightarrow{\partial}}{\partial\bar{\theta}} + \frac{1}{2}\gamma^\mu\theta\frac{\partial}{\partial x_+^\mu}\right), \qquad P_R\mathbb{D} = P_R\frac{\overrightarrow{\partial}}{\partial\bar{\theta}}. \tag{14.109}$$

This simplifies the constraint (14.107) and implies that a chiral superfield has an expansion of the form

$$\Phi(x_+,\theta) = Z(x_+) + \frac{1}{\sqrt{2}}\bar{\theta}P_L\chi(x_+) + \frac{1}{4}\bar{\theta}P_L\theta F(x_+). \tag{14.110}$$

Numerical factors were chosen so that the superspace operator $\bar{\epsilon}P_L\mathbb{Q} + \bar{\epsilon}P_R\mathbb{Q}$ gives the component transformation rules (6.15).

Exercise 14.13 *Verify these transformation rules. Use (3.72) to rearrange products of the θ and the fact that $\bar{\theta}\gamma_{\mu\nu}\theta = 0$ due to the symmetry properties.*

Gauge multiplets are defined as real superfields Φ on which a 'supergauge' transformation acts via

$$\Phi \to \Phi + \mathrm{i}\left(\Lambda - \bar{\Lambda}\right). \tag{14.111}$$

The gauge parameters are the components of the chiral superfield Λ. To obtain the component transformations of the gauge multiplet in the Wess–Zumino gauge, one chooses the parameters so that

$$C = 0, \qquad \zeta = 0, \qquad \mathcal{H} = 0. \tag{14.112}$$

This fixes all components of the superfield Λ except for one real field, which remains as the conventional gauge parameter of an abelian gauge potential.

One nice feature of the superspace formalism is that the product of two superfields is a superfield. Multiplication of superfields is equivalent to multiplication in the multiplet calculus but usually simpler. The following exercise contains an example.

Exercise 14.14 *If Φ is a chiral superfield given by the expansion (14.110), show that the expansion of Φ^2 is*

$$\begin{aligned}\Phi(x_+,\theta)^2 = {} & Z(x_+)^2 + \sqrt{2}\bar{\theta}P_L Z(x_+)\chi(x_+) \\ & + \tfrac{1}{4}\bar{\theta}P_L\theta\left(2z(x_+)F(x_+) - \bar{\chi}(x_+)P_L\chi(x_+)\right).\end{aligned} \tag{14.113}$$

Check that this result is compatible with the result of Ex. 14.3.

With similar manipulations one also obtains one of the main formulas of this chapter: (14.15).

Action integrals in the superspace formalism include integration over the θ variables. One form involves the integral $\int \mathrm{d}^4x\, \mathrm{d}^4\theta\, \Phi(x,\theta)$ where Φ is a real superfield, usually composite. By definition integration over Grassmann variables is equivalent to differentiation.

Using the component expansion (14.100), one can see that the θ integral 'selects' the θ^4 component, and we get

$$\int \mathrm{d}^4x\,\mathrm{d}^4\theta\,\Phi(x,\theta) = \tfrac{1}{8}\int \mathrm{d}^4x\,D(x). \tag{14.114}$$

This agrees, except for a numerical factor, with the D-term action in (14.12), which we used in the component approach. If Φ is a chiral superfield, then one can form the action integral $\int \mathrm{d}^4x\,\mathrm{d}^2P_L\theta\,\Phi(x,\theta)$. Using (14.110), one can see that this isolates the F-component. The result is proportional to the F-term action of (14.12).

In $\mathcal{N}=1$, $D=4$ global supersymmetry, superfields give natural and elegant analogues of the multiplet calculus constructions presented in the main part of this chapter. There is a well-developed superspace approach to $\mathcal{N}=1$, $D=4$ supergravity (see [126, 127, 128, 129, 130] or the books [120, 121, 122, 123]) in which the geometry of superspace is described by frame and torsion superfields. However, these fields satisfy many constraints which make it complicated to derive physical results. This is why we chose the multiplet calculus as the principal method used in this book. Multiplets are defined by the properties of their lowest components and then completed by enforcing the supersymmetry algebra.

Appendix 14B Appendix: Covariant supersymmetry transformations

Note that the supersymmetry transformations are not covariant under Z reparametrizations. Indeed, we get

$$\delta P_L\chi'^{\alpha} = \frac{\partial Z'^{\alpha}}{\partial Z^{\beta}}\delta P_L\chi^{\beta} + \frac{\partial^2 Z'^{\alpha}}{\partial Z^{\gamma}\partial Z^{\beta}}P_L\chi^{\beta}\delta Z^{\gamma}. \tag{14.115}$$

However, one can define a 'covariant transformation'

$$\hat{\delta} P_L\chi^{\alpha} = \delta P_L\chi^{\alpha} + \Gamma^{\alpha}_{\beta\gamma}P_L\chi^{\gamma}\delta Z^{\beta}. \tag{14.116}$$

This transforms covariantly in the sense that

$$\hat{\delta} P_L\chi'^{\alpha} = \frac{\partial Z'^{\alpha}}{\partial Z^{\beta}}\hat{\delta} P_L\chi^{\beta}. \tag{14.117}$$

Exercise 14.15 *Prove this fact using the transformation of a Levi-Civita connection that was obtained in (7.103).*

Exercise 14.16 *Check that the commutator of covariant transformations on a coordinate vector such as χ^{α} is*

$$\left[\hat{\delta}_1,\hat{\delta}_2\right]P_L\chi^{\alpha} = \hat{\delta}_3 P_L\chi^{\alpha} + (\delta_1 Z^c)(\delta_2 Z^d)R_{cd}{}^{\alpha}{}_{\beta}P_L\chi^{\beta}, \tag{14.118}$$

where δ_3 is the transformation in the commutator $[\delta_1,\delta_2]$, and where c, d stand for indices that can be holomorphic or anti-holomorphic.

The previous considerations apply to any transformation, not just supersymmetry, on quantities that behave under reparametrizations as in (14.23). Replacing δ with a spacetime derivative, (14.116) reduces to (14.21), and this explains the use of this covariant derivative in writing the action. Note that the covariant SUSY transformation (14.116) can be simply expressed in terms of $\tilde{F}^\alpha$ defined in (14.26):

$$\hat{\delta} P_L \chi^\alpha = \frac{1}{\sqrt{2}} P_L (\partial\!\!\!/ Z^\alpha + \tilde{F}^\alpha)\epsilon. \tag{14.119}$$

We can understand the definition (14.116) as follows. In the main text, we considered Z^α and χ^α (and the auxiliary F^α) as the basis of independent fields in the chiral multiplet. One could take a basis with other fermion fields $\chi'^\alpha(Z, \chi)$. A covariant formalism should allow such a change of basis. Thus, in a covariant formalism, the fermions can be functions of the scalars. A full transformation $\hat{\delta}\chi^\alpha$ then involves the transformation of these scalars, and could take the form

$$\hat{\delta} P_L \chi^\alpha = \delta P_L \chi^\alpha + \frac{\partial P_L \chi^\alpha}{\partial Z^\beta} \delta Z^\beta. \tag{14.120}$$

However, the expression $\partial\chi^\alpha/\partial Z^\beta$ only makes sense if one compares different bases. In the same way, the expression $\Gamma^\alpha_{\beta\gamma}\chi^\gamma$ is not covariant. The only covariant object is the covariant derivative. A covariant version of (14.120) is

$$\hat{\delta} P_L \chi^\alpha = \delta P_L \chi^\alpha + \left(\frac{\partial P_L \chi^\alpha}{\partial Z^\beta} + \Gamma^\alpha_{\beta\gamma} P_L \chi^\gamma \right) \delta Z^\beta. \tag{14.121}$$

This reduces to the covariant transformation (14.116) for the basis that we use where $\partial P_L \chi^\alpha/\partial Z^\beta = 0$, i.e. when the fermions and bosons are considered as independent fields.

Using the covariant transformation as in (14.116), the non-covariant expression (14.34) also becomes covariant:

$$\hat{\delta}(\theta) P_L \chi^\alpha = \left(\nabla_\beta k_A{}^\alpha(Z)\right) P_L \chi^\beta. \tag{14.122}$$

PART V

SUPERCONFORMAL CONSTRUCTION OF SUPERGRAVITY THEORIES

15 Gravity as a conformal gauge theory

The main goal of the next three chapters is to study the formulation and structure of $\mathcal{N} = 1,\ D = 4$ supergravity, discussed in Ch. 9, coupled to gauge and chiral multiplets. These 'matter multiplets' were introduced at the level of global supersymmetry in Ch. 6, and discussed more generally in Ch. 14. Matter-coupled supergravity theories are rather complicated, and the simplest systematic approach is the 'superconformal method'. This method makes maximal use of the ideas of symmetry. Coleman and Mandula [29] showed that the maximal spacetime symmetry of a non-trivial field theory is the conformal symmetry. The supersymmetric extension of conformal symmetry is the superconformal algebra introduced by Haag, Łopuszański, and Sohnius [31]. This algebra contains P_a, M_{ab}, and Q_α of the Poincaré supersymmetry algebra plus additional generators (which we introduce and explain in Ch. 16). The superconformal method is based on early work on the understanding of supergravity from the superconformal algebra [131, 132, 133, 134]. Superconformal methods can also be used in superspace; see the review [135].

To apply the superconformal method one first formulates gauge theories of the superconformal algebra using the algebraic approach to gauge theories introduced in Ch. 11 with the modifications for gravity discussed in Sec. 11.3. These theories contain extra fields that are important in the formulation but are then eliminated to obtain the desired gauge theories of the Poincaré supersymmetry subalgebra. Some extra fields are eliminated using curvature constraints similar to the constraint of Ex. 11.10, which expresses the Lorentz connection $\omega_\mu{}^{ab}$ in terms of $e_\mu{}^a$ and ψ_μ. Other unwanted fields are called compensating fields, which are eliminated by gauge fixing the extra symmetries.

The theories that result from this process are the desired matter-coupled supergravity theories with local Poincaré supersymmetry. The other symmetries are not visible in the final result. Thus, we stress that it is not our purpose to consider conformal invariant theories. The conformal symmetry is just used as a tool to construct matter-coupled Poincaré supergravity and gain insight into its structure.

In this chapter we begin our program in a situation of less technical complexity. We show how conventional Einstein gravity emerges from a theory that is initially invariant under the bosonic conformal algebra. This algebra, which is isomorphic to $\mathfrak{so}(4,2)$, contains P_a, M_{ab} plus the generators K_a of special conformal transformation and dilatations D. The conformal gauge multiplet is then coupled to a real scalar compensating field. After imposing curvature constraints and gauge fixing, we are left with conventional bosonic gravity with the usual Hilbert action.

Beginning in Sec. 15.2 we define the conformal algebra, and discuss the transformation rules of matter and gauge fields, and the constraints needed to obtain the action describing the coupling of the conformal gauge multiplet to a real scalar. The simple action that

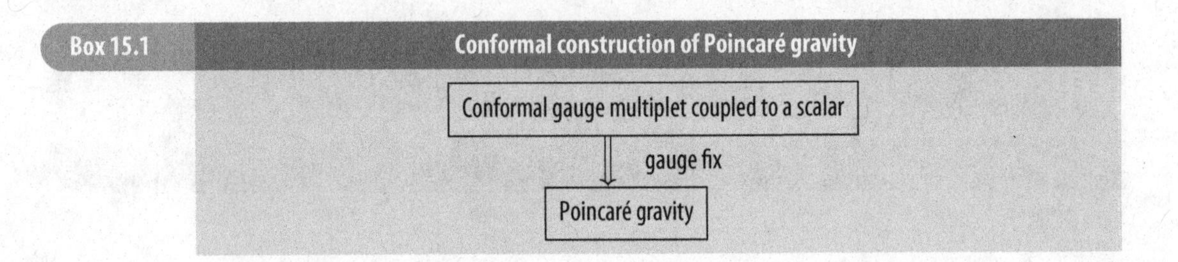

appears before the final gauge fixing in this case was already encountered in Ex. 8.7. So in the next section we begin by reviewing that result and stating the general strategy of the conformal approach. We follow the scheme of Box 15.1.

15.1 The strategy

The Lagrangian of Ex. 8.7 contains the spacetime metric $g_{\mu\nu}(x)$ and a real scalar field $\phi(x)$. It is invariant under conventional general coordinate transformations and local scale transformations, which are often called Weyl transformations. We specialize to $D = 4$ and use $\lambda_{\rm D}(x)$ to denote the gauge parameter of Weyl transformations. The Lagrangian[1] and scale transformations are

$$\mathcal{L} = \sqrt{g}\left[\tfrac{1}{2}(\partial_\mu\phi)(\partial^\mu\phi) + \tfrac{1}{12}R\phi^2\right], \tag{15.1}$$

$$\delta\phi = \lambda_{\rm D}\phi\,, \qquad \delta g_{\mu\nu} = -2\lambda_{\rm D}g_{\mu\nu}\,. \tag{15.2}$$

Readers who have not yet worked through Ex. 8.7 are again invited to demonstrate that the action (15.1) is invariant under (15.2). The proof requires a compact expression for $\delta R_{\mu\nu}$ obtained using (7.105) and (7.125).

As in most gauge theory actions, gauge fixing is needed for many physical applications. A compensating field is one that can be immediately eliminated by gauge fixing, and it is easy to see that the condition[2]

$$\phi = \frac{\sqrt{6}}{\kappa} \tag{15.3}$$

locks the scalar field at a fixed value, at which the action becomes the standard Hilbert action for gravity (8.1), namely

$$\mathcal{L} = \frac{1}{2\kappa^2}\sqrt{g}R\,, \tag{15.4}$$

which depends only on the physical metric $g_{\mu\nu}(x)$.

[1] We choose the opposite sign from Ex. 8.7 because our purpose now is to obtain a physical theory of gravity in which the Hilbert action must have positive sign.

[2] Alternatively, gauge fixing can be interpreted as a redefinition of the field variables so that only one field still transforms under the corresponding transformations. Then, the invariance is expressed as the absence of this field from the action. In this case we would use $g'_{\mu\nu} = (\kappa^2/6)g_{\mu\nu}\phi^2$ as dilatation invariant metric. One can check that this redefinition also leads to (15.4) in terms of the new field.

15.2 The conformal algebra

Conformal symmetry is initially defined as the symmetry of Minkowski space that preserves the 'angle' between any two vectors x^μ, y^μ. Specifically, conformal transformations are coordinate transformations that leave the quantity $x \cdot y/\sqrt{x \cdot x\, y \cdot y}$ invariant. Infinitesimal transformations are determined by vector fields ξ^μ that induce a change in the metric $\eta_{\mu\nu}$ by a scale factor. Such vector fields are called conformal Killing vectors. They are solutions of the 'conformal Killing equation'

$$\partial_\mu \xi_\nu + \partial_\nu \xi_\mu - \frac{2}{D}\eta_{\mu\nu}\partial_\rho \xi^\rho = 0 \,. \tag{15.5}$$

In two dimensions, this has an infinite set of solutions, which define the Virasoro algebra of string theory. We concentrate on $D > 2$, where the conformal algebra is finite-dimensional. Indeed, the solutions of (15.5) are

$$\xi^\mu(x) = a^\mu + \lambda^{\mu\nu}x_\nu + \lambda_{\rm D} x^\mu + (x^2\lambda_{\rm K}^\mu - 2x^\mu x \cdot \lambda_{\rm K}) \,. \tag{15.6}$$

They contain the familiar D translation and $D(D-1)/2$ Lorentz parameters a^μ and $\lambda^{\mu\nu}$ plus $D+1$ new parameters $\lambda_{\rm D}$ and $\lambda_{\rm K}^\mu$ for scale and special conformal transformations. In total the conformal group contains $(D+1)(D+2)/2$ parameters.

Exercise 15.1 *Show that the transformations above preserve angles. To do so, show that the transform of the line element* $\mathrm{d}s^2 = \mathrm{d}x^\mu \eta_{\mu\nu}\mathrm{d}x^\nu$ *is proportional to itself, i.e.* $\delta\, \mathrm{d}s^2 \sim \mathrm{d}s^2$. *You should find*

$$\begin{aligned} \delta\, \mathrm{d}x^\mu &= \lambda^{\mu\nu}\mathrm{d}x_\nu + \lambda_{\rm D}\mathrm{d}x^\mu + 2x \cdot \mathrm{d}x\, \lambda_{\rm K}^\mu - 2\mathrm{d}x^\mu x \cdot \lambda_{\rm K} - 2x^\mu \mathrm{d}x \cdot \lambda_{\rm K} \,, \\ \delta\, \mathrm{d}s^2 &= 2\Lambda_{\rm D}\mathrm{d}s^2 \,, \qquad \Lambda_{\rm D}(x) \equiv \lambda_{\rm D} - 2x \cdot \lambda_{\rm K} \,. \end{aligned} \tag{15.7}$$

The full set of conformal transformations can be expressed in terms of generators as in (11.1), i.e.

$$\delta(\epsilon) = a^\mu P_\mu + \tfrac{1}{2}\lambda^{\mu\nu}M_{[\mu\nu]} + \lambda_{\rm D} D + \lambda_{\rm K}^\mu K_\mu \,. \tag{15.8}$$

From Lie brackets of the various transformations in (15.6), one can obtain the conformal algebra. The non-vanishing commutators are

$$\begin{aligned} [M_{\mu\nu}, M_{\rho\sigma}] &= 4\eta_{[\mu[\rho}M_{\sigma]\nu]} = \eta_{\mu\rho}M_{\sigma\nu} - \eta_{\nu\rho}M_{\sigma\mu} - \eta_{\mu\sigma}M_{\rho\nu} + \eta_{\nu\sigma}M_{\rho\mu} \,, \\ [P_\mu, M_{\nu\rho}] &= 2\eta_{\mu[\nu}P_{\rho]} \,, \qquad [K_\mu, M_{\nu\rho}] = 2\eta_{\mu[\nu}K_{\rho]} \,, \\ [P_\mu, K_\nu] &= 2(\eta_{\mu\nu}D + M_{\mu\nu}) \,, \\ [D, P_\mu] &= P_\mu \,, \qquad [D, K_\mu] = -K_\mu \,. \end{aligned} \tag{15.9}$$

This is the SO$(D,2)$ algebra. Indeed, one can define (with $\hat\mu, \hat\nu = 0, \ldots, D, D+1$)

$$M^{\hat\mu\hat\nu} = \begin{pmatrix} M^{\mu\nu} & \frac{1}{2}(P^\mu - K^\mu) & \frac{1}{2}(P^\mu + K^\mu) \\ -\frac{1}{2}(P^\nu - K^\nu) & 0 & -D \\ -\frac{1}{2}(P^\nu + K^\nu) & D & 0 \end{pmatrix} . \tag{15.10}$$

Indices are raised by the diagonal metric

$$\hat{\eta} = \mathrm{diag}(-1, 1, ..., 1, -1)\,. \tag{15.11}$$

15.3 Conformal transformations on fields

In the previous section we identified the conformal group as the transformations of flat spacetime that preserve angles. Now we will implement the algebra (15.9) as a set of transformations of a system of fields. In this section we first consider global conformal transformations on a set of fields $\phi^i(x)$, where i enumerates the fields. Then we extend the definition to local (i.e. gauge) conformal transformations. In the next section we will introduce gauge fields. In many cases, the index i includes component indices of the representation of the Lorentz group in which the field ϕ^i transforms, for example the spinor or vector representation.

In Ch. 1 we saw that Lorentz transformations have an 'orbital part', which acts on the spacetime point x^μ, plus a 'spin part' or 'intrinsic part', which acts on representation indices. Dilatations and special conformal transformations also have 'orbital parts' and may have 'intrinsic parts' which we denote by $k_{\mathrm{D}}{}^i(\phi)$ or $k_\mu{}^i(\phi)$. We absorb the orbital parts of all generators in the conformal Killing vector $\xi^\mu(x)$ of (15.6) and write the action of a generic conformal transformation (in flat spacetime) as

$$\begin{aligned}\delta_C\phi^i(x) &= \xi^\mu(x)\partial_\mu\phi^i(x) - \tfrac{1}{2}\Lambda_M^{\mu\nu}(x)\, m_{[\mu\nu]}{}^i{}_j\phi^j(x)\\ &\quad + \Lambda_{\mathrm{D}}(x)\, k_{\mathrm{D}}{}^i(\phi)(x) + \lambda_{\mathrm{K}}^\mu k_\mu{}^i(\phi)(x)\,,\end{aligned} \tag{15.12}$$

where

$$\begin{aligned}\Lambda_{M\,\mu\nu}(x) &= \partial_{[\nu}\xi_{\mu]} = \lambda_{\mu\nu} - 4x_{[\mu}\lambda_{\mathrm{K}\,\nu]}\,,\\ \Lambda_{\mathrm{D}}(x) &= (1/D)\partial_\rho\xi^\rho = \lambda_{\mathrm{D}} - 2x\cdot\lambda_{\mathrm{K}}\,.\end{aligned} \tag{15.13}$$

Intrinsic special conformal transformations contain terms that act as x-dependent Lorentz and scale transformations. It is convenient to absorb these terms in the parameters $\Lambda_M(x)$ and $\Lambda_{\mathrm{D}}(x)$ in (15.13). These are x-dependent combinations of the independent constant parameters a^μ, $\lambda_{\mu\nu}$, λ_{D} and λ_{K} of global conformal transformations. We must still specify $m_{[\mu\nu]}{}^i{}_j$, $k_{\mathrm{D}}{}^i(\phi)$ and $k_\mu{}^i(\phi)$ for the various fields we deal with. This will be done below.

Exercise 15.2 *Consider a field ϕ with intrinsic dilatation transformation $k_{\mathrm{D}}(\phi) = w\,\phi$, determined by a real number w, which is called the 'Weyl weight'. Use (15.12) to calculate the dilatation transformation of $\partial_\mu\phi(x)$ and show that this behaves as a field with Weyl weight $w+1$.*

Exercise 15.3 *Prove that the Maxwell action in four-dimensional Minkowski spacetime is invariant under dilatations and special conformal transformations. This is a basic result in relativistic field theory which will allow the reader to explore how the various definitions made in this chapter are used in practice. The main equation needed is (15.12) for the* *gauge potential A_μ. This has Weyl weight 1, which means that $k_{D\,\mu} = A_\mu$. There are no extra special conformal transformations, so $k_\nu{}^\mu = 0$. However, Λ_M in (15.13) includes a term involving λ_K, so one needs the Lorentz transformation $m_{\{\mu\nu\}\rho}{}^\sigma = 2\eta_{\rho[\mu}\delta^\sigma_{\nu]}$. As an intermediate step show that $F_{\mu\nu}$ transforms as*

$$\delta F_{\mu\nu} = \xi^\rho \partial_\rho F_{\mu\nu} + 2\Lambda_{[\mu}{}^\rho F_{\nu]\rho} + 2\Lambda_D F_{\mu\nu}\,, \tag{15.14}$$

indicating that $F_{\mu\nu}$ has Weyl weight 2. To show that the action is invariant, you need the value of Λ_D in terms of ξ^ρ as in (15.13).

Next, we take the first steps towards *gauged conformal symmetry*, which is our final goal. With a theory of gravity in view we no longer restrict the discussion to Minkowski space, but assume that the fields $\phi^i(x)$ 'live on' a dynamic spacetime manifold, determined by the frame field $e^a_\mu(x)$. Recall from Sec. 11.3 that in theories of gravity it is common to use frame indices for fields and for transformation parameters and generators. Therefore, we will now switch from coordinate indices $\mu, \nu, \ldots$ to frame indices $a, b, \ldots$. Fields with frame indices only are called 'world scalars' or 'coordinate scalars'. Further, the symmetry parameters we introduce below are now gauge parameters; they are arbitrary functions of the coordinates.

In Sec. 11.3 we saw that it was necessary to separate translations from the other 'standard transformations' of an algebra and replace translations by covariant general coordinate transformations defined for various fields in Sec. 11.3.2. As in (11.56) we will regard $\xi^a(x) = e^a_\mu \xi^\mu(x)$ as the independent parameter of these covariant general coordinate transformations. 'Standard gauge transformations' of the conformal algebra are expressed as

$$\delta(\epsilon) = \epsilon^A T_A = \tfrac{1}{2}\lambda^{ab}(x) M_{[ab]} + \lambda_D(x) D + \lambda_K^a(x) K_a\,. \tag{15.15}$$

As explained at the beginning of Sec. 11.3.1, we absorb all the individual transformations of (15.6) into the single parameter $\xi^\mu(x)$. Similarly we now absorb the terms in (15.13) into gauge parameters $\lambda^{ab}(x)$ for Lorentz rotations and $\lambda_D(x)$ for local dilatations. This rearrangement is a change of basis in the conformal algebra, for which the generators M_{ab}, D and K_a of the standard subalgebra satisfy the commutation relations of (15.9).

Local conformal transformations for fields that are 'coordinate scalars' can thus be written as the following general expression:

$$\begin{aligned}\delta_C \phi^i(x) &= \xi^a(x) \mathcal{D}_a \phi^i(x) - \tfrac{1}{2}\lambda^{ab}(x)\, m_{\{ab\}}{}^i{}_j \phi^j(x) \\ &\quad + \lambda_D(x)\, k_D{}^i(\phi)(x) + \lambda_K^a(x) k_a{}^i(\phi)(x)\,.\end{aligned} \tag{15.16}$$

The first term is the covariant general coordinate transformation defined in (11.61). We now discuss the specific form of the other terms.

(i) **Lorentz transformations** are encoded in the matrix $(m_{\{ab\}})^i{}_j$. For scalars, this vanishes. For spinors, i and j include spinor indices (which are usually not written explicitly) and we have

$$m_{\{ab\}} = \tfrac{1}{2}\gamma_{ab} \qquad \Rightarrow \qquad \delta\Psi = -\tfrac{1}{4}\lambda^{ab}\gamma_{ab}\Psi\,. \tag{15.17}$$

For vectors we have

$$m_{\{ab\}}{}^c{}_d = 2\delta^c_{[a}\eta_{b]d} \qquad \Rightarrow \qquad \delta V^a = -\lambda^{ab} V_b\,. \tag{15.18}$$

On any other field, the transformation is determined by its frame indices and spinor indices.

(ii) Dilatations are specified by $k^i_{\rm D}(\phi)$. The fields considered in this chapter are simply scaled by a constant, i.e.

$$k^i_{\rm D} = w\,\phi^i \qquad \Rightarrow \qquad \delta\phi^i = w\,\phi^i\lambda_{\rm D}\,. \tag{15.19}$$

The real number w is called the 'Weyl weight' of the field ϕ^i. In (15.2), for example, the Weyl weight of ϕ is 1, and the Weyl weight of the metric $g_{\mu\nu}$ is -2.

A more general form of the Weyl transformation is needed for scalar fields of the nonlinear σ-model because (15.19) is not reparametrization invariant. We will introduce this in Sec. 15.7, using the concept of a 'closed homothetic Killing vector'.

(iii) Special conformal transformations are defined by a function $k_a{}^i(\phi)$. We consider a general expression of the form

$$\delta\phi^i = \lambda^a_{\rm K} k_a{}^i(\phi) = \lambda^a_{\rm K}\chi_a{}^i\,, \tag{15.20}$$

in which $\chi_a{}^i$ might be another field of the system or a derivative of another field. However, (15.9) tells us that a special conformal transformation decreases the Weyl weight from w to $w-1$, i.e. the Weyl weight of $\chi_a{}^i$ has to be one lower than the Weyl weight of ϕ^i. Often, there is no field in the multiplet that has the required lower Weyl weight. In that case the field ϕ^i has vanishing special conformal transformation. In fact, $k_a{}^i(\phi)$ does vanish for many fields, but we will find in Sec. 15.4 that several gauge fields do have special conformal transformations.

Although we described the transformations of coordinate scalars here, the discussion also applies to fields such as Yang–Mills gauge fields and conformal gauge fields (which will be treated below). The only change required is to replace the first term in (15.16) by the appropriate covariant general coordinate transformation described in Sec. 11.3.2.

Exercise 15.4 *Prove first that the Weyl weight of the frame field $e_\mu{}^a$ is -1. Then check that if ϕ has Weyl weight w, then $D_a\phi \equiv e_a{}^\mu(\partial_\mu - w\,b_\mu)\phi$ has Weyl weight $w+1$, where b_μ is the gauge field transforming under dilatations as $\delta b_\mu = \partial_\mu\lambda_D$. Compare with the result that you obtained for global conformal transformations in Ex. 15.2.*

To summarize the discussion of this section, we repeat that the conformal algebra contains spatial translations and Lorentz transformations (i.e. the Poincaré algebra) and, in addition, dilatations and special coordinate transformations. For gauge conformal transformations of world scalar fields $\phi^i(x)$, translations are treated as covariant general coordinate transformations and Lorentz transformations are, as usual, determined by the frame indices of the field. Dilatations are specified by the Weyl weight of the field. Fields with intrinsic special conformal transformations will appear in the next section.

The attentive reader may wonder whether we cannot omit the special conformal transformations entirely. Indeed, the algebra (15.9) closes if K_a is omitted. However, in the local conformal theory we will use special conformal transformations to remove the gauge field of dilatations. Indeed, we do not want this field to remain in the final Poincaré theory.

15.4 The gauge fields and constraints

We now introduce gauge fields for the conformal algebra. The following table lists the generators, the corresponding gauge parameters, and the gauge fields.

P_a	M_{ab}	D	K_a
ξ^a	λ^{ab}	$\lambda_{\rm D}$	$\lambda_{\rm K}^a$
e_μ^a	$\omega_\mu{}^{ab}$	b_μ	$f_\mu{}^a$

This will lead us to a gravity theory. As usual, the translations P_a, which are gauged by e_μ^a, play a special role and are distinguished from the 'standard transformations' generated by M_{ab}, D, K_a. The 'standard transformations' of gauge fields, obtained from (11.24), are given by

$$\begin{aligned}
\delta e_\mu{}^a &= -\lambda^{ab} e_{\mu b} - \lambda_{\rm D} e_\mu{}^a\,,\\
\delta\omega_\mu{}^{ab} &= \partial_\mu \lambda^{ab} + 2\omega_{\mu c}{}^{[a}\lambda^{b]c} - 4\lambda_{\rm K}^{[a} e_\mu{}^{b]}\,,\\
\delta b_\mu &= \partial_\mu \lambda_{\rm D} + 2\lambda_{\rm K}^a e_{\mu a}\,,\\
\delta f_\mu{}^a &= \partial_\mu \lambda_{\rm K}^a - b_\mu \lambda_{\rm K}^a + \omega_\mu{}^{ab}\lambda_{{\rm K}b} - \lambda^{ab} f_{\mu b} + \lambda_{\rm D} f_\mu{}^a\,.
\end{aligned} \tag{15.21}$$

Covariant general coordinate transformations, specified in (11.61), should be included to complete the result. The Lorentz transformations are determined by the index structure of the various fields, and the dilatation transformations can be described by saying that the vielbein has Weyl weight -1, the field $f_\mu{}^a$ has Weyl weight 1, and the others have Weyl weight 0. Furthermore, we see here that all these gauge fields, except the vielbein, have intrinsic special conformal transformations.

Exercise 15.5 *Readers may wish to show that the commutator of two transformations (15.21) realizes the conformal algebra (15.9); see (11.17).*

However, as discussed in the introduction to this chapter, we do not want all these gauge fields to describe new physical degrees of freedom. The only independent field in Einstein gravity is the frame field. The others are determined by imposing constraints or by gauge fixing as we now proceed to discuss.

First, as already seen in Sec. 11.3.2, the spin connection $\omega_\mu{}^{ab}$ is a composite field, determined by a constraint on the translation curvature $R_{\mu\nu}(P^a)$. We will impose the same constraint here. However, the algebra has changed from Poincaré to conformal, so we will find a new value for $\omega_\mu{}^{ab}$. Indeed, there is one new term in the P-curvature, related to the commutator $[D, P_a] = P_a$. This leads to the new constraint

$$R_{\mu\nu}(P^a) \equiv 2\left(\partial_{[\mu} + b_{[\mu}\right) e^a_{\nu]} + 2\omega_{[\mu}{}^{ab} e_{\nu]b} = 0\,. \tag{15.22}$$

Solving for the spin connection, we find that its value is modified compared to the usual formula obtained from (7.93):

$$\omega_\mu{}^{ab}(e,b) = \omega_\mu{}^{ab}(e) + 2e_\mu{}^{[a} e^{b]\nu} b_\nu\,, \qquad \omega_\mu{}^{ab}(e) = 2e^{\nu[a}\partial_{[\mu} e_{\nu]}{}^{b]} - e^{\nu[a} e^{b]\sigma} e_{\mu c}\partial_\nu e_\sigma{}^c\,. \tag{15.23}$$

Exercise 15.6 *Check that the transformations of the spin connection in (15.21) are consistent with its value in (15.23) using the transformations of the independent fields $e_\mu{}^a$ and b_μ.*

Second, we add a new constraint to fix the gauge field of special conformal transformations. The following constraint allows us to solve for f^a_μ algebraically:

$$e^\nu_b R_{\mu\nu}(M^{ab}) = 0, \qquad R_{\mu\nu}(M^{ab}) = R_{\mu\nu}{}^{ab} + 8 f_{[\mu}{}^{[a} e_{\nu]}{}^{b]}\,, \tag{15.24}$$

where we have split the curvature tensor of the Lorentz rotations into a term involving f^a_μ and the rest. The latter, $R_{\mu\nu}{}^{ab}$, is the expression found for the Poincaré algebra[3] and is given in (7.115). This allows us to solve the constraint for f^a_μ in terms of the Ricci tensor, obtaining

$$2(D-2) f^a_\mu = -R_\mu{}^a + \frac{1}{2(D-1)} e^a_\mu R\,, \qquad R_{\mu\nu} \equiv R_{\rho\mu}(M^{ab}) e_a{}^\rho e_{\nu b}\,, \qquad R = R_\mu{}^\mu\,. \tag{15.25}$$

After these two constraints, the independent fields that remain are the vielbein $e_\mu{}^a$ and the gauge field of dilatations b_μ. We will impose a condition on b_μ to fix the gauge of special conformal transformations. Indeed, the last term of the transformation of b_μ in (15.21) can be written as $\delta_K(\lambda_K) b_\mu = 2\lambda_{K\mu}$. This implies that the value of this field can be changed at will by a special conformal transformation. Therefore, we gauge-fix the special conformal transformations by fixing $b_\mu \equiv 0$:

$$K\text{-gauge:} \quad b_\mu = 0\,. \tag{15.26}$$

After this choice, only the vielbein remains as an independent field, and the special conformal transformations are no longer independent transformations. In fact, (15.21) shows that the gauge condition (15.26) is preserved if

$$\lambda_{K\mu} = -\tfrac{1}{2}\partial_\mu \lambda_D\,. \tag{15.27}$$

This type of relation, imposed by the preservation of a gauge condition, is called a *decomposition law*. It expresses a 'gauge-fixed' symmetry parameter in terms of gauge parameters of symmetries that remain in the theory.

[3] In fact, it includes b_μ hidden in the expression for $\omega_\mu{}^{ab}$. Below we fix the gauge so that b_μ vanishes, and this difference will disappear.

15.5 The action

Finally, we will derive the action (15.1) within the framework of gauged conformal symmetry. It is the action obtained by coupling the conformal gauge fields to a real scalar. We will make excellent use of the tricks that we learned in Sec. 11.3.4 for handling covariant quantities.

We begin with the scalar field whose gauge transformations are general coordinate transformations and Weyl transformations, the latter given by the usual scaling,

$$\delta\phi = w\,\lambda_{\rm D}\phi\,, \tag{15.28}$$

in which the Weyl weight w is a parameter to be determined. Our aim is first to construct a conformal invariant action, and then to perform the gauge fixing. Hence, in the first step, we do not yet impose (15.26), and b_μ is still an independent field.

To construct an action, we will first construct the conformal d'Alembertian

$$\Box^C\phi \equiv \eta^{ab}\mathcal{D}_b\mathcal{D}_a\phi\,. \tag{15.29}$$

Observe that we define it using covariant derivatives $\mathcal{D}_a$, which obey simple transformation rules. So let us compute (15.29) in two steps. First, consider

$$\mathcal{D}_a\phi = e_a{}^\mu\left(\partial_\mu - w\,b_\mu\right)\phi\,. \tag{15.30}$$

We know from Sec. 11.3.4 that this is a covariant quantity, i.e. its transformation does not involve derivatives of a gauge parameter. We have also learned that to calculate its transformation, we can restrict attention to terms in the transformation laws of the gauge fields that are not the derivatives of the parameters and do not include gauge fields of standard transformations. In other words, we can delete all terms that contain any gauge field other than e^a_μ. This reduces the relevant part of (15.21) to

$$\begin{aligned}
\delta e_\mu{}^a &= -\lambda^{ab}e_{\mu b} - \lambda_{\rm D}e_\mu{}^a\,,\\
\delta\omega_\mu{}^{ab} &= \ldots - 4\lambda_{\rm K}^{[a}e_\mu{}^{b]}\,,\\
\delta b_\mu &= \ldots + 2\lambda_{{\rm K}a}e_\mu{}^a\,,\\
\delta f_\mu{}^a &= \ldots\,,
\end{aligned} \tag{15.31}$$

where $\ldots$ denote terms that can be omitted by the lemma on covariant derivatives. Thus, we find the variation

$$\delta\mathcal{D}_a\phi = (w+1)\lambda_{\rm D}\mathcal{D}_a\phi - \lambda_a{}^b\mathcal{D}_b\phi - 2w\lambda_{{\rm K}a}\phi\,. \tag{15.32}$$

This implies that (15.29) is

$$\Box^C\phi \equiv \eta^{ab}\mathcal{D}_b\mathcal{D}_a\phi = e^{a\mu}\left(\partial_\mu\mathcal{D}_a\phi - (w+1)b_\mu\mathcal{D}_a\phi + \omega_{\mu\,ab}\mathcal{D}^b\phi + 2wf_{\mu a}\phi\right)\,. \tag{15.33}$$

To calculate the transformations of this quantity, we again need only consider (15.31). The transformations of the frame field ensure that $\Box^C\phi$ is a Lorentz scalar, and has Weyl

Box 15.2 **Signs of the kinetic terms**

To obtain gravity with positive kinetic term, one has to start with a compensating scalar with negative kinetic term.

weight $w+2$. Only the special conformal transformations of the field b_μ and $\omega_\mu{}^{ab}$ have to be considered, and we find

$$\delta\Box^C\phi = (w+2)\lambda_D\Box^C\phi + (2D-4w-4)\lambda_K^a\mathcal{D}_a\phi\,. \tag{15.34}$$

The last term drops if we choose

$$w = \tfrac{1}{2}D - 1\,, \tag{15.35}$$

which means, for example, Weyl weight $w=1$ for $D=4$. The Weyl weight of the vielbein determinant e is $-D$, since it is the product of D factors e_μ^a. Therefore, with w specified in (15.35), the quantity $e\phi\Box^C\phi$ is invariant under local dilatations and can be used as the invariant Lagrangian we need!

Thus we consider the action

$$I = -\frac{1}{2}\int \mathrm{d}^D x\, e\phi\Box^C\phi\,. \tag{15.36}$$

This action is conformal invariant. Now we use the gauge condition (15.26) to simplify (15.33). Further we insert the explicit form of f_μ^a given in (15.25), specifically

$$e_a^\mu f_\mu^a = -\frac{1}{4(D-1)}R\,. \tag{15.37}$$

The action can then be written as

$$\begin{aligned} I &= -\frac{1}{2}\int \mathrm{d}^D x\, e\phi\left(D^a(e_a^\nu\partial_\nu\phi) - \frac{D-2}{4(D-1)}R\phi\right) \\ &= \int \mathrm{d}^D x\, e\left(\frac{1}{2}g^{\mu\nu}(\partial_\mu\phi)(\partial_\nu\phi) + \frac{D-2}{8(D-1)}R\phi^2\right), \end{aligned} \tag{15.38}$$

where D_a is the Lorentz covariant derivative, and the last line follows after integration by parts. We thus have obtained the conformal action of a scalar. For $D=4$ it agrees with the action (15.1), which began our discussion. The final gauge fixing

$$\phi = \sqrt{\frac{4(D-1)}{(D-2)\kappa}} \tag{15.39}$$

gives the Hilbert action of gravity in D dimensions.

15.6 Recapitulation

We have seen how the method of conformal calculus works in this simple example, and we have also introduced some useful tricks and tools of gauge theories in the presence of local translations.

The local conformal algebra was realized on two independent gauge fields e_μ^a and b_μ and two dependent fields $\omega_\mu{}^{ab}$ and f_μ^a, whose values were determined by constraints. We introduced the scalar ϕ and constructed a conformal invariant action. Then we fixed the gauge of the superfluous gauge symmetries, first setting the field b_μ to zero, which breaks special conformal transformations. Finally, the scalar compensating field ϕ is fixed by (15.39), breaking local dilatations. The action that remains is just the Hilbert action.

We add a final note concerning the sign of scalar field kinetic term in (15.1) or (15.38). Readers may have noticed that it is the 'wrong' or unphysical sign. The wrong sign is acceptable here, because $\phi(x)$ is *not* a physical field. Indeed the wrong sign is needed, because the final action (15.1) or (15.38) describes only the spin-2 graviton with the conventional positive kinetic term required for gravity. It is a general feature of the conformal and superconformal methods that the kinetic term of the scalar compensator field appears with unphysical sign.

15.7 Homothetic Killing vectors

In this section we discuss the implementation of conformal symmetry in systems of scalars ϕ^i whose dynamics is described by a nonlinear σ-model with target space metric $g_{ij}(\phi)$. Conformal symmetry, whether global or gauged, strongly constrains this metric and the vector fields that generate the symmetries. The vector fields are homothetic Killing vectors or special cases called closed and exact homothetic Killing vectors. The properties of these vectors are defined below. These concepts play an important role in the development of matter-coupled supergravity theories in Ch. 17. The target space in those theories is a Kähler manifold.

As mentioned in Sec. 15.3, the dilatation Killing vector (15.19) cannot always be used when the target space geometry is non-trivial. Instead the intrinsic part of dilatations is given by $\delta\phi^i = \lambda_{\rm D} k_{\rm D}{}^i(\phi)$, where $k_{\rm D}{}^i$ is a *homothetic Killing vector*. The dilatation Killing vector also determines the action of special conformal transformations. For global symmetries this happens because $\lambda_{\rm K}$ appears in $\Lambda_{\rm D}(x)$ of (15.13). For local conformal symmetry it happens because the dilatation gauge field b_μ transforms under special conformal transformations; see (15.21).

Global special conformal transformations on scalars are given by

$$\delta\phi^i = (x^2\lambda_{\rm K}^\mu - 2x^\mu x\cdot\lambda_{\rm K})\partial_\mu\phi^i - 2x^\mu\lambda_{{\rm K}\mu}k_{\rm D}{}^i(\phi)\,. \tag{15.40}$$

This equation was obtained from (15.12) by dropping all symmetry parameters except $\lambda_{\rm K}^\mu$ and also assuming that $k_\mu{}^i(\phi) = 0$. This assumption is certainly valid if there are no scalars of lower Weyl weight in the theory. The nonlinear σ-model is then invariant if $k_{\rm D}{}^i$ is a *closed homothetic Killing vector* [136], as readers will show in an exercise below. The appropriate local extension of the nonlinear σ-model is also invariant under special conformal transformations if $k_{\rm D}{}^i$ is a closed homothety. This will also be shown in an exercise. Note that $k_{\rm D}{}^i$ is a vector field on the target space that is needed to realize conformal symmetry in the nonlinear σ-model field theory.

We now define the vector fields discussed above for general target space metric $g_{ij}(\phi)$, where the ϕ^i are real coordinates. A Killing vector k is a vector whose Lie derivative annihilates the metric: $\mathcal{L}_k g_{ij} = 0$. A conformal Killing vector is one for which

$$\text{conformal Killing vector: } \mathcal{L}_k g_{ij} = \nabla_i k_j + \nabla_j k_i = w(\phi)\, g_{ij}\,, \tag{15.41}$$

where ∇ contains the Levi-Civita connection for the metric, and $w(\phi)$ can be an arbitrary scalar function on the manifold. Note that a conformal Killing vector is thus *not* a Killing vector. Then one defines [137, 138]

$$\text{homothetic Killing vector: } \mathcal{L}_k g_{ij} = \nabla_i k_j + \nabla_j k_i = 2\, w\, g_{ij}\,, \tag{15.42}$$

where w is a real number. Finally a closed homothetic Killing vector is one for which the 1-form $k_i \mathrm{d}\phi^i = k^j g_{ij} \mathrm{d}\phi^i$ is closed. Thus also $\partial_i k_j - \partial_j k_i = 0$ and we can write

$$\text{closed homothetic Killing vector: } \nabla_i k_j = w\, g_{ij}\,. \tag{15.43}$$

Observe that a manifold has a closed homothetic Killing vector if and only if there exists (locally) a function $\tilde{\kappa}(\phi)$ that satisfies

$$\nabla_i \partial_j \tilde{\kappa} = g_{ij}\,. \tag{15.44}$$

Exercise 15.7 *Prove the integrability condition for a closed homothetic Killing vector*

$$R_{ijk\ell} k^\ell = 0\,. \tag{15.45}$$

This condition strongly restricts the target spaces which support special conformal symmetry.

In explicit computations it is convenient to use the alternative form of (15.43), which depends only on the Levi-Civita connection on the target space,

$$\partial_i k_{\rm D}{}^j + \Gamma_{ik}{}^j k_{\rm D}{}^k = w\, \delta_i^j\,. \tag{15.46}$$

As in (15.19), the parameter w is called the Weyl weight of the fields ϕ^i. When $\Gamma_{ij}{}^k = 0$, the unique solution is the simple form in (15.19).

Exercise 15.8 *In this example we consider the nonlinear σ-model in flat Minkowski space. Global dilatation and special conformal transformations of the scalars are defined by*

$$\delta\phi^i = \xi^\mu \partial_\mu \phi^i + \Lambda_{\rm D}(x) k_{\rm D}{}^i(\phi)\,. \tag{15.47}$$

It follows from (15.5) and (15.13) that the quantities $\xi^\mu(x)$ *and* $\Lambda_{\rm D}(x)$ *are related by* $\partial_{(\mu}\xi_{\nu)} = \eta_{\mu\nu}\Lambda_{\rm D}(x)$. *Check that*

$$\mathcal{L} = -\tfrac{1}{2}g_{ij}(\phi)\partial_\mu\phi^i\,\partial^\mu\phi^j \tag{15.48}$$

transforms as

$$\delta\mathcal{L} = \partial_\mu(\xi^\mu\mathcal{L}) + (-D+2+2w)\mathcal{L}\Lambda_{\rm D}(x) - k_{{\rm D}i}\partial_\mu\phi^i\partial^\mu\Lambda_{\rm D}(x) \tag{15.49}$$

if and only if ${k_{\rm D}}^i$ *are homothetic Killing vectors with weight* w. *Then, with the choice of weights (15.35) there is dilatation invariance, i.e. invariance of the action under the transformations with* $\Lambda_{\rm D}(x) = \lambda_{\rm D}$. *On the other hand, show that if* $k_{{\rm D}i}(\phi) = \partial_i\tilde{k}(\phi)$, *the action is also invariant for* $\Lambda_{\rm D}(x) = -2x\cdot\lambda_{\rm K}$.

Exercise 15.9 *We now couple the* σ*-model to gravity and consider its* local *spacetime symmetries, in particular dilatations and special conformal transformations with local parameters* $\lambda_{\rm D}(x)$ *and* $\lambda_{\rm K}^a(x)$. *The local field transformations are*

$$\begin{aligned} \delta\phi^i &= \lambda_{\rm D}{k_{\rm D}}^i(\phi)\,, \qquad \delta e_\mu{}^a = -\lambda_{\rm D}e_\mu{}^a\,, \qquad \delta b_\mu = \partial_\mu\lambda_{\rm D} + 2\lambda_{\rm K}^a e_{\mu a}\,,\\ \delta f_a{}^a &= 2\lambda_{\rm D}f_a{}^a + e_a^\mu\left(\partial_\mu - b_\mu\right)\lambda_{\rm K}^a + e_\mu{}^a\omega_\mu{}^{ab}(e,b)\lambda_{{\rm K}b}\,. \end{aligned} \tag{15.50}$$

Calculate the transformations of the Lagrangian

$$\begin{aligned} L &= -\frac{1}{2}e\,g_{ij}(\phi)D_a\phi^i\,D^a\phi^j + e\frac{1}{w}f_a{}^a\,{k_{\rm D}}^i g_{ij}{k_{\rm D}}^j\,,\\ D_a\phi^i &= e_a{}^\mu\left(\partial_\mu\phi^i - b_\mu{k_{\rm D}}^i\right), \end{aligned} \tag{15.51}$$

in which $f_a{}^a$ *is defined in (15.37). Hint: use the methods of Sec. 11.3.4 to check that*

$$\delta D_a\phi^i = \lambda_D\left(D_a\phi^i + \partial_j{k_{\rm D}}^i D_a\phi^j\right) - 2{k_{\rm D}}^i\lambda_{{\rm K}a}\,. \tag{15.52}$$

You will find that if ${k_{\rm D}}^i$ *are homothetic Killing vectors both terms of the action are separately invariant under local dilatations. For the special conformal transformations, you need the equation*

$$D_a\left({k_{\rm D}}^i g_{ij}{k_{\rm D}}^j\right) = 2w\,k_{{\rm D}i}D_a\phi^i \tag{15.53}$$

to prove that the action is invariant. This equation is based on (15.46).

Exercise 15.10 *The Lagrangian*

$$\mathcal{L} = \left(1 + \frac{\phi^1}{\phi^2}\right)(\partial_\mu\phi^1)(\partial^\mu\phi^2) \tag{15.54}$$

defines a σ*-model whose only non-zero metric component is* $g_{12} = (\phi^1+\phi^2)/\phi^2$. *Calculate the Levi-Civita connections*

$$\Gamma^1_{11} = \frac{1}{\phi^1+\phi^2}\,, \qquad \Gamma^2_{22} = -\frac{\phi^1/\phi^2}{\phi^1+\phi^2}\,. \tag{15.55}$$

Show that the solutions of the conformal Killing equations

$$\partial_{(i}k_{j)D} - \Gamma^k_{ij}k_{kD} = wg_{ij} \tag{15.56}$$

are

$$k_{\mathrm{D}}^{1} = w\,\phi^{1}\,, \qquad k_{\mathrm{D}}^{2} = w\,\phi^{2}\,. \tag{15.57}$$

Check that this does not solve the equation (15.46). Hence, this model is invariant under global dilatations, but not under global special conformal transformations.

Exercise 15.11 *In rigorous differential geometry a closed homothetic Killing vector is defined as a vector k with the property that $\nabla_Y k = Y$, for an arbitrary vector Y. Note that $\nabla_Y = Y^i \nabla_i$. Show that this definition is equivalent to (15.43).*

16 The conformal approach to pure $\mathcal{N} = 1$ supergravity

In this chapter we extend the previous construction to the superconformal algebra. We couple the gauge fields of this algebra to a chiral multiplet of $\mathcal{N} = 1$, $D = 4$ SUSY, which is called the compensating multiplet. This multiplet replaces the scalar compensator field of the previous chapter. After appropriate constraints and gauge fixing, the final product of the construction will be the pure $\mathcal{N} = 1$, $D = 4$ discussed in Ch. 9, but now with auxiliary fields and a superalgebra that closes off-shell. Readers may wish to reread the introductory discussion of the method at the beginning of Ch. 15. The scheme of Box 15.1 is now replaced by that of Box 16.1.

16.1 Ingredients

To carry out the construction, we first discuss the superconformal algebra, and then introduce gauge fields and constraints to reduce the number of fields in the Weyl multiplet as we did for gravity. Then we consider the superconformal transformations of a chiral multiplet. This will prepare us to construct the action in Sec. 16.2.

16.1.1 Superconformal algebra

The superconformal algebra is a superalgebra that contains the conformal algebra SO$(D, 2)$ in its Lie subalgebra. In general the generators of a superalgebra can be considered to be matrix elements in a vector space with both commuting and anti-commuting components. The matrices take the form

$$\begin{pmatrix} \text{conformal algebra} & Q, S \\ Q, S & R\text{-symmetry} \end{pmatrix}. \tag{16.1}$$

A systematic analysis of superconformal algebras has been performed by Nahm [58]. For $\mathcal{N} = 1$ in four dimensions the relevant superalgebra[1] is SU(2, 2|1). The generators of the bosonic subalgebra SU(2, 2) × U(1) appear in the diagonal blocks in (16.1). The SU(2, 2) subalgebra is the conformal algebra, which contains the generators P_a, M_{ab}, K_a and D, which we are familiar with. SU(2, 2) is the covering group of SO(4, 2). This means that it has the same algebra, but allows fermionic representations, which we obviously need for supersymmetry.

[1] Appendix B.3 contains more information on these superalgebras, but it is not needed to follow the discussion in this chapter.

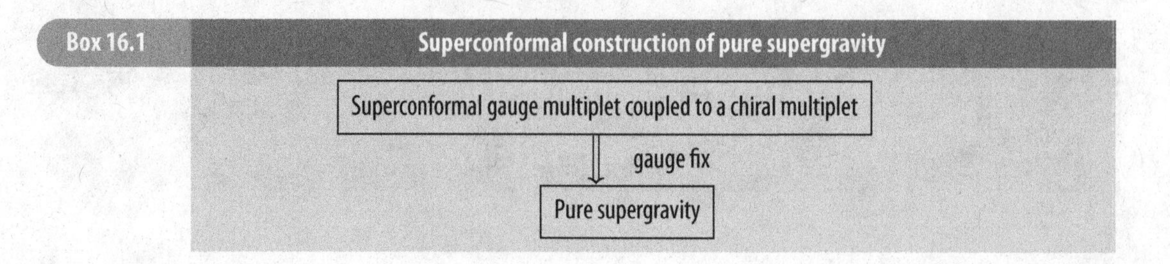

The U(1) symmetry is an R-symmetry, the symmetry that we first encountered in Sec. 6.2.1 and discussed more generally in Sec. 12.2. As in (12.10), we use T to denote the R-symmetry generator of the superconformal algebra. Recall that the T_R symmetry of Sec. 6.2.1 is an optional symmetry in global $\mathcal{N} = 1$ SUSY. It is present if the superpotential has R-charge 2, but not otherwise. The superconformal T-symmetry acts somewhat differently. Since it is part of the superconformal algebra, it is gauged in the superconformal action we obtain in Sec. 16.2. However, only general coordinate transformations, Q_α and M_{ab}, are gauged in the physical supergravity theories that are the goal of our construction. So the T-symmetry will be one of the extra symmetries, which will be gauge fixed at the end of our work. (The general matter-coupled theories we construct in Ch. 17 do, however, have the option, depending on details of the model, of a global or even a gauged T_R symmetry. This issue will be discussed in Ch. 19.)

The fermionic generators are the conventional Poincaré supercharge Q_α plus the conformal supercharge S_α, which is quite different. Its 'square' is the special conformal generator K_a. Both K_a and S_α are extra symmetries that will be gauge fixed. We point out that all spinors considered in this chapter are Majorana spinors.

The commutators of the conformal algebra are specified in (15.9), and these charges all commute with the U(1) generator T. The commutators of M_{ab} with the fermionic generators are

$$[M_{ab}, Q] = -\tfrac{1}{2}\gamma_{ab} Q\,, \qquad [M_{ab}, S] = -\tfrac{1}{2}\gamma_{ab} S\,, \tag{16.2}$$

which is the just the statement that they are Lorentz spinors.

The dilatations provide a 5-grading of the superconformal algebra. By this, we mean that the generators can be ordered in five groups according to their Weyl weight:

$$\begin{aligned} 1 &: P_a \\ \tfrac{1}{2} &: Q \\ 0 &: D\,,\ M_{ab}\,,\ T \\ -\tfrac{1}{2} &: S \\ -1 &: K_a\,. \end{aligned} \tag{16.3}$$

These weights determine all commutators involving D, for example

$$[D, Q] = \tfrac{1}{2} Q\,, \qquad [D, S] = -\tfrac{1}{2} S\,. \tag{16.4}$$

As we discussed above, T is an R-symmetry. We saw in (12.10) that this means that it rotates the supercharges. Since we discuss $\mathcal{N}=1$, the only choice is a chiral transformation, specifically

$$[T, Q] = -\tfrac{3}{2}\mathrm{i}\gamma_* Q, \qquad [T, S] = \tfrac{3}{2}\mathrm{i}\gamma_* S. \tag{16.5}$$

This means that $P_L Q$ and $P_L S$ have opposite U(1) charge. The value of this charge, which we have chosen to be 3/2, is arbitrary. Physical results for $\mathcal{N}=1$ supergravity and its matter couplings do not depend on this choice.[2] (Note that the normalization of T_R in (6.4) is different from T.)

The other non-vanishing commutators between bosonic and fermionic generators are

$$[K_a, Q] = \gamma_a S, \qquad [P_a, S] = \gamma_a Q. \tag{16.6}$$

Finally, the anti-commutators of the fermionic generators must be specified. We include spinor indices for clarity.[3] We have

$$\begin{aligned}
\{Q_\alpha, Q^\beta\} &= -\tfrac{1}{2}(\gamma^a)_\alpha{}^\beta P_a, \qquad \{S_\alpha, S^\beta\} = -\tfrac{1}{2}(\gamma^a)_\alpha{}^\beta K_a, \\
\{Q_\alpha, S^\beta\} &= -\tfrac{1}{2}\delta_\alpha{}^\beta D - \tfrac{1}{4}(\gamma^{ab})_\alpha{}^\beta M_{ab} + \tfrac{1}{2}\mathrm{i}(\gamma_*)_\alpha{}^\beta T.
\end{aligned} \tag{16.7}$$

Exercise 16.1 *Check the super-Jacobi identity*

$$\left[\{Q_\alpha, Q^\beta\}, S_\beta\right] = \left[Q_\alpha, \{Q^\beta, S_\beta\}\right] + \left[Q^\beta, \{Q_\alpha, S_\beta\}\right] = 0. \tag{16.8}$$

In this question, we have written a sum over the spinor index β, which simplifies the calculation.

16.1.2 Gauge fields, transformations, and curvatures

As in Sec. 15.4, we begin by assigning parameters and gauge fields to all generators of the superalgebra; see Table 16.1. As was done in Ch. 11, the generators are divided into two classes: the translations P_a, which will be represented as covariant general coordinate transformations; and the remaining 'standard superconformal transformations'. The latter are denoted collectively by

$$\delta = \epsilon^A T_A = \tfrac{1}{2}\lambda^{ab} M_{[ab]} + \lambda_{\rm D} D + \lambda_{\rm K}^a K_a + \lambda_T T + \epsilon^\alpha Q_\alpha + \eta^\alpha S_\alpha. \tag{16.9}$$

With spinor indices omitted, the last two terms can be written as $\bar\epsilon Q + \bar\eta S$.

Table 16.1 Parameters and gauge fields for the superconformal $\mathcal{N}=1$ algebra.

P_a	M_{ab}	D	K_a	T	Q	S
ξ^a	λ^{ab}	$\lambda_{\rm D}$	$\lambda_{\rm K}^a$	λ_T	ϵ	η
e_μ^a	$\omega_\mu{}^{ab}$	b_μ	$f_\mu{}^a$	A_μ	ψ_μ	ϕ_μ

[2] We take a different normalization from most papers in the literature of $\mathcal{N}=1$ superconformal supergravity, which will turn out to be useful below. The factor 3 originates in the superconformal algebra for $\mathcal{N}$-extended supersymmetry, where factors $4-\mathcal{N}$ appear [132].

[3] We use the notation of Sec. 3.2.2, e.g. $\bar{Q}^\alpha = Q^\alpha = \mathcal{C}^{\alpha\beta} Q_\beta$.

As in Ch. 11 the commutators of the generators define structure constants of the superalgebra and these determine the transformation laws of the gauge fields; see (11.24). The coordinate transformations (cgct) of these gauge fields, with parameter $\xi^\mu = e^\mu_a \xi^a$, follow from the rule (11.65). For the 'standard' transformations we use (11.24), in which the sum over B includes translations, which are gauged by e^a_μ. As we saw in the previous chapter, the important terms in practical calculations are the contributions where B is the translation (see (15.31)). Therefore, we underline these terms in the following result:

$$
\begin{aligned}
\delta e_\mu{}^a &= \underline{-\lambda^a{}_b e_\mu{}^b - \lambda_{\rm D} e_\mu{}^a} + \tfrac{1}{2}\bar\epsilon\gamma^a\psi_\mu\,,\\
\delta\omega_\mu{}^{ab} &= 2\lambda^{[a}{}_c\omega_\mu{}^{b]c} \underline{-4\lambda_{\rm K}^{[a} e_\mu{}^{b]}} + \tfrac{1}{2}\bar\epsilon\gamma^{ab}\phi_\mu + \tfrac{1}{2}\bar\eta\gamma^{ab}\psi_\mu\,,\\
\delta b_\mu &= \partial_\mu\lambda_{\rm D} \underline{+2\lambda_{\rm K}^a e_{\mu a}} + \tfrac{1}{2}\bar\epsilon\phi_\mu - \tfrac{1}{2}\bar\eta\psi_\mu\,,\\
\delta f_\mu{}^a &= -\lambda^a{}_b f_\mu{}^b + \partial_\mu\lambda_{\rm K}^a - b_\mu\lambda_{\rm K}^a + \omega_\mu{}^{ab}\lambda_{{\rm K}b} + \lambda_{\rm D} f_\mu{}^a + \tfrac{1}{2}\bar\eta\gamma^a\phi_\mu\,,\\
\delta A_\mu &= \partial_\mu\lambda_T - \tfrac{1}{2}{\rm i}\bar\epsilon\gamma_*\phi_\mu + \tfrac{1}{2}{\rm i}\bar\eta\gamma_*\psi_\mu\,,\\
\delta\psi_\mu &= -\tfrac{1}{4}\lambda^{ab}\gamma_{ab}\psi_\mu + \left(\partial_\mu + \tfrac{1}{2}b_\mu + \tfrac{1}{4}\omega_\mu{}^{ab}\gamma_{ab} - \tfrac{3}{2}{\rm i}A_\mu\gamma_*\right)\epsilon \underline{-\gamma_a e^a_\mu\eta}\\
&\quad -\tfrac{1}{2}\lambda_{\rm D}\psi_\mu + \tfrac{3}{2}{\rm i}\lambda_T\gamma_*\psi_\mu\,,\\
\delta\phi_\mu &= -\tfrac{1}{4}\lambda^{ab}\gamma_{ab}\phi_\mu + \left(\partial_\mu - \tfrac{1}{2}b_\mu + \tfrac{1}{4}\omega_\mu{}^{ab}\gamma_{ab} + \tfrac{3}{2}{\rm i}A_\mu\gamma_*\right)\eta - \gamma_a f_\mu{}^a\epsilon\\
&\quad +\tfrac{1}{2}\lambda_{\rm D}\phi_\mu - \tfrac{3}{2}{\rm i}\lambda_T\gamma_*\phi_\mu + \lambda_{\rm K}^a\gamma_a\psi_\mu\,.
\end{aligned}
\tag{16.10}
$$

The Lorentz transformations obtained above are the usual ones expected from the frame and spinor indices of the various fields, and the dilatations reflect the Weyl weight of the fields.

Exercise 16.2 *Verify the b_μ term in the supersymmetry transformation of ψ_μ above. Some gymnastics with spinor indices is needed. The relevant commutator of (16.4) is $[D, Q_\alpha] = \frac{1}{2}\delta_\alpha{}^\beta Q_\beta$. Now use (11.24) to obtain the term $\delta\psi^\alpha_\mu = \frac{1}{2}\epsilon^\alpha b_\mu$. This gives the transformation of $\bar\psi_\mu$, since the spinor index is up. Lowering the index is straightforward and gives the transformation written in (16.10). Now try another term in (16.10), e.g. the Q-supersymmetry transformation of ϕ_μ, to practice your skills.*

Similarly, using (11.41), we can write down the curvatures of all generators:

$$
\begin{aligned}
R_{\mu\nu}(P^a) &= 2\left(\partial_{[\mu} + b_{[\mu}\right) e^a_{\nu]} + 2\omega_{[\mu}{}^{ab} e_{\nu]b} - \tfrac{1}{2}\bar\psi_\mu\gamma^a\psi_\nu\,,\\
R_{\mu\nu}(M^{ab}) &= 2\partial_{[\mu}\omega_{\nu]}{}^{ab} + 2\omega_{[\mu}{}^a{}_c\omega_{\nu]}{}^{cb} + 8 f_{[\mu}^{[a} e_{\nu]}^{b]} - \bar\psi_{[\mu}\gamma^{ab}\phi_{\nu]}\,,\\
R_{\mu\nu}(D) &= 2\partial_{[\mu}b_{\nu]} - 4 f^a_{[\mu}e_{\nu]a} - \bar\psi_{[\mu}\phi_{\nu]}\,,\\
R_{\mu\nu}(K^a) &= 2\left(\partial_{[\mu} - b_{[\mu}\right) f^a_{\nu]} + 2\omega_{[\mu}{}^{ab} f_{\nu]b} - \tfrac{1}{2}\bar\phi_\mu\gamma^a\phi_\nu\,,\\
R_{\mu\nu}(T) &= 2\partial_{[\mu}A_{\nu]} + {\rm i}\bar\psi_{[\mu}\gamma_*\phi_{\nu]}\,,\\
R_{\mu\nu}(Q) &= 2\left(\partial_{[\mu} + \tfrac{1}{2}b_{[\mu} - \tfrac{3}{2}{\rm i}A_{[\mu}\gamma_* + \tfrac{1}{4}\omega_{[\mu}{}^{ab}\gamma_{ab}\right)\psi_{\nu]} - 2\gamma_{[\mu}\phi_{\nu]}\,,\\
R_{\mu\nu}(S) &= 2\left(\partial_{[\mu} - \tfrac{1}{2}b_{[\mu} + \tfrac{3}{2}{\rm i}A_{[\mu}\gamma_* + \tfrac{1}{4}\omega_{[\mu}{}^{ab}\gamma_{ab}\right)\phi_{\nu]} - 2\gamma_a f^a_{[\mu}\psi_{\nu]}\,.
\end{aligned}
\tag{16.11}
$$

Weyl multiplet **Box 16.2**

The independent fields of the Weyl multiplets are $e_\mu{}^a$, b_μ, A_μ and ψ_μ. The other gauge fields in Table 16.1 are functions of these, determined by constraints.

16.1.3 Constraints

In the previous chapter, we imposed two 'conventional' constraints, i.e. constraints that could be solved algebraically for a gauge field: (15.22) was solved to determine the spin connection $\omega_\mu{}^{ab}$, and (15.24) was solved for $f_\mu{}^a$. Therefore, the remaining *independent* fields of the bosonic Weyl multiplet were $e_\mu{}^a$ and b_μ. In this section we will impose similar constraints, which will also determine the gauge field of S-supersymmetry, ϕ_μ. Therefore, the independent fields of the Weyl multiplet will be $e_\mu{}^a$, b_μ, A_μ and ψ_μ. The transformation laws of these fields are shown in (16.10). (Readers may wish to skip the explanation of the constraints and just note the results given below for the dependent gauge fields $\omega_\mu{}^{ab}(e, b, \psi)$, $f_\mu{}^a$ and ϕ_μ. See (16.13) and (16.24) and see (16.25) and (16.23) for the notation.)

The three constraints needed to reduce the superconformal gauge multiplet to the independent fields listed above are

$$\begin{aligned}
&R_{\mu\nu}(P^a) = 0\,,\\
&e_b^\nu \widehat{R}_{\mu\nu}(M^{ab}) = 0\,,\\
&\gamma^\mu R_{\mu\nu}(Q) = 0\,.
\end{aligned} \tag{16.12}$$

The second constraint contains a modification of the curvature $R(M)$ in (16.11) that we will discuss below. The purpose is to make the constraint invariant under Q-supersymmetry. The modification is not strictly necessary but it is convenient. A change in a constraint redefines the field that we solve for.

The new constraint involves $R(Q)$. To motivate the specific choice above, note that ϕ_μ appears linearly in the last term of $R_{\mu\nu}(Q)$ in (16.11). The γ-trace of $R_{\mu\nu}(Q)$ has the same number of components as the field ϕ_μ, which we want to solve for, so there is a unique solution of the constraint.

Let us elaborate on the way in which the constraints are used. These constraints generalize the second order formalism in gravity, and we already applied such constraints in the previous chapter in Sec. 15.4. For supersymmetric theories it is necessary to have a good understanding of their significance. In the second order formalism of general relativity, $\omega_\mu{}^{ab}$ is defined as a function of the frame field. It is not an independent field. In the same way, here $\omega_\mu{}^{ab}$, $f_\mu{}^a$ and ϕ_μ are now 'dependent fields' (or 'composite fields'). They are to be understood as functions of the independent fields e_μ^a, b_μ, A_μ and ψ_μ that are obtained by solving the constraints (16.12).

Why do we need these constraints? Before we impose the constraints, we have an algebra that is closed on all the gauge fields of the Weyl multiplet, but translations do not act as they should. We gave an example in the context of Poincaré supergravity at the end of Sec. 11.1.2 where the gravitino is invariant under translations. We saw in Sec. 11.3 that the gauge translations should be replaced by covariant general coordinate transformations, and that this desired result is obtained for the frame field by imposing the first constraint of (16.12). A careful analysis [139] shows that the other independent fields transform as covariant general coordinate transformations after all three constraints (16.12) are imposed.[4]

We now study the solutions of the constraints and the modified transformation rules. Let's look at the first constraint. From (16.11), we see that $R_{\mu\nu}(P^a)$ in the superconformal algebra is very similar to the curvature of its Poincaré SUSY subalgebra, which was given in Ex. 11.10 of Ch. 11. The only difference is the appearance of the gauge field b_μ, which can be viewed as an added term in $\omega_\mu{}^{ab}$, exactly as in (15.23). These observations help us to obtain the solution of the first constraint, which is

$$\omega_\mu{}^{ab}(e,b,\psi) = \omega_\mu{}^{ab}(e,b) + \tfrac{1}{2}\bar{\psi}_\mu\gamma^{[a}\psi^{b]} + \tfrac{1}{4}\bar{\psi}^a\gamma_\mu\psi^b\,, \tag{16.13}$$

where the first term is given in (15.23). The quadratic term in ψ^a is the gravitino torsion term that already occurred in our discussion of $\mathcal{N} = 1$, $D = 4$ supergravity in (9.21).

How should we now determine the transformations of the dependent field? Is the constraint invariant under supersymmetry or other transformations? To answer these questions, we symbolically represent the constraint as $C(\omega,\phi) = 0$, where ω stands here for any of the dependent fields, and ϕ for the independent ones. All the constraints are chosen to be solvable, i.e. $\delta C/\delta\omega$ is invertible. A solution to the constraint means that there is a function $\omega(\phi)$ such that

$$C\left(\omega(\phi),\phi\right) \equiv 0 \tag{16.14}$$

for all ϕ. Hence we can take a derivative, and find

$$\frac{\partial C}{\partial\omega}\frac{\partial\omega}{\partial\phi} + \frac{\partial C}{\partial\phi} = 0\,. \tag{16.15}$$

Since ω is now to be considered as shorthand for $\omega(\phi)$, its transformations are

$$\delta\omega = \frac{\partial\omega}{\partial\phi}\delta\phi\,. \tag{16.16}$$

With (16.15) this implies immediately that $\delta C = 0$ without any further calculation. This shows also how we could compute the transformations of the dependent fields. We just use (16.16). This thus means for the spin connection that we compute $\delta\omega$ by using the expression (16.13) and inserting the transformations of the independent fields e, b and ψ. This is conceptually simple but in practice involves a lot of work.

A second way to compute the transformations needs a bit more thought, but is easier in practice. We can split the transformations of the dependent fields into the sum of their 'old' transformations, as dictated by the gauge theory algebra, plus extra transformations. The structure is then

$$\delta\omega = \delta_{\rm gauge}\omega + \delta_{\mathcal{M}}\omega\,, \tag{16.17}$$

[4] This analysis uses the modified transformation laws that we will discuss below and the formula for a covariant general coordinate transformation in (11.73).

where the latter notation is inspired by (11.69). Then the full transformation of the constraint is

$$0 = \delta C = \delta_{\text{gauge}} C + \delta_{\mathcal{M}} C = \delta_{\text{gauge}} C + \frac{\partial C}{\partial \omega} \delta_{\mathcal{M}} \omega \,. \tag{16.18}$$

The first term is easy to calculate since the constraint is written in terms of curvatures, for which we have the rule (11.42), and the invertibility of $\partial C / \partial \omega$ allows us then to obtain the new parts of the transformation of the dependent fields.

Let us apply this now for the first constraint. The rule (11.42), applied to the curvature of the translation, leads to (using $R_{\mu\nu}(P^a) = 0$ on the right-hand side)

$$\delta_{\text{gauge}} R_{\mu\nu}(P^a) = \tfrac{1}{2} \bar{\epsilon} \gamma^a R_{\mu\nu}(Q) \,. \tag{16.19}$$

Thus, (16.18) takes the form

$$0 = \tfrac{1}{2} \bar{\epsilon} \gamma^a R_{\mu\nu}(Q) + \frac{\partial}{\partial \omega_\rho{}^{bc}} R_{\mu\nu}(P^a) \delta_{\mathcal{M}} \omega_\rho{}^{bc} \,. \tag{16.20}$$

This can be solved for $\delta_{\mathcal{M}} \omega$ by the usual trick: writing the equation three times with different indices, adding two and subtracting one (see Ex. 7.26). One obtains

$$\begin{aligned} \delta_Q(\epsilon) \omega_\mu{}^{ab}(e, b, \psi) &= \delta_{\text{gauge}}(\epsilon) \omega_\mu{}^{ab} + \delta_{\mathcal{M}}(\epsilon) \omega_\mu{}^{ab} \\ &= \tfrac{1}{2} \bar{\epsilon} \gamma^{ab} \phi_\mu - \tfrac{1}{2} \bar{\epsilon} \gamma^{[a} R_\mu{}^{b]}(Q) - \tfrac{1}{4} \bar{\epsilon} \gamma_\mu R^{ab}(Q) \,. \end{aligned} \tag{16.21}$$

Although not essential, it is possible to simplify this expression by making use of a consequence of the third constraint, which is equivalent to $\gamma_{[\mu} R_{\nu\rho]}(Q) = 0$.

Exercise 16.3 *It is an interesting exercise in γ-algebra to show this equivalence, and use it to obtain the simplified transformation*

$$\delta_Q(\epsilon) \omega_\mu{}^{ab}(e, b, \psi) = \tfrac{1}{2} \bar{\epsilon} \gamma^{ab} \phi_\mu - \tfrac{1}{2} \bar{\epsilon} \gamma_\mu R^{ab}(Q) \,. \tag{16.22}$$

The new supersymmetry transformation of the spin connection is of the general type discussed in (11.69). The terms called $\mathcal{M}$ in that expression are exactly the $R^{ab}(Q)$ terms in (16.21). The $\mathcal{M}$ terms also modify the curvature for Lorentz transformations according to (11.72):

$$\begin{aligned} \hat{R}_{\mu\nu}(M^{ab}) &= \hat{R}_{\mu\nu}{}^{ab} + 8 f_{[\mu}^{[a} e_{\nu]}^{b]} \,, \\ \hat{R}_{\mu\nu}{}^{ab} &= r_{\mu\nu}{}^{ab} + \bar{\psi}_{[\mu} \gamma^{[a} R_{\nu]}{}^{b]}(Q) + \tfrac{1}{2} \bar{\psi}_{[\mu} \gamma_{\nu]} R^{ab}(Q) \,, \\ r_{\mu\nu}{}^{ab} &= 2 \partial_{[\mu} \omega_{\nu]}{}^{ab} + 2 \omega_{[\mu}{}^{a}{}_{c} \omega_{\nu]}{}^{cb} - \bar{\psi}_{[\mu} \gamma^{ab} \phi_{\nu]} \,. \end{aligned} \tag{16.23}$$

As in (15.24), we have separated the terms containing the dependent gauge field $f_\mu{}^a$. It is the modified curvature $\widehat{R}_{\mu\nu}(M^{ab})$, rather than $R_{\mu\nu}(M^{ab})$, that we use in the second line of (16.12). This gives a constraint that transforms covariantly.

It is now straightforward to solve the second and third constraints of (16.12) for the gauge fields $f_\mu{}^a$ and ϕ_μ:

$$\begin{aligned} f_\mu{}^a &= -\tfrac{1}{4} \hat{R}_\mu{}^a + \tfrac{1}{24} e_\mu{}^a \hat{R} \,, \\ \phi_\mu &= -\tfrac{1}{2} \gamma^\nu R'_{\mu\nu}(Q) + \tfrac{1}{12} \gamma_\mu \gamma^{ab} R'_{ab}(Q) \,. \end{aligned} \tag{16.24}$$

Here, we use Ricci tensors defined from the covariant $\hat{R}_{\mu\nu}{}^{ab}$ in (16.23), and we have also split the supersymmetry curvature by omitting terms containing the gauge field ϕ_μ. These curvatures and Ricci tensors in (16.24) are given by

$$\begin{aligned}
&\hat{R} = e_a^\mu \hat{R}_\mu{}^a\,, \qquad \hat{R}_\mu{}^a = \hat{R}_{\mu\nu}{}^{ab} e_b^\nu,\\
&R_{\mu\nu}(Q) = R'_{\mu\nu}(Q) - 2\gamma_{[\mu}\phi_{\nu]}\,,\\
&R'_{\mu\nu}(Q) = 2\left(\partial_{[\mu} + \tfrac{1}{2}b_{[\mu} - \tfrac{3}{2}\mathrm{i}A_{[\mu}\gamma_* + \tfrac{1}{4}\omega_{[\mu}{}^{ab}\gamma_{ab}\right)\psi_{\nu]}\,.
\end{aligned} \tag{16.25}$$

For future use, here are three more formulas that follow from these definitions:

$$\begin{aligned}
f_\mu{}^\mu &= -\tfrac{1}{12}\hat{R} = -\tfrac{1}{12}\left(R(\omega) - \bar{\psi}_a\gamma^{ab}\phi_b\right),\\
\gamma^\mu\phi_\mu &= -\tfrac{1}{6}\gamma^{\mu\nu}R'_{\mu\nu}(Q)\,,\\
\gamma^{ab}\phi_b &= -\tfrac{1}{4}\gamma^{abc}R'_{bc}(Q)\,.
\end{aligned} \tag{16.26}$$

Here, $R(\omega)$ is the usual Ricci scalar (defined from the $\partial\omega$ and $\omega\omega$ terms only) in which ω is now the solution (16.13).

16.1.4 Superconformal transformation rules of a chiral multiplet

The next step in the strategy sketched in Box 15.1 is to study the superconformal properties of a chiral multiplet. In global SUSY the Q-supersymmetry transformation rules of a general chiral multiplet, either elementary or composite, were derived in Sec. 6.2; see (6.15). They are valid in supergravity with the following changes:

1. The transformation parameter is now an arbitrary function $\epsilon(x)$.
2. The derivatives $\partial_\mu Z$ and $\partial_\mu P_L\chi$ in (6.15) must be replaced by superconformal covariant derivatives defined as in (11.58).

This replacement is an example of the well-known 'covariantization' procedure. The resulting expressions transform properly under local superconformal transformations. However, the explicit form of the superconformal covariant derivatives is not completely known because we still have to find the dilatation and other superconformal transformations. Let's get started.

First we define how dilatations act, which means that we must determine the appropriate Weyl weights for the fields.

Exercise 16.4 *Suppose that the field Z has Weyl weight w. Consider the commutator $[D,Q]$ in (16.4) applied to the field Z. Show that this implies that $P_L\chi$ has Weyl weight $w+\frac{1}{2}$.*

After this exercise, convince yourself that, owing to the structure of (16.4), the Q-supersymmetry parameter ϵ can be viewed as a quantity that carries Weyl weight $-1/2$, and the S-supersymmetry parameter η as a quantity with Weyl weight $1/2$. Hence the supersymmetry rules imply that when Z has Weyl weight w, the fermion $P_L\chi$ has Weyl weight $w+\frac{1}{2}$, and the auxiliary field F has Weyl weight $w+1$.

A similar discussion applies to T-transformations. We define a chiral weight c for the complex scalar Z by the transformation rule

$$\delta_T Z = \mathrm{i}c\lambda_T Z\,, \tag{16.27}$$

and similarly for all other fields. The commutator (16.5) then implies that $P_L\epsilon$ (or $\bar{\epsilon}P_L$) has chiral weight $3/2$, while $P_R\epsilon$ has chiral weight $-3/2$. Similarly $P_L\eta$ has chiral weight $-3/2$, and $P_R\eta$ has chiral weight $3/2$.

Next, we show that the Weyl weights of the fields and the parameters require that Z is invariant under S-supersymmetry. Indeed, since η has weight $+1/2$, Z can only transform into a field of Weyl weight $w - \frac{1}{2}$. There is no such field in the multiplet. Further, the result of Ex. 15.4 implies that a covariant derivative $\mathcal{D}_a$ also carries Weyl weight $+1$. So derivatives of fields produce fields of even higher Weyl weight. Thus, the only possibility is that $\delta_S Z = 0$.

This observation allows us to make a simple calculation to determine a relation between the Weyl weight and the chiral weight. We will check the commutator of Q- and S-supersymmetry on Z. The anti-commutation relations (16.7) become commutators when the Grassmann transformation parameters are included, as we learned in Sec. 11.1.1; see (11.12). We write the result here for the three relevant commutators:

$$\begin{aligned}
&\left[\delta_Q(\epsilon_1),\delta_Q(\epsilon_2)\right] = \delta_{\rm cgct}\left(\xi^\mu = \tfrac{1}{2}\bar{\epsilon}_2\gamma^\mu\epsilon_1\right),\\
&[\delta_S(\eta_1),\delta_S(\eta_2)] = \delta_K\left(\lambda_{\rm K}^a = \tfrac{1}{2}\bar{\eta}_2\gamma^a\eta_1\right),\\
&\left[\delta_S(\eta),\delta_Q(\epsilon)\right] = \delta_{\rm D}\left(\lambda_{\rm D} = \tfrac{1}{2}\bar{\epsilon}\eta\right) + \delta_M\left(\lambda^{ab} = \tfrac{1}{4}\bar{\epsilon}\gamma^{ab}\eta\right) + \mathrm{i}\delta_T\left(\lambda_T = -\tfrac{1}{2}\bar{\epsilon}\gamma_*\eta\right).
\end{aligned} \tag{16.28}$$

First note that the second commutator tells us that $\delta_K Z = 0$. This fact also follows from the Weyl weight argument above applied to special conformal symmetry, which assigns Weyl weight $+1$ to λ_K. So a chiral multiplet scalar is also invariant under a special conformal transformation. Now proceed to the main task, which involves the third commutator. Since we already learned that Z is invariant under S-supersymmetry, we can write

$$\left[\delta_S(\eta),\delta_Q(\epsilon)\right]Z = \delta_S(\eta)\delta_Q(\epsilon)Z = \frac{1}{\sqrt{2}}\bar{\epsilon}P_L\delta_S(\eta)\chi\,. \tag{16.29}$$

Note that $\bar{\epsilon}P_R$ does not appear in the result. Consider then what we should obtain according to (16.28). Z is a scalar, so the δ_M term vanishes. Only the δ_D and δ_T terms remain, and these are determined by the Weyl and chiral weights of Z. If we split $\bar{\epsilon} = \bar{\epsilon}P_L + \bar{\epsilon}P_R$ and isolate the $\bar{\epsilon}P_R$ terms, then the absence of such terms in (16.29) implies that

$$0 = \tfrac{1}{2}wZ - \tfrac{1}{2}cZ \qquad \Longrightarrow \qquad c = w\,. \tag{16.30}$$

The chiral weight is thus determined by the Weyl weight in any chiral multiplet. On the other hand, when we consider the left-chiral terms in this calculation, we obtain

$$\frac{1}{\sqrt{2}}\bar{\epsilon}P_L\delta_S(\eta)\chi = \frac{1}{2}w\,Z\,\bar{\epsilon}P_L\eta + \frac{1}{2}w\,Z\,\bar{\epsilon}P_L\eta \quad\Rightarrow\quad \delta_S(\eta)P_L\chi = \sqrt{2}w\,ZP_L\eta\,. \tag{16.31}$$

By checking the same commutator of Q- and S-supersymmetry on the fermion $P_L\chi$, one obtains the S-supersymmetry transformation of the auxiliary field, as the following exercise shows.

Exercise 16.5 *Consider the commutator $[\delta_S, \delta_Q]$ on the fermion χ. It should produce the combination of symmetries on the right-hand side of the last equation of (16.28). These transformations are determined by the Weyl and R-symmetry weights of χ, as explained above. We already know all transformations needed to compute $\delta_Q(\epsilon)\delta_S(\eta)\chi$. Then we must compute $\delta_S(\eta)\delta_Q(\epsilon)\chi$. The Q-supersymmetry transformation of χ involves $\mathcal{D}_\mu Z$ and F. Thus one needs the S-supersymmetry transformation of $\mathcal{D}_\mu Z$ to determine $\delta_S(\eta)F$. Since Z is invariant, the lemma on covariant derivatives implies that the S-supersymmetry transformation only results from the underlined terms in (16.10). This gives*

$$\delta_S(\eta)\mathcal{D}_a Z = -\frac{1}{\sqrt{2}}(\delta_S(\eta)\overline{\psi}_a)P_L\chi = -\frac{1}{\sqrt{2}}\bar{\eta}\gamma_a P_L\chi\,. \tag{16.32}$$

Finally you must combine these results, using a Fierz rearrangement, to obtain $\delta_S(\eta)F = \sqrt{2}(1-w)\bar{\eta}P_L\chi$.

Finally we can write the complete superconformal transformation of the chiral multiplet (to simplify we omit general coordinate and Lorentz transformations):

$$\begin{aligned}
\delta Z &= [w\lambda_{\rm D} + {\rm i}\, w\,\lambda_T]Z + \frac{1}{\sqrt{2}}\bar{\epsilon}P_L\chi\,,\\
\delta P_L\chi &= \left[\left(w+\frac{1}{2}\right)\lambda_{\rm D} + \left(w-\frac{3}{2}\right){\rm i}\lambda_T\right]P_L\chi + \frac{1}{\sqrt{2}}P_L\left(\not{\mathcal{D}}Z+F\right)\epsilon\\
&\quad + \sqrt{2}wZP_L\eta\,,\\
\delta F &= [(w+1)\,\lambda_{\rm D} + (w-3)\,{\rm i}\lambda_T]\,F + \frac{1}{\sqrt{2}}\bar{\epsilon}\,\not{\mathcal{D}}P_L\chi + \sqrt{2}(1-w)\bar{\eta}P_L\chi\,.
\end{aligned} \tag{16.33}$$

Observe that the T-transformations are the same as the $\mathrm{U}(1)_R$ symmetry transformations (6.28), except for normalization.[5] Since this symmetry is part of the superconformal algebra, any superconformal invariant action will also be T-symmetric, even locally T-symmetric. Now the superconformal covariant derivatives follow easily by applying the gauge transformation laws 'stepwise'. By inspection of δZ we obtain $D_a Z$ and by inspection of $\delta P_L\chi$ we obtain $D_a P_L\chi$:

$$\begin{aligned}
\mathcal{D}_a Z &= e_a^\mu\left(\partial_\mu Z - w\,b_\mu Z - {\rm i}w\,A_\mu Z - \frac{1}{\sqrt{2}}\bar{\psi}_\mu P_L\chi\right),\\
\mathcal{D}_a P_L\chi &= e_a^\mu P_L\left[\left(\partial_\mu + \frac{1}{4}\omega_\mu{}^{bc}\gamma_{bc} - \left(w+\frac{1}{2}\right)b_\mu - \left(w-\frac{3}{2}\right){\rm i}A_\mu\right)\chi\right.\\
&\quad \left. -\frac{1}{\sqrt{2}}\left(\not{\mathcal{D}}Z+F\right)\psi_\mu - \sqrt{2}w\,Z\phi_\mu\right].
\end{aligned} \tag{16.34}$$

[5] The relation between the parameters here and in Sec. 6.2.1 is $\lambda_T = \frac{3}{2}\rho$. The transformations of the chiral multiplet then agree if we identify $r = \frac{2}{3}w$.

In global supersymmetry, the highest component, F, transforms into a total derivative, and is therefore an appropriate quantity to use in an action. In a gravity theory, we might expect something like $e\,F$ as the Lagrangian. Since e carries Weyl weight -4, the field F should carry weight $+4$. In view of (16.33) this means $w = 3$. One notices immediately that F is then invariant under the T-transformations.

However, $\int \mathrm{d}^4x\, e\, F$ is not invariant in the superconformal setting because of the form of the $\delta_Q F$ and $\delta_S F$ terms in (16.33). For example, after a partial integration the variation contains a term with $\int \mathrm{d}^4x\, e\, \partial_\mu \bar{\epsilon}\, \gamma^\mu P_L \chi$. As we will now discuss, the modified integral[6]

$$S_F = \int \mathrm{d}^4x\, e\, \mathrm{Re}\left[F + \frac{1}{\sqrt{2}}\bar{\psi}_\mu \gamma^\mu P_L \chi + \frac{1}{2} Z \bar{\psi}_\mu \gamma^{\mu\nu} P_R \psi_\nu\right] \tag{16.35}$$

does give an S and Q invariant action for a chiral multiplet with $w = 3$. Both the real and imaginary parts of the quantity in brackets give invariant integrals. We will use the real part as our basic action. Indeed, this formula is a master formula that will be used many times.

Exercise 16.6 *Check S-supersymmetry of (16.35). This is a very simple calculation that determines the constants in (16.35). It is much simpler than checking the Q-supersymmetry.*

It is more difficult to check the Q-supersymmetry, but this can be done using the Noether method, which we now outline for the present purpose. The key is to notice that the $\delta \bar{\psi}_\mu = \partial_\mu \bar{\epsilon}$ variation of the second term cancels the non-invariant $\partial_\mu \bar{\epsilon}$ term mentioned above. However, there are other $\bar{\psi}_\mu \partial_\nu \epsilon$ variations of the second term that are canceled by adding the third term.

One more important remark. A chiral multiplet is determined by its lowest component both in global SUSY, as we saw in Ch. 14, and in the superconformal setting of this chapter. This means that if one has a field that transforms like Z in (16.33) with χ replaced by an arbitrary spinor, then the superconformal algebra determines that this χ should transform as the χ in (16.33) for some undetermined function F. The latter then transforms as the F in (16.33). We will use this observation in the next section.

16.2 The action

16.2.1 Superconformal action of the chiral multiplet

Our goal now is to construct the superconformal generalization of the action (15.36). This requires a scalar compensator field with Weyl weight $w = 1$ (in four-dimensional spacetime) and the conformal d'Alembertian of this scalar. It is reasonable to anticipate that the

[6] This action assumes that there is no boundary in spacetime. Recently an action formula in the presence of a boundary has been constructed [140].

scalar compensator in the superconformal extension is the lowest component Z of a chiral multiplet and that this scalar should again have Weyl weight $w = 1$.

Thus we will start with a superconformal chiral multiplet $(Z,\ P_L\chi,\ F)$ with $w = 1$ and construct the action using superconformal extensions of two exercises in global SUSY:

1. Given an anti-chiral multiplet $(\bar{Z},\ P_R\chi,\ \bar{F})$, Ex. 6.8 shows that $(Z' = \bar{F},\ P_L\chi' = \partial\!\!\!/ P_R\chi,\ F' = \Box\bar{Z})$ are components of a chiral multiplet. It is obvious that the Weyl weight will be $w_F = 2$.
2. Imitating Ex. 14.3, we will consider the product multiplet with components $(Z\bar{F},\ P_L(Z\partial\!\!\!/\chi + \bar{F}\chi),\ F\bar{F} + Z\Box\bar{Z} - \bar{\chi}P_L\partial\!\!\!/\chi)$. The lowest component has Weyl weight $w = 1 + 2 = 3$ and the top component has Weyl weight $w = 4$.
3. The action formula (16.35) gives a locally superconformal invariant action for a chiral multiplet whose F component has Weyl weight $w = 4$. We will take this chiral multiplet to be the local superconformal version of the product multiplet of step 2 above. The action we need is then obtained by inserting the several components of the product multiplet in (16.35).

The first step is to obtain the superconformal version of the chiral multiplet whose lowest component is $\bar{F}$. We observe that δF in (16.33) simplifies when $w = 1$. The η term cancels so δF is quite similar to $\delta\bar{Z}'$ for a scalar Z' with $w_{Z'} = 2$. From the complex conjugate we identify $P_L\chi' = \not{\mathcal{D}} P_R\chi$. With some work one can then use the Q-supersymmetry transform of $P_L\chi'$ and find the top F' component of the multiplet. The conclusion is that

$$Z' = \bar{F}\,, \qquad P_L\chi' = \not{\mathcal{D}} P_R\chi\,, \qquad F' = \Box^C\bar{Z} \tag{16.36}$$

is a superconformal chiral multiplet[7] with $w = 2$. Some advice on the last step of this construction is given below. We first discuss the superconformal derivatives that appear in (16.36). The derivatives $\mathcal{D}_a Z$ and $\mathcal{D}_a P_R\chi$ can be obtained from (16.34) for the relevant Weyl weight $w = 1$. The explicit form of the superconformal covariant d'Alembertian is

$$\begin{aligned}\Box^C Z = e^{a\mu}\Big(&\partial_\mu\mathcal{D}_a Z - 2b_\mu\mathcal{D}_a Z + \omega_{\mu ab}\mathcal{D}^b Z + 2f_{\mu a}Z - \mathrm{i}A_\mu\mathcal{D}_a Z \\ &-\frac{1}{\sqrt{2}}\bar{\psi}_\mu P_L\mathcal{D}_a\chi + \frac{1}{\sqrt{2}}\bar{\phi}_\mu\gamma_a P_L\chi\Big)\,.\end{aligned} \tag{16.37}$$

Note: To check the Q-supersymmetry of $P_L\chi'$, one needs the transformation of the covariant derivative $\gamma^a\mathcal{D}_a P_R\chi$. Since there are no terms with ϵ underlined in (16.10), the application of the lemma on covariant derivatives (see Sec. 11A.1) leads immediately to $\not{\mathcal{D}} P_R\left(\not{\mathcal{D}}\bar{Z} + \bar{F}\right)\epsilon$. The last term corresponds to the $\not{\mathcal{D}} Z'$ term in (16.33). In the product $\not{\mathcal{D}}\not{\mathcal{D}} = \Box^C + \gamma^{ab}\mathcal{D}_a\mathcal{D}_b$, we can neglect the second term, since we know that there is no γ^{ab} in the final result for the transformation of the second component of a chiral multiplet in (16.33). The final remark of Sec. 16.1.4 leads to this conclusion.

If you really want to go through the full calculation without using this argument, you need the curvature constraints, and consequences thereof like $R_{\mu\nu}(D) = -\mathrm{i}\tilde{R}_{\mu\nu}(T)$, where the tilde denotes the dual of the antisymmetric tensor. Such equations can be proven from the curvature constraints (16.12) using Bianchi identities.

[7] In superspace, this corresponds to the superfield $\bar{D}\bar{D}\bar{\Phi}$.

The product of the chiral multiplet $(Z, P_L\chi, F)$ and the multiplet (16.36) is a chiral multiplet with Weyl weight 3 with components[8]

$$\begin{aligned} Z(Z\bar{F}) &= Z\bar{F}\,, \\ P_L\chi(Z\bar{F}) &= P_L\chi\bar{F} + Z\not{D}P_R\chi\,, \\ F(Z\bar{F}) &= F\bar{F} + Z\Box^C\bar{Z} - \bar{\chi}P_L\not{D}\chi\,. \end{aligned} \tag{16.38}$$

It is already clear that the last component leads indeed to the simple action (6.17) in global supersymmetry. Moreover we see that the scalars appear with the superconformal extension of the d'Alembertian expected from the bosonic theory in Sec. 15.5. Now we substitute these results in (16.35) and change the sign:

$$\begin{aligned} S_{\rm SG} = -\int \mathrm{d}^4x\, e\, \mathrm{Re}\Big[& F\bar{F} + Z\Box^C\bar{Z} - \bar{\chi}P_L\not{D}\chi \\ & + \frac{1}{\sqrt{2}}\bar{\psi}_\mu\gamma^\mu\left(P_L\chi\bar{F} + Z\not{D}P_R\chi\right) + \frac{1}{2}Z\bar{F}\bar{\psi}_\mu\gamma^{\mu\nu}P_R\psi_\nu\Big]\,. \end{aligned} \tag{16.39}$$

Many simplifications occur if we insert the explicit forms of the covariant derivatives. For example, it is easy to show that all F-dependent terms cancel except the leading $F\bar{F}$ term.

16.2.2 Gauge fixing

The next step in our strategy is gauge fixing the superconformal symmetries that are not in the super-Poincaré algebra, namely K_a, D, T and S-supersymmetry. For the special conformal transformation we can still choose the gauge condition (15.26) used previously. The generalization of the dilatation gauge choice (15.3) is to fix the modulus of Z. But we must now fix the T-symmetry also, so we can fix Z completely by the condition

$$D\text{-gauge and } T\text{-gauge:} \quad Z = \frac{\sqrt{3}}{\kappa}\,. \tag{16.40}$$

Finally, the obvious choice for S-supersymmetry is

$$S\text{-gauge:} \quad \chi = 0\,. \tag{16.41}$$

This does not leave a lot from the expressions (16.34) and (16.37). The only remaining terms are

$$\begin{aligned} \mathcal{D}_a Z &= -\mathrm{i}A_a Z\,, \\ \mathcal{D}_a P_L\chi &= e_a^\mu P_L\left[\frac{1}{\sqrt{2}}\left(\mathrm{i}\not{A}Z - F\right)\psi_\mu - \sqrt{2}Z\phi_\mu\right], \\ \Box^C Z &= -\mathrm{i}e^{a\mu}\left(\partial_\mu A_a Z + \omega_{\mu ab}A^b Z\right) + 2f_\mu{}^\mu Z - A^a A_a Z - \frac{1}{\sqrt{2}}\bar{\psi}^a P_L\mathcal{D}_a\chi\,. \end{aligned} \tag{16.42}$$

[8] The notation $Z(Z\bar{F})$ means the 'Z component of the product multiplet', etc.

After substitution in (16.39) and some simplification,[9] the action becomes

$$S_{\rm SG} = -\int {\rm d}^4x\, e\left[F\bar{F} + \frac{3}{\kappa^2}\left(2f_\mu{}^\mu - A^a A_a - \tfrac{1}{2}\bar{\psi}_\mu\gamma^{\mu\nu}\phi_\nu + \tfrac{1}{4}{\rm i}\bar{\psi}_\mu\gamma_*\gamma^{\mu\nu\rho}A_\rho\psi_\nu\right)\right]. \tag{16.43}$$

16.2.3 The result

Finally, after use of the constraints (16.24), specifically the relations (16.26) derived from the constraints, we obtain

$$S_{\rm SG} = \int {\rm d}^4x\, e\left[\frac{1}{2\kappa^2}\left(R(\omega(e,\psi)) - \bar{\psi}_\mu\gamma^{\mu\nu\rho}\left(\partial_\nu + \tfrac{1}{4}\omega_\nu{}^{ab}(e,\psi)\gamma_{ab}\right)\psi_\rho \right.\right.$$
$$\left.\left. + 6A^a A_a\right) - F\bar{F}\right]. \tag{16.44}$$

When the gravitino torsion terms are split off from the spin connection, one finds exactly the extra 4-gravitino terms of $\mathcal{N}=1$, $D=4$ supergravity, as the reader was invited to check in Ex. 9.4.

The complex fields F and the real vector A_a are auxiliary fields, which can be eliminated from the action using their rather trivial field equations. In this section we used a conformal chiral multiplet to construct the supergravity action, and that led us to these auxiliary fields, which are referred to in the literature as the 'old minimal' set of auxiliary fields [141, 142, 143]. Starting with action formulas for other multiplets leads to supergravity with different sets of auxiliary fields, such as the 'new minimal set' [144] or the 'non-minimal set' [145, 146]. Furthermore there are the 16 + 16 sets [147, 148, 149]. We will not discuss these other options here since they do not lead to different physical actions [150].

Note the minus sign that we introduced in (16.43). This sign is needed to produce the action $S_{\rm SG}$ with positive graviton kinetic energy. A similar minus sign was discussed at the end of Sec. 15.6. This sign is an important feature in the structure of supergravity theories. It implies that the kinetic energy of the (unphysical) compensating scalar is negative, and it is the origin of the appearance of non-compact symmetry groups and of negative terms in the scalar potential.

We also want the local supersymmetry transformations, under which the action (16.44) is invariant. The vierbein transformation is unaltered from (16.10), but we write it again to assemble the key results of this chapter in one place:

$$\delta e_\mu{}^a = \tfrac{1}{2}\bar{\epsilon}\gamma^a\psi_\mu\,. \tag{16.45}$$

For the gravitino we have to take the gauge fixing conditions into account. Most important is (16.41), which is not invariant under Q-supersymmetry by itself, but only under

[9] There are simplifications in the projection to the Re part in (16.35), which eliminates several A_μ terms. Actually, the eliminated terms combine into a total derivative, so it is not necessary to use the Re part in this application of (16.35).

a combination of S- and Q-supersymmetry. Indeed, taking into account the other gauge conditions, the consistency of this gauge choice under the transformations (16.33) leads to

$$P_L\eta = \frac{1}{2}\mathrm{i}P_L \not{A}\epsilon - \frac{\kappa}{2\sqrt{3}} F P_L\epsilon\,. \tag{16.46}$$

This is what we called a decomposition law at the end of Sec. 15.4. Inserting this into (16.10) leads, for example, to the Poincaré supersymmetry transformation of the gravitino:

$$\delta P_L\psi_\mu = P_L\left(\partial_\mu + \frac{1}{4}\omega_\mu{}^{ab}\gamma_{ab} - \frac{3}{2}\mathrm{i}A_\mu + \frac{1}{2}\mathrm{i}\gamma_\mu\not{A} + \frac{\kappa}{2\sqrt{3}}\gamma_\mu\bar{F}\right)\epsilon\,. \tag{16.47}$$

The remaining fields are the auxiliary A_a and complex F. We obtain the transformation law of A_a from (16.10). The λ_T is fixed to zero, but for the S-supersymmetry parameter we should in principle use the decomposition law (16.46). However, the lemma on covariant derivatives can again be used to avoid a difficult calculation. This term is proportional to ψ_μ, which is a gauge field in Poincaré supergravity. So we know that such terms need not be calculated explicitly; rather they are automatically included in covariantizations. So, we just need only consider the ϵ term and using (16.24) we obtain

$$\delta A_\mu = -\tfrac{1}{4}\mathrm{i}\bar{\epsilon}\gamma_*\left(-\gamma^\nu\widehat{R}_{\mu\nu}(Q) + \tfrac{1}{6}\gamma_\mu\gamma^{ab}\widehat{R}_{ab}(Q)\right)\,. \tag{16.48}$$

We introduce here $\widehat{R}_{\mu\nu}(Q)$ as a covariant curvature for the super-Poincaré algebra; see (11.72). This means that it should take into account the transformations of the gravitino in (16.47) that do not involve gauge fields. The expression for the left-handed projection is therefore

$$P_L\widehat{R}_{\mu\nu}(Q) = 2P_L\left(\partial_{[\mu} + \frac{1}{4}\omega_{[\mu}{}^{ab}\gamma_{ab} - \frac{3}{2}\mathrm{i}A_{[\mu} + \frac{1}{2}\mathrm{i}\gamma_{[\mu}\not{A} + \frac{\kappa}{2\sqrt{3}}\gamma_{[\mu}\bar{F}\right)\psi_{\nu]}\,. \tag{16.49}$$

For the auxiliary field F, we start from (16.33) with $w = 1$, and we can use the simplification of $\mathcal{D}_a\chi$ in (16.42). Again, the ψ_μ terms can be omitted by using the same covariantization, and with (16.26) one obtains

$$\delta F = \frac{\sqrt{3}}{6\kappa}\bar{\epsilon}\,P_R\gamma^{\mu\nu}\widehat{R}_{\mu\nu}(Q)\,. \tag{16.50}$$

With these transformations, the local SUSY algebra closes without use of the field equations.

With the present formalism it is easy to construct extensions. Since we started with a chiral multiplet, the obvious extension is to add a superpotential as discussed in Secs. 6.2 and 14.2.1. However, we must now ensure that the interaction is conformal.

Exercise 16.7 *Check that the interaction (6.18) can only be conformal invariant for $W(Z) = cZ^3$, where c is a constant.*

In the framework of this section, the superpotential term in the action is then obtained by starting with a chiral multiplet with lowest component $W(Z)$ and Weyl weight $w = 3$. We need only substitute the components of this multiplet in (16.35). Using the previous gauge conditions, and the specific $W(Z) = cZ^3$ with real c (and real Z after the gauge condition), one finds the very simple result

$$S_c = \int \mathrm{d}^4x\, e \left[3cZ^2 \operatorname{Re} F + \tfrac{1}{4}cZ^3 \bar{\psi}_\mu \gamma^{\mu\nu} \psi_\nu\right]. \tag{16.51}$$

We now consider the theory whose total action is $S_{SG}+S_c$. The auxiliary field A_a in (16.44) still vanishes, but the field equation for F gives the value

$$F = -\frac{3}{2}cZ^2 = -\frac{9c}{2\kappa^2}. \tag{16.52}$$

Using this value, the action becomes

$$S_{SG} + S_c = \int \mathrm{d}^4x\, e \frac{1}{2\kappa^2}\Big[R(\omega(e,\psi)) - \Lambda \tag{16.53}$$
$$- \bar{\psi}_\mu \gamma^{\mu\nu\rho}\left(\partial_\nu + \tfrac{1}{4}\omega_\nu{}^{ab}(e,\psi)\gamma_{ab}\right)\psi_\rho + m_{3/2}\bar{\psi}_\mu\gamma^{\mu\nu}\psi_\nu\Big].$$

We defined

$$m_{3/2} = \frac{1}{L} = \frac{1}{2}\kappa^2 cZ^3 = \frac{3^{3/2}\,c}{2\kappa}, \qquad \Lambda = -\frac{81\,c^2}{2\kappa^2} = -6m_{3/2}^2. \tag{16.54}$$

We thus reobtain in this way the anti-de Sitter supergravity discussed in Sec. 9.6, with parameter L given in (16.54). The constant c thus sets the value of the cosmological constant.

Exercise 16.8 *Owing to (16.52) and (16.46), the S-supersymmetry contributes to the supersymmetry transformation of the gravitino. Check that this leads to (9.52).*

Exercise 16.9 *Use the decomposition law (16.46), the equations of motion of the auxiliary fields, and the superconformal algebra (16.28) to obtain that the supersymmetry algebra of the preserved ϵ supersymmetries contains a Lorentz transformation with parameter*

$$\lambda^{ab} = \frac{1}{4L}\bar{\epsilon}_2\gamma^{ab}\epsilon_1. \tag{16.55}$$

Check that this agrees with (12.17).

17 Construction of the matter-coupled $\mathcal{N} = 1$ supergravity

We are now ready to construct the general action of $\mathcal{N} = 1$ supergravity coupled to chiral and vector multiplets,[1] using conformal and superconformal methods. Again there is a chiral compensator multiplet, which plays a special role. It is needed to apply superconformal symmetry, but its scalar and spinor components are subsequently gauge fixed and do not correspond to particles in the final supergravity theory.

For matter-coupled theories we start with $n+1$ chiral multiplets whose components will be denoted by X^I, Ω^I and F^I, $I = 0, 1, \ldots, n$. One of these is the compensating multiplet. We reserve the previous notation z^α, χ^α, F^α, $\alpha = 1, \ldots, n$, for the physical chiral multiplets which will be obtained from the larger set by gauge fixing. We will apply our previous knowledge of the general couplings of chiral multiplets in global supersymmetry from Ch. 14 and the information on Kähler geometry from Ch. 13. This information will be combined with the important new element of conformal symmetry.

Two Kähler manifolds appear in the construction. The first of these is the embedding manifold, which has complex dimension $n + 1$ and local coordinates X^I, $\bar{X}^{\bar{I}}$. Its Kähler metric has signature $(- + \ldots +)$; the negative direction corresponds to the scalar compensator. This metric is strongly constrained by the property of conformal symmetry. The initial action we construct incorporates the structure of a superconformal extension of the nonlinear σ-model on the embedding manifold. As in the previous chapter we then fix the gauge of all symmetries that are not part of the final supergravity theory that is our goal. Gauge fixing eliminates the fields of the compensator multiplet (except for an auxiliary field). The n chiral multiplets that remain are physical and they have a different geometric structure. They are coordinate fields of a projective Kähler manifold (also called a Kähler–Hodge manifold) that has metric signature $(+ + \ldots +)$. The differential geometry of the projection of the final Kähler manifold from the embedding manifold[2] will be discussed in Sec. 17.3 and in Appendix 17A. Our strategy is illustrated in Box 17.1.

Gauge multiplets $A_\mu{}^A$, λ^A, D^A (sometimes called vector multiplets) are also important. Their basic interactions in global supersymmetry are already conformal invariant, so it is relatively simple to include them in supergravity. We also incorporate a general superpotential. The final product of this chapter is then the general coupling of $\mathcal{N} = 1$ gauge and chiral multiplets to supergravity.

[1] We do not treat tensor (or linear) multiplets. The superconformal approach to these multiplets is discussed in [150]. An alternative approach based on 'Kähler superspace geometry' can be found in the review [151].

[2] The mathematics of projective Kähler manifolds allows more general signature relations between the embedding manifold and its projection. In one of the examples considered in Sec. 17.3 the embedding manifold has positive definite signature.

Box 17.1 Superconformal construction of chiral multiplet couplings to supergravity

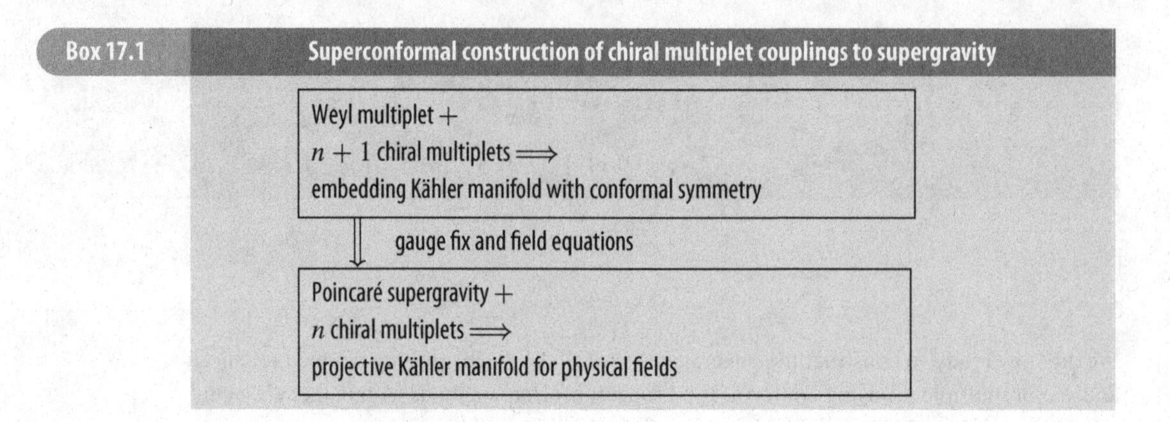

Our treatment of matter couplings follows [152] (which incorporates earlier work on the chiral multiplet in [153]) and uses the improved superconformal methods of [154]. Further details can be found in [139, 155, 156].

17.1 Superconformal tensor calculus

The general actions of global supersymmetry studied in Ch. 14 are invariant under the Poincaré supersymmetry algebra (14.1). In this section we extend those constructions to incorporate invariance under the superconformal algebra discussed in Sec. 16.1.1. This requires that we clearly specify the Weyl weight of the multiplets we work with, namely chiral multiplets, gauge multiplets and real multiplets. Therefore we must define the superconformal transformations of the component fields of these multiplets. This was already done for chiral multiplets in Sec. 16.1.4, although we now need to include the case when they transform under a gauged internal symmetry. We will also consider gauge multiplets and real multiplets. We recommend that readers consult Ch. 14 as we work out the various superconformal extensions.

17.1.1 The superconformal gauge multiplet

By definition a superconformal gauge multiplet contains a gauge vector potential, called A_μ^A in (6.49), together with its Q-supersymmetry partners λ^A, D^A. The gauged internal symmetry transformations that act on these fields are not part of the superconformal group; so they must commute with the dilatations and the R-symmetry. This implies that the Weyl weight and chiral weight of the vector potential vanish. As we explained in Sec. 16.1.4 for the chiral multiplet, this fixes the Weyl weights of the other fields, namely $w_\lambda = 3/2$, $w_D = 2$. Furthermore, $A_\mu{}^A$ and D^A are real, so their chiral weights vanish, and the chiral weight of $P_L\lambda^A$ is $c = 3/2$. It then turns out that nothing more has to be done to specify the superconformal transformations. There are no extra S- or K-transformations for the fields of this multiplet. So we can take over the rules of global supersymmetry and covariantize derivatives. The result is therefore (we omit general coordinate and Lorentz transformations)[3]

$$
\begin{aligned}
\delta A_\mu^A &= -\tfrac{1}{2}\bar{\epsilon}\gamma_\mu\lambda^A\,,\\
\delta\lambda^A &= \tfrac{3}{2}\left[\lambda_D + i\gamma_*\lambda_T\right]\lambda^A + \left[\tfrac{1}{4}\gamma^{ab}\widehat{F}_{ab}{}^A + \tfrac{1}{2}i\gamma_* D^A\right]\epsilon\,,\\
\delta D^A &= 2\lambda_D D^A + \tfrac{1}{2}i\,\bar{\epsilon}\gamma_*\gamma^\mu \mathcal{D}_\mu\lambda^A\,,\\
\widehat{F}_{ab}{}^A &= e_a{}^\mu e_b{}^\nu\left(2\partial_{[\mu}A_{\nu]}{}^A + f_{BC}{}^A A_\mu{}^B A_\nu{}^C + \bar{\psi}_{[\mu}\gamma_{\nu]}\lambda^A\right),\\
\mathcal{D}_\mu\lambda^A &\equiv \left(\partial_\mu - \tfrac{3}{2}b_\mu + \tfrac{1}{4}\omega_\mu{}^{ab}\gamma_{ab} - \tfrac{3}{2}i\gamma_* A_\mu\right)\lambda^A + \lambda^C A_\mu{}^B f_{BC}{}^A\\
&\quad -\left[\tfrac{1}{4}\gamma^{ab}\widehat{F}_{ab}{}^A + \tfrac{1}{2}i\gamma_* D^A\right]\psi_\mu\,.
\end{aligned}
\tag{17.1}
$$

The covariant field strength $\widehat{F}_{ab}{}^A$ appears here, as explained in Sec. 11.3 and also considered in Appendix 11A.

17.1.2 The superconformal real multiplet

To include the real multiplet in the superconformal framework, we must first determine the Weyl and chiral weights. As above, the leading field C is real. It cannot undergo a complex transformation, and thus the chiral weight vanishes. On the other hand, the Weyl weight of C is an arbitrary parameter w. The Weyl weights of the other fields are determined by imposing $[D, Q] = \frac{1}{2}Q$. Rather than presenting the Weyl and chiral transformations of all these fields separately, we refer to Table 17.1 where the weights of all the fields that we need are collected. The weights of the Weyl and chiral multiplets correspond to the transformations already given in (16.10) and (16.33).

The Q-supersymmetry transformations are a rather easy generalization of those for global supersymmetry in (14.10). A check of the algebra, as done in Sec. 16.1.4 for the chiral multiplet, now leads to extra S-supersymmetry and special conformal transformations for some of the fields. The results for the superconformal transformations (omitting covariant general coordinate transformations, Lorentz transformations, dilatations, and T) of the real multiplet are then

$$
\begin{aligned}
\delta C &= \tfrac{1}{2}i\bar{\epsilon}\gamma_*\zeta\,,\\
\delta P_L\zeta &= \tfrac{1}{2}P_L\left(i\mathcal{H} - \not{B} - i\not{\mathcal{D}}C\right)\epsilon - iwCP_L\eta\,,\\
\delta\mathcal{H} &= -i\bar{\epsilon}P_R\left(\lambda + \not{\mathcal{D}}\zeta\right) + (w-2)i\bar{\eta}P_L\zeta\,,\\
\delta B_a &= -\tfrac{1}{2}\bar{\epsilon}\left(\gamma_a\lambda + \mathcal{D}_a\zeta\right) + \tfrac{1}{2}(1+w)\bar{\eta}\gamma_a\zeta\,,
\end{aligned}
$$

[3] Note that A_μ without upper index A is the gauge field of R-symmetry introduced in Sec. 16.1.2. Furthermore, the Lorentz connection which appears is $\omega_\mu^{ab}(e, b, \psi)$ of (16.13). Modified Lorentz connections are used later in the chapter and indicated explicitly.

Table 17.1 Weyl and chiral weights for fields in superconformal $\mathcal{N}=1$ multiplets. The corresponding transformations of the fields are $\delta\phi = (w\lambda_D + ic\lambda_T)\phi$. (The Weyl weights of chiral and real multiplets are independent. We use a common parameter w for simplicity.)

	Weyl multiplet							Chiral multiplet		
	$e_\mu{}^a$	$P_L\psi_\mu$	b_μ	$\omega_\mu{}^{ab}$	A_μ	$P_L\phi_\mu$	$f_\mu{}^a$	X^I	Ω^I	F^I
w	-1	$-\frac{1}{2}$	0	0	0	$\frac{1}{2}$	1	w	$w+\frac{1}{2}$	$w+1$
c	0	$\frac{3}{2}$	0	0	0	$-\frac{3}{2}$	0	w	$w-\frac{3}{2}$	$w-3$

	Real multiplet						Gauge multiplet		
	C	$P_L\zeta$	$\mathcal{H}$	B_a	$P_R\lambda$	D	A_μ^A	$P_R\lambda^A$	D^A
w	w	$w+\frac{1}{2}$	$w+1$	$w+1$	$w+\frac{3}{2}$	$w+2$	0	$\frac{3}{2}$	2
c	0	$-\frac{3}{2}$	-3	0	$-\frac{3}{2}$	0	0	$-\frac{3}{2}$	0

$$
\begin{aligned}
\delta P_R\lambda &= P_R\left[\tfrac{1}{4}\gamma^{ab}F_{ab} - \mathrm{i}\tfrac{1}{2}D\right]\epsilon + \tfrac{1}{2}wP_R\left[\mathrm{i}\mathcal{H} - \not{B} + \mathrm{i}\not{\mathcal{D}}C\right]\eta - w\lambda\!\!\!/_{\mathrm{K}}P_L\zeta\,,\\
\delta D &= \tfrac{1}{2}\mathrm{i}\,\bar\epsilon\gamma_*\not{\mathcal{D}}\lambda + \mathrm{i}w\bar\eta\left(\lambda + \tfrac{1}{2}\not{\mathcal{D}}\zeta\right) - 2w\lambda_{\mathrm{K}}^a\mathcal{D}_aC,\\
F_{ab} &\equiv 2\mathcal{D}_{[a}B_{b]} + \varepsilon_{abcd}\mathcal{D}^c\mathcal{D}^dC\,.
\end{aligned}
\tag{17.2}
$$

The covariant derivatives $\mathcal{D}_\mu$ are covariant under all symmetries of the superconformal algebra. Since the vector is not a gauge field, we write expressions for B_a rather than B_μ. Then the right-hand side of (17.2) involves only covariant quantities (see Sec. 11.3).

In Sec. 14.1.2 we saw that, in global supersymmetry, real multiplets can be reduced to (elementary) gauge multiplets. This was related to the Wess–Zumino gauge. The same reduction in the superconformal setting requires that B_μ must carry Weyl weight 0, since it is a gauge field. Hence, $B_a = e_a^\mu B_\mu$ must have Weyl weight 1, since it involves the inverse frame field. Table 17.1 tells us that this can only be done for a real multiplet of Weyl weight $w=0$. Note that many transformation laws simplify in this case. However, we will also need (composite) real multiplets with $w=2$ in order to construct superconformal actions, as we will discuss below.

17.1.3 Gauge transformations of superconformal chiral multiplets

We now extend the treatment of superconformal chiral matter multiplets, discussed in Sec. 16.1.4, to multiplets which transform in a representation of a gauged internal symmetry group.[4] This is similar to what was done for global supersymmetry in Sec. 14.4.2, but we must now consider the requirements of superconformal symmetry.

Gauge symmetries must be compatible with the conformal structure, and this means that gauge and superconformal transformations must commute. For scalar fields X^I the non-trivial requirement is evident in (14.33). The Killing vector $k_A{}^I(X)$ must carry the same Weyl weight w as the chiral multiplet headed by X^I. The rest is determined by the rules of global supersymmetry and the result (16.33) for the 'gauge singlet' superconformal chiral multiplet. We can thus write the superconformal Q- and S-supersymmetry transformations as (with the notation that $\Omega^I = P_L\Omega^I$ and $\Omega^{\bar{I}} = P_R\Omega^{\bar{I}}$, to be clarified in the next paragraph)

$$
\begin{aligned}
\delta X^I &= \frac{1}{\sqrt{2}}\bar{\epsilon}\Omega^I ,\\
\delta\Omega^I &= \frac{1}{\sqrt{2}}P_L\left(\not{D}X^I + F^I\right)\epsilon + \sqrt{2}wX^I P_L\eta ,\\
\delta F^I &= \frac{1}{\sqrt{2}}\bar{\epsilon}\not{D}\Omega^I + \bar{\epsilon}P_R\lambda^A k_A{}^I(X) + \sqrt{2}(1-w)\bar{\eta}\Omega^I .
\end{aligned} \tag{17.3}
$$

These transformations are the extension of (16.33) to include gauge symmetries, so that gauge connection terms are now added to the previous superconformal covariant derivatives. These derivatives are written below in (17.20). Note that the extra λ^A-term is also consistent with the superconformal weights.

We now use a new notation for the spinors of chiral and anti-chiral multiplets, so let us be clear about this. We omit the explicit projector P_L for spinors Ω^I, but it is to be understood that the projector is present in all expressions. Thus Ω^I should be viewed as a four-component Dirac spinor that is *invariant* under the P_L projection. In a Weyl representation it takes the form of the left-hand spinor in (3.95). Given a chiral multiplet X^I, Ω^I, F^I, we write the conjugate anti-chiral multiplet as $\bar{X}^{\bar{I}}$, $\Omega^{\bar{I}}$, $\bar{F}^{\bar{I}}$. The bosonic components are related to those of the chiral multiplet by complex conjugation, e.g. $\bar{X}^{\bar{I}} = (X^I)^*$. The fermion component $\Omega^{\bar{I}}$ is a four-component spinor invariant under right-handed projection. It is the charge conjugate spinor of Ω^I; see (3.74). In a Weyl representation it takes the form of the right-hand spinor in (3.95), with components related to the complex conjugates of those of Ω^I. As explained in the last paragraph of Sec. 3.4.2, one can define a Majorana spinor from the left-handed and right-handed projections, in this case $\Omega^I + \Omega^{\bar{I}}$. In a Weyl representation the third and fourth components, which are components of $\Omega^{\bar{I}}$, are related to the complex conjugates of the first two components as in (3.93). The projector P_R is understood to be present in all expressions involving $\Omega^{\bar{I}}$. Conjugate spinors for both Ω^I and $\Omega^{\bar{I}}$ are defined as in (3.50) using the charge conjugation matrix. Again projectors are implicitly present. Thus $\bar{\Omega}^I = (\Omega^I)^T C = \bar{\Omega}^I P_L$ and $\bar{\Omega}^{\bar{I}} = (\Omega^{\bar{I}})^T C = \bar{\Omega}^{\bar{I}} P_R$.

[4] We will use the term 'gauge symmetry' to refer to these internal symmetries, although superconformal transformations are also gauged.

17.1.4 Invariant actions

In global supersymmetry, invariant actions are built directly from the highest components of chiral and real multiplets. However, as discussed in the previous chapter, for a superconformal chiral multiplet we need to use the action formula (16.35) rather than just the F-component. Further, F must belong to a chiral multiplet with Weyl weight 3. This ensures that F has Weyl weight 4, compensating the Weyl weight of the determinant of the frame field. Arguing in the same way for the real multiplet, it is clear that we need a real multiplet with $w=2$. Its D-component will then have Weyl weight 4, and we can hope to find a superconformal extension of the general global SUSY action S_D of (14.12). It turns out that we can simplify the work by first noticing that, for $w=2$, the field $\mathcal{H}$ does not transform under S-supersymmetry, and thus transforms as the lowest component of an anti-chiral multiplet. Our strategy is to identify the components of the conjugate chiral multiplet and simply insert them in the action formula (16.35).

In fact, one can use the transformations (17.2) with $w=2$ to verify that the following fields form a chiral multiplet[5]

$$Z=-\overline{\mathcal{H}}\,,\qquad P_L\chi=-\sqrt{2}\mathrm{i}P_L\left(\lambda+\not{D}\zeta\right),\qquad F=D+\mathcal{D}^a\mathcal{D}_aC+\mathrm{i}\mathcal{D}^aB_a\,. \quad (17.4)$$

The fields on the right-hand sides are those of the real multiplet. We insert Z, $P_L\chi$ and F in (16.35) to obtain the desired action formula for a real multiplet with Weyl weight $w=2$ [157, 158]:

$$S_D=\int \mathrm{d}^4x\, e\left[D+\mathcal{D}^a\mathcal{D}_aC+\tfrac{1}{2}\left(\mathrm{i}\bar{\psi}\cdot\gamma P_R(\lambda+\not{D}\zeta)-\tfrac{1}{4}\bar{\psi}_\mu P_L\gamma^{\mu\nu}\psi_\nu\mathcal{H}+\text{h.c.}\right)\right]. \quad (17.5)$$

The contraction $\bar{\psi}\cdot\gamma$ stands for $\bar{\psi}^\mu\gamma_\mu$. The covariant derivatives in (17.5) are similar to those of (16.34) and (16.37), but now for a real multiplet with $w=2$. According to the transformation rules in (17.2), they are given by

$$\begin{aligned}
\mathcal{D}_aC&=\partial_aC-2b_aC-\tfrac{1}{2}\mathrm{i}\bar{\psi}_a\gamma_*\zeta\,,\\
\mathcal{D}_aP_L\zeta&=\left(\partial_a+\tfrac{1}{4}\omega_a{}^{bc}\gamma_{bc}-\tfrac{5}{2}b_a+\tfrac{3}{2}\mathrm{i}A_a\right)\zeta-\tfrac{1}{2}P_L\left(\mathrm{i}\mathcal{H}-\not{B}-\mathrm{i}\not{D}C\right)\psi_a\\
&\quad+2\mathrm{i}CP_L\phi_a\,, \qquad (17.6)\\
\square^CC&=\partial^a\mathcal{D}_aC-3b^a\mathcal{D}_aC+\omega_a{}^{ab}\mathcal{D}_bC+4C\,f_a{}^a-\tfrac{1}{2}\mathrm{i}\bar{\psi}^a\gamma_*\mathcal{D}_a\zeta+\tfrac{1}{2}\mathrm{i}\bar{\phi}\cdot\gamma\gamma_*\zeta\,.
\end{aligned}$$

Here $\square^CC=\mathcal{D}^a\mathcal{D}_aC$. Its bosonic terms were already determined in (15.33).

The next step is to use the expressions for the dependent gauge fields (16.26) and (16.25), which introduces the graviton and gravitino curvatures $R(\omega)$ and $R'(Q)$. After further manipulations discussed in Appendix 17B we find the action formula

$$\begin{aligned}
S_D=\int \mathrm{d}^4x\, e\Big[&D-\tfrac{1}{2}\bar{\psi}\cdot\gamma\mathrm{i}\gamma_*\lambda-\tfrac{1}{3}C\,R(\omega)+\tfrac{1}{6}\left(C\,\bar{\psi}_\mu\gamma^{\mu\rho\sigma}-\mathrm{i}\bar{\zeta}\gamma^{\rho\sigma}\gamma_*\right)R'_{\rho\sigma}(Q)\\
&+\tfrac{1}{4}\varepsilon^{abcd}\bar{\psi}_a\gamma_b\psi_c\left(B_d-\tfrac{1}{2}\bar{\psi}_d\zeta\right)\Big]. \qquad (17.7)
\end{aligned}$$

[5] Although we do not discuss the antisymmetric tensor multiplet in this book, we remark that such a multiplet is defined as a real multiplet with the constraint that the fields of this chiral multiplet vanish.

This S_D is the superconformal extension of the D-term action of (14.12) for global supersymmetry, and we will use it to obtain general superconformal actions.

17.2 Construction of the action

We now have all the ingredients needed to extend the procedure of Ch. 14 to construct general actions of chiral and gauge multiplets coupled to supergravity. The actions will depend on three general functions: the real Kähler potential $N(X,\bar{X})$ of the embedding manifold, a holomorphic superpotential $\mathcal{W}(X)$, and a gauge kinetic function $f_{AB}(X)$. In particular the Kähler potential will be identified with the lowest component of the real multiplet by

$$C = \tfrac{1}{2}N(X,\bar{X}), \tag{17.8}$$

as was done in Sec. 14.2.2 for the Kähler potential $K(Z,\bar{Z})$.

However the important new ingredient here is that the requirement of conformal symmetry must be imposed on the three functions above. This is quite straightforward for $\mathcal{W}$ and f_{AB}, but more substantive for N. We begin the discussion by determining a convenient choice of the conformal weights of the scalars X^I.

17.2.1 Conformal weights

The theories we wish to construct contain the Weyl multiplet, gauge multiplets, and chiral multiplets. Conformal weights for the first two cases are unique, as discussed for the Weyl multiplet in Ch. 15 and in Sec. 17.1.1 for gauge multiplets. For chiral multiplets in a superconformal setting the Weyl weights are not yet determined. One cannot take the Weyl weight of all chiral multiplets equal to zero, because the C field in the action formula (17.7) must have Weyl weight $w=2$, and this would be inconsistent with (17.8). Once there is at least one scalar with non-zero Weyl weight, say X^0, the weights of the other fields X^I depend on the choice of coordinates for the Kähler manifold. Indeed, suppose for example that X^0 has Weyl weight 1, and X^1 has weight w. Then one can redefine $X'^1 \equiv (X^0)^{w'-w}X^1$ such that the primed field has Weyl weight w'. The field X^0 must be non-vanishing so that the transformation above is non-degenerate. We already saw in the simplest example of pure supergravity that the compensating multiplet has a non-vanishing scalar; see (16.40). Thus, the choice of Weyl weight of the multiplets is arbitrary. We will take them all of Weyl weight 1, which makes the construction symmetric and easier. At this point we have $n+1$ chiral multiplets on essentially equal footing. The 'reference' multiplet headed by X^0 is not necessarily the compensator. In Sec. 17.3.3 we will discuss how to isolate the compensator and physical multiplets.

17.2.2 Superconformal invariant action (ungauged)

We can now extend the different constructions of Sec. 14.2 to conformal supergravity. There, we first looked at the superpotential. To use the superconformal chiral action (16.35) we must require that the superpotential, now denoted by $\mathcal{W}(X)$, has Weyl weight 3. This is expressed by the equation

$$X^I \mathcal{W}_I = 3\mathcal{W}(X), \qquad \mathcal{W}_I \equiv \frac{\partial}{\partial X^I}\mathcal{W}(X). \tag{17.9}$$

Assuming that X^0 is non-zero, one can solve this by

$$\mathcal{W}(X) = (X^0)^3 W\left(\frac{X^\alpha}{X^0}\right), \qquad \alpha = 1,\ldots,n, \tag{17.10}$$

where W is an arbitrary holomorphic function of the n ratios. But we only need the equation (17.9).

Next, we consider the kinetic terms of the chiral multiplets. In global supersymmetry they are determined by a real function $K(Z,\bar{Z})$ using the D-action formula. In the superconformal tensor calculus this must be a function of Weyl weight 2, which we indicate by $N(X,\bar{X})$. Since it must also be real (or, equivalently, invariant under the superconformal U(1)) it is homogeneous of first degree in both X and $\bar{X}$. This requirement is expressed by

$$N(\lambda X, \bar{\lambda}\bar{X}) = (\lambda\bar{\lambda})N(X,\bar{X}), \tag{17.11}$$

for any complex number λ. Examples are

$$N = a_{I\bar{J}} X^I \bar{X}^{\bar{J}}, \qquad N = \mathrm{i}\frac{\bar{X}^0 (X^1)^3}{(X^0)^2} - 3\mathrm{i}\frac{(X^1)^2\bar{X}^1}{X^0} - \mathrm{i}\frac{X^0(\bar{X}^1)^3}{(\bar{X}^0)^2} + 3\mathrm{i}\frac{(\bar{X}^1)^2 X^1}{\bar{X}^0},$$
$$N = \mathrm{i}\bar{X}^0\sqrt{(X^1)^3(X^0)^{-1}} + 3\mathrm{i}\bar{X}^1\sqrt{X^0X^1} - \mathrm{i}X^0\sqrt{(\bar{X}^1)^3(\bar{X}^0)^{-1}} - 3\mathrm{i}X^1\sqrt{\bar{X}^0\bar{X}^1}. \tag{17.12}$$

These examples show that many structures are possible.[6]

The homogeneity properties can be written as differential equations. Indicating the derivatives with respect to X^I as N_I, and the derivatives with respect to $\bar{X}^{\bar{I}}$ as $N_{\bar{I}}$, these are

$$N(X,\bar{X}) = X^I N_I = \bar{X}^{\bar{I}} N_{\bar{I}} = X^I \bar{X}^{\bar{J}} N_{I\bar{J}}, \qquad N_I = N_{I\bar{J}}\bar{X}^{\bar{J}}, \tag{17.13}$$
$$X^I N_{IJ} = 0, \qquad N_{IJ\bar{K}}\bar{X}^{\bar{K}} = N_{IJ}, \qquad X^K N_{KI\bar{J}} = 0.$$

Exercise 17.1 *Prove all these equations from the first one. Although the embedding manifold is a Kähler manifold, it is curious that the required conformal properties, expressed in (17.13), do not allow Kähler transformations $N(X,\bar{X}) \to N(X,\bar{X}) + f(X) + \bar{f}(\bar{X})$.*

In global supersymmetry the kinetic terms of gauge multiplets were obtained by applying the F-term action formula to the composite chiral multiplet (14.17). In conformal

[6] The choices are inspired by some symmetric spaces that occur in $\mathcal{N}=2$ extended supergravity.

supergravity we require that its top component has Weyl weight 3. Since $\bar{\lambda}^A P_L \lambda^B$ has Weyl weight 3, the kinetic matrix $f_{AB}(X)$ must have Weyl weight 0, i.e.

$$X^I f_{AB\,I} = 0\,, \qquad f_{AB\,I} \equiv \frac{\partial}{\partial X^I} f_{AB}(X)\,. \tag{17.14}$$

With the three ingredients discussed in this section we can write an action formula which is the conformal extension of the three terms in (14.29).[7] The total action can be written as

$$\mathcal{L} = [N(X,\bar{X})]_D + [\mathcal{W}(X)]_F + \left[f_{AB}(X)\bar{\lambda}^A P_L \lambda^B\right]_F\,, \tag{17.15}$$

where the subscript F refers to the formula (16.35), while the subscript D refers to (17.7) (up to normalization factors, which are taken as in Ch. 14). Each of the three terms is superconformal invariant. Their detailed form will be presented in the next section after we include gauged internal symmetry, as in the discussion of Sec. 14.4 for global supersymmetry.

The content of this Lagrangian can be understood from the analogous terms in (14.28) and (14.29). For example, the first term includes the kinetic Lagrangian for the complex scalars X^I, with the Kähler metric

$$G_{I\bar{J}} = N_{I\bar{J}}\,, \tag{17.16}$$

derived from the Kähler potential N. The full Lagrangian consists of the results of (14.29) completed with terms involving the fields of the Weyl multiplet, as we will see in (17.19). One example of the full Lagrangian already appeared in (16.39), which can be obtained by $N = -X^0\bar{X}^{\bar{0}}$. The variable Z of that chapter is now denoted by X^0. Note that the Kähler metric in this case is negative. In general, we must always use a Kähler metric with one negative direction related to the sign discussed at the end of Sec. 15.6.

17.2.3 Gauged superconformal supergravity

We start with the gauge transformations of global supersymmetry; see Sec. 14.4.1. In the present notation, the symmetry transformations of the scalars and fermions of chiral multiplets are thus

$$\delta X^I = \theta^A k_A{}^I\,, \qquad \delta\Omega^I = \theta^A \partial_J k_A{}^I \Omega^J\,. \tag{17.17}$$

As noted in Sec. 17.1.3, the main new ingredient needed for conformal supersymmetry is that the holomorphic Killing vectors $k_A{}^I$ must have the same Weyl weight as the field X^I, thus $w = 1$. In other words, they should be homogeneous of first degree in the X^I. In global SUSY the Kähler potential need not be invariant under gauge transformations; instead it

[7] We now use the D-term action formula (17.7) to construct the chiral kinetic action rather than the F-term used in Ch. 16. In the limit of global SUSY, the Lagrangians differ by a total derivative.

may change by a Kähler transformation; see (13.57). However, we saw in Ex. 17.1 that Kähler transformations of $N(X,\bar{X})$ are forbidden. So N must be gauge invariant. As in global supersymmetry, the superpotential $\mathcal{W}$ must be gauge invariant.

Exercise 17.2 *Show that these statements lead to the following equations:*

$$\begin{aligned}
&\partial_{\bar{J}} k_A{}^I = 0\,, \qquad X^J\partial_J k_A{}^I = k_A{}^I\,,\\
&N_I k_A{}^I + N_{\bar{I}} k_A{}^{\bar{I}} = 0\,;\\
&\mathcal{P}_A = \tfrac{1}{2}\mathrm{i}\left(N_I k_A{}^I - N_{\bar{I}} k_A{}^{\bar{I}}\right) = \mathrm{i}N_I k_A{}^I = -\mathrm{i}N_{\bar{I}} k_A{}^{\bar{I}}\,, \qquad \partial_{\bar{I}}\mathcal{P}_A = \mathrm{i}N_{J\bar{I}} k_A{}^J\,,\\
&\mathcal{W}_I k_A{}^I = 0\,.
\end{aligned} \tag{17.18}$$

The real quantities $\mathcal{P}_A$ are the moment maps associated with holomorphic symmetries of Kähler manifolds, as discussed in Sec. 13.4.1. Note that $\mathcal{P}_A$ has (Weyl, chiral) weight (2, 0), and can thus not be changed by adding constants.

We now present the full expressions for the three superconformal Lagrangians of (17.15). The D-terms are obtained from (17.7) and the F-terms from (16.35), using components that are covariantizations of those found for global supersymmetry in Sec. 14.2. This leads to

$$\begin{aligned}
[N]_D e^{-1} = {} & N_{I\bar{J}}\left(-\mathcal{D}_\mu X^I \mathcal{D}^\mu \bar{X}^{\bar{J}} - \frac{1}{2}\bar{\Omega}^I \not{\mathcal{D}}\Omega^{\bar{J}} - \frac{1}{2}\bar{\Omega}^{\bar{J}}\not{\mathcal{D}}\Omega^I + F^I \bar{F}^{\bar{J}}\right)\\
& + \frac{1}{2}\left[N_{IJ\bar{K}}\left(-\bar{\Omega}^I\Omega^J \bar{F}^{\bar{K}} + \bar{\Omega}^I(\not{\mathcal{D}}X^J)\Omega^{\bar{K}}\right) + \text{h.c.}\right]\\
& + \frac{1}{4}N_{IJ\bar{K}\bar{L}}\,\bar{\Omega}^I\Omega^J\bar{\Omega}^{\bar{K}}\Omega^{\bar{L}}\\
& - \mathrm{i}N_I k_A{}^I D^A - \sqrt{2}\bar{\lambda}^A N_{I\bar{J}}\left(\Omega^I k_A{}^{\bar{J}} + \Omega^{\bar{J}} k_A{}^I\right)\\
& + \left[\frac{1}{2\sqrt{2}}\bar{\psi}\cdot\gamma\left(N_{I\bar{J}}F^I\Omega^{\bar{J}} - N_{I\bar{J}}\not{\mathcal{D}}\bar{X}^{\bar{J}}\Omega^I - \frac{1}{2}N_{IJ\bar{K}}\Omega^{\bar{K}}\bar{\Omega}^I\Omega^J + N_I P_R\lambda^A k_A{}^I\right)\right.\\
& \left. + \frac{1}{8}\mathrm{i}\varepsilon^{\mu\nu\rho\sigma}\bar{\psi}_\mu\gamma_\nu\psi_\rho\left(N_I \mathcal{D}_\sigma X^I + \frac{1}{2}N_{I\bar{J}}\bar{\Omega}^I\gamma_\sigma\Omega^{\bar{J}} + \frac{1}{\sqrt{2}}N_I\bar{\psi}_\sigma\Omega^I\right) + \text{h.c.}\right]\\
& + \frac{1}{6}N\left(-R(\omega) + \tfrac{1}{2}\bar{\psi}_\mu\gamma^{\mu\nu\rho}R'_{\nu\rho}(Q)\right) - \frac{1}{6\sqrt{2}}\left(N_I\bar{\Omega}^I + N_{\bar{I}}\bar{\Omega}^{\bar{I}}\right)\gamma^{\mu\nu}R'_{\mu\nu}(Q)\,,
\end{aligned}$$

$$[\mathcal{W}]_F e^{-1} = \mathcal{W}_I F^I - \tfrac{1}{2}\mathcal{W}_{IJ}\bar{\Omega}^I\Omega^J + \mathcal{W}_I\bar{\psi}\cdot\gamma\Omega^I + \tfrac{1}{2}\mathcal{W}\bar{\psi}_\mu P_R\gamma^{\mu\nu}\psi_\nu + \text{h.c.}\,,$$

$$\begin{aligned}
\left[f_{AB}\bar{\lambda}^A P_L\lambda^B\right]_F e^{-1} = {} & -\frac{1}{4}f_{AB}\left[2\bar{\lambda}^A P_L\not{\mathcal{D}}\lambda^B + \widehat{F}^{-A}_{\mu\nu}\widehat{F}^{\mu\nu-B} - D^A D^B\right.\\
& \left. + \frac{1}{4}\bar{\psi}\cdot\gamma P_L\left(\frac{1}{2}\gamma^{\mu\nu}\widehat{F}^{-A}_{\mu\nu} - \mathrm{i}D^A\right)\lambda^B - \frac{1}{8}\bar{\psi}_\mu\gamma^{\mu\nu}P_R\psi_\nu\bar{\lambda}^A P_L\lambda^B\right]\\
& + \frac{1}{2\sqrt{2}}f_{AB\,I}\bar{\Omega}^I\left(-\frac{1}{2}\gamma^{\mu\nu}\widehat{F}^{-A}_{\mu\nu} + \mathrm{i}D^A\right)\lambda^B\\
& + \left[-\frac{1}{4}f_{AB\,I}\left(F^I + \frac{1}{\sqrt{2}}\bar{\psi}\cdot\gamma\Omega^I\right) + \frac{1}{8}f_{AB\,IJ}\bar{\Omega}^I\Omega^J\right]\bar{\lambda}^A P_L\lambda^B + \text{h.c.}
\end{aligned} \tag{17.19}$$

Terms that do not depend on gauge fields of the superconformal algebra are the same as in (14.57), but the covariant derivatives $\mathcal{D}_\mu$ are the full superconformal derivatives. See (17.1) where $\widehat{F}^A_{\mu\nu}$ and $\mathcal{D}_\mu\lambda^A$ are written. Derivatives of the chiral multiplet fields are given by

$$\begin{aligned}
\mathcal{D}_\mu X^I &= (\partial_\mu - b_\mu - \mathrm{i}\mathcal{A}_\mu)\, X^I - \frac{1}{\sqrt{2}}\bar\psi_\mu\Omega^I - A^A_\mu k_A{}^I\,,\\
\mathcal{D}_\mu\Omega^I &= \left(\partial_\mu - \frac{3}{2}b_\mu + \frac{1}{4}\omega_\mu{}^{ab}\gamma_{ab} + \frac{1}{2}\mathrm{i}\mathcal{A}_\mu\right)\Omega^I\\
&\quad - \frac{1}{\sqrt{2}}P_L\left(\not{\mathcal{D}}X^I + F^I\right)\psi_\mu - \sqrt{2}X^I P_L\phi_\mu - A^A_\mu\Omega^J\partial_J k_A{}^I\,. \qquad (17.20)
\end{aligned}$$

Note that $R'_{\mu\nu}(Q)$ in $[N]_D$ is given in (16.25).

The Lagrangians of (17.19) are very complicated. However, many cancelations occur in terms with gravitinos when the various covariantizations are written in detail. For example, the D^A term in the second line of the gauge kinetic action is canceled by the $D^A\psi_\mu$ term in the covariant derivative of the gaugino. There are also simplifications in the four-fermion terms. A detailed check of these requires many Fierz identities and an immense work that we cannot ask a reader to do (or the authors to write down). Therefore we note that the result of all this work can be found in [155]. We will use this result in the formulas below, but postpone recording the new form of the Lagrangians until auxiliary fields are eliminated.

17.2.4 Elimination of auxiliary fields

We now eliminate auxiliary fields, still maintaining the superconformal invariance. The auxiliary fields are the complex conjugate pair F^I, $\bar F^{\bar I}$ plus the real fields D^A and $\mathcal{A}_\mu$. The values of these auxiliary fields are

$$\begin{aligned}
-\bar F^{\bar J}N_{I\bar J} &= \mathcal{W}_I - \tfrac{1}{2}N_{I\bar J\bar K}\bar\Omega^{\bar J}\Omega^{\bar K} - \tfrac{1}{4}f_{AB\,I}\bar\lambda^A P_L\lambda^B\,,\\
(\mathrm{Re}\, f_{AB})D^B &= \mathcal{P}_A + \mathcal{P}^{\mathrm{F}}_A\,,\\
\mathcal{P}^{\mathrm{F}}_A &= -\mathrm{i}\frac{1}{2\sqrt{2}}f_{AB\,I}\bar\Omega^I\lambda^B + \mathrm{i}\frac{1}{2\sqrt{2}}f_{AB\,\bar I}\bar\Omega^{\bar I}\lambda^B\,,\\
\mathcal{A}_\mu &= A_\mu + \mathcal{A}^{\mathrm{F}}_\mu\,,\\
A_\mu &= \mathrm{i}\frac{1}{2N}\left[N_{\bar I}\left(\partial_\mu\bar X^{\bar I} - A^A_\mu k_A{}^{\bar I}\right) - N_I\left(\partial_\mu X^I - A^A_\mu k_A{}^I\right)\right]\\
&= \mathrm{i}\frac{1}{2N}\left[N_{\bar I}\,\partial_\mu\bar X^{\bar I} - N_I\,\partial_\mu X^I\right] + \frac{1}{N}A^A_\mu\mathcal{P}_A,\\
\mathcal{A}^{\mathrm{F}}_\mu &\equiv \mathrm{i}\frac{1}{4N}\left[\sqrt{2}\bar\psi_\mu\left(N_I\Omega^I - N_{\bar I}\Omega^{\bar I}\right) + N_{I\bar J}\bar\Omega^I\gamma_\mu\Omega^{\bar J}\right.\\
&\qquad\left. + \frac{3}{2}(\mathrm{Re}\, f_{AB})\bar\lambda^A\gamma_\mu\gamma_*\lambda^B\right]. \qquad (17.21)
\end{aligned}$$

We split the values of D^A and $\mathcal{A}_\mu$ into bosonic and fermionic terms. The moment map $\mathcal{P}_A$ appears as the bosonic part of D^A.

Simplifications occur when the values of the auxiliary fields are inserted in the Lagrangian. The combined three terms of (17.19) can then be rewritten as[8]

$$
\begin{aligned}
e^{-1}\mathcal{L} = {}& \tfrac{1}{6}N\left[-R(e,b) + \bar{\psi}_\mu R^\mu + e^{-1}\partial_\mu(e\,\bar{\psi}\cdot\gamma\psi^\mu) - \mathcal{L}_{\rm SG,torsion}\right] \\
& - N_{I\bar{J}}\left[D^\mu X^I\, D_\mu \bar{X}^{\bar{J}} + \tfrac{1}{2}\bar{\Omega}^I \not{D}^{(0)}\Omega^{\bar{J}} + \tfrac{1}{2}\bar{\Omega}^{\bar{J}} \not{D}^{(0)}\Omega^I\right] \\
& + (\mathrm{Re}\, f_{AB})\left[-\tfrac{1}{4}F_{\mu\nu}^A F^{\mu\nu\, B} - \tfrac{1}{2}\bar{\lambda}^A \not{D}^{(0)}\lambda^B\right] \\
& + \tfrac{1}{4}\mathrm{i}\left[(\mathrm{Im}\, f_{AB})\, F_{\mu\nu}^A \tilde{F}^{\mu\nu\, B} + (D_\mu\, \mathrm{Im}\, f_{AB})\,\bar{\lambda}^A\gamma_*\gamma^\mu\lambda^B\right] \\
& - N_{I\bar{J}}F^I\bar{F}^{\bar{J}} - \tfrac{1}{2}(\mathrm{Re}\, f_{AB})D^A D^B \\
& + \left\{\tfrac{1}{2}\mathcal{W}\bar{\psi}_\mu P_R\gamma^{\mu\nu}\psi_\nu - \tfrac{1}{2}\mathcal{W}_{IJ}\bar{\Omega}^I\Omega^J - \sqrt{2}N_{I\bar{J}}k_A{}^I\bar{\lambda}^A\Omega^{\bar{J}} + \text{h.c.}\right\} \\
& + \tfrac{1}{8}(\mathrm{Re}\, f_{AB})\bar{\psi}_\mu\gamma^{ab}\left(F_{ab}^A + \widehat{F}_{ab}^A\right)\gamma^\mu\lambda^B + \left\{\frac{1}{\sqrt{2}}N_{I\bar{J}}\bar{\psi}_\mu \not{D}\bar{X}^{\bar{J}}\gamma^\mu\Omega^I + \text{h.c.}\right\} \\
& + \left\{\frac{1}{2}N_{IJ\bar{K}}\bar{\Omega}^I\gamma^\mu\Omega^{\bar{K}}D_\mu X^J - \frac{1}{4\sqrt{2}}f_{AB\,I}\bar{\Omega}^I\gamma^{ab}\widehat{F}_{ab}^A\lambda^B\right. \\
& - \frac{2}{3\sqrt{2}}N_I\bar{\Omega}^I\gamma^{\mu\nu}D_\mu\psi_\nu + \bar{\psi}\cdot\gamma P_L\left(\frac{1}{2}\mathrm{i}\mathcal{P}_A\lambda^A + \frac{1}{\sqrt{2}}\mathcal{W}_I\Omega^I\right) \\
& \left. - \frac{1}{4\sqrt{2}}f_{AB\,I}\bar{\psi}\cdot\gamma\Omega^I\bar{\lambda}^A P_L\lambda^B + \frac{1}{8}f_{AB\,IJ}\bar{\Omega}^I\Omega^J\bar{\lambda}^A P_L\lambda^B + \text{h.c.}\right\} \\
& + \tfrac{1}{16}\mathrm{i}\,e^{-1}\varepsilon^{\mu\nu\rho\sigma}\bar{\psi}_\mu\gamma_\nu\psi_\rho\left(N_{I\bar{J}}\bar{\Omega}^{\bar{J}}\gamma_\sigma\Omega^I + \tfrac{1}{2}\,\mathrm{Re}\, f_{AB}\bar{\lambda}^A\gamma_*\gamma_\sigma\lambda^B\right) \\
& - \tfrac{1}{2}N_{I\bar{J}}\bar{\psi}_\mu\Omega^{\bar{J}}\,\bar{\psi}^\mu\Omega^I \\
& + \tfrac{1}{4}N_{IJ\bar{K}\bar{L}}\bar{\Omega}^I\Omega^J\,\bar{\Omega}^{\bar{K}}\Omega^{\bar{L}} + N(A_\mu^{\rm F})^2 .
\end{aligned}
\tag{17.22}
$$

Although much work remains before we obtain the final physical action, the result (17.22) is an important milestone along the way. So it is useful to pause and discuss the general structure and some details. In this action the auxiliary fields have the values from (17.21). The torsion terms in the first line of (17.22) are those that we already found in (9.19). They appear as in Sec. 9.2 because we have extracted the torsion from the spin connection. Thus, $R(e,b)$ is again defined with the spin connection $\omega_\mu{}^{ab}(e,b)$. Similarly, other fermion terms are extracted from covariant derivatives D_μ, so that the T connection involves only the bosonic term $\mathcal{A}_\mu$. Thus, explicitly,

$$
\begin{aligned}
D_\mu X^I &= \partial_\mu X^I - b_\mu X^I - A_\mu^A k_A{}^I - \mathrm{i}\mathcal{A}_\mu X^I , \\
D_\mu^{(0)}\Omega^I &= \left(\partial_\mu - \tfrac{3}{2}b_\mu + \tfrac{1}{4}\omega_\mu{}^{ab}(e,b)\gamma_{ab} + \tfrac{1}{2}\mathrm{i}\mathcal{A}_\mu\right)\Omega^I - A_\mu^A\partial_J k_A{}^I\,\Omega^J , \\
D_\mu^{(0)}\lambda^A &= \left(\partial_\mu - \tfrac{3}{2}b_\mu + \tfrac{1}{4}\omega_\mu{}^{ab}(e,b)\gamma_{ab} - \tfrac{3}{2}\mathrm{i}\mathcal{A}_\mu\gamma_*\right)\lambda^A - A_\mu^C\lambda^B f_{BC}{}^A .
\end{aligned}
\tag{17.23}
$$

[8] Note that, after elimination of auxiliary fields, the action remains superconformal invariant, but the commutator algebra of two supersymmetries contains new transformations of the superconformal algebra plus terms that vanish when the field equations are satisfied. It is an example of an 'open algebra', as discussed in Sec. 11.1.3.

We also defined in a similar way

$$R^\mu \equiv \gamma^{\mu\rho\sigma}\left(\partial_\rho + \tfrac{1}{2}b_\rho + \tfrac{1}{4}\omega_\rho{}^{ab}(e,b)\gamma_{ab} - \tfrac{3}{2}\mathrm{i}\mathcal{A}_\rho\gamma_*\right)\psi_\sigma\,, \tag{17.24}$$

while $D_\mu\psi_\nu$ contains also ψ-torsion in the spin connection:

$$D_\mu\psi_\nu = \left(\partial_\mu + \tfrac{1}{2}b_\mu + \tfrac{1}{4}\omega_\mu{}^{ab}\gamma_{ab} - \tfrac{3}{2}\mathrm{i}\mathcal{A}_\mu\gamma_*\right)\psi_\nu\,. \tag{17.25}$$

To arrive at some of the simplifications, observe that

$$N_I D_\mu X^I = N_{\bar I} D_\mu \bar X^{\bar I} = \tfrac{1}{2}\left(N_I\partial_\mu X^I + N_{\bar I}\partial_\mu\bar X^{\bar I} - 2b_\mu N\right) = \tfrac{1}{2}(\partial_\mu - 2b_\mu)N\,. \tag{17.26}$$

Exercise 17.3 *Several terms in (17.22) can be written in a Kähler covariant form. Observe that the term with $\bar\Omega^I\gamma^\mu\Omega^{\bar K}D_\mu X^J$ can be included in the kinetic terms of Ω^I using a modified covariant derivative*

$$\hat D_\mu^{(0)}\Omega^I = D_\mu^{(0)}\Omega^I + \Gamma^I_{JK}\Omega^K D_\mu X^J\,. \tag{17.27}$$

This uses the connection components of the embedding Kähler manifold (see (13.19)), and is similar to the formula (14.58) for global supersymmetry. Further, notice that the second term in the expression for the auxiliary field F^I in (17.21) can be written as

$$F^I = \ldots + \tfrac{1}{2}\Gamma^I_{JK}\bar\Omega^J\Omega^K + \ldots\,. \tag{17.28}$$

Check then that the $F\bar F$ term leads to covariantizations of other terms, such that the actions contain

$$\begin{aligned} e^{-1}\mathcal{L} = \ldots &+ \left\{-\tfrac{1}{2}\nabla_I W_J\bar\Omega^I\Omega^J + \tfrac{1}{8}\nabla_I f_{AB\,J}\bar\Omega^I\Omega^J\bar\lambda^A P_L\lambda^B + \text{h.c.}\right\} \\ &+ \tfrac{1}{4}R_{I\bar K J\bar L}\bar\Omega^I\Omega^J\bar\Omega^{\bar K}\Omega^{\bar L}\,, \end{aligned} \tag{17.29}$$

where (13.22) can be used for the last term.

Using the results of the exercise and other rewritings, we can split the superconformal action into several parts:

$$\begin{aligned} e^{-1}\mathcal{L} = {}&\tfrac{1}{6}N\left[-R(e,b) + \bar\psi_\mu R^\mu + e^{-1}\partial_\mu(e\bar\psi\cdot\gamma\psi^\mu)\right] \\ &+ \mathcal{L}_0 + \mathcal{L}_{1/2} + \mathcal{L}_1 - V + \mathcal{L}_m + \mathcal{L}_{\text{mix}} + \mathcal{L}_d + \mathcal{L}_{4\text{f}}, \end{aligned} \tag{17.30}$$

where

$$\begin{aligned} \mathcal{L}_0 &= -G_{I\bar J}D^\mu X^I D_\mu\bar X^{\bar J}\,, \\ \mathcal{L}_{1/2} &= -\tfrac{1}{2}G_{I\bar J}\left[\bar\Omega^I\hat{\not{D}}^{(0)}\Omega^{\bar J} + \bar\Omega^{\bar J}\hat{\not{D}}^{(0)}\Omega^I\right], \\ \mathcal{L}_1 &= (\operatorname{Re} f_{AB})\left[-\tfrac{1}{4}F_{\mu\nu}^A F^{\mu\nu\,B} - \tfrac{1}{2}\bar\lambda^A\not{D}^{(0)}\lambda^B\right] \\ &\quad + \tfrac{1}{4}\mathrm{i}\left[(\operatorname{Im} f_{AB})\,F_{\mu\nu}^A\tilde F^{\mu\nu\,B} + (D_\mu\operatorname{Im} f_{AB})\,\bar\lambda^A\gamma_*\gamma^\mu\lambda^B\right], \\ V &= V_F + V_D = G^{I\bar J}W_I\overline{W}_{\bar J} + \tfrac{1}{2}(\operatorname{Re} f)^{-1\,AB}\mathcal{P}_A\mathcal{P}_B\,, \end{aligned}$$

$$
\begin{aligned}
\mathcal{L}_m &= \tfrac{1}{2}W\bar\psi_\mu P_R\gamma^{\mu\nu}\psi_\nu - \tfrac{1}{2}\nabla_I W_J\bar\Omega^I\Omega^J + \tfrac{1}{4}G^{I\bar J}\overline{W}_{\bar J} f_{ABI}\bar\lambda^A P_L\lambda^B \\
&\quad +\sqrt{2}\mathrm{i}\left[-\partial_I\mathcal{P}_A + \tfrac{1}{4}f_{ABI}(\mathrm{Re}\, f)^{-1\,BC}\mathcal{P}_C\right]\bar\lambda^A\Omega^I + \text{h.c.}\,, \\
\mathcal{L}_{\text{mix}} &= \bar\psi\cdot\gamma P_L\left(\frac{1}{2}\mathrm{i}\mathcal{P}_A\lambda^A + \frac{1}{\sqrt{2}}W_I\Omega^I\right) + \text{h.c.}\,, \\
\mathcal{L}_d &= \frac{1}{8}(\mathrm{Re}\, f_{AB})\bar\psi_\mu\gamma^{ab}\left(F_{ab}^A + \widehat{F}_{ab}^A\right)\gamma^\mu\lambda^B + \frac{1}{\sqrt{2}}\left\{G_{I\bar J}\bar\psi_\mu \not{D}X^{\bar J}\gamma^\mu\Omega^I\right. \\
&\quad \left. -\tfrac{1}{4}f_{ABI}\bar\Omega^I\gamma^{ab}\widehat{F}_{ab}^A\lambda^B - \tfrac{2}{3}N_I\bar\Omega^I\gamma^{\mu\nu}D_\mu\psi_\nu + \text{h.c.}\right\}, \\
\mathcal{L}_{4\text{f}} &= -\tfrac{1}{6}N\mathcal{L}_{\text{SG,torsion}} \\
&\quad +\left\{-\frac{1}{4\sqrt{2}}f_{ABI}\bar\psi\cdot\gamma\Omega^I\bar\lambda^A P_L\lambda^B + \frac{1}{8}\nabla_I f_{ABJ}\bar\Omega^I\Omega^J\bar\lambda^A P_L\lambda^B + \text{h.c.}\right\} \\
&\quad +\tfrac{1}{16}\mathrm{i}\, e^{-1}\varepsilon^{\mu\nu\rho\sigma}\bar\psi_\mu\gamma_\nu\psi_\rho\left(\bar\Omega^{\bar J}\gamma_\sigma\Omega^I + \tfrac{1}{2}\,\mathrm{Re}\, f_{AB}\bar\lambda^A\gamma_*\gamma_\sigma\lambda^B\right) \\
&\quad -\tfrac{1}{2}G_{I\bar J}\bar\psi_\mu\Omega^{\bar J}\,\bar\psi^\mu\Omega^I \\
&\quad +\tfrac{1}{4}R_{I\bar K J\bar L}\bar\Omega^I\Omega^J\,\bar\Omega^{\bar K}\Omega^{\bar L} - \tfrac{1}{16}G^{I\bar J}f_{ABI}\bar\lambda^A P_L\lambda^B\,\bar f_{CD\bar J}\bar\lambda^C P_R\lambda^D \\
&\quad +\tfrac{1}{16}(\mathrm{Re}\, f)^{-1\,AB}\left(f_{ACI}\bar\Omega^I - \bar f_{AC\bar I}\bar\Omega^{\bar I}\right)\lambda^C\left(f_{BDJ}\bar\Omega^J - \bar f_{BD\bar J}\bar\Omega^{\bar J}\right)\lambda^D \\
&\quad +N(A_\mu^{\text{F}})^2\,.
\end{aligned}
\tag{17.31}
$$

The expression (17.30) has the following structure. The first line is $N(X,\bar X)$ times the pure supergravity action (which includes also the four-fermion torsion terms). $\mathcal{L}_0$ and $\mathcal{L}_{1/2}$ contain the kinetic terms of the chiral multiplets. $\mathcal{L}_1$ contains the standard kinetic terms of the gauge multiplets. V is the scalar potential. $\mathcal{L}_m$ contains fermion mass terms, while $\mathcal{L}_{\text{mix}}$ contains terms that identify the Goldstino, as we will discuss in Sec. 19.1.1. In the first line of $\mathcal{L}_d$, the gravitino is contracted with the Noether supercurrent for the gauge and chiral multiplets.[9] The second line contains other derivative interactions between the fields (its last term will disappear after gauge fixing of S-supersymmetry). $\mathcal{L}_{4\text{f}}$ contain only four-fermion interactions. Finally, the entire structure is invariant under holomorphic changes of coordinates X^I on the embedding Kähler manifold as discussed in Ch. 13. As discussed in Ch. 14, these induce transformations among entire chiral multiplets X^I, Ω^I, F^I.

The two steps that remain in the long march towards the physical theory are the gauge fixing of the superconformal symmetries that are not part of the Poincaré supersymmetry algebra and the specification of physical chiral multiplets. The special conformal K, dilatation D, conformal supersymmetry S and chiral transformations T must be fixed. In the next section we discuss gauge fixing conditions for K, D and S. It is convenient to defer gauge fixing for T until we have chosen physical coordinates later in the chapter.

Exercise 17.4 *Check that the scalar potential can be obtained from (14.68) using the Q-supersymmetry. The gravitino does not contribute to this calculation, since its Q-supersymmetry transformation in (16.10) does not involve a shift with scalars.*

[9] Note that the structure $F+\hat F$ (with $\hat F$ defined in (17.1)) was already encountered in $D=11$ supergravity; see (10.26).

17.2.5 Partial gauge fixing

We now follow steps similar to those of Sec. 16.2.2. First, we apply the K-gauge condition (15.26). Thus we set $b_\mu = 0$ in all equations. As in Ch. 15, we choose the dilatation gauge condition so that the standard Hilbert action appears. Since the product NR appears in $[N]_D$ in (17.19) and in the first line of (17.30), the appropriate choice is [154]

$$D\text{-gauge:} \quad N = -\frac{3}{\kappa^2}\,. \tag{17.32}$$

This choice inserts the dimensionful gravitational coupling constant in the action, which was scale invariant up to this point. Observe that another argument for this gauge choice is that it eliminates mixed kinetic terms of the graviton and scalars.

Similarly, the S-supersymmetry gauge condition will be chosen to eliminate kinetic mixing terms between the gravitino and other fermions. These are present in the last term of $[N]_D$ or of $\mathcal{L}_d$ in (17.31). Therefore we take

$$S\text{-gauge:} \quad N_I \Omega^I = 0\,, \tag{17.33}$$

which also implies that $N_{\bar{I}} \Omega^{\bar{I}} = 0$. This is a good S-gauge condition since its S-transformation is proportional to $N_I X^I = N$. The gauge transformations of the condition (17.32) determine the decomposition law for the dilatations. Since the Q-SUSY transformation of (17.32) is (17.33), we just get $\lambda_{\rm D} = 0$.

There are modest simplifications in the action formula (17.30) when the gauge conditions above are used. We recognize the pure supergravity action in the first line. The remaining terms involve matter multiplets, but the $n+1$ not-yet-physical chiral multiplets headed by the X^I are still in the game. We need to select n physical scalars z^α from the larger set. These fields can be interpreted as the complex coordinates of a particular type of Kähler manifold called a projective Kähler manifold or more commonly a Kähler–Hodge manifold. In the next section we will explain projective Kähler geometry in some detail. The Kähler–Hodge condition is discussed in Sec. 17.5.1, and in more detail in Appendix 17A.

17.3 Projective Kähler manifolds

In the conformal methodology of this chapter, the chiral fields X^I of supergravity are initially coordinates of a Kähler manifold that we have called the embedding manifold. Dilatations and T-transformations act as symmetries of this manifold. In Sec. 17.2.5 we discussed schematically how the gauge fixing of these symmetries leads to a projective

Kähler manifold whose coordinates are the physical scalars Z^α of matter-coupled supergravity.

In this section we discuss projective Kähler manifolds from a more general mathematical viewpoint. A projective manifold is obtained from its embedding manifold by identifying points which lie on the orbit of the vector field that generates a symmetry of the latter space. In the application to supergravity this symmetry is the complex transformation that combines the dilatation and T-transformation. Identification of points on the symmetry orbit is essentially equivalent to gauge fixing the symmetry. The motivation for the more general viewpoint is that similar constructions apply to vector multiplets in $\mathcal{N} = 2$ supergravity and also determine the structure of hypermultiplets in four-, five-, and six-dimensional supergravity (see Ch. 20).

We will find that projective Kähler manifolds require that the Kähler potential N of the embedding space has the homogeneity properties of (17.13). Hence, the scalar terms of matter-coupled supergravities are those of a projective Kähler manifold. Our discussion of how the Kähler structure of the projective manifold emerges from its embedding space follows the treatment of [159]. Later, in Sec. 17.5.1, we will discuss how the Kähler geometry of the embedding manifold leads to the Kähler–Hodge condition on the projective manifold.

The signature of metric of the embedding manifold does not play a significant role in our discussion. So the results apply to the embedding spaces for $\mathcal{N} = 1$ supergravity in which the metric has a negative direction as well as to Euclidean signature metrics as in the example in the next section.

17.3.1 The example of CP^n

The first example of the construction of projective Kähler manifolds was already discussed in Sec. 13.3. The coordinates of the embedding manifold are the X^I, and we now consider the scaling operation used to define CP^n. In this case, the scaling is the simple operation (13.33). The embedding metric, extracted from (13.35), is flat and takes the simple form

$$\mathrm{d}s^2 = \mathrm{d}X^I G_{I\bar{J}} \mathrm{d}X^{\bar{J}} , \qquad G_{I\bar{J}} = \delta_{I\bar{J}} . \tag{17.34}$$

This is a Kähler manifold with Kähler potential

$$N = X^I \delta_{I\bar{J}} X^{\bar{J}} , \tag{17.35}$$

which satisfies the requirements of (17.13).

A scale transformation $X^I = \rho X'^I$ leads to

$$\mathrm{d}s^2 = |\rho|^2 \mathrm{d}X'^I \delta_{I\bar{J}} \mathrm{d}X^{\bar{J}} = \mathrm{d}X'^I \gamma'_{I\bar{J}} \mathrm{d}X'^{\bar{J}} , \qquad \gamma'_{I\bar{J}} = |\rho|^2 \delta_{I\bar{J}} . \tag{17.36}$$

The modulus of the ρ-transformation scales the metric. Thus it differs from the symmetries that we discussed in Sec. 7.12, which left the metric invariant. The metric is invariant under the phase part of ρ which is a normal symmetry.

Let us examine the infinitesimal form of the scale transformation to find the vector fields that are its generators. We use the infinitesimal parameters defined by[10]

$$\rho = \exp(-\lambda_{\rm D} - {\rm i}\lambda_T)\,. \tag{17.37}$$

Then we can write the variation $\delta X = X' - X$ as

$$\delta X^I = \lambda_{\rm D} k_{\rm D}^I + \lambda_T k_T^I\,, \qquad k_{\rm D}^I = X^I\,, \qquad k_T^I = {\rm i}X^I\,. \tag{17.38}$$

Note that these transformation vectors are holomorphic. As in (7.140) we now have

$$\begin{aligned}
\delta G_{I\bar J} &= -\left(\lambda_{\rm D}\mathcal{L}_{k_{\rm D}} + \lambda_T \mathcal{L}_{k_T}\right) G_{I\bar J}\,,\\
\mathcal{L}_k G_{I\bar J} &= k^K\partial_K G_{I\bar J} + k^{\bar K}\partial_{\bar K} G_{I\bar J} + \partial_I k^K G_{K\bar J} + \partial_{\bar J} k^{\bar K} G_{I\bar K}\,,\\
\mathcal{L}_{k_{\rm D}} G_{I\bar J} &= 2G_{I\bar J}\,, \qquad \mathcal{L}_{k_T} G_{I\bar J} = 0\,.
\end{aligned} \tag{17.39}$$

Thus k_T is a Killing vector, but $k_{\rm D}$ is not. In fact, $k_{\rm D}$ is a closed homothetic Killing vector, a concept introduced in Sec. 15.7.

Exercise 17.5 *Show that the scaling relations of (17.39) are also valid for any Kähler metric (17.16) whose Kähler potential has the homogeneity properties of (17.13).*

17.3.2 Dilatations and holomorphic homothetic Killing vectors

In Ch. 15 we studied the conformal algebra and determined how fields transform under dilatations and special conformal transformations. The dilatations are determined by a vector ${k_{\rm D}}^i$. It also induces special conformal transformations of covariant derivatives of the fields; see (15.52).

In Sec. 15.7, we discussed the transformations of a set of scalar fields $\phi^i(x)$ that are coordinates of the target space for a nonlinear σ-model with metric tensor $g_{ij}(\phi)$. It was explained that the vector ${k_{\rm D}}^i$ must be a closed homothetic Killing vector so that the full conformal group is realized on the scalar sector. A closed homothetic Killing vector satisfies $\nabla_i {k_{\rm D}}^j = \delta_i^j$. Locally it can be expressed as the gradient of a scalar, i.e. $k_{{\rm D}i} = \partial_i \tilde{k}$. We have assigned the Weyl weight $w = (D-2)/2 \to 1$ appropriate to $D = 4$ spacetime dimensions.

In this section we extend these considerations to the case of interest in matter-coupled $\mathcal{N} = 1$ supergravity where the target space of the scalar fields is a Kähler manifold. We begin with the real description of such a manifold. Given a closed homothetic Killing vector ${k_{\rm D}}^i$, one can use the complex structure ${J_j}^i$ to construct a Killing vector k_T:

$$(k_T)^i \equiv (k_{\rm D})^j {J_j}^i\,. \tag{17.40}$$

Indeed, owing to (13.29) and (15.43) we have

$$\nabla_i\left((k_{\rm D})^k J_{kj}\right) = J_{ij}\,. \tag{17.41}$$

Since J_{ij} is antisymmetric, k_T is a Killing vector.

[10] The parameters λ_D and λ_T are analogues of the corresponding gauge parameters of the superconformal algebra.

We now introduce a holomorphic basis of coordinates X^I, $\bar{X}^{\bar{I}}$, in which the hermitian metric is $G_{I\bar{J}}$. Then, we define a holomorphic homothetic Killing vector as a holomorphic vector $h = h^I(X)\partial/\partial X^I$ that satisfies

$$\text{holomorphic homothetic Killing vector: } \mathcal{L}_h G_{I\bar{J}} = G_{K\bar{J}} \nabla_I h^K = 2\, G_{I\bar{J}}\,. \tag{17.42}$$

It is easy to see that the real and imaginary parts of h are the real vectors $k_{\rm D}$ and k_T we have discussed above; specifically we define

$$\begin{aligned} k_{\rm D} &= \tfrac{1}{2}(h+\bar{h})\,, & k_{\rm D}^I &= \tfrac{1}{2}h^I\,, & k_{\rm D}^{\bar{I}} &= \tfrac{1}{2}\bar{h}^{\bar{I}}\,,\\ k_T &= \tfrac{1}{2}{\rm i}(h-\bar{h})\,, & k_T^I &= \tfrac{1}{2}{\rm i}h^I\,, & k_T^{\bar{I}} &= -\tfrac{1}{2}{\rm i}\bar{h}^{\bar{I}}\,. \end{aligned} \tag{17.43}$$

It then follows that $k_{\rm D}$ is a closed homothetic Killing vector, while k_T is the Killing vector (17.40).

Exercise 17.6 *The Kähler potential in conformal supergravity satisfies the conditions of (17.13). Show that (15.44) is satisfied by the choice $\tilde{\kappa} = N$. In the complex coordinate basis, (15.44) becomes the pair of equations $\partial_I\partial_{\bar{J}}\tilde{\kappa} = G_{I\bar{J}}$ and $\nabla_I\partial_J\tilde{\kappa} = 0$.*

Exercise 17.7 *For the flat embedding space of (17.34) check that the vector $h = 2\,X^I\partial_I$ is a holomorphic homothetic Killing vector. Then the vectors of (17.43) agree with (17.38). Show that the same h is also a holomorphic homothetic Killing vector for the metric (17.16) of conformal supergravity. You need $\nabla_I X^J = \delta_I^J$, which follows from (13.19) with the properties (17.13).*

17.3.3 The projective parametrization

The homothetic Killing vector $h(X)$ defines a holomorphic map of the embedding manifold into itself. The map defines the action of the Lie group of combined dilatations and T-transformations. We assume that this group acts freely. This means that two distinct points are related by at most one transformation of the group. The map then defines a foliation of the embedding manifold. The projective Kähler manifold is defined by identifying all points that are related by a group transformation.

We will now show that these properties of h^I imply that there is a parametrization in which the embedding manifold has Kähler potential N satisfying (17.11). This shows that embedding manifolds are the same as the manifolds we use to describe superconformal chiral multiplets. In Sec. 17.5.1 we will show that the projective manifolds are the Kähler–Hodge manifolds.

The Frobenius theorem allows us to choose holomorphic coordinates y and z^α such that $\partial/\partial y$ points in the direction of the holomorphic homothetic Killing vector. Thus,

$$h = 2\,y\frac{\partial}{\partial y}\,. \tag{17.44}$$

The factor 2 is included for convenience (determining the normalization of y). The complex lines of fixed z^α and varying y are the fibers of a fibration.

We now require that the $n+1$ coordinates X^I, heretofore an arbitrary holomorphic set, have the property of homogeneous coordinates. This means that they are defined by the set y, z^α through functions $Z^I(z)$:

$$X^I = y\, Z^I(z)\,. \tag{17.45}$$

We do not specify the $(n+1)$ functions Z^I of the base space coordinates z^α, so that we keep the freedom of arbitrary coordinates on the base. The Z^I must be non-degenerate in the sense that the $(n+1)\times(n+1)$ matrix

$$\begin{pmatrix} Z^I \\ \partial_\alpha Z^I \end{pmatrix} \tag{17.46}$$

should have rank $n+1$. There are many ways to choose the Z^I. One simple choice, labeling the I index from 0 to n, can be

$$Z^0 = 1\,, \qquad Z^\alpha = z^\alpha\,. \tag{17.47}$$

In the application to supergravity we use X^I to denote the scalar fields in the embedding space. The physical scalar fields are z^α. With n chiral multiplets, α will run over $1,\ldots,n$, while I runs over $0,1,\ldots,n$.

As we now demonstrate, it is in homogeneous coordinates that the Kähler potential N of the embedding space has the scaling properties of (17.11), which are needed to couple chiral multiplets to supergravity. First note that one can use the chain rule to rewrite the holomorphic homothetic Killing vector as

$$h = 2X^I\frac{\partial}{\partial X^I} \qquad \text{or} \qquad h^I = 2X^I\,. \tag{17.48}$$

The condition (17.42) then reduces to

$$G_{K\bar J}\Gamma^K_{IL}X^L = X^L\frac{\partial}{\partial X^L}G_{I\bar J} = 0\,. \tag{17.49}$$

The Kähler metric is the second derivative of a Kähler potential, which we denote now by N',

$$G_{I\bar J} = \frac{\partial}{\partial X^I}\frac{\partial}{\partial \bar X^{\bar J}}N'\,, \tag{17.50}$$

but (17.49) implies that

$$\frac{\partial}{\partial X^I}\frac{\partial}{\partial \bar X^{\bar J}}\left(X^K\frac{\partial}{\partial X^K}N' - N'\right) = 0\,. \tag{17.51}$$

By a Kähler transformation, which we can always perform, this condition and its complex conjugate show that we can define a function N such that

$$X^I\frac{\partial}{\partial X^I}N = N\,, \qquad \bar X^{\bar I}\frac{\partial}{\partial \bar X^{\bar I}}N = N\,. \tag{17.52}$$

This states that N is homogeneous of first degree in X^I, and in $\bar{X}^{\bar{I}}$, and can be identified with the function in (17.11).

The real and imaginary parts of the holomorphic homothetic Killing vector h generate the dilatations and T-transformations of the superconformal algebra. Specifically, $k_{\rm D}$ and k_T of (17.43) induce the following transformation of the homogeneous coordinates:

$$\delta X^I = \lambda_{\rm D} X^I + {\rm i}\lambda_T X^I\,, \qquad \delta \bar{X}^{\bar{I}} = \lambda_{\rm D} \bar{X}^{\bar{I}} - {\rm i}\lambda_T \bar{X}^{\bar{I}}\,, \tag{17.53}$$

in which we have included parameters $\lambda_{\rm D}$ and λ_T. Since $h = 2y\,{\rm d}/{\rm d}y$, the complex scale transformations (17.53) are represented on y and z^α as

$$\delta y = (\lambda_{\rm D} + {\rm i}\lambda_T)\,y\,, \qquad \delta \bar{y} = (\lambda_{\rm D} - {\rm i}\lambda_T)\,\bar{y}\,, \qquad \delta z^\alpha = \delta \bar{z}^{\bar{\alpha}} = 0\,. \tag{17.54}$$

We now want to write an action for the scalar fields $X^I(x)$ coupled to a general spacetime metric $g_{\mu\nu}(x)$ that is invariant under *local* versions of these transformations, i.e. with arbitrary functions $\lambda_{\rm D}(x)$, $\lambda_T(x)$. The gauging of the T-transformation will be implemented through an (auxiliary) U(1) gauge field $\mathcal{A}_\mu$, as was done in Sec. 13.3. The action will correspond quite precisely to the scalar kinetic and gravity terms of the Lagrangian (17.30) (with $b_\mu = 0$). Our purpose is to make contact between the scalar sector of $\mathcal{N}=1$ supergravity and the mathematical theory of projective Kähler manifolds.

The desired action is

$$S = \int {\rm d}^4x\,\mathcal{L} = -\int {\rm d}^4x\, e\left[\tfrac{1}{6} N R(e) + G_{I\bar{J}}\, g^{\mu\nu}\,(\partial_\mu - {\rm i}\mathcal{A}_\mu)\,X^I\,(\partial_\nu + {\rm i}\mathcal{A}_\nu)\,\bar{X}^{\bar{J}}\right]. \tag{17.55}$$

We assign the transformation rule $\delta\mathcal{A}_\mu = \partial_\mu\lambda_T$, so that S is obviously invariant under T-transformations. Invariance under local dilatations follows from an interesting extension of Ex. 8.7.

Exercise 17.8 *Prove the local scale invariance of S. This is related to Ex. 15.9, but with $b_\mu = 0$. This gauge choice is permitted because of the local special conformal symmetry of the Lagrangian (15.51).*

The Euler–Lagrange equation for $\mathcal{A}_\mu$ is algebraic, and its solution is

$$\begin{aligned} 2\,{\rm i}\mathcal{A}_\mu &= -\frac{1}{N}\left(X^I G_{I\bar{J}}\partial_\mu \bar{X}^{\bar{J}} - \partial_\mu X^I G_{I\bar{J}} \bar{X}^{\bar{J}}\right) \\ &= -\frac{1}{N}\left(\partial_\mu \bar{X}^{\bar{J}} N_{\bar{J}} - \partial_\mu X^I N_I\right). \end{aligned} \tag{17.56}$$

Note that this agrees with the supergravity auxiliary field $\mathcal{A}_\mu$ in (17.21) when all fields except scalars are dropped. Inserting this result, the covariant derivative in (17.55) can be written as

$$(\partial_\mu - {\rm i}\mathcal{A}_\mu)\,X^I = \partial_\mu X^I - X^I \partial_\mu X^K \frac{\partial}{\partial X^K}\ln N + \frac{1}{2} X^I \partial_\mu \ln N\,. \tag{17.57}$$

The contraction of the first two terms with N_I vanishes. This simplifies the rewriting of the Lagrangian, which takes the form[11]

$$\mathcal{L} = -\frac{1}{4N}(\partial_\mu N)(\partial^\mu N) - N(\partial_\mu X^I)(\partial^\mu \bar{X}^{\bar{J}})\frac{\partial}{\partial X^I}\frac{\partial}{\partial \bar{X}^{\bar{J}}}\ln N\,. \tag{17.58}$$

By virtue of its construction, the action S is invariant, after restoration of the $NR(e)$ term, under the local transformation (17.53).

Exercise 17.9 *Consider the Lagrangian (17.58) in the flat space limit in which the spacetime metric becomes $g_{\mu\nu} \to \eta_{\mu\nu}$. Show that, for a flat embedding space with Kähler potential $N \to X^I \delta_{I\bar{J}} X^{\bar{J}}$, the Lagrangian agrees with the Lagrangian (13.37) for the CP^n model (after the constraint (13.34) is imposed).*

17.3.4 The Kähler cone

We will now show that the geometry of the embedding manifold is a generalized cone. We remind the reader that a cone, with opening angle 2α, embedded in three-dimensional Euclidean space has the induced metric

$$\mathrm{d}s^2 = \mathrm{d}r^2 + r^2(\mathrm{d}x^2 + \mathrm{d}y^2)\,, \qquad x^2 + y^2 = \sin^2\alpha\,. \tag{17.59}$$

We will use this to identify the target space geometry associated with the embedding space Lagrangian (17.58). The cone is drawn in Fig. 17.1.

We use the basis of fields $\{y, z^\alpha\}$, which are related to the homogeneous coordinates by (17.45). The X^I, and hence the Z^I, are zero modes of

$$\frac{\partial}{\partial X^I}\frac{\partial}{\partial \bar{X}^{\bar{J}}}\ln N = \frac{G_{I\bar{J}}}{N} - \frac{G_{I\bar{K}}\bar{X}^{\bar{K}} G_{\bar{J}L} X^L}{N^2}\,, \tag{17.60}$$

so the spacetime gradient $\partial_\mu y$ does not survive in the second term of (17.58). The Lagrangian can then be written as

$$\mathcal{L} = -\frac{1}{4N}(\partial_\mu N)(\partial^\mu N) - N(\partial_\mu z^\alpha)(\partial^\mu \bar{z}^{\bar{\beta}})\frac{\partial}{\partial z^\alpha}\frac{\partial}{\partial \bar{z}^{\bar{\beta}}}\ln\left[Z^I(z) G_{I\bar{J}} \bar{Z}^{\bar{J}}(\bar{z})\right]. \tag{17.61}$$

To visualize the geometry we can regard N as a radial coordinate. If the embedding metric has positive signature, as in the CP^n example of Sec. 17.3.1, then N will be positive. We then define $N = r^2$ and rewrite (17.61) as

$$\mathcal{L} = -\partial_\mu r \partial^\mu r - r^2 g_{\alpha\bar{\beta}}(z,\bar{z})\partial_\mu z^\alpha \partial^\mu \bar{z}^{\bar{\beta}}\,. \tag{17.62}$$

This looks like the metric of a cone, whose 'cross-section' is an n-dimensional Kähler manifold with metric $g_{\alpha\bar{\beta}}(z,\bar{z})$ implicitly defined by comparing the second terms of (17.62) and (17.61).

As we have already discussed, embedding metrics for chiral multiplets in supergravity have Lorentzian signature $(-++\ldots+)$. The Kähler potential N must be negative so that

[11] We omit the $NR(e)$ term henceforth because it is inessential to our main purpose, which is to explain projective Kähler manifolds.

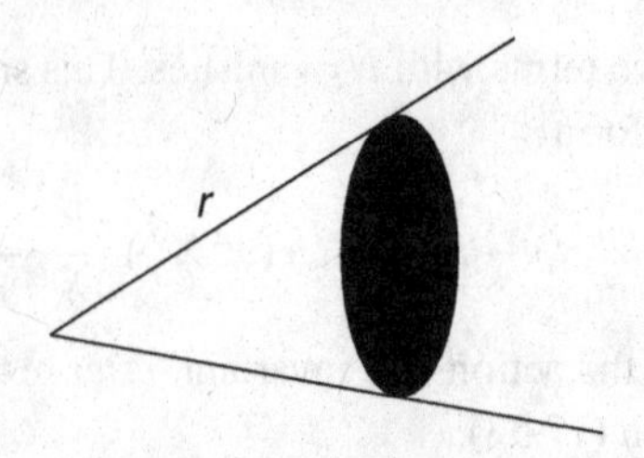

Fig. 17.1 Conical geometry of the $(n + 1)$-dimensional embedding space.

the gravity Lagrangian $NR(e, b)$ in (17.30) has the proper physical sign. In this case we would define $N = -r^2$, and the $\partial_\mu r \partial^\mu r$ term of (17.63) would have opposite sign:

$$\mathcal{L} = \partial_\mu r \partial^\mu r - r^2 g_{\alpha\bar{\beta}}(z, \bar{z}) \partial_\mu z^\alpha \partial^\mu \bar{z}^{\bar{\beta}} , \tag{17.63}$$

where $g_{\alpha\bar{\beta}}$ is defined appropriately (see below). The geometry is still that of a Kähler cone which may be compared with a Lorentzian version of (17.59).

In differential geometry, a Kähler manifold with a homothetic Killing vector is called a *Sasakian manifold*. The Killing vector generating the U(1) symmetry is called the Reeb vector field.

17.3.5 The projection

It is clear that when the embedding Kähler manifold has complex dimension $n + 1$, the projective Kähler manifold has dimension n. It is the base manifold of a bundle with fibers defined by the action of the complex scaling.

We proceed in much the same way as in Sec. 13.3, and impose a constraint that selects one representative in each set of scale equivalent configurations. Since N transforms under the scaling as

$$\delta N = 2\,\lambda_{\rm D}\, N , \tag{17.64}$$

it is appropriate to impose the constraint directly on N. We fix the dilatation gauge by requiring that N is constant, viz.

$$N(X, \bar{X}) = -a . \tag{17.65}$$

The value of a depends on which theory is being considered. In the example of Sec. 13.3 we have chosen $a = -1$. For $\mathcal{N} = 1$ supergravity we noted in (17.32) that the appropriate choice is $a = 3\kappa^{-2}$, where κ is the gravitational coupling constant. For Kähler manifolds that appear in $\mathcal{N} = 2$ supergravity, the appropriate choice will be $a = \kappa^{-2}$. We must also fix the T-gauge, which we will do at the end of Sec. 17.3.6.

We choose the n physical scalars z^α as coordinates on the resulting projective manifold. In fact the magnitude of the scaling coordinate y is determined in terms of the z^α and $\bar{z}^{\bar{\alpha}}$ by the constraint

$$y\bar{y} = -a\left[Z^I(z)G_{I\bar{J}}\bar{Z}^{\bar{J}}(\bar{z})\right]^{-1}. \tag{17.66}$$

Since $\mathcal{N}$ is constant the Lagrangian of the physical scalars is just the second term of (17.61). The target space metric that appears in that term can be obtained from the following Kähler potential for the projective manifold:

$$\mathcal{K}(z,\bar{z}) = -a\ln\left[-a^{-1}\,Z^I(z)G_{I\bar{J}}\bar{Z}^{\bar{J}}(\bar{z})\right] = a\ln y\bar{y}. \tag{17.67}$$

Thus, after gauge fixing the dilatations, our Lagrangian takes the form of a nonlinear σ-model on an n-dimensional Kähler manifold, namely

$$\mathcal{L} = -g_{\alpha\bar{\beta}}\partial_\mu z^\alpha\partial^\mu\bar{z}^{\bar{\beta}} = -\partial_\alpha\partial_{\bar{\beta}}\mathcal{K}(z,\bar{z})\partial_\mu z^\alpha\partial^\mu\bar{z}^{\bar{\beta}}. \tag{17.68}$$

From now on ∂_α is considered to be a derivative on the projective manifold, while y in (17.67) is no longer an independent variable. Note that the 'holomorphic' variables X^I of the embedding manifold are not holomorphic in the complex structure defined on the projective Kähler manifold, since y as a solution of (17.66) is a function of both z and $\bar{z}$.

Exercise 17.10 *Show that the homogeneity conditions (17.13) lead to*

$$\partial_\alpha\mathcal{K} = -a\frac{y}{N}N_I\partial_\alpha Z^I, \tag{17.69}$$

and that the value of $\mathcal{A}_\mu$ from (17.56) can be written as

$$\mathrm{i}\,\mathcal{A}_\mu = \tfrac{1}{2}a^{-1}\left(\partial_\mu\bar{z}^{\bar{\alpha}}\partial_{\bar{\alpha}}\mathcal{K} - \partial_\mu z^\alpha\partial_\alpha\mathcal{K}\right) + \tfrac{1}{2}\partial_\mu\ln(y/\bar{y}). \tag{17.70}$$

The last term is 'pure gauge'.

17.3.6 Kähler transformations

In Ex. 17.1 readers showed that the homogeneity properties (17.13) of the Kähler potential $N(X,\bar{X})$ in homogeneous coordinates do not permit Kähler transformations of N. However, Kähler transformations of the Kähler potential $\mathcal{K}(z,\bar{z})$ of the projected space are allowed and play an important role in the structure of the physical Lagrangians of $\mathcal{N}=1$ supergravity. The origin of these transformations lies in the definition (17.45). In fact, the variable y is not uniquely defined. We may consider redefinitions

$$y' = y\,\mathrm{e}^{f(z)/a}, \qquad Z'^I = Z^I\,\mathrm{e}^{-f(z)/a}. \tag{17.71}$$

If the Z'^I are used in (17.67), the Kähler potential is modified and reads

$$\mathcal{K}'(z,\bar{z}) = \mathcal{K}(z,\bar{z}) + f(z) + \bar{f}(\bar{z}). \tag{17.72}$$

This is precisely a Kähler transformation; see (13.18).

If a quantity transforms under Kähler transformations, it is useful to include a new term, called the Kähler connection, in its covariant derivative. Consider, for example, the functions $Z^I(z)$ and $\bar{Z}^{\bar{I}}(\bar{z})$. We define their covariant derivatives by

$$\begin{aligned} \nabla_\alpha Z^I &\equiv \partial_\alpha Z^I + a^{-1}(\partial_\alpha \mathcal{K})\, Z^I\,, & \overline{\nabla}_{\bar\alpha} Z^I &\equiv \partial_{\bar\alpha} Z^I = 0\,, \\ \overline{\nabla}_{\bar\alpha} \bar{Z}^{\bar I} &\equiv \partial_{\bar\alpha} \bar{Z}^{\bar I} + a^{-1}(\partial_{\bar\alpha} \mathcal{K})\, \bar{Z}^{\bar I}\,, & \nabla_\alpha \bar{Z}^{\bar I} &\equiv \partial_\alpha \bar{Z}^{\bar I} = 0\,. \end{aligned} \tag{17.73}$$

Viewing $f(z)$ and $\bar f(\bar z)$ as independent gauge parameters, it is easy to see that all covariant derivatives in (17.73) transform with the factors $\mathrm{e}^{-f/a}$ or $\mathrm{e}^{-\bar f/a}$ under combined Kähler transformations of Z^I, $\bar Z^{\bar I}$ and $\mathcal{K}$. Therefore we can identify $\partial_\alpha \mathcal{K}$ and $\partial_{\bar\alpha}\mathcal{K}$ as the 'gauge fields' for holomorphic and anti-holomorphic Kähler transformations, respectively. When a quantity transforms under complex coordinate transformations on the projective manifold, i.e. when it has α or $\bar\alpha$ indices, one must also include the metric connection of (13.19) in its covariant derivatives.

We now define the Kähler covariant derivative on more general functions $V(z,\bar z)$, which transform under Kähler transformations as

$$V'(z,\bar z) = V(z,\bar z)\exp\left[-a^{-1}\left[w_+ f(z) + w_- \bar f(\bar z)\right]\right]. \tag{17.74}$$

We then say that these functions have Kähler weights (w_+, w_-). The scalar functions $Z^I(z)$ and $\bar Z^{\bar I}(\bar z)$ are examples of this with weights $(w_+, w_-) = (1,0)$ and $(0,1)$, respectively. The variable y has weights $(-1,0)$. Functions on the embedding space do not transform under Kähler transformations. For example a scalar function $\mathcal{V}(X,\bar X)$ on the embedding space that transforms under $\lambda_{\rm D}$ and λ_T as

$$\delta\mathcal{V} = w_+(\lambda_{\rm D} + \mathrm{i}\lambda_T)\mathcal{V} + w_-(\lambda_{\rm D} - \mathrm{i}\lambda_T)\mathcal{V} \tag{17.75}$$

can be expressed in terms of y, z^α and their complex conjugates as

$$\mathcal{V}(X,\bar X) = y^{w_+}\bar y^{w_-} V(z,\bar z)\,. \tag{17.76}$$

Then $V(z,\bar z)$ is inert under conformal scalings but transforms under Kähler transformations as in (17.74). The covariant derivatives that generalize (17.73) are

$$\begin{aligned} \nabla_\alpha V(z,\bar z) &= \partial_\alpha V(z,\bar z) + w_+ a^{-1}(\partial_\alpha \mathcal{K}) V(z,\bar z)\,, \\ \overline{\nabla}_{\bar\alpha} V(z,\bar z) &= \partial_{\bar\alpha} V(z,\bar z) + w_- a^{-1}(\partial_{\bar\alpha} \mathcal{K}) V(z,\bar z)\,. \end{aligned} \tag{17.77}$$

Exercise 17.11 *Prove that, under Kähler transformations, $\nabla_\alpha V$ is a covariant object like (17.74) with the same weights as V itself.*

The commutator of holomorphic Kähler covariant derivatives $[\nabla_\alpha, \nabla_\beta]$ vanishes. However, the mixed commutator is not zero. In fact[12]

$$\left[\nabla_\alpha, \overline{\nabla}_{\bar\beta}\right] V(z, \bar z) = a^{-1}(w_- - w_+) g_{\alpha\bar\beta} V(z, \bar z) . \qquad (17.78)$$

The general definition (11.41) of curvature shows that there is curvature associated with the imaginary part of the Kähler transformations in projective Kähler manifolds. Indeed, this curvature is the fundamental 2-form (13.14) of the Kähler manifold, which is significant for the definition of Kähler–Hodge manifolds in Appendix 17A.2.

We can also define the covariant derivatives in spacetime, using

$$\nabla_\mu = (\partial_\mu z^\alpha)\nabla_\alpha + (\partial_\mu \bar z^{\bar\alpha})\overline{\nabla}_{\bar\alpha} . \qquad (17.79)$$

After complete gauge fixing, y becomes a function of z and $\bar z$. However, at this point we have only fixed the modulus of y via (17.67), while its phase is the T-gauge degree of freedom. Therefore, we will add the connection for T-transformations to the definition of ∇_μ for quantities such as y and $\bar y$ which carry chiral weight $c \neq 0$. For example, y has Kähler weight $(-1, 0)$, and chiral weight 1, so we define

$$\nabla_\mu y \equiv \left[\partial_\mu - \mathrm{i}\mathcal{A}_\mu - a^{-1}(\partial_\mu z^\alpha)\partial_\alpha \mathcal{K}\right] y = 0 . \qquad (17.80)$$

It can be seen that this covariant derivative vanishes when the specific form of (17.70) and a spacetime derivative of (17.67) is substituted. After any choice of the T-gauge, y can be expressed as a function of z and $\bar z$. Then we can split $\mathcal{A}_\mu$, (17.70), into terms with $\partial_\mu z^\alpha$ and $\partial_\mu \bar z^{\bar\alpha}$. The condition (17.80) then splits into the two relations

$$\nabla_\alpha y = \overline{\nabla}_{\bar\alpha} y = 0 . \qquad (17.81)$$

The explicit form of the quantities $\nabla_\alpha y$ and $\nabla_{\bar\alpha} y$ is not determined until the T-gauge condition is specified, but one can use the fact that they vanish before this is done. This facilitates many calculations.

Here are some exercises to give the reader practice in working with the quantities we have introduced, and which are used later to rewrite the superconformal action in terms of physical fields.

Exercise 17.12 *Using the fact that X does not transform under Kähler transformations, show that the derivative that enters in (17.55) is*

$$\left(\partial_\mu - \mathrm{i}\mathcal{A}_\mu\right) X^I = \nabla_\mu X^I = (\nabla_\mu y) Z^I + y\,\nabla_\mu Z^I = y\,(\partial_\mu z^\alpha)\nabla_\alpha Z^I . \qquad (17.82)$$

You will have to combine information from (17.80) and (17.73).

[12] For vectors and tensors with Kähler indices α or $\bar\alpha$, one must also include the curvature tensor as in (13.24).

Exercise 17.13 *Next consider the scalar function N that is invariant under chiral and Kähler transformations, and fixed to a constant value by the dilatation gauge condition. Apply first* ∇_α *and then* $\overline{\nabla}_{\bar\beta}$ *to obtain*

$$N_I \nabla_\alpha Z^I = 0\,, \qquad N_I = \bar{y}\, G_{I\bar{J}} \bar{Z}^{\bar{J}}\,,$$
$$g_{\alpha\bar\beta} = y\bar{y} \nabla_\alpha Z^I G_{I\bar{J}} \overline{\nabla}_{\bar\beta} \bar{Z}^{\bar{J}}\,. \tag{17.83}$$

Use the homogeneity properties (17.13), the definition of homogeneous coordinates (17.45) and the commutator (17.78).

Exercise 17.14 *Combine previous results and (17.66) into the matrix equation*

$$\begin{pmatrix} -a & 0 \\ 0 & g_{\alpha\bar\beta} \end{pmatrix} = y\bar{y} \begin{pmatrix} Z^I \\ \nabla_\alpha Z^I \end{pmatrix} G_{I\bar{J}} \left(\bar{Z}^{\bar{J}} \quad \overline{\nabla}_{\bar\beta} \bar{Z}^{\bar{J}} \right). \tag{17.84}$$

The $(n+1) \times (n+1)$ *matrices on both sides of this equation must be invertible so that scalar kinetic terms are non-degenerate. Observe that this requirement implies the non-degeneracy of (17.46). Compute the inverse of* $G_{I\bar{J}}$ *to find*

$$G^{I\bar{J}} = y\bar{y} \left(-a^{-1} Z^I \bar{Z}^{\bar{J}} + g^{\alpha\bar\beta} \nabla_\alpha Z^I \, \overline{\nabla}_{\bar\beta} \bar{Z}^{\bar{J}} \right). \tag{17.85}$$

This result will be used in Sec. 17.4.2 below.

Exercise 17.15 *Furthermore, by applying* ∇_β *to the second row of the matrix equation above, you can derive*[13]

$$\nabla_\beta \nabla_\alpha Z^I = -y\, G^{I\bar{L}} N_{JK\bar{L}} \nabla_\alpha Z^J \nabla_\beta Z^K = -y\, \Gamma^I_{JK} \nabla_\alpha Z^J \nabla_\beta Z^K\,. \tag{17.86}$$

This equation can be used to calculate the curvature of the projective manifold. The curvature may be defined by the Ricci identity for the vector $\nabla_\alpha Z^I$, *which reads*

$$\left[\overline{\nabla}_{\bar\alpha}, \nabla_\beta\right] \nabla_\alpha Z^I = a^{-1} g_{\beta\bar\alpha} \nabla_\alpha Z^I + R_{\bar\alpha\beta\alpha}{}^\gamma \nabla_\gamma Z^I\,. \tag{17.87}$$

Apply $y\bar{y} \overline{\nabla}_{\bar\beta} \bar{Z}^{\bar{J}} G_{I\bar{J}} \overline{\nabla}_{\bar\alpha}$ *to both sides of (17.86) and use (17.87) to obtain*

$$R_{\alpha\bar\alpha\beta\bar\beta} - 2a^{-1} g_{\bar\alpha(\alpha} g_{\beta)\bar\beta} = (y\bar{y})^2 R_{I\bar{I}J\bar{J}} \nabla_\alpha Z^I \nabla_\beta Z^J \overline{\nabla}_{\bar\alpha} \bar{Z}^{\bar{I}} \overline{\nabla}_{\bar\beta} \bar{Z}^{\bar{J}}\,. \tag{17.88}$$

Observe that, owing to the homogeneity conditions, Z^I *is a zero eigenvector of the curvature tensor; see (15.45). So the derivatives on the right-hand side can be replaced by ordinary derivatives, and this formula gives the curvature of the projective manifold in terms of the pull-back of the curvature of the embedding manifold.*

[13] One might define also a covariant derivative that takes the connection in the embedding manifold into account, similar to (17.27), i.e. $\hat\nabla_\beta \nabla_\alpha Z^I = \nabla_\beta \nabla_\alpha Z^I + \Gamma^I_{JK} \nabla_\alpha Z^K \nabla_\beta (y Z^J)$. Then this equation reads as $\hat\nabla_\beta \nabla_\alpha Z^I = 0$. Note that, owing to $\Gamma^I_{JK} Z^J = 0$, this modification has no effect on $\hat\nabla_\alpha Z^I = \nabla_\alpha Z^I$.

17.3.7 Physical fermions

In order to define the physical bosons, we changed from the conformal basis $\{X^I\}$ to the basis $\{y, z^\alpha\}$. The modulus of y is fixed by the dilatation gauge condition, and its phase corresponds to the remaining gauge freedom of T-transformations. We now make a similar change of basis from the conformal fermions $\{\Omega^I\}$ to a new basis $\{\chi^0, \chi^\alpha\}$. We will define this basis so that the S-gauge condition (17.33) implies that $\chi^0 = 0$. The fields χ^α then remain as the physical fermions. They are partners of the bosons z^α under the conventional chiral multiplet transformations

$$\delta z^\alpha = \frac{1}{\sqrt{2}}\bar{\epsilon}\chi^\alpha . \tag{17.89}$$

We again use the implicit chiral notation explained below (17.3), i.e. $P_L\chi^\alpha = \chi^\alpha$ and $P_R\chi^{\bar{\alpha}} = \chi^{\bar{\alpha}}$.

We express the $(n+1)$ fields Ω^I in terms of the χ^α and an extra χ^0 by

$$\Omega^I = y\chi^0 Z^I + y\chi^\alpha\nabla_\alpha Z^I = y\begin{pmatrix} Z^I & \nabla_\alpha Z^I \end{pmatrix}\begin{pmatrix}\chi^0\\ \chi^\alpha\end{pmatrix} . \tag{17.90}$$

The covariant derivative ensures uniform behavior under Kähler transformations. Note also that the vectors Z^I and $\nabla_\alpha Z^I$ are orthogonal vectors on the embedding space; see (17.83). Using the invertibility of the matrices in Ex. 17.14, we find that

$$\begin{pmatrix}\chi^0\\ \chi^\alpha\end{pmatrix} = \bar{y}\begin{pmatrix} -a^{-1} & 0\\ 0 & g^{\alpha\bar{\beta}}\end{pmatrix}\begin{pmatrix}\bar{Z}^{\bar{J}}\\ \bar{\nabla}_{\bar{\beta}}\bar{Z}^{\bar{J}}\end{pmatrix} G_{\bar{J}I}\Omega^I . \tag{17.91}$$

Explicitly,

$$\begin{aligned}\chi^0 &= -a^{-1}\bar{y}\bar{Z}^{\bar{J}}G_{I\bar{J}}\Omega^I = -a^{-1}N_I\Omega^I ,\\ \chi^\alpha &= \bar{y}\,\Omega^I G_{I\bar{J}}g^{\alpha\bar{\beta}}\bar{\nabla}_{\bar{\beta}}\bar{Z}^{\bar{J}} .\end{aligned} \tag{17.92}$$

Hence the S-gauge condition can be written as $\chi^0 = 0$.

Exercise 17.16 *Start with the transformation rule $\delta X^I = \bar{\epsilon}\Omega^I/\sqrt{2}$. Use the defining property (17.45) of homogeneous coordinates on the left-hand side and the relation (17.90) on the right-hand side (with $\chi^0 = 0$). Write these as*

$$Z^I\delta y + y\left[\nabla_\alpha - a^{-1}(\partial_\alpha\mathcal{K})\right]Z^I\delta z^\alpha = \frac{1}{\sqrt{2}}y\bar{\epsilon}\chi^\alpha\nabla_\alpha Z^I . \tag{17.93}$$

Deduce the transformation rule (17.89) and also the accompanying transformation

$$\delta y = \frac{1}{\sqrt{2}a}y\bar{\epsilon}\chi^\alpha\partial_\alpha\mathcal{K} , \qquad \delta\bar{y} = \frac{1}{\sqrt{2}a}\bar{y}\bar{\epsilon}\chi^{\bar{\alpha}}\partial_{\bar{\alpha}}\mathcal{K} . \tag{17.94}$$

17.3.8 Symmetries of projective Kähler manifolds

In Sec. 13.4.1 it was shown that Killing symmetries of Kähler manifolds must respect the metric and the complex structure. There is still another important structure in the embedding manifold, namely the holomorphic homothetic Killing vector defined in (17.42). To preserve the structure of the homothety, the Killing vector $k_A{}^I\,\partial/\partial X^I$ of a symmetry must commute with $h = 2X^I\,\partial/\partial X^I$. This is the requirement that Killing vectors have Weyl weight 1, as noted at the beginning of Sec. 17.2.3. As also noted there, the Kähler potential N must be invariant.

In the coordinates y and z^α related to the X^I in (17.45), commutation with $h = 2y\,\mathrm{d}/\mathrm{d}y$ implies that

$$\delta z^\alpha = \theta^A k_A{}^\alpha(z)\,, \qquad \delta y = a^{-1}\theta^A y r_A(z)\,. \tag{17.95}$$

Note that $r_A(z)$ can be considered to be the component of the Killing vector in the direction of y. The Killing vector condition $\nabla_I k_{\bar{J}} + \nabla_{\bar{J}} k_I = 0$ implies that $\nabla_\alpha k_{\bar{\beta}} + \nabla_{\bar{\beta}} k_\alpha = 0$. Thus each Killing vector k_A^α, $k_A^{\bar{\alpha}}$ is a holomorphic Killing vector of the projected metric $g_{\alpha\bar{\beta}}$. However, owing to the relation (17.67), which follows from the fixing of dilatations in (17.65), the Kähler potential $\mathcal{K}$ is not always invariant. Rather it transforms as

$$\delta\mathcal{K} = \theta^A \mathcal{L}_{k_A}\mathcal{K} = \theta^A\left[r_A(z) + \bar{r}_A(\bar{z})\right]\,, \tag{17.96}$$

which is a Kähler transformation. The moment map is given as in (13.59) by

$$\mathcal{P}_A = \mathrm{i}\left(k_A{}^\alpha \partial_\alpha \mathcal{K} - r_A\right) = -\mathrm{i}\left(k_A{}^{\bar{\alpha}}\partial_{\bar{\alpha}}\mathcal{K} - \bar{r}_A\right)\,. \tag{17.97}$$

Exercise 17.17 *Prove the following relation between the Killing vectors in the embedding space and those in the projective manifold:*

$$k_A{}^I = y\left[k_A{}^\alpha \nabla_\alpha Z^I + \mathrm{i}a^{-1}\mathcal{P}_A Z^I\right]\,. \tag{17.98}$$

To do this, use (17.95) to obtain the coordinate variations δX^I. Then use (17.97) and (17.73).

In Sec. 14.4.3 we introduced the Fayet–Iliopoulos constants, in global supersymmetry as real constants in $\mathcal{P}_A$. We now see that they originate in supergravity from imaginary constants in r_A. They are therefore phase transformations of the compensating scalar y. We will discuss this further in Sec. 19.5.

Exercise 17.18 *The commutator of two gauge transformations (17.95) on the field y must satisfy the gauge algebra. Show that this requirement can be written as*

$$f_{AB}{}^C r_C = k_A{}^\alpha \partial_\alpha r_B - (A \leftrightarrow B)\,. \tag{17.99}$$

Use (17.97) and (13.60) to show that this leads to equivariance relation (13.62).

The composite connection $\mathcal{A}_\mu$, given before gauging in (17.70), now includes gauge potentials. One can use the result of Ex. 17.17 to rewrite (17.21) in terms of physical fields as

$$\begin{aligned}\mathcal{A}_\mu &= \tfrac{1}{2}\mathrm{i}a^{-1}\left(\partial_\mu z^\alpha \partial_\alpha \mathcal{K} - \partial_\mu \bar{z}^{\bar{\alpha}}\partial_{\bar{\alpha}}\mathcal{K}\right) - \tfrac{1}{2}\mathrm{i}\partial_\mu \ln(y/\bar{y}) - a^{-1}A_\mu^A \mathcal{P}_A \\ &= \tfrac{1}{2}\mathrm{i}a^{-1}\left(\hat{\partial}_\mu z^\alpha \partial_\alpha \mathcal{K} - \hat{\partial}_\mu \bar{z}^{\bar{\alpha}}\partial_{\bar{\alpha}}\mathcal{K}\right) - \tfrac{1}{2}\mathrm{i}\hat{\partial}_\mu \ln(y/\bar{y})\,,\end{aligned} \tag{17.100}$$

in which we used (17.97), and the definitions

$$\hat{\partial}_\mu z^\alpha \equiv \partial_\mu z^\alpha - A_\mu^A k_A{}^\alpha\,, \qquad \hat{\partial}_\mu y \equiv \partial_\mu y - a^{-1}A_\mu^A r_A\, y\,. \tag{17.101}$$

With these formulas, we can gauge covariantize the ∇_μ derivative (17.79). When gauge internal symmetries are present, we will write it as

$$\nabla_\mu = (\hat{\partial}_\mu z^\alpha)\nabla_\alpha + (\hat{\partial}_\mu \bar{z}^{\bar{\alpha}})\nabla_{\bar{\alpha}}\,. \tag{17.102}$$

The important fact is that inclusion of the gauge connection does not spoil (17.80).

Exercise 17.19 *Check that the inclusion of the gauge connection in $\mathcal{A}$ as in (17.100) is still consistent with $\nabla_\mu y = 0$.*

17.3.9 T-gauge and decomposition laws

The selection of one representative among complex scale invariant configurations in the embedding space requires gauge fixing for both dilatation and T-transformations. We have already used the dilatation gauge choice (17.65) extensively. For T-transformations, the simple condition

$$T\text{-gauge:}\quad y = \bar{y}\,, \qquad \text{i.e.} \qquad y = \mathrm{e}^{\mathcal{K}/(2a)} \tag{17.103}$$

is convenient. The last equality follows from (17.67).

This is the last gauge condition needed to reduce the superconformal symmetries to the super-Poincaré algebra. Now we determine the combination of the superconformal symmetries that remain after the gauge fixing. We first fix the gauge of special conformal transformations by setting $b_\mu = 0$. Considering (16.10) this determines the relation

$$2\lambda_{\mathrm{K}\mu} = -\partial_\mu \lambda_{\mathrm{D}} - \tfrac{1}{2}\bar{\epsilon}\phi_\mu + \tfrac{1}{2}\bar{\eta}\psi_\mu\,, \tag{17.104}$$

which is the extension of (15.27) to the $\mathcal{N}=1$ theory. It is a ‘decomposition law’ for the K-symmetry.

Then we imposed the dilatation gauge (17.32). This is invariant under all other gauge symmetries apart from dilatations itself. For this to occur, it was essential that the S-gauge (17.33) was chosen. Hence, this simply leads to

$$\lambda_{\rm D} = 0\,. \tag{17.105}$$

The transformations of the condition (17.103) are less trivial, since the condition is not invariant under gauge transformations (see (17.95)) or under supersymmetry. Furthermore it is not invariant under the Kähler reparametrizations (17.71). To obtain the decomposition law we write

$$\begin{aligned}\delta(y-\bar{y}) &= {\rm i}\lambda_T(y+\bar{y}) + \frac{1}{a}\theta^A(yr_A - \bar{y}\bar{r}_A) \\ &\quad + \frac{1}{a\sqrt{2}}\bar{\epsilon}\left(y\chi^\alpha\partial_\alpha\mathcal{K} - \bar{y}\chi^{\bar{\alpha}}\partial_{\bar{\alpha}}\mathcal{K}\right) + \frac{1}{a}\left[yf(z) - \bar{y}\bar{f}(\bar{z})\right] = 0\,.\end{aligned} \tag{17.106}$$

Using $y=\bar{y}$, we solve for λ_T and obtain

$$\lambda_T = \frac{1}{2a}{\rm i}\theta^A(r_A - \bar{r}_A) + \frac{1}{2a\sqrt{2}}{\rm i}\bar{\epsilon}\left(\chi^\alpha\partial_\alpha\mathcal{K} - \chi^{\bar{\alpha}}\partial_{\bar{\alpha}}\mathcal{K}\right) + \frac{\rm i}{2a}\left[f(z) - \bar{f}(\bar{z})\right]. \tag{17.107}$$

This shows that a T-transformation must be included in the effective symmetry transformation of any field that transforms under T in the embedding space. For example, the effective gauge symmetry transformation in the projected manifold, $\delta_{\rm proj}$, is related to the gauge transformation in the embedding manifold, $\delta_{\rm emb}$, by

$$\delta_{\rm proj}(\theta) = \delta_{\rm emb}(\theta) + \delta_T(\lambda_T(\theta))\,. \tag{17.108}$$

The new terms from λ_T have no effect on the scalars z^α, since they are invariant under the T-transformation, but the fermions do transform. The definition (17.92) leads to the T-transformation

$$\delta_T\chi^\alpha = -\tfrac{3}{2}{\rm i}\lambda_T\chi^\alpha\,. \tag{17.109}$$

Thus, after the gauge fixing, fermions will transform under Kähler transformations, i.e.

$$\delta_{\rm Kahler}\chi^\alpha = \tfrac{3}{2}{\rm i}\,a^{-1}({\rm Im}\,f)\chi^\alpha\,. \tag{17.110}$$

Exercise 17.20 *Prove that for the gauge transformations and supersymmetry we can write*

$$a\lambda_T = \theta^A\mathcal{P}_A + \tfrac{1}{2}{\rm i}\left(\delta z^\alpha\partial_\alpha\mathcal{K} - \delta\bar{z}^{\bar{\alpha}}\partial_{\bar{\alpha}}\mathcal{K}\right). \tag{17.111}$$

Exercise 17.21 *Show that the rule (17.107) implies that an object with chiral weight c obtains Kähler weights $(w_+, w_-) = (c/2, -c/2)$. Deduce from this that the Kähler weights of χ^α are $(-3/4, 3/4)$ (and opposite for $\chi^{\bar{\alpha}}$).*

We draw readers' attention to the fact that the mixing of gauge symmetries and T-symmetry means that gauge symmetries no longer commute with supersymmetry in supergravity when $r_A \neq 0$. In global supersymmetry, there is a clear distinction between gauge

Box 17.2

Decomposition of T-symmetry

After gauge fixing, the T-transformations act as gauge transformations, supersymmetries and Kähler transformations. Therefore, the gauge group in the super-Poincaré theory can act partly as a gauged R-symmetry, following the scheme:

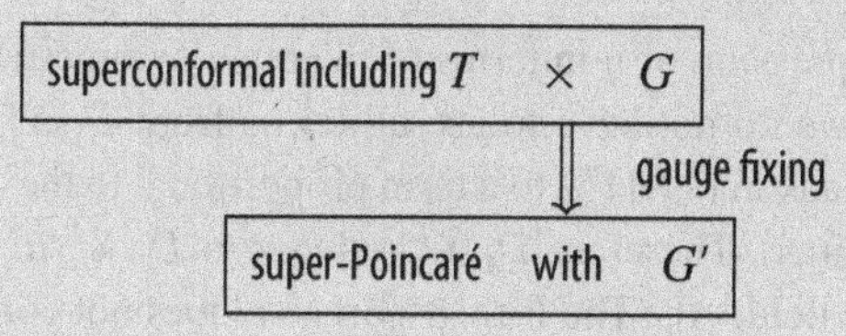

where G' is G with a mixing of superconformal T-symmetries.

symmetries that commute with supersymmetry, and R-symmetry, which does not; see Sec. 12.2. In the superconformal setup, this structure is maintained. The T-symmetry, which does not commute with supersymmetry, is part of the superconformal group. The symmetries gauged by the vectors A_μ^A commute with supersymmetry in the superconformal theory, but not necessarily in the super-Poincaré theory; see Box 17.2.

The gauge field $\mathcal{A}_\mu$ was introduced as the gauge field of T-transformations, i.e. $\mathcal{A}_\mu \to \mathcal{A}_\mu + \partial_\mu \lambda_T$, and it thus acts now as gauge field for the imaginary part of Kähler transformations:

$$\delta_{\text{Kahler}}\mathcal{A}_\mu = -a^{-1}\partial_\mu \operatorname{Im} f\,. \tag{17.112}$$

Using (17.101) and (17.103), its value in (17.100) can be rewritten as

$$\mathcal{A}_\mu = \tfrac{1}{2}\mathrm{i}a^{-1}\left(\hat{\partial}_\mu z^\alpha \partial_\alpha \mathcal{K} - \hat{\partial}_\mu \bar{z}^{\bar{\alpha}} \partial_{\bar{\alpha}} \mathcal{K}\right) + \tfrac{1}{2}\mathrm{i}a^{-1}A_\mu^A(r_A - \bar{r}_A). \tag{17.113}$$

One may wonder now whether the covariant derivatives introduced previously are still consistent. The connection $\mathcal{A}_\mu$ was initially the gauge field for T-transformations. The superconformal covariant derivatives with the value of $\mathcal{A}_\mu$ in (17.113) contain exactly the new connection terms that must appear due to the modified transformations. Hence the covariant derivatives constructed either before or after T-gauge fixing lead to identical results.

Exercise 17.22 *Check that, owing to the decomposition law (17.107), the Kähler weights of y change from the previous $(w_+, w_-) = (-1, 0)$ to $(-\frac{1}{2}, -\frac{1}{2})$. Show that the covariant derivative of (17.80) becomes*

$$\nabla_\mu y = \left[\hat{\partial}_\mu - \tfrac{1}{2}a^{-1}(\hat{\partial}_\mu \mathcal{K})\right] y = 0\,, \tag{17.114}$$

after the value of the connection in (17.113) is inserted. Show that this derivative incorporates the modified Kähler weights. Use the value of y from (17.103) and extract the $\partial_\mu z^\alpha$ and $\partial_\mu \bar{z}^{\bar{\alpha}}$ 'components' of $\nabla_\mu y$ to find an explicit example of the covariant derivatives of (17.81).

Finally, we consider the S-gauge condition (17.33). It transforms to a multiple of itself under dilatations, T and Lorentz transformations. In the exercise below readers are asked to check that it is also invariant under gauge transformations (17.17). It is not invariant under the Q- and S-supersymmetry transformations of (17.3). Indeed the variation of (17.33) gives

$$0 = N_I P_L \left(\not{D} X^I + F^I \right) \epsilon - 2a\, P_L \eta + \Omega^I \bar{\epsilon} \left(N_{IJ} \Omega^J + N_{I\bar{J}} \Omega^{\bar{J}} \right) . \tag{17.115}$$

This leads to the expression of η in terms of the other symmetries. Various simplifications are possible. First, we do not have to use all terms from $\mathcal{D}_\mu X^I$. The ψ_μ dependent term written in (17.20) leads in (17.115) to a term proportional to the S-gauge condition (17.33) and can thus be omitted. Therefore $\mathcal{D}_\mu X^I$ reduces to $D_\mu X^I$ in (17.23) and the fermionic part of the auxiliary field: $A_\mu^{\rm F}$. The former part also does not contribute due to (17.26) and the dilatation condition that puts N equal to a constant. Thus the only part of the $\mathcal{D}_\mu X^I$ that contributes is $-{\rm i} A_\mu^{\rm F} \gamma^\mu \epsilon N$. Then we use the field equations (17.21). The ψ_μ terms of $A_\mu^{\rm F}$ vanish again due to the S-gauge condition. For the auxiliary F^I one uses also the homogeneity equations (17.13) and (17.9). Finally, a Fierz transformation (3.72) simplifies the decomposition law to

$$2\kappa^{-2} P_L \eta = -\overline{\mathcal{W}} P_L \epsilon + \gamma_a P_R \epsilon \left(\tfrac{1}{4} N_{I\bar{J}} \bar{\Omega}^I \gamma^a \Omega^{\bar{J}} + \tfrac{1}{8} \operatorname{Re} f_{AB} \bar{\lambda}^A \gamma^a \gamma_* \lambda^B \right) . \tag{17.116}$$

Exercise 17.23 *Check that the gauge fixing condition for S-supersymmetry is invariant under gauge transformations by using a derivative of the second equation of (17.18).*

17.3.10 An explicit example: SU(1, 1)/U(1) model

The projective manifold

We consider a model of supergravity with one physical chiral multiplet. In the conformal approach we must include the compensating multiplet also. Thus the embedding space has two homogeneous coordinates X^0, X^1. We choose the parametrization

$$X^0 = y\,, \qquad X^1 = yz\,, \qquad \text{i.e.} \qquad Z^0 = 1\,, \qquad Z^1 = z, \tag{17.117}$$

so that the physical scalar is $z = z^1$.

We consider the following quadratic Kähler potential in the embedding space:

$$N = {\rm i} a (\bar{X}^0 X^1 - X^0 \bar{X}^1) . \tag{17.118}$$

The dilatation constraint $N = -a = -2ay\bar{y} \operatorname{Im} z$ selects the upper half-plane as the domain of z. Using (17.67) and the value $a = 3\kappa^{-2}$ for $\mathcal{N} = 1$ supergravity, we find the Kähler potential

$$\mathcal{K} = -3\kappa^{-2} \ln\left[-{\rm i}(z - \bar{z})\right] . \tag{17.119}$$

The Kähler metric is

$$g_{1\bar{1}} = -3\kappa^{-2} \frac{1}{(z - \bar{z})^2} . \tag{17.120}$$

This is the metric of the Poincaré plane, which we studied at the end of Sec. 7.12, although the scale is different. One may check with the formulas of Sec. 13.1 that the scalar curvature is $R = 2g^{1\bar{1}}R_{1\bar{1}} = -\frac{4}{3}\kappa^2$. This value is independent of the normalization of (17.118).

Exercise 17.24 *Consider the modified Kähler potential*

$$N = aX^0\bar{X}^0\left(\mathrm{i}\frac{X^1}{X^0} - \mathrm{i}\frac{\bar{X}^1}{\bar{X}^0}\right)^b . \tag{17.121}$$

Verify that the Kähler metric and the scalar curvature of the projective manifold become

$$g_{1\bar{1}} = -3b\kappa^{-2}\frac{1}{(z-\bar{z})^2}, \qquad R = -\frac{4}{3b}\kappa^2 . \tag{17.122}$$

The lesson is that the modified embedding metric gives us the freedom of an arbitrary scale for the projective manifold.

Isometry group

It is clear that the symmetries that preserve the form of N in (17.118) are given by the $\mathrm{U}(1) \times \mathrm{SL}(2, \mathbb{R})$ transformations:

$$\delta X^I = -\theta^A t_A{}^I{}_J X^J . \tag{17.123}$$

We choose the basis of matrices

$$t_0 = \begin{pmatrix} \mathrm{i} & 0 \\ 0 & \mathrm{i} \end{pmatrix}, \qquad t_1 = \begin{pmatrix} 0 & 0 \\ -1 & 0 \end{pmatrix}, \qquad t_2 = \frac{1}{2}\begin{pmatrix} 1 & 0 \\ 0 & -1 \end{pmatrix}, \qquad t_3 = \begin{pmatrix} 0 & 1 \\ 0 & 0 \end{pmatrix}, \tag{17.124}$$

which act on the vector $\begin{pmatrix} X^0 \\ X^1 \end{pmatrix}$. This leads to the Lie algebra of (7.158) plus the commuting generator t_0.

When we use the parametrization of the projective manifold, we find

$$\delta y = \delta X^0 = \frac{1}{a} y\,\theta^A r_A , \qquad \delta z = \frac{\delta X^1}{X^0} - z\frac{\delta X^0}{X^0} = \theta^A k_A{}^z , \tag{17.125}$$

which explicitly leads to

$$\theta^A r_A = a\left(-\mathrm{i}\theta^0 - \tfrac{1}{2}\theta^2 - \theta^3 z\right), \qquad \theta^A k_A{}^z = \theta^1 + \theta^2 z + \theta^3 z^2 . \tag{17.126}$$

Note that $k_A{}^z$ are the same Killing vectors as found in (7.157).

We will discuss the gauging of (part of) the isometry group in Sec. 19.5.3.

17.4 From conformal to Poincaré supergravity

In the previous section, we discussed how the kinetic terms of chiral multiplets define a projective Kähler manifold. We used the parametrization (17.45), with y constrained by (17.66), to introduce the coordinates z^α that describe physical scalars. Analogously we introduced the physical fermions by the equation

$$\Omega^I = y\,\chi^\alpha \nabla_\alpha Z^I . \qquad (17.127)$$

In this section, we will reexpress the conformal Lagrangian (17.30) in terms of these physical fields. To obtain the correct normalization of the gravity Lagrangian $\mathcal{L} = eR/(2\kappa^2)$ in (17.30), we choose the values

$$a = 3\kappa^{-2}, \qquad y = \exp\left(\tfrac{1}{6}\kappa^2\mathcal{K}\right) . \qquad (17.128)$$

17.4.1 The superpotential

Our first task is to develop the physical form of the scalar potential of $\mathcal{N}=1$ supergravity, which is obtained from a superpotential on the embedding manifold. In (17.9) we saw that the superpotential must be a homogeneous holomorphic function of third degree. In the coordinates y and z^α it takes the form

$$\mathcal{W} = y^3 W(z) = \exp\left(\tfrac{1}{2}\kappa^2\mathcal{K}\right) W(z) . \qquad (17.129)$$

The holomorphic function $W(z)$ is the superpotential of the Poincaré supergravity theory.

In the terminology of (17.76) the equation above indicates that the superpotential transforms under the Kähler transformations (17.71) as a function with weights $(w_+, w_-) = (3,0)$. Thus, its Kähler covariant derivative is

$$\nabla_\alpha W(z) = \partial_\alpha W(z) + \kappa^2(\partial_\alpha\mathcal{K})W(z) . \qquad (17.130)$$

The homogeneity conditions lead to relations between the derivatives of $\mathcal{W}$ and W:

$$\begin{aligned} y\mathcal{W}_I Z^I &= \mathcal{W}_I X^I = 3\mathcal{W} = 3y^3 W(z) , \\ y\mathcal{W}_I \nabla_\alpha Z^I &= \nabla_\alpha \mathcal{W} = y^3 \nabla_\alpha W . \end{aligned} \qquad (17.131)$$

W and $\overline{W}$ combine with the Kähler potential to form the Kähler-invariant function $\mathcal{G}(z,\bar z)$:

$$\mathcal{G} = \kappa^2\mathcal{K} + \log(\kappa^6 W\overline{W}) , \qquad (17.132)$$

which is often used in the literature.

The reader can check the gauge transformation properties of the superpotential in the next exercise. These imply that the superpotential is *not* invariant if $r_A \neq 0$. This is the case, for example, if the theory includes Fayet–Iliopoulos constants.

Exercise 17.25 *The gauge invariance condition for the superconformal $\mathcal{W}$ takes the simple form given in (17.18). Use the expression (17.98) that you derived in Ex. 17.17, and (17.131) to rewrite (17.18) as*

$$k_A{}^\alpha \nabla_\alpha W + \mathrm{i}\kappa^2 \mathcal{P}_A W = W_\alpha k_A{}^\alpha + \kappa^2 r_A\, W = 0 . \qquad (17.133)$$

Exercise 17.26 *Show that (17.133) implies that a* constant *superpotential must* vanish *if the theory contains gauged symmetries with $r_A \neq 0$. Show that the superpotential must also vanish if it depends only on scalars that do not transform under a symmetry, i.e. $k_A{}^\alpha = 0$ but $r_A \neq 0$.*

17.4.2 The potential

We now consider the scalar potential of a general $\mathcal{N} = 1$ supergravity theory. As in global supersymmetry, it contains an F-term and a D-term. The potential $V = V_F + V_D$ appears in the superconformal formulation in (17.31). As in global supersymmetry V_D is non-negative, since $\operatorname{Re} f_{AB}$ must be a positive quadratic form. Otherwise gauge field kinetic terms would not be positive. However, the situation is different for V_F because the compensating multiplet in the conformal approach to supergravity has negative norm in the embedding space metric $G_{I\bar{J}}$. Thus V_F is not intrinsically positive, in marked distinction to the situation in global supersymmetry.

The inverse of $G_{I\bar{J}}$ in the basis of physical fields was obtained in (17.85), and this expression clearly exhibits the rank-1 negative mode. We combine this with (17.131) and use (17.128) to express V_F as

$$V_F = e^{\kappa^2 \mathcal{K}} \left(-3\kappa^2 W \overline{W} + \nabla_\alpha W g^{\alpha\bar{\beta}} \overline{\nabla}_{\bar{\beta}} \overline{W} \right) . \tag{17.134}$$

This form clearly displays the negative contribution from supergravity to the potential.

Let's discuss the D-term part of the potential briefly. Our discussion of the symmetries of a projective Kähler manifold led to the expression (17.97) for the moment map of a holomorphic Killing vector. This expression agrees with the form obtained in Sec. 13.4.1 for the general Kähler manifolds that appear as target spaces in global supersymmetry. Thus V_D takes the same form in global and local $\mathcal{N} = 1$ SUSY.

The net result of this section is that the full potential for $\mathcal{N} = 1$ supergravity can be split into negative and positive definite parts, viz.

$$V = V_- + V_+ , \tag{17.135}$$

$$V_- = -3\kappa^2 e^{\kappa^2 \mathcal{K}} W \overline{W} , \qquad V_+ = e^{\kappa^2 \mathcal{K}} \nabla_\alpha W g^{\alpha\bar{\beta}} \overline{\nabla}_{\bar{\beta}} \overline{W} + \tfrac{1}{2} (\operatorname{Re} f)^{-1\,AB} \mathcal{P}_A \mathcal{P}_B .$$

17.4.3 Fermion terms

The gravitino and gaugino kinetic terms of the conformal action (17.30) already involve the fields needed in the Poincaré description of supergravity. However, for the fermions of the chiral multiplets, we have to use (17.127). It was noted in (17.109) that the physical fermion χ^α has T-charge $-3/2$. Further, χ^α transforms as a tangent vector V^α of the projective manifold under gauge transformations and holomorphic reparametrizations. These considerations require that its physical kinetic term involves the covariant derivative

$$D_\mu^{(0)}\chi^\alpha = \left(\partial_\mu + \tfrac14\omega_\mu{}^{ab}(e)\gamma_{ab} + \tfrac32\mathrm{i}\mathcal{A}_\mu\right)\chi^\alpha - A_\mu^A\frac{\partial k_A{}^\alpha(z)}{\partial z^\beta}\chi^\beta + \Gamma^\alpha_{\beta\gamma}\chi^\gamma\hat\partial_\mu z^\beta. \quad (17.136)$$

The same derivative should emerge when $\hat D_\mu^{(0)}\Omega^I$, (17.27), is evaluated using (17.127) and indeed it does. The methods of Sec. 17.3.6 can be used to obtain

$$\begin{aligned}\hat D_\mu^{(0)}\Omega^I &= D_\mu^{(0)}\left(y\chi^\alpha\nabla_\alpha Z^I\right) + y^2\Gamma^I_{JK}\chi^\alpha\nabla_\alpha Z^K\nabla_\beta Z^J\hat\partial_\mu z^\beta\\ &= y\left(D_\mu^{(0)}\chi^\alpha\right)\nabla_\alpha Z^I + y\chi^\alpha\left(\hat\partial_\mu z^\beta\nabla_\beta + \hat\partial_\mu\bar z^{\bar\beta}\overline\nabla_{\bar\beta}\right)\nabla_\alpha Z^I\\ &\quad + y^2\Gamma^I_{JK}\chi^\alpha\nabla_\alpha Z^K\nabla_\beta Z^J\hat\partial_\mu z^\beta\\ &= y\left(D_\mu^{(0)}\chi^\alpha\right)\nabla_\alpha Z^I + \tfrac13\kappa^2 y\chi^\alpha\hat\partial_\mu\bar z^{\bar\beta}g_{\alpha\bar\beta}Z^I\,. \end{aligned} \quad (17.137)$$

One needs (17.86) and (17.78) to obtain this result. The second term does not contribute in $\mathcal{L}_{1/2}$, (17.31), since it then produces a term proportional to $\overline\nabla_{\bar\alpha}\bar Z^{\bar J}G_{I\bar J}Z^I = 0$; see (17.84). Finally, the latter equation also implies that $\mathcal{L}_{1/2}$ reduces to the physical kinetic term

$$\mathcal{L}_{1/2} = -\tfrac12 g_{\alpha\bar\beta}\left[\bar\chi^\alpha\not{D}^{(0)}\chi^{\bar\beta} + \bar\chi^{\bar\beta}\not{D}^{(0)}\chi^\alpha\right]. \quad (17.138)$$

Exercise 17.27 *Show, using (17.110) and (17.112), that the Lagrangian (17.138) is invariant under Kähler transformations.*

Similar manipulations can be used for other terms in the action. For example the mass terms of χ^α are obtained by first taking another covariant derivative of (17.131) using again (17.86), which then implies

$$y^2\nabla_J\mathcal{W}_I\nabla_\alpha Z^I\nabla_\beta Z^J = y^3\nabla_\beta\nabla_\alpha W\,. \quad (17.139)$$

This can be directly used in the second term of $\mathcal{L}_m$ in (17.31). For the mass terms involving λ, we first need an equation for the derivative of f_{AB}:

$$f_{AB\,\alpha} = \nabla_\alpha f_{AB} = y\, f_{AB\,I}\nabla_\alpha Z^I\,. \quad (17.140)$$

For the $\lambda\lambda$ term we use the expression for $G^{I\bar J}$ in (17.85) and the degree zero homogeneity of f_{AB}; i.e. $f_{AB\,I}Z^I = 0$. Then, (17.131) can be used to translate

$$G^{I\bar J}\overline{\mathcal{W}}_{\bar J}f_{ABI} = \bar y^3 g^{\alpha\bar\beta}f_{AB\,\alpha}\overline\nabla_{\bar\beta}\overline W\,. \quad (17.141)$$

For the mixed $\lambda\chi$ mass terms, we use the same equation as (17.140) and its analogue with f_{AB} replaced with $\mathcal{P}_A$.

For the four-fermion terms we use similar manipulations, with frequent use of (17.84). For $\nabla_I f_{AB\,J}$ we repeat the method used in (17.139). Owing to the gauge condition (17.33), the fermionic part of the auxiliary field A_μ simplifies from (17.21) and becomes

$$A_\mu^{\mathrm F} = \mathrm{i}\kappa^2\left[\tfrac{1}{12}g_{\alpha\bar\beta}\bar\chi^\alpha\gamma_\mu\chi^{\bar\beta} - \tfrac18(\mathrm{Re}\, f_{AB})\bar\lambda^A\gamma_\mu\gamma_*\lambda^B\right]. \quad (17.142)$$

The square of the first term combines (after a Fierz transformation) with the curvature term, where (17.88) is now convenient. This leads to a term

$$\tfrac14\left(R_{\alpha\bar\gamma\beta\bar\delta} - \tfrac12\kappa^2 g_{\alpha\bar\gamma}g_{\beta\bar\delta}\right)\bar\chi^\alpha\chi^\beta\bar\chi^{\bar\gamma}\chi^{\bar\delta}\,. \quad (17.143)$$

The results of all these manipulations will be incorporated in the final physical action in Sec. 18.1.

17.5 Review and preview

Let's review the main steps of this chapter. The general $\mathcal{N} = 1$ supergravity theory was constructed as a gauge theory of the superconformal algebra. In this theory the gauge and chiral 'matter multiplets' of global SUSY are coupled to the Weyl multiplet, which contains the gauge fields for the algebra. The action and transformation for the Weyl multiplet were derived in Ch. 16. In Sec. 17.1 we discussed the gauge transformations of the matter multiplets. In Sec. 17.2 we constructed the action and eliminated auxiliary fields. At this point we presented the action (17.30), which is invariant under the full superconformal algebra. Specifically it is invariant under the transformation rules of (15.21), (17.3) and (17.1) with auxiliary fields replaced by the solution of their equations of motion. In Sec. 17.2.5 we discussed the gauge fixing of those superconformal symmetries (except the T-symmetry) which are not part of the Poincaré SUSY algebra.

The dynamics of the chiral multiplets is a major concern. It is basic to supersymmetry in four spacetime dimensions that this dynamics requires a Kähler manifold. We introduced the Kähler potential $N(Z, \bar{Z})$ of the $(n+1)$-dimensional embedding manifold. Superconformal symmetry requires that $N(Z, \bar{Z})$ satisfies the homogeneity conditions (17.11). In Sec. 17.3 we studied the geometry and symmetries of projective Kähler manifolds, and we showed that the embedding metric satisfies the conditions of projective Kähler geometry. We imposed the gauge condition (17.65) and used projective coordinates (17.45) to introduce the fields z^α, χ^α of the n physical chiral multiplets of the final Poincaré SUSY theory. The Kähler potential $\mathcal{K}(z, \bar{z})$ in (17.67) was obtained by projection from the embedding space. In Sec. 17.4.2 we discussed how the scalar potential $V(z, \bar{z})$ of the final theory inherits properties from projective Kähler geometry. One consequence is that the supergravity potential $V(z, \bar{z})$ is not positive definite as it is in global SUSY.

The full physical Lagrangian can be obtained by incorporating the results above in the Lagrangian (17.30), using Kähler covariant derivatives and final steps discussed in Sec. 17.4.3. The physical Lagrangian is presented in the next chapter. It is invariant under transformation rules that we also present there. They are obtained by incorporating results of this chapter in the transformation rules of Sec. 17.1. An important feature of the final transformation rules is the decomposition laws of Sec. 17.3.9, which are a byproduct of the gauge fixing procedure.

The superconformal approach and projective Kähler geometry determine many features of the final Lagrangian and ensure its invariance under all local symmetries of the Poincaré SUSY algebra and non-abelian gauge symmetry. However, once the final structure is found, one can set much of the initial approach aside.

In particular, one can use any Kähler manifold, with potential $\mathcal{K}(z, \bar{z})$ as the target space of the physical scalars.[14] Let us consider how this comes about. We started with a real function $N(X, \bar{X})$ of $(n+1)$ complex variables X^I that must satisfy the homogeneity relations (17.11). The kinetic terms were transformed to the form (17.61). After gauge fixing the physical Kähler potential was defined in (17.67) and we found the Lagrangian (17.68). The latter depends only on the Kähler potential $\mathcal{K}(z, \bar{z})$, a real function of n complex variables

[14] If the manifold has non-trivial 2-cycles it must satisfy the Kähler–Hodge condition to be discussed below.

z^α, and we need only the latter to write down the action. In fact, the space of *homogeneous* functions $N(X,\bar{X})$ of $(n+1)$ variables is equivalent to the space of arbitrary functions $\mathcal{K}(z,\bar{z})$. Indeed, in the variables y and z^α, homogeneity requires that N is the product $y\bar{y}$ times a function of z and $\bar{z}$. With the notations of the Kähler potential that we adopted, we can write this relation as

$$N=(-a)\,y\bar{y}\exp(-\mathcal{K}/a)\,. \tag{17.144}$$

Any homogeneous N must take this form for some function $\mathcal{K}(z,\bar{z})$, and on the other hand, any function $\mathcal{K}(z,\bar{z})$ determines in this way a function N, written here as a function of $\{y,z^\alpha\}$ and its conjugates. The transformation from $\{y,z^\alpha\}$ to $\{X^I\}$ depends on the choice of the functions $Z^I(z)$, but it is invertible since the matrix (17.46) must be invertible. The function N in (17.144) has the required homogeneity properties and gives a Kähler metric $G_{I\bar{J}}$ in the embedding space of signature $(-+++\ldots)$. This fact is evident from (17.84). We will illustrate the use of (17.144) in Sec. 17.5.2, but we first discuss a global issue that has been neglected so far.

Exercise 17.28 *Consider the flat Kähler metric $g_{\alpha\bar{\beta}}=\delta_{\alpha\bar{\beta}}$ in a model with one physical scalar z^1. Show that this is produced by a Kähler potential in the embedding manifold that is*

$$N=(-a)X^0\bar{X}^0\exp\left[-\frac{1}{a}\frac{X^1\bar{X}^1}{X^0\bar{X}^0}\right], \tag{17.145}$$

if we also take the standard choice (17.47) for the embedding Z^I.

17.5.1 Projective and Kähler–Hodge manifolds

Our construction involves a natural U(1) fiber bundle with connection $\mathcal{A}_\mu$ given in (17.56). In this section we show that its curvature, the field strength $\mathcal{F}_{\mu\nu}\equiv\partial_\mu\mathcal{A}_\nu-\partial_\nu\mathcal{A}_\mu$, is the pull-back of the fundamental Kähler form (13.14) on the projective space. In Appendix 17A we discuss the relation of the quantization condition for integrals of $\mathcal{F}_{\mu\nu}$ and the Kähler–Hodge condition (17.165). When applied here this relation shows that the projective manifolds required for $\mathcal{N}=1$ supergravity are Kähler–Hodge manifolds.

The curvature of $\mathcal{A}_\mu$ in (17.70) is

$$\mathcal{F}_{\mu\nu}=-\frac{2\mathrm{i}}{a}\partial_\alpha\partial_{\bar{\beta}}\mathcal{K}\left(\partial_{[\mu}z^\alpha\partial_{\nu]}\bar{z}^{\bar{\beta}}\right). \tag{17.146}$$

The argument of Sec. 17A.1 shows that the proper definition of the symmetry requires the quantization of integrals of the curvature of the U(1) connection over closed 2-surfaces in spacetime, i.e.

$$\iint\frac{1}{2}\mathcal{F}_{\mu\nu}\mathrm{d}x^\mu\wedge\mathrm{d}x^\nu=\frac{2\pi n}{q}. \tag{17.147}$$

According to the discussion in Sec. 17A.1 the constant q is the minimal coefficient of the coupling of $\mathcal{A}_\mu$ to physical fields in the theory in which the symmetry appears. The coupling to the gaugino and gravitino[15] in (17.23) and (17.24) leads to $q = 3/2$.

Now note that a solution of the equations of motion of the Lagrangian (17.68) specifies a map from spacetime into the projective Kähler target space determined by the functions $z^\alpha(x)$. A closed 2-sphere in spacetime maps into a 2-cycle in the target space. By (17.146) the spacetime integral (17.147) is equal to an integral of the Kähler form over that cycle. It then takes the quantized value

$$\iint \frac{1}{2}\mathcal{F}_{\mu\nu}\mathrm{d}x^\mu \wedge \mathrm{d}x^\nu = -\mathrm{i}\frac{1}{a}\iint g_{\alpha\bar{\beta}}\mathrm{d}z^\alpha \wedge \mathrm{d}\bar{z}^{\bar{\beta}} = \frac{1}{2a}\iint \Omega = \frac{2\pi n}{q}. \tag{17.148}$$

In the last integral we identified the fundamental 2-form (13.14). The final integral in (17.148) then agrees with the quantization condition for Kähler–Hodge manifolds (17.165).

The argument above shows that the projective Kähler manifolds of matter-coupled $\mathcal{N} = 1$ supergravity are Kähler–Hodge manifolds. This construction of Kähler–Hodge manifolds is probably exhaustive. The *Kodaira embedding theorem* (see e.g. p. 181 of [160]) says that a compact Kähler–Hodge manifold can be embedded in a projective space. We are not aware of a generalization of the theorem to non-compact spaces, though this is probable, so that any Kähler–Hodge manifold can appear as a target manifold for chiral multiplets in supergravity.

These quantization issues that lead us to Kähler–Hodge manifolds are related to the fact that the T-symmetry, U(1) in this case, has a non-trivial action on the scalar manifold. Specifically, the U(1) generates Kähler transformations after gauge fixing, as we showed in Sec. 17.3.9. This is a general phenomenon of supergravity theories. The analogous symmetry (usually called an R-symmetry) always acts on the scalar manifolds, as we mentioned in Sec. 12.5; see Table 12.3. This does not occur for scalar manifolds in theories with global supersymmetry. That is why we did not find a Kähler–Hodge condition for the global supersymmetry σ-models of Ch. 14. The coupling of the connection $\mathcal{A}_\mu$ vanishes in the global limit when the gravitational coupling $\kappa = 0$.

17.5.2 Compact manifolds

The scalar manifolds of higher $\mathcal{N} > 1$ supergravity are always non-compact. See Sec. 12.5 for an overview, or [161] for a general argument. However, for $\mathcal{N} = 1$ compact scalar manifolds are possible. We now consider the compact manifold CP^1 as a target space for a supergravity theory with one physical chiral multiplet. We will use this to illustrate the application of (17.144) and then discuss the corresponding Kähler–Hodge condition.

15 The homogeneous fermionic fields Ω^I have smaller coupling, but we saw in Sec. 17.3.7 that the physical fermions obtained from them in (17.91) also have the same coupling $q = 3/2$; see (17.109).

The Kähler potential, metric, scalar curvature and Kähler form of CP^1 are (with an arbitrary scale b)

$$\mathcal{K}=b\ln(1+z\bar{z}),\qquad g_{1\bar{1}}=\frac{b}{(1+z\bar{z})^2},\qquad R=\frac{4}{b},\qquad \Omega=-2\mathrm{i}\frac{b}{(1+z\bar{z})^2}\mathrm{d}z\wedge\mathrm{d}\bar{z}. \tag{17.149}$$

In Sec. 13.3 we obtained this metric from a flat positive signature embedding space, but this is not suitable for application to supergravity, where we need an indefinite metric so that the graviton has positive kinetic energy. We choose $Z^0=1$, $Z^1=z$, and express (17.144) in terms of $X^I=yZ^I$. The embedding space Kähler potential for the CP^1 model is then

$$N=-a\left(X^0\bar{X}^0+X^1\bar{X}^1\right)^{-b/a}\left(X^0\bar{X}^0\right)^{1+b/a}. \tag{17.150}$$

If we impose the constraint $N=-a$ and apply the projection discussed in Sec. 17.3.5, we reproduce the data of (17.149).

Exercise 17.29 *Check that with $a,b>0$, the matrix of second derivatives of N has indefinite signature (it is sufficient to calculate it at the point $z=0$).*

Now we consider the quantization issue. To apply Sec. 17.5.1, we have to consider the couplings of the gauge field $\mathcal{A}_\mu$ to fermions. As mentioned above, this leads to $q=3/2$. Using polar coordinates, you can prove that

$$\int \mathrm{d}z\wedge\mathrm{d}\bar{z}\,(2\mathrm{i})\frac{1}{(1+z\bar{z})^2}=4\pi. \tag{17.151}$$

Thus we find from (17.148) that

$$\frac{3b}{2a}=\frac{b\kappa^2}{2}=n. \tag{17.152}$$

This is the 'quantization of the Newton's constant' in terms of b [162].

All CP^n spaces have closed homology 2-cycles. Hence, the Kähler–Hodge condition is relevant in all cases.

Appendix 17A Kähler–Hodge manifolds

Kähler manifolds that can be coupled consistently to $\mathcal{N}=1$ supergravity must satisfy the additional requirements that define what mathematicians call Kähler–Hodge manifolds [163, 164]. In this section we discuss these conditions, which are related to the more familiar subject of the mathematics of magnetic monopoles in U(1) gauge theories (see e.g. [24] for a review on magnetic monopoles). We briefly discuss the situation of magnetic monopoles in Sec. 17A.1 and then Kähler–Hodge manifolds in Sec. 17A.2.

The common feature of both cases is the idea that a U(1) gauge potential A_μ transforms as the connection of a fiber bundle. This means that the potentials $A_\mu(x)$ and $A'_\mu(x')$ in overlapping coordinate neighborhoods of a manifold M are related by

$$A'_\mu(x') = \frac{\partial x^\nu}{\partial x'^\mu} A_\nu(x) + \partial_\mu \theta(x)\,, \tag{17.153}$$

in which the covariant vector transformation rule is combined with a gauge transformation. The sections of the bundle are fields $\psi(x)$ with electric charge q, which transform as $\psi(x) \to \mathrm{e}^{\mathrm{i}q\theta(x)}\psi(x)$. The phase factors $\mathrm{e}^{\mathrm{i}q\theta(x)}$ are elements of the U(1) structure group of the bundle.

There is one important global issue in the application of gauge theories in general, and which we will need for the Kähler manifolds of supergravity. In a gauge theory, the gauge potential $A_\mu(x)$ need not be a continuous function over the whole manifold. Instead, in a cover of the manifold by coordinate charts, there is a continuous function $A_\mu(x)$ in each chart, but in an overlap of two charts M and M', the potentials A_μ and A'_μ differ by the gauge transformation $\partial_\mu\theta(x)$, where θ is defined on the overlap. If the manifold has a non-trivial structure, these gauge changes must act consistently on charged fields. This means that a charged field must return to its original value if one traverses a closed path on the manifold. In mathematical terms this is the cocycle condition for gauge transformations ϕ_{ji} from a patch i to a patch j. These gauge transformations act on charged fields by multiplication with $\exp(\mathrm{i}q\theta_{ji})$. In the intersection of three patches we must have

$$\phi_{kj}\phi_{ji} = \phi_{ki}\,. \tag{17.154}$$

In some cases it might not be possible to find such consistent globally defined field, i.e. 'a well-defined bundle'. This is, for example, not obvious for a magnetic monopole. The requirement that the system of a charged particle and a monopole is globally well defined leads to the Dirac quantization condition. We review this here, taking care of factors that will be important for the Kähler transformations too.

17A.1 Dirac quantization condition

The Maxwell equations involve only the field strengths. The gauge field appears in the coupling of a charged particle to electromagnetism. The field equations that charged particles satisfy contain the covariant derivative

$$D_\mu\psi = (\partial_\mu - \mathrm{i}\,q\,A_\mu)\,\psi\,, \tag{17.155}$$

where q is the charge of the particle described by ψ. The field equations will be invariant if a gauge transformation of A_μ is combined with a transformation of the phase of ψ in the following way:

$$\begin{aligned} \psi &\to \mathrm{e}^{\mathrm{i}q\theta}\psi\,, \qquad A_\mu \to A_\mu + \partial_\mu\theta\,,\\ D_\mu\psi &\to \mathrm{e}^{\mathrm{i}q\theta}D_\mu\psi. \end{aligned} \tag{17.156}$$

This shows that $\mathrm{e}^{\mathrm{i}q\theta}$ is the quantity that should be uniquely defined at any spacetime point. Hence, when we follow the evolution of θ over a closed path, the initial and final values of $q\theta$ should differ only by $2\pi n$, where $n \in \mathbb{Z}$. As alluded to before, it might not be possible to define A_μ everywhere with one continuous function. The closed path may be in the overlap of two regions with different values of A_μ related as in (17.156). Consider therefore that we describe a closed 2-surface built from two open 2-surfaces with the path

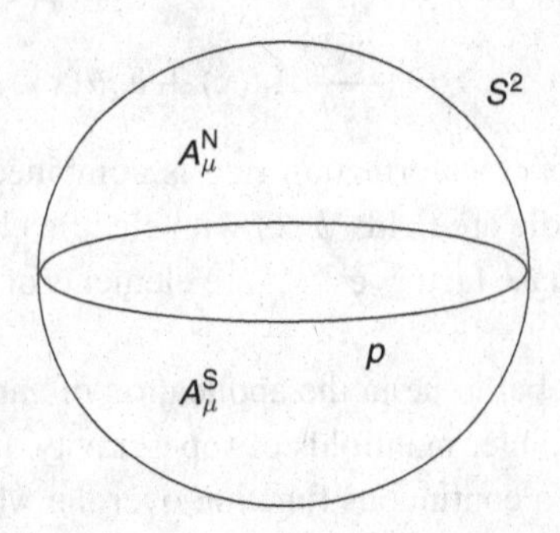

Fig. 17.2 The closed path as border of two regions with different gauge choices for A_μ.

p as border (see Fig. 17.2). In each of these regions there are well-defined gauge fields, $A_\mu^{\rm N}$ and $A_\mu^{\rm S}$, respectively, leading to the same field strength $F_{\mu\nu}$. We use Stokes' theorem to do the integral:

$$\iint_{S^2} \tfrac{1}{2} F_{\mu\nu} {\rm d}x^\mu \wedge {\rm d}x^\nu = \left(\iint_N + \iint_S\right) \tfrac{1}{2} F_{\mu\nu} {\rm d}x^\mu \wedge {\rm d}x^\nu = \int_p (A_\mu^{\rm N} - A_\mu^{\rm S}){\rm d}x^\mu = \int_p {\rm d}\theta\,. \tag{17.157}$$

The last term must give $2\pi n/q$, so we find the consistency condition

$$q \iint \tfrac{1}{2} F_{\mu\nu} {\rm d}x^\mu \wedge {\rm d}x^\nu = 2\pi n\,. \tag{17.158}$$

Consider as an example a magnetic monopole solution

$$\vec{B} = p \frac{\vec{r}}{4\pi r^3}\,, \tag{17.159}$$

where the magnetic field is defined as $F_{ij} = \varepsilon_{ijk} B^k$. The origin is a singular point, and we thus consider the integral of the field strength on a sphere surrounding it. In spherical coordinates, $\frac{1}{2}\varepsilon_{ijk} x^k {\rm d}x^i {\rm d}x^j = r^3 \sin\theta\, {\rm d}\theta\, {\rm d}\phi$. The integral (17.158) then gives

$$\frac{q\,p}{4\pi} \int_{S^2} \sin\theta\, {\rm d}\theta\, {\rm d}\phi = q\,p = 2\pi n\,, \tag{17.160}$$

which is the Dirac quantization condition between the electric charges and magnetic charges of particles.

17A.2 Kähler–Hodge manifolds

In supergravity a similar situation occurs. The relevant invariance that must be well defined is the Kähler symmetry (13.18). In supergravity, fermions transform under this symmetry, and there is therefore an extra condition that the Kähler transformations should be well defined on intersections of coordinate charts in the projective manifold [162, 165].

The quantities $\partial_\alpha \mathcal{K}$ and $\partial_{\bar\alpha} \mathcal{K}$ transform with the derivatives of the parameters f and $\bar f$, and act thus as the gauge fields for these transformations:

$$\delta\partial_\alpha\mathcal{K} = \partial_\alpha f = 2\mathrm{i}\partial_\alpha \operatorname{Im} f\,,$$
$$\delta\partial_{\bar\alpha}\mathcal{K} = \partial_{\bar\alpha}\bar f = -2\mathrm{i}\partial_{\bar\alpha} \operatorname{Im} f\,. \tag{17.161}$$

The last equations follow from the holomorphy of f. Note that if we were to write the transformations for $\operatorname{Re} f$, the gauge fields would be $\partial_a\mathcal{K}$, which is a total derivative, and hence its curvature vanishes. For the imaginary part, however, the gauge field is

$$\omega_a = \{\omega_\alpha = -\tfrac{1}{2}\mathrm{i}\partial_\alpha\mathcal{K},\ \omega_{\bar\alpha} = \tfrac{1}{2}\mathrm{i}\partial_{\bar\alpha}\mathcal{K}\} = -\tfrac{1}{2}J_a{}^b\partial_b\mathcal{K}\,, \tag{17.162}$$

which transforms as $\delta\omega_a = \partial_a \operatorname{Im} f$. It has non-vanishing curvature

$$R^{(2)} = \tfrac{1}{2}\left(\partial_a\omega_b - \partial_b\omega_a\right)\mathrm{d}z^a\wedge\mathrm{d}z^b = \left(\partial_\alpha\omega_{\bar\beta} - \partial_{\bar\beta}\omega_\alpha\right)\mathrm{d}z^\alpha\wedge\mathrm{d}\bar z^{\bar\beta}$$
$$= \mathrm{i}\partial_\alpha\partial_{\bar\beta}\mathcal{K}\,\mathrm{d}z^\alpha\wedge\mathrm{d}\bar z^{\bar\beta} = -\tfrac{1}{2}\Omega\,, \tag{17.163}$$

where Ω was introduced in (13.14). We have seen in Sec. 17.3.9 that the imaginary part of f remains as the U(1) gauge transformation T; see (17.107). This transformation acts on the fermions, and is thus indeed the relevant transformation for the issues that we consider here.

In the discussion of the monopoles we saw that the quantization condition depends on the coupling of the gauge field to charged particles. If there are fields that couple to ω_a with a covariant derivative of the form[16]

$$\mathcal{D}_a = \partial_a - \mathrm{i}q'\,\omega_a\,, \tag{17.164}$$

the argument of the Dirac quantization condition says that

$$q'\iint\Omega = 4\pi\,n\,. \tag{17.165}$$

As in the case of magnetic monopoles, all 'Kähler charges' must be a multiple of a minimum charge q' when the manifold has non-contractible 2-cycles. It is this minimal charge that gives the strongest condition on allowed Kähler metrics. Kähler metrics that satisfy such a condition are called Kähler–Hodge manifolds.

To compare the result to (17.152) note that the coupling

$$q\mathrm{i}A_\mu = -\tfrac{1}{2}qa^{-1}\partial_\mu z^\alpha\partial_\alpha\mathcal{K} + \ldots = -\tfrac{1}{3}\mathrm{i}q\kappa^2\partial_\mu z^\alpha\omega_\alpha + \ldots \tag{17.166}$$

leads to $q' = \frac{1}{3}q\kappa^2 = \frac{1}{2}\kappa^2$.

Let us express this in a more mathematical fashion. The form defining the Kähler manifold is Ω, called the Kähler form. It is a closed 2-form. Hence, it defines a cohomology that is called the Kähler class of the manifold. All metrics that differ by an exact form are in the same class. Therefore an integral over a closed 2-surface gives a characteristic number for this class.

The Kähler transformations act on other objects as U(1) transformations, which means that they define a line bundle. In a line bundle there is a distinguished Chern class called

[16] Here q' takes the role of q in (17.155), but has dimension $[M]^{-2}$, since the Kähler potential, and hence ω_a, has dimension $[M]^2$. See Sec. 18.3.1 for more information about the engineering dimensions of different quantities.

the first Chern class. The first Chern class means the set of curvature 2-forms defined up to addition of an exact differential form. These 2-forms are the $q'R^{(2)}$, where $R^{(2)}$ is defined in (17.163) and the factor q' in (17.164). The Chern class of any U(1) bundle belongs to the integral cohomology class, which means that by proper normalization the integrals over closed curves give integers. This is the consistency condition that we discussed above. The equation (17.163) says that this cohomology is equal to the Kähler class. The equality of the Kähler class to the Chern class of a U(1) line bundle is the mathematical definition of a *Kähler–Hodge manifold* (also called *Kähler manifold of restricted type*) [163, 164].

Definition: A Kähler manifold $\mathcal{M}$ is a Kähler–Hodge manifold if and only if there exists a line bundle $L \to \mathcal{M}$ such that $c_1(L) = [\mathcal{K}]$ where c_1 denotes the first Chern class and $[\mathcal{K}]$ denotes the Kähler class.

If several line bundles exist, having different U(1) charges that are multiples of a basic unit, then the most stringent condition is given by a bundle having this unit charge.

It has recently been remarked [166] that one can consider topological non-trivial target spaces and restrict the configuration space to a sector with a fixed topological charge k. In that case the quantization condition changes in the sense that n is replaced by n/k.

Appendix 17B Steps in the derivation of (17.7)

In this appendix we present more details of the derivation of (17.7), which is a master formula for the construction of the action. This derivation illustrates also the use of some formulas that have been derived earlier. We insert the expression for the covariant derivatives and covariant d'Alembertian (17.6) in (17.5). We need then some expressions for the dependent fields. For example, (16.13) and (16.26) lead to

$$\omega_a{}^{ab} = \omega_a{}^{ab}(e) + 3b^b + \tfrac{1}{2}\bar{\psi}\cdot\gamma\psi^b\,, \qquad f_a{}^a = -\tfrac{1}{12}\widehat{R}\,. \tag{17.167}$$

After deleting a total derivative (using the torsionless connection), we thus get

$$\begin{aligned} S_D = &\int \mathrm{d}^4x\, e\,\Big[D + \tfrac{1}{2}\bar{\psi}\cdot\gamma\psi^b\mathcal{D}_bC - \tfrac{1}{3}C\,\widehat{R} + \tfrac{1}{2}\mathrm{i}\bar{\phi}\cdot\gamma\gamma_*\zeta \\ &+\Big(\tfrac{1}{2}\mathrm{i}\bar{\psi}\cdot\gamma P_R\lambda + \tfrac{1}{2}\mathrm{i}\bar{\psi}_a\gamma^{ab}\mathcal{D}_bP_L\zeta - \tfrac{1}{4}\bar{\psi}_\mu P_L\gamma^{\mu\nu}\psi_\nu\mathcal{H} + \text{h.c.}\Big)\Big] \qquad (17.168)\\ = &\int \mathrm{d}^4x\, e\,\Big[D + \tfrac{1}{2}\bar{\psi}\cdot\gamma\psi^b\mathcal{D}_bC - \tfrac{1}{3}C\,R(\omega) + \tfrac{1}{2}\mathrm{i}\bar{\phi}\cdot\gamma\gamma_*\zeta - \tfrac{2}{3}C\,\bar{\psi}_a\gamma^{ab}\phi_b \\ &+\Big(\tfrac{1}{2}\mathrm{i}\bar{\psi}\cdot\gamma P_R\lambda + \tfrac{1}{2}\mathrm{i}\bar{\psi}_a\gamma^{ab}D_bP_L\zeta + \tfrac{1}{4}\mathrm{i}\bar{\psi}_\mu P_L\gamma^{\mu\nu}(\not{B} + \mathrm{i}\not{\mathcal{D}}C)\psi_\nu + \text{h.c.}\Big)\Big]\,. \end{aligned}$$

In the second expression, we used the value of $\widehat{R}$ from (16.26), and we decomposed the covariant derivative $\mathcal{D}_a\zeta$ from (17.6) into its linear part in ζ, i.e. $D_a\zeta$ and the remainder. We then use the symmetry and reality properties of spinor bilinears and (3.43). We further also use the last of (16.26) to obtain

$$S_D = \int d^4x\, e \Big[D - \tfrac{1}{2}i\bar{\psi}\cdot\gamma\gamma_*\lambda - \tfrac{1}{3}C\,R(\omega) + \tfrac{1}{2}i\bar{\phi}\cdot\gamma\gamma_*\zeta + \tfrac{1}{6}C\bar{\psi}_a\gamma^{abc}R'_{bc}(Q)$$
$$+\tfrac{1}{2}i\bar{\psi}_a\gamma^{ab}\gamma_* D_b\zeta + \tfrac{1}{4}\varepsilon^{abcd}\bar{\psi}_a\gamma_d\psi_b B_c\Big]\,. \qquad (17.169)$$

After use of the second equation of (16.26), we are very close to our goal, which is (17.7). What remains to be proven is that

$$\tfrac{1}{2}i\bar{\psi}_a\gamma^{ab}\gamma_* D_b\zeta = -\tfrac{1}{4}i\bar{\zeta}\gamma^{\rho\sigma}\gamma_* R'_{\rho\sigma}(Q) - \tfrac{1}{8}\varepsilon^{abcd}\bar{\psi}_a\gamma_b\psi_c\bar{\psi}_d\zeta + \text{total derivative}\,. \quad (17.170)$$

The remainder of the appendix is the proof of this equation.

The leading term with $(\partial_b + \frac{1}{4}\omega_{bcd}\gamma^{cd})\zeta$ is equal to the leading derivative term in $R'_{\rho\sigma}(Q)$ after integration by parts. To complete the job we must consider the torsion. The explicit b_μ terms in the covariant derivative and in $R'_{\rho\sigma}(Q)$ must be combined with those in (15.23), and then can be checked to agree on both sides of (17.170). It is more difficult to handle the ψ^3 terms for which we must show that

$$i\omega_a{}^{cd}(\psi)\left[\gamma_{cd}, \gamma^{ab}\right]\gamma_*\psi_b = \varepsilon^{abcd}\psi_d\bar{\psi}_a\gamma_b\psi_c\,, \qquad (17.171)$$

where $\omega(\psi)$ indicates the torsion terms only. The left-hand side is

$$\begin{aligned} 8i\gamma_c{}^b\gamma_*\omega_{[a}{}^{ca}(\psi)\psi_{b]} &= -4\gamma_{de}\varepsilon_c{}^{bde}\omega_{[a}{}^{ca}(\psi)\psi_{b]} \\ &= \gamma_{de}\varepsilon^{cbde}\psi_b\bar{\psi}_a\gamma^a\psi_c - \tfrac{1}{2}\gamma_{de}\varepsilon^{cbde}\psi_a\bar{\psi}_b\gamma^a\psi_c \\ &= \tfrac{5}{2}\gamma_{[de}\varepsilon^{cbde}\psi_b\bar{\psi}_a\gamma^a\psi_{c]} + \gamma_{da}\varepsilon^{cbde}\psi_b\bar{\psi}_e\gamma^a\psi_c\,. \end{aligned} \quad (17.172)$$

To understand the step to the last line, one has to write out the antisymmetrization in the lower five indices. For that, one just has to distinguish the five possible positions of the index a, since in the upper indices there is already a factor ε^{cbde}. If a goes on a gravitino, we find the terms that are in the second line of (17.172) (two of the three terms are equal, owing to the antisymmetry of the gravitino bilinear). When a sits on the leading γ-matrix, it gives the term that we compensate for in the last line.

The first term is antisymmetric in the five lower indices and thus vanishes (note the Schouten identity (3.11)). Finally, we write

$$\gamma_{da}\varepsilon^{cbde}\psi_b\bar{\psi}_e\gamma^a\psi_c = \gamma_d\gamma_a\varepsilon^{cbde}\psi_b\bar{\psi}_e\gamma^a\psi_c - \varepsilon^{cbae}\psi_b\bar{\psi}_e\gamma_a\psi_c\,. \qquad (17.173)$$

The first term vanishes due to the Fierz identity of (3.68). After relabeling indices the second term agrees with the right-hand side of (17.171). This finishes the proof.

PART VI

$\mathcal{N} = 1$ SUPERGRAVITY ACTIONS AND APPLICATIONS

The physical $\mathcal{N} = 1$ matter-coupled supergravity

18

In the previous chapter we applied superconformal methods and the mathematics of projective Kähler manifolds to derive the ingredients of the general $\mathcal{N} = 1$ matter-coupled supergravity theory. In this chapter we assemble these ingredients and present the action and transformation rules in a self-contained form suitable for applications. The superconformal methodology provides many insights into the intricate structures that appear in the final theory, but, like the scaffolding for a new building, it is needed for construction but can be removed after human use begins.

Each of the elements of the general $\mathcal{N} = 1$ global SUSY theory of Ch. 14 appears in the final supergravity theory equipped with the extra structure required for invariance under local SUSY with arbitrary spinor transformation parameters $\epsilon(x)$. We begin by reviewing those elements or inputs.

1. The choice of a set of chiral multiplets, z^α, χ^α, enumerated by the index α, whose kinetic terms are determined by the Kähler potential $\mathcal{K}(z, \bar{z})$. The metric of the scalar target space is invariant under a Kähler transformation, viz.

$$g_{\alpha\bar{\beta}} = \partial_\alpha \partial_{\bar{\beta}} \mathcal{K}(z, \bar{z}) , \qquad \mathcal{K}(z, \bar{z}) \to \mathcal{K}'(z, \bar{z}) = \mathcal{K}(z, \bar{z}) + f(z) + \bar{f}(\bar{z}) . \qquad (18.1)$$

   Other elements of the theory also transform so that the overall structure is Kähler invariant.

2. The choice of a set of gauge multiplets, $A_\mu{}^A$, λ^A, enumerated by the index A. Gauge multiplet kinetic terms are determined by holomorphic functions $f_{AB}(z) = f_{BA}(z)$.

3. One can choose a holomorphic superpotential of the scalar fields, $W(z)$. The scalar potential $V(z, \bar{z})$, fermion mass terms and couplings depend on $W(z)$ and its complex conjugate. $W(z)$ is a scalar under coordinate transformations on the target space, and it transforms as $W(z) \to \mathrm{e}^{-\kappa^2 f(z)} W(z)$ under Kähler transformations.

4. One can choose a Lie group with structure constants $f_{AB}{}^C$ as the gauge group G of the theory provided that $\mathcal{K}$, f_{AB} and W obey suitable conditions. The generators of the group are associated with (a subset of) the gauge vectors A_μ^A and gauge parameters $\theta^A(x)$. The gauge transformations of chiral multiplet scalars involve holomorphic Killing vectors, i.e. $\delta z^\alpha = \theta^A k_A^\alpha(z)$, and the latter are determined from real scalar moment maps $\mathcal{P}_A(z, \bar{z})$ which satisfy

$$k_A{}^\alpha(z) = -\mathrm{i} g^{\alpha\bar{\beta}} \partial_{\bar{\beta}} \mathcal{P}_A(z, \bar{z}) , \qquad \nabla_\alpha \partial_\beta \mathcal{P}_A(z, \bar{z}) = 0; \qquad (18.2)$$

see Sec. 13.4.1. The covariant derivative ∇_α contains the Levi-Civita connection of the Kähler metric in (18.1). The moment maps must also form a closed algebra, which is the equivariance relation (13.62):

$$k_A{}^\alpha g_{\alpha\bar\beta} k_B^{\bar\beta} - k_B^\alpha g_{\alpha\bar\beta} k_A{}^{\bar\beta} = \mathrm{i} f_{AB}{}^C \mathcal{P}_C . \tag{18.3}$$

The Kähler potential $\mathcal{K}$ need not be invariant, but can change by a Kähler transformation

$$\delta\mathcal{K} = \theta^A \left[r_A(z) + \bar r_A(\bar z) \right] , \qquad r_A = k_A{}^\alpha \partial_\alpha \mathcal{K} + \mathrm{i}\mathcal{P}_A . \tag{18.4}$$

The kinetic matrix $f_{AB}(z)$ must be compatible with the gauge group. We require that it transform in the bi-adjoint representation of G, as written[1] in (14.56) with $C_{AB,C} = 0$. Furthermore the gauge transformation of the superpotential should satisfy (17.133),

$$W_\alpha k_A{}^\alpha = -\kappa^2 r_A W . \tag{18.5}$$

18.1 The physical action

The physical action of the general $\mathcal{N} = 1$ supergravity theory is the spacetime integral of a Lagrangian that is obtained from (17.30) by incorporating the many steps of the previous chapter. It takes the form[2]

$$\begin{aligned}
e^{-1}\mathcal{L} = {}& \frac{1}{2\kappa^2}\left[R(e) - \bar\psi_\mu R^\mu \right] \\
& - g_{\alpha\bar\beta}\left[\hat\partial_\mu z^\alpha \hat\partial^\mu \bar z^{\bar\beta} + \tfrac12 \bar\chi^\alpha \not{D}^{(0)} \chi^{\bar\beta} + \tfrac12 \bar\chi^{\bar\beta} \not{D}^{(0)} \chi^\alpha \right] - V \\
& + (\operatorname{Re} f_{AB})\left[-\tfrac14 F_{\mu\nu}^A F^{\mu\nu\, B} - \tfrac12 \bar\lambda^A \not{D}^{(0)} \lambda^B \right] \\
& + \tfrac14 \mathrm{i}\left[(\operatorname{Im} f_{AB}) F_{\mu\nu}^A \tilde F^{\mu\nu\, B} + (\hat\partial_\mu \operatorname{Im} f_{AB}) \bar\lambda^A \gamma_* \gamma^\mu \lambda^B \right] \\
& + \tfrac18 (\operatorname{Re} f_{AB}) \bar\psi_\mu \gamma^{ab} \left(F_{ab}^A + \widehat F_{ab}^A \right) \gamma^\mu \lambda^B + \left[\frac{1}{\sqrt2} g_{\alpha\bar\beta} \bar\psi_\mu \hat{\not\partial} \bar z^{\bar\beta} \gamma^\mu \chi^\alpha + \text{h.c.} \right] \\
& + \left[\frac{1}{4\sqrt2} f_{AB\,\alpha} \bar\lambda^A \gamma^{ab} \widehat F_{ab}^B \chi^\alpha + \text{h.c.} \right] + \mathcal{L}_m + \mathcal{L}_{\text{mix}} + \mathcal{L}_{4\text{f}} .
\end{aligned} \tag{18.6}$$

The first line contains the graviton and gravitino kinetic terms with R_μ defined from (17.24) as

$$R^\mu \equiv \gamma^{\mu\rho\sigma} \left(\partial_\rho + \tfrac14 \omega_\rho{}^{ab}(e) \gamma_{ab} - \tfrac32 \mathrm{i} \mathcal{A}_\rho \gamma_* \right) \psi_\sigma . \tag{18.7}$$

The second line includes kinetic terms for chiral multiplets with covariant derivatives defined by

[1] The generalization [109] to $C_{AB,C} \neq 0$ (with a Chern–Simons term) is omitted for simplicity.

[2] Chiral multiplet spinors χ^α are to be understood as the projection with $P_L = \frac12(1+\gamma_*)$ of a four-component Majorana spinor, while $\chi^{\bar\alpha}$ is the projection with $P_R = \frac12(1-\gamma_*)$. The gravitino field ψ_μ and the gauginos λ^A are Majorana spinors. Conjugate spinors are defined in (3.50).

$$\hat{\partial}_\mu z^\alpha = \partial_\mu z^\alpha - A_\mu^A k_A{}^\alpha\,, \tag{18.8}$$

$$D_\mu^{(0)}\chi^\alpha = \left(\partial_\mu + \tfrac{1}{4}\omega_\mu{}^{ab}(e)\gamma_{ab} + \tfrac{3}{2}\mathrm{i}\mathcal{A}_\mu\right)\chi^\alpha - A_\mu^A \frac{\partial k_A{}^\alpha(z)}{\partial z^\beta}\chi^\beta + \Gamma^\alpha_{\beta\gamma}\chi^\gamma\hat{\partial}_\mu z^\beta\,.$$

The scalar covariant derivative contains only the Yang–Mills connection. The spinor derivative contains the Yang–Mills gauge and Christoffel connection $\Gamma^\alpha_{\beta\gamma}$ terms already present in global supersymmetry, plus Lorentz spin connection and Kähler connection as in (17.113):

$$\begin{aligned}
\Gamma^\alpha_{\beta\gamma} &= g^{\alpha\bar\delta}\partial_\beta g_{\gamma\bar\delta}\,,\\
\mathcal{A}_\mu &= \tfrac{1}{6}\mathrm{i}\kappa^2\left(\partial_\mu z^\alpha\partial_\alpha\mathcal{K} - \partial_\mu\bar z^{\bar\alpha}\partial_{\bar\alpha}\mathcal{K}\right) - \tfrac{1}{3}\kappa^2 A_\mu^A\mathcal{P}_A\\
&= \tfrac{1}{6}\mathrm{i}\kappa^2\left[\hat\partial_\mu z^\alpha\partial_\alpha\mathcal{K} - \hat\partial_\mu\bar z^{\bar\alpha}\partial_{\bar\alpha}\mathcal{K} + A_\mu^A(r_A - \bar r_A)\right].
\end{aligned} \tag{18.9}$$

The Kähler connection transforms under gauge and Kähler transformations as

$$\delta\mathcal{A}_\mu = -\tfrac{1}{3}\kappa^2\partial_\mu\left(\theta^A\,\mathrm{Im}\,r_A + \mathrm{Im}\,f\right). \tag{18.10}$$

The second line also includes the scalar potential V of (17.135):

$$\begin{aligned}
V &= V_- + V_+\,,\\
V_- &= -3\kappa^2 e^{\kappa^2\mathcal{K}}W\overline{W}\,, \qquad V_+ = e^{\kappa^2\mathcal{K}}\nabla_\alpha W g^{\alpha\bar\beta}\overline{\nabla}_{\bar\beta}\overline{W} + \tfrac{1}{2}\left(\mathrm{Re}\,f\right)^{-1\,AB}\mathcal{P}_A\mathcal{P}_B\,.
\end{aligned} \tag{18.11}$$

Note that W appears with its Kähler covariant derivative

$$\nabla_\alpha W(z) = \partial_\alpha W(z) + \kappa^2(\partial_\alpha\mathcal{K})W(z)\,. \tag{18.12}$$

The third and fourth lines contain the gravitational extension of the gauge multiplet kinetic terms. Their form is determined by global SUSY; see (14.57). The derivative $D_\mu^{(0)}\lambda^A$ is defined in (17.23):

$$D_\mu^{(0)}\lambda^A = \left(\partial_\mu + \tfrac{1}{4}\omega_\mu{}^{ab}(e)\gamma_{ab} - \tfrac{3}{2}\mathrm{i}\mathcal{A}_\mu\gamma_*\right)\lambda^A - A_\mu^C\lambda^B f_{BC}{}^A\,. \tag{18.13}$$

The fifth line contains terms of the form $\bar\psi_\mu\mathcal{J}^\mu$ where $\mathcal{J}^\mu$ is the basic supercurrent of the gauge (see (6.10)) and chiral (see (6.27)) multiplets. It is a general feature of Lagrangians with gauged symmetries that a gauge field appears contracted with the Noether current of the global limit of the theory. However, as the reader now can see, much more is required to gauge global SUSY!

The supercovariant gauge curvature $\widehat{F}_{ab}$ is given in (17.1):

$$\widehat{F}_{ab}{}^A = e_a{}^\mu e_b{}^\nu\left(2\partial_{[\mu}A_{\nu]}{}^A + f_{BC}{}^A A_\mu{}^B A_\nu{}^C + \bar\psi_{[\mu}\gamma_{\nu]}\lambda^A\right). \tag{18.14}$$

It contains a fermion bilinear, which ensures that it transforms under local SUSY without derivatives of $\epsilon(x)$. This was shown in Ex. 11.14; see (11.88). It is well known that a symmetry transforms the Euler–Lagrange equations for the various fields of a theory into one another. This is true in a gauge theory only if these equations of motion transform without derivatives of the gauge parameters. The structure $(F + \hat{F})$ as well as other terms in (18.6) ensure that the supergravity equations of motion are supercovariant. This was shown in detail in Ch. 10.

Let's discuss the last line of (18.6). The first term[3] is already part of the global SUSY Lagrangian (14.29). The last line also includes bilinear mass and mixing terms, plus quartic couplings for fermion fields. The fermion mass Lagrangian is given by

$$\begin{aligned}\mathcal{L}_m = {} & \tfrac{1}{2} m_{3/2} \bar{\psi}_\mu P_R \gamma^{\mu\nu} \psi_\nu \\ & - \tfrac{1}{2} m_{\alpha\beta} \bar{\chi}^\alpha \chi^\beta - m_{\alpha A} \bar{\chi}^\alpha \lambda^A - \tfrac{1}{2} m_{AB} \bar{\lambda}^A P_L \lambda^B + \text{h.c.}\end{aligned} \tag{18.15}$$

The (complex) gravitino mass parameter[4] and spin-1/2 mass matrices are

$$m_{3/2} = \kappa^2 e^{\kappa^2 \mathcal{K}/2} W\,, \tag{18.16}$$

$$\begin{aligned} m_{\alpha\beta} &= e^{\kappa^2\mathcal{K}/2} \nabla_\alpha \nabla_\beta W \equiv e^{\kappa^2\mathcal{K}/2}\left[\partial_\alpha + \left(\kappa^2 \partial_\alpha \mathcal{K}\right)\right] \nabla_\beta W - e^{\kappa^2\mathcal{K}/2} \Gamma^\gamma_{\alpha\beta} \nabla_\gamma W\,, \\ m_{\alpha A} &= i\sqrt{2}\left[\partial_\alpha \mathcal{P}_A - \tfrac{1}{4} f_{AB\,\alpha} \left(\operatorname{Re} f\right)^{-1\,BC} \mathcal{P}_C\right] = m_{A\alpha}\,, \\ m_{AB} &= -\tfrac{1}{2} e^{\kappa^2\mathcal{K}/2} f_{AB\,\alpha} g^{\alpha\bar{\beta}} \overline{\nabla}_{\bar{\beta}} \overline{W}\,. \end{aligned} \tag{18.17}$$

Terms that mix the gravitino and spin-1/2 fermions are included in

$$\mathcal{L}_{\text{mix}} = \bar{\psi} \cdot \gamma \left[\frac{1}{2} i P_L \lambda^A \mathcal{P}_A + \frac{1}{\sqrt{2}} \chi^\alpha e^{\kappa^2\mathcal{K}/2} \nabla_\alpha W\right] + \text{h.c.} \tag{18.18}$$

Finally the four-fermion terms contain the gravitino torsion terms $\mathcal{L}_{\text{SG,torsion}}$ in (9.19) plus other terms. The full expression is

$$\begin{aligned} \mathcal{L}_{4\text{f}} = {} & \tfrac{1}{2}\kappa^{-2} \mathcal{L}_{\text{SG,torsion}} \\ & + \left\{ -\frac{1}{4\sqrt{2}} f_{AB\,\alpha} \bar{\psi} \cdot \gamma \chi^\alpha \bar{\lambda}^A P_L \lambda^B + \frac{1}{8} (\nabla_\alpha f_{AB\,\beta}) \bar{\chi}^\alpha \chi^\beta \bar{\lambda}^A P_L \lambda^B + \text{h.c.} \right\} \\ & + \tfrac{1}{16} i e^{-1} \varepsilon^{\mu\nu\rho\sigma} \bar{\psi}_\mu \gamma_\nu \psi_\rho \left(\tfrac{1}{2} \operatorname{Re} f_{AB} \bar{\lambda}^A \gamma_* \gamma_\sigma \lambda^B + g_{\alpha\bar{\beta}} \bar{\chi}^{\bar{\beta}} \gamma_\sigma \chi^\alpha \right) - \tfrac{1}{2} g_{\alpha\bar{\beta}} \bar{\psi}_\mu \chi^{\bar{\beta}} \bar{\psi}^\mu \chi^\alpha \\ & + \tfrac{1}{4} \left(R_{\alpha\bar{\gamma}\beta\bar{\delta}} - \tfrac{1}{2}\kappa^2 g_{\alpha\bar{\gamma}} g_{\beta\bar{\delta}} \right) \bar{\chi}^\alpha \chi^\beta \bar{\chi}^{\bar{\gamma}} \chi^{\bar{\delta}} \\ & + \tfrac{3}{64} \kappa^2 \left[(\operatorname{Re} f_{AB}) \bar{\lambda}^A \gamma_\mu \gamma_* \lambda^B \right]^2 - \tfrac{1}{16} f_{AB\,\alpha} \bar{\lambda}^A P_L \lambda^B g^{\alpha\bar{\beta}} \bar{f}_{CD\,\bar{\beta}} \bar{\lambda}^C P_R \lambda^D \\ & + \tfrac{1}{16} (\operatorname{Re} f)^{-1\,AB} \left(f_{AC\,\alpha} \bar{\chi}^\alpha - \bar{f}_{AC\,\bar{\alpha}} \bar{\chi}^{\bar{\alpha}} \right) \lambda^C \left(f_{BD\,\beta} \bar{\chi}^\beta - \bar{f}_{BD\,\bar{\beta}} \bar{\chi}^{\bar{\beta}} \right) \lambda^D \\ & - \tfrac{1}{4} \kappa^2 g_{\alpha\bar{\beta}} (\operatorname{Re} f_{AB}) \bar{\chi}^\alpha \lambda^A \bar{\chi}^{\bar{\beta}} \lambda^B\,. \end{aligned} \tag{18.19}$$

Exercise 18.1 *Derive the following equation for the gradient of the potential:*

$$\partial_\alpha V = e^{\kappa^2\mathcal{K}/2}\left(-2\overline{m}_{3/2} \nabla_\alpha W + m_{\alpha\beta} g^{\beta\bar{\beta}} \overline{\nabla}_{\bar{\beta}} \overline{W}\right) - \frac{i}{\sqrt{2}} \mathcal{P}_A (\operatorname{Re} f)^{-1\,AB} m_{\alpha B}\,. \tag{18.20}$$

This is an important equation since the condition $\partial_\alpha V = 0$ determines the possible vacuum states of the theory. Commutators of Kähler covariant derivatives, as in (17.78), are helpful in the proof.

[3] Observe that the F^B_{ab} can be written as F^{-B}_{ab} due to the chirality of χ^α; see Ex. 4.7.

[4] The generation of a physical mass for the gravitino will be discussed in Sec. 19.1.1.

Exercise 18.2 *Derive that the F-term potential can be rewritten in terms of the Kähler invariant function $\mathcal{G}(z,\bar{z})$, (17.132), as*

$$V_F = \kappa^{-4}\mathrm{e}^{\mathcal{G}}\left(\mathcal{G}_\alpha \mathcal{G}^{\alpha\bar{\beta}}\mathcal{G}_{\bar{\beta}} - 3\right), \tag{18.21}$$

where $\mathcal{G}_\alpha = \partial_\alpha \mathcal{G}, \ldots$ and $\mathcal{G}^{\alpha\bar{\beta}}$ is the inverse of $\mathcal{G}_{\alpha\bar{\beta}} = \kappa^2 g_{\alpha\bar{\beta}}$.

18.2 Transformation rules

The superconformal approach of Chs. 16 and 17 guarantees that the action integral of the Lagrangian (18.6) is invariant under transformation rules of local supersymmetry. These rules are obtained from the transformations of the superconformal algebra given in (16.10), (17.1) and (17.3). One must reexpress (17.3) in terms of the physical chiral multiplet fields, and incorporate the gauge fixing conditions used in Ch. 17 and the residual decomposition laws of Sec. 17.3.9. This procedure leads to the physical form of the transformation rules:

$$\begin{aligned}
\delta e_\mu^a &= \tfrac{1}{2}\bar{\epsilon}\gamma^a\psi_\mu\,,\\
\delta P_L\psi_\mu &= \left(\partial_\mu + \tfrac{1}{4}\omega_\mu{}^{ab}(e)\gamma_{ab} - \tfrac{3}{2}\mathrm{i}\mathcal{A}_\mu\right)P_L\epsilon + \tfrac{1}{2}\kappa^2\gamma_\mu \mathrm{e}^{\kappa^2\mathcal{K}/2}W P_R\epsilon\\
&\quad + \tfrac{1}{4}\kappa^2 P_L\psi_\mu\theta^A(\bar{r}_A - r_A) + \text{cubic in fermions},\\
\delta z^\alpha &= \frac{1}{\sqrt{2}}\bar{\epsilon}\chi^\alpha\,,\\
\delta\chi^\alpha &= \frac{1}{\sqrt{2}}P_L\left(\hat{\not\partial} z^\alpha - \mathrm{e}^{\kappa^2\mathcal{K}/2}g^{\alpha\bar{\beta}}\overline{\nabla}_{\bar{\beta}}\overline{W}\right)\epsilon\\
&\quad + \theta^A\left[\frac{\partial k_A{}^\alpha}{\partial z^\beta}\chi^\beta + \frac{1}{4}\kappa^2(r_A - \bar{r}_A)\chi^\alpha\right] + \text{cubic in fermions},\\
\delta A_\mu^A &= -\tfrac{1}{2}\bar{\epsilon}\gamma_\mu\lambda^A + \partial_\mu\theta^A + \theta^C A_\mu^B f_{BC}{}^A\,,\\
\delta\lambda^A &= +\left[\tfrac{1}{4}\gamma^{\mu\nu}F_{\mu\nu}{}^A + \tfrac{1}{2}\mathrm{i}\gamma_*(\operatorname{Re} f)^{-1\,AB}\mathcal{P}_B\right]\epsilon\\
&\quad + \theta^B\left[\lambda^C f_{CB}{}^A + \tfrac{1}{4}\kappa^2\gamma_*(\bar{r}_B - r_B)\lambda^A\right] + \text{cubic in fermions}.
\end{aligned} \tag{18.22}$$

The phrase 'cubic in fermions' means terms that are quadratic in fermion fields multiplied by ϵ. These terms are omitted because they are not needed in most applications. Observe that the gauge transformations of the fermions include a term proportional to $\operatorname{Im} r_A$ due to the decomposition law (17.107). Fermion fields also transform under Kähler transformations,

$$\begin{aligned}
&P_L\psi_\mu \to \mathrm{e}^{-\mathrm{i}\kappa^2(\operatorname{Im} f)/2}P_L\psi_\mu\,,\\
&P_L\lambda^A \to \mathrm{e}^{-\mathrm{i}\kappa^2(\operatorname{Im} f)/2}P_L\lambda^A\,, \qquad \chi^\alpha \to \mathrm{e}^{\mathrm{i}\kappa^2(\operatorname{Im} f)/2}\chi^\alpha\,.
\end{aligned} \tag{18.23}$$

In Sec. 14.5.2 we showed that the scalar potential in global SUSY can be written as a sum of squares of the scalar parts of the transformation laws of the fermion fields χ^α and λ^A. This structure is valid for the supergravity potential (18.11) if one includes the scalar part of the gravitino transformation.[5] Specifically, the scalar parts of the transformation rules ('fermion shifts') are defined by

$$\delta_s \chi^\alpha = -\frac{1}{\sqrt{2}} e^{\kappa^2 \mathcal{K}/2} g^{\alpha\bar\beta} \overline{\nabla}_{\bar\beta} \overline{W} , \qquad \delta_s P_L \lambda^A = \frac{1}{2} \mathrm{i} (\operatorname{Re} f)^{-1\,AB} \mathcal{P}_B ,$$
$$\delta_s P_L \psi_\mu = \tfrac{1}{2} \kappa^2 e^{\kappa^2 \mathcal{K}/2} W \gamma_\mu . \qquad (18.24)$$

The potential (17.135) can then be expressed as the 'quadratic form'

$$V = -\tfrac{1}{2}(D-2)(D-1)\kappa^2 e^{\kappa^2 \mathcal{K}} |W|^2 \qquad (18.25)$$
$$+2 \left(\delta_s P_L \chi^\alpha\right) g_{\alpha\bar\beta} \left(\delta_s P_R \chi^{\bar\beta}\right) + 2 \left(\delta_s P_L \lambda^A\right) \operatorname{Re} f_{AB} \left(\delta_s P_R \lambda^B\right) .$$

We are interested in the case $D=4$, but the result is much more general. It is valid for supergravity with scalars in higher dimensions. In this case the coefficient of the gravitino term arises from the contraction with the gravitino kinetic term $\gamma_\mu \gamma^{\mu\rho\sigma} \gamma_\sigma \partial_\rho = (D-2)(D-1)\partial\!\!\!/$. A similar form is also valid for extended $\mathcal{N}>1$ supergravity (using any one of the $\mathcal{N}$ supersymmetries) [167, 168, 169]. The derivation of the potential using fermion shifts (18.25) is explained in [170].

18.3 Further remarks

18.3.1 Engineering dimensions

Scalar fields z^α can appear nonlinearly in the Lagrangian, and it is most convenient to take them dimensionless. The kinetic term $g_{\alpha\bar\beta} \partial_\mu z^\alpha \partial^\mu \bar z^{\bar\beta}$ implies that $g_{\alpha\bar\beta}$ must have mass dimension 2, so the Kähler potential $\mathcal{K}(z,\bar z)$ carries dimensions of mass squared, $[M]^2$. In particular models where the Kähler potential is homogeneous, as in the simple but important case $\mathcal{K} = \kappa^{-2} z^\alpha \delta_{\alpha\bar\beta} \bar z^{\bar\beta}$, one can redefine $z' = \kappa^{-1} z$ to avoid factors κ in the kinetic term. The dimensions of various quantities are

$$\mathcal{K}\colon [M]^2 , \qquad g_{\alpha\bar\beta}\colon [M]^2 , \qquad W\colon [M]^3 , \qquad \kappa\colon [M]^{-1} ,$$
$$A_\mu\colon [M], \qquad r_A\colon [M]^2 , \qquad \mathcal{P}_A\colon [M]^2 , \qquad f_{AB}\colon [M]^0 . \qquad (18.26)$$

18.3.2 Rigid or global limit

The normalizations of the independent functions $\mathcal{K}$, W, f_{AB}, $\mathcal{P}_A$ or r_A have been chosen so that the global supersymmetry limit is obtained by taking $\kappa = 0$ after rescaling $\psi_\mu \to \kappa \psi_\mu$.

[5] Note that in the superconformal framework the gravitino shift is absent, but the method still works; see Ex. 17.4.

18.3.3 Quantum effects and global symmetries

The structure of the *classical* supergravity theory was discussed in this chapter. Perturbative quantum corrections can impose further restrictions on this structure. These restrictions include anomaly cancelation and quantization of black hole charges.

A further restriction on supergravity models occurs when one requires the quantum mechanical consistency of black hole solutions. A folk theorem says that the final theory must not have any continuous global (i.e. ungauged) symmetries [34, 171, 172].

19 Applications of $\mathcal{N} = 1$ supergravity

In this chapter we want to discuss some aspects of $\mathcal{N} = 1$ supergravity that are important for applications. This will also improve our understanding of the structure of supergravity theories.

19.1 Supersymmetry breaking and the super-BEH effect

We discussed supersymmetry breaking in global supersymmetry in Sec. 14.5. The main result is that supersymmetry is broken when at least one auxiliary field does not vanish at the minimum of the potential. Since the potential is positive definite, this implies that the vacuum energy is positive. A massless fermion, the Goldstino, is characteristic of spontaneously broken global SUSY.

We expect two main changes in this scenario in supergravity. Since supersymmetry is a gauge symmetry, a mechanism similar to the Brout–Englert–Higgs (BEH) effect in conventional gauge theories should give mass to the gauge field, namely to the gravitino, and the Goldstone fermion should disappear from the set of physical fields. This mechanism is called the super-BEH effect. Secondly, we saw in Sec. 17.4.2 that the potential in supergravity also has a negative contribution and thus the result $V > 0$ for supersymmetry-breaking vacua can be circumvented.

The latter is good news. In a theory with gravity, a positive vacuum expectation value of V means that there is a positive cosmological constant. In most models this is far too big for any realistic cosmology. On the other hand, without the contribution of matter, the potential (17.135) has only the negative term V_-. In that case, we have the situation discussed at the end of Ch. 16, where we encountered 'anti-de Sitter' gravity. This scenario is also excluded by cosmology. It is thus fortunate that matter-coupled supergravity allows a delicate balance between matter contributions to V_+ and the supergravity terms of V_- such that the resulting cosmological constant is close to zero but positive.

19.1.1 Goldstino and the super-BEH effect

We discussed the Goldstino in global supersymmetry in Sec. 14.5.4, and derived the expression (14.89). There are corrections to this result in supergravity that can be obtained from the terms that mix the gravitino with other fermions in (18.18). We thus write[1]

[1] In superconformal variables it has the same form as in global supersymmetry, namely $P_L \upsilon = -\frac{1}{\sqrt{2}}\mathcal{W}_I \Omega^I - \frac{1}{2}\mathrm{i}\mathcal{P}_A P_L \lambda^A$.

$$\mathcal{L}_{\text{mix}} = -\bar{\psi}\cdot\gamma P_L \upsilon + \text{h.c.}\,,$$
$$P_L\upsilon = -\frac{1}{\sqrt{2}}\chi^\alpha e^{\kappa^2\mathcal{K}/2}\nabla_\alpha W - \frac{1}{2}\mathrm{i}P_L\lambda^A\mathcal{P}_A\,. \qquad (19.1)$$

Another useful form, similar to (14.90), involves the fermion shifts:

$$P_L\upsilon = \chi^\alpha\delta_{\text{s}}\chi_\alpha + P_L\lambda^A\delta_{\text{s}}P_R\lambda_A\,,$$
$$\delta_{\text{s}}\chi_\alpha = g_{\alpha\bar{\beta}}\delta_{\text{s}}\chi^{\bar{\beta}} = -\frac{1}{\sqrt{2}}e^{\kappa^2\mathcal{K}/2}\nabla_\alpha W\,,$$
$$\delta_{\text{s}}P_R\lambda_A = (\text{Re}\,f)_{AB}\delta_{\text{s}}P_R\lambda^B = -\frac{1}{2}\mathrm{i}\mathcal{P}_A\,. \qquad (19.2)$$

The supersymmetry transformation of the Goldstino is illuminating. One can check, using the transformations (18.22), and the value of the positive part of the potential (17.135), that

$$\delta P_L\upsilon = \tfrac{1}{2}V_+P_L\epsilon + \ldots\,, \qquad (19.3)$$

where the omitted terms involve vectors, derivatives of scalars, or are cubic in fermions. Since $V_+ > 0$ in a SUSY-breaking vacuum, the Goldstino undergoes a shift under supersymmetry transformations. One can then impose the convenient supersymmetry gauge condition

$$\text{supersymmetry gauge:}\quad \upsilon = 0\,, \qquad (19.4)$$

which eliminates υ from the theory.

This gauge choice eliminates the mixing of the gravitino with other fermions. The remaining gravitino mass term (18.15) has the same form found in the massive Rarita–Schwinger action in (5.48). Thus, in Minkowski space we can interpret $|m_{3/2}|$ as the physical mass of the gravitino, produced by the super-BEH effect.

Exercise 19.1 *When scalars are constant and other terms in the action are neglected, the gravitino terms become*

$$e^{-1}\mathcal{L} = \frac{1}{2\kappa^2}\bar{\psi}_\mu\left[-\gamma^{\mu\nu\rho}\partial_\nu + \left(m_{3/2}P_R + \overline{m}_{3/2}P_L\right)\gamma^{\mu\rho}\right]\psi_\rho - \bar{\psi}_\mu\gamma^\mu\upsilon\,. \qquad (19.5)$$

The purpose of this exercise is to diagonalize this Lagrangian by introducing the massive gravitino field

$$P_L\Psi_\mu = P_L\psi_\mu - \frac{2\kappa^2}{3|m_{3/2}|^2}\partial_\mu P_L\upsilon - \frac{\kappa^2}{3\overline{m}_{3/2}}\gamma_\mu P_R\upsilon\,. \qquad (19.6)$$

Check first that the additional terms correspond to a supersymmetry transformation with appropriately chosen parameter. Then check that the Lagrangian can be rewritten as

$$e^{-1}\mathcal{L} = \frac{1}{2\kappa^2}\bar{\Psi}_\mu\left[-\gamma^{\mu\nu\rho}\partial_\nu + \left(m_{3/2}P_R + \overline{m}_{3/2}P_L\right)\gamma^{\mu\rho}\right]\Psi_\rho$$
$$+\frac{\kappa^2}{3|m_{3/2}|^2}\left[\bar{\upsilon}\partial\!\!\!/\upsilon + 2\bar{\upsilon}\left(m_{3/2}P_R + \overline{m}_{3/2}P_L\right)\upsilon\right]. \qquad (19.7)$$

Consider a model with one chiral multiplet and no gauge multiplets. Show that the kinetic term of the Goldstino in (19.7) cancels the kinetic term of the chiral fermion if the potential vanishes, i.e.

$$\nabla_1 W \overline{\nabla}_{\bar{1}}\overline{W} = 3\kappa^2 g_{1\bar{1}}|W|^2. \qquad (19.8)$$

The elimination of the Goldstino leads to new contributions to the fermion mass matrices (18.17), which can be obtained by inserting the expression (19.2) in the last term in (19.7). The new contributions are

$$m^{(\upsilon)}_{\alpha\beta} = -\frac{4\kappa^2}{3m_{3/2}}\left(\delta_s\chi_\alpha\right)\left(\delta_s\chi_\beta\right),$$
$$m^{(\upsilon)}_{\alpha A} = -\frac{4\kappa^2}{3m_{3/2}}\left(\delta_s\chi_\alpha\right)\left(\delta_s P_R\lambda_A\right),$$
$$m^{(\upsilon)}_{AB} = -\frac{4\kappa^2}{3m_{3/2}}\left(\delta_s P_R\lambda_A\right)\left(\delta_s P_R\lambda_B\right). \qquad (19.9)$$

The full mass matrices are thus

$$m^{g}_{\alpha\beta} = m_{\alpha\beta} + m^{(\upsilon)}_{\alpha\beta}, \qquad m^{g}_{\alpha A} = m_{\alpha A} + m^{(\upsilon)}_{\alpha A}, \qquad m^{g}_{AB} = m_{AB} + m^{(\upsilon)}_{AB}. \qquad (19.10)$$

The derivative of the potential (18.20) and the gauge invariance condition on the superpotential (17.133) can be written as

$$m_{\alpha\beta}\delta_s\chi^\beta + m_{\alpha B}\delta_s P_L\lambda^B - 2\overline{m}_{3/2}\delta_s\chi_\alpha = -\frac{1}{\sqrt{2}}\partial_\alpha V$$
$$m_{\beta A}\delta_s\chi^\beta - 2\overline{m}_{3/2}\delta_s P_R\lambda_A = 0. \qquad (19.11)$$

Combining this with (19.9) and using (18.25) (and the gravitino mass from (18.16)) leads to

$$m^{g}_{\alpha\beta}\delta_s\chi^\beta + m^{g}_{\alpha A}\delta_s P_L\lambda^B + \frac{2\kappa^2}{3m_{3/2}}V\,\delta_s\chi_\alpha = -\frac{1}{\sqrt{2}}\partial_\alpha V,$$
$$m^{g}_{A\beta}\delta_s\chi^\beta + m^{g}_{AB}\delta_s P_L\lambda^B + \frac{2\kappa^2}{3m_{3/2}}V\,\delta_s P_R\lambda_A = 0. \qquad (19.12)$$

Goldstino in supergravity **Box 19.1**

The fermion mass matrix of the previous chapter has a zero mode. In global SUSY the zero mode is the Goldstino, a massless physical particle. In supergravity the zero mode corresponds to the reduction of rank of the full mass matrix m^g because the Goldstino disappears.

In the vacuum ($\partial_\alpha V = 0$) and for zero cosmological constant ($\langle V\rangle = 0$), these conditions show that the full mass matrix has a zero eigenvector. In Sec. 14.5.4 we used similar equations to find a zero eigenvector of the fermion mass matrix in global SUSY. The mathematics is similar but the interpretation is different; see Box 19.1.

19.1.2 Extension to cosmological solutions

In cosmological applications of supergravity, scalar fields need not be constant. Typically they depend on the time. There are then new mixing terms of the gravitino and chiral fermions which involve derivatives of the scalars. The relevant term in (18.6) can be written as

$$\begin{aligned}\mathcal{L} &= \frac{1}{\sqrt{2}} g_{\alpha\bar\beta}\bar\psi_\mu \hat{\not\partial}\bar z^{\bar\beta}\gamma^\mu\chi^\alpha + \text{h.c.}\\ &= \bar\psi\cdot\gamma\frac{1}{\sqrt{2}} g_{\alpha\bar\beta}\hat{\not\partial} z^\alpha\chi^{\bar\beta} + \sqrt{2} g_{\alpha\bar\beta}\bar\psi_\mu\gamma^{\nu\mu}\hat\partial_\nu\bar z^{\bar\beta}\chi^\alpha + \text{h.c.}\end{aligned} \tag{19.13}$$

The first term can be interpreted as the extra contribution,

$$P_L\upsilon' = -\frac{1}{\sqrt{2}} g_{\alpha\bar\beta}\hat{\not\partial} z^\alpha\chi^{\bar\beta}, \tag{19.14}$$

to the Goldstino, while the second term includes mixing of a different type. For cosmological solutions the Goldstino is the sum $\upsilon+\upsilon'$, and it is useful to consider the supersymmetry transformations of the two parts:

$$\begin{aligned}\delta P_L\upsilon &= -\tfrac{1}{2}\mathrm{e}^{\kappa^2\mathcal{K}/2}\nabla_\alpha W\not\partial z^\alpha P_R\epsilon + \tfrac{1}{2}V_+ P_L\epsilon + \dots,\\ \delta P_L\upsilon' &= +\tfrac{1}{2}\mathrm{e}^{\kappa^2\mathcal{K}/2}\nabla_\alpha W\not\partial z^\alpha P_R\epsilon - \tfrac{1}{2}\not\partial z^\alpha g_{\alpha\bar\beta}\not\partial\bar z^{\bar\beta} P_L\epsilon + \dots,\end{aligned} \tag{19.15}$$

where ... stands for terms with vector fields and terms cubic in fermions. In the transformation of the total Goldstino, the first terms cancel, while the others combine to

$$\delta P_L(\upsilon+\upsilon') = \tfrac{1}{2}\left(-\not\partial z^\alpha g_{\alpha\bar\beta}\not\partial\bar z^{\bar\beta} + V_+\right) P_L\epsilon + \dots. \tag{19.16}$$

The new term compared to (19.3) is positive for cosmological solutions with only time dependent scalars. Therefore, the variation of this Goldstino is non-vanishing for non-trivial solutions that are of interest, and methods similar to those of the previous section can be applied [155].

19.1.3 Mass sum rules in supergravity

In Sec. 14.5.5 we studied the mass supertrace sum rules, which give useful information on the mass spectra of particles when global supersymmetry is broken. In this section we discuss the modified sum rules which contain similar information when supersymmetry is broken in supergravity [152, 118, 173].

We first calculate the trace of the scalar masses using the methods of Sec. 14.5.5. We must take into account the modified form of the potential in (17.135) and the modification of commutators of covariant derivatives. The latter imply that $\nabla_\alpha \overline{\nabla}_{\bar{\alpha}} \overline{W} = \kappa^2 g_{\alpha\bar{\alpha}} \overline{W}$. As in the discussion of global supersymmetry we restrict to $f_{AB} = \delta_{AB}$ (see [118] for a more general result). The main steps are the following:

$$
\begin{aligned}
\partial_\alpha V &= \mathrm{e}^{\kappa^2\mathcal{K}}\left(-2\kappa^2\overline{W}\nabla_\alpha W + \nabla_\alpha\nabla_\beta W\,\overline{\nabla}^\beta\overline{W}\right) + \mathcal{P}^A\partial_\alpha\mathcal{P}_A\,,\\
\partial_\alpha\partial_{\bar{\alpha}} V &= \mathrm{e}^{\kappa^2\mathcal{K}}\Big(-2\kappa^4 g_{\alpha\bar{\alpha}}W\overline{W} - \kappa^2\nabla_\alpha W\overline{\nabla}_{\bar{\alpha}}\overline{W} + \kappa^2 g_{\alpha\bar{\alpha}}\nabla_\beta W\,\overline{\nabla}^\beta\overline{W}\\
&\quad - R_{\alpha\bar{\alpha}}{}^{\bar{\beta}\beta}\nabla_\beta W\,\overline{\nabla}_{\bar{\beta}}\overline{W} + \nabla_\alpha\nabla_\beta W\,\overline{\nabla}_{\bar{\alpha}}\overline{\nabla}^\beta\overline{W}\Big)\\
&\quad + \partial_\alpha\mathcal{P}^A\partial_{\bar{\alpha}}\mathcal{P}_A + \mathcal{P}^A\partial_{\bar{\alpha}}\partial_\alpha\mathcal{P}_A\,,\\
\operatorname{tr}\mathcal{M}_0^2 &= 2\mathrm{e}^{\kappa^2\mathcal{K}}\Big(-2n\kappa^4 W\overline{W} + (n-1)\kappa^2\nabla_\alpha W\overline{\nabla}^\alpha\overline{W} + \nabla_\alpha\nabla_\beta W\overline{\nabla}^\alpha\overline{\nabla}^\beta\overline{W}\\
&\quad + R^{\bar{\beta}\beta}\nabla_\beta W\overline{\nabla}_{\bar{\beta}}\overline{W}\Big) + 2\partial_\alpha\mathcal{P}_A\,\partial^\alpha\mathcal{P}^A + 2\mathcal{P}^A\partial^\alpha\partial_\alpha\mathcal{P}_A\,.
\end{aligned}
\tag{19.17}
$$

For the spin-1/2 contribution, we have

$$
\begin{aligned}
\operatorname{tr}\mathcal{M}_{1/2}^2 &= m^{\mathrm{g}}_{\alpha\beta}\overline{m}^{\mathrm{g}\alpha\beta} + 2m^{\mathrm{g}}_{\alpha A}\overline{m}^{\mathrm{g}\alpha A} + m^{\mathrm{g}}_{AB}\overline{m}^{\mathrm{g}AB}\\
&= m_{\alpha\beta}\overline{m}^{\alpha\beta} + 2m_{\alpha A}\overline{m}^{\alpha A} - \frac{8\kappa^2}{3}V_+ + \frac{4\kappa^4}{9|m_{3/2}|^2}V_+^2\,,
\end{aligned}
\tag{19.18}
$$

where we used (19.11) with $\partial_\alpha V = 0$ to treat the Goldstino addition (19.9) to the original mass matrices. Using the results of (18.17),

$$
m_{\alpha\beta} = \mathrm{e}^{\kappa^2\mathcal{K}/2}\nabla_\alpha\nabla_\beta W\,,\qquad m_{\alpha A} = \mathrm{i}\sqrt{2}\partial_\alpha\mathcal{P}_A\,,
\tag{19.19}
$$

we find similar terms to those in (19.17). The mass terms of the spin-1 fields are the same as in global SUSY (see (14.98)):

$$
\operatorname{tr}\mathcal{M}_1^2 = 2\partial_\alpha\mathcal{P}_A\partial^\alpha\mathcal{P}^A\,,
\tag{19.20}
$$

while the gravitino mass is $|m_{3/2}|$. We now impose vanishing cosmological term, i.e. $V = -3\kappa^{-2}|m_{3/2}|^2 + V_+ = 0$, and find the new result:

$$
\begin{aligned}
\operatorname{Supertr}\mathcal{M}^2 &\equiv \sum_J(-)^{2J}(2J+1)m_J^2\\
&= \operatorname{tr}\mathcal{M}_0^2 - 2\operatorname{tr}\mathcal{M}_{1/2}^2 + 3\operatorname{tr}\mathcal{M}_1^2 - 4|m_{3/2}|^2
\end{aligned}
$$

$$= (n-1)\left(2|m_{3/2}|^2 - \kappa^2 \mathcal{P}^A \mathcal{P}_A\right) + 2\mathrm{e}^{\kappa^2 \mathcal{K}} R^{\alpha\bar{\alpha}} \nabla_\alpha W \overline{\nabla}_{\bar{\alpha}} \overline{W}$$
$$+ 2\mathrm{i} D^A \nabla_\alpha k_A^\alpha \,. \tag{19.21}$$

In the last term we rewrote $\mathcal{P}^A \partial^\alpha \partial_\alpha \mathcal{P}_A$ as in (14.99). The interesting term is the first one. It implies that on average scalars can get masses of the order of the gravitino mass, while the fermion masses remain low. The sum rule will be explored in examples later in this chapter.

19.2 The gravity mediation scenario

In this section we explore some features of the gravity mediation scenario for SUSY breaking in phenomenological applications of supergravity. The scenario has both successes and difficulties in particle phenomenology. We discuss it because it is a scenario in which the supergravity modifications of global supersymmetry are very important. The chiral multiplets of a model of this type are divided into two sectors, a hidden sector containing scalars z^α and an observable sector with scalars labeled Φ^i. The observable sector also contains gauge multiplets. We ignore them since SUSY breaking is determined by the chiral multiplet scalars.

The Kähler potential and superpotential are sums of two decoupled terms. We take a flat Kähler potential for simplicity:

$$\mathcal{K} = \sum_\alpha z^\alpha \bar{z}^\alpha + \sum_i \Phi^i \bar{\Phi}^i \,, \qquad W = W_\mathrm{h}(z^\alpha) + W_\mathrm{o}(\Phi^i)\,. \tag{19.22}$$

In the limit $\kappa \to 0$ in which gravitational effects can be neglected, we simply have a model with two decoupled sectors with a scalar potential of the type in (14.67) (schematically with derivatives indicated by $'$)

$$V = |W'_\mathrm{h}|^2 + |W'_\mathrm{o}|^2. \tag{19.23}$$

When the coupling to supergravity is included, the two sectors are coupled in a scalar potential of the form (17.135), given by

$$V = \mathrm{e}^{\kappa^2(|z|^2+|\Phi|^2)} \left[\left| W'_\mathrm{h} + \kappa^2 \bar{z}(W_\mathrm{h} + W_\mathrm{o}) \right|^2 + \left| W'_\mathrm{o} + \kappa^2 \bar{\Phi}(W_\mathrm{h} + W_\mathrm{o}) \right|^2 \right.$$
$$\left. - 3\kappa^2 \left| W_\mathrm{h} + W_\mathrm{o} \right|^2 \right]. \tag{19.24}$$

The hidden sector superpotential W_h contains couplings that carry a high scale, at or near the Planck mass $m_\mathrm{P} = 1/\kappa = 2.4 \times 10^{18}$ GeV; see (A.11). It is constructed so that SUSY is broken in this sector. The potential is minimized at $V = 0$, so the cosmological constant vanishes. Since hidden sector terms dominate, it is a good approximation to minimize the terms in V that are quadratic in W_h. The vacuum conditions that state that there is supersymmetry breaking (see (18.24)), and that the potential vanishes are

Box 19.2 Gravity mediation scenario

In this scenario chiral multiplets occur in two sectors, a hidden sector at a high mass scale and the observable sector. The two sectors are decoupled when $\kappa \to 0$, but they are coupled by supergravity in the full theory. SUSY breaking is determined by the potential in the hidden sector. This leads to soft SUSY-breaking terms for the fields of the observable sector.

$$F_{\rm h} \equiv {\rm e}^{\kappa^2\mathcal{K}/2}\left(W_{\rm h}' + \kappa^2 \bar{z} W_{\rm h}\right) \neq 0\,, \qquad |F_{\rm h}| = \sqrt{3}{\rm e}^{\kappa^2\mathcal{K}/2}\kappa |W|\,. \tag{19.25}$$

Typically the vacuum values of the hidden sector z^α are of order $z_0 = m_{\rm P} = 1/\kappa$. Since the two sectors are coupled in the full potential (19.24), hidden sector SUSY breaking is communicated to the observable sector. We now show in a simple model that such effects are important at the energy scales probed by the LHC accelerator.

19.2.1 The Polónyi model of the hidden sector

We study the earliest and simplest model of the hidden sector, the Polónyi model [174]. It consists of a single chiral multiplet z, χ with superpotential

$$W_{\rm h}(z) = \mu m_{\rm P}(z+\beta)\,. \tag{19.26}$$

For phenomenological reasons, the value of the parameter μ is in the range 100 GeV to 1 TeV while β is of order $m_{\rm P}$. We take μ and β to be real. In this section we discuss the determination of the vacuum in the model; in the next section we sketch some consequences in a toy model of the observable sector. Our discussion is similar to that of [34].

To simplify the algebra we work temporarily in units where $\kappa = 1/m_{\rm P} = 1$ and use dimensional analysis to restore physical dimensions in the final results. The vacuum conditions require that the hidden sector potential and its gradient simultaneously vanish:

$$V_{\rm h} = \mu^2 {\rm e}^{z\bar{z}}\left(|1+\bar{z}(z+\beta)|^2 - 3|z+\beta|^2\right) = 0\,, \tag{19.27}$$
$$\partial_{\bar{z}} V_{\rm h} = zV + \mu^2 {\rm e}^{z\bar{z}}\left\{(z+\beta)\left[-2+z(\bar{z}+\beta)\right] + z\left[1+\bar{z}(z+\beta)\right]\right\} = 0\,.$$

Let us look for a real solution z_0. The condition $V_{\rm h}=0$ becomes

$$(z_0^2 + z_0\beta) + 1 = \sqrt{3}(z_0+\beta)\,. \tag{19.28}$$

This can be used to rewrite the $\partial_{\bar{z}} V_{\rm h} = 0$ condition in the factored form

$$(z_0+\beta)\left(2z_0+\beta-\sqrt{3}\right) = 0\,. \tag{19.29}$$

There is a common solution of the two previous equations, with the vacuum expectation value $z_0 = (\sqrt{3}-\beta)/2$, provided that the coupling β takes one of the two values $\beta = \pm 2 - \sqrt{3}$. This is an example of the 'fine tuning' that is necessary to obtain vanishing cosmological constant.

It is interesting to check whether there is a SUSY-preserving vacuum solution with $V_h(z_0) < 0$. Otherwise, a locally stable SUSY-breaking vacuum could tunnel to this state. The condition for a SUSY-preserving vacuum is

$$\nabla_z W_h = \partial_z W_h + \mathcal{K}_z W_h = \mu\left[1 + \bar{z}(z+\beta)\right] = 0\,. \tag{19.30}$$

The last form is relevant to the Polónyi model. Since β is real, any root of this equation must be real, i.e. $z \to x$. To avoid a SUSY vacuum we must ensure that the roots

$$x = (-\beta \pm \sqrt{\beta^2 - 4})/2 \tag{19.31}$$

are complex. This requires the inequality $|\beta| < 2$. Of the two possible values of β found above, only $\beta = 2 - \sqrt{3}$ satisfies this.

Summary. If β is tuned to the value $\beta = (2-\sqrt{3})m_\text{P}$, then the Polónyi model has a stable, SUSY-breaking vacuum with vanishing cosmological constant and $z_0 = (\sqrt{3}-1)m_\text{P}$. The Goldstino field of the model is described by the first term in (19.1). Since $\nabla_z W_h(z_0) \neq 0$, the gauge condition (19.4) can be imposed to eliminate the Goldstino from the theory. We are left with the gravitino. Its mass is given in (18.16):

$$m_{3/2} = \kappa^2 e^{\kappa^2|z_0|^2/2} W_h(z_0) = \mu e^{2-\sqrt{3}}\,. \tag{19.32}$$

Exercise 19.2 *Check that the two solutions $\kappa z = \sqrt{3} \mp 1$ lead to*

$$\begin{aligned} &\langle W\rangle = \pm\mu\kappa^{-2}\,, \qquad \langle\nabla_z W\rangle = \sqrt{3}\mu\kappa^{-1}\,, \qquad \langle\nabla_z\nabla_z W\rangle = \pm 2\mu\,,\\ &m_{11} = 2m_{3/2}\,, \end{aligned} \tag{19.33}$$

where $m_{3/2}$ is $\pm$ the value given in (19.32). Show that the fermion mass matrix m^g of (19.10) vanishes. This is what we expect in a supergravity model with only one spin-1/2 fermion and broken supersymmetry.

Exercise 19.3 *Consider an extension of the Polónyi model with several scalar fields, all with minimal Kähler potential, $\mathcal{K} = \bar{z}^{\bar{\alpha}} z^\alpha \delta_{\alpha\bar{\alpha}}$, but with the same superpotential W_h in (19.26), where the scalar z is denoted as z^1. Obtain that*

$$\partial_\alpha \partial_{\bar{\alpha}} V = m_{3/2}^2 \left(\delta_{\alpha\bar{\alpha}} + \delta_\alpha^1 \delta_{\bar{\alpha}}^{\bar{1}}\right). \tag{19.34}$$

This shows that the sum of the squared masses of the two real scalars with label '1' is $4m_{3/2}^2$ (in fact the real and imaginary parts have different masses; see [153, 175]) while the other scalars each have mass $m_{3/2}$. The spin-1/2 fields are massless, taking into account the Goldstino correction as shown in Ex. 19.2. Check that this is consistent with the mass formula (19.21).

19.2.2 Soft SUSY breaking in the observable sector

In realistic applications of gravity mediation, the scalar fields of the observable sector are partners of the quarks and leptons of the standard model or scalars that mediate electroweak symmetry breaking. Although this structure is very important, we will be content here to work with a generic observable superpotential of the form

$$W_{\rm o}(\Phi^i) = m_{ij}\Phi^i\Phi^j + g_{ijk}\Phi^i\Phi^j\Phi^k\,. \tag{19.35}$$

To explore the 'phenomenology' of this toy model we need to study the full potential (19.24), which couples the hidden and observable sectors. The main steps are as follows:

1. Rewrite V dropping quadratic terms in $W_{\rm h}$ that have canceled due to the vacuum condition $V_{\rm h}(z_0) = 0$.
2. We take a limit in which we keep all order $\kappa z_0 \approx 1$ effects on low energy physics induced by the coupling to the hidden sector, but drop all Planck scale effects which are negligible at low energy. This is limit $m_{\rm p} \to \infty$, $F_{\rm h} \to \infty$, with $m_{3/2}$ fixed.

In this limit we have

$$\begin{aligned} W_{\rm h} &= \mu\kappa^{-2}\,, & m_{3/2} &= \mu {\rm e}^{\kappa^2\mathcal{K}_{\rm h}(z_0)/2}\,,\\ \nabla_z W_{\rm h} &= \sqrt{3}\mu\kappa^{-1}\,, & \nabla_z W_0 &= \kappa^2\bar z_0 W_0\,,\\ \nabla_i W_{\rm h} &= \kappa^2 W_{\rm h}\bar\Phi^i = \mu\bar\Phi^i\,, & \nabla_i W_0 &= \partial_i W_0\,. \end{aligned} \tag{19.36}$$

Here μ is a scale defined in the first line, and agrees with the one introduced for the Polónyi model. This leads to the scalar potential

$$V(z_0,\Phi^i) = {\rm e}^{\kappa^2\mathcal{K}_{\rm h}(z_0)}\sum_i\Big[|\partial_i W_{\rm o}|^2 + \mu^2|\Phi^i|^2 + \mu(\Phi^i\partial_i W_{\rm o} + AW_{\rm o} + {\rm h.c.})\Big]\,. \tag{19.37}$$

In the Polónyi model, we have ${\rm e}^{\kappa^2\mathcal{K}_{\rm h}(z_0)} = {\rm e}^{4-2\sqrt{3}}$ and $A \equiv \sqrt{3}\kappa\bar z_0 - 3 = -\sqrt{3}$.

Let us discuss the physical implications of this result. The first term is the scalar potential expected in global supersymmetry multiplied by a factor of order 1. The net coefficient of the second term is $m_{3/2}^2$, while that of the third term differs from $m_{3/2}$ by an order 1 factor. These terms clearly break SUSY and originate from the coupling to supergravity. In the gravity mediation scenario, $m_{3/2}$ is usually adjusted to be near the scale of electroweak symmetry breaking, 100 GeV $< m_{3/2} <$ 1 TeV. Naive analysis suggests that gravitational effects on elementary particles become important only near the Planck scale, but we see that they actually appear at much lower energy due to the Kähler structure of matter-coupled supergravity.

The structure of the SUSY-breaking terms in (19.37) is interesting. To discuss it we remind readers that one important motivation to explore supersymmetric explanations of physics beyond the standard model is that there are no quadratic divergences in global supersymmetry. The stability of the Higgs mass is thus improved. However, in the MSSM (minimal supersymmetric standard model) and its generalizations, the couplings of a supersymmetric theory are not adequate for phenomenology. One must include additional terms which explicitly break SUSY *softly*. This means that they do not reintroduce quadratic divergences. Such soft breaking terms were classified by Girardello and Grisaru [176]. For a superpotential of maximum degree 3 such as (19.35), the breaking terms are exactly the allowed scalar terms of [176]. The cubic terms are especially significant since only holomorphic $\Phi^i\Phi^j\Phi^k$ monomials (or their conjugates) occur in (19.37), and the analysis of [176] indicates that only these are soft. It is an attractive feature of gravity mediation

Box 19.3

No-scale models

No-scale models avoid two awkward features of gravity mediation. The first is the blatant fine tuning needed to achieve vanishing cosmological constant at the classical level. The second is that the gravitino mass scale, which is 'put in by hand' in gravity mediation, is determined dynamically. Classically there are flat directions, and the value of $m_{3/2}$ is undetermined.

that the soft breaking terms that are arbitrarily added to make the MSSM viable are automatically generated. For further information, see Ch. 6 of [34], Ch. 31 of [30] or the reviews [177, 178].

19.3 No-scale models

The essential idea of no-scale models is given in Box 19.3. They are cleverly constructed so that the classical potential (with D-terms from gauge multiplets ignored) vanishes identically for all values of the scalar fields. It is said to have 'flat directions'. Yet SUSY is formally broken everywhere on the scalar manifold. This is indicated by the fact that $m_{3/2}$ is non-vanishing although undetermined classically. The properties of the actual vacuum and the value of $m_{3/2}$ then emerge from quantum corrections to the classical approximation. The flat directions are useful in applications to cosmology. Furthermore, no-scale models commonly occur in string phenomenology. Our discussion below is limited to the construction of classical potentials with no-scale properties. We refer readers to the original literature [179, 180] or the review [181] for more information.

Consider a general chiral multiplet model with fields z^α, χ^α. We have Kähler potential $\mathcal{K}(z,\bar{z})$, target space metric $g_{\alpha\bar{\beta}} = \partial_\alpha\partial_{\bar{\beta}}\mathcal{K}$, and superpotential $W(z)$. The scalar potential given in (18.11) was rewritten in (18.21) in terms of the Kähler invariant function $\mathcal{G}$ of (17.132). This is equivalent to a Kähler transformation $\mathcal{K} \to \mathcal{K} + f(z) + \bar{f}(\bar{z})$ with $f(z) = \kappa^{-2}\ln W$. It is precisely the Kähler transformation that sends the superpotential $W \to \mathrm{e}^{-\kappa^2 f} W \equiv 1$. The overall Kähler invariance of $\mathcal{N} = 1$ supergravity ensures that the original theory with 'input data' $\mathcal{K}$, W is physically equivalent to the theory with new data $\mathcal{G}$, $W \equiv 1$ (and with fermion fields transformed as in (18.22)). Thus we can reexpress the gravitino mass term (18.15) and Goldstino field (19.1) as

$$\mathcal{L}_{3/2} = \tfrac{1}{2} m_{3/2}\bar{\psi}_\mu\gamma^{\mu\nu}\psi_\nu\,, \qquad m_{3/2} = \kappa^{-1}\mathrm{e}^{\mathcal{G}/2}\,,$$
$$P_L v = -\frac{1}{\sqrt{2}}\kappa^{-3}\chi^\alpha \mathrm{e}^{\mathcal{G}/2}\partial_\alpha\mathcal{G}\,. \tag{19.38}$$

Thus we see that SUSY is broken if $\partial_\alpha\mathcal{G} \neq 0$ at the minimum of the potential. The Goldstino can then be removed from the theory by the gauge condition (19.4) and a non-vanishing gravitino mass is generated.

What remains to be investigated is how to choose $\mathcal{G}(z,\bar{z})$ so that the potential vanishes identically. This will be true provided that $\mathcal{G}$ satisfies the nonlinear partial differential equation

$$\mathcal{G}^{\alpha\bar{\beta}}\partial_\alpha\mathcal{G}\,\partial_{\bar{\beta}}\mathcal{G}=3\,. \tag{19.39}$$

The vanishing of the cosmological constant is thus enforced by a geometric condition on the Kähler metric.

Exercise 19.4 *Here is a guided exercise to find the target space that satisfies this equation when there is only one chiral scalar z. Show that in this case (19.39) can be rewritten as*

$$\partial_z\partial_{\bar{z}}\mathcal{G}=\tfrac{1}{3}\partial_z\mathcal{G}\partial_{\bar{z}}\mathcal{G}\,. \tag{19.40}$$

Express $\mathcal{G}=c\ln H$ where the constant c and the real function H are to be determined. Substitute this in (19.40) to find that the choice $c=-3$ implies that H satisfies $\partial_z\partial_{\bar{z}}H=0$ with solution $\mathcal{G}=-3\ln\left[h(z)+\bar{h}(\bar{z})\right]$. One can then redefine the coordinate z and express the Kähler potential and metric as

$$\mathcal{G}(z,\bar{z})=-3\ln(z+\bar{z})\,,\qquad \mathrm{d}s^2=\frac{3}{(z+\bar{z})^2}\mathrm{d}z\mathrm{d}\bar{z}\,. \tag{19.41}$$

After a further change of coordinate $z\to -\mathrm{i}z$ one can recognize this as the special case of the Poincaré plane which appeared in Sec. 17.3.10.

Realistic implementations of the no-scale principle must incorporate chiral multiplets that can be interpreted as observable sector fields whose mass spectrum can be identified with the spectrum of the standard model particles plus their SUSY partners. There turns out to be considerable freedom to add additional chiral scalars ϕ^i to the previous model. Given any real function $h(\phi,\bar{\phi})$, the potential V of (18.21) satisfies the flatness condition (19.39) for

$$\mathcal{G}=-3\ln\left[z^1+\bar{z}^{\bar{1}}-h(z^i,\bar{z}^{\bar{\imath}})\right]\,,\qquad i=2,\ldots,n\,. \tag{19.42}$$

The scalar z is now indicated by z^1 and the ϕ^i are renamed as z^i. The derivatives satisfy

$$\mathcal{G}_1=-3\mathrm{e}^{\mathcal{G}/3}\,,\qquad \mathcal{G}_i=-h_i\mathcal{G}_1\,. \tag{19.43}$$

One can then write the metric as

$$\mathcal{G}_{\alpha\bar{\beta}}=\kappa^2 g_{\alpha\bar{\beta}}=3\mathrm{e}^{2\mathcal{G}/3}\begin{pmatrix}1 & 0\\ -h_i & -h_{i\bar{k}}\end{pmatrix}\begin{pmatrix}1 & 0\\ 0 & \mathrm{e}^{-\mathcal{G}/3}h^{\bar{k}\ell}\end{pmatrix}\begin{pmatrix}1 & -h_{\bar{\jmath}}\\ 0 & -h_{\ell\bar{\jmath}}\end{pmatrix}. \tag{19.44}$$

One proves from this that $\mathcal{G}_\alpha\mathcal{G}^{\alpha\bar{\beta}}=-\mathrm{e}^{-\mathcal{G}/3}\delta_{\bar{1}}^{\bar{\beta}}$. Contraction with $\mathcal{G}_{\bar{\beta}}$ then produces the condition (19.39), which shows that V vanishes identically. Thus the vacuum configuration of the fields is not determined by classical analysis, and there are n flat directions or moduli. Yet SUSY is spontaneously broken.

Phenomenological no-scale models contain both chiral and gauge multiplets. The full potential then contains F-terms and D-terms:

$$V = \kappa^{-4} e^{\mathcal{G}} \left(\mathcal{G}_\alpha \mathcal{G}^{\alpha\bar{\beta}} \mathcal{G}_{\bar{\beta}} - 3 \right) + \tfrac{1}{2} (\mathrm{Re}\, f)^{-1\,AB} \mathcal{P}_A \mathcal{P}_B \,. \tag{19.45}$$

The no-scale property means that $\mathcal{G}$ is chosen so that the F-term part vanishes. The dynamics of the vacuum is then determined by both the D-terms and perturbative quantum effects; see [181].

Exercise 19.5 *Check that (19.44) leads to the Kähler curvature*

$$R_{\alpha\bar{\beta}} = -\tfrac{1}{3}(n+1)\mathcal{G}_{\alpha\bar{\beta}} - \partial_\alpha \partial_{\bar{\beta}} \log \det h_{i\bar{j}} \,. \tag{19.46}$$

The second term does not contribute to the mass sum rule (19.21), owing to the multiplication of $R_{\alpha\bar{\beta}}$ by $g^{\bar{\beta}\gamma}\nabla_\gamma W$, which is proportional to $\mathcal{G}^{\bar{\beta}\gamma}\mathcal{G}_\gamma$ and thus only non-zero in the direction $\bar{\beta} = 1$. Check that, without vector multiplets, the sum rule leads to

$$\mathrm{tr}\, \mathcal{M}_0^2 - 2\, \mathrm{tr}\, \mathcal{M}_{1/2}^2 = 0 \,. \tag{19.47}$$

This means that the expected gravitino contribution cancels and that, on average, there is no splitting between the masses of scalars and their spinor partners.

The no-scale models are most important for applications in cosmology. The main virtues of supergravity for applications in cosmology, and especially for models of inflation, are twofold. First, there is the abundance of flat directions. After SUSY breaking these are lifted, but they still have a gentle slope without excessive fine tuning. Second, there is the possibility to cancel the positive term in the potential by the negative one in (18.11), or something very close, leading to a realistic value of the Hubble constant $H^2 = \frac{1}{3}\kappa^2 V$, where V is the value of the potential. Useful references are Ch. 8 of [182] and [34].

19.4 Supersymmetry and anti-de Sitter space

Supersymmetric solutions of the supergravity occur only in Minkowski space and in anti-de Sitter space. Such solutions are characterized by the vanishing of the supersymmetry transformations of the fermions. Hence, (18.25) shows immediately that in such solutions we can only have a negative (or zero) term depending on whether W vanishes or not. If $W \neq 0$ then the vanishing of the supersymmetry variation of the gravitino in (18.22) has to be compensated by a non-zero $\omega_\mu{}^{ab}$, hence a curved space. We saw in the simplest example in Sec. 9.6 that this occurs in an AdS manifold, and that the non-vanishing value of the potential is related to the AdS curvature and to the mass parameter of the gravitino. In a curved space the meaning of the mass parameters in the action is different from the one in Minkowski space.

In Minkowski space the parameter m^2 in the field equation for a scalar field,

$$\Box\phi - m^2\phi = 0 \,, \tag{19.48}$$

Box 19.4 **Supersymmetric solutions of $\mathcal{N}=1$ supergravity**

Supersymmetric solutions with constant scalar and vector fields can occur in a spacetime background that is Minkowski or anti-de Sitter. The corresponding ground states are stable.

should be positive. If this is not the case, the field has tachyonic modes, which implies that the system is unstable. In an anti-de Sitter background, stability with respect to fluctuations of the scalar fields is maintained for negative values of the parameter m^2 as long as

$$m^2 > \frac{D-1}{2(D-2)}\kappa^2 V \overset{D=4}{\longrightarrow} \frac{3}{4}\kappa^2 V\,. \tag{19.49}$$

This is usually called the Breitenlohner–Freedman (BF) bound [183, 184, 185]. In Ch. 23, we sketch a derivation from the viewpoint of unitary representations of the AdS$_D$ isometry group SO$(D-2,2)$..

The parameter m^2 in a field equation of the type (19.48) is determined by the ratio of the second derivative of the potential with respect to the metric occurring in the kinetic terms. Hence, for an arbitrary scalar fluctuation u^α and hermitian kinetic terms:

$$m^2(u,\bar u) = \frac{2u^\alpha \bar u^{\bar\beta}\partial_\alpha\partial_{\bar\beta}V + u^\alpha u^\beta \partial_\alpha\partial_\beta V + \bar u^{\bar\alpha}\bar u^{\bar\beta}\partial_{\bar\alpha}\partial_{\bar\beta}V}{2u^\gamma \bar u^{\bar\gamma} g_{\gamma\bar\gamma}}\,. \tag{19.50}$$

We now prove that, in $\mathcal{N}=1$ supergravity, this satisfies the bound (19.49) for any supersymmetric solution. For the latter, the vanishing of the transformations of the physical fermions lead to $\nabla_\alpha W = 0$ and $\mathcal{P}_A = 0$. This drastically simplifies the $\partial_\alpha\partial_{\bar\alpha}V$ expression in (19.17), and leads also to a simple expression for $\partial_\alpha\partial_\beta V$. Inserting these in (19.50) gives

$$\begin{aligned} m^2(u,\bar u) &= \frac{1}{2u^\gamma \bar u_\gamma}\Big[\frac{1}{2}e^{\kappa^2\mathcal{K}}\left(2u^\alpha\nabla_\alpha\nabla_\beta W - \kappa^2 W\bar u_\beta\right)\left(2\bar u^{\bar\alpha}\overline{\nabla}_{\bar\alpha}\overline{\nabla}^\beta\overline{W} - \kappa^2\overline{W}u^\beta\right) \\ &\qquad - \frac{9}{2}e^{\kappa^2\mathcal{K}}\kappa^4 W\overline{W}u^\alpha\bar u_\alpha + \left(u^\alpha\partial_\alpha\mathcal{P}_A + \bar u^{\bar\alpha}\partial_{\bar\alpha}\mathcal{P}_A\right)^2\Big] \\ &\geq -\frac{9}{4}e^{\kappa^2\mathcal{K}}\kappa^4 W\overline{W} = \frac{3}{4}\kappa^2 V\,. \end{aligned} \tag{19.51}$$

Hence, supersymmetric ground states are stable. This result turns out to be more general than for the $\mathcal{N}=1$ four-dimensional supergravities that we considered here [186, 187]. It is due to the general form of the potential in supergravity theories.

19.5 *R*-symmetry and Fayet–Iliopoulos terms

The $\mathcal{N}=1$ superconformal algebra contains the U(1) transformation called T in Sec. 16.1.1. It becomes the R-symmetry of the Poincaré supersymmetry algebra of the physical theory. We retain the name T because its normalization differs from that in Sec. 12.2.

R-symmetry part of gauge symmetries in supergravity **Box 19.5**

Any gauged symmetry acts partly as a gauged R-symmetry, whose form is determined by the moment map and the Kähler connection, as expressed by the imaginary part of r_A.

In this section we examine its role in physical $\mathcal{N}=1$ supergravity. The key feature is that it is gauged by a composite gauge field, so we have gauged R-symmetry.

19.5.1 The R-gauge field and transformations

The T gauge field in the superconformal Weyl multiplet is $\mathcal{A}_\mu$. It is an auxiliary field. After use of the equations of motion, $\mathcal{A}_\mu$ takes the form[2] given in (18.9) with $a = 3\kappa^{-2}$:

$$
\begin{aligned}
a\mathcal{A}_\mu &= -\omega_\alpha \partial_\mu z^\alpha - \omega_{\bar\alpha}\partial_\mu \bar z^{\bar\alpha} - A_\mu^A \mathcal{P}_A\,, \qquad \omega_\alpha = -\tfrac{1}{2}\mathrm{i}\partial_\alpha \mathcal{K}\,, \qquad \omega_{\bar\alpha} = \tfrac{1}{2}\mathrm{i}\partial_{\bar\alpha}\mathcal{K}\,,\\
a\mathcal{A}_\mu &= -\omega_\alpha \hat\partial_\mu z^\alpha - \omega_{\bar\alpha}\hat\partial_\mu \bar z^{\bar\alpha} + \tfrac{1}{2}\mathrm{i}A_\mu^A (r_A - \bar r_A)\,.
\end{aligned} \tag{19.52}
$$

We recognize the structure in (19.52):

1. In the top line, the first two terms are the pull-back to spacetime of the 'Kähler connection'.
2. The third term contains products of the physical vector fields A_μ^A and their moment maps $\mathcal{P}_A$.
3. In the next line the gauge field terms are regrouped into covariant $\hat\partial$ derivatives plus the imaginary part of r_A.

After gauge fixing of the superconformal theory, a decomposition law leads to the form (17.107) of the T-gauge parameter. This form shows that T is not an independent symmetry of the physical theory. Instead there is an effective T-gauge symmetry whose parameters include those of Yang–Mills, supersymmetry, and Kähler transformations. The first two are spacetime gauge symmetries (supersymmetry and Yang–Mills gauge symmetries) and the last one is a σ-model symmetry. For the first two, we can write, using (17.111) and (19.52),

$$
a\lambda_T = -\omega_\alpha \delta z^\alpha - \omega_{\bar\alpha}\delta \bar z^{\bar\alpha} - \theta^A \mathcal{P}_A\,. \tag{19.53}
$$

The implication for Yang–Mills symmetries is the important statement in Box 19.5. The contribution of (19.53) to supersymmetry transformations is included in (18.22) in the terms cubic in the fermions. Bosons in this physical basis are inert under the T-symmetry.

The last term in (17.107) concerns Kähler transformations. These are not gauge symmetries of the theory, but determine an invariance of the action under reparametrizations of the data that specify the theory, i.e. the Kähler potential and the superpotential. The rule (17.107) implies that the fermions have to be reparametrized accordingly.

Exercise 19.6 *Check that the gravitino mass parameter in (18.16) changes under Kähler transformations to*

[2] We concentrate here on the bosonic terms.

$$m'_{3/2} = m_{3/2} \exp\left(\tfrac{1}{2}\kappa^2(\bar{f}(\bar{z}) - f(z))\right), \qquad (19.54)$$

and that (18.23) specifies the transformation of the gravitino such that its mass term in the action is invariant. Note that the physical gravitino mass $|m_{3/2}|$ is also invariant.

19.5.2 Fayet–Iliopoulos terms

The essential elements of both global supersymmetric gauge theories and 'gauged supergravities', that is supergravity theories with Yang–Mills symmetries, are the moment maps $\mathcal{P}_A$ which determine the holomorphic Killing vectors $k_A^\alpha(z)$ of the gauged isometries of the Kähler manifold (see item 4 on p. 385). Moment maps are defined only up to additive real constants, because one can add arbitrary imaginary constants $\mathrm{i}\xi_A$ to the $r_A(z)$ defined by the gauge transformation of the Kähler potential. Thus we have

$$\mathcal{P}_A \to \mathcal{P}_A + \xi_A\,, \qquad r_A \to r_A + \mathrm{i}\xi_A\,. \qquad (19.55)$$

The arbitrary ξ_A are called Fayet–Iliopoulos (FI) constants.

In global supersymmetry, an important restriction comes from the equivariance relation; see (13.62) or (18.3). This condition shows that the $\mathcal{P}_A$ of non-abelian symmetries are unique. Arbitrary ξ_A are possible only for generators that are not in the derived algebra, say U(1) symmetries. In supergravity the equivariance condition was derived in Ex. 17.18 by requiring that the gauge algebra is satisfied on the compensating scalar y. Further, the FI constants affect the gauge invariance condition (17.133) of the superpotential in supergravity. In physical applications the FI terms facilitate vacua with non-vanishing D-term contribution to the scalar potential.

The FI terms can also be understood from the viewpoint of a symmetry of the embedding superconformal theory. The Killing vector k_T, (17.40), can be reinterpreted as a gauge symmetry that can be coupled to one of the gauge fields. We now indicate it with index 0, i.e. $\delta X = \mathrm{i}\theta_0 X$, which can be expressed as

$$k_0 = \mathrm{i}X^I\frac{\partial}{\partial X^I} = \mathrm{i}y\frac{\partial}{\partial y}\,. \qquad (19.56)$$

The last expression in terms of the coordinates y and z^α shows that the symmetry acts only on y, hence the components $k_0{}^\alpha$ vanish. However, it has $r_0 = 3\mathrm{i}\kappa^{-2}$, as follows from (17.95). Hence, it commutes with the other symmetries.

Therefore, we may add this '0'-symmetry to any U(1). Changing the generator $k_A \to k_A + \frac{1}{3}\kappa^2\xi_A k_0$ changes the moment map as in (19.55).

The general ideas of this section are not restricted to $\mathcal{N}=1$ supergravity. Similar structures with moment maps for gauged R-symmetry and FI terms exist also for $\mathcal{N}=2$ supergravity in four, five or six dimensions, as we will see in Ch. 20, and especially in Sec. 20.4.5.

19.5.3 An example with non-minimal Kähler potential

We again consider the example of Sec. 17.3.10, with superconformal Kähler potential (17.118). The symmetries of the target space were labelled '0, 1, 2, 3'. The symmetry

'0' is the one discussed in (19.56). The corresponding Killing vector has no effect on the coordinate z as shown in the second expression in (17.126), but there is a non-trivial $r_A(z)$. In this section we wish to study the effects of a gauged U(1) symmetry with Fayet–Iliopoulos term.

We can choose which symmetry to gauge. We take one gauge multiplet and couple it to the Killing vector

$$k = k_3 - \tfrac{1}{3}\kappa^2\xi k_0 \,. \tag{19.57}$$

We thus have

$$\delta z = \theta k^z = \theta z^2 \,, \qquad r(z) = -3\kappa^{-2}z + \mathrm{i}\xi \,. \tag{19.58}$$

To make the model more interesting, we add the superpotential,

$$W(z) = z^3 \exp \frac{\mathrm{i}\kappa^2\xi}{z} \,, \tag{19.59}$$

chosen such that it transforms under the symmetry $\delta z = \theta z^2$ as

$$\delta W(z) = W(z)(3z - \mathrm{i}\kappa^2\xi) = -\kappa^2 r(z)\, W(z) \,, \tag{19.60}$$

which is required according to (18.5).

Calculating the F-potential we find that the negative term V_- in (18.11) cancels the positive part for $\xi = 0$, and the result is

$$V_F = \kappa^4\xi \mathrm{e}^{\kappa^2\mathcal{K}} W\bar{W} \left[2\mathrm{i}\left(\frac{1}{\bar{z}} - \frac{1}{z}\right) - \frac{1}{3}\kappa^2\xi\left(\frac{1}{\bar{z}} - \frac{1}{z}\right)^2 \right] . \tag{19.61}$$

With standard kinetic terms for the gauge field, the D-term potential is

$$V_D = \frac{1}{2}\mathcal{P}^2 = \frac{1}{2}\left(\xi - 3\mathrm{i}\kappa^{-2}\frac{z\bar{z}}{z-\bar{z}}\right)^2 . \tag{19.62}$$

This potential is invariant under the transformation of the scalars as in (19.58). The gauge transformation of the fermions is composed of the part as in global supersymmetry and the part where it acts as an R-symmetry due to the remnant of the T-symmetry (18.22), for example

$$\delta P_L\chi = \theta\left[\partial_z k^z + \tfrac{1}{4}\kappa^2(r - \bar{r})\right] P_L\chi = \theta\left[2z - \tfrac{3}{4}(z - \bar{z}) - \tfrac{1}{2}\mathrm{i}\kappa^2\xi\right] P_L\chi \,. \tag{19.63}$$

In a general model with FI terms, the chiral transformation depends on the FI constants. If the target space is topological non-trivial, the quantization condition of Secs. 17.5.1 and 17A implies that the FI constants are quantized. This has been discussed in [166, 188].

It has been argued [171, 189, 166] that theories with FI constants cannot be obtained by starting from a global supersymmetric theory with FI constant and adding the supergravity fields order by order in κ without adding more dynamical fields, unless the theory has an exact global symmetry.

Exercise 19.7 *Check that for $\xi = 0$ the Kähler invariant function is of the form (19.42) after the redefinition $z^1 = \mathrm{i}z^{-1}$. This explains why (19.61) vanishes for $\xi = 0$.*

PART VII

EXTENDED $\mathcal{N} = 2$ SUPERGRAVITY

20 Construction of the matter-coupled $\mathcal{N} = 2$ supergravity

In the previous chapters we used the superconformal method to construct couplings of matter multiplets to $\mathcal{N} = 1$, $D = 4$ supergravity, and discussed the resulting action and some of its applications. In this chapter, we will move on to the first extended supersymmetry: $\mathcal{N} = 2$. The $\mathcal{N} = 2$ supergravity theories in four dimensions have many applications, for example to Calabi–Yau compactifications of superstring theory and to studies of black hole solutions. The main mathematical feature of $\mathcal{N} = 2$ supersymmetry and supergravity is the requirement of new scalar field target spaces, known by the name special geometry.

The term $\mathcal{N} = 2$ supersymmetry refers to the theories with eight real supercharges listed in Table 12.2. The maximum spacetime dimension is six, where this theory is also described as $(1, 0)$ supergravity, since it is based on symplectic Majorana–Weyl spinors, which are spinors of one chirality. We discuss the main multiplets of the six- and five-dimensional theories at the level of global supersymmetry. The five-dimensional theories will help explain the structure of special geometry. However, our focus is on $D = 4$, $\mathcal{N} = 2$, and the main goal is the derivation of the action and transformation laws of the matter-coupled supergravity theories with vector and hypermultiplets.

This chapter is rather long. In the next chapter, we will write the results in an independent way for readers who are only interested in understanding the action and transformation laws for applications.

In Sec. 20.1 we start with global supersymmetry for the six-, five- and four-dimensional theories. We discuss gauge multiplets and hypermultiplets. The formulation in $D = 6$ is simplest, but readers can confine their attention to the discussion of $D = 4$. The final result of this section is the action of coupled vector and hypermultiplets obtained by gauging an isometry group of the hypermultiplet scalar metric.

In the same way as was done for $\mathcal{N} = 1$, we construct $\mathcal{N} = 2$ supergravity theories by gauging the superconformal algebra. This method was the original approach to special Kähler geometry [94], and it nicely elucidates the symplectic nature of this geometric structure. In Sec. 20.2 we discuss the conformal structure and its gauging. For the matter multiplets and their actions we restrict to $D = 4$. We choose the partial gauge fixing conditions that will lead to the super-Poincaré theory, and eliminate the auxiliary fields. The final action in that section is already the full matter-coupled theory but it is expressed in terms of the fields of the 'embedding manifold' as done for $\mathcal{N} = 1$. It is not yet in the most convenient variables for the super-Poincaré theory.

Special geometry as a mathematical structure will be discussed in Sec. 20.3. We discuss the general family of these manifolds, and then the properties of very special real geometry (for gauge multiplets in $D = 5$), special Kähler geometry and quaternionic-Kähler geometry. The subsection on special Kähler geometry is especially important, but quaternionic-Kähler geometry is also used in the description of the Poincaré theory in $D = 4$ supergravity.

In Sec. 20.4 we will explain the final steps to obtain the super-Poincaré theory for $\mathcal{N} = 2$ in four dimensions.

Appendices 20A and 20B explain some calculational tools. Appendix 20C gives the mathematical definition of rigid special Kähler geometry.

We end this introduction with remarks on other approaches. First, there are superspace approaches to $\mathcal{N} = 2$ supersymmetry. The fact that there are central charges in the $\mathcal{N} = 2$ superalgebra (see Sec. 12.6.2) complicates the superspace approaches. To deal with them, the concepts of harmonic [190, 191] or projective [192, 193, 194, 195, 196] superspaces were developed. In the same context the superconformal approach, originally developed by Howe [197, 198], is important. A recent account of this is given in [199]. The group manifold approach has been explained in detail in [200]. The resulting action was given in [201]. Other extensions of the theories discussed here are obtained by adding the tensor or vector–tensor multiplets [202, 203]. They may give a useful reformulation of vector or hypermultiplets.

20.1 Global supersymmetry

20.1.1 Gauge multiplets for $D = 6$

$\mathcal{N} = 2$ gauge multiplets are simplest in six dimensions, where they are similar to the $\mathcal{N} = 1$ gauge multiplets in four dimensions: the physical fields are gauge vectors A_μ^I and 'gaugini' λ^{iI} in the adjoint representation of the gauge group.[1] The extra $i = 1, 2$ index reflects that they are symplectic Weyl spinors. The counting of degrees of freedom gives a first indication that fields can form a multiplet. From Table 6.2 we see that a gauge vector has five off-shell and four on-shell degrees of freedom. The spinors have eight off-shell and four on-shell degrees of freedom. Hence, the on-shell counting already fits. Off-shell we need three auxiliary bosonic degrees of freedom, which we will denote as $\vec{Y}^I$. They are the analogues of the D^A auxiliary field of the $\mathcal{N} = 1$, $D = 4$ gauge multiplets. Their value after field equations will be related to a quaternionic structure of the isometries of the matter sector, induced by the SU(2) R-symmetry group. This fits with the fact that $\vec{Y}^I$ are triplets for each gauge multiplet, rather than the singlets D^A. With all this information, it should not be difficult to guess the form of the transformation laws. The only difficulty is the use of notation with the indices i, related to the symplectic Majorana spinors of $D = 6$.

In Sec. 3.3.2 it was explained that symplectic Majorana spinors are chiral and satisfy a reality condition that depends on the antisymmetric tensor ε_{ij}. As a result, the following identities hold for the gauginos and supersymmetry parameters, ϵ^i,

[1] The gauge fields are denoted here with an index I. Therefore, gauge transformation parameters are denoted by θ^I rather than the θ^A used in earlier chapters.

$$\lambda^i = P_L\lambda^i\,, \qquad \bar\lambda^i = \bar\lambda^i P_R\,, \tag{20.1}$$
$$\lambda_i = (\lambda^i)^C\,, \qquad \lambda^i = -(\lambda_i)^C\,, \tag{20.2}$$
$$\lambda_i = \lambda^j\varepsilon_{ji}\,, \qquad \lambda^i = \varepsilon^{ij}\lambda_j\,, \tag{20.3}$$

where one replaces λ with ϵ for the supersymmetry parameters. The fact that the second equation has P_R where the first has P_L follows from (3.56). The second line is consistent with the result of Ex. 3.33 with $t_1 = 1$, and the third line follows the NW–SE rule for raising and lowering of indices.

The triplet of auxiliary fields can be written as a real vector $\vec{Y}$ or as a symmetric tensor $Y^{ij} = Y^{ji}$. The relation is explained in Appendix 20A.

The form of the supersymmetry transformation laws for the fields in the gauge multiplet are fixed by the index structure of the fields. The numerical coefficients are determined by checking the supersymmetry algebra. The transformation laws

$$\begin{aligned}
\delta A_\mu^I &= \tfrac{1}{2}\bar\epsilon^i\gamma_\mu\lambda_i^I\,,\\
\delta\lambda^{iI} &= -\tfrac{1}{4}\gamma^{\mu\nu}F_{\mu\nu}{}^I\epsilon^i - Y^{ijI}\epsilon_j\,,\\
\delta Y^{ijI} &= -\tfrac{1}{2}\bar\epsilon^{(i}\not{D}\lambda^{j)I}
\end{aligned} \tag{20.4}$$

realize the algebra

$$[\delta(\epsilon_1),\delta(\epsilon_2)] = \delta_{\rm cgct}(\xi^\mu)\,, \qquad \xi^\mu = \tfrac{1}{2}\bar\epsilon_2^i\gamma^\mu\epsilon_{1i}\,. \tag{20.5}$$

The right-hand side is a covariant translation, i.e. the combination of a translation with a gauge transformation acting as the covariant general coordinate transformation discussed in Sec. 11.3.2, for the case of constant parameters.

The action that is invariant under these transformations is

$$S = \int \mathrm{d}^6x\,\delta_{IJ}\left[-\tfrac{1}{4}F_{\mu\nu}^I F^{\mu\nu\,J} - \tfrac{1}{2}\bar\lambda^{iI}\not{D}\lambda_i{}^J + 2\vec{Y}^I\cdot\vec{Y}^J\right], \tag{20.6}$$

where δ_{IJ} is a Cartan–Killing metric for the gauge group. Note that for invariance in the case of non-abelian gauge groups, one needs the result of Ex. 3.27 and the extension of (3.67) to $D = 6$ discussed in footnote 3 there.

20.1.2 Gauge multiplets for $D = 5$

The five-dimensional gauge multiplets are the dimensional reduction of the $D = 6$ ones. The fermions are again symplectic fermions, hence (20.3) and (20.2) still apply. However, there is no chirality since the spacetime dimension is odd. The main new ingredient is that the gauge vector of six dimensions gives rise to a gauge vector in five dimensions plus a scalar (see Sec. 5.3.3, restricting to the $k = 0$ sector, where A_{D0} is the new scalar field). We will denote this extra scalar by σ^I.

The transformation laws of the gauge multiplet in five dimensions can be obtained from those of $D=6$ by dimensional reduction. In Appendix 20B the technical aspects of the reduction of spinors are explained. The result is[2]

$$\begin{aligned}
\delta A_\mu^I &= \tfrac{1}{2}\bar{\epsilon}^i \gamma_\mu \lambda_i{}^I\,,\\
\delta\sigma^I &= \tfrac{1}{2}\mathrm{i}\bar{\epsilon}^i \lambda_i{}^I\,,\\
\delta\lambda^{iI} &= -\tfrac{1}{4}\gamma^{\mu\nu} F_{\mu\nu}{}^I \epsilon^i - \tfrac{1}{2}\mathrm{i}\not{D}\sigma^I \epsilon^i - Y^{ij\,I}\epsilon_j\,,\\
\delta\vec{Y}^I &= -\tfrac{1}{4}\vec{\tau}_{ij}\bar{\epsilon}^i\left[\not{D}\lambda^{jI} + \mathrm{i} f_{JK}{}^I \sigma^J \lambda^{jK}\right].
\end{aligned} \tag{20.7}$$

The gauge part in the $\delta_{\rm cgct}$ in (20.5) also contains A_μ, which is the origin of a new term in five dimensions, coming from the last component of the $D=6$ gauge field, which is here σ^I:

$$[\delta(\epsilon_1), \delta(\epsilon_2)] = \delta_{\rm cgct}(\xi^\mu) + \delta_G(\theta^I)\,, \qquad \theta^I = -\tfrac{1}{2}\mathrm{i}\sigma^I \bar{\epsilon}_2{}^i \epsilon_{1i} = -\tfrac{1}{2}\mathrm{i}\sigma^I \bar{\epsilon}_2{}^i \epsilon_1{}^j \varepsilon_{ji}\,. \tag{20.8}$$

That last term resembles a central charge, as discussed for four dimensions in Sec. 12.6.2: it appears in the commutator of two supersymmetries with a factor $\bar{\epsilon}_2{}^i \epsilon_1{}^j \varepsilon_{ji}$. Comparing to (12.23), Z is here $\sigma^I T_I$, where T_I is the action of the gauge symmetries. Note, however, that $\sigma^I T_I$ is not yet an operator that commutes with all the other symmetries, i.e. it is not central. However, when solutions to the field equations of the theory are considered, the value of σ^I is often fixed to a number. In this case, the $\sigma^I T_I$ term in the algebra of the symmetries of the supergravity action can result in a central charge in the algebra of preserved symmetries of the solutions. Therefore, we can call it a central charge-like term.

Exercise 20.1 *Check that such a central charge-like term would not be possible in six dimensions, not only due to the fact that there is no scalar, but also due to the chirality properties of the spinors.*

We could write the analogue of the action (20.6) at this point. However, we are going to present another action that will be more important in the context of supergravity. It makes use of a Chern–Simons term, similar to the one that we encountered in 11 dimensions: $A \wedge F \wedge F$, where A is a 3-form. In five dimensions such a term can be constructed using the 1-form gauge field. Owing to the requirement of gauge invariance, such a term can only be multiplied with constants, as can be seen in a calculation similar to the one in (10.25). Hence, we will introduce a constant three-index symmetric tensor C_{IJK} to contract with the gauge indices of the $A^I \wedge F^J \wedge F^K$ term. For non-abelian gauge multiplets, it has to be 'gauge invariant', i.e.

$$f_{I(J}{}^M C_{KL)M} = 0\,. \tag{20.9}$$

For example, the 'd-symbols'

$$\mathrm{tr}(t_A\{t_B, t_C\}) = d_{ABC}\,, \tag{20.10}$$

satisfy these relations.

It is an important aspect of $D=5$ supersymmetric theories that tensor multiplets can lead to more general gauged supergravities than vector multiplets, as we already mentioned

[2] Concerning the appearance of factors of i in these formulas, remember Ex. 3.36.

Gauge multiplets with global $\mathcal{N}=2$, $D=5$ supersymmetry Box 20.1

The action that has superconformal symmetry depends on a symmetric tensor C_{IJK}. The gauge multiplets contain real scalars σ^I in the adjoint of the gauge group, whose kinetic terms define (rigid) 'very special real geometry'.

in Sec. 12.3.2. As a free field, antisymmetric tensors are equivalent to vector fields in $D=5$; see Sec. 7.8. However, vector gauge fields should transform in the adjoint representation, which is not necessary for tensors. Therefore, more general possibilities than (20.9) are possible when including vector–tensor multiplets. However, we will omit them here and refer the interested reader to the literature [204, 205, 206, 207].

The complete Lagrangian is

$$\begin{aligned}\mathcal{L}_{5v} = C_{IJK}\Big[&\Big(-\tfrac{1}{4}F^I_{\mu\nu}F^{\mu\nu\,J} - \tfrac{1}{2}\bar{\lambda}^{iI}\not{D}\lambda_i{}^J - \tfrac{1}{2}D_\mu\sigma^I D^\mu\sigma^J + 2\vec{Y}^I\cdot\vec{Y}^J\Big)\sigma^K\\ &-\tfrac{1}{24}\varepsilon^{\mu\nu\rho\sigma\tau}A^I_\mu\Big[F^J_{\nu\rho}F^K_{\sigma\tau} + f_{LM}{}^J A^L_\nu A^M_\rho\Big(-\tfrac{1}{2}F^K_{\sigma\tau} + \tfrac{1}{10}f_{NP}{}^K A^N_\sigma A^P_\tau\Big)\Big]\\ &-\tfrac{1}{8}\mathrm{i}\bar{\lambda}^{iI}\gamma^{\mu\nu}F^J_{\mu\nu}\lambda_i{}^K - \tfrac{1}{2}\mathrm{i}\bar{\lambda}^{iI}\lambda^{jJ}\vec{\tau}_{ij}\cdot\vec{Y}^K + \tfrac{1}{4}\mathrm{i}\sigma^I\sigma^J\bar{\lambda}^{iL}\lambda_i{}^M f_{LM}{}^K\Big].\end{aligned} \tag{20.11}$$

We used the convention $\gamma^{\mu\nu\rho\sigma\tau} = \mathrm{i}\varepsilon^{\mu\nu\rho\sigma\tau}$, i.e. the lower sign in (3.41). For physical applications, as in supergravity, it will be important that σ^I acquires a non-zero value such that the kinetic terms are non-degenerate. Note that, owing to (20.8), this also implies that there is a central charge.

Exercise 20.2 *Check the gauge invariance of the non-abelian Chern–Simons term.*

A remarkable fact is that the Lagrangian (20.11) is superconformal invariant. Therefore, it can be used in a superconformal construction of $D=5$ supergravity, for which we refer to the literature [208, 209, 206, 210, 207]. The kinetic terms of the scalars define a (rigid) very special real manifold, which we will discuss in Sec. 20.3.2.

20.1.3 Gauge multiplets for $D=4$

Now we reduce to four dimensions. The same mechanism that generated the scalar in the reduction from six to five dimensions now generates a second scalar. Therefore, the gauge multiplets of $\mathcal{N}=2$, $D=4$ contain complex scalars $X^I = \frac{1}{2}(A^I_4 - \mathrm{i}\sigma^I)$. We denote the fermion SU(2) doublets as $\Omega_i{}^I$. They correspond to the λ^{iI} in $D=5$, but we give it a different name since the spinors are not symplectic in $D=4$. We work again with chiral spinors in $D=4$, using the notation introduced in Sec. 6A, so that the position of the index i on the fermions indicates their chirality. Whether upper indices indicate left- or right-chiral spinors is chosen independently for each type of spinor separately to make the formulas simpler. We choose here

$$\Omega_i{}^I = P_L\Omega_i{}^I\,, \qquad \Omega^{iI} = P_R\Omega^{iI}\,, \qquad \epsilon^i = P_L\epsilon^i\,, \qquad \epsilon_i = P_R\epsilon_i\,. \tag{20.12}$$

The relation with the spinors of $D=5$ is given in Sec. 20B.2.

The supersymmetry transformations are

$$\begin{aligned}
\delta X^I &= \tfrac{1}{2}\bar{\epsilon}^i\Omega_i^I\,,\\
\delta\Omega_i^I &= \not{D}X^I\epsilon_i + \tfrac{1}{4}\gamma^{\mu\nu}F_{\mu\nu}{}^I\varepsilon_{ij}\epsilon^j + Y_{ij}{}^I\epsilon^j + X^J\bar{X}^K f_{JK}{}^I\varepsilon_{ij}\epsilon^j\,,\\
\delta A_\mu^I &= \tfrac{1}{2}\varepsilon^{ij}\bar{\epsilon}_i\gamma_\mu\Omega_j{}^I + \text{h.c.}\,,\\
\delta\vec{Y}^I &= \tfrac{1}{4}\vec{\tau}^{ij}\bar{\epsilon}_i\not{D}\Omega_j^I - \tfrac{1}{2}f_{JK}{}^I\vec{\tau}_i{}^j\bar{\epsilon}_jX^J\Omega^{iK} + \text{h.c.}\,,
\end{aligned} \tag{20.13}$$

where we included the case of non-abelian gauge multiplets [211], in which all the fields transform in the adjoint of the gauge group.

The supersymmetry transformations realize the algebra

$$\begin{aligned}
&[\delta(\epsilon_1),\delta(\epsilon_2)] = \delta_{\rm cgct}(\xi^\mu) + \delta_G(\theta^I)\,,\\
&\qquad \xi^\mu = \tfrac{1}{2}\bar{\epsilon}_2^i\gamma^\mu\epsilon_{1i} + \text{h.c.}\,, \qquad \theta^I = X^I\varepsilon^{ij}\bar{\epsilon}_{2i}\epsilon_{1j} + \bar{X}^I\varepsilon_{ij}\bar{\epsilon}_2^i\epsilon_1^j\,.
\end{aligned} \tag{20.14}$$

Similar to the case of $D=5$, the additional terms contained in $\delta_G(\theta^I)$ can lead to a central charge for solutions when $X^I\neq 0$.

Exercise 20.3 *Check the supersymmetry algebra both on X^I and on $A_\mu{}^I$ to verify the appearance of the central charge-like term.*

Observe that the supersymmetry transformation of X^I is chiral: $\epsilon_i = P_R\epsilon_i$ does not appear. This is very similar to the transformation of the complex scalar in chiral multiplets of $\mathcal{N}=1$ supersymmetry. In fact, we used this property to define chiral multiplets and the same can be done here. However, for $\mathcal{N}=2$, a chiral multiplet is reducible [212, 213, 214], i.e. the supersymmetry algebra closes on a subset of the fields of the chiral multiplet. A full chiral multiplet has $16+16$ components; a reduced one has $8+8$ components. The gauge multiplet is such a reduced chiral multiplet.[3] This fact is important to understand the construction of the actions for gauge multiplets.

The construction of an action is very similar to the procedure in $\mathcal{N}=1$ supergravity in Sec. 14.2. An arbitrary holomorphic function $F(X)$ determines another $\mathcal{N}=2$ chiral multiplet. It is not a constrained one: it has $16+16$ components. The last component of the latter defines an invariant action, very similar to the procedure of the F-terms and D-terms in $\mathcal{N}=1$; see (14.12). Hence, we find an action that depends on this holomorphic function, which is called the *prepotential*. This Lagrangian is

$$\begin{aligned}
\mathcal{L}_{4v} = {}& \mathrm{i}F_{IJ}D_\mu X^I D^\mu\bar{X}^J + \tfrac{1}{4}\mathrm{i}F_{IJ}F_{\mu\nu}{}^{-I}F^{-\mu\nu J} + \tfrac{1}{2}\mathrm{i}F_{IJ}\bar{\Omega}_i^I\not{D}\Omega^{iJ} - \mathrm{i}F_{IJ}\vec{Y}^I\cdot\vec{Y}^J\\
&+ \tfrac{1}{4}\mathrm{i}F_{IJK}\vec{Y}^I\cdot\vec{\tau}^{ij}\bar{\Omega}_i^J\Omega_j^K\\
&- \tfrac{1}{16}\mathrm{i}F_{IJK}\varepsilon^{ij}\bar{\Omega}_i^I\gamma^{\mu\nu}F_{\mu\nu}{}^{-J}\Omega_j^K + \tfrac{1}{48}\mathrm{i}F_{IJKL}\bar{\Omega}_i^I\Omega_\ell^J\bar{\Omega}_j^K\Omega_k^L\varepsilon^{ij}\varepsilon^{k\ell}
\end{aligned}$$

[3] You can compare this with the real multiplet in $\mathcal{N}=1$ built from a chiral multiplet with $C=\operatorname{Im}Z$, which is used as supergauge transformations in (14.11): its B_μ component is a total derivative of a scalar and its λ and D components vanish.

Gauge multiplets with global $\mathcal{N} = 2$, $D = 4$ supersymmetry **Box 20.2**

The kinetic terms are determined by a holomorphic prepotential function $F(X)$. The latter determines the Kähler potential of the manifold of the scalars. The Kähler manifolds that are of this type are called rigid special Kähler manifolds.

$$
\begin{aligned}
&+\tfrac{1}{2}\mathrm{i}F_I f_{JK}{}^I \bar{\Omega}^{iJ}\Omega^{jK}\varepsilon_{ij} - \tfrac{1}{2}\mathrm{i}F_{IJ} f_{KL}{}^I \bar{X}^K \bar{\Omega}_i^J \Omega_j^L \varepsilon^{ij} \\
&-\mathrm{i}F_I f_{JK}{}^I f_{LM}{}^J \bar{X}^K \bar{X}^L X^M + \text{h.c.}
\end{aligned} \tag{20.15}
$$

Here F_I, $F_{IJ}, \ldots$ denote the derivatives of $F(X)$. The $F_{\mu\nu}{}^{-I}$ are the anti-self-dual field strengths (see Sec. 4.2.1). Let us first consider the action without gauging.

The reader will recognize that the kinetic terms of the scalars X^I are of the hermitian form. In fact, the scalars form a Kähler manifold, with Kähler potential

$$
K = \mathrm{i}X^I \bar{F}_I(\bar{X}) - \mathrm{i}\bar{X}^I F_I(X) . \tag{20.16}
$$

Exercise 20.4 *Check that this Kähler potential leads to the kinetic terms of the scalars in (20.15). Check that a quadratic term in F with real coefficients, i.e. $F = a_{IJ}X^I X^J$ with real a_{IJ}, does not contribute to the Kähler metric.*

The kinetic terms of scalars, gauge fields and gauginos are determined from the imaginary part of the second derivatives. The manifold of the scalars is called a *rigid special Kähler manifold*. It is 'special' in the sense that the Kähler potential is not an arbitrary real function of X and $\bar{X}$, but of the form (20.16) where $F(X)$ is an arbitrary holomorphic function. We will discuss this from the geometric point of view in Sec. 20.3.3 and Appendix 20C.

The simplest action is obtained by choosing the prepotential $F(X) = \mathrm{i}\delta_{IJ}X^I X^J$. Then $\mathrm{i}F_{IJ} = -\delta_{IJ}$, and if the algebra is abelian, we just have the first line of (20.15). On the other hand, for $F(X) = -\frac{1}{6}C_{IJK}X^I X^J X^K$, this action is the reduction of the five-dimensional one in (20.11).

In the case of gauging, the covariant derivatives are the gauge covariant expressions. We assumed here that F is a gauge invariant function, i.e.

$$
\delta_G F \equiv F_I \theta^K X^J f_{JK}{}^I = 0 , \tag{20.17}
$$

although more generally one may allow a quadratic variation of F with real coefficients [215, 216]:

$$
\delta_G F = -\theta^K C_{K,IJ} X^I X^J , \tag{20.18}
$$

where $C_{K,IJ}$ are real constants[4]

[4] Do not confuse these coefficients $C_{I,JK}$ with the coefficients C_{IJK} that appear in the $D = 5$ gauge multiplets.

$$C_{K,IJ} = -f_{K(I}{}^L F_{J)L} = -f_{K(I}{}^L \bar{F}_{J)L} \,. \tag{20.19}$$

These constants enter in an additional Chern–Simons term that should be added to the action (which we will write below for the theory with supergravity). Such a possibility exists because quadratic terms in F with real coefficients lead only to total derivatives in (20.15); see Ex. 20.4.

20.1.4 Hypermultiplets

Hypermultiplets [213, 214] are the analogues of the chiral multiplets of $\mathcal{N}=1$ supersymmetry. They contain four scalars and two spin-1/2 fields. Note that the off-shell number of degrees of freedom do not match. To match them, we would have to introduce four auxiliary fields. This can be done, but (for non-trivial hypermultiplet couplings) involves a more complicated procedure using off-shell central charges.

Since dimensional reduction for scalars and spin-1/2 fermions leads to the same type of particles in lower dimensions, the properties of the hypermultiplets do not depend on whether we consider $D=6$, $D=5$ or $D=4$ (or even $D=3$). In practice, there is a technical difference since the four on-shell (or eight off-shell) degrees of freedom are captured in symplectic Weyl, symplectic or Majorana spinors, respectively.

We consider a set of n_H hypermultiplets. The real scalars are denoted as q^X, with $X=1,\ldots,4n_H$. We will see that they are coordinates on a manifold with a hypercomplex structure, which is Kähler in the case of global supersymmetry, and is then called a 'hyper-Kähler manifold'. For supergravity the scalars span a quaternionic manifold. We will explain these notions in Sec. 20.3. The $2n_H$ fermions are indicated by $\zeta^{\mathcal{A}}$, $\mathcal{A}=1,\ldots,2n_H$. These fields are connected by a frame field $f^X{}_{i\mathcal{A}}(q)$, which can be seen as a $4n_H\times 4n_H$ invertible matrix. The inverse is written as $f^{i\mathcal{A}}{}_X(q)$, i.e.

$$f^{i\mathcal{A}}{}_Y f^X{}_{i\mathcal{A}} = \delta^X_Y \,, \qquad f^{i\mathcal{A}}{}_X f^X{}_{j\mathcal{B}} = \delta^i_j \delta^{\mathcal{A}}_{\mathcal{B}} \,. \tag{20.20}$$

The frame field satisfies a reality condition

$$\left(f^{i\mathcal{A}}{}_X\right)^* = f^{j\mathcal{B}}{}_X \varepsilon_{ji} \rho_{\mathcal{B}\bar{\mathcal{A}}} \,, \qquad \left(f^X{}_{i\mathcal{A}}\right)^* = \varepsilon^{ij} \rho^{\bar{\mathcal{A}}\mathcal{B}} f^X{}_{j\mathcal{B}} \,, \tag{20.21}$$

in terms of a non-degenerate tensor $\rho_{\mathcal{A}\bar{\mathcal{B}}}$ that satisfies

$$\rho_{\mathcal{A}\bar{\mathcal{B}}} \rho^{\bar{\mathcal{B}}\mathcal{C}} = -\delta^{\mathcal{C}}_{\mathcal{A}} \,, \qquad \rho^{\bar{\mathcal{A}}\mathcal{B}} = \left(\rho_{\mathcal{A}\bar{\mathcal{B}}}\right)^* \,. \tag{20.22}$$

The above conditions have as consequence that

$$2 f^{i\mathcal{A}}{}_X f^Y{}_{j\mathcal{A}} = \delta^Y_X \delta^i_j + \vec{\tau}_j{}^i \cdot \vec{J}_X{}^Y \,, \qquad \vec{J}_X{}^Y = \left(\vec{J}_X{}^Y\right)^* = -f^{i\mathcal{A}}{}_X f^Y{}_{j\mathcal{A}} \vec{\tau}_i{}^j \,, \tag{20.23}$$

where the latter are almost quaternionic structures. This means that they satisfy the quaternion algebra: for any vectors $\vec{A}$ and $\vec{B}$,

$$\vec{A}\cdot\vec{J}_X{}^Z\,\vec{B}\cdot\vec{J}_Z{}^Y = -\delta_X{}^Y\vec{A}\cdot\vec{B} + (\vec{A}\times\vec{B})\cdot\vec{J}_X{}^Y\,. \tag{20.24}$$

The frame field is covariantly constant using a connection related to the indices $\mathcal{A}$, and a torsionless connection related to the indices X:

$$\begin{aligned}\nabla_Y f^X{}_{i\mathcal{A}} &\equiv \partial_Y f^X{}_{i\mathcal{A}} - \omega_{Y\mathcal{A}}{}^{\mathcal{B}}(q)f^X{}_{i\mathcal{B}} + \Gamma^X_{YZ}(q)f^Z{}_{i\mathcal{A}} = 0\,,\\ \nabla_Y f^{i\mathcal{A}}{}_X &\equiv \partial_Y f^{i\mathcal{A}}{}_X + f^{i\mathcal{B}}{}_X\omega_{Y\mathcal{B}}{}^{\mathcal{A}}(q) - \Gamma^Z_{YX}(q)f^{i\mathcal{A}}{}_Z = 0\,,\end{aligned} \tag{20.25}$$

where the second equation follows from the first using (20.21). We can solve these equations for $\omega_{X\mathcal{A}}{}^{\mathcal{B}}$,

$$\omega_{X\mathcal{A}}{}^{\mathcal{B}} = \frac{1}{2}f^{i\mathcal{B}}{}_Y\left(\partial_X f^Y{}_{i\mathcal{A}} + \Gamma^Y_{XZ}f^Z{}_{i\mathcal{A}}\right), \tag{20.26}$$

such that the independent connection is $\Gamma_{XY}{}^Z$. The latter is the unique connection on the manifold of the scalars q^X that 'preserves' the complex structures in (20.23):

$$\nabla_Z\vec{J}_X{}^Y \equiv \partial_Z\vec{J}_X{}^Y - \Gamma_{ZX}{}^U\vec{J}_U{}^Y + \Gamma_{ZU}{}^Y\vec{J}_X{}^U = 0\,. \tag{20.27}$$

A connection satisfying this property is called the 'Obata' connection [217]. Below, we will also recognize $\Gamma_{XY}{}^Z$ as the Christoffel connection of a metric. With condition (20.27) the almost quaternionic structure is promoted to a 'quaternionic structure'. The complex conjugate of $\omega_{X\mathcal{A}}{}^{\mathcal{B}}$ is

$$\left(\omega_{X\mathcal{A}}{}^{\mathcal{B}}\right)^* \equiv \bar{\omega}_X{}^{\bar{\mathcal{A}}}{}_{\bar{\mathcal{B}}} = -\rho^{\bar{\mathcal{A}}\mathcal{C}}\omega_{X\mathcal{C}}{}^{\mathcal{D}}\rho_{\mathcal{D}\bar{\mathcal{B}}}\,. \tag{20.28}$$

The matrix $\rho_{\mathcal{A}\bar{\mathcal{B}}}$ used in these reality conditions, which should also be covariantly constant, enters also in the definition of the symplectic Majorana spinors of hypermultiplets in five and six dimensions.[5] The spinors of the $D=6$ hypermultiplets are moreover right-handed chiral. For $D=4$ on the other hand, charge conjugation raises or lowers the indices on the fermions (and changes the chirality):

$$\begin{aligned}D=6:&\quad (\zeta^{\mathcal{A}})^C = \zeta^{\mathcal{B}}\rho_{\mathcal{B}\bar{\mathcal{A}}}\,, &\quad \zeta^{\mathcal{A}} = P_R\zeta^{\mathcal{A}}\,,\\ D=5:&\quad (\zeta^{\mathcal{A}})^C = \zeta^{\mathcal{B}}\rho_{\mathcal{B}\bar{\mathcal{A}}}\,, &\\ D=4:&\quad (\zeta^{\mathcal{A}})^C = \zeta_{\bar{\mathcal{A}}} = P_R\zeta_{\bar{\mathcal{A}}}\,, &\quad (\zeta_{\bar{\mathcal{A}}})^C = \zeta^{\mathcal{A}} = P_L\zeta^{\mathcal{A}}\,.\end{aligned} \tag{20.29}$$

The supersymmetry transformation laws are simplest for $D=6$:

$$\begin{aligned}D=6:\quad \delta q^X &= \bar{\epsilon}^i\zeta^{\mathcal{A}}f^X{}_{i\mathcal{A}}\,,\\ \delta\zeta^{\mathcal{A}} &= \tfrac{1}{2}f^{i\mathcal{A}}{}_X\not{D}q^X\epsilon_i - \zeta^{\mathcal{B}}\omega_{X\mathcal{B}}{}^{\mathcal{A}}\delta q^X\,.\end{aligned} \tag{20.30}$$

Notice that the last term is necessary for covariance under a change of basis of the spinors $\zeta^{\mathcal{A}}$ and drops out of the definition of covariant transformations as explained in Appendix

[5] Note that we defined symplectic Majorana spinors in (3.86) using an antisymmetric ε_{ij}. However, the condition (20.22) suffices.

14B. Furthermore, we allow that the hypermultiplet transforms also under a gauge group, using the covariant derivative

$$D_\mu q^X = \partial_\mu q^X - A_\mu{}^I k_I{}^X . \tag{20.31}$$

We will discuss the properties of the Killing vectors later. The supersymmetry algebra (20.5) is satisfied on the q^X fields. Calculating the commutator of two transformations on the fermions leads to a difference with the right-hand side of (20.5) that looks like an equation of motion (despite the fact that we did not yet discuss an action).

The step to five dimensions is easy. Owing to the fact that the charge conjugation satisfies different properties, we have to include some factors i to respect reality conditions. Furthermore, the scalar σ^I of the gauge multiplet enters (as sixth component of the gauge vector of $D=6$), such that the supersymmetry transformation laws are

$$\begin{aligned} D=5: \qquad \delta q^X &= -\mathrm{i}\bar\epsilon^i \zeta^{\mathcal{A}} f^X{}_{i\mathcal{A}} , \\ \delta\zeta^{\mathcal{A}} &= \tfrac12 \mathrm{i} f^{i\mathcal{A}}{}_X \not{D} q^X \epsilon_i - \zeta^{\mathcal{B}} \omega_{X\mathcal{B}}{}^{\mathcal{A}} \delta q^X + \tfrac12 \sigma^I k_I{}^X f^{i\mathcal{A}}{}_X \epsilon_i . \end{aligned} \tag{20.32}$$

For $D=4$, the transformation laws are[6] [95, 219]

$$\begin{aligned} D=4: \qquad \delta q^X &= -\mathrm{i}\bar\epsilon^i \zeta^{\mathcal{A}} f^X{}_{i\mathcal{A}} + \mathrm{i}\varepsilon^{ij} \rho^{\bar{\mathcal{A}}\mathcal{B}} \bar\epsilon_i \zeta_{\bar{\mathcal{A}}} f^X{}_{j\mathcal{B}} , \\ \delta\zeta^{\mathcal{A}} &= \tfrac12 \mathrm{i} f^{i\mathcal{A}}{}_X \not{D} q^X \epsilon_i - \zeta^{\mathcal{B}} \omega_{X\mathcal{B}}{}^{\mathcal{A}} \delta q^X + \mathrm{i}\bar X^I k_I{}^X f^{i\mathcal{A}}{}_X \varepsilon_{ij} \epsilon^j . \end{aligned} \tag{20.33}$$

Observe that the transformations of the scalars in $D=6$ and $D=5$ are real due to the properties of the spinors. For $D=4$ the dimensional reduction according to (20.214) leads to a second term that is the complex conjugate of the first one using (20.21).

To construct an action, we need a metric. The first new ingredient is a (hermitian) metric in tangent space $d^{\bar{\mathcal{A}}}{}_{\mathcal{B}} = (d^{\bar{\mathcal{B}}}{}_{\mathcal{A}})^*$. For applications in global supersymmetry, we can assume $d^{\bar{\mathcal{A}}}{}_{\mathcal{B}} = \delta^{\bar{\mathcal{A}}}{}_{\mathcal{B}}$. However, for the superconformal construction of supergravity we will also be interested in a metric with non-positive definite signature. This tensor $d^{\bar{\mathcal{A}}}{}_{\mathcal{B}}$ should respect the quaternionic structure, which can most simply be expressed as the antisymmetry of a tensor $C_{\mathcal{A}\mathcal{B}}$:

$$C_{\mathcal{A}\mathcal{B}} = -C_{\mathcal{B}\mathcal{A}} , \qquad C_{\mathcal{A}\mathcal{B}} \equiv \rho_{\mathcal{A}\bar{\mathcal{C}}} d^{\bar{\mathcal{C}}}{}_{\mathcal{B}} . \tag{20.34}$$

It has been proven in [215], using the theorems of [220], that at any point one can choose a basis such that

$$\rho_{\mathcal{A}\bar{\mathcal{B}}} = \begin{pmatrix} 0 & \mathbb{1}_{n_H} \\ -\mathbb{1}_{n_H} & 0 \end{pmatrix} , \tag{20.35}$$

and at the same time

$$d^{\bar{\mathcal{A}}}{}_{\mathcal{B}} = \begin{pmatrix} \mathbb{1}_p & & & \\ & -\mathbb{1}_q & & \\ & & \mathbb{1}_p & \\ & & & -\mathbb{1}_q \end{pmatrix} , \qquad p+q = n_H . \tag{20.36}$$

[6] The formulation for $D=4$ used here, which is similar to the one for $D=5$, has been obtained in [218].

The metric of the scalars is defined from the metric $d^{\bar{\mathcal{A}}}{}_{\mathcal{B}}$ and the frame fields:

$$g_{XY} = (f^{i\bar{\mathcal{A}}}{}_X)^* d^{\bar{\mathcal{A}}}{}_{\mathcal{B}} f^{i\mathcal{B}}{}_Y = f^{i\mathcal{A}}{}_X \varepsilon_{ij} C_{\mathcal{AB}} f^{j\mathcal{B}}{}_Y . \tag{20.37}$$

One further needs the covariant constancy of $C_{\mathcal{AB}}$:

$$\nabla_X C_{\mathcal{AB}} \equiv \partial_X C_{\mathcal{AB}} + 2\omega_{X[\mathcal{A}}{}^{\mathcal{C}} C_{\mathcal{B}]\mathcal{C}} = 0 . \tag{20.38}$$

This implies the covariant constancy of g_{XY}, and implies therefore also that $\Gamma^X_{YZ}(q)$ is the Christoffel connection of the metric.

We use the antisymmetric tensor $C_{\mathcal{AB}}$ to raise and lower $\mathcal{A}$ indices in NW–SE convention. For an arbitrary $\phi^{\mathcal{A}}$:

$$\phi^{\mathcal{A}} = C^{\mathcal{AB}}\phi_{\mathcal{B}} , \qquad \phi_{\mathcal{A}} = \phi^{\mathcal{B}} C_{\mathcal{BA}} , \qquad C^{\mathcal{AC}} C_{\mathcal{BC}} = \delta^{\mathcal{A}}_{\mathcal{B}} . \tag{20.39}$$

The frame fields with upper X indices were defined in (20.20) as the inverse of those with lower X indices. This is consistent with raising and lowering indices with the metric:

$$f_{Xi\mathcal{A}} = g_{XY} f^Y{}_{i\mathcal{A}} = f^{j\mathcal{B}}{}_X \varepsilon_{ji} C_{\mathcal{BA}} . \tag{20.40}$$

Exercise 20.5 *Check the reality condition*

$$C^{\bar{\mathcal{A}}\bar{\mathcal{B}}} \equiv (C_{\mathcal{AB}})^* = d^{\bar{\mathcal{A}}}{}_{\mathcal{C}} d^{\bar{\mathcal{B}}}{}_{\mathcal{D}} C^{\mathcal{CD}} = \rho^{\bar{\mathcal{A}}\mathcal{C}} d^{\bar{\mathcal{B}}}{}_{\mathcal{C}} , \tag{20.41}$$

and that the raising and lowering conventions imply that $d^{\bar{\mathcal{A}}}{}_{\mathcal{B}} = -\rho^{\bar{\mathcal{A}}}{}_{\mathcal{B}}$.

We will discuss the geometry of these manifolds in Sec. 20.3.4, but we need curvature relations to present the action. The curvature is defined in the usual way:

$$\mathcal{R}_{XY\mathcal{B}}{}^{\mathcal{A}} \equiv 2\partial_{[X}\omega_{Y]\mathcal{B}}{}^{\mathcal{A}} + 2\omega_{[X|\mathcal{C}|}{}^{\mathcal{A}}\omega_{Y]\mathcal{B}}{}^{\mathcal{C}} . \tag{20.42}$$

It satisfies reality equations similar to (20.28). The integrability condition of the covariant constancy conditions of $C_{\mathcal{AB}}$ and $d^{\bar{\mathcal{A}}}{}_{\mathcal{B}}$ implies relations on the curvature:

$$\begin{aligned} 0 &= [\nabla_X, \nabla_Y] C_{\mathcal{AB}} = 2\mathcal{R}_{XY[\mathcal{A}}{}^{\mathcal{C}} C_{\mathcal{B}]\mathcal{C}} = -2\mathcal{R}_{XY[\mathcal{AB}]} , \\ 0 &= [\nabla_X, \nabla_Y] d^{\bar{\mathcal{A}}}{}_{\mathcal{B}} = -\left(\mathcal{R}_{XY\mathcal{A}}{}^{\mathcal{C}}\right)^* d^{\bar{\mathcal{C}}}{}_{\mathcal{B}} - \mathcal{R}_{XY\mathcal{B}}{}^{\mathcal{C}} d^{\bar{\mathcal{A}}}{}_{\mathcal{C}} . \end{aligned} \tag{20.43}$$

These say that the curvature matrices preserve the antisymmetric metric $C_{\mathcal{AB}}$ and the metric $d^{\bar{\mathcal{A}}}{}_{\mathcal{B}}$, and that they are therefore in the algebra $\mathfrak{usp}(2p, 2q)$; see (B.2). This is expressed by the statement that the 'structure group' is USp$(2p, 2q)$.

The integrability condition on the covariant constant frame fields (20.25) leads to a relation between the curvature defined by Γ^Z_{XY} and the curvature in (20.42), as is the case for the similar spacetime curvatures. Further use of the cyclicity properties of the curvatures that we studied in (7.10) leads to an expression in terms of a tensor $W_{\mathcal{ABCD}}$:

$$\begin{aligned} R_{XY}{}^W{}_Z = f^W{}_{i\mathcal{A}} f^{i\mathcal{B}}{}_Z \mathcal{R}_{XY\mathcal{B}}{}^{\mathcal{A}} &= \tfrac{1}{2} f^{\mathcal{A}i}{}_X f_i{}^{\mathcal{B}}{}_Y f^{k\mathcal{C}}{}_Z f^W{}_{k\mathcal{D}} W_{\mathcal{ABC}}{}^{\mathcal{D}} , \\ W_{\mathcal{ABCD}} \equiv f^{Xi}{}_{\mathcal{A}} f^Y{}_{i\mathcal{B}} \mathcal{R}_{XY\mathcal{CD}} &= \tfrac{1}{2} f^{Xi}{}_{\mathcal{A}} f^Y{}_{i\mathcal{B}} f^{Zk}{}_{\mathcal{C}} f^W{}_{k\mathcal{D}} R_{XYZW} . \end{aligned} \tag{20.44}$$

Box 20.3 **Hypermultiplets in global $\mathcal{N}=2$ supersymmetry**

The kinetic terms of the hypermultiplets are governed by a metric that respects three complex structures. The corresponding scalar manifold is a hyper-Kähler manifold, which has vanishing Ricci tensor. Gauging is determined by a triplet of moment maps, defining triholomorphic Killing vectors.

The tensor $W_{\mathcal{ABCD}} \equiv W_{\mathcal{ABC}}{}^{\mathcal{E}} C_{\mathcal{ED}}$ is symmetric in its four indices. This will be consistent with its appearance in the four-fermion term of the action, similar to the appearance of the curvature term in the kinetic action of the $\mathcal{N}=1$ chiral multiplet (14.29). Moreover, it implies that the manifold is *Ricci flat*:

$$R_{YZ} = R_{XY}{}^{X}{}_{Z} = 0\,. \tag{20.45}$$

The Lagrangians for the different dimensions are

$$\begin{aligned}
\mathcal{L}_6 = \mathcal{L}_5 &= -\tfrac{1}{2} g_{XY}\partial_\mu q^X \partial^\mu q^Y + \bar\zeta_{\mathcal{A}} \nabla\!\!\!\!/\, \zeta^{\mathcal{A}} - \tfrac{1}{4} W_{\mathcal{ABCD}} \bar\zeta^{\mathcal{A}} \zeta^{\mathcal{B}} \bar\zeta^{\mathcal{C}} \zeta^{\mathcal{D}}\,,\\
\mathcal{L}_4 = -\tfrac{1}{2} g_{XY}\partial_\mu q^X \partial^\mu q^Y &- d^{\bar{\mathcal{A}}}{}_{\mathcal{B}} \left(\bar\zeta_{\bar{\mathcal{A}}} \nabla\!\!\!\!/\, \zeta^{\mathcal{B}} + \bar\zeta^{\mathcal{B}} \nabla\!\!\!\!/\, \zeta_{\bar{\mathcal{A}}}\right)\\
&+ \tfrac{1}{2} W_{\mathcal{AB}}{}^{\mathcal{EF}} d^{\bar{\mathcal{C}}}{}_{\mathcal{E}} d^{\bar{\mathcal{D}}}{}_{\mathcal{F}} \bar\zeta_{\bar{\mathcal{C}}} \zeta_{\bar{\mathcal{D}}} \bar\zeta^{\mathcal{A}} \zeta^{\mathcal{B}}\,,
\end{aligned} \tag{20.46}$$

where $\nabla_\mu \zeta^{\mathcal{A}} = \partial_\mu \zeta^{\mathcal{A}} + \partial_\mu q^X \omega_{X\mathcal{B}}{}^{\mathcal{A}} \zeta^{\mathcal{B}}$.

Exercise 20.6 *For the simplest hypermultiplet, $n_H = 1$, we can take $\rho_{\mathcal{A}\bar{\mathcal{B}}} = \varepsilon_{\mathcal{AB}}$. Check that the reality constraints allow the solution where the only non-vanishing elements of $f^{i\mathcal{A}}{}_X$ are*

$$\begin{aligned}
f^{12}{}_1 = f^{21}{}_1 = \mathrm{i}\,, &\qquad -f^{12}{}_2 = f^{21}{}_2 = 1\,,\\
f^{11}{}_3 = -f^{22}{}_3 = \mathrm{i}\,, &\qquad f^{11}{}_4 = f^{22}{}_4 = 1\,.
\end{aligned} \tag{20.47}$$

Choose $d^{\bar{\mathcal{A}}}{}_{\mathcal{B}}$ such that $g_{XY} = \delta_{XY}$. Since these are all constants, $\Gamma^Z_{XY} = \omega_{X\mathcal{A}}{}^{\mathcal{B}} = W_{\mathcal{ABCD}} = 0$. Calculate the complex structure, and with the notation $\alpha = 1,2,3$, for three components of $\vec{J}$ as well as for $X = 1,2,3$, find that

$$(J^\alpha)_\beta{}^\gamma = -\varepsilon_{\alpha\beta\gamma}\,, \qquad (J^\alpha)_\beta{}^4 = -(J^\alpha)_4{}^\beta = \delta^\alpha_\beta\,. \tag{20.48}$$

20.1.5 Gauged hypermultiplets

We now consider the gauging of symmetries in the action of the hypermultiplet. This leads to interactions of the gauge multiplet and the hypermultiplet, and is the analogue of the interaction of chiral and gauge multiplets in $\mathcal{N}=1$; see Sec. 6.3.2. It will turn out that this mechanism will also allow us to describe massive hypermultiplets. To describe massive hypermultiplets we need central charges. In (20.8) and (20.14) we saw that central charges are obtained by gauge symmetries with non-vanishing scalars in the gauge multiplet.

Symmetries are defined by Killing vectors of the metric g_{XY}, denoted by $k_I{}^X$. For Kähler manifolds, we saw that these Killing vectors should be holomorphic, which means (in the present notation) that $\nabla_X k_I{}^Y$ is a matrix that must commute with the complex structure; see (13.47). The requirement for hypermultiplets is that the Killing vectors should be *triholomorphic*, meaning that this matrix must commute with the three complex structures (20.23):

$$\left(\nabla_X k_I{}^Y\right) \vec{J}_Y{}^Z = \vec{J}_X{}^Y \left(\nabla_Y k_I{}^Z\right) . \tag{20.49}$$

For the Kähler manifolds, this requirement implied that the holomorphic Killing vectors are related to a moment map as in (13.51). Here we obtain that they are related to a triplet of moment maps:

$$\partial_X \vec{P}_I = \vec{J}_X{}^Y k_{IY} . \tag{20.50}$$

In the case that the Killing vectors satisfy a non-abelian algebra, we also obtain that supersymmetry implies an 'equivariance relation', which is the analogue of (13.62):

$$k_I{}^X \vec{J}_{XY} k_J{}^Y = f_{IJ}{}^K \vec{P}_K . \tag{20.51}$$

Exercise 20.7 *For the flat hyper-Kähler manifold of Ex. 20.6, $k_X = \lambda_{XY} q^Y$ for any antisymmetric λ_{XY} defines* SO(4) *isometries. Check that the complex structure (20.48) is anti-self-dual using the antisymmetric tensor: $\varepsilon_{1234} = 1$. Check then that any self-dual tensor λ_{XY} defines a triholomorphic symmetry. Hint: use (4.39). This implies that, from the* SO(4) = SU(2) × SU(2) *symmetries, the complex structures sit in one factor, and the triholomorphic isometries are those that are in the other factor.*

Exercise 20.8 *Continue with the same example, and use the triholomorphic Killing vector*

$$k^1 = q^2, \qquad k^2 = -q^1, \qquad k^3 = q^4, \qquad k^4 = -q^3 . \tag{20.52}$$

Check that upon changing the signs of k^3 and k^4, it becomes a non-triholomorphic Killing vector. Check that the moment maps corresponding to (20.52) are

$$\begin{aligned} P^1 &= -q^1 q^3 - q^2 q^4 , \\ P^2 &= q^1 q^4 - q^2 q^3 , \\ P^3 &= \tfrac{1}{2}\left[(q^1)^2 + (q^2)^2 - (q^3)^2 - (q^4)^2\right] . \end{aligned} \tag{20.53}$$

The transformations of the fermions under the gauge group are governed by a (q-dependent) matrix[7]

$$t_{I\mathcal{A}}{}^{\mathcal{B}} \equiv \tfrac{1}{2} f^Y{}_{i\mathcal{A}} \nabla_Y k_I{}^X f^{i\mathcal{B}}{}_X . \tag{20.54}$$

[7] Under covariant transformations as in (14.122) the gauge transformation is $\hat{\delta}_G \zeta^{\mathcal{A}} = \theta^I \zeta^{\mathcal{B}} t_{I\mathcal{B}}{}^{\mathcal{A}}$.

These matrices satisfy the commutation relations

$$[t_I, t_J]_{\mathcal{B}}{}^{\mathcal{A}} = f_{IJ}{}^K t_{K\mathcal{B}}{}^{\mathcal{A}} - k_I^X k_J^Y \mathcal{R}_{XY\mathcal{B}}{}^{\mathcal{A}} . \tag{20.55}$$

The properties

$$\begin{aligned} &t_I{}^{\mathcal{A}\mathcal{B}} \equiv C^{\mathcal{A}\mathcal{C}} t_{I\mathcal{C}}{}^{\mathcal{B}} = t_I{}^{\mathcal{B}\mathcal{A}} , \\ &\bar{t}_I{}^{\bar{\mathcal{A}}}{}_{\bar{\mathcal{B}}} \equiv (t_{I\mathcal{A}}{}^{\mathcal{B}})^* = -\rho^{\bar{\mathcal{A}}\mathcal{D}} t_{I\mathcal{D}}{}^{\mathcal{C}} \rho_{\mathcal{C}\bar{\mathcal{B}}} , \qquad \bar{t}_I{}^{\bar{\mathcal{A}}}{}_{\bar{\mathcal{C}}} d^{\bar{\mathcal{C}}}{}_{\mathcal{B}} = -d^{\bar{\mathcal{A}}}{}_{\mathcal{C}} t_{I\mathcal{B}}{}^{\mathcal{C}} \end{aligned} \tag{20.56}$$

are similar to the properties of the curvature tensor (20.43), and imply that these matrices are elements of $\mathfrak{usp}(2n_H - 2p, 2p)$.

With these ingredients we can write the actions of the gauged hypermultiplets:

$$\begin{aligned} \mathcal{L}_6 = &-\tfrac{1}{2} g_{XY} D_\mu q^X D^\mu q^Y + \bar{\zeta}_{\mathcal{A}} \hat{\not{D}} \zeta^{\mathcal{A}} - \tfrac{1}{4} W_{\mathcal{A}\mathcal{B}\mathcal{C}\mathcal{D}} \bar{\zeta}^{\mathcal{A}} \zeta^{\mathcal{B}} \bar{\zeta}^{\mathcal{C}} \zeta^{\mathcal{D}} \\ &- 2k_I^X f^{i\mathcal{A}}{}_X \bar{\zeta}_{\mathcal{A}} \lambda_i{}^I + 2\vec{P}_I \cdot \vec{Y}^I , \\ \mathcal{L}_5 = &-\tfrac{1}{2} g_{XY} D_\mu q^X D^\mu q^Y + \bar{\zeta}_{\mathcal{A}} \hat{\not{D}} \zeta^{\mathcal{A}} - \tfrac{1}{4} W_{\mathcal{A}\mathcal{B}\mathcal{C}\mathcal{D}} \bar{\zeta}^{\mathcal{A}} \zeta^{\mathcal{B}} \bar{\zeta}^{\mathcal{C}} \zeta^{\mathcal{D}} \\ &- 2\mathrm{i} k_I^X f^{i\mathcal{A}}{}_X \bar{\zeta}_{\mathcal{A}} \lambda_i{}^I + \mathrm{i}\sigma^I t_{I\mathcal{B}}{}^{\mathcal{A}} \bar{\zeta}_{\mathcal{A}} \zeta^{\mathcal{B}} + 2\vec{P}_I \cdot \vec{Y}^I - \tfrac{1}{2} \sigma^I \sigma^J k_I^X k_{JX} , \\ \mathcal{L}_4 = &-\tfrac{1}{2} g_{XY} D_\mu q^X D^\mu q^Y - \left(\bar{\zeta}_{\bar{\mathcal{A}}} \hat{\not{D}} \zeta^{\mathcal{B}} d^{\bar{\mathcal{A}}}{}_{\mathcal{B}} + \text{h.c.} \right) + \tfrac{1}{2} W_{\mathcal{A}\mathcal{B}}{}^{\mathcal{E}\mathcal{F}} d^{\bar{\mathcal{C}}}{}_{\mathcal{E}} d^{\bar{\mathcal{D}}}{}_{\mathcal{F}} \bar{\zeta}_{\bar{\mathcal{C}}} \zeta_{\bar{\mathcal{D}}} \bar{\zeta}^{\mathcal{A}} \zeta^{\mathcal{B}} \\ &+ \left(2X^I t_{I\mathcal{A}\mathcal{B}} \bar{\zeta}^{\mathcal{A}} \zeta^{\mathcal{B}} + 2\mathrm{i} f_X^{i\mathcal{A}} k_I^X \bar{\zeta}_{\bar{\mathcal{B}}} \Omega^{jI} \varepsilon_{ij} d^{\bar{\mathcal{B}}}{}_{\mathcal{A}} + \text{h.c.} \right) \\ &+ 2\vec{P}_I \cdot \vec{Y}^I - 2\bar{X}^I X^J k_I{}^X k_{JX} . \end{aligned} \tag{20.57}$$

The second line in $\mathcal{L}_6$, and corresponding lines in $\mathcal{L}_5$ and $\mathcal{L}_4$, are similar to S_{coupling} in (6.59). The covariant derivatives read

$$\begin{aligned} D_\mu q^X &= \partial_\mu q^X - A_\mu^I k_I^X , \\ \hat{D}_\mu \zeta^{\mathcal{A}} &= \partial_\mu \zeta^{\mathcal{A}} + \partial_\mu q^X \omega_{X\mathcal{B}}{}^{\mathcal{A}} \zeta^{\mathcal{B}} - A_\mu^I t_{I\mathcal{B}}{}^{\mathcal{A}} \zeta^{\mathcal{B}} . \end{aligned} \tag{20.58}$$

Exercise 20.9 *Check that, for the gauging (20.52), $t_{I\mathcal{A}}{}^{\mathcal{B}} = \mathrm{i}(\sigma_3)_{\mathcal{A}}{}^{\mathcal{B}}$. Calculate the action for $D = 4$, and use for the gauge multiplet $X^I = m$ with the other fields equal to zero. You should find*

$$\mathcal{L}_4 = -\tfrac{1}{2} \partial_\mu q^X \partial^\mu q^X - \bar{\zeta}_{\mathcal{A}} \not{\partial} \zeta^{\mathcal{A}} + 2\mathrm{i}m \left(\bar{\zeta}_1 \zeta_2 - \bar{\zeta}^1 \zeta^2 \right) - 2m^2 q^X q^X . \tag{20.59}$$

The last exercise, which could as well be done for $D = 5$, shows how a massive hypermultiplet is obtained. Remark that a gauge multiplet with first scalar equal to a constant and all other fields equal to zero is consistent with the transformations of abelian gauge multiplets.

This finishes the overview of $\mathcal{N} = 2$ global supersymmetry. The final actions in $D = 4, 5, 6$ are the sums of those of the gauge multiplets and of the hypermultiplets.

Box 20.4 **Superconformal construction of matter-coupled $\mathcal{N} = 2$ supergravity**

Weyl multiplet +
$n_V + 1$ vector multiplets and $n_H + 1$ hypermultiplets $\Rightarrow$
embedding rigid special Kähler manifold and
hyper-Kähler manifold with conformal symmetry

$\Downarrow$ gauge fix and field equations

Poincaré supergravity +
n_V vector multiplets and n_H hypermultiplets $\Rightarrow$
special Kähler manifold and
quaternionic-Kähler manifold for physical fields

20.2 $\mathcal{N} = 2$ superconformal calculus

The construction of the full supergravity theories for $\mathcal{N} = 2$ uses the same superconformal methods as in Ch. 17 for $\mathcal{N} = 1$. In this case there will be two compensating multiplets: a gauge multiplet and a hypermultiplet. Only one field of these compensating multiplets is physical: the vector of the gauge multiplet. It combines with the graviton and gravitini, which are parts of the Weyl multiplet, to make the pure $\mathcal{N} = 2$ Poincaré supergravity multiplet. The Weyl multiplet gauges the superconformal symmetries. Therefore, we first have to study the $\mathcal{N} = 2$ superconformal algebra and its gauging, and then define the matter multiplets as representations of that algebra. When this is done, and a corresponding action is found, we can perform the gauge fixing that leads to the matter-coupled $\mathcal{N} = 2$ supergravity.

The general scheme of matter couplings is shown in Box 20.4.

20.2.1 The superconformal algebra

The superconformal algebra contains as bosonic part the direct sum of the conformal algebra and the R-symmetry. As can be seen from Table 12.1, the R-symmetry group is $\mathrm{USp}(2) = \mathrm{SU}(2)$ for $D = 6$ and $D = 5$, and $\mathrm{SU}(2) \times \mathrm{U}(1)$ for $D = 4$. There are 16 real supercharges: the eight generators Q of ordinary supersymmetry, and the eight special supersymmetries S. The simple superalgebras that have this content are

$$D = 6\colon \mathrm{OSp}(8^*|2)\,, \qquad D = 5\colon F^2(4)\,, \qquad D = 4\colon \mathrm{SU}(2,2|2)\,. \tag{20.60}$$

Despite the different names, for which the reader can consult Appendix B.3, these algebras are very similar. The conformal algebra can be found in (15.9). For the three generators of the $\mathfrak{su}(2)$ algebra, we use the triplet notation $\vec{U}$ or the traceless 2×2 notation $U_i{}^j$, which are related as in (20.207). The $\mathfrak{su}(2)$ algebra can be written as

$$[U_i{}^j, U_k{}^\ell] = \delta_i{}^\ell U_k{}^j - \delta_k{}^j U_i{}^\ell , \tag{20.61}$$

while in the triplet notation it can be written as $[U^1, U^2] = U^3, \ldots$. For $D=4$ we denote the generator of the U(1) part of the R-symmetry group as T, as we did for $\mathcal{N}=1$. This T commutes with all the other bosonic generators.

The fermionic generators are symplectic Majorana–Weyl spinors for $D=6$ with chirality[8]

$$D=6: \qquad Q^i = P_R Q^i = -\gamma_* Q^i , \qquad S^i = P_L S^i = \gamma_* S^i . \tag{20.62}$$

For $D=4$ we use the chiral notation as in (6.82), i.e. Q_i are left-handed. For the special supersymmetries, the opposite convention is used:

$$D=4: \qquad S^i = P_L S^i = \gamma_* S^i , \qquad S_i = P_R S_i = -\gamma_* S_i . \tag{20.63}$$

Commutators of the bosonic generators with the Q- and S-supersymmetries reflect that they are spinors, doublets under $\mathfrak{su}(2)$, and for $D=4$, that they carry a chiral weight. The $[K,Q]$ and $[P,S]$ commutators have a different form in different dimensions due to the difference in reality and chirality conditions:

$$\begin{aligned}
&[M_{ab}, Q^i_\alpha] = -\tfrac12(\gamma_{ab}Q^i)_\alpha , && [M_{ab}, S^i_\alpha] = -\tfrac12(\gamma_{ab}S^i)_\alpha ,\\
&[D, Q_\alpha{}^i] = \tfrac12 Q_\alpha{}^i , && [D, S_\alpha{}^i] = -\tfrac12 S_\alpha{}^i ,\\
&[U_i{}^j, Q_\alpha{}^k] = \delta_i{}^k Q^j_\alpha - \tfrac12\delta_i{}^j Q_\alpha{}^k , && [U_i{}^j, S_\alpha{}^k] = \delta_i{}^k S^j_\alpha - \tfrac12\delta_i{}^j S_\alpha{}^k ,\\
&[U_i{}^j, Q_{\alpha k}] = -\delta_k{}^j Q_{\alpha i} + \tfrac12\delta_i{}^j Q_{\alpha k} , && [U_i{}^j, S_{\alpha k}] = \delta_k{}^j S_{\alpha i} - \tfrac12\delta_i{}^j S_{\alpha k} ,\\
D=4:\ &[T, Q_\alpha{}^i] = \mathrm{i}\tfrac12 Q_\alpha{}^i , && [T, S_\alpha{}^i] = \mathrm{i}\tfrac12 S_\alpha{}^i ,\\
D=4:\ &[K_a, Q^i_\alpha] = (\gamma_a S^i)_\alpha , && [P_a, S_\alpha{}^i] = (\gamma_a Q^i)_\alpha ,\\
D=5:\ &[K_a, Q^i_\alpha] = \mathrm{i}(\gamma_a S^i)_\alpha , && [P_a, S_\alpha{}^i] = -\mathrm{i}(\gamma_a Q^i)_\alpha ,\\
D=6:\ &[K_a, Q^i_\alpha] = -(\gamma_a S^i)_\alpha , && [P_a, S_\alpha{}^i] = -(\gamma_a Q^i)_\alpha .
\end{aligned} \tag{20.64}$$

The anti-commutation relations between the fermionic generators are

$$\begin{aligned}
&\{Q_{i\alpha}, Q^{j\beta}\} = -\tfrac12\delta_i{}^j(\gamma^a)_\alpha{}^\beta P_a , \qquad \{S_{i\alpha}, S^{j\beta}\} = -\tfrac12\delta_i{}^j(\gamma^a)_\alpha{}^\beta K_a ,\\
D=4:\ &\{Q_\alpha{}^i, Q_j{}^\beta\} = -\tfrac12\delta_j{}^i(\gamma^a)_\alpha{}^\beta P_a , \qquad \{S_\alpha{}^i, S_j{}^\beta\} = -\tfrac12\delta_j{}^i(\gamma^a)_\alpha{}^\beta K_a ,\\
D=4:\ &\{Q_\alpha{}^i, Q^{j\beta}\} = 0 , \qquad \{S_\alpha{}^i, S^{j\beta}\} = 0 , \qquad \{Q_\alpha{}^i, S^{j\beta}\} = 0 ,\\
D=4:\ &\{Q_{i\alpha}, S^{j\beta}\} = -\tfrac12\delta_i{}^j\delta_\alpha{}^\beta D - \tfrac14\delta_i{}^j(\gamma^{ab})_\alpha{}^\beta M_{ab} + \mathrm{i}\tfrac12\delta_i{}^j\delta_\alpha{}^\beta T - \delta_\alpha{}^\beta U_i{}^j ,\\
D=4:\ &\{Q_\alpha{}^i, S_j{}^\beta\} = -\tfrac12\delta_j{}^i\delta_\alpha{}^\beta D - \tfrac14\delta_j{}^i(\gamma^{ab})_\alpha{}^\beta M_{ab} - \mathrm{i}\tfrac12\delta_j{}^i\delta_\alpha{}^\beta T + \delta_\alpha{}^\beta U_j{}^i ,\\
D=5:\ &\{Q_{i\alpha}, S^{j\beta}\} = -\mathrm{i}\tfrac12\left(\delta_i{}^j\delta_\alpha{}^\beta D + \tfrac12\delta_i{}^j(\gamma^{ab})_\alpha{}^\beta M_{ab} + 3\delta_\alpha{}^\beta U_i{}^j\right) ,\\
D=6:\ &\{Q_{i\alpha}, S^{j\beta}\} = \tfrac12\left(\delta_i{}^j\delta_\alpha{}^\beta D + \tfrac12\delta_i{}^j(\gamma^{ab})_\alpha{}^\beta M_{ab} + 4\delta_\alpha{}^\beta U_i{}^j\right) .
\end{aligned} \tag{20.65}$$

For readability of the formulas, we omitted from the right-hand side P_L and P_R projection matrices for $D=4$ and $D=6$, which follow from the chirality properties of the generators on the left-hand side.

Exercise 20.10 *Check that the anti-commutation relations are consistent with symmetry properties of anti-commutators and with reality properties of the different types of spinors.*

[8] Q_i and S_i have the same chirality since we raise and lower i indices in $D=6$ with ε_{ij}.

20.2.2 Gauging of the superconformal algebra

All the standard rules of gauge theories can be used to define the transformations of gauge fields, the curvatures and covariant derivatives, as we have extensively discussed in earlier chapters. We will restrict ourselves in the explicit formulas from now on to $D = 4$. For $D = 5$ [221, 208, 209, 206] and $D = 6$ [222] we refer to the literature.

We assign parameters and gauge fields to all the symmetries as shown in Table 20.1. The total set of 'standard superconformal transformations', analogous to (16.9), is

$$\begin{aligned}\delta = \epsilon^A T_A = \tfrac{1}{2}\lambda^{ab} M_{[ab]} + \lambda_{\mathrm{D}} D + \lambda_{\mathrm{K}}^a K_a + \lambda_i{}^j U_j{}^i + \lambda_T T \\ + \bar{\epsilon}^i Q_i + \bar{\epsilon}_i Q^i + \bar{\epsilon}^i S_i + \bar{\epsilon}_i S^i \,. \end{aligned} \tag{20.66}$$

The spinors are split into left- and right-handed chiral components.

Then, in the same way as for $\mathcal{N} = 1$ in Sec. 16.1.3, we impose constraints such that $\omega_\mu{}^{ab}$, $f_\mu{}^a$ and $\phi_{i\mu}$ become composite fields. For $\mathcal{N} = 1$ this leads to a multiplet of the remaining gauge fields with $8 + 8$ off-shell degrees of freedom, called the Weyl multiplet. However, this counting does not work out for $\mathcal{N} = 2$. The present gauge fields, after subtraction of the composite fields and gauge degrees of freedom, do not lead to an equal number of bosonic and fermionic degrees of freedom. Hence, the general theorem of Sec. 6B implies that we need extra auxiliary fields to close the superconformal algebra in a way in which all fields undergo general coordinate transformations. The solution is not unique. We present the 'standard Weyl multiplet', which is the one that is mostly used. It has 24+24 off-shell degrees of freedom. The fields and their number of degrees of freedom are given in Table 20.2. Apart from the gauge fields, there is a real scalar D, an antisymmetric tensor T_{ab} and a spinor χ^i. They will be auxiliary fields in the action that we will construct. Their role in Poincaré supergravity will be clarified in Secs. 20.2.5 and 20.2.6. The real antisymmetric tensor can be written as a complex (anti-)self-dual tensor T_{ab}^-. This is more convenient, especially for the U(1) R-symmetry transformations, which act on these as complex scalings.

The dependent gauge fields are determined by constraints as in Sec. 16.1.3; see (16.12). The first of these constraints, which puts $R(P) = 0$, is universal. We have used it already in the general treatment of gravity theories in Sec. 11.3. The exact form of the others is a matter of convenience, and different choices amount to field redefinitions. We use the constraints [223]

$$\begin{aligned}&\gamma^b \widehat{R}_{ba}(Q^i) + \tfrac{3}{2}\gamma_a \chi^i = 0\,, \\ &\widehat{R}_{ac}(M^{bc}) + \mathrm{i}\widetilde{\widehat{R}}_a{}^b(A) + \tfrac{1}{4} T_{ca}^- T^{+bc} + \tfrac{3}{2}\delta_a{}^b D = 0\,. \end{aligned} \tag{20.67}$$

Table 20.1 Parameters and gauge fields for the superconformal $\mathcal{N} = 2$ algebras.

P_a	M_{ab}	D	K_a	$U_i{}^j$	T	Q^i	S^i
ξ^a	λ^{ab}	λ_{D}	λ_{K}^a	$\lambda_j{}^i$	λ_T	ϵ_i	η_i
e_μ^a	$\omega_\mu{}^{ab}$	b_μ	$f_\mu{}^a$	$V_{\mu j}{}^i$	A_μ	$\psi_{i\mu}$	$\phi_{i\mu}$

Table 20.2 Number of components in the fields of the standard Weyl multiplet, with the number of components of the fields with gauge transformations subtracted. The next columns contain the Weyl weight w, the chiral weight c (with $\delta\phi = \mathrm{i}c\phi\lambda_T$), and the chirality for the fermions. The last column mentions the gauge transformations that have been used to reduce the number of degrees of freedom in the counting.

field	#	w	c	L/R	gauge subtracted
$e_\mu{}^a$	5	-1	0		$P^a,\ M_{ab},\ D$
b_μ	0	0	0		$D,\ K^a$
$\omega_\mu{}^{ab}$	composite	0	0		
$f_\mu{}^a$	composite	1	0		
$V_{\mu i}{}^j$	9	0	0		SU(2)
A_μ	3	0	0		U(1)
T_{ab}^-	6	1	-1		
D	1	2	0		
$\psi_\mu{}^i$	16	$-\frac{1}{2}$	$\frac{1}{2}$	L	$Q_i,\ S_i$
$\phi_\mu{}^i$	composite	$+\frac{1}{2}$	$\frac{1}{2}$	R	
χ^i	8	$\frac{3}{2}$	$\frac{1}{2}$	L	

Here $\widehat{R}_{ba}(Q^i)$ and $\widehat{R}_{ac}(M^{bc})$ are the modified curvatures following the principles of Sec. 11.3.3. We will give explicit expressions below. The coefficients in these constraints are chosen such that they are invariant under S-supersymmetry (transformation rules are given below).

The dilatation and T-transformations of the fields follow from the weights mentioned in Table 20.2. The SU(2) transformations are fixed by the position of indices i, for example

$$\delta_{\mathrm{SU(2)}}\chi^i = \chi^j\lambda_j{}^i\,,\qquad \delta_{\mathrm{SU(2)}}\chi_i = -\lambda_i{}^j\chi_j\,. \tag{20.68}$$

Exercise 20.11 *Check that the rules (20.68) are consistent with raising and lowering indices by complex conjugation. Check also that an object with contracted i, j indices, like $\bar{\epsilon}_i\chi^i$, is invariant.*

The Lorentz transformations follow from the a, b indices or the spinorial character, and the only special conformal transformation of the elementary fields is the shift of b_μ as in (15.21): $\delta b_\mu = 2\lambda_{K\mu}$. Then, the only transformations that we still have to specify are the Q- and S-supersymmetries. They are

$$\begin{aligned}
\delta e_\mu{}^a &= \tfrac12\bar\epsilon^i\gamma^a\psi_{\mu i} + \text{h.c.}\,,\\
\delta b_\mu &= \tfrac12\bar\epsilon^i\phi_{\mu i} - \tfrac12\bar\eta^i\psi_{\mu i} - \tfrac38\bar\epsilon^i\gamma_\mu\chi_i + \text{h.c.}\,,\\
\delta A_\mu &= -\tfrac12\mathrm{i}\bar\epsilon^i\phi_{\mu i} - \tfrac12\mathrm{i}\bar\eta^i\psi_{\mu i} - \tfrac38\mathrm{i}\bar\epsilon^i\gamma_\mu\chi_i + \text{h.c.}\,,\\
\delta V_{\mu i}{}^j &= -\bar\epsilon_i\phi_\mu^j - \bar\eta_i\psi_\mu^j + \tfrac34\bar\epsilon_i\gamma_\mu\chi^j + \bar\epsilon^j\phi_{\mu i} + \bar\eta^j\psi_{\mu i} - \tfrac34\bar\epsilon^j\gamma_\mu\chi_i\\
&\quad -\tfrac12\delta_i{}^j\left(-\bar\epsilon_k\phi_\mu^k - \bar\eta_k\psi_\mu^k + \tfrac34\bar\epsilon_k\gamma_\mu\chi^k + \bar\epsilon^k\phi_{\mu k} + \bar\eta^k\psi_{\mu k} - \tfrac34\bar\epsilon^k\gamma_\mu\chi_k\right),
\end{aligned}$$

$$\begin{aligned}
\delta\psi_\mu^i &= \left(\partial_\mu + \tfrac{1}{2}b_\mu + \tfrac{1}{4}\gamma^{ab}\omega_{\mu ab} - \tfrac{1}{2}\mathrm{i}A_\mu\right)\epsilon^i + V_\mu{}^i{}_j\epsilon^j - \tfrac{1}{16}\gamma^{ab}T^-_{ab}\varepsilon^{ij}\gamma_\mu\epsilon_j - \gamma_\mu\eta^i ,\\
\delta T^-_{ab} &= 2\bar\epsilon^i\widehat{R}_{ab}(Q^j)\varepsilon_{ij} ,\\
\delta\chi^i &= \tfrac{1}{2}D\epsilon^i + \tfrac{1}{6}\gamma^{ab}\left[-\tfrac{1}{4}\not{D}T^-_{ab}\varepsilon^{ij}\epsilon_j - \widehat{R}_{ab}(U_j{}^i)\epsilon^j + \mathrm{i}\widehat{R}_{ab}(T)\epsilon^i + \tfrac{1}{2}T^-_{ab}\varepsilon^{ij}\eta_j\right] ,\\
\delta D &= \tfrac{1}{2}\bar\epsilon^i\not{D}\chi_i + \text{h.c.}
\end{aligned} \tag{20.69}$$

The transformations of the gauge fields have various non-gauge terms. In the terminology of Sec. 11.3, these are the $\mathcal{M}$-terms, which, due to the methods of Sec. 11.3.4, are the most relevant terms for calculations.

In (20.69) appear covariant curvatures for Q-supersymmetry, for SU(2) and for U(1), which can be split into the pure gauge parts ($r_{\mu\nu}$; see Sec. 11.3.3) and covariantizations:

$$\begin{aligned}
\widehat{R}_{\mu\nu}(Q^i) &= \widehat{R}'_{\mu\nu}(Q^i) - 2\gamma_{[\mu}\phi^i_{\nu]} ,\\
\widehat{R}'_{\mu\nu}(Q^i) &= r_{\mu\nu}(Q^i) - \tfrac{1}{8}\gamma^{ab}T^-_{ab}\varepsilon^{ij}\gamma_{[\mu}\psi_{\nu]j} ,\\
\widehat{R}_{\mu\nu}(U_i{}^j) &= r_{\mu\nu}(U_i{}^j) + \tfrac{3}{2}\bar\psi_{[\nu i}\gamma_{\mu]}\chi^j - \tfrac{3}{2}\bar\psi^j_{[\nu}\gamma_{\mu]}\chi_i - \tfrac{3}{4}\delta_i{}^j\bar\psi_{[\nu k}\gamma_{\mu]}\chi^k + \tfrac{3}{4}\delta_i{}^j\bar\psi^k_{[\nu}\gamma_{\mu]}\chi_k ,\\
\widehat{R}_{\mu\nu}(T) &= r_{\mu\nu}(T) - \tfrac{3}{4}\mathrm{i}\bar\psi^i_{[\nu}\gamma_{\mu]}\chi_i + \tfrac{3}{4}\mathrm{i}\bar\psi_{[\nu i}\gamma_{\mu]}\chi^i .
\end{aligned} \tag{20.70}$$

The spin connection $\omega_\mu{}^{ab}$ is the one that we also found in (16.13) where the gravitino bilinears have to be summed over the two copies (and over the chiralities). The other composite fields are modified with respect to the $\mathcal{N}=1$ expressions in (16.24) due to the appearance of the auxiliary fields. The expressions in terms of the physical fields are

$$\begin{aligned}
\omega_\mu{}^{ab} &= \omega_\mu{}^{ab}(e,b) + \tfrac{1}{2}\left[\bar\psi^i_\mu\gamma^{[a}\psi^{b]}_i + \bar\psi_{\mu i}\gamma^{[a}\psi^{b]i} + \bar\psi^{[a}_i\gamma_\mu\psi^{b]i}\right] ,\\
f_\mu{}^a &= -\tfrac{1}{4}\hat R_\mu{}^a + \tfrac{1}{24}e_\mu{}^a\hat R + \tfrac{1}{4}\mathrm{i}\tilde{\widehat{R}}_\mu{}^a(T) - \tfrac{1}{16}T^-_{c\mu}T^{+ac} - \tfrac{1}{8}e_\mu{}^a D ,\\
\phi^i_\mu &= -\tfrac{1}{2}\gamma^\nu R'_{\mu\nu}(Q^i) + \tfrac{1}{12}\gamma_\mu\gamma^{ab}R'_{ab}(Q^i) + \tfrac{1}{4}\gamma_\mu\chi^i .
\end{aligned} \tag{20.71}$$

Here appears the dual of the U(1) curvature, and the Ricci tensor corresponding to the covariant Lorentz curvature after subtraction of appropriate terms:

$$\widehat{R}_{\mu\nu}{}^{ab} = r_{\mu\nu}{}^{ab} + \left[\bar\psi^i_{[\mu}\gamma_{\nu]}\left(\tfrac{3}{4}\gamma^{ab}\chi_i + \widehat{R}^{ab}(Q_i)\right) + \tfrac{1}{2}\bar\psi^i_\mu\psi^j_\nu\varepsilon_{ij} + \text{h.c.}\right] . \tag{20.72}$$

The removed terms are those that are used for solving the constraints, as in (16.23) and (16.25); $r_{\mu\nu}{}^{ab}$ is the $\mathcal{N}=1$ expression in (16.23). The relevant parts of the transformations of the composite fields are the non-gauge parts ($\mathcal{M}$ in (11.69)):

$$\begin{aligned}
\delta\omega_\mu{}^{ab} &= \ldots - \tfrac{3}{8}\bar\epsilon^i\gamma_\mu\gamma^{ab}\chi_i - \tfrac{1}{2}\bar\epsilon^i\gamma_\mu\widehat{R}^{ab}(Q_i) ,\\
\delta f^a_\mu &= \ldots - \tfrac{3}{16}e_\mu{}^a\bar\epsilon^i\not{D}\chi_i + \tfrac{1}{4}\bar\epsilon^i\gamma_\mu D_b\widehat{R}^{ba}(Q_i) + \text{h.c.} ,\\
\delta\phi^i_\mu &= \ldots - \tfrac{1}{32}\not{D}T^-_{ab}\gamma^{ab}\gamma_\mu\varepsilon^{ij}\epsilon_j - \tfrac{1}{8}\widehat{R}_{ab}(U_j{}^i)\gamma^{ab}\gamma_\mu\epsilon^j - \tfrac{1}{8}\mathrm{i}\widehat{R}_{ab}(T)\gamma^{ab}\gamma_\mu\epsilon^i .
\end{aligned} \tag{20.73}$$

Table 20.3 Fields of superconformal matter multiplets, with number of real off-shell or on-shell degrees of freedom, and chirality. For each multiplet, the bosonic fields are given above the line, and the fermionic fields below.

field	#	w	c	L/R
		Off-shell gauge multiplet		
X^I	2	1	1	
A_μ^I	3	0	0	
$\vec{Y}^I$	3	2	0	
Ω_i^I	8	3/2	1/2	L
		On-shell hypermultiplet		
q^X	4	1	0	
ζ^A	4	3/2	−1/2	L

The algebra after imposing the constraints is modified to a soft algebra in which the auxiliary matter fields appear in the structure functions:

$$[\delta_Q(\epsilon_1), \delta_Q(\epsilon_2)] = \delta_{\text{cgct}}\left(\xi_3^a\right) + \delta_M\left(\lambda_3^{ab}\right) + \delta_K\left(\Lambda_{3K}^a\right) + \delta_S\left(\eta_3^i\right), \tag{20.74}$$

where the associated parameters are given by

$$\begin{aligned}
\xi_3^a &= \tfrac{1}{2}\bar{\epsilon}_2^i\gamma^a\epsilon_{1i} + \text{h.c.}, \\
\lambda_3^{ab} &= \tfrac{1}{4}\bar{\epsilon}_1^i\epsilon_2^j\, T^{+ab}\varepsilon_{ij} + \text{h.c.}, \\
\Lambda_{3K}^a &= \tfrac{1}{8}\bar{\epsilon}_1^i\epsilon_2^j\, D_b T^{+ba}\varepsilon_{ij} + \tfrac{3}{16}\bar{\epsilon}_2^i\gamma^a\epsilon_{1i}\, D + \text{h.c.}, \\
\eta_3^i &= \tfrac{3}{4}\,\bar{\epsilon}_{[1}^i\epsilon_{2]}^j\,\chi_j\,.
\end{aligned} \tag{20.75}$$

We will see below how these extra terms give rise to central charges when the fields appearing in the structure functions are given non-zero vacuum expectation values. We already saw that, in the presence of gauge multiplets with their gauge transformations, there appear similar terms involving the matter scalars of the vector multiplet; see (20.14).

20.2.3 Conformal matter multiplets

We will restrict ourselves to the off-shell gauge multiplets and the on-shell hypermultiplet. Chiral, linear and tensor multiplets are also useful, but we will not discuss them in this book.

The global supersymmetric form of these multiplets have been discussed in Sec. 20.1. When we make the step to the gauged superconformal algebra, the main extra ingredients are the weights of the fields under the dilatation and U(1) part of the R-symmetry. Table 20.3 is the analogue of Table 20.2 for these matter multiplets. None of these fields transforms under special conformal symmetry. The Q- and S-supersymmetry transformations of the gauge multiplet are

$$
\begin{aligned}
\delta X^I &= \tfrac{1}{2}\bar{\epsilon}^i\Omega_i^I\,,\\
\delta\Omega_i^I &= \not{\mathcal{D}}X^I\epsilon_i + \tfrac{1}{4}\gamma^{ab}\mathcal{F}_{ab}{}^I\varepsilon_{ij}\epsilon^j + Y_{ij}{}^I\epsilon^j + X^J\bar{X}^K f_{JK}{}^I\varepsilon_{ij}\epsilon^j + 2X^I\eta_i\,,\\
\delta A_\mu^I &= \tfrac{1}{2}\varepsilon^{ij}\bar{\epsilon}_i\gamma_\mu\Omega_j{}^I + \varepsilon^{ij}\bar{\epsilon}_i\psi_{\mu j}X^I + \text{h.c.}\,,\\
\delta\vec{Y}^I &= \tfrac{1}{4}\vec{\tau}^{ij}\bar{\epsilon}_i\not{\mathcal{D}}\Omega_j^I - \tfrac{1}{2}f_{JK}{}^I\vec{\tau}_i{}^j\bar{\epsilon}_jX^J\Omega^{iK} + \text{h.c.}
\end{aligned}
\tag{20.76}
$$

Here we use the quantities

$$
\begin{aligned}
\mathcal{F}_{ab}^{I-} &\equiv \widehat{F}_{ab}^{I-} - \tfrac{1}{2}\bar{X}^I T_{ab}^-\,,\\
\widehat{F}_{\mu\nu}{}^I &= F_{\mu\nu}{}^I + \left(-\varepsilon_{ij}\bar{\psi}^i_{[\mu}\gamma_{\nu]}\Omega^{Ij} - \varepsilon_{ij}\bar{\psi}^i_\mu\psi^j_\nu\bar{X}^I + \text{h.c.}\right),\\
F_{\mu\nu}{}^I &= \partial_\mu A_\nu{}^I - \partial_\nu A_\mu{}^I + A_\mu{}^J A_\nu{}^K f_{JK}{}^I\,,
\end{aligned}
\tag{20.77}
$$

and the complete superconformal derivatives

$$
\begin{aligned}
\mathcal{D}_\mu X^I &= D_\mu X^I - \tfrac{1}{2}\bar{\psi}^i_\mu\Omega_i^I\,,\\
D_\mu X^I &= \left(\partial_\mu - b_\mu - \mathrm{i}A_\mu\right)X^I + A_\mu{}^J X^K f_{JK}{}^I\,,\\
\mathcal{D}_\mu\Omega_i^I &= D_\mu\Omega_i^I - \not{\mathcal{D}}X^I\psi_{\mu i} - \tfrac{1}{4}\gamma^{ab}\mathcal{F}_{ab}^{I-}\varepsilon_{ij}\psi^j_\mu\\
&\quad -Y_{ij}^I\psi^j_\mu - X^J\bar{X}^K f_{JK}{}^I\varepsilon_{ij}\psi^j_\mu - 2X^I\phi_{\mu i}\,,
\end{aligned}
\tag{20.78}
$$

$$
D_\mu\Omega_i^I = \left(\partial_\mu + \tfrac{1}{4}\omega_\mu{}^{ab}\gamma_{ab} - \tfrac{3}{2}b_\mu - \tfrac{1}{2}\mathrm{i}A_\mu\right)\Omega_i + V_{\mu i}{}^j\Omega_j{}^I + A_\mu^J\Omega_i^K f_{JK}{}^I\,.
$$

For the hypermultiplets, the conformal structure is determined [137, 224] from a closed homothetic Killing vector $k_{\mathrm{D}}{}^X$, which generates the dilatations. We saw in (17.40) that, in the presence of a complex structure, the homothetic Killing vector gives rise to a Killing vector. Since we have here three complex structures we can construct a triplet of Killing vectors:

$$
\vec{k}^X = \tfrac{1}{2}k_{\mathrm{D}}{}^Y\vec{J}_Y{}^X\,. \tag{20.79}
$$

They determine SU(2) transformations of the superconformal algebra. The dilatations and R-symmetry transformations of the fields of the hypermultiplet are

$$
\begin{aligned}
\delta q^X &= \lambda_{\mathrm{D}}k_{\mathrm{D}}{}^X - 2\vec{\lambda}\cdot\vec{k}^X\,,\\
\delta\zeta^{\mathcal{A}} &= \left(\tfrac{3}{2}\lambda_{\mathrm{D}} - \tfrac{1}{2}\mathrm{i}\lambda_T\right)\zeta^{\mathcal{A}} - \zeta^{\mathcal{B}}\omega_{X\mathcal{B}}{}^{\mathcal{A}}\delta q^X\,,\\
\delta\zeta_{\bar{\mathcal{A}}} &= \left(\tfrac{3}{2}\lambda_{\mathrm{D}} + \tfrac{1}{2}\mathrm{i}\lambda_T\right)\zeta_{\bar{\mathcal{A}}} - \zeta_{\bar{\mathcal{B}}}\bar{\omega}_X{}^{\bar{\mathcal{B}}}{}_{\bar{\mathcal{A}}}\delta q^X\,.
\end{aligned}
\tag{20.80}
$$

Observe that the scalars are inert under U(1), but transform under the SU(2), while the reverse is true for the spinors, at least when we neglect the last term (see ‘covariant transformations’ in Sec. 14B). The Q-supersymmetry transformations are the straightforward superconformal covariantizations of (20.33). From dimensional arguments it follows that

q must be inert under S-supersymmetry, and that ζ can transform to q. Imposing the superconformal algebra on q one finds

$$\delta_S \zeta^{\mathcal{A}} = \mathrm{i} f^{i\mathcal{A}}{}_X k_{\mathrm{D}}{}^X \eta_i \,. \tag{20.81}$$

Exercise 20.12 *Check that (20.64) and (20.66) imply that the commutator of* SU(2) *and* Q*-supersymmetry in terms of parameters is*

$$[\delta(\lambda), \delta(\epsilon)] = \delta(\epsilon')\,, \qquad \epsilon'_i = \lambda_i{}^j \epsilon_j\,, \qquad \epsilon'^i = -\epsilon^j \lambda_j{}^i\,. \tag{20.82}$$

Check that this is satisfied on the scalars. Hint: covariantize the derivatives and use the covariant constancy of the complex structures $\vec{J}_X{}^Y$, $\nabla_X k_{\mathrm{D}}{}^Y = \delta_X^Y$ *and*

$$\vec{\tau}_i{}^j \cdot \vec{J}_Y{}^X f^Y{}_{k\mathcal{A}} = 2\delta_k^j f^X{}_{i\mathcal{A}} - \delta_i^j f^X{}_{k\mathcal{A}}\,, \tag{20.83}$$

which can be proven from the definitions (20.20) and (20.23). If you would like a simpler exercise, you can check the commutator of SU(2) *with* S*-supersymmetry on* $\zeta^{\mathcal{A}}$*, and the result should be analogous to (20.82).*

Rather than repeating all the transformations, we give here the covariant derivatives (which encode the transformations):

$$\begin{aligned}
\mathcal{D}_\mu q^X &= D_\mu q^X + \mathrm{i}\bar\psi^i_\mu \zeta^{\mathcal{A}} f^X{}_{i\mathcal{A}} - \mathrm{i}\varepsilon^{ij}\rho^{\bar{\mathcal{A}}\mathcal{B}}\bar\psi_{\mu i}\zeta_{\bar{\mathcal{A}}} f^X{}_{j\mathcal{B}}\,,\\
D_\mu q^X &= \partial_\mu q^X - b_\mu k_{\mathrm{D}}{}^X + 2\vec{V}\cdot\vec{k}^X - A_\mu{}^I k_I{}^X\,,\\
\widehat{\mathcal{D}}_\mu \zeta^{\mathcal{A}} &= \hat{D}_\mu \zeta^{\mathcal{A}} - \tfrac{1}{2}\mathrm{i} f^{i\mathcal{A}}{}_X \not{\mathcal{D}} q^X \psi_{\mu i} - \mathrm{i}\bar{X}^I k_I{}^X f^{i\mathcal{A}}{}_X \varepsilon_{ij}\psi^j_\mu - \mathrm{i} f^{i\mathcal{A}}{}_X k_{\mathrm{D}}{}^X \phi_{\mu i}\,,\\
\hat{D}_\mu \zeta^{\mathcal{A}} &= \left(\partial_\mu + \tfrac{1}{4}\omega_\mu{}^{ab}\gamma_{ab} - \tfrac{3}{2}b_\mu + \tfrac{1}{2}\mathrm{i}A_\mu\right)\zeta^{\mathcal{A}} - A^I_\mu t_{I\mathcal{B}}{}^{\mathcal{A}}\zeta^{\mathcal{B}} + \partial_\mu q^X \omega_{X\mathcal{B}}{}^{\mathcal{A}}\zeta^{\mathcal{B}}\,.
\end{aligned} \tag{20.84}$$

Note that the hatted covariant derivatives are also covariant for target space transformations, and that the $\partial_\mu q^X$ in the last term should not be covariantized to obtain this covariant expression $\widehat{\mathcal{D}}_\mu \zeta^{\mathcal{A}}$.

The local superconformal algebra (20.74) is realized on the scalars, with the addition of the central-charge-like term as in (20.14) assuming that the hypermultiplets transform under gauge transformations as discussed in Sec. 20.1.5. However, the algebra does not close on the spinors. The non-closure should be proportional to field equations. This fixes the action up to the proportionality functions, which are the components of the metric $d^{\bar{\mathcal{A}}}{}_{\mathcal{B}}$ mentioned in Sec. 20.1.4.

20.2.4 Superconformal actions

The superconformal actions of gauge multiplets and hypermultiplets are the covariantizations of (20.15) and (20.57). The new requirement is the invariance under dilatations, after which all the other symmetries follow. Considering the weights in Table 20.3, you can use any term of (20.15) to determine that the requirement for the dilatational weight of the prepotential $F(X)$ is 2:

$$X^I \partial_I F(X) = 2F(X) . \tag{20.85}$$

This corresponds to zero weight for F_{IJ}, and we define

$$N_{IJ} \equiv 2\,\mathrm{Im}\,F_{IJ} = -\mathrm{i}F_{IJ} + \mathrm{i}\bar{F}_{IJ} , \qquad N \equiv N_{IJ} X^I \bar{X}^J . \tag{20.86}$$

With the correspondence of dilatational and chiral weight of X it follows that F_{IJ} has also zero chiral weight, such that the formula for N_{IJ} is consistent in this respect. Therefore, N has the same weights as the $N(X, \bar{X})$ in Sec. 17.2. There is only one difference in notation. For $\mathcal{N}=2$ there are no barred and unbarred I indices. The quantity that is denoted by $N_{I\bar{J}}$ in Ch. 17 is here N_{IJ}, since for $\mathcal{N}=2$

$$\frac{\partial}{\partial X^I}\frac{\partial}{\partial \bar{X}^J} N = N_{IJ} . \tag{20.87}$$

To prove this relation, one has to use the homogeneity of F, i.e.

$$F_{IJK} X^K = 0 . \tag{20.88}$$

With this understanding all the relations (17.13) are valid, replacing $N_{I\bar{J}}$ by N_{IJ}, while the $N_{IJ} \equiv \partial_I \partial_J N$ of (17.13) is here equal to $-\mathrm{i}F_{IJK}\bar{X}^K$.

As was the case there, the Lagrangian can be simplified after use of the constraints and extracting terms from covariant derivatives [215]:

$$\begin{aligned}
e^{-1}\mathcal{L}_g = & -\tfrac{1}{6} N\, R - N\, D - N_{IJ} D_\mu X^I D^\mu \bar{X}^J + N_{IJ} \vec{Y}^I \cdot \vec{Y}^J \\
& + N_{IJ} f_{KL}{}^I \bar{X}^K X^L f_{MN}{}^J \bar{X}^M X^N \\
& + \Big\{ -\tfrac{1}{4}\mathrm{i}\bar{F}_{IJ} \widehat{F}^{+I}_{\mu\nu} \widehat{F}^{+\mu\nu J} - \tfrac{1}{16} N_{IJ} X^I X^J T^{+}_{ab} T^{+ab} + \tfrac{1}{4} N_{IJ} X^I \widehat{F}^{+J}_{ab} T^{+ab} \\
& - \tfrac{1}{4} N_{IJ} \bar{\Omega}^{iI} \not{D} \Omega_i^J + \tfrac{1}{6} N\, \bar{\psi}_{i\mu} \gamma^{\mu\nu\rho} D_\nu \psi_\rho^i \\
& - \tfrac{1}{2} N\, \bar{\psi}_{ia} \gamma^a \chi^i + N_{IJ} X^I \bar{\Omega}^{iJ} \chi_i - \tfrac{1}{3} N_{IJ} X^J \bar{\Omega}^{iI} \gamma^{\mu\nu} D_\mu \psi_{\nu i} \\
& + \tfrac{1}{2} N_{IJ} \bar{\psi}^i_\mu \not{D} \bar{X}^I \gamma^\mu \Omega_i^J + \tfrac{1}{4} N_{IJ} \bar{X}^I \bar{\psi}_{ai} \gamma^{abc} \psi_b^i D_c X^J \\
& + \tfrac{1}{2}\mathrm{i}\bar{F}_{IJ} \varepsilon_{ij} \left(\bar{\Omega}^{iI} \gamma_\mu - \bar{X}^I \bar{\psi}^i_\mu \right) \psi^j_\nu \tilde{\hat{F}}^{\mu\nu J} - \tfrac{1}{16}\mathrm{i} F_{IJK} \bar{\Omega}_i^I \gamma^{\mu\nu} \Omega_j^J \varepsilon^{ij} F^{-K}_{\mu\nu} \\
& + \tfrac{1}{2} N_{IJ} \bar{\Omega}_i^I f_{KL}{}^J \left(\Omega_j^L + \gamma^a \psi_{aj} X^L \right) \bar{X}^K \varepsilon^{ij} \\
& + \left(\tfrac{1}{12} N\, \bar{\psi}_i^a \psi_j^b - \tfrac{1}{6} N_{IJ} \bar{X}^I \bar{\Omega}_i^J \gamma^a \psi_j^b + \tfrac{1}{32}\mathrm{i} F_{IJK} \bar{\Omega}_i^I \gamma^{ab} \Omega_j^J \bar{X}^K \right) T^{-}_{ab} \varepsilon^{ij} \\
& - \tfrac{1}{4}\mathrm{i} F_{IJK} D_\mu X^I \bar{\Omega}_i^J \gamma^\mu \Omega^{iK} + \tfrac{1}{4}\mathrm{i} F_{IJK} Y^{ijI} \bar{\Omega}_i^J \Omega_j^K + \text{h.c.} \Big\} \\
& + \text{four-fermion terms.}
\end{aligned} \tag{20.89}$$

Notations for the covariant derivatives are in (20.78), and

$$D_\mu \psi_{\nu i} = \left(\partial_\mu + \tfrac{1}{4}\omega_\mu{}^{ab}\gamma_{ab} + \tfrac{1}{2} b_\mu + \tfrac{1}{2}\mathrm{i} A_\mu \right) \psi_{\nu i} + V_{\mu i}{}^j \psi_{\nu j} . \tag{20.90}$$

For the coupling of the hypermultiplet to the gauge multiplets in the presence of the superconformal algebra, the new requirement is that the gauge transformations should commute with the dilatational transformations:

$$0 = k_{\rm D}{}^Y \partial_Y k_I{}^X - k_I{}^Y \partial_Y k_{\rm D}{}^X = k_{\rm D}{}^Y \nabla_Y k_I{}^X - k_I{}^X \,. \tag{20.91}$$

This implies that they also commute with the SU(2) transformations generated by (20.79). The moment maps must have Weyl weight $D-2=2$ in order that the $\vec{P}_I \cdot \vec{Y}^I$ terms in (20.57) have Weyl weight $D=4$. This fixes their solution of (20.50) to

$$\vec{P}_I = \vec{k}_X k_I{}^X \,. \tag{20.92}$$

Exercise 20.13 *Prove that this solves (20.50), by using (20.79), (15.46) with $w=1$, (20.49) and (20.91).*

The superconformal hypermultiplet action with gauged isometries is

$$\begin{aligned}
e^{-1}\mathcal{L}_h = {}& -\tfrac{1}{12}k_{\rm D}^2 R + \tfrac14 k_{\rm D}^2 D - \tfrac12 g_{XY} D_\mu q^X D^\mu q^Y - 2\bar{X}^I X^J k_I{}^X k_{JX} + 2\vec{P}_I\cdot\vec{Y}^I \\
&+\Big\{ -\bar{\zeta}_{\bar{\mathcal{A}}}\widehat{\not{D}}\zeta^{\mathcal{B}} d^{\bar{\mathcal{A}}}{}_{\mathcal{B}} + \tfrac{1}{12}k_{\rm D}^2 \bar{\psi}_{i\mu}\gamma^{\mu\nu\rho}D_\nu\psi_\rho^i \\
&+\tfrac18 k_{\rm D}^2\bar{\psi}_{ia}\gamma^a\chi^i - 2{\rm i}d^{\bar{\mathcal{A}}}{}_{\mathcal{B}}A^{i\mathcal{B}}\bar{\zeta}_{\bar{\mathcal{A}}}\chi_i \\
&+{\rm i}\bar{\zeta}_{\bar{\mathcal{A}}}\gamma^a \not{D} q^X\psi_{ai} f^{i\mathcal{B}}{}_X d^{\bar{\mathcal{A}}}{}_{\mathcal{B}} - \tfrac13 {\rm i} d^{\bar{\mathcal{A}}}{}_{\mathcal{B}}A^{i\mathcal{B}}\bar{\zeta}_{\bar{\mathcal{A}}}\gamma^{\mu\nu}D_\mu\psi_{\nu i} \\
&+\left(\tfrac{1}{12}{\rm i}d^{\bar{\mathcal{A}}}{}_{\mathcal{B}}A^{i\mathcal{B}}\bar{\zeta}_{\bar{\mathcal{A}}}\gamma_a\psi_b^j - \tfrac{1}{48}k_{\rm D}^2\bar{\psi}_a^i\psi_b^j\right)T^{+ab}\varepsilon_{ij} \\
&-\tfrac18\bar{\zeta}_{\bar{\mathcal{A}}}\gamma^{ab}T_{ab}^+\zeta_{\bar{\mathcal{B}}}C^{\bar{\mathcal{A}}\bar{\mathcal{B}}} + 2{\rm i}\bar{X}^I k_I^X\bar{\zeta}_{\bar{\mathcal{A}}}\gamma^a\psi_a^j\varepsilon_{ij}d^{\bar{\mathcal{A}}}{}_{\mathcal{B}}f^{i\mathcal{B}}{}_X \\
&+2X^I\bar{\zeta}^{\mathcal{A}}\zeta^{\mathcal{B}}t_{I\mathcal{A}\mathcal{B}} + 2{\rm i}k_I^X f_X^{i\mathcal{A}}\bar{\zeta}_{\bar{\mathcal{B}}}\Omega^{jI}\varepsilon_{ij}d^{\bar{\mathcal{B}}}{}_{\mathcal{A}} \\
&+\tfrac12\bar{\psi}_{aj}\gamma^a\Omega_i^I P_I{}^{ij} + \tfrac12\bar{X}^I\bar{\psi}_a^i\gamma^{ab}\psi_b^j P_{Iij} + {\rm h.c.}\Big\} \\
&-\tfrac12\bar{\psi}_a^i\gamma^{abc}\psi_{bj}D_c q^X\vec{k}_X\cdot\vec{\tau}_i{}^j + \text{four-fermion terms,}
\end{aligned} \tag{20.93}$$

where the covariant derivatives can be found in (20.84),

$$A^{i\mathcal{A}} \equiv f^{i\mathcal{A}}{}_X k_{\rm D}{}^X\,, \qquad P_{Iij} \equiv P_{Ii}{}^k\varepsilon_{kj}\,, \qquad P_I{}^{ij} \equiv \varepsilon^{ik}P_{Ik}{}^j\,. \tag{20.94}$$

20.2.5 Partial gauge fixing

$\mathcal{N}=2$ Poincaré supergravity can be obtained by the same procedure that we discussed extensively for $\mathcal{N}=1$. We construct a superconformal invariant action with more matter multiplets than those that appear in the Poincaré theory. For $\mathcal{N}=1$, we could suffice with one compensating chiral multiplet. For $\mathcal{N}=2$ it turns out that we need two compensating multiplets. Different choices have been discussed in the literature, leading to different sets of auxiliary fields [225, 226, 227, 228, 229, 230, 231, 222, 232, 233, 210, 207]. We choose here one compensating gauge multiplet and one compensating hypermultiplet,[9] since this gives the most insight into the structure of the physical theory using the methods of projective manifolds. We thus consider the Lagrangian that is the sum of (20.89) and (20.93):

$$\mathcal{L} = \mathcal{L}_g + \mathcal{L}_h \,. \tag{20.95}$$

We first impose, as in the previously discussed theories, the gauge

$$K\text{-gauge:} \quad b_\mu = 0 \,. \tag{20.96}$$

We saw in Ch. 17 that the geometry for the scalars of the matter multiplet in supergravity is obtained as a projective manifold from the conformal invariant embedding manifold. However, how can one make such a projection for both gauge and hypermultiplets since there is only one dilatation generator in the superconformal algebra? Here the auxiliary fields of the Weyl multiplets come to the rescue. In fact, the first two terms of (20.89) and of (20.93) show what happens. If we combine these terms, we have

$$e^{-1}\mathcal{L} = \left(-\tfrac{1}{6}N - \tfrac{1}{12}k_{\mathrm{D}}^2\right) R + \left(-N + \tfrac{1}{4}k_{\mathrm{D}}^2\right) D + \dots \,. \tag{20.97}$$

Thus, the field equation of the auxiliary field D puts $N = \frac{1}{4}k_{\mathrm{D}}^2$ and we use as dilatational gauge

$$D\text{-gauge:} \quad -\tfrac{1}{6}N - \tfrac{1}{12}k_{\mathrm{D}}^2 = \tfrac{1}{2}\kappa^{-2} \,. \tag{20.98}$$

Combining these equations leads to

$$N = -\kappa^{-2} \,, \qquad k_{\mathrm{D}}^2 = -4\kappa^{-2} \,. \tag{20.99}$$

These give the main projection equations for special Kähler manifolds (gauge multiplets) and quaternionic-Kähler manifolds (hypermultiplets), whose geometries we will discuss in Sec. 20.3. The first equation is of the form (17.65) with $a = \kappa^{-2}$. A similar constant in the construction of quaternionic-Kähler manifolds is fixed by the second equation, as we will show below in (20.180).

The two minus signs in (20.99) imply that in both embedding manifolds there should be multiplets with negative kinetic terms. Since the physical multiplets should have all positive kinetic terms, the signature should be for both $(- + \dots +)$, where the $-$ stands for a full multiplet, i.e. a complex scalar for the gauge multiplets and a quaternion for the hypermultiplets.

The conditions (20.99) only fix the norm of the complex or quaternionic scalar in the compensating multiplet. Here, the structure of the superconformal algebra turns out to be exactly what one needs. The R-symmetry part of the superconformal algebra is $\mathrm{U}(1) \times \mathrm{SU}(2)$. In Sec. 20.2.3 we have seen that the scalars of the gauge multiplets transform under the U(1) factor and are inert under SU(2), while the opposite holds for the scalars of the

[9] The hypermultiplet is considered on-shell, which is sufficient to obtain the physical theories. To discuss the full off-shell structure, one should add four real auxiliary fields to this multiplet, which appear as central charges in the transformations of the scalars q^X [234, 235].

hypermultiplets. Hence, the U(1) gauge fixing can be used to fix the phase of the complex scalars of the compensating gauge multiplet and the SU(2) gauge fixing can fix the three phases of the quaternions of the compensating hypermultiplet.

How does it work for the fermionic sector? In the $\mathcal{N}=1$ theory, the spinor of the chiral multiplet was fixed by S-supersymmetry. For $\mathcal{N}=2$, we have an extra condition since the Weyl multiplet contains an auxiliary field χ_i. Some terms with χ_i cancel between (20.89) and (20.93) due to (20.99), and we are left with the field equation

$$N_{IJ}X^I\bar{\Omega}^{iJ} - 2\mathrm{i}d^{\bar{A}}{}_{\mathcal{B}}A^{i\mathcal{B}}\zeta_{\bar{A}} = 0\,. \tag{20.100}$$

On the other hand, we choose a gauge fixing to eliminate the mixed kinetic terms of the gravitino with the matter spinor fields:

$$S\text{-gauge:}\quad N_{IJ}X^J\Omega^{iI} + \mathrm{i}d^{\bar{A}}{}_{\mathcal{B}}A^{i\mathcal{B}}\zeta_{\bar{A}} = 0\,. \tag{20.101}$$

Combining the gauge fixing with the field equation (20.100) gives

$$N_{IJ}X^J\Omega^{iI} = 0\,,\qquad d^{\bar{A}}{}_{\mathcal{B}}A^{i\mathcal{B}}\zeta_{\bar{A}} = 0\,. \tag{20.102}$$

Thus, the fields of the compensating hypermultiplet disappear by this combination of gauge fixings and field equations, while for the gauge multiplet only the gauge field and the auxiliary field $\vec{Y}$ remain. The gauge field joins the physical fields of the Weyl multiplet as the 'graviphoton': the spin-1 field that should be in the super-Poincaré multiplet according to Table 6.1.

20.2.6 Elimination of auxiliary fields

Figure 20.1 (inspired by [218]) gives a schematic overview of the fields involved in the superconformal and in the super-Poincaré theory. To clarify the procedure, we split the vector and hypermultiplets into the compensating and the physical ones. For the compensating hypermultiplet we use q^0 to indicate the norm of the quaternion, q^α for its phases and ζ^i is the corresponding fermion doublet. The physical quantities are written as $\{q^u, \zeta^A\}$. The fields of the compensating gauge multiplet are denoted by $\{y, \chi^{i0}, A^0_\mu, \vec{Y}^0\}$, and the remaining physical ones are indicated by an index α.

We now discuss the role of the auxiliary fields that were not yet discussed in Sec. 20.2.5. The fields A_μ and $V_{\mu i}{}^j$ appeared in the Weyl multiplet as gauge fields of the $\mathrm{U}(1)\times\mathrm{SU}(2)$ R-symmetry. We have seen in detail in Sec. 17.3 how the field A_μ serves a role as an auxiliary field to produce the action of the projective Kähler manifold. All this remains true for the special Kähler manifolds. The only difference is the value of the parameter a, and the replacement of $G_{I\bar{J}}$ in that section with N_{IJ} of (20.86). Similarly $V_{\mu i}{}^j$ plays a

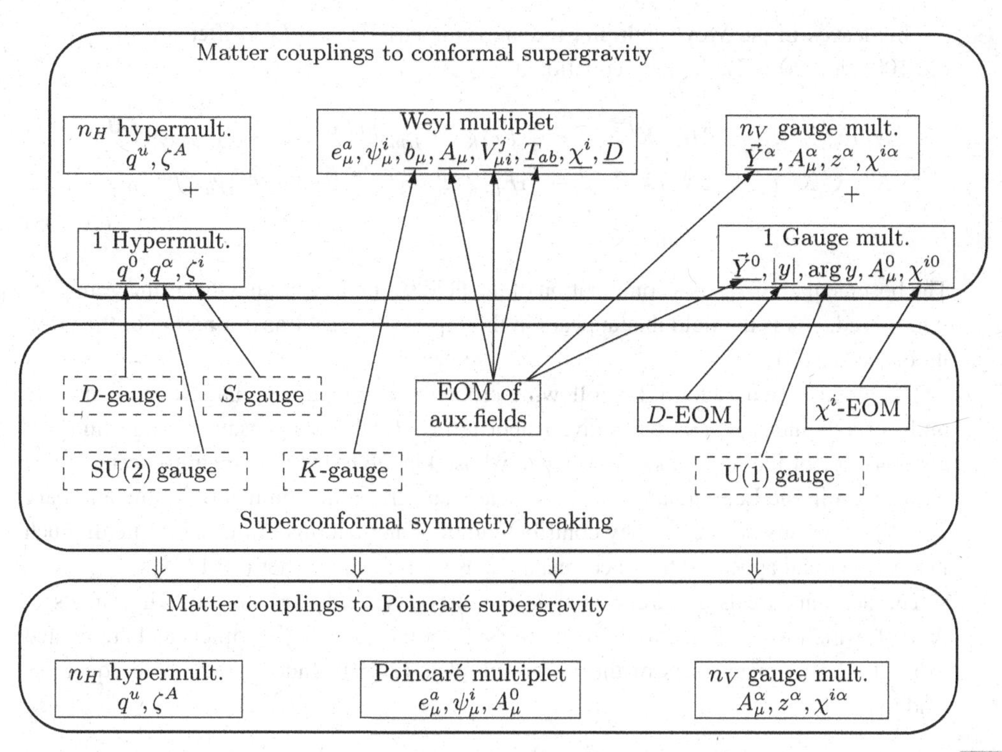

Fig. 20.1 Summary of gauge fixing and auxiliary fields. The underlined terms of the upper box will be eliminated. The method that is used is indicated in the middle box, where EOM stands for elimination by an equation of motion. Other fields are eliminated by gauge conditions. This leads to the field content indicated in the lower box.

role as an auxiliary field to produce the action for the quaternionic-Kähler manifold. The values of these fields are[10]

$$\begin{aligned}
A_\mu &= \mathcal{A}_\mu + A_\mu^{\mathrm{F}}\,,\\
&\quad \mathcal{A}_\mu = -\tfrac{1}{2}\mathrm{i}\kappa^2 N_{IJ} X^I \overset{\leftrightarrow}{\hat{\partial}}_\mu \bar{X}^J\,,\\
&\quad A_\mu^{\mathrm{F}} = -\tfrac{1}{8}\mathrm{i}\kappa^2 N_{IJ}\bar{\Omega}^{iI}\gamma_\mu\Omega_i^J + \tfrac{1}{2}\mathrm{i}\kappa^2\bar{\zeta}_{\bar{\mathcal{A}}}\gamma_\mu\zeta^{\mathcal{B}} d^{\bar{\mathcal{A}}}{}_{\mathcal{B}}\,,\\
\vec{V}_\mu &= \vec{\mathcal{V}}_\mu + \vec{V}_\mu^{\mathrm{F}},\\
&\quad \vec{\mathcal{V}}_\mu = \tfrac{1}{2}\kappa^2\vec{k}_X\hat{\partial}_\mu q^X = \tfrac{1}{2}\kappa^2\left(\vec{k}_X\partial_\mu q^X - A_\mu^I\vec{P}_I\right),\\
&\quad \vec{V}_\mu^{\mathrm{F}} = \tfrac{1}{8}\kappa^2 N_{IJ}\bar{\Omega}^{iI}\gamma_\mu\Omega_j^J\vec{\tau}_i{}^j\,,
\end{aligned} \tag{20.103}$$

where we used

$$\begin{aligned}
\hat{\partial}_\mu X^I &= \partial_\mu X^I + A_\mu{}^J X^K f_{JK}{}^I\,,\\
\hat{\partial}_\mu q^X &= \partial_\mu q^X - A_\mu{}^I k_I{}^X\,.
\end{aligned} \tag{20.104}$$

[10] If you want to check this, a useful equation is $(\vec{k}_X\cdot\vec{\tau}_j{}^i)(\vec{k}^X\cdot\vec{\tau}_\ell{}^k) = \varepsilon^{ik}\varepsilon_{j\ell}$. This equation is also necessary in order to show that the gravitino does not contribute to the $\vec{V}_\mu^{\mathrm{F}}$ connection, in the same way as it does not contribute to the A_μ^{F}.

This leaves in the Weyl multiplet the antisymmetric tensor T_{ab}. After use of (20.99), (20.100) and (20.102), the field equation of T_{ab} is

$$\begin{aligned} N_{IJ}X^IX^JT^+_{ab} &= 2N_{IJ}X^I\widehat{F}^{+J}_{ab} - \tfrac{1}{4}\mathrm{i}\bar{F}_{IJK}\bar{\Omega}^{iI}\gamma_{ab}\Omega^{jJ}X^K\varepsilon_{ij} - \bar{\zeta}_{\bar{\mathcal{A}}}\gamma_{ab}\zeta_{\bar{\mathcal{B}}}C^{\bar{\mathcal{A}}\bar{\mathcal{B}}}\,,\\ N_{IJ}\bar{X}^I\bar{X}^JT^-_{ab} &= 2N_{IJ}\bar{X}^I\widehat{F}^{-J}_{ab} + \tfrac{1}{4}\mathrm{i}F_{IJK}\bar{\Omega}^I_i\gamma_{ab}\Omega^J_j\bar{X}^K\varepsilon^{ij} - \bar{\zeta}^{\mathcal{A}}\gamma_{ab}\zeta^{\mathcal{B}}C_{\mathcal{A}\mathcal{B}}\,. \end{aligned} \tag{20.105}$$

The bosonic part of T_{ab} is a combination of the field strengths F^I_{ab}, and this is the graviphoton. Indeed, T_{ab} is the field that appears in the supersymmetry transformation of the gravitino; see (20.69).

You might have thought of the following problem. We mentioned that the compensating multiplets should appear with negative kinetic terms. But the compensating gauge multiplet contains one of the physical spin-1 fields, whose kinetic terms thus seem to occur with a wrong sign in the Lagrangian. The terms that result from the elimination of the auxiliary field T_{ab} cure this problem. They contribute to the kinetic terms of the spin-1 fields, such that they will all appear with the correct sign, as we will show after (20.137).

The auxiliary fields $\vec{Y}^I$ are similar to the auxiliary fields D^A of gauge multiplets in $\mathcal{N}=1$, which were related to moment maps. Here they are SU(2) triplets and are related to the triplet moment maps of the hypermultiplets (20.50). Indeed, their field equations lead to

$$N_{IJ}\vec{Y}^J = -\vec{P}_I - \tfrac{1}{8}\left(\mathrm{i}F_{IJK}\bar{\Omega}^J_i\Omega^K_j\vec{\tau}^{ij} + \text{h.c.}\right). \tag{20.106}$$

The reader may wonder what happened to the U(1) moment map of the special Kähler manifold. The gauge transformations of the field X^I are determined by the fact that they transform in the adjoint representation of the gauge group, being supersymmetry partners of the gauge fields. Hence these Killing vectors are

$$k_J{}^I = X^K f_{KJ}{}^I\,. \tag{20.107}$$

The indices A in (17.18) that denote all the gauge vectors coincide in this case with the indices I. The moment map that corresponds to this transformation, which we indicate by P^0, can be written in various ways:

$$\begin{aligned} P^0_I &= -\mathrm{i}N_{IJ}f_{KL}{}^JX^K\bar{X}^L = \mathrm{i}N_{JL}f_{KI}{}^JX^K\bar{X}^L = -\mathrm{i}N_{JK}f_{LI}{}^JX^K\bar{X}^L\\ &= f_{IJ}{}^K(X^J\bar{F}_K + \bar{X}^JF_K) + 2C_{I,JK}X^J\bar{X}^K = -\langle T_IV,\bar{V}\rangle\,. \end{aligned} \tag{20.108}$$

The first line implies that the potential-like term on the second line of (20.89) can be written as $-N^{-1|IJ}P^0_IP^0_J$. In the second line appears the tensor $C_{I,JK}$ defined in (20.18).

Exercise 20.14 *Note that (20.19) implies*

$$f_{K(I}{}^LN_{J)L} = 0\,, \tag{20.109}$$

and use this to prove the equalities in the first line of (20.108).

20.2.7 Complete action

For writing the complete action, we use the gauge conditions, write out the $\hat{F}_{\mu\nu}{}^I$ from (20.77), and use Ex. 17.3, with (17.27).[11] This leads to (adding also the Chern–Simons term, related to the coefficients $C_{I,JK}$ in (20.18))

$$
\begin{aligned}
e^{-1}\mathcal{L} = {} & \kappa^{-2}\left(\tfrac{1}{2}R - \bar{\psi}_{i\mu}\gamma^{\mu\nu\rho}D_\nu\psi^i_\rho\right) - N_{IJ}D_\mu X^I D^\mu \bar{X}^J - \tfrac{1}{2}g_{XY}D_\mu q^X D^\mu q^Y \\
& - N_{IJ}\vec{Y}^I\cdot\vec{Y}^J - N^{-1|IJ}P^0_I P^0_J - 2\bar{X}^I X^J k_I{}^X k_{JX} \\
& + \tfrac{2}{3}C_{I,JK}e^{-1}\varepsilon^{\mu\nu\rho\sigma}A_\mu{}^I A_\nu{}^J\left(\partial_\rho A_\sigma{}^K + \tfrac{3}{8}f_{LM}{}^K A_\rho{}^L A_\sigma{}^M\right) \\
& + \tfrac{1}{2}\bar{\psi}_{ai}\gamma^{abc}\psi^j_b\left(\delta^i_j N_{IJ}\bar{X}^I D_c X^J + D_c q^X \vec{k}_X\cdot\vec{\tau}_j{}^i\right) \\
& + \Big\{-\tfrac{1}{4}\mathrm{i}\bar{F}_{IJ}F^{+I}_{\mu\nu}F^{+\mu\nu J} + \tfrac{1}{16}N_{IJ}X^I X^J T^+_{ab}T^{+ab} \\
& - \tfrac{1}{4}N_{IJ}\bar{\Omega}^{iI}\hat{\not{D}}\Omega^J_i - \bar{\zeta}_{\bar{\mathcal{A}}}\hat{\not{D}}\zeta^{\mathcal{B}}d^{\bar{\mathcal{A}}}{}_{\mathcal{B}} \\
& + \tfrac{1}{2}N_{IJ}\bar{\psi}_{ia}\not{D}X^I\gamma^a\Omega^{iJ} + \mathrm{i}\bar{\psi}_{ia}\not{D}q^X\gamma^a\zeta_{\bar{\mathcal{A}}}f^{i\mathcal{B}}{}_X d^{\bar{\mathcal{A}}}{}_{\mathcal{B}} \\
& - \tfrac{1}{2}N_{IJ}\varepsilon_{ij}\left(\bar{\Omega}^{iI}\gamma_\mu - \bar{X}^I\bar{\psi}^i_\mu\right)\psi^j_\nu F^{-\mu\nu J} - \tfrac{1}{16}\mathrm{i}F_{IJK}\bar{\Omega}^I_i\gamma^{\mu\nu}\Omega^J_j\varepsilon^{ij}F^{-K}_{\mu\nu} \\
& + \tfrac{1}{2}N_{IJ}\bar{\Omega}^I_i f_{KL}{}^J\left(\Omega^L_j + \gamma^a\psi_{aj}X^L\right)\bar{X}^K\varepsilon^{ij} \\
& + 2X^I\bar{\zeta}^{\mathcal{A}}\zeta^{\mathcal{B}}t_{I\mathcal{A}\mathcal{B}} + 2\mathrm{i}k^X_I f^{i\mathcal{B}}{}_X\varepsilon_{ij}d^{\bar{\mathcal{A}}}{}_{\mathcal{B}}\bar{\zeta}_{\bar{\mathcal{A}}}\left(\Omega^{jI} + \gamma^a\psi^j_a\bar{X}^I\right) \\
& + \tfrac{1}{2}\bar{\psi}_{aj}\gamma^a\Omega^I_i P_I{}^{ij} + \tfrac{1}{2}\bar{X}^I\bar{\psi}^i_a\gamma^{ab}\psi^j_b P_{Iij} + \text{h.c.}\Big\} \\
& + \text{four-fermion terms.}
\end{aligned}
\tag{20.110}
$$

We have used the field equations to determine the U(1) and SU(2) connections. The auxiliary fields are no longer to be considered as independent fields, but stand for their values defined by the field equations (20.103), (20.105) and (20.106). The covariant derivatives are (omitting all contributions that gives rise to four-fermion terms)

$$
\begin{aligned}
D_\mu X^I &= \left(\partial_\mu - \mathrm{i}\mathcal{A}_\mu\right)X^I + A_\mu{}^J X^K f_{JK}{}^I\,, \\
D_\mu q^X &= \partial_\mu q^X + 2\vec{\mathcal{V}}_\mu\cdot\vec{k}^X - A_\mu{}^I k_I{}^X\,, \\
D_\mu\Omega^I_i &= \left(\partial_\mu + \tfrac{1}{4}\omega_\mu{}^{ab}(e)\gamma_{ab} - \tfrac{1}{2}\mathrm{i}\mathcal{A}_\mu\right)\Omega_i + \mathcal{V}_{\mu i}{}^j\Omega_j{}^I + A^J_\mu\Omega^K_i f_{JK}{}^I\,, \\
\hat{D}_\mu\zeta^{\mathcal{A}} &= \left(\partial_\mu + \tfrac{1}{4}\omega_\mu{}^{ab}(e)\gamma_{ab} + \tfrac{1}{2}\mathrm{i}\mathcal{A}_\mu\right)\zeta^{\mathcal{A}} - A^I_\mu t_{I\mathcal{B}}{}^{\mathcal{A}}\zeta^{\mathcal{B}} + \partial_\mu q^X\omega_{X\mathcal{B}}{}^{\mathcal{A}}\zeta^{\mathcal{B}}\,, \\
D_\mu\psi_{\nu i} &= \left(\partial_\mu + \tfrac{1}{4}\omega_\mu{}^{ab}(e)\gamma_{ab} + \tfrac{1}{2}\mathrm{i}\mathcal{A}_\mu\right)\psi_{\nu i} + \mathcal{V}_{\mu i}{}^j\psi_{\nu j}\,,
\end{aligned}
\tag{20.111}
$$

where $\mathcal{A}_\mu$ and $\mathcal{V}_{\mu i}{}^j$ are the bosonic parts of the auxiliary R-symmetry gauge fields given in (20.103) (with the usual translation between triplet and doublet notation; see (20.207)).

[11] Since we do not write four-fermion terms here, we can omit the $^{(0)}$ indication for the covariant derivative, which is used in Ch. 17 to indicate that ψ-torsion terms are omitted.

20.2.8 $D = 5$ and $D = 6$, $\mathcal{N} = 2$ supergravities

We do not discuss in detail the $D = 5$ and $D = 6$ theories, but the main procedure is very similar in these cases: one has to use two compensating multiplets. For $D = 5$ one can again use a compensating gauge multiplet and a hypermultiplet. The difference from $D = 4$ is that the gauge multiplet has only a real scalar. This fits remarkably well with the fact that the R-symmetry group contains only SU(2), and no U(1), which we used in $D = 4$ for fixing the phase of the scalar field of the gauge multiplet. The full $D = 5$ theory has been written down in [207], for a large part based on previous work, e.g. [93, 204, 205].

For $D = 6$ one of the compensating multiplets is a tensor multiplet containing on the bosonic side a real scalar, which can be used for the dilatational gauge fixing, and a self-dual antisymmetric tensor. The latter combines with the anti-self-dual tensor T^- of the Weyl multiplet to make the field strength of a physical 2-form field in the super-Poincaré multiplet [222, 236, 237, 238]. The general super-Poincaré theory is given in [239]. It builds on earlier work, e.g. [240, 241, 242].

20.3 Special geometry

As mentioned in Sec. 12.5 and at the beginning of this chapter, the scalar sectors of $\mathcal{N} = 2$ supersymmetry and supergravity define *special geometries*. As was mentioned in Sec. 12.5, the R-symmetry group plays an important role for the geometries of the scalars in supergravity. The R-symmetry group is SU(2) for $D = 5$ and 6, and SU(2)× U(1) for $D = 4$. The SU(2) group acts on the scalars of the hypermultiplets. This leads to the three complex structures of the corresponding manifolds. The U(1) factor acts on the complex scalars of the gauge multiplet in $D = 4$, whose manifold therefore inherits one complex structure. This manifold is a Kähler manifold, but with 'special' properties, which are related to the symplectic duality transformations of the vector fields to which the gauge multiplet scalars are related by supersymmetry.

20.3.1 The family of special manifolds

There are three types of special manifolds which we shall discuss, related to the real scalars of gauge multiplets in $D = 5$, the complex scalars of $D = 4$ gauge multiplets and the quaternionic scalars of hypermultiplets. Furthermore, these spaces are not completely independent, and we additionally discuss the relation between these special manifolds. Since there are no scalars in the gauge multiplets of $D = 6$, there is no geometry in that case.

We furthermore consider the manifolds of the scalars in global supersymmetry and in supergravity. We have seen in Sec. 17.3 that the manifold of the scalars in $\mathcal{N} = 1$ supergravity can be understood as a projective manifold. To define this manifold, we first imposed conformal symmetry on the global supersymmetry theory, and included one multiplet that contains 'compensating fields', i.e. fields that are not physical; instead they are

Table 20.4 Geometries from supersymmetric theories with eight real supercharges, and the connections provided by the **r**-map and the **c**-map.

	$D = 5$ gauge multiplets	$D = 4$ gauge multiplets	hypermultiplets
rigid (affine)	affine very special real	affine special Kähler	hyper-Kähler
local (projective)	(projective) very special real	(projective) special Kähler	quaternionic-Kähler

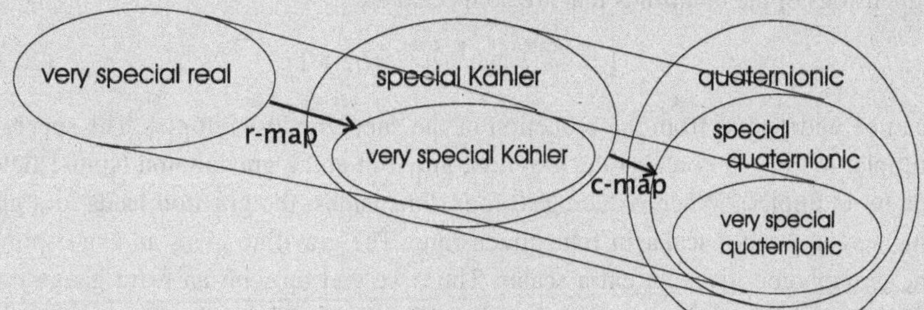

very special real
special Kähler
quaternionic
r-map
very special Kähler
special quaternionic
c-map
very special quaternionic

gauge degrees of freedom for the superconformal theory. The projection from the conformal invariant theory to the physical one lead to the manifolds that appear in supergravity. In Sec. 20.2.5 we saw already that similar projections can be defined for $\mathcal{N} = 2$. In view of this construction the geometries that are related to global supersymmetry have been called 'affine' in the mathematics literature [243, 244] and 'rigid' in the physics literature, while those for supergravity are called 'projective'.[12] This leads to the scheme in the upper part of Table 20.4. Note that the systematic nomenclature is not followed for the hypermultiplets, since what could have been called 'affine quaternionic-Kähler manifolds' were already named *hyper-Kähler manifolds* by mathematicians.[13]

The lower part of Table 20.4 represents a relation between these manifolds that can be understood from dimensional reduction. The manifolds of supergravity theories with gauge multiplets in $D = 5$ are called very special real manifolds. When these theories are dimensionally reduced to $D = 4$, they give rise to theories with special Kähler geometry. This map between a very special real manifold and a special Kähler manifold is called the **r**-map [245]. A theory with gauge multiplets in $D = 4$ can be dimensionally reduced to $D = 3$. As discussed in Sec. 7.8, see especially (7.77), a vector field ($p = 1$) in $D = 3$ can be transformed into a scalar. Therefore, the dimensional reduction of the bosonic part of gauge multiplets can be written in terms of scalars only, and

[12] When we do not specify and, for example, talk about 'special Kähler' geometry, we refer to the geometry of the scalars in supergravity.

[13] Hyper-Kähler manifolds can be seen as a special case of 'quaternionic-Kähler manifolds' with vanishing SU(2) curvature. We will restrict ourselves to the quaternionic-Kähler manifolds of negative scalar curvature, as those are the only ones that appear in supergravity with positive kinetic energies for all fields. For manifolds with continuous isometries, this implies that they are non-compact.

thus they become hypermultiplets. This provides a map between vector multiplets and hypermultiplets, which is the **c**-map (named for Calabi) [246]. A scalar action looks the same in any dimension, and therefore the **c**-map can be considered also as a map between the special Kähler geometry of vector multiplets in $D = 4$ and the geometry of hypermultiplets. The latter map has also an interpretation from T-duality between IIA and IIB superstring theories, but we will not discuss this here.

The **r**-map thus maps a real geometry into a complex one, and the **c**-map connects a complex geometry to a quaternionic one. But in fact these maps also increase the number of multiplets. Denoting by n the complex dimension of the special Kähler manifolds, the real dimensions of the manifolds that are connected are

$$n - 1 \overset{\mathbf{r}-\mathrm{map}}{\longrightarrow} 2n \overset{\mathbf{c}-\mathrm{map}}{\longrightarrow} 4(n+1)\,. \tag{20.112}$$

This can be understood from the reduction of the supergravity multiplet. The supergravity multiplet in $D = 5$ contains the graviton, gravitini and a graviphoton (spin-1 field of the gravity multiplet). When reducing to four dimensions, the graviton leads to a graviton plus a vector and a scalar in four dimensions. The gravitino gives an extra spin 1/2, and the graviphoton gives an extra scalar. Thus, we end up with an extra gauge multiplet. The same happens in the reduction from $D = 4$ to $D = 3$, where the graviton gives rise to a vector and a scalar, and also the graviphoton gives a vector and a scalar. Both vectors are duality transformed into scalars, and thus we have an extra quaternionic scalar.

The dimensional reduction, and thus the **r**-map and the **c**-map, lead only to a specific subclass of special Kähler and quaternionic-Kähler manifolds. This induces a terminology for subclasses of special Kähler and quaternionic-Kähler manifolds. The image of the very special real manifolds under the **r**-map defines the very special Kähler manifolds, a non-trivial subset of special Kähler manifolds. The **c**-map defines in the same way the 'special quaternionic-Kähler' manifolds as a non-trivial subset of quaternionic-Kähler manifolds, and in an obvious way also very special quaternionic-Kähler manifolds; see the lower part of Table 20.4.

20.3.2 Very special real geometry

The real manifolds that appear in five dimensions [93] are called very special real manifolds. In Sec. 20.1.2 we discussed the couplings of gauge multiplets in global $D = 5$ supersymmetry, enumerated by indices $I = 1, \ldots, n$ and determined by a symmetric tensor C_{IJK}. The scalar fields of these multiplets are real fields σ^I, and from (20.11) we extract the metric of the scalar manifold

$$g_{IJ} = C_{IJK}\sigma^K\,. \tag{20.113}$$

Exercise 20.15 *Prove that the curvature tensor corresponding to this metric is* $R_{IJKL} = -\frac{1}{2}g^{MN}C_{NI[K}C_{L]JN}$.

This metric has a homothetic Killing vector. Indeed, (15.46) is satisfied for $k_{\mathrm{D}}^I = \sigma^I$ and $w = (D-2)/2 = 3/2$. Therefore, we can define the projective manifold, by defining a

Table 20.5 Symmetric very special manifolds. The real dimensions of real, Kähler and quaternionic manifolds are $(n-1)$, $2n$ and $4(n+1)$, respectively.

q, P	n	real	Kähler	quaternionic
$-1, 0$	2	$\mathrm{SO}(1,1)$	$\left[\frac{\mathrm{SU}(1,1)}{\mathrm{U}(1)}\right]^2$	$\frac{\mathrm{SO}(3,4)}{(\mathrm{SU}(2))^3}$
$-1, P$	$2+P$	$\frac{\mathrm{SO}(P+1,1)}{\mathrm{SO}(P+1)}$		
$0, P$	$3+P$	$\mathrm{SO}(1,1)\otimes\frac{\mathrm{SO}(P+1,1)}{\mathrm{SO}(P+1)}$	$\frac{\mathrm{SU}(1,1)}{\mathrm{U}(1)}\otimes\frac{\mathrm{SO}(P+2,2)}{\mathrm{SO}(P+2)\otimes\mathrm{SO}(2)}$	$\frac{\mathrm{SO}(P+4,4)}{\mathrm{SO}(P+4)\otimes\mathrm{SO}(4)}$
$1, 1$	6	$\frac{\mathrm{SL}(3,\mathbb{R})}{\mathrm{SO}(3)}$	$\frac{\mathrm{Sp}(6)}{\mathrm{U}(3)}$	$\frac{F_{4,4}}{\mathrm{USp}(6)\otimes\mathrm{SU}(2)}$
$2, 1$	9	$\frac{\mathrm{SL}(3,\mathbb{C})}{\mathrm{SU}(3)}$	$\frac{\mathrm{SU}(3,3)}{\mathrm{SU}(3)\otimes\mathrm{SU}(3)\otimes\mathrm{U}(1)}$	$\frac{E_{6,2}}{\mathrm{SU}(6)\otimes\mathrm{SU}(2)}$
$4, 1$	15	$\frac{\mathrm{SU}^*(6)}{\mathrm{USp}(6)}$	$\frac{\mathrm{SO}^*(12)}{\mathrm{SU}(6)\otimes\mathrm{U}(1)}$	$\frac{E_{7,-5}}{\mathrm{SO}(12)\otimes\mathrm{SU}(2)}$
$8, 1$	27	$\frac{E_{6,-26}}{F_4}$	$\frac{E_{7,-25}}{E_6\otimes\mathrm{U}(1)}$	$\frac{E_{8,-24}}{E_7\otimes\mathrm{SU}(2)}$

fixed value for a 'radius'. Using the conformal methods for $D=5$, one finds, similar to (20.99), together with $k_{\mathrm{D}}^2=-9\kappa^{-2}$, [207]

$$C_{IJK}\sigma^I\sigma^J\sigma^K=-3\kappa^{-2}\,. \tag{20.114}$$

We choose coordinates ϕ^x on the $(n-1)$-dimensional submanifold (20.114). In order to convert to the standard normalizations in the literature [93, 204, 205, 207], we rescale the C_{IJK} symbol and the gauge multiplet scalars as follows:

$$h^I\equiv\kappa\sqrt{\frac{2}{3}}\sigma^I\,,\qquad \mathcal{C}_{IJK}\equiv-\frac{1}{2\kappa}\sqrt{\frac{3}{2}}C_{IJK}\,, \tag{20.115}$$

such that

$$\mathcal{C}_{IJK}h^Ih^Jh^K=1\,. \tag{20.116}$$

The metric on the scalar manifold is then

$$g_{xy}=-3(\partial_xh^I)(\partial_yh^J)\mathcal{C}_{IJK}h^K\,. \tag{20.117}$$

In general, very special real manifolds are not homogeneous manifolds. The manifold turns out to be a homogeneous manifold for certain choices of the C tensor [245]. These manifolds have been called $L(q,P)$ for $q=-1,0,1,2,\ldots$ and $P=0,1,2,\ldots$, and for $q=4m$ there is an extra possibility $L(q,P,\dot{P})=L(q,\dot{P},P)$. For some specific values of q and P they are symmetric spaces, and their dimensional reduction according to Table 20.4 gives rise, up to the exceptional case $L(-1,P)$, to symmetric very special Kähler and very special quaternionic manifolds; see Table 20.5.

20.3.3 Special Kähler geometry

Gauge multiplets in $D=4$, $\mathcal{N}=2$ have complex scalars, and the kinetic terms determine Kähler manifolds, which are of a special type. The restricted class of Kähler metrics that occur in the scalar sector of $\mathcal{N}=2$, $D=4$ gauge multiplets are called special Kähler manifolds [91]. We saw already in (20.16) that for global supersymmetry the Kähler potential is obtained from a prepotential $F(X)$. However, the X^I are specific coordinates, and (20.16) is not preserved when we change to other coordinates. We should look for a coordinate independent characterization of special Kähler geometry. Definitions of such manifolds independent of supergravity were given in [247, 243, 244, 159].

The main ingredient of the geometry of gauge multiplets in four dimensions is the connection to the gauge fields. We saw in Sec. 4.2 that the symmetries of the field equations and Bianchi identities of these gauge fields define a symplectic group. The matrix that enters in the kinetic terms, which was called f_{AB} in Sec. 4.2.4, depends on the scalars of the gauge multiplets. Therefore, the rule (4.74) determines the action of symplectic transformations on the scalars. This extra ingredient characterizes special Kähler geometry as a subclass of Kähler manifolds. They are the Kähler manifolds in which the scalars 'feel' the symplectic transformations.

Rigid special Kähler geometry

We will first discuss special Kähler geometry in the context of *global supersymmetry* (i.e. 'affine' or 'rigid' special Kähler manifolds), and will then use the projection method to define special geometry as it is known in supergravity. In Appendix 20C we give a mathematical definition, which essentially demands that it is a Kähler manifold with an extra connection that preserves a symplectic structure and has zero curvature. We discuss here the resulting structure.

A rigid special Kähler manifold is a Kähler manifold with holomorphic coordinates z^α, and 'holomorphic symplectic sections'. A symplectic section means that at any point of the manifold there are $2n$ quantities $X^I(z)$ and $F_I(z)$ that transform as a vector under the symplectic transformations discussed in Sec. 4.2.4 (there presented with indices A). Here $I=1,\ldots,n$ and $\alpha=1,\ldots,n$. The z^α behave as arbitrary coordinates under holomorphic coordinate transformations. The X^I and F_I should be vectors under the symplectic transformations. Here F_I are the lower components of such a symplectic vector, and not the derivative of a scalar function F. But we will soon find that under regular assumptions it is of gradient form locally in any patch.

The supersymmetry transformations in the gauge multiplets connect the vectors to the fermions, and the latter to the scalars, in such a way that the symplectic transformations of the vectors are connected to those on the scalar sector by the correspondence

$$V=\begin{pmatrix} X^I \\ F_I \end{pmatrix} \Longleftrightarrow \begin{pmatrix} F^I_{\mu\nu} \\ G_{\mu\nu I} \end{pmatrix}. \tag{20.118}$$

Symplectic inner products between two vectors, denoted by $\langle\cdot,\cdot\rangle$, are symplectic invariants that are defined using the antisymmetric matrix Ω in (4.77), for example

$$\langle V,\bar{V}\rangle \equiv X^I\bar{F}_I - F_I\bar{X}^I = \begin{pmatrix} X^I & F_I \end{pmatrix}\begin{pmatrix} 0 & \mathbb{1} \\ -\mathbb{1} & 0 \end{pmatrix}\begin{pmatrix} \bar{X}^I \\ \bar{F}_I \end{pmatrix}. \tag{20.119}$$

Special Kähler geometry can be defined as a Kähler manifold with metric $g_{\alpha\bar\beta}$ based on a symplectic vector $V(z)$ such that

$$\langle \partial_\alpha V, \partial_\beta V \rangle = 0\,,$$
$$g_{\alpha\bar\beta} = \mathrm{i}\partial_\alpha \partial_{\bar\beta} \langle V, \bar V \rangle = \mathrm{i}\langle \partial_\alpha V, \partial_{\bar\beta} \bar V \rangle\,. \tag{20.120}$$

The Kähler potential is thus the symplectic invariant expression $\mathrm{i}\langle V, \bar V\rangle$. The X^I are the scalars that we used in the previous sections.

If X^I are independent variables, i.e. if $\partial_\alpha X^I$ is invertible, then the first condition in (20.120) is the integrability condition for the (local) existence of a holomorphic function $F(X)$, called the *prepotential*, with

$$F_I(X) = \frac{\partial F(X)}{\partial X^I}\,. \tag{20.121}$$

The function $F(X)$ is an arbitrary holomorphic function for rigid special Kähler geometry. For the conformal theory, we have already seen in (20.85) that it must be homogeneous of second order, and this requirement will then also appear when we discuss supergravity.

If the metric is positive definite, $\partial_\alpha X^I$ should be invertible [247] and thus a prepotential can be defined. Then the metric is

$$g_{\alpha\bar\beta} = N_{IJ}\partial_\alpha X^I \partial_{\bar\beta} \bar X^J\,, \qquad N_{IJ} \equiv 2\,\mathrm{Im}\,F_{IJ} = -\mathrm{i}F_{IJ} + \mathrm{i}\bar F_{IJ}\,. \tag{20.122}$$

This metric is the one that can be extracted from the action (20.15). We thus identify $F(X)$ as the holomorphic function which has been used to construct that supersymmetric action. We remind the reader that we used $F(X)$ in Sec. 20.1.3 as the first component of a chiral multiplet whose highest component gives the action. In superspace language, this means that the action is $\int \mathrm{d}^4x\, \mathrm{d}^8\theta\, F$+h.c. However, the last steps above hold only in a local frame. $F(X)$ is in general not a well-defined function over the manifold. As a global definition, we should consider the symplectic formulation in which $(X^I,\ F_I)$ are local components of a symplectic bundle. When we go from one patch to another, the symplectic vectors in both patches should be related by symplectic transformations [247, 243, 159].

The situation is quite different if the metric is not positive definite. This case is of course important for supergravity, since the condition (20.99) implies that there should be negative contributions to the metric N_{IJ}, due to the compensating multiplet with negative signature. In that case, the function $X^I(z)$ may in some instances not be invertible, and we cannot go beyond (20.120), i.e. there may not be a prepotential such that (20.121) holds [248]. Hence, the symplectic formulation is necessary to understand the structure of special geometry.

The kinetic terms of the gauge fields are also determined by special geometry. The relevant complex matrix was called f_{AB} in Sec. 4.2.4 and in the treatment of $\mathcal{N} = 1$, but the standard notation for extended supersymmetry is $\mathcal{N}_{IJ} \equiv -\mathrm{i}\bar f_{IJ}$, as we mentioned already in (12.11). We thus write the spin-1 kinetic terms as

Box 20.5 Special Kähler geometry

In special Kähler geometry the symplectic symmetry $\mathrm{Sp}(2(n_V+1))$ of the gauge vectors in the multiplet acts also in the scalars that live in a complex n_V dimensional Kähler manifold. The geometry can be described in terms of a prepotential $F(X)$ that must be homogeneous of degree 2 in the (n_V+1) complex variables X^I.

$$\mathcal{L}_1 = \tfrac{1}{4}e(\mathrm{Im}\,\mathcal{N}_{IJ})F^I_{\mu\nu}F^{\mu\nu J} - \tfrac{1}{8}(\mathrm{Re}\,\mathcal{N}_{IJ})\varepsilon^{\mu\nu\rho\sigma}F^I_{\mu\nu}F^J_{\rho\sigma} = \tfrac{1}{2}e\,\mathrm{Im}\left(\mathcal{N}_{IJ}F^{+I}_{\mu\nu}F^{+\mu\nu J}\right). \tag{20.123}$$

The symplectic transformation on this matrix is

$$\mathcal{N}' = (C + D\mathcal{N})(A + B\mathcal{N})^{-1}, \qquad \mathcal{N}_{IJ} \equiv -\mathrm{i}\bar{f}_{IJ}. \tag{20.124}$$

Comparing the general form of the action (4.78) with (20.15) identifies $\mathcal{N}_{IJ} = \bar{F}_{IJ}$. A symplectic invariant definition is

$$\overline{\mathcal{N}}_{IJ}\partial_\alpha X^J = \partial_\alpha F_I. \tag{20.125}$$

This defines $\mathcal{N}_{IJ}$ for an invertible $\partial_\alpha X^I$. For supergravity, where this matrix may not be invertible, we will need an extra equation.

We finally draw the attention of the reader to the formulas (20.223) and (20.236), which give the curvature of rigid special Kähler manifolds.

Projective special Kähler manifolds

To define projective special Kähler geometry, we have to start with the definition of the dilatation operator. In the context of the superconformal calculus, we found already that the X^I have Weyl weight 1, and that a prepotential F should have weight 2, such that F_I also has weight 1.

Exercise 20.16 *Check that for any Kähler manifold the condition for a closed homothetic Killing vector (15.46) using a holomorphic and an anti-holomorphic index states that k^α should be holomorphic. Check that this condition (with $w=1$) for two holomorphic indices is equivalent to the condition $\partial_\alpha(k^\beta g_{\beta\bar{\beta}}) = g_{\alpha\bar{\beta}}$. Use then the metric (20.120) and show that it is solved when*

$$k^\alpha_{\mathrm{D}}\partial_\alpha V(z) = V(z). \tag{20.126}$$

Prove finally that with (20.121) this reproduces the statements of dilatational transformations of X^I and the homogeneity of the prepotential F.

As mentioned before, we can thus use the properties of the projective Kähler manifolds discussed in Sec. 17.3 with $a = \kappa^{-2}$ for supergravity, as follows from (20.99). We split the coordinates similar to (17.45) and can do this for the whole symplectic section:

$$V = y\, v(z)\,, \qquad v(z) \equiv \begin{pmatrix} Z^I(z) \\ F_I(z) \end{pmatrix}, \tag{20.127}$$

where v is a symplectic vector with $2(n_V+1)$ components.[14] The holomorphic coordinates are as for $\mathcal{N} = 1$: $\{y, z^\alpha\}$, $\alpha = 1, \ldots, n_V$. Hence this set of $n_V + 1$ coordinates of the embedding manifold were in the presentation of the rigid Kähler manifold denoted by z^α. The condition (20.126) and the definition of y as the coordinate in the direction of the holomorphic homothetic Killing vector imply $F_I(y,z) = y\, F_I(z)$. In the case that they are determined from a prepotential, we have

$$F_I(y,z) = F_I(X(y,z)) = F_I(y\, Z(z)) = y F_I(Z(z))\,. \tag{20.128}$$

Owing to the homogeneity, we have in fact $F(X) = y^2 F(Z)$, and $F_I(Z) = \partial F(Z)/\partial Z^I$. Since these functions $F(X)$ and $F(Z)$ have the same functional forms, we do not introduce different names. Note that, according to (20.118), V transforms under symplectic transformations in the same way as the gauge field strengths. When we consider symplectic transformations that do not leave y invariant, v does not transform in the same way.

Following the steps of Sec. 17.3.5, we obtain here (with $a = \kappa^{-2}$ and taking the U(1) gauge $y = \bar{y}$)

$$y = \mathrm{e}^{\kappa^2 \mathcal{K}(z,\bar{z})/2}\,, \qquad \mathrm{e}^{-\kappa^2 \mathcal{K}(z,\bar{z})} = -\mathrm{i}\kappa^2 \langle v, \bar{v}\rangle\,. \tag{20.129}$$

We then define the Kähler transformations and introduce Kähler covariant derivatives as in Sec. 17.3.6. They can be applied to the full symplectic section, and the covariant derivatives (17.73) are

$$\begin{aligned} \nabla_\alpha v &\equiv \partial_\alpha v + \kappa^2 (\partial_\alpha \mathcal{K}) v\,, & \overline{\nabla}_{\bar\alpha} v &\equiv 0\,, \\ \overline{\nabla}_{\bar\alpha} \bar{v} &\equiv \partial_{\bar\alpha} \bar{v} + \kappa^2 (\partial_{\bar\alpha} \mathcal{K}) \bar{v}\,, & \nabla_\alpha \bar{v} &\equiv 0\,. \end{aligned} \tag{20.130}$$

Remember from Sec. 17.3.6 that $\nabla_\alpha y = \nabla_\alpha \bar{y} = 0$ as well, before or after the T gauge fixing. We can also write

$$\nabla_\alpha X^I = \partial_\alpha X^I + \tfrac{1}{2}\kappa^2 (\partial_\alpha \mathcal{K}) X^I = y \nabla_\alpha Z^I\,, \qquad \nabla_\alpha V = y \nabla_\alpha v\,, \tag{20.131}$$

which allows us to write down the Kähler and symplectic covariant form of the metric

$$g_{\alpha\bar\beta} = \mathrm{i}\,\langle \nabla_\alpha V, \overline{\nabla}_{\bar\beta} \bar{V}\rangle = \mathrm{i}\, y\bar{y}\, \langle \nabla_\alpha v, \overline{\nabla}_{\bar\beta} \bar{v}\rangle\,. \tag{20.132}$$

It is also important to realize the vanishing of the symplectic holomorphic inner products

$$\langle v, \nabla_\alpha v\rangle = \langle \nabla_\alpha v, \nabla_\beta v\rangle = 0\,. \tag{20.133}$$

The vanishing of the first expression implies the second one (while the inverse statement is valid for $n > 1$), and follows from the homogeneity of first order of $F_I(X)$.

[14] We use n_V for the complex dimension of the projective manifold, and thus the rigid special Kähler manifold has dimension $n_V + 1$.

The matrix methods introduced in Ex. 17.14 become even more important in the present case. However, the expressions there have to be completed to symplectic covariant expressions. The invertibility of the metric implies [247] that the $2(n_V+1)\times(n_V+1)$ matrix

$$(v \quad \nabla_\alpha v) = \begin{pmatrix} Z^I(z) & \nabla_\alpha Z^I(z) \\ F_I(z) & \nabla_\alpha F_I(z) \end{pmatrix} \tag{20.134}$$

has rank (n_V+1). However, the upper $(n_V+1)\times(n_V+1)$ matrix could be non-invertible. If it is invertible, then the prepotential exists. Indeed, this upper part corresponds to $\{\partial_y X^I, \partial_\alpha X^I\}$ for the embedding manifold, whose invertibility was the condition for the existence of the prepotential. In either case, if the metric is invertible one can prove that

$$(\bar{Z}^I(z) \quad \nabla_\alpha Z^I(z)) \tag{20.135}$$

is invertible. This invertible matrix appears in the expression

$$\bar{\mathcal{N}}_{IJ} = \big(\bar{F}_I(\bar{z}) \quad \nabla_\alpha F_I(z)\big)\big(\bar{Z}^J(\bar{z}) \quad \nabla_\alpha Z^J(z)\big)^{-1}, \tag{20.136}$$

which defines the kinetic matrix of the vector fields. When we discuss the Poincaré supergravity theory, we will see that this is the matrix that appears in the spin-1 action (20.123). As previously mentioned, the definition (20.136) is independent of the existence of the prepotential. If F_I is derivable from a prepotential, (20.136) can be written as

$$\mathcal{N}_{IJ}(z,\bar{z}) = \bar{F}_{IJ}(\bar{z}) + \mathrm{i}\frac{N_{IK}Z^K(z)N_{JL}Z^L(z)}{N_{MN}Z^M(z)Z^N(z)} = \bar{F}_{IJ} + \mathrm{i}\frac{N_{IK}X^K N_{JL}X^L}{N_{MN}X^M X^N}. \tag{20.137}$$

Exercise 20.17 *Show that inserting the bosonic part of (20.105) in (20.89) leads to (20.123) with $\mathcal{N}_{IJ}$ as in (20.137).*

The positivity of the kinetic energy of the vectors follows from the $(n_V+1)\times(n_V+1)$ matrix equation

$$\begin{aligned}\begin{pmatrix} \kappa^{-2} & 0 \\ 0 & g_{\alpha\bar{\beta}} \end{pmatrix} &= \mathrm{i}(y\bar{y})\begin{pmatrix} \langle \bar{v}, v\rangle & \langle \bar{v}, \overline{\nabla}_{\bar{\beta}} v\rangle \\ \langle \nabla_\alpha v, v\rangle & \langle \nabla_\alpha v, \overline{\nabla}_{\bar{\beta}}\bar{v}\rangle \end{pmatrix} \\ &= -2(y\bar{y})\begin{pmatrix} \bar{Z}^I \\ \nabla_\alpha Z^I \end{pmatrix} \operatorname{Im}\mathcal{N}_{IJ}\,(Z^J \quad \overline{\nabla}_{\bar{\beta}}\bar{Z}^J) .\end{aligned} \tag{20.138}$$

It shows that $\operatorname{Im}\mathcal{N}_{IJ}$ is negative definite if the metric is positive definite.[15]

The curvature of a special Kähler manifold can be obtained from the general method in (17.88) in terms of the curvature of the embedding manifold, which is derived in Appendix 20C; see (20.223). This leads to

$$R_{\alpha\bar{\alpha}\beta\bar{\beta}} = 2\kappa^2 g_{\bar{\alpha}(\alpha}g_{\beta)\bar{\beta}} - C_{\alpha\beta\gamma}g^{\gamma\bar{\gamma}}\bar{C}_{\bar{\alpha}\bar{\beta}\bar{\gamma}}\,. \tag{20.139}$$

[15] In fact it is also used here that κ^2 is positive, which means in supergravity that the positivity of the kinetic terms of the spin-1 fields follows from the positive definiteness of the kinetic terms of the spin-2 and of the spin-0 fields.

The tensor $C_{\alpha\beta\gamma}$ can be written as

$$C_{\alpha\beta\gamma} = \mathrm{i}\,\langle \nabla_\alpha \nabla_\beta V, \nabla_\gamma V\rangle = \mathrm{i}F_{IJK}\nabla_\alpha X^I \nabla_\beta X^J \nabla_\gamma X^K\,, \tag{20.140}$$

where the last expression can only be used when the prepotential $F(X)$ exists.

The examples in the following exercises will be used in Ch. 22.

Exercise 20.18 *Consider the prepotential*[16]

$$F(X) = -\mathrm{i}\kappa^{-2}X^0X^1\,. \tag{20.141}$$

This is a model with $n_V = 1$*, i.e. with one coordinate* $z = z^1$*. One has to choose a parametrization to be used in the upper part of (20.127). Choose* $Z^0 = 1$ *and* $Z^1 = -\mathrm{i}z$ *(in order to use the same variable* z *as in Chs. 7 and 22). Find the holomorphic symplectic vector*

$$v = \begin{pmatrix} Z^0 \\ Z^1 \\ F_0 \\ F_1 \end{pmatrix} = \begin{pmatrix} 1 \\ -\mathrm{i}z \\ -\kappa^{-2}z \\ -\mathrm{i}\kappa^{-2} \end{pmatrix}. \tag{20.142}$$

Obtain the Kähler potential directly from (20.129), and find the metric in the scalar and vector sector:

$$\mathrm{e}^{-\kappa^2\mathcal{K}} = 4\,\mathrm{Im}\,z\,, \qquad g_{z\bar z} = \partial_z\partial_{\bar z}\mathcal{K} = (2\kappa\,\mathrm{Im}\,z)^{-2}\,, \qquad \mathcal{N} = \kappa^{-2}\begin{pmatrix} -z & 0 \\ 0 & 1/z \end{pmatrix}. \tag{20.143}$$

We remark that the domain of positivity for both metrics (i.e. $g_{z\bar z} > 0$ *and* $\mathrm{Im}\,\mathcal{N}_{IJ} < 0$*) is* $\mathrm{Im}\,z > 0$.

Exercise 20.19 *Perform a symplectic transformation on the vector of the previous example:*

$$V' = \mathcal{S}V = \begin{pmatrix} 1 & 0 & 0 & 0 \\ 0 & 0 & 0 & -\kappa^2 \\ 0 & 0 & 1 & 0 \\ 0 & \kappa^{-2} & 0 & 0 \end{pmatrix} y v = y\begin{pmatrix} 1 \\ \mathrm{i} \\ -\kappa^{-2}z \\ -\mathrm{i}\kappa^{-2}z \end{pmatrix}. \tag{20.144}$$

After this transformation, z *does not appear any more in* (X'^0, X'^1)*, the upper two components of the symplectic vector* V'*. Hence,* $\partial_\alpha X^I = 0$*, and thus the symplectic vector cannot be obtained from a prepotential. The Kähler metric is a symplectic invariant, and is still as in (20.143). Compute the kinetic matrix of the vector sector in two ways: (1) from (20.136) and (2) using (20.124):*

$$\mathcal{N}' = (C + D\mathcal{N})(A + B\mathcal{N})^{-1} = -\kappa^{-2}z\mathbb{1}\,. \tag{20.145}$$

[16] The factor i in the overall normalization is necessary to get a non-trivial model; see Ex. 20.4. The factor κ^{-2} is a normalization chosen for convenience in Ch. 22.

Symmetries of special Kähler manifolds

The symmetries of a special Kähler manifold are a subset of the symplectic transformations. They are those that are consistent with the chosen symplectic vectors. Infinitesimal symplectic transformations are matrices of the form (4.76) that are close to the unit matrix, i.e.

$$\mathcal{S} = \mathbb{1} + T\,, \qquad T = \begin{pmatrix} a & b \\ c & -a^T \end{pmatrix}, \qquad b = b^T\,, \qquad c = c^T\,. \tag{20.146}$$

We thus look for infinitesimal transformations δy and δz^α that are consistent with the transformation of $V = y\,v(z)$. Following (17.95) we write $\delta y = \kappa^2 y\, r(z)$, and the condition is then

$$\delta V = T\,V\,, \qquad \delta v(z) = T\,v(z) - \kappa^2 r(z)\,v(z)\,. \tag{20.147}$$

For a good understanding, one has to consider an example as in the following exercise.

Exercise 20.20 *Assume arbitrary δz, and compare the transformation of the symplectic vector (20.142) with the formula (20.147). Find out which δz and $r(z)$ are possible for a real constant T matrix of the form (20.146). Do then the same for (20.144). You should find that in both cases they are of the form*

$$r = \theta^\Sigma r_\Sigma = \kappa^{-2}\left(\mathrm{i}\theta^0 - \tfrac{1}{2}\theta^2 - \theta^3 z\right), \qquad \delta z = \theta^\Sigma k_\Sigma{}^z = \theta^1 + \theta^2 z + \theta^3 z^2\,. \tag{20.148}$$

These are the Killing vectors of $\mathfrak{su}(1,1)$ that we found already in (7.157), and θ^0 is the parameter of a commuting abelian (R-symmetry) generator. The symplectic embeddings are different in the two cases, such that in the first case there is only one perturbative symmetry (a symmetry such that $b = 0$) while in the second case there are three perturbative symmetries.[17] *The solution that you should find for the case (20.144) is*

$$T = \begin{pmatrix} a & b \\ c & -a^T \end{pmatrix} = \begin{pmatrix} -\frac{1}{2}\theta^2 & \theta^0 & \kappa^2\theta^3 & 0 \\ -\theta^0 & -\frac{1}{2}\theta^2 & 0 & \kappa^2\theta^3 \\ -\kappa^{-2}\theta^1 & 0 & \frac{1}{2}\theta^2 & \theta^0 \\ 0 & -\kappa^{-2}\theta^1 & -\theta^0 & \frac{1}{2}\theta^2 \end{pmatrix}. \tag{20.149}$$

Exercise 20.21 *Write the left-hand side of (20.147) as $k^\alpha \partial_\alpha v$ and prove from this, using $T = \theta^\Sigma T_\Sigma$, the analogue of (17.98):*

$$T_\Sigma\, v = k_\Sigma{}^\alpha \nabla_\alpha v + \mathrm{i}\kappa^2 P^0_\Sigma v\,, \qquad P^0_\Sigma = -\mathrm{e}^{\kappa^2\mathcal{K}}\langle T_\Sigma v, \bar v\rangle\,. \tag{20.150}$$

P^0_Σ is the moment map (20.108) for this case.

The special Kähler manifolds that are the dimensional reduction of the very special real manifolds (i.e. very special Kähler manifolds; see Table 20.4) have a description in terms of a prepotential of the form

[17] The example that we used here appears in a reduction to $\mathcal{N}=2$ of two versions of $\mathcal{N}=4$ supergravity, known respectively as the 'SO(4) formulation' [249, 250, 251] and the 'SU(4) formulation' of pure $\mathcal{N}=4$ supergravity [252].

$$F(X) = \frac{C_{ABC}X^A X^B X^C}{X^0}, \tag{20.151}$$

where $\{X^I\} = \{X^0, X^A\}$, and the C_{ABC} are the coefficients used in Sec. 20.3.2, there denoted by C_{IJK}. In particular, this includes all the Kähler manifolds in Table 20.5. The only other symmetric special Kähler manifolds [253] are the dimensional reductions of pure $D = 5$ supergravity, leading to SU(1, 1)/U(1), and (the non-compact versions of) the complex projective spaces

$$\frac{\mathrm{SU}(n_V, 1)}{\mathrm{SU}(n_V) \times \mathrm{U}(1)}. \tag{20.152}$$

The latter can be obtained by the prepotential

$$F(X) = \mathrm{i}\left(-X^0X^0 + X^AX^A\right). \tag{20.153}$$

One can see that the list of symmetric spaces is consistent with (13.69), where all irreducible symmetric Kähler spaces were listed. However, for general Kähler spaces, one can take products of two Kähler spaces to make a new one, which is not the case for special Kähler manifolds. The only special Kähler manifolds [254] that are product manifolds are those in the (0, P) Kähler entry of Table 20.5. Since those manifolds are important in many applications of special geometry, we will consider them in more detail. They are the symmetric spaces

$$\frac{\mathrm{SU}(1,1)}{\mathrm{U}(1)} \otimes \frac{\mathrm{SO}(n_V - 1, 2)}{\mathrm{SO}(n_V - 1) \otimes \mathrm{SO}(2)}. \tag{20.154}$$

These manifolds occur in the classical limit of the compactified heterotic string, where the first factor contains the dilaton-axion.

Exercise 20.22 *Special Kähler models of type $L(0, n_V - 3)$ can be described by the prepotential*

$$F(X) = \frac{1}{2X^0} X^1 X^r X^t \eta_{rt}, \tag{20.155}$$

where $\{X^I\} = \{X^0, X^1\, X^r\}$, with $r = 2, \ldots, n_V$, and η_{rt} is a constant diagonal metric with signature $(+, -, \ldots, -)$. By choosing suitable coordinates, S and z^r, obtain the symplectic section

$$v = \begin{pmatrix} 1 \\ S \\ z^r \\ -\frac{1}{2} S \eta_{rs} z^r z^s \\ \frac{1}{2}\eta_{rs} z^r z^s \\ S\eta_{rs} z^s \end{pmatrix}. \tag{20.156}$$

In this parametrization only an SO($n_V - 2$) subgroup of SO($n_V - 1, 2$) is linearly realized (acting on the X^r). This is surprising from a string compactification point of view. The full SO($n_V - 1, 2$) should be a perturbative symmetry, as realized in the $\mathcal{N} = 4$ theory described in [255, 256]. We can, however, obtain a linear SO($n_V - 1, 2$) symmetry after a symplectic transformation [248], as clarified in the following exercise.

Exercise 20.23 *Consider the symplectic transformation on the components* $\{Z^0, Z^1, F_0, F_1\}$ *of (20.156) (leaving invariant* Z^r *and* F_r*), generated by*

$$\begin{pmatrix} -\frac{1}{2} & 0 & 0 & 1 \\ \frac{1}{2} & 0 & 0 & 1 \\ 0 & -\frac{1}{2} & -1 & 0 \\ 0 & -\frac{1}{2} & 1 & 0 \end{pmatrix} . \tag{20.157}$$

Prove first that this is a symplectic transformation, and show then that after this transformation

$$\begin{aligned} Z'^0 &= -\tfrac{1}{2}(1 - \eta_{ij} z^i z^j) , \\ Z'^1 &= +\tfrac{1}{2}(1 + \eta_{ij} z^i z^j) , \\ Z'^i &= z^i , \\ F'_I &= S\, \eta_{IJ} Z'^J , \end{aligned} \tag{20.158}$$

such that $Z'^I \eta_{IJ} Z'^J = 0$*, with* $\eta_{IJ} = \text{diag}(+ - + - \ldots -)$*. Show that this symplectic section is consistent with symmetry transformations of the form (20.147) with*

$$a^I{}_J = -d_J{}^I = \Lambda^I{}_J , \qquad b = c = 0 , \qquad -\kappa^2 r(z) = \left(\Lambda^0{}_J - \Lambda^1{}_J\right) Z'^J , \tag{20.159}$$

where $\Lambda^I{}_J$ *are parameters of* $\text{SO}(2, n_V - 1)$ *transformations, i.e.* $\eta_{IJ}\Lambda^J{}_K = -\eta_{KJ}\Lambda^J{}_I$.

The result (20.158) is similar to (20.144) in that the variable S *does not appear in the upper part. Therefore, also here the second formulation cannot be derived from a prepotential* F*. This important example illustrates the usefulness of formulations without a prepotential.*

The case $n_V = 3$ of (20.154) has some special properties. Since $\text{SO}(2,2) = \text{SU}(1,1) \times \text{SU}(1,1)$, this model is

$$\left(\frac{\text{SU}(1,1)}{\text{U}(1)}\right)^3 . \tag{20.160}$$

It is called the STU-model, giving the names S, T and U to the three complex scalar fields, and is often used for studies of black holes and for applications in string theory.

20.3.4 Hyper-Kähler and quaternionic-Kähler manifolds

Hyper-Kähler manifolds

We introduced hypermultiplets in Sec. 20.1.4. We found that they have a multiple of four real scalars, which we denoted by q^X. Their supersymmetry transformations involve 'frame fields', such that squares of the latter lead not only to the unit matrix, but also to hypercomplex structures; see (20.23). A hypercomplex structure is a triple $\vec{J}_X{}^Y$ of complex structures with the relation $J^1 J^2 = J^3$. The geometries of manifolds with complex structures have been discussed in the physics literature in [257] because of their importance in the context of two-dimensional field theories.

We assume that the manifold has also a metric structure, as we need to build an action, with Levi-Civita connection ∇. If the hypercomplex structure is covariantly constant with this connection, i.e. (20.27) holds, then we have a *hyper-Kähler manifold.*

Exercise 20.24 *Did you check that the complex structures (20.48) satisfy the algebra of a hypercomplex structure? Consider now*

$$(J^\alpha)_\beta{}^\gamma = -\varepsilon_{\alpha\beta\gamma}\,, \qquad (J^\alpha)_\beta{}^4 = \mathrm{e}^{-h}\delta^\alpha_\beta\,, \qquad (J^\alpha)_4{}^\beta = -\mathrm{e}^h\delta^\alpha_\beta\,, \qquad h \equiv q^4\,. \tag{20.161}$$

Check that the algebra is still satisfied. Consider the metric

$$g_{\alpha\beta} = \mathrm{e}^{-2h}\delta_{\alpha\beta}\,, \qquad g_{\alpha 4} = 0\,, \qquad g_{44} = 1\,. \tag{20.162}$$

The non-zero Christoffel connection coefficients are $\Gamma^4_{\alpha\beta} = \mathrm{e}^{-2h}\delta_{\alpha\beta}$ *and* $\Gamma^\alpha_{4\beta} = \Gamma^\alpha_{\beta 4} = -\delta^\alpha_\beta$. *Check that* ∇_4 *annihilates the complex structures, but the covariant derivatives in the first three directions of* q^X *are*

$$\nabla_\beta (J^\alpha)_X{}^Y = \varepsilon^{\alpha\beta\gamma}\mathrm{e}^{-h}(J^\gamma)_X{}^Y\,. \tag{20.163}$$

Hence the manifold is not hyper-Kähler. We will find that it is quaternionic-Kähler.

Quaternionic-Kähler manifolds

We will first discuss the geometry of quaternionic-Kähler manifolds. Then we will obtain them as projective manifolds from hyper-Kähler embedding spaces. We will use the notation q^u for the coordinates and use indices A for the flat basis. Hence, the fermionic indices in supergravity will be denoted by A. The basic equations of Sec. 20.1.4 are for these coordinates

$$\begin{aligned}
&f^{iA}{}_v f^u{}_{iA} = \delta^u_v\,, \qquad f^{iA}{}_u f^u{}_{jB} = \delta^i_j\delta^A_B\,,\\
&\left(f^{iA}{}_u\right)^* = f^{jB}{}_u\varepsilon_{ji}\rho_{B\bar{A}}\,, \qquad \left(f^u{}_{iA}\right)^* = \varepsilon^{ij}\rho^{\bar{A}B}f^u{}_{jB}\,,\\
&\rho_{A\bar{B}}\rho^{\bar{B}C} = -\delta^C_A\,, \qquad \rho^{\bar{A}B} = \left(\rho_{A\bar{B}}\right)^*\,,\\
&2f^{iA}{}_u f^v{}_{jA} = \delta^v_u\delta^i_j + \vec{\tau}_j{}^i\cdot\vec{J}_u{}^v\,, \qquad \vec{J}_u{}^v = (\vec{J}_u{}^v)^* = -f^{iA}{}_u f^v{}_{jA}\vec{\tau}_i{}^j\,,\\
&h_{uv} = f^{iA}{}_u\varepsilon_{ij}C_{AB}f^{jB}{}_v\,,
\end{aligned} \tag{20.164}$$

where h_{uv} is the metric. What distinguishes quaternionic-Kähler manifolds from hyper-Kähler manifolds is that the hypercomplex structures are only covariantly constant up to a rotation between them:

$$\widetilde{\nabla}_w\vec{J}_u{}^v \equiv \nabla_w\vec{J}_u{}^v + 2\vec{\omega}_w\times\vec{J}_u{}^v = 0\,, \tag{20.165}$$

where $\vec{\omega}_u(q)$ is a triplet of functions on the manifold. Quaternionic-Kähler manifolds entered physics in [95], and [258] contains a lot of interesting properties. Reviews of the geometries of quaternionic manifolds can be found in [259, 170, 206, 260].

Box 20.6 Quaternionic-Kähler manifolds

Quaternionic-Kähler manifolds are Einstein manifolds with holonomy group $\mathrm{USp}(2n_H)\times\mathrm{SU}(2)$. The $\mathrm{SU}(2)$ factor mixes the three complex structures.

In (20.44) we saw that the metric curvature of hyper-Kähler manifolds depends on the curvature of the $\mathrm{USp}(2n_H)=\mathrm{U}(n_H,\mathbb{H})$ holonomy group.[18] To investigate the curvature of quaternionic-Kähler manifolds, we start from the covariant constancy of frame fields. The condition (20.25) is modified to

$$\tilde{\nabla}_v f^{iA}{}_u=\partial_v f^{iA}{}_u+f^{iB}{}_u\omega_{vB}{}^A+f^{jA}{}_u\omega_{vj}{}^i-\Gamma^w_{vu}f^{iA}{}_w=0\,, \qquad (20.166)$$

where again we switch to the notation $\omega_i{}^j$ for the triplet $\vec{\omega}$. This condition gives the $\mathrm{USp}(2n_H)$ and $\mathrm{SU}(2)$ connections given the Christoffel connection of the metric as

$$\omega_{uj}{}^i\delta_B{}^A+\omega_{uB}{}^A\delta_j{}^i=-f^v_{jB}\left(\partial_u f^{iA}{}_v-\Gamma^w_{uv}f^{iA}{}_w\right). \qquad (20.167)$$

If the left-hand side were an arbitrary matrix $\Omega_{u\,jB}{}^{iA}$, the holonomy group would be $\mathrm{GL}(4n_H)$. This form is the expression of the fact that the holonomy group is restricted[19] to $\mathrm{USp}(2n_H)\times\mathrm{SU}(2)$. The curvature relation is obtained as integrability condition of (20.166):

$$\begin{aligned}
R_{uv}{}^w{}_x&=f^w{}_{iA}f^{iB}{}_x\mathcal{R}_{uvB}{}^A-\vec{J}_x{}^w\cdot\vec{\mathcal{R}}_{uv}\,,\\
R_{uv}{}^w{}_x&\equiv 2\partial_{[u}\Gamma^w_{v]x}+2\Gamma^w_{y[u}\Gamma^y_{v]x}\,,\\
\mathcal{R}_{uvB}{}^A&\equiv 2\partial_{[u}\omega_{v]B}{}^A+2\omega_{[u|C|}{}^A\omega_{v]B}{}^C\,,\\
\vec{\mathcal{R}}_{uv}&\equiv 2\partial_{[u}\vec{\omega}_{v]}+2\vec{\omega}_u\times\vec{\omega}_v\,.
\end{aligned} \qquad (20.168)$$

Quaternionic-Kähler manifolds turn out to be Einstein,[20] i.e.

$$R_{uv}=\frac{1}{4n_H}h_{uv}R\,. \qquad (20.169)$$

A further property is that the $\mathrm{SU}(2)$ curvature is proportional to the complex structures (with an index lowered using the metric):

$$\vec{\mathcal{R}}_{uv}=\tfrac{1}{2}\nu\vec{J}_{uv}\,,\qquad \nu\equiv\frac{1}{4n_H(n_H+2)}R\,. \qquad (20.170)$$

[18] The Ambrose–Singer theorem [261] states that the Lie algebra of the restricted holonomy group of the frame bundle coincides with the algebra generated by the curvature.

[19] Actually it is $\mathrm{USp}(2n_H)\times\mathrm{SU}(2)/\mathbb{Z}_2$, but since we consider mainly the algebra rather than the group, we will neglect this difference here.

[20] For $n_H>1$ this can be proven, but for $n_H=1$ the properties below are often included in the definition of quaternionic-Kähler manifolds [262], which from the physical point of view is appropriate as they follow also from supersymmetry requirements.

Here ν is a dimensionful constant, which in supergravity is related to the Planck mass: $\nu = -\kappa^2$. We will show below how this results from the projective construction of quaternionic-Kähler manifolds using the superconformal calculus. The negative value for supergravity implies that the manifold is non-compact (if there is at least one isometry). Hyper-Kähler manifolds are those where the SU(2) curvature is zero, i.e. $\nu = 0$, and these are thus Ricci flat, as mentioned before.

There is a relevant curvature relation, which splits the curvature of any quaternionic-Kähler manifold into the curvature of the (non-compact version of the) quaternionic projective space of the same dimension,

$$\mathbb{H}P_{n_H} = \frac{\mathrm{USp}(2n_H, 2)}{\mathrm{USp}(2n_H) \times \mathrm{SU}(2)}, \tag{20.171}$$

and a remaining part,

$$\begin{aligned} R_{uvwx} &= \nu(R^{\mathbb{H}P_{n_H}})_{uvwx} + \tfrac{1}{2}(f^{iA}{}_u f^{jB}{}_v \varepsilon_{ij})(f^{kC}{}_w f^{\ell D}{}_x \varepsilon_{k\ell})\mathcal{W}_{ABCD}, \\ \left(R^{\mathbb{H}P_{n_H}}\right)_{uvwx} &\equiv \tfrac{1}{2} h_{w[u} h_{v]x} + \tfrac{1}{2}\vec{J}_{uv} \cdot \vec{J}_{wx} - \tfrac{1}{2}\vec{J}_{w[u} \cdot \vec{J}_{v]x}, \end{aligned} \tag{20.172}$$

with $\mathcal{W}_{ABCD}$ completely symmetric. The $\mathcal{W}$-dependent part ('Weyl curvature') is Ricci flat, while the first part is completely determined by the Ricci tensor ('Ricci curvature'). That part gives the Ricci tensor (20.169) and the value of the scalar curvature in terms of ν in (20.170).

Exercise 20.25 *Calculate the curvature for the manifold in Ex. 20.24, and see that it agrees with (20.172) for $\nu = -1$ and $\mathcal{W}_{ABCD} = 0$. Hence this manifold is $\mathbb{H}P_1$. Check also (20.170). You will have to deduce from (20.163) that $\vec{\omega} \equiv \vec{\omega}_u \mathrm{d}q^u = -\frac{1}{2}\mathrm{e}^{-h}\mathrm{d}\vec{q}$.*

Isometries

The symmetries of hyper-Kähler manifolds were identified in Sec. 20.1.5 as the isometries that are triholomorphic, which was the condition (20.49). For these isometries we found that there is a triplet moment map $\vec{P}_I$. For quaternionic-Kähler manifolds the isometries are 'quaternionic', which means that

$$\left(\nabla_u k_I{}^v\right) \vec{J}_v{}^w - \vec{J}_u{}^v \left(\nabla_v k_I{}^w\right) = \nu \vec{J}_u{}^w \times \vec{P}_I. \tag{20.173}$$

This implies again the existence of moment maps with formulas as for hyper-Kähler manifolds up to 'covariantization' with the SU(2) connection:

$$\begin{aligned} &\tilde{\nabla}_u \vec{P}_I \equiv \partial_u \vec{P}_I + 2\vec{\omega}_u \times \vec{P}_I = \vec{J}_{uv} k_I{}^v, \\ &\nabla_u k_I{}^v = -\vec{R}_u{}^v \cdot \vec{P}_I + f_u^{iB} f^v{}_{iA} t_{IB}{}^A, \qquad t_{IA}{}^B \equiv \tfrac{1}{2} f^v{}_{iA} \nabla_v k_I{}^u f^{iB}{}_u, \\ &2n_H \nu \vec{P}_I = \vec{J}_u{}^v \nabla_v k_I{}^u. \end{aligned} \tag{20.174}$$

All these equations are true for hyper-Kähler manifolds when $\vec{\omega} = \vec{R} = \nu = 0$.

As we have seen before for other similar cases, there is one condition for the non-abelian Killing vectors: the equivariance relation. It generalizes (20.51),

$$k_I{}^u \vec{J}_{uv} k_J{}^v = f_{IJ}{}^K \vec{P}_K + \nu \vec{P}_I \times \vec{P}_J \,. \tag{20.175}$$

Exercise 20.26 *Show that the isometries of the metric defined in (20.162) can be parametrized by* $\{\theta^I\} = \{\theta^X, \theta^{[XY]}\}$, *such that (*$\alpha$ *indices are lowered by* $\delta_{\alpha\beta}$*)*

$$\begin{aligned}
\theta^I k_I^X &= \theta^\alpha k_\alpha{}^X + \theta^4 k_4{}^X + \tfrac{1}{2}\theta^{[\alpha\beta]} k_{[\alpha\beta]}{}^X + \theta^{4\alpha} k_{[4\alpha]}{}^X \,,\\
\theta^I k_I{}^\alpha &= \theta^\alpha + \theta^\alpha{}_\beta q^\beta + \theta^4 q_\alpha + q^\beta q^\beta \theta^{[4\alpha]} - 2 q^\alpha q^\beta \theta^{[4\beta]} + \mathrm{e}^{2h}\theta^{[4\alpha]} \,,\\
\theta^I k_I{}^4 &= \theta^4 - 2 q^\alpha \theta^{[4\alpha]} \,,\\
\theta^I P_I{}^\alpha &= \varepsilon_{\alpha\beta\gamma}\left(\tfrac{1}{2}\theta^{[\beta\gamma]} + 2\theta^{[4\beta]} q^\gamma\right) - 2\mathrm{e}^h \theta^{4\alpha} + \mathrm{e}^{-h}\theta^I k_I^\alpha \,.
\end{aligned} \tag{20.176}$$

To solve the Killing vector equations, you might compare with (15.5)–(15.6). In this way, you can recognize the algebra $\mathfrak{so}(4,1)$ *for the (Euclidean) conformal group in three dimensions. With (B.7) you can identify this group with* $\mathfrak{usp}(2,2)$. *With more energy, you can explicitly check (20.173).*

Quaternionic-Kähler manifolds as projected hyper-Kähler manifolds

We now turn to the construction of quaternionic-Kähler manifolds as projective hyper-Kähler manifolds [263, 206, 207, 260]. We start from hyper-Kähler cones. That means that it is a hyper-Kähler manifold that possesses a closed homothetic Killing vector as defined in Sec. 15.7. The construction is produced by the superconformal methods in Sec. 20.2. We saw in (20.79) that the dilatation with the complex structures defines an SU(2) factor in the isometry group. The homothetic Killing vector and the three SU(2) Killing vectors define four directions in the hyper-Kähler manifold that are to be eliminated to define the projective manifold. We thus have to define suitable coordinates between the $4(n_H+1)$ components of q^X to make the projection. We do this by defining four coordinates, indicated by q^0 and q^α, to run along the four directions defined by the homothetic Killing vector and the SU(2) Killing vectors:

$$\begin{aligned}
&\{q^X\} = \{q^0, q^\alpha, q^u\}\,, \\
&k_{\mathrm{D}}{}^X = \{2wq^0, 0, 0, \ldots, 0\}\,, \qquad \vec{k}^X = \frac{1}{2w} k_{\mathrm{D}}{}^Y \vec{J}_Y{}^X = \{0, \vec{k}^\alpha, 0, \ldots, 0, 0\}\,.
\end{aligned} \tag{20.177}$$

The other $4n_H$ coordinates q^u are (using the Frobenius theorem) defined such that these four vectors have no components in the u-directions.

The projective manifold is defined by a fixed value of the homothetic Killing vector and for the coordinates q^α. These are the conditions in (20.99) and the SU(2) gauge conditions in the superconformal construction of supergravity. After appropriate definitions of the variables,[21] the metric can be brought into a cone-light form as in Sec. 17.3.4, i.e.

$$g_{XY}\mathrm{d}q^X\mathrm{d}q^Y = -\frac{1}{q^0}\mathrm{d}q^0\mathrm{d}q^0 + h_{uv}\mathrm{d}q^u\mathrm{d}q^v - \frac{1}{q^0}\vec{k}_X\cdot\vec{k}_Y\mathrm{d}q^X\mathrm{d}q^Y\,,$$
$$\vec{k}_X\mathrm{d}q^X = \vec{k}_\alpha\mathrm{d}q^\alpha + \vec{k}_u\mathrm{d}q^u\,. \tag{20.178}$$

Note that although $\vec{k}^X$ has only three non-zero components according to (20.177), $\vec{k}_X$ has components in the α- and u-directions. This implies a relation between g_{uv} and h_{uv},

$$h_{uv} = g_{uv} - \frac{1}{q^0}\vec{k}_u\cdot\vec{k}_v\,, \tag{20.179}$$

which can be seen with the superconformal methods from the contribution of the auxiliary field $\vec{\mathcal{V}}_\mu$ (20.103).

The chosen value for q^0 defines the parameter ν mentioned in (20.170):

$$k_{\mathrm{D}}{}^X g_{XY} k_{\mathrm{D}}{}^Y = -4w^2 q^0 = 4w^2\frac{1}{\nu}\,. \tag{20.180}$$

For supergravity we use $w = (D-2)/2$, which is $w = 1$ for $D = 4$, such that comparison with (20.99) leads to

$$\nu = -\kappa^2 = -\frac{1}{q^0}\,. \tag{20.181}$$

The choice (20.177) implies that $\vec{J}_0{}^X$ has only components in the $X = \alpha$ directions, and one can obtain more information by analyzing the defining relations of the complex structures, (20.24), with various types of indices. One finds (using there the index value $X = 0$) that also $\vec{J}_\alpha{}^v = 0$, and this implies for all $\vec{A}$ and $\vec{B}$ that

$$\vec{A}\cdot\vec{J}_u{}^Z\vec{B}\cdot\vec{J}_Z{}^v = \vec{A}\cdot\vec{J}_u{}^w\vec{B}\cdot\vec{J}_w{}^v\,. \tag{20.182}$$

Hence, the components of the complex structures in the u-directions, i.e. $\vec{J}_u{}^v$, define a good hypercomplex structure on the projective manifold. However, its covariant derivative is no longer zero using the Levi-Civita connection of the projected metric h_{uv}. After analyzing the defining relations, one finds that it satisfies (20.165) with

$$\vec{\omega}_u = -\frac{1}{2q^0}\vec{k}_u\,. \tag{20.183}$$

This proves that the projective manifold is a quaternionic-Kähler space.

[21] More details are given in [207, 260].

Exercise 20.27 *Prove from the defining relations that*

$$\vec{J}_0{}^\alpha = \frac{1}{q^0}\vec{k}^\alpha\,,\qquad \vec{J}_\alpha{}^0 = \vec{k}_\alpha\,,\qquad -q^0\vec{k}^\alpha\cdot\vec{k}_\beta = \delta^\alpha_\beta\,,$$
$$\vec{A}\cdot\vec{k}^X\vec{k}_X\cdot\vec{B} = \vec{A}\cdot\vec{k}^\alpha\vec{k}_\alpha\cdot\vec{B} = -q^0\vec{A}\cdot\vec{B}\,. \tag{20.184}$$

As we did for the projective Kähler manifolds, we demand that symmetries of the projective manifold respect the closed homothetic Killing vector; see (20.91). The components of these Killing vectors in the projection are

$$k_I{}^X = \{k_I{}^0 = 0\,, k_I{}^\alpha = \vec{k}^\alpha\cdot\vec{r}_I, k_I{}^u\}\,, \tag{20.185}$$

with

$$\vec{r}_I = -2\vec{\omega}_u k_I{}^u + \nu\vec{P}_I\,,\qquad \vec{P}_I = \vec{k}_X k_I{}^X\,. \tag{20.186}$$

Note that this is very similar to the relation (17.97), where the role of the SU(2) gauge vectors $\vec{\omega}_u$ is played by $\mathrm{i}\partial_\alpha\mathcal{K}$, i.e. the gauge field for T-symmetry.

The indices of the local frame in the embedding manifold, $\mathcal{A}$, are split into $\{\mathcal{A}\} = \{i, A\}$, with $i = 1, 2$ and $A = 1, \ldots, 2n_H$, such that the symplectic metric $C_{\mathcal{AB}}$ in the embedding manifold has components $C_{ij} = \varepsilon_{ij}$, $C_{iA} = 0$ and C_{AB}, and $f^{i\mathcal{A}}{}_0$ is split into [260]

$$f^{ij}{}_0 = \mathrm{i}\,\varepsilon^{ij}\sqrt{\tfrac{1}{2}q^0}\,,\qquad f^{iA}{}_0 = 0\,. \tag{20.187}$$

The factor i takes care of the minus sign in g_{00}; see (20.178), using (20.37).

It turns out that the curvature tensors of the embedding manifold and the projective manifold are related by the fact that the components of the tensor (20.44) in the A-direction are the components of $\mathcal{W}$ in (20.172), i.e. $W_{ABCD} = \mathcal{W}_{ABCD}$. Also the frame fields $f^u{}_{iA}$ are identified with the corresponding components of the frame field in the embedding metric.

The simplest example starts from flat space in the embedding manifold, i.e. $W_{ABCD} = 0$. Since then also $\mathcal{W}_{ABCD} = 0$, (20.172) implies that the projective manifold is $\mathbb{H}P_{n_H}$. Other homogeneous spaces can be obtained from the **c**-map, i.e. they are special quaternionic. First there are the (very special) quaternionic spaces $L(q, P)$ discussed in Sec. 20.3.2. There are no other homogeneous, non-symmetric quaternionic-Kähler manifolds. Some of them, mentioned in Table 20.5, are symmetric. Other symmetric spaces are collected in Table 20.6, where SG_4 and SG_5 denote pure supergravity in four or five dimensions, and the dimensional reductions thereof.

A final, but important, remark is that one can add the Lagrangians for two hypermultiplets for global supersymmetry, and join them in one action. The corresponding mathematical statement is that a product of two hyper-Kähler manifolds is hyper-Kähler. However, this is not so for supergravity, since the projection is only applied to the full hyper-Kähler cone. A product of two quaternionic manifolds is not quaternionic-Kähler. This is can be seen from (20.170), which is a nonlinear relation on different multiplets. It is a typical feature of supergravity, where elimination of auxiliary fields mixes multiplets.

Table 20.6 Non-very special symmetric manifolds.

	n	real	Kähler	quaternionic
$L(-3,P)$	P			$\frac{\mathrm{USp}(2P+2,2)}{\mathrm{USp}(2P+2)\otimes\mathrm{SU}(2)}$
SG_4	0		SG	$\frac{\mathrm{U}(1,2)}{\mathrm{U}(1)\otimes\mathrm{U}(2)}$
$L(-2,P)$	$1+P$		$\frac{\mathrm{U}(P+1,1)}{\mathrm{U}(P+1)\otimes\mathrm{U}(1)}$	$\frac{\mathrm{SU}(P+2,2)}{\mathrm{SU}(P+2)\otimes\mathrm{SU}(2)\otimes\mathrm{U}(1)}$
SG_5	1	SG	$\frac{\mathrm{SU}(1,1)}{\mathrm{U}(1)}$	$\frac{G_2}{\mathrm{SU}(2)\otimes\mathrm{SU}(2)}$

20.4 From conformal to Poincaré supergravity

In the final section of this chapter, we explain the steps that are needed for rewriting (20.110) in terms of the variables of Poincaré supergravity. The final result will be given in the next chapter.

20.4.1 Kinetic terms of the bosons

We use the conditions (20.99) to fix the modulus of the scalars of the compensating gauge multiplet and hypermultiplet. The superconformal U(1) gauge and SU(2) gauge condition are reality conditions for these scalars. In the gauge multiplet this is (17.103) and leads to the projective Kähler metric as in Sec. 17.3. As SU(2) gauge condition, we demand the vanishing of the phases of the first quaternion, i.e. $q^\alpha = 0$, $\alpha = 1,2,3$, in (20.177). After the elimination of the auxiliary fields $\vec{\mathcal{V}}_\mu$, the kinetic terms of the scalars q^u of the hypermultiplets have the metric (20.179).

In the covariant derivatives of these physical scalars, the U(1), respectively SU(2), connections are absent. Indeed, the physical scalars z^α do not transform under the U(1) transformation, and q^u do not transform under the SU(2); see (20.177). Therefore the covariant derivative D_μ reduces to $\hat{\partial}_\mu z^\alpha$ as in (17.101), and $D_\mu q^u$ reduces to $\hat{\partial}_\mu q^u$ in (20.104).

The kinetic terms of the gauge fields have the standard form (20.123); see Ex. 20.17.

20.4.2 Identities of special Kähler geometry

There are several identities between the matrices $g_{\alpha\bar{\beta}}$, N_{IJ} and $\mathcal{N}_{IJ}$ that are important to write the action and transformation laws in convenient ways. One important relation is the $(n_V+1)\times(n_V+1)$ matrix equation (20.138). The reader is invited to derive others in the following three exercises.

Exercise 20.28 *Prove that*

$$F_I(X) = \mathcal{N}_{IJ} X^J \,, \tag{20.188}$$

from the explicit expression in the presence of a prepotential (20.137) as well as from the definition of $\mathcal{N}_{IJ}$ in (20.136). Check also the following useful relations:

$$\begin{aligned}\operatorname{Im}\mathcal{N}_{IJ}\bar{X}^J &= -\frac{\kappa^{-2}}{2}\frac{N_{IJ}X^J}{N_{LM}X^LX^M}\,,\\ (N)^{-1|IJ} &= -\tfrac{1}{2}(\operatorname{Im}\mathcal{N})^{-1|IJ} - \kappa^2\left(X^I\bar{X}^J + \bar{X}^I X^J\right),\end{aligned} \tag{20.189}$$

where, for example, $(N)^{-1|IJ}$ are the components of the inverse of N_{IJ}.

Exercise 20.29 *Prove the following consequences of (20.138):*

$$\begin{aligned}-\tfrac{1}{2}(\operatorname{Im}\mathcal{N})^{-1|IJ} &= \left(\nabla_\alpha X^I g^{\alpha\bar\beta}\bar\nabla_{\bar\beta}\bar{X}^J + \kappa^2\bar{X}^I X^J\right),\\ (N)^{-1|IJ} &= \left(\nabla_\alpha X^I g^{\alpha\bar\beta}\bar\nabla_{\bar\beta}\bar{X}^J - \kappa^2 X^I\bar{X}^J\right).\end{aligned} \tag{20.190}$$

Exercise 20.30 *Prove the equations:*

$$\begin{aligned}X^I(\operatorname{Im}\mathcal{N}_{IJ})X^J &= \frac{1}{2\kappa^4\bar{X}^I N_{IJ}\bar{X}^J}\,,\\ C_{\alpha\beta\gamma}g^{\gamma\bar\delta}\bar\nabla_{\bar\delta}\bar{X}^I(\operatorname{Im}\mathcal{N}_{IJ})X^J &= -\frac{\mathrm{i}\kappa^{-2}}{2\bar{X}^I N_{IJ}\bar{X}^J}F_{KLM}\nabla_\alpha X^K\nabla_\beta X^L\bar{X}^M\,,\\ \nabla_\beta\nabla_\alpha X^I = \mathrm{i}\,N^{-1|IL}F_{JKL}\nabla_\alpha X^J\nabla_\beta X^K &= C_{\alpha\beta\gamma}g^{\gamma\bar\delta}\bar\nabla_{\bar\delta}\bar{X}^I\,.\end{aligned} \tag{20.191}$$

20.4.3 The potential

The remaining bosonic part of the action is the scalar potential. We start from the second line in (20.110), which agrees with the rule (14.68) using the transformations in (20.76) and in (20.33). Using the bosonic part of (20.106), this potential is

$$V = N^{-1|IJ}\vec{P}_I\cdot\vec{P}_J + N^{-1|IJ}P_I^0P_J^0 + 2\bar{X}^I X^J k_I{}^X k_J{}^Y g_{XY}\,. \tag{20.192}$$

Now we transform this to the projective space. We start with the last term, and using the metric (20.178) we find

$$k_I{}^X k_J{}^Y g_{XY} = k_I{}^u k_J{}^v h_{uv} - \kappa^2\vec{k}_X\cdot\vec{k}_Y k_I{}^X k_J{}^Y = k_I{}^u k_J{}^v h_{uv} - \kappa^2\vec{P}_I\cdot\vec{P}_J\,, \tag{20.193}$$

where in the last equality we used (20.92). Using (20.189) and $X^I P_I^0 = 0$ we obtain

$$V = \left[-\tfrac{1}{2}(\operatorname{Im}\mathcal{N})^{-1|IJ} - 4\kappa^2 X^I\bar{X}^J\right]\vec{P}_I\cdot\vec{P}_J + 2\bar{X}^I X^J k_I{}^u k_J{}^v h_{uv} + N^{-1|IJ}P_I^0P_J^0\,. \tag{20.194}$$

20.4.4 Physical fermions and other terms

We introduced the projection from the fermions of the superconformal theory to fermions in the super-Poincaré theory for $\mathcal{N}=1$ in Sec. 17.3.7. The same ideas can be used to project the fermions of the gauge multiplets of $\mathcal{N}=2$. We write

$$\Omega_i^I = \chi_i^0 X^I + \chi_i^\alpha \nabla_\alpha X^I , \qquad \Omega^{iI} = \chi^{i0}\bar{X}^I + \chi^{i\bar{\alpha}}\overline{\nabla}_{\bar{\alpha}}\bar{X}^I . \tag{20.195}$$

The gauge fixing (20.102) leads as in $\mathcal{N} = 1$ to $\chi^{i0} = 0$, and one can apply Sec. 17.4.3 to obtain the kinetic terms as in (17.138).

For the fermions of the hypermultiplets, we split the $\mathcal{A}$-indices into those involved in the compensating multiplet and those that remain in the Poincaré theory:

$$\left\{\zeta^{\mathcal{A}}\right\} = \left\{\zeta^i, \zeta^A\right\} . \tag{20.196}$$

Owing to (20.177) and (20.187), $A^{i\mathcal{A}}$, defined in (20.94), has only components in the directions where $\mathcal{A}$ is i, and (20.102) implies that $\zeta^i = 0$. Hence, the physical components are ζ^A. With these definitions the next steps are similar to Sec. 17.4 for $\mathcal{N} = 1$.

The fourth line of (20.110) disappears after the gauge fixings. For the term with $D_c X^J$, this follows from (17.26). Further, it follows from (20.184) that $D_\mu q^X \vec{k}_X = \hat{\partial}_\mu q^X \vec{k}_X - 2\kappa^{-2}\vec{V}_\mu$, which vanishes due to (20.103).

Making extensive use of the special geometry relations in (20.189) and (20.191), one can write the solution for the graviphoton field strength (20.105) as

$$T_{ab}^- = -\kappa^2 \operatorname{Im}\mathcal{N}_{IJ} X^J \left[4\widehat{F}_{ab}^{-J} + \tfrac{1}{2} C_{\alpha\beta\gamma} g^{\gamma\bar{\delta}} \overline{\nabla}_{\bar{\delta}} \bar{X}^I \bar{\chi}_i^\alpha \gamma_{ab} \chi_j^\beta \varepsilon^{ij} + 2\kappa^2 X^I \bar{\zeta}^A \gamma_{ab} \zeta^B C_{AB}\right] . \tag{20.197}$$

This expression appears squared in the action (20.110), and thus generates also two-fermion terms. These combine with other terms in (20.110) to the following form of the coupling of fermions and gauge field strengths (called 'Pauli terms'):

$$\begin{aligned} e^{-1}\mathcal{L}_{\text{Pauli}} &= F_{ab}^{-I} \operatorname{Im}\mathcal{N}_{IJ} Q^{ab-J} + \text{h.c.}, \\ Q^{ab-J} &\equiv \overline{\nabla}_{\bar{\alpha}} \bar{X}^J \left(\tfrac{1}{8} g^{\beta\bar{\alpha}} C_{\beta\gamma\delta} \bar{\chi}_i^\gamma \gamma_{ab} \chi_j^\delta \varepsilon^{ij} + \bar{\chi}^{\bar{\alpha}i} \gamma^a \psi^{bj} \varepsilon_{ij}\right) \\ &\quad + X^J \left(\bar{\psi}_i^a \psi_j^b \varepsilon^{ij} + \tfrac{1}{2}\kappa^2 \bar{\zeta}^A \gamma^{ab} \zeta^B C_{AB}\right) . \end{aligned} \tag{20.198}$$

The reader who has the courage to go through the detailed calculations will notice that diverse miraculous cancelations occur to obtain (20.198). This formula has the special features that (1) F_{ab}^{-I} is at the end always multiplied with $\operatorname{Im}\mathcal{N}_{IJ}$ and (2) Q^{ab-J} is proportional to X^J or to $\overline{\nabla}_{\bar{\alpha}}\bar{X}^J$, but contains no terms with $\bar{X}^J$ or $\nabla_\alpha X^J$. These features are not accidents, but follow from the symplectic symmetry, which we will discuss in Sec. 21.2.4.

20.4.5 Supersymmetry and gauge transformations

As was done for $\mathcal{N} = 1$ supergravity in Sec. 18.2, we obtain the transformation laws of the action by considering the combination of the superconformal transformations that are conserved by the partial gauge fixing of Sec. 20.2.5. The superconformal transformations from which we have to start are those of the Weyl multiplet, (20.69), the vector multiplets,

(20.76), and the hypermultiplets, (20.33) and (20.81). Restricting ourselves to the physical fields, the Q- and S-supersymmetries are[22]

$$\begin{aligned}
\delta e_\mu{}^a &= \tfrac{1}{2}\bar\epsilon^i\gamma^a\psi_{\mu i} + \text{h.c.}\,,\\
\delta\psi_\mu^i &= \left(\partial_\mu + \tfrac{1}{4}\gamma^{ab}\omega_{\mu ab} - \tfrac{1}{2}\mathrm{i}A_\mu\right)\epsilon^i + V_\mu{}^i{}_j\epsilon^j - \tfrac{1}{16}\gamma^{ab}T_{ab}^-\varepsilon^{ij}\gamma_\mu\epsilon_j - \gamma_\mu\eta^i\,,\\
\delta X^I &= \tfrac{1}{2}\bar\epsilon^i\Omega_i^I\,,\\
\delta\Omega_i^I &= \not{D}X^I\epsilon_i + \tfrac{1}{4}\gamma^{ab}\mathcal{F}_{ab}{}^I\varepsilon_{ij}\epsilon^j + Y_{ij}{}^I\epsilon^j + X^J\bar X^K f_{JK}{}^I\varepsilon_{ij}\epsilon^j + 2X^I\eta_i\,,\\
\delta A_\mu^I &= \tfrac{1}{2}\varepsilon^{ij}\bar\epsilon_i\gamma_\mu\Omega_j{}^I + \varepsilon^{ij}\bar\epsilon_i\psi_{\mu j}X^I + \text{h.c.}\,,\\
\delta q^u &= -\mathrm{i}\bar\epsilon^i\zeta^A f^u{}_{iA} + \mathrm{i}\varepsilon^{ij}\rho^{\bar A B}\bar\epsilon_i\zeta_{\bar A} f^u{}_{jB}\,,\\
\delta\zeta^A &= \tfrac{1}{2}\mathrm{i}f^{iA}{}_u\hat{\not\partial}q^u\epsilon_i - \zeta^B\omega_{uB}{}^A\delta q^u + \mathrm{i}\bar X^I k_I{}^u f^{iA}{}_u\varepsilon_{ij}\epsilon^j + \mathrm{i}f^{iA}{}_u k_{\mathrm{D}}{}^u\eta_i\,,
\end{aligned}\tag{20.199}$$

while the relevant bosonic symmetry transformations are the T and SU(2) R-symmetries and the Yang–Mills gauge symmetries:

$$\begin{aligned}
\delta\psi_\mu^i &= \tfrac{1}{2}\mathrm{i}\lambda_T\psi_\mu^i + \psi_\mu^j\lambda_j{}^i\,,\\
\delta X^I &= \mathrm{i}\lambda_T X^I + \theta^J X^K f_{KJ}{}^I\,,\\
\delta\Omega_i^I &= \tfrac{1}{2}\mathrm{i}\lambda_T\Omega_i^I - \lambda_i{}^j\Omega_j^I + \theta^J\Omega_i^K f_{KJ}{}^I\,,\\
\delta A_\mu^I &= \partial_\mu\theta^I + \theta^J A_\mu^K f_{KJ}{}^I\,,\\
\delta q^u &= \theta^I k_I{}^u\,,\\
\delta\zeta^A &= -\tfrac{1}{2}\mathrm{i}\lambda_T\zeta^A + \theta^I\zeta^B t_{IB}{}^A - \zeta^B\omega_{uB}{}^A\delta q^u\,.
\end{aligned}\tag{20.200}$$

We have already used a few elementary results of the partial gauge fixing. We have put $b_\mu = 0$, and since none of these fields transforms under special conformal transformations, we do not need the corresponding decomposition law of $\lambda_{\mathrm{K}\mu}$. As in Sec. 18.2, $\lambda_D = 0$ after the partial gauge fixing, and therefore we have not mentioned the dilatations here.

For the gauge fixing of the T-symmetry, and the corresponding decomposition law for λ_T, we can use the results of Sec. 17.3.9. The corresponding discussion of gauged R-symmetry in Sec. 19.5 can be directly applied here, using $a = \kappa^{-2}$, and replacing $\mathcal{P}_A$ with P_I^0.

The general ideas can also be applied for the SU(2) part of the R-symmetry group, using now the scalars of the hypermultiplet.[23] As mentioned in Sec. 20.4.1, the SU(2) gauge fixing is performed by annihilating the phases of the quaternion of the compensating hypermultiplet, i.e. $q^\alpha = 0$. The preservation of that condition is the equation

$$\delta q^\alpha = -2\vec\lambda\cdot\vec k^\alpha + \theta^I k_I{}^\alpha + \left[-\mathrm{i}\bar\epsilon^i\zeta^A f^\alpha{}_{iA} + \text{h.c.}\right] = 0\,, \qquad f^\alpha{}_{iA} = -2\vec\omega_u\cdot\vec k^\alpha f^u{}_{iA}\,. \tag{20.201}$$

The last relation is obtained by a careful examination of the conditions that the frame fields should satisfy in the basis with (20.177) and (20.187) [260]. Using also (20.185) and (20.186), one obtains as decomposition law

[22] The terms $-\zeta^B\omega_{uB}{}^A\delta q^u$, which one finds in any transformation of the ζ^A in many equations of this section, and already appeared from (20.33), can be understood from the general principles of covariant transformations in Sec. 14B.

[23] The σ-model symmetries that correspond to the Kähler transformations are rotations of the complex structures, discussed shortly in Sec. 21.5.

$$\vec{\lambda} = -\vec{\omega}_u \delta q^u - \tfrac{1}{2}\kappa^2 \theta^I \vec{P}_I \,. \tag{20.202}$$

Thus the gauge transformations of the fields have now a part that acts as an SU(2) R-symmetry transformation as well as the T-transformation discussed in Sec. 17.3.9. Note the similarity between (19.53) and (20.202).

Finally, we need the decomposition law for the S-supersymmetry. We can calculate the variation of any of the equations in (20.102). Using $N_{IJ}X^J\Omega^{iI} = 0$, many steps are similar to the way in which we obtained (17.116). A useful relation, due to the field equation (20.105), is

$$N_{IJ}\bar{X}^J \mathcal{F}^{-I}_{ab} = -\tfrac{1}{8}\mathrm{i}F_{IJK}\bar{\Omega}^I_i \gamma_{ab}\Omega^J_j \bar{X}^K \varepsilon^{ij} + \tfrac{1}{2}C_{AB}\bar{\zeta}^A\gamma^{ab}\zeta^B \,. \tag{20.203}$$

Using also the field equation (20.106), we obtain

$$\begin{aligned}\kappa^{-2}\eta_i = {} & -\tfrac{1}{2}\bar{X}^I P_{Iij}\epsilon^j + \tfrac{1}{8}\gamma^a\epsilon_j g_{\alpha\bar{\beta}}\bar{\chi}^\alpha_i \gamma_a \chi^{j\bar{\beta}} + \tfrac{1}{16}\gamma^{ab}\epsilon^j\varepsilon_{ij}C_{AB}\bar{\zeta}^A\gamma^{ab}\zeta^B \\ & + \gamma^a\epsilon_i\left[-\tfrac{1}{16}g_{\alpha\bar{\beta}}\bar{\chi}^\alpha_j\gamma_a\chi^{j\bar{\beta}} + \tfrac{1}{4}\bar{\zeta}^A\gamma_a\zeta_{\bar{B}}d^{\bar{B}}{}_A\right] .\end{aligned} \tag{20.204}$$

Using these decomposition rules, the transformation laws of Poincaré supersymmetry can be written down. The explicit result will be given in Sec. 21.3.2. To check the results one needs the formula

$$G^-_{ab\,\bar{\alpha}} \equiv -\tfrac{1}{2}\nabla_{\bar{\alpha}}\bar{X}^I N_{IJ}\mathcal{F}^{-J}_{ab} = \nabla_{\bar{\alpha}}\bar{X}^I \,\mathrm{Im}\,\mathcal{N}_{IJ}F^{-J}_{ab} + \text{fermionic terms}, \tag{20.205}$$

which follows from the definition (20.77) and the field equation (20.105).

Appendix 20A SU(2) conventions and triplets

Remember that we use the NW–SE raising and lowering convention for SU(2) indices as in (20.3). This implies for any two doublets A, B that $A^iB_i = -A_iB^i$.

For any antisymmetric tensor A^{ij} one can write $A^{ij} = \tfrac{1}{2}\varepsilon^{ij}A_k{}^k$.

To convert SU(2) triplets into 2×2 matrices, we use the anti-hermitian $\vec{\tau}$, related to the three Pauli matrices $\vec{\sigma}_i{}^j$ in (2.3) as

$$\vec{\tau}_i{}^j = \mathrm{i}\vec{\sigma}_i{}^j \,. \tag{20.206}$$

We then define the triplets as the traceless anti-hermitian matrices

$$Y_i{}^j \equiv \vec{\tau}_i{}^j\cdot\vec{Y}\,, \qquad \vec{Y} = -\tfrac{1}{2}\vec{\tau}_i{}^j Y_j{}^i\,, \qquad \lambda_i{}^j Y_j{}^i = -2\vec{\lambda}\cdot\vec{Y}\,. \tag{20.207}$$

We can also raise indices using (20.3), which converts (20.206) into

$$\vec{\tau}^{ij} = \varepsilon^{ik}\vec{\tau}_k{}^j = \vec{\tau}^{ji} = (\mathrm{i}\sigma_3, -\mathbb{1}, -\mathrm{i}\sigma_1) = \left(\vec{\tau}_{ij}\right)^* . \tag{20.208}$$

With indices at equal height they are a basis of the 2×2 symmetric matrices. We make the correspondence between real symmetric matrices and triplets by

$$Y^{ij} = \vec{\tau}^{ij}\cdot\vec{Y} = (Y_{ij})^*\,, \qquad \vec{Y} = \tfrac{1}{2}\vec{\tau}^{ij}Y_{ij} = \tfrac{1}{2}\vec{\tau}_{ij}Y^{ij}\,. \tag{20.209}$$

Useful translation formulas for the symmetric tensors are

$$\begin{aligned}\vec{\tau}_{ij}\cdot\vec{\tau}^{k\ell} &= \delta_i^k\delta_j^\ell + \delta_j^k\delta_i^\ell\,,\\ A_{ij}B^{jk} &= \delta_i^k\vec{A}\cdot\vec{B} + (\vec{A}\times\vec{B})\cdot\vec{\tau}_i{}^k\,, \qquad \text{i.e.}\quad A_{ij}B^{ij} = 2\vec{A}\cdot\vec{B}\,.\end{aligned} \tag{20.210}$$

Appendix 20B Dimensional reduction $6\to 5\to 4$

In the treatment of global supersymmetry in this chapter, we reduced formulas from $D=6$ to $D=5$ and $D=4$. In this appendix, we show the relations among these spinors. Let us first remind the reader that the gamma algebras have different properties in different dimensions. The coefficients that appear in the Majorana flip relation (3.51) are

$$\begin{aligned} D=6:&\quad t_0=t_3=-1\,, &\quad t_1=t_2=+1\,,\\ D=5:&\quad t_2=t_3=-1\,, &\quad t_0=t_1=+1\,,\\ D=4:&\quad t_1=t_2=-1\,, &\quad t_0=t_3=+1\,.\end{aligned} \tag{20.211}$$

This is relevant for the transposition of spinor bilinears, and has consequences for calculation rules as (3.77), (3.79) and (3.55).

20B.1 Reducing from $D=6\ \to\ D=5$

For spinors that are left-handed chiral in $D=6$ we identify the $D=5$ spinor field with the one in $D=6$, i.e. $\epsilon^i\to\epsilon^i$. For the right-handed ones of $D=6$ on the other hand, $\zeta^{\mathcal{A}}\to -\mathrm{i}\zeta^{\mathcal{A}}$, and then $\bar\zeta^{\mathcal{A}}\to\mathrm{i}\bar\zeta^{\mathcal{A}}$.

The dimensional reduction of the last γ-matrix is done by $\gamma_5\lambda^i\to\mathrm{i}\lambda^i$ for a spinor λ^i that is left-handed in $D=6$, while for a spinor like $\zeta^{\mathcal{A}}$ that is right-handed in $D=6$ we replace $\gamma_5\zeta^{\mathcal{A}}\to-\mathrm{i}\zeta^{\mathcal{A}}$.

20B.2 Reducing from $D=5\ \to\ D=4$

Spinors in $D=5$ and $D=4$ are quite different. Therefore the formulas are more involved for this reduction. The last γ-matrix of $D=5$, γ_4, is identified with γ_* in $D=4$. The charge conjugation matrix of $D=5$ is $C\gamma_*$ in $D=4$. This is relevant for the translation of the barred spinors. It also implies that charge conjugation of a spinor in five dimensions is $-\gamma_*$ times the charge conjugate of four dimensions.

These rules lead to the following identifications. For a spinor that is left-handed in four dimensions, $\chi^i = P_L\chi^i$, and that is the dimensional reduction of a spinor ϵ^i in $D = 5$, we have

$$\begin{aligned} \epsilon^i &\to \chi^i + \varepsilon^{ij}\chi_j\,, & \bar{\epsilon}^i &\to \bar{\chi}^i - \varepsilon^{ij}\bar{\chi}_j\,,\\ \epsilon_i &\to \chi^j\varepsilon_{ji} + \chi_i\,, & \bar{\epsilon}_i &\to \bar{\chi}^j\varepsilon_{ji} - \bar{\chi}_i\,. \end{aligned} \qquad (20.212)$$

For a spinor that is right-handed in four dimensions, $\Omega^i = P_R\Omega^i$, and that is the reduction of a spinor λ^i in $D = 5$, we have

$$\begin{aligned} \lambda^i &\to -\Omega^i + \varepsilon^{ij}\Omega_j\,, & \bar{\lambda}^i &\to \bar{\Omega}^i + \varepsilon^{ij}\bar{\Omega}_j\,,\\ \lambda_i &\to -\Omega^j\varepsilon_{ji} + \Omega_i\,, & \bar{\lambda}_i &\to \bar{\Omega}^j\varepsilon_{ji} + \bar{\Omega}_i\,. \end{aligned} \qquad (20.213)$$

Some care is needed for the hypermultiplet, where the complex conjugation condition is determined by the tensor $\rho_{\mathcal{A}\bar{\mathcal{B}}}$ in $D = 5$. The translation formulas are

$$\begin{aligned} \zeta^{\mathcal{A}} &\to \zeta^{\mathcal{A}} - \zeta_{\bar{\mathcal{B}}}\rho^{\bar{\mathcal{B}}\mathcal{A}}\,, & \bar{\zeta}^{\mathcal{A}} &\to \bar{\zeta}^{\mathcal{A}} + \bar{\zeta}_{\bar{\mathcal{B}}}\rho^{\bar{\mathcal{B}}\mathcal{A}}\,,\\ \zeta_{\mathcal{A}} &\to \zeta^{\mathcal{B}}C_{\mathcal{B}\mathcal{A}} + \zeta_{\bar{\mathcal{B}}}d^{\bar{\mathcal{B}}}{}_{\mathcal{A}}\,, & \bar{\zeta}_{\mathcal{A}} &\to \bar{\zeta}^{\mathcal{B}}C_{\mathcal{B}\mathcal{A}} - \bar{\zeta}_{\bar{\mathcal{B}}}d^{\bar{\mathcal{B}}}{}_{\mathcal{A}}\,. \end{aligned} \qquad (20.214)$$

Appendix 20C Definition of rigid special Kähler geometry

The definition of rigid (or affine) special Kähler geometry can be shortly formulated as follows [243, 159]:

> A rigid special Kähler manifold *is a Kähler manifold with Kähler form* Ω *and complex structure* J *such that there is a real, flat, torsion-free, symplectic connection* $\nabla_{\rm s}$*, whose connection components we denote by* $A_{ij}{}^k$*. The covariant differential* $\mathrm{d}_{\nabla_{\rm s}}$ *built from this connection annihilates* J *seen as a 1-form:*
>
> $$\mathrm{d}_{\nabla_{\rm s}}\left[(\mathrm{d}\phi^i)J_i{}^j\right] \equiv \mathrm{d}\phi^k\wedge\left[\partial_k\delta^j_\ell + A_{k\ell}{}^j\right]\left[(\mathrm{d}\phi^i)J_i{}^\ell\right] = 0\,, \qquad i,j = 1,\ldots,2n\,, \qquad (20.215)$$
>
> *where* ϕ^i *denote the (real) coordinates of the manifold.*

Now we analyze this definition. The symplectic structure in a Kähler manifold can be obtained from (13.32) with the antisymmetric J_{ij}. Holomorphic 1-forms are of the form

$$\pi^i \equiv \tfrac{1}{2}(\mathrm{d}\phi^j)\left(\delta_j{}^i - \mathrm{i}J_j{}^i\right)\,, \qquad \pi^i J_i{}^j = \mathrm{i}\pi^i\,. \qquad (20.216)$$

The definition involves another connection $\nabla_{\rm s}$. It is mentioned that it is flat, i.e. its curvature tensor vanishes. Thus, these conditions say that the real coefficients A^k_{ij} satisfy

$$\begin{aligned} &\text{torsion free:} && A_{ij}{}^k = A_{ji}{}^k\,, \quad \text{i.e.} \quad \mathrm{d}_{\nabla_{\rm s}}\mathrm{d}\phi^i = A^i_{jk}\mathrm{d}\phi^j\wedge\mathrm{d}\phi^k = 0\,,\\ &\text{symplectic:} && \partial_i J_{jk} + 2A_{i[j}{}^\ell J_{k]\ell} = 0\,, \quad \text{i.e.} \quad \nabla_{\rm s}\Omega = 0\,,\\ &\text{flat:} && \partial_{[i}A_{j]k}{}^\ell + A_{m[i}{}^\ell A_{j]k}{}^m = 0\,,\\ &(20.215): && \partial_{[i}J_{j]}{}^k + A_{\ell[i}{}^k J_{j]}{}^\ell = 0\,. \end{aligned} \qquad (20.217)$$

Exercise 20.31 *In order to make these concepts concrete, consider an example. Use two real coordinates x and y, and the metric*

$$ds^2 = (1+y)(dx^2+dy^2)\,, \qquad -\Gamma^y_{xx} = \Gamma^x_{xy} = \Gamma^y_{yy} = \tfrac{1}{2}(1+y)^{-1}\,. \tag{20.218}$$

Check that those Γ-coefficients are the only non-zero entries of the Christoffel symbols of this metric. The manifold is Kähler with the definition $J_x{}^y = -J_y{}^x = 1$, for which you should check (13.29). Prove that the conditions (20.217) are satisfied for another connection, which has as the only non-zero entries

$$A_{yy}{}^y = -A_{xx}{}^y = (1+y)^{-1}\,. \tag{20.219}$$

Check with (7.103) that a redefinition

$$x' = x\,, \qquad y' = -y + \tfrac{1}{2}(x^2 - y^2)\,, \tag{20.220}$$

leads to zero coefficients $A_{ij}{}^k$, and to $J_{x'y'} = -1$.

The main properties of the manifold can be seen from comparing two sets of coordinates:

1. The complex coordinates z^α and $\bar{z}^{\bar\alpha}$ that we used for a Kähler manifold in Sec. 13.1, in which the complex structure has the form (13.28).
2. Flat real components. Owing to the flatness of the connection, one can choose a frame with $A^k_{ij} = 0$, and J_{ij} is the standard form for an antisymmetric tensor.

The equations (20.217) in the first set of coordinates imply (using $J_{ij} = J_i{}^k g_{kj}$) that the non-vanishing connection coefficients are

$$A^\gamma_{\alpha\beta} = \Gamma^\gamma_{\alpha\beta}\,, \qquad A^{\bar\gamma}_{\alpha\beta} = g^{\bar\gamma\gamma} C_{\alpha\beta\gamma}(z)\,, \tag{20.221}$$

and their complex conjugates, where $\Gamma^\gamma_{\alpha\beta}$ are the Levi-Civita connections (13.19) for the Kähler metric and $C_{\alpha\beta\gamma}(z)$ is a completely symmetric tensor, with holomorphic components satisfying

$$\nabla_{[\delta} C_{\alpha]\beta\gamma} \equiv \partial_{[\delta} C_{\alpha]\beta\gamma} - \Gamma^\epsilon_{\beta[\delta} C_{\alpha]\epsilon\gamma} - \Gamma^\epsilon_{\gamma[\delta} C_{\alpha]\beta\epsilon} = 0\,. \tag{20.222}$$

The flatness condition can be split into the contributions from Γ and those from C, which leads to

$$R^\delta{}_{\beta\alpha\bar\gamma}(\Gamma) = C_{\alpha\beta\epsilon} g^{\epsilon\bar\epsilon} \bar{C}_{\bar\epsilon\bar\beta\bar\delta} g^{\bar\delta\delta}\,, \tag{20.223}$$

where the left-hand side is the metric curvature.

We denote the $2n$ coordinates of the second mentioned frame as

$$\{\phi^i\} = \{x^I,\, f_I\}\,, \qquad I = 1, \ldots, n\,, \tag{20.224}$$

such that

$$\Omega = -J_{ij} \mathrm{d}\phi^i \wedge \mathrm{d}\phi^j = 2\mathrm{d}x^I \wedge \mathrm{d}f_I\,. \tag{20.225}$$

Note that in a frame where J_{ij} is constant, $J_i{}^j = J_{ik} g^{kj}$ is in general not constant, but (20.215) implies that $\mathrm{d}\phi^j J_j{}^i$ is closed in a frame with zero coefficients $A_{ij}{}^k$. Hence the $2n$ holomorphic 1-forms $\{\pi^I, \pi_I\}$ are closed, and thus locally exact:

$$\pi^I = \mathrm{d}X^I\,, \qquad \pi_I = \mathrm{d}F_I\,. \tag{20.226}$$

The definition of π^i in (20.216) implies that, in terms of the holomorphic coordinates, they depend only on z^α. Obviously, one can repeat the same for the projections to antiholomorphic quantities, and this leads to

$$\mathrm{d}x^I = \mathrm{d}X^I(z) + \mathrm{d}\bar{X}^I(\bar{z}), \qquad \mathrm{d}f_I = \mathrm{d}F_I(z) + \mathrm{d}\bar{F}_I(\bar{z}). \tag{20.227}$$

Note that the X^I and F_I are complex, while x^I and f_I have been defined as real coordinates.

Exercise 20.32 *Continue with the two-dimensional example of Ex. 20.31, where the variables in the flat frame are x^1 and f_1. Identify the latter as x and y', and obtain*

$$\begin{aligned} \pi^x &= \tfrac{1}{2}\mathrm{d}z, & \pi^y &= -\tfrac{1}{2}\mathrm{i}\mathrm{d}z, & z &\equiv x + \mathrm{i}y, \\ \pi^1 &= \tfrac{1}{2}\mathrm{d}z, & \pi_1 &= \tfrac{1}{2}(\mathrm{i} + z)\mathrm{d}z. \end{aligned} \tag{20.228}$$

Calculate X^1 and F_1, express F_1 as a function of X^1 and obtain the relation

$$F_1 = \frac{\partial}{\partial X^1}\left[\frac{1}{2}\mathrm{i}(X^1)^2 + \frac{1}{3}(X^1)^3\right]. \tag{20.229}$$

We now compare (13.14) and (20.225):

$$\Omega = -2\mathrm{i}g_{\alpha\bar{\beta}}\mathrm{d}z^\alpha \wedge \mathrm{d}\bar{z}^{\bar{\beta}} = \mathrm{d}X^I \wedge \mathrm{d}F_I + \mathrm{d}X^I \wedge \mathrm{d}\bar{F}_I + \mathrm{d}\bar{X}^I \wedge \mathrm{d}F_I + \mathrm{d}\bar{X}^I \wedge \mathrm{d}\bar{F}_I. \tag{20.230}$$

The first term in the last expression is quadratic in holomorphic variables, and should thus vanish by itself:

$$\mathrm{d}X^I(z) \wedge \mathrm{d}F_I(z) = 0. \tag{20.231}$$

The same holds for the last term, which is the complex conjugate of (20.231). This equation is the first line of (20.120), and also the second line follows from the remaining terms in (20.230).

If $\partial_\alpha X^I$ is invertible, $X(z)$ can be inverted to obtain $z(X)$. Then also F_I can be considered as a function of the X^I as $F_I(X) = F_I(z(X))$. Therefore we have in that case

$$\mathrm{d}F_I(z) = \frac{\partial F_I(X)}{\partial X^J}\mathrm{d}X^J(z), \tag{20.232}$$

and (20.231) is the integrability condition for the existence of a holomorphic function $F(X)$, called the *prepotential*, with (20.121) [247]. Then the metric is the expression given in (20.122). If this metric is positive definite it follows from (20.120) that indeed $\partial_\alpha X^I$ is invertible, and a prepotential exists. This gives all the ingredients used in Sec. 20.3.3 to describe special Kähler geometry.

Exercise 20.33 *Use the result (20.229) to prove that for that example*

$$g_{z\bar{z}} = \tfrac{1}{2} + \tfrac{1}{4}\mathrm{i}(\bar{z} - z), \tag{20.233}$$

such that $\mathrm{d}s^2 = 2g_{z\bar{z}}\mathrm{d}z\mathrm{d}\bar{z}$ agrees with (20.218).

Exercise 20.34 *Note from (20.216) that* $\mathrm{d}\phi^i = \pi^i + \bar{\pi}^i$. *Use this to derive the matrix for changing basis from* $\{x^I, f_I\}$ *to* $\{z^\alpha, \bar{z}^{\bar{\alpha}}\}$, *assuming the invertibility of* $\partial_\alpha X^I$:

$$\begin{aligned} \partial_\alpha x^I &= \partial_\alpha X^I\,, & \partial_\alpha f_I &= \partial_\alpha X^J F_{JI}\,,\\ \partial_{\bar{\alpha}} x^I &= \partial_{\bar{\alpha}} \bar{X}^I\,, & \partial_{\bar{\alpha}} f_I &= \partial_{\bar{\alpha}} \bar{X}^J \bar{F}_{JI}\,. \end{aligned} \tag{20.234}$$

Using (20.122), find also the inverse transformations:

$$\begin{aligned} \frac{\partial z^\alpha}{\partial x^I} &= \mathrm{i} g^{\alpha\bar{\beta}} \partial_{\bar{\beta}} \bar{X}^J\, \bar{F}_{IJ}\,, & \frac{\partial \bar{z}^{\bar{\alpha}}}{\partial x^I} &= -\mathrm{i} g^{\bar{\alpha}\beta} \partial_\beta X^J\, F_{IJ}\,,\\ \frac{\partial z^\alpha}{\partial f_I} &= -\mathrm{i} g^{\alpha\bar{\beta}} \partial_{\bar{\beta}} \bar{X}^I\,, & \frac{\partial \bar{z}^{\bar{\alpha}}}{\partial f_I} &= \mathrm{i} g^{\bar{\alpha}\beta} \partial_\beta X^I\,. \end{aligned} \tag{20.235}$$

Since the connection coefficients $A^i{}_{jk}$ *are zero in the basis* $\{x^I, f_I\}$, *one can use (7.103) to find the connection coefficients in the complex basis. Establish then (20.221) with*

$$C_{\alpha\beta\gamma} = \mathrm{i} F_{IJK}\, \partial_\alpha X^I\, \partial_\beta X^J\, \partial_\gamma X^K\,. \tag{20.236}$$

21 The physical $\mathcal{N} = 2$ matter-coupled supergravity

The previous chapter contained a detailed discussion of $\mathcal{N} = 2$ supersymmetry and supergravity. We explained how the supergravity theory can be obtained from a covariantization of actions of global supersymmetry with superconformal symmetry. In this chapter, we will write the results in such a way that they are accessible to readers who skipped the previous chapter, but want to know the results in view of applications.

We first repeat the essential properties of the bosonic part in Sec. 21.1. We devote a separate section, Sec. 21.2, to the symplectic formulation, which is particularly useful for both the central charges of black hole solutions and the moduli spaces of Calabi–Yau manifolds, as is shown in later parts of this chapter; see Sec. 21.4. The action and transformation laws of the couplings of gauge and hypermultiplets in $D = 4, \mathcal{N} = 2$ supergravity are fully discussed in Sec. 21.3. After the applications in Sec. 21.4, a final section contains remarks on Fayet–Iliopoulos terms, σ-model symmetries and engineering dimensions of the objects in this chapter.

21.1 The bosonic sector

We summarize here the main features of the coupling of $\mathcal{N} = 2$, $D = 4$ supergravity to n_V vector (gauge) multiplets, and n_H hypermultiplets. Apart from the graviton field, the bosonic sector contains $n_V + 1$ gauge fields (containing the 'graviphoton') A_μ^I, $I = 0, 1, \ldots, n_V$, and scalars.

21.1.1 The basic (ungauged) $\mathcal{N} = 2$, $D = 4$ matter-coupled supergravity

The manifold of the scalars is a direct product of two target spaces determined by the scalars in gauge multiplets and scalars in hypermultiplets, respectively. The n_V complex scalars z^α, $\bar{z}^{\bar{\alpha}}$ of the gauge multiplets define a special Kähler manifold. The real scalars of the hypermultiplets q^u, $u = 1, \ldots, 4n_H$, define an n_H-dimensional quaternionic-Kähler manifold. The kinetic terms of the gauge vectors depend only on the scalars of the special Kähler manifold via a matrix $\mathcal{N}_{IJ}(z, \bar{z})$ defined below.

A *special Kähler manifold* is a Kähler manifold that possesses a symplectic structure $\mathrm{Sp}(2(n_V + 1))$. The latter arises because $\mathcal{N} = 2$ supersymmetry connects the scalars to the $n_V + 1$ gauge vectors, which exhibit duality transformations. The symplectic structure is clearly visible in the 'embedding Kähler manifold', which is used for the superconformal description. We postpone the full symplectic formulation to Sec. 21.2, and first consider the formulation that is most similar to the $\mathcal{N}=1$ supergravity as in Sec. 17.3. In that formulation the scalars are described by n_V+1 complex fields X^I, which are constrained such that

$$N = N_{IJ}X^I\bar{X}^J = -a = -\kappa^{-2}, \qquad N_{IJ} \equiv 2\,\mathrm{Im}\,F_{IJ}. \tag{21.1}$$

Here F_{IJ} is the second derivative of a holomorphic function $F(X)$, homogeneous of second degree in X^I, called the prepotential. This differs from $\mathcal{N}=1$ supergravity where $N(X,\bar{X})$ can be an arbitrary homogeneous function of first degree in X and in $\bar{X}$. We can apply what we learned in Ch. 17 by replacing there $G_{I\bar{J}} \to N_{IJ}$, and taking $a=\kappa^{-2}$. There we mentioned that it is convenient to parametrize $X^I = y\,Z^I(z)$, and determine y by the constraint (21.1) and a gauge choice of the U(1) R-symmetry group $y=\bar{y}$. We then get a Kähler potential as in (17.67):

$$\mathcal{K} = -\kappa^{-2}\ln\left(-\kappa^2 N_{IJ}Z^I\bar{Z}^J\right), \qquad y = \mathrm{e}^{\kappa^2\mathcal{K}(z,\bar{z})/2}. \tag{21.2}$$

We often use covariant derivatives

$$\nabla_\alpha X^I = \partial_\alpha X^I + \tfrac{1}{2}\kappa^2(\partial_\alpha\mathcal{K})X^I = y\nabla_\alpha Z^I, \qquad \nabla_\alpha Z^I = \partial_\alpha Z^I + \kappa^2(\partial_\alpha\mathcal{K})Z^I. \tag{21.3}$$

The kinetic terms of the gauge fields are

$$\mathcal{L}_1 = -\tfrac{1}{4}\mathrm{i}e\mathcal{N}_{IJ}F^{+I}_{\mu\nu}F^{+\mu\nu J} + \text{h.c.} = \tfrac{1}{4}e(\mathrm{Im}\,\mathcal{N}_{IJ})F^I_{\mu\nu}F^{\mu\nu J} - \tfrac{1}{8}(\mathrm{Re}\,\mathcal{N}_{IJ})\varepsilon^{\mu\nu\rho\sigma}F^I_{\mu\nu}F^J_{\rho\sigma}, \tag{21.4}$$

where[1]

$$\mathcal{N}_{IJ}(z,\bar{z}) = \bar{F}_{IJ}(\bar{z}) + \mathrm{i}\frac{N_{IK}Z^K(z)N_{JL}Z^L(z)}{N_{MN}Z^M(z)Z^N(z)} = \bar{F}_{IJ} + \mathrm{i}\frac{N_{IK}X^KN_{JL}X^L}{N_{MN}X^MX^N}. \tag{21.5}$$

A *quaternionic-Kähler manifold* is a manifold with metric h_{uv} and a triplet of complex structures $\vec{J}_u{}^v$, whose Levi-Civita covariant derivatives, determined by the metric h_{uv}, rotate them into themselves:

$$\widetilde{\nabla}_w\vec{J}_u{}^v \equiv \nabla_w\vec{J}_u{}^v + 2\,\vec{\omega}_w\times\vec{J}_u{}^v = 0. \tag{21.6}$$

This defines an SU(2) connection, $\vec{\omega}_u$, whose curvature is proportional to the 2-form defined by the complex structures, and the manifold is Einstein, related to the same proportionality constant:

$$\vec{\mathcal{R}}_{uv} = \tfrac{1}{2}\nu\vec{J}_{uv}, \qquad R_{uv} = (n_H+2)\nu h_{uv}, \qquad \nu = -\kappa^2. \tag{21.7}$$

The constant ν is arbitrary for general quaternionic-Kähler manifolds, but is determined to be $-\kappa^2$ in supergravity, fixing the scalar curvature to $-4n_H(n_H+2)\kappa^2$. Another important

[1] Owing to the homogeneity of F, $F_{IJ}(X) = F_{IJ}(Z)$, where the former is the second derivative of $F(X)$ with respect to X and the latter is the second derivative of $F(Z)$ with respect to Z.

property is that the product of two quaternionic-Kähler manifolds is not quaternionic-Kähler. For more detail see Sec. 20.3.4.

In a supersymmetric theory, we have to write the metric h_{uv} in terms of frame fields $f^{iA}{}_u$, where $A = 1, \ldots, 2n_H$, and considering (iA) together, the frame field is a $4n_H \times 4n_H$ matrix. The metric is

$$h_{uv} = f^{iA}{}_u \varepsilon_{ij} C_{AB} f^{jB}{}_v, \tag{21.8}$$

where C_{AB} is an invertible antisymmetric matrix, which is used for raising and lowering indices in the familiar NW–SE convention. The frame fields satisfy a reality condition[2]

$$\left(f^{iA}{}_u\right)^* = f^{jB}{}_u \varepsilon_{ji} C_{BA}, \qquad \left(f^u{}_{iA}\right)^* = \varepsilon^{ij} C^{AB} f^u{}_{jB}, \tag{21.9}$$

where

$$C_{AB} C^{BC} = -\delta_A^C, \qquad C^{AB} = (C_{AB})^* . \tag{21.10}$$

The frame fields are covariantly constant using connections on every index:

$$\widetilde{\nabla}_v f^{iA}{}_u \equiv \partial_v f^{iA}{}_u + f^{jA}{}_u \omega_{vj}{}^i + f^{iB}{}_u \omega_{vB}{}^A - \Gamma^w_{vu} f^{iA}{}_w = 0. \tag{21.11}$$

Notice that the connection $\vec{\omega}_u$ is written here as a traceless anti-hermitian $\omega_{ui}{}^j$. The correspondence is explained in Sec. 20A.

21.1.2 The gauged supergravities

The gauging involves the choice of the gauge group, i.e. the structure constants $f_{IJ}{}^K$, and the action of the gauge generators on gauge multiplet and hypermultiplet scalars. The latter is encoded in the 'Killing vectors' $k_I{}^\alpha$ and $k_I{}^u$. Since there are $n_V + 1$ gauge vectors, this is the maximum dimension of the gauge group. The scalars in the embedding manifold, X^I, should be in the adjoint of the gauge group, and the prepotential should be invariant up to a quadratic function with real coefficients $C_{I,JK}$:

$$\delta_G F \equiv F_I \theta^K X^J f_{JK}{}^I = -\theta^K C_{K,IJ} X^I X^J . \tag{21.12}$$

The transformations on the scalars z^α are then, using (17.98), encoded in

$$\delta z^\alpha = \theta^J k_J{}^\alpha, \qquad k_J{}^\alpha \nabla_\alpha Z^I = Z^K f_{KJ}{}^I - \mathrm{i}\kappa^2 Z^I P_J^0, \tag{21.13}$$

[2] In terms of the tensors of the previous chapter, (21.9) involves another tensor $\rho_{A\bar{B}}$ such that $C_{AB} = \rho_{A\bar{C}} d^{\bar{C}}{}_B$, with $d^{\bar{A}}{}_B$ a hermitian matrix, which determines the kinetic terms of the physical hypermultiplets. Since we are now only interested in positive kinetic terms, we take a basis where $d^{\bar{A}}{}_B = \delta^{\bar{A}}{}_B$. In this way there is no longer any difference between barred $\bar{A}$ and unbarred A indices.

where the (real) moment map P_I^0 is

$$P_I^0 = -\mathrm{i}N_{IJ}f_{KL}{}^J X^K \bar{X}^L = \mathrm{i}N_{JL}f_{KI}{}^J X^K \bar{X}^L = -\mathrm{i}N_{JK}f_{LI}{}^J X^K \bar{X}^L. \tag{21.14}$$

Note that this implies that $X^I P_I^0 = 0$.

The gauge group can also act as part of the isometry group of the hypermultiplets. The action is determined by triplet moment maps $\vec{P}_I$, which should satisfy the equivariance condition

$$2n_H \nu \vec{P}_I = \vec{J}_u{}^v \nabla_v k_I{}^u, \qquad k_I{}^u \vec{J}_{uv} k_J{}^v = f_{IJ}{}^K \vec{P}_K + \nu \vec{P}_I \times \vec{P}_J. \tag{21.15}$$

The scalar potential is then

$$V = \left[-\tfrac{1}{2}(\operatorname{Im}\mathcal{N})^{-1|IJ} - 4\kappa^2 X^I \bar{X}^J\right] \vec{P}_I \cdot \vec{P}_J + 2\bar{X}^I X^J k_I{}^u k_J{}^v h_{uv} + N^{-1|IJ} P_I^0 P_J^0. \tag{21.16}$$

All terms of the potential originate from gauging.[3] They are thus consequences of the 'deformations' (Sec. 12.4.2). The only negative term in the potential is $-4\kappa^2 X^I \bar{X}^J \vec{P}_I \cdot \vec{P}_J$. We will derive alternative forms of the potential in Ex. 21.1. But for that it is useful to have a symplectic formalism.

21.2 The symplectic formulation

We derived the $\mathcal{N}=2$ matter-coupled supergravity starting from a prepotential $F(X)$ to construct the action (20.89). However, the result no longer refers to the explicit prepotential. It is expressed in quantities that fit in a general treatment of special Kähler geometry based on the symplectic group. The final formulas are more generally applicable than the way in which they are derived.

The definition of special Kähler geometry is based on the symplectic group of duality transformations. Though any special Kähler geometry can be written with a prepotential, it may be useful to make a duality transformation to another symplectic frame in which the prepotential does not exist. A few examples were already given in Sec. 20.3.3, and below we will further discuss the usefulness of the symplectic covariant formulation.

21.2.1 Symplectic definition

The basic ingredients are symplectic vectors, related to those that we encountered for abelian gauge fields and gauge field strengths in Sec. 4.2.4 (apart from the redefinition that the f_{AB} there is denoted by $-\mathrm{i}\overline{\mathcal{N}}_{IJ}$ here). The basic symplectic vector V for the scalars contains X^I in the upper parts and F_I in the lower part. They are both of the form y times

[3] Though $\vec{P}$ may be non-zero when there is only a compensating hypermultiplet. These are then the Fayet–Iliopoulos terms; see Sec. 21.5.

a function of z, but F_I is not necessarily a function of X^I as is the case in the formulation with a prepotential:

$$V = \begin{pmatrix} X^I \\ F_I \end{pmatrix} = y\, v, \qquad v = \begin{pmatrix} Z^I(z) \\ F_I(z) \end{pmatrix}. \tag{21.17}$$

The symplectic definition of special Kähler manifold that we discussed in Sec. 20.3.3 states that a special Kähler manifold is determined by a symplectic vector such that

$$\langle V, \nabla_\alpha V\rangle = 0, \qquad \nabla_\alpha V = y\nabla_\alpha v, \qquad \nabla_\alpha v \equiv \partial_\alpha v + (\partial_\alpha \mathcal{K})v, \tag{21.18}$$

where the symplectic inner product has been used, e.g.

$$\langle V, \bar{V}\rangle \equiv X^I \bar{F}_I - F_I \bar{X}^I = \begin{pmatrix} X^I & F_I \end{pmatrix} \begin{pmatrix} 0 & \mathbb{1} \\ -\mathbb{1} & 0 \end{pmatrix} \begin{pmatrix} \bar{X}^I \\ \bar{F}_I \end{pmatrix}. \tag{21.19}$$

The Kähler potential $\mathcal{K}$ is determined by

$$\langle V, \bar{V}\rangle = \mathrm{i}\kappa^{-2}, \qquad (y\bar{y})^{-1} = \mathrm{e}^{-\kappa^2 \mathcal{K}(z,\bar{z})} = -\mathrm{i}\kappa^2 \langle v, \bar{v}\rangle, \tag{21.20}$$

and determines the metric in the usual way:

$$g_{\alpha\bar{\beta}} = \partial_\alpha \partial_{\bar{\beta}} \mathcal{K} = \mathrm{i}\langle \nabla_\alpha V, \overline{\nabla}_{\bar{\beta}} \bar{V}\rangle. \tag{21.21}$$

The matrix $\mathcal{N}_{IJ}$ is in the symplectic formulation defined as

$$\overline{\mathcal{N}}_{IJ} = \begin{pmatrix} \bar{F}_I & \nabla_\alpha F_I \end{pmatrix} \begin{pmatrix} \bar{X}^J & \nabla_\alpha X^J \end{pmatrix}^{-1}. \tag{21.22}$$

The invertibility of the metric together with non-singularity of the kinetic term of the graviton is equivalent to the invertibility of the $(n_V + 1) \times (n_V + 1)$ matrix $\begin{pmatrix} \bar{X}^J & \nabla_\alpha X^J \end{pmatrix}$.

The symplectic vectors have their lower part related to the upper part by $\mathcal{N}$ or $\overline{\mathcal{N}}$, such that the transformations are consistent using (20.124). We have encountered the following symplectic vectors:

$$\begin{array}{lll} V = \begin{pmatrix} X^I \\ F_I \end{pmatrix}, & \overline{\nabla}_{\bar{\alpha}} \bar{V} = \begin{pmatrix} \overline{\nabla}_{\bar{\alpha}} \bar{X}^I \\ \overline{\nabla}_{\bar{\alpha}} \bar{F}_I \end{pmatrix}, & \begin{pmatrix} F^{+I}_{\mu\nu} \\ G^{+}_{\mu\nu I} \end{pmatrix}, \\ F_I = \mathcal{N}_{IJ} X^J, & \overline{\nabla}_{\bar{\alpha}} \bar{F}_I = \mathcal{N}_{IJ} \overline{\nabla}_{\bar{\alpha}} \bar{X}^J, & G^{+}_{\mu\nu I} = \mathcal{N}_{IJ} F^{+J}_{\mu\nu}, \\ & & \\ \bar{V} = \begin{pmatrix} \bar{X}^I \\ \bar{F}_I \end{pmatrix}, & \nabla_\alpha V = \begin{pmatrix} \nabla_\alpha X^I \\ \nabla_\alpha F_I \end{pmatrix}, & \begin{pmatrix} F^{-I}_{\mu\nu} \\ G^{-}_{\mu\nu I} \end{pmatrix}, \\ \bar{F}_I = \overline{\mathcal{N}}_{IJ} \bar{X}^J, & \nabla_\alpha F_I = \overline{\mathcal{N}}_{IJ} \nabla_\alpha X^J, & G^{-}_{\mu\nu I} = \overline{\mathcal{N}}_{IJ} F^{-J}_{\mu\nu}. \end{array} \tag{21.23}$$

The symplectic inner product between two vectors of the same row in (21.23) vanishes automatically. Between one of the upper row, and one of the lower row, the symplectic inner product leads to a product using $\operatorname{Im} \mathcal{N}_{IJ}$ between the upper parts, for example

$$\begin{aligned} \left\langle \begin{pmatrix} \bar{X}^I \\ \bar{F}_I \end{pmatrix}, \begin{pmatrix} F^{+I}_{\mu\nu} \\ G^{+}_{\mu\nu I} \end{pmatrix} \right\rangle &= \bar{X}^I G^{+}_{I\,ab} - \bar{F}_I F^{+I}_{ab} = \bar{X}^I \mathcal{N}_{IJ} F^{+J}_{ab} - \bar{X}^I \overline{\mathcal{N}}_{IJ} F^{+J}_{ab} \\ &= 2\mathrm{i}\bar{X}^I \operatorname{Im} \mathcal{N}_{IJ} F^{+J}_{ab}. \end{aligned} \tag{21.24}$$

21.2.2 Comparison of symplectic and prepotential formulation

If the matrix

$$(X^I \quad \nabla_\alpha X^I) \tag{21.25}$$

is invertible, then the condition (21.18) is solved by $F_I = \partial_I F(X)$, where the prepotential $F(X)$ is a homogeneous function of second degree in X. In that case the Kähler potential is given by (21.2), and (21.22) agrees with (21.5).

One can always make a symplectic transformation to a frame with invertible matrix (21.25) [247], and thus with a prepotential. Hence, one could wonder why we need other frames.

A symplectic transformation involves, as discussed in Sec. 4.2.2, a mixing of electric and magnetic fields. Since the conventional gauging of symmetries is done with electric fields, the choice of frames is important for the gauging. Hence, the choice between symplectic frames is important for:

1. **The definition of a prepotential.** A prepotential can only be defined if (21.25) is invertible.
2. **The gauging.** As discussed in Sec. 20.3.3, the symmetries of special Kähler geometry are embedded in the symplectic group. They can only be gauged if the upper right part of the symplectic transformation matrix vanishes, $B = 0$ in (4.71). Such symmetries are said to be 'of electric type'.

For any gauging one can find a basis such that the symmetries are all of electric type. On the other hand, as mentioned, one can also always find a symplectic frame in which (21.25) is invertible. However, both choices cannot always be taken simultaneously. If one wants a basis with a prepotential, then not all gaugeable symmetries are of electric type. On the other hand, when the gaugings are of electric type, then in some cases one needs the formalism without a prepotential. In this book, we have chosen to restrict gaugings to electric type. There does exist a completely symplectic invariant formulation where the gauge symmetries can be 'magnetic', i.e. the submatrix B of (4.71) is not zero. This uses the formalism of embedding tensors [82, 83, 264, 84, 265], where the usual electric gauge fields $A_\mu{}^I$ are accompanied by magnetic gauge fields $A_{\mu I}$. However, this involves a more complicated formalism with more fields and gauge invariances, which we will not treat here.

21.2.3 Gauge transformations and symplectic vectors

In general, a subgroup of the isometry group can be gauged. We found in Sec. 20.3.3 that the isometry group is embedded in the symplectic group. Hence, any gauge transformation acts on the symplectic vectors as $\delta V = TV$, where T is an infinitesimal symplectic transformation (see (20.146)). On the other hand, since these symplectic transformations should also act on the field strengths, the upper part of these transformations should also be consistent with (4.93).[4] Therefore the symplectic matrix T should be of the form

[4] This is the statement that can be avoided by the use of magnetic gauge fields, as in the embedding tensor formalism.

$$T = \theta^K T_K, \qquad T_K = \begin{pmatrix} -f_{KJ}{}^I & 0 \\ 2C_{K,IJ} & f_{KI}{}^J \end{pmatrix}. \tag{21.26}$$

In the lower part, $C_{K,IJ}$ is symmetric in the last two indices and these are the constants mentioned (in the presence of a prepotential) in (20.18). In order that the matrices T_I satisfy the gauge algebra, we have to require that

$$\begin{aligned} C_{(I,JK)} &= 0, \\ f_{KL}{}^M C_{M,IJ} &= 2 f_{J[K}{}^M C_{L],IM} + 2 f_{I[K}{}^M C_{L],JM}. \end{aligned} \tag{21.27}$$

In the prepotential formulation, this is satisfied by (20.19).

The moment map (21.14) can also be written in a symplectic form as $P_I^0 = -\langle T_I V, \bar{V}\rangle$; see (20.108). This leads also to a rewriting of the scalar potential, as the reader can check with the following exercise.

Exercise 21.1 *First prove with (21.26), (21.27) and (20.108) that*

$$\bar{X}^I X^J \langle T_I V, T_J \bar{V}\rangle = -\mathrm{i} N^{-1|IJ} P_I^0 P_J^0. \tag{21.28}$$

Use then (20.150) and $X^I P_I^0 = \bar{X}^I P_I^0 = 0$ *to prove that*

$$\mathrm{e}^{\kappa^2 \mathcal{K}} \bar{X}^I X^J \langle T_I v, T_J \bar{v}\rangle = -\mathrm{i} \bar{X}^I X^J k_I{}^\alpha k_J{}^{\bar{\beta}} g_{\alpha\bar{\beta}}. \tag{21.29}$$

This implies that (21.16) can be written as

$$V = -\tfrac{1}{2}(\mathrm{Im}\,\mathcal{N})^{-1|IJ} \vec{P}_I \cdot \vec{P}_J + \bar{X}^I X^J \left(-4\kappa^2 \vec{P}_I \cdot \vec{P}_J + 2k_I{}^u k_J{}^v h_{uv} + k_I{}^\alpha k_J{}^{\bar{\beta}} g_{\alpha\bar{\beta}}\right). \tag{21.30}$$

An alternative derivation consists in using (17.83), (21.13) and (20.108) to show that

$$X^I k_I{}^{\bar{\beta}} g_{\alpha\bar{\beta}} = -\mathrm{i} P_I^0 \nabla_\alpha X^I. \tag{21.31}$$

21.2.4 Physical fermions and duality

The physical fermions of the gauge multiplets will be denoted by $\chi_i^\alpha = P_L \chi_i^\alpha$ and those of the hypermultiplets by $\zeta^A = P_L \zeta^A$, $A = 1, \ldots, 2n_H$ (and their right-handed components $\chi^{i\bar{\alpha}}$ and ζ_A).

The Pauli terms describe the coupling of gauge field strengths and fermions:

$$\begin{aligned} e^{-1}\mathcal{L}_{\text{Pauli}} &= \tfrac{1}{2}\mathrm{i} F_{ab}^{-I} G_{I\,\text{ferm}}^{ab-} + \text{h.c.}, \qquad G_{I\,\text{ferm}}^{ab-} \equiv -2\mathrm{i}\,\mathrm{Im}\,\mathcal{N}_{IJ} Q^{ab-J}, \\ Q^{ab-J} &\equiv \nabla_{\bar{\alpha}} \bar{X}^J \left(\tfrac{1}{8} g^{\beta\bar{\alpha}} C_{\beta\gamma\delta} \bar{\chi}_i^\gamma \gamma_{ab} \chi_j^\delta \varepsilon^{ij} + \bar{\chi}^{\bar{\alpha} i} \gamma^a \psi^{bj} \varepsilon_{ij}\right) \\ &\quad + X^J \left(\bar{\psi}_i^a \psi_j^b \varepsilon^{ij} + \tfrac{1}{2}\kappa^2 \bar{\zeta}^A \gamma^{ab} \zeta^B C_{AB}\right). \end{aligned} \tag{21.32}$$

The quantity $G_{I\,\text{ferm}}^{ab-}$ is, according to (4.68), the fermionic part of G_I^{ab-}. The structure of (21.32) with $\mathrm{Im}\,\mathcal{N}_{IJ}$ is similar to (21.24), and thus we can identify

$$\begin{pmatrix} Q^{ab-I} \\ \mathcal{N}_{IJ} Q^{ab-J} \end{pmatrix} \tag{21.33}$$

as a symplectic vector of the form of the upper row in (21.23). The definition in (21.32) shows that it is built from the first two symplectic vectors in (21.23). More detail on the duality transformations of the fermionic part of this action can be found in [248], where also the four-fermion terms have been analyzed in this context.

21.3 Action and transformation laws

21.3.1 Final action

We can now write the full action. It has been derived by rewriting (20.110), obtained from superconformal methods, in terms of the 'Poincaré variables'. The same action has been obtained using another method, called the 'group manifold approach', in [201]. The action is

$$
\begin{aligned}
e^{-1}\mathcal{L} = {} & \kappa^{-2}\left(\tfrac{1}{2}R - \bar\psi_{i\mu}\gamma^{\mu\nu\rho}D_\nu\psi_\rho^i\right) - g_{\alpha\bar\beta}\hat\partial_\mu z^\alpha\hat\partial^\mu\bar z^{\bar\beta} - \tfrac{1}{2}h_{uv}\hat\partial_\mu q^u\hat\partial^\mu q^v - V \\
& + \tfrac{2}{3}C_{I,JK}e^{-1}\varepsilon^{\mu\nu\rho\sigma}A_\mu{}^I A_\nu{}^J\left(\partial_\rho A_\sigma{}^K + \tfrac{3}{8}f_{LM}{}^K A_\rho{}^L A_\sigma{}^M\right) \\
& + \Big\{-\tfrac{1}{4}\mathrm{i}\mathcal{N}_{IJ}F_{\mu\nu}^{+I}F^{+\mu\nu J} + F_{\mu\nu}^{+I}\,\mathrm{Im}\,\mathcal{N}_{IJ}Q^{\mu\nu+J} \\
& - \tfrac{1}{4}g_{\alpha\bar\beta}\bar\chi_i^\alpha\not{D}\chi^{i\bar\beta} - \bar\zeta_A\hat{\not{D}}\zeta^A \\
& + \tfrac{1}{2}g_{\alpha\bar\beta}\bar\psi_{ia}\hat{\not\partial}z^\alpha\gamma^a\chi^{i\bar\beta} - \mathrm{i}\bar\psi_a^i\hat{\not\partial}q^u\gamma^a\zeta^A f_{uiA} + \text{h.c.}\Big\} \\
& + \mathcal{L}_m - \bar\psi_i\cdot\gamma v^i + \text{four-fermion terms.}
\end{aligned} \tag{21.34}
$$

The covariant derivatives are

$$
\begin{aligned}
\hat\partial_\mu z^\alpha &= \partial_\mu z^\alpha - A_\mu{}^I k_I{}^\alpha, \qquad \hat\partial_\mu q^u = \partial_\mu q^u - A_\mu{}^I k_I{}^u, \\
D_\mu\psi_{\nu i} &= \left(\partial_\mu + \tfrac{1}{4}\omega_\mu{}^{ab}(e)\gamma_{ab} + \tfrac{1}{2}\mathrm{i}\mathcal{A}_\mu\right)\psi_{\nu i} + \mathcal{V}_{\mu i}{}^j\psi_{\nu j}, \\
D_\mu\chi_i^\alpha &= \left(\partial_\mu + \tfrac{1}{4}\omega_\mu{}^{ab}(e)\gamma_{ab} + \tfrac{1}{2}\mathrm{i}\mathcal{A}_\mu\right)\chi_i^\alpha + \mathcal{V}_{\mu i}{}^j\chi_j^\alpha - A_\mu^I\chi_i^\beta\partial_\beta k_I{}^\alpha + \Gamma^\alpha_{\beta\gamma}\chi_i^\gamma\hat\partial_\mu z^\beta, \\
\hat D_\mu\zeta^A &= \left(\partial_\mu + \tfrac{1}{4}\omega_\mu{}^{ab}(e)\gamma_{ab} + \tfrac{1}{2}\mathrm{i}\mathcal{A}_\mu\right)\zeta^A - A_\mu^I t_{IB}{}^A\zeta^B + \partial_\mu q^u\omega_{uB}{}^A\zeta^B.
\end{aligned} \tag{21.35}
$$

Here appear the (bosonic part of the) effective U(1) and SU(2) composite gauge fields:[5]

$$
\begin{aligned}
\mathcal{A}_\mu &= -\kappa^2\omega_\alpha\partial_\mu z^\alpha - \kappa^2\omega_{\bar\alpha}\partial_\mu\bar z^{\bar\alpha} - \kappa^2 A_\mu^I P_I^0 = -\kappa^2\omega_\alpha\hat\partial_\mu z^\alpha + \tfrac{1}{2}\mathrm{i}A_\mu^I r_I + \text{h.c.}, \\
\vec{\mathcal{V}}_\mu &= -\vec\omega_u\partial_\mu q^u - \tfrac{1}{2}\kappa^2 A_\mu^I\vec P_I = -\vec\omega_u\hat\partial_\mu q^u + \tfrac{1}{2}A_\mu^I\vec r_I.
\end{aligned} \tag{21.36}
$$

[5] The first one is the rewriting of (19.52). In the second one, $\vec\omega_u$ is also of order κ^2 in the sense that this connection is a supergravity effect, which is absent in global supersymmetry.

Remember that $\omega_\alpha = -\frac{1}{2}\mathrm{i}\partial_\alpha\mathcal{K}$, while $\vec{\omega}_u$ and $\omega_{uB}{}^A$, which appears in (21.35), are the connections mentioned in (21.6) and (21.11).

The gauge transformation of the fermions of the hypermultiplets are determined by the matrix

$$t_{IA}{}^B \equiv \tfrac{1}{2} f^v{}_{iA}\nabla_v k_I{}^u f^{iB}{}_u. \tag{21.37}$$

The potential V, determined by the gauging, is given in (21.16) or (21.30). We will still give another expression in (21.46). The latter will be based on the 'fermion shifts', i.e. the non-derivative part of the supersymmetry transformations of the fermions, which we will discuss below. One finds that also the mass terms of the fermions in (21.34) are determined by the gauging:

$$\mathcal{L}_m = \tfrac{1}{2}S_{ij}\bar{\psi}^i_\mu\gamma^{\mu\nu}\psi^j_\nu - \tfrac{1}{2}m^{ij}{}_{\alpha\beta}\bar{\chi}^\alpha_i\chi^\beta_j - m_{i\bar{\alpha}}{}^A\bar{\chi}^{i\bar{\alpha}}\zeta_A - \tfrac{1}{2}m_{AB}\bar{\zeta}^A\zeta^B + \text{h.c.}, \tag{21.38}$$

where the following matrices have been used:[6]

$$\begin{aligned}
S_{ij} &= P_{Iij}\bar{X}^I,\\
m^{ij}{}_{\alpha\beta} &= \tfrac{1}{2}P_I^{ij}C_{\alpha\beta\gamma}g^{\gamma\bar{\delta}}\overline{\nabla}_{\bar{\delta}}\bar{X}^I + \varepsilon^{ij}\nabla_\alpha X^I k_I{}^{\bar{\gamma}}g_{\beta\bar{\gamma}},\\
m_{i\bar{\alpha}}^A &= 2\mathrm{i}k_I^u\varepsilon_{ij}f^{jA}{}_u\overline{\nabla}_{\bar{\alpha}}\bar{X}^I,\\
m_{AB} &= -4X^I t_{IAB}.
\end{aligned} \tag{21.39}$$

Finally, the Goldstino in the last line of (21.34) is

$$\begin{aligned}
\upsilon^i &= \tfrac{1}{2}W_\alpha{}^{ij}\chi_j^\alpha + 2N^i{}_A\zeta^A,\\
W_\alpha{}^{ij} &\equiv \left(\mathrm{i}\varepsilon^{ij}P_I^0 - P_I{}^{ij}\right)\nabla_\alpha X^I = \varepsilon^{ij}g_{\alpha\bar{\beta}}k_I{}^{\bar{\beta}}X^I + P_I{}^{ij}\nabla_\alpha X^I,\\
N^i{}_A &\equiv \mathrm{i}f^{iB}{}_u k_I{}^u X^I C_{BA}.
\end{aligned} \tag{21.40}$$

21.3.2 Supersymmetry transformations

The supersymmetry transformations of the bosonic fields are

$$\delta e_\mu{}^a = \tfrac{1}{2}\bar{\epsilon}^i\gamma^a\psi_{\mu i} + \text{h.c.}, \qquad \delta z^\alpha = \tfrac{1}{2}\bar{\epsilon}^i\chi_i^\alpha, \tag{21.41}$$
$$\delta A_\mu{}^I = \tfrac{1}{2}\varepsilon^{ij}\bar{\epsilon}_i\gamma_\mu\chi_j^\alpha\nabla_\alpha X^I + \varepsilon^{ij}\bar{\epsilon}_i\psi_{\mu j}X^I + \text{h.c.}, \qquad \delta q^u = -\mathrm{i}\bar{\epsilon}^i\zeta^A f^u{}_{iA} + \text{h.c.}$$

We will write here the bosonic part of the transformations of the fermions, since these are the important terms in the search for solutions of the field equations and their preserved supersymmetries (see Sec. 12.6):

[6] The triplet moment maps $\vec{P}_I$ are written as symmetric tensors P_{Iij}. Also this correspondence is explained in Sec. 20A.

$$\begin{aligned}
\delta\psi_\mu^i &= \left(\partial_\mu + \tfrac{1}{4}\omega_\mu{}^{ab}\gamma_{ab} - \tfrac{1}{2}\mathrm{i}\mathcal{A}_\mu\right)\epsilon^i + \mathcal{V}_\mu{}^i{}_j\epsilon^j - \tfrac{1}{16}\gamma^{ab}T^-_{ab}\varepsilon^{ij}\gamma_\mu\epsilon_j + \tfrac{1}{2}\kappa^2\gamma_\mu S^{ij}\epsilon_j,\\
\delta\chi_i^\alpha &= \hat{\partial\!\!\!/} z^\alpha\epsilon_i - \tfrac{1}{2}G^{-\alpha}_{ab}\gamma^{ab}\varepsilon_{ij}\epsilon^j + g^{\alpha\bar\beta}\overline{W}_{\bar\beta ji}\epsilon^j,\\
\delta\zeta^A &= \tfrac{1}{2}\mathrm{i}f^{iA}{}_u\hat{\partial\!\!\!/} q^u\epsilon_i - \zeta^B\omega_{uB}{}^A\delta q^u + \bar{N}_i{}^A\epsilon^i.
\end{aligned} \tag{21.42}$$

In the fermion shifts appear the complex conjugates of the expressions in (21.40):

$$\begin{aligned}
\overline{W}_{\bar\alpha ij} &= \left(-\mathrm{i}\varepsilon_{ij}P_I^0 - P_{Iij}\right)\overline{\nabla}_{\bar\alpha}\bar{X}^I = \varepsilon_{ij}g_{\beta\bar\alpha}k_I{}^\beta\bar{X}^I + P_{Iij}\overline{\nabla}_{\bar\alpha}\bar{X}^I,\\
\bar{N}_i{}^A &\equiv -\mathrm{i}\varepsilon_{ij}f^{jA}{}_u k_I{}^u\bar{X}^I.
\end{aligned} \tag{21.43}$$

In the transformation of the gravitino appears the field strength T_{ab}, which is therefore called the graviphoton field strength. Its bosonic part is

$$T^+_{ab}\big|_{\rm bos} = -4\kappa^2\bar{X}^I\,\mathrm{Im}\,\mathcal{N}_{IJ}F^{+J}_{ab} = 2\mathrm{i}\kappa^2\left(\bar{X}^I G^+_{I\,ab} - \bar{F}_I F^{+I}_{ab}\right). \tag{21.44}$$

The graviphoton is thus a scalar field dependent combination of the field strengths of the (n_V+1) gauge fields. It is proportional to the symplectic inner product (21.24). This expression can thus also be used in the absence of a prepotential as in the example of Ex. 20.19, which we will use for black holes. If in the last expression we replace F^{+I}_{ab} and $G^+_{I\,ab}$ with F^{-I}_{ab} and $G^-_{I\,ab}$, this expression would vanish (being an inner product of two symplectic vectors in the same row in (21.23)). Therefore, we can use this result for T^+_{ab} omitting the + indications in the last expression (as will be relevant for central charges in Sec. 21.4.2).

In the transformation of the gauginos appears another symplectic invariant:

$$G^{-\alpha}_{ab} = g^{\alpha\bar\beta}\overline{\nabla}_{\bar\beta}\bar{X}^I\,\mathrm{Im}\,\mathcal{N}_{IJ}F^{-J}_{ab}. \tag{21.45}$$

The identity (14.68), which connects the potential to the fermion shifts, is valid for each of the supersymmetries, and leads to zero when multiplying fermion shifts related to different supersymmetries. The identity in this case is

$$-3\kappa^2 S^{ik}S_{jk} + W_\alpha{}^{ik}g^{\alpha\bar\beta}\overline{W}_{\bar\beta jk} + 4N^i{}_A\bar{N}_j{}^A = \delta^i_j V. \tag{21.46}$$

Exercise 21.2 *Prove this identity using (20.210), (20.190), (20.108), (20.37) and (20.164) and, for the (ij) non-diagonal part, also the equivariance condition (20.175).*

Exercise 21.3 *Recover the pure $\mathcal{N}=2$ supergravity, i.e. $n_V=n_H=0$. The prepotential should be homogeneous of second degree in the compensating field X^0. Check that with*

$$F = -\tfrac{1}{2}\mathrm{i}(X^0)^2, \tag{21.47}$$

the gauge fixing leads to $X^0 = (\sqrt{2}\kappa)^{-1}$, *and one gets* $\mathcal{N}_{00} = -\mathrm{i}$, *such that the action contains the standard kinetic terms for one vector field, as is clear from (21.4). Obtain then also that*

$$T_{ab} = 2\sqrt{2}\kappa\, F_{ab}, \tag{21.48}$$

and thus the supersymmetry transformation

$$\delta\psi^i_\mu = \left(\partial_\mu + \tfrac{1}{4}\gamma^{ab}\omega_{\mu ab}\right)\epsilon^i - \tfrac{1}{8}\sqrt{2}\kappa\gamma^{ab}F_{ab}\varepsilon^{ij}\gamma_\mu\epsilon_j. \tag{21.49}$$

Exercise 21.4 *Gauged pure* $\mathcal{N} = 2$ *supergravity can be obtained by considering the previous model with non-zero moment maps. The latter are then Fayet–Iliopoulos constants. Compare the potential (21.16) for this case with (9.51) and identify* $\vec{P}\cdot\vec{P} = 2/(\kappa L)^2$. *The reader can get further practice with these formulas by comparing (9.52) with (21.42), putting* $\vec{P}$ *in an arbitrary direction in* SU(2) *space.*

21.4 Applications

21.4.1 Partial supersymmetry breaking

One important application is partial supersymmetry breaking of $\mathcal{N} = 2$ to $\mathcal{N} = 1$. The no-go theorem of [266, 169] stated that it is not possible to find a vacuum that preserves one of the two supersymmetries in Minkowski space. This statement was obtained in special Kähler geometry in the presence of a prepotential. The preservation of one supersymmetry means that there should be one spinor ϵ_1^i for which the transformations (21.42) vanish. For a Minkowski vacuum the vector fields vanish, and the scalars are constants in spacetime. Hence we are left with the conditions

$$S_{ij}\epsilon_1^j = 0, \qquad \overline{W}_{\bar{\alpha}ji}\epsilon_1^j = 0, \qquad \bar{N}_i{}^A\epsilon_1^i = 0. \tag{21.50}$$

Using the explicit forms of S_{ij} in (21.39), $\overline{W}_{\bar{\alpha}ji}$ in (21.43), and $\bar{X}^I P_I^0 = 0$, the first two conditions are

$$\left(\mathrm{i}\varepsilon_{ji}P_I^0 + P_{Iij}\right)\bar{X}^I\epsilon_1^j = 0, \qquad \left(\mathrm{i}\varepsilon_{ji}P_I^0 + P_{Iij}\right)\overline{\nabla}_{\bar{\alpha}}\bar{X}^I\epsilon_1^j = 0. \tag{21.51}$$

When there is a prepotential, the matrix (21.25) (or its complex conjugate) is invertible. Hence we get that, for every I, the 2×2 matrix in brackets should have a zero eigenvalue. Using the reality of P_I^0 and $\vec{P}_I$, it follows that these should be zero. If that is the case, (21.51) is also valid with ϵ_1^i replaced by any other ϵ_2^i. Hence both supersymmetries are preserved.[7]

[7] To make the argument complete, one has to argue that also $\bar{N}_i{}^A\epsilon_2^i = 0$. This can be deduced from the previous results and the fact that we look for a solution in Minkowski space. In that case $S_{ij} = 0$ and applying (21.46) on ϵ_2^j gives the required result.

The way out has been found in [267] after the formalism without a prepotential had been introduced [248]. In this case, (21.25) is not necessarily invertible, and hence one cannot conclude from (21.51) that the matrix in brackets vanishes for all I. Ferrara, Girardello and Porrati [267] in fact constructed an explicit model with partial supersymmetry breaking in Minkowski space. It can be easier to study such models anyway in the presence of a prepotential, using a symplectic frame where all gauge symmetries are not of electric type (thus using magnetic vectors $A_{\mu I}$). This has been done in [268, 269].

21.4.2 Field strengths and central charges

The symplectic expression (21.44) for the gravitino field strength is useful in the context of central charges. In the previous chapter, we have seen that $T_{\mu\nu}$ appears in the soft algebra of local supersymmetry; see (20.74) and (20.75). Comparing this with (12.23), we see that it appears in the form of a central charge term in the algebra. The exact identification depends on the type of solution that one considers. In Sec. 22.5, we will consider black hole solutions with electric and magnetic charges. Therefore it is useful to define central charges at infinity as

$$\begin{aligned} \mathcal{Z} &\equiv -\tfrac{1}{2}\mathrm{i}\kappa^{-2}\int T^-_{\mu\nu}\mathrm{d}x^\mu \wedge \mathrm{d}x^\nu = -\tfrac{1}{2}\int \left(X^I G_{I\,\mu\nu} - F_I F^I_{\mu\nu}\right)\mathrm{d}x^\mu \wedge \mathrm{d}x^\nu \\ &= 2\left(X^I q_I - F_I p^I\right), \end{aligned} \tag{21.52}$$

where we assumed that the values of the scalar fields are constants on the space surface on which we integrate. We furthermore made use of the definitions (7.65) of electric and magnetic charges. This quantity is complex if the value of X^I and F_I for the solution is complex.

Note that the covariant derivatives of these quantities are related to the symplectic quantity that we found in the supersymmetry transformation of the gauginos:

$$\begin{aligned} \nabla_\alpha \mathcal{Z} &= 2\left(\nabla_\alpha X^I q_I - \nabla_\alpha F_I p^I\right) = -\int \left(\nabla_\alpha X^I G_{I\,\mu\nu} - \nabla_\alpha F_I F^I_{\mu\nu}\right)\mathrm{d}x^\mu \wedge \mathrm{d}x^\nu \\ &= -\int \left(\nabla_\alpha X^I G^+_{I\,\mu\nu} - \nabla_\alpha F_I F^{+I}_{\mu\nu}\right)\mathrm{d}x^\mu \wedge \mathrm{d}x^\nu \\ &= -2\mathrm{i}\int \nabla_\alpha X^I\, \mathrm{Im}\,\mathcal{N}_{IJ}\, F^{+J}_{\mu\nu}\,\mathrm{d}x^\mu \wedge \mathrm{d}x^\nu = -2\mathrm{i}\int G_{\mu\nu\alpha}\,\mathrm{d}x^\mu \wedge \mathrm{d}x^\nu . \end{aligned} \tag{21.53}$$

Using the symplectic notation, the values of $\mathcal{Z}$ and its derivatives are

$$\mathcal{Z} = 2\langle V, \Gamma\rangle, \qquad \nabla_\alpha \mathcal{Z} = 2\langle \nabla_\alpha V, \Gamma\rangle, \qquad \Gamma \equiv \begin{pmatrix} p^I \\ q_I \end{pmatrix}, \qquad V \equiv \begin{pmatrix} X^I \\ F_I \end{pmatrix}. \tag{21.54}$$

21.4.3 Moduli spaces of Calabi–Yau manifolds

The symplectic formulation of special geometry is very useful for describing the geometry of moduli spaces of Calabi–Yau manifolds of complex dimension 3, which are the principal candidates for internal manifolds in superstring theories. A Calabi–Yau manifold is a Kähler manifold with vanishing first Chern class. We will restrict ourselves to some aspects of these manifolds that relate to special Kähler geometry. Many books on string theory contain an introduction to Calabi–Yau manifolds.

An important role is played by the (p, q) forms on the manifold (hence $p + q \leq 6$). There is one (cohomologically) unique $(3, 0)$ form, denoted by $\Omega^{(3,0)}$, that characterizes the Calabi–Yau manifold. The dimension of the cohomology space of (p, q) forms is indicated by h^{pq}, and the 'Hodge diamond' gives these numbers in a table:

$$\begin{array}{ccccccc} & & & h^{00}=1 & & & \\ & & 0 & & 0 & & \\ & 0 & & h^{11}=m & & 0 & \\ h^{30}=1 & & h^{21}=n & & h^{12}=n & & h^{03}=1 \\ & 0 & & h^{22}=m & & 0 & \\ & & 0 & & 0 & & \\ & & & h^{33}=1 & & & \end{array} \tag{21.55}$$

Dual to the cohomology group is the homology group of cycles. Since the above table says that there are $2(n+1)$ non-trivial 3-forms, there are also $2(n+1)$ non-trivial cycles, which we indicate as c_Λ, $\Lambda = 1, \ldots, 2(n+1)$. One can order them such that their intersection matrix is of the form

$$Q_{\Lambda\Sigma} \equiv c_\Lambda \cap c_\Sigma = \begin{pmatrix} 0 & \mathbb{1} \\ -\mathbb{1} & 0 \end{pmatrix}. \tag{21.56}$$

The possible deformations of such manifolds are called the moduli. The geometry of deformations of the complex structure of Calabi–Yau 3-folds is a special Kähler geometry. This moduli space is of dimension $n = h^{21}$. The coordinates of this space are the fields z^α of special Kähler geometry.

The symplectic vectors are identified as vectors of integrals of 3-forms over the $2(n+1)$ 3-cycles:

$$v = \int_{c_\Lambda} \Omega^{(3,0)}, \qquad \nabla_\alpha v = \int_{c_\Lambda} \Omega_\alpha^{(2,1)}. \tag{21.57}$$

Here $\Omega_\alpha^{(2,1)}$ is a basis of the $(2, 1)$ forms, determined by the choice of basis for z^α. That these moduli spaces give rise to special Kähler geometry was found in [270, 271, 272, 273, 274, 246].

The defining equations of special Kähler geometry are automatically satisfied, due to an integral theorem, containing sums over 3-cycles:

$$\int_{c_\Lambda} \omega_1 \cdot Q_{\Lambda\Sigma} \cdot \int_{c_\Sigma} \omega_2 = \int_{CY} \omega_1 \wedge \omega_2. \tag{21.58}$$

For example, one can see how the crucial equation (20.133) or (21.18) is realized:

$$\begin{aligned}\langle \nabla_\alpha v, \nabla_\beta v\rangle &= \int_{C\Lambda} \Omega_\alpha^{(2,1)} \cdot Q_{\Lambda\Sigma} \cdot \int_{C\Sigma} \Omega_\beta^{(2,1)} \\ &= \int_{CY} \Omega_\alpha^{(2,1)} \wedge \Omega_\beta^{(2,1)} = 0. \end{aligned} \tag{21.59}$$

The last expression vanishes since there are only three independent holomorphic coordinates on the Calabi–Yau space.

In this context, the symplectic transformations correspond to changes of the basis of the cycles used to construct the symplectic vectors, maintaining the intersection matrix (21.56).

21.5 Remarks

21.5.1 Fayet–Iliopoulos terms

If the quaternionic-Kähler manifold is not empty, equations (21.15) determine the moment maps. However, in the case that $n_H = 0$, in which there is no physical hypermultiplet, the moment maps can still be non-vanishing. In the superconformal context, they then describe the gauge transformations of the compensating hypermultiplet, which we need in any case for consistency of the field equations. These moment maps can then be non-vanishing constants, which are the FI constants. Hence, FI terms exist in $\mathcal{N}=2$ supergravity only in the case $n_H = 0$. They enter, for example, in the construction of gauged pure $\mathcal{N}=2$ supergravity, as shown in Ex. 21.4, or in theories with coupled abelian gauge multiplets.

21.5.2 σ-model symmetries

Kähler manifolds have Kähler transformations, which appeared first in (13.57). In the conformal context they originate from the non-uniqueness of the split of the variables X^I in y and Z^I; see (17.71). The pull-back to spacetime of the gauge field of these transformations, ω_α, is the composite gauge field of the T part of the R-symmetry group; see the first line of (21.36).

Similarly, quaternionic manifolds have such symmetries with gauge field $\vec{\omega}_u$. They originate from the rotations between the basis of the three complex structures: $\delta(\vec{\ell})\vec{J}_u{}^v = \vec{\ell}\times\vec{J}_u{}^v$. The composite gauge field of the SU(2) part of the R-symmetry group, $\vec{\mathcal{V}}_\mu$, is the pull-back to spacetime of $\vec{\omega}_u$; see the second line of (21.36).

21.5.3 Engineering dimensions

Similar to the choices in $\mathcal{N}=1$ (see Sec. 18.3.1), the scalars are dimensionless, but the prepotential should be chosen such that it has mass dimension 2; see e.g. (20.141). In that way the Kähler potential has mass dimension 2. Also the metric of the quaternionic space should have mass dimension 2. With these definitions $\mathcal{N}_{IJ}$ has dimension $[M]^2$, and the field strengths $F^I_{\mu\nu}$ have dimension $[M]$. Therefore magnetic charges proportional to $\int F$ over a two-dimensional surface have dimension of a length, while electric charges proportional to $\int \mathcal{N} F$ have dimension $[M]$. Symplectic transformations are then not dimensionless; see (20.144) and (20.149).

In particular models where the Kähler potential is homogeneous (see e.g. (21.47)), one may use X^I fields with dimension $[M]$, such that the second derivative of $F(X)$, and hence $\mathcal{N}$, has dimension 0. In this setting, $F_{\mu\nu}$ has dimension 2, leading to dimensionless electric and magnetic charges. Alternative definitions are used in Sec. 22.5 below.

PART VIII

CLASSICAL SOLUTIONS AND THE AdS/CFT CORRESPONDENCE

Classical solutions of gravity and supergravity 22

Many spacetimes that are solutions of the gravitational field equations, with and without matter sources, have appeared in applications of supergravity in various spacetime dimensions. Supergravity theories always include both fermions and bosons, but the classical solutions are configurations of the bosonic fields only, which satisfy the Euler–Lagrange equations of the action with all fermion fields vanishing. Fermions play a vital role in determining the structure of a supergravity theory, but they do not appear directly in the classical limit.

In this chapter we select and discuss a few simple and useful solutions that have interesting applications in supergravity or which illustrate some of the general ideas. We contrast solutions obtained directly from the Einstein field equations with the special but quite large and interesting class of Bogomol'nyi–Prasad–Sommerfield (BPS) solutions. In a supersymmetry or supergravity theory, the term BPS designates a solution of the theory that is invariant under a subalgebra of the SUSY algebra of the action that contains at least one fermionic generator. The BPS solution can be obtained by solving first order differential equations associated with the fermion transformation rules of the theory. For each preserved supersymmetry there is a Killing spinor and a subset of the Killing vectors of the geometry can be obtained from them. This central concept will be explained in detail in Sec. 22.2.

Many subjects cannot be treated here as thoroughly as they are discussed in texts devoted to general relativity. Other important subjects must be omitted entirely. These include Penrose diagrams, the global structure of black hole geometries and black hole thermodynamics.

22.1 Some solutions of the field equations

22.1.1 Prelude: frames and connections on spheres

Many useful solutions have spherical symmetry. In this case it is desirable to formulate an ansatz for the fields involved that incorporates the symmetry. This reduces the number of essential variables needed to two, a radial variable r and a time variable t. If the full solution of interest is a D-dimensional spacetime, the remaining variables can be taken as angles on the unit sphere S^{D-2}, which is a submanifold of that spacetime.

We will need the following facts about the differential geometry of S^n:

(i) There are frame 1-forms $\bar{e}^a$, $a = 1, \dots, n$, such that the line element of S^n can be written as

$$d\Omega_n^2 = \delta_{ab}\bar{e}^a\bar{e}^b \,. \tag{22.1}$$

(ii) There are torsion-free connection 1-forms $\bar{\omega}^{ab}$, which satisfy the Cartan structure equation (7.81) (with $T^a = 0$).

(iii) S^n is a maximally symmetric space, so the curvature 2-forms satisfy

$$\bar{\rho}^{ab} = \bar{e}^a \wedge \bar{e}^b \,. \tag{22.2}$$

For most of the discussion below, the facts above are sufficient to find the complete spherically symmetric solution of interest. However, it is occasionally useful to have an explicit set of frame and connection forms, and we now define a set based on hyperspherical coordinates.

It is very well known that the unit sphere S^2 is the surface $(x^1)^2 + (x^2)^2 + (x^3)^2 = 1$ embedded in flat Euclidean space $\mathbb{R}^3$ with Cartesian coordinates x^i, $i = 1, 2, 3$. Intrinsic angular coordinates θ^1, θ^2 with range $0 \le \theta^1 \le 2\pi, 0 \le \theta^2 \le \pi$ are introduced via[1] $x^1 = \sin\theta^2 \sin\theta^1$, $x^2 = \sin\theta^2 \cos\theta^1$, $x^3 = \cos\theta^2$. Substituting these in the Euclidean line element gives the intrinsic metric:

$$\begin{aligned} d\Omega_2^2 &= (dx^1)^2 + (dx^2)^2 + (dx^3)^2 \\ &= (d\theta^2)^2 + \sin^2\theta^2 (d\theta^1)^2 \,. \end{aligned} \tag{22.3}$$

Proceeding as discussed in Ch. 7, one can define the frame 1-forms

$$\bar{e}^2 = d\theta^2 \,, \qquad \bar{e}^1 = \sin\theta^2 d\theta^1 \,, \tag{22.4}$$

and then use (7.81) to obtain $\bar{\omega}^{12} = \cos\theta^2 d\theta^1$. We follow the recursive method used in [275] to extend this construction to S^n, which we view as the surface $\sum_{i=1}^{n+1}(x^i_{(n)})^2 = 1$ embedded in flat Euclidean $\mathbb{R}^{n+1}$. We express the Cartesian coordinates $x^i_{(n)}$ in terms of the principal angle θ^n and the coordinates $x^i_{(n-1)}$ of S^{n-1} by the recursive formula

$$x^{n+1}_{(n)} = \cos\theta^n \,, \qquad x^a_{(n)} = \sin\theta^n x^a_{(n-1)} \,, \qquad a \le n \,, \qquad 0 \le \theta^n \le \pi \,. \tag{22.5}$$

This implicitly defines angular coordinates θ^a, $a = 1, \dots, n$, which are called 'hyperspherical coordinates'. The angular range is $0 \le \theta^1 < 2\pi$ and $0 \le \theta^a \le \pi$ for $a > 1$. The line element is

$$d\Omega_n^2 = (d\theta^n)^2 + \sin^2\theta^n d\Omega_{n-1}^2 \,. \tag{22.6}$$

[1] Instead of the usual coordinates x, y, z, we choose $x^1 = y, x^2 = x$ and $x^3 = z$ to prepare for the extension to arbitrary n-spheres below.

Frame and connection forms are

$$\bar{e}^n_{(n)} = \mathrm{d}\theta^n\,, \qquad \bar{e}^a_{(n)} = \sin\theta^n \bar{e}^a_{(n-1)}\,, \qquad a \leq n-1\,,$$
$$\bar{\omega}^{ab}_{(n)} = \bar{\omega}^{ab}_{(n-1)}\,, \qquad \bar{\omega}^{an}_{(n)} = \cos\theta^n \bar{e}^a_{(n-1)}\,. \tag{22.7}$$

More explicitly for the n-sphere:

$$\bar{e}^a = \left(\prod_{j=a+1}^{n} \sin\theta^j\right) \mathrm{d}\theta^a\,, \qquad a \leq n\,,$$
$$\bar{\omega}^{ab} = \cos\theta^b \left(\prod_{j=a+1}^{b-1} \sin\theta^j\right) \mathrm{d}\theta^a\,, \qquad 1 \leq a < b \leq n\,. \tag{22.8}$$

Exercise 22.1 *Show that the curvature 2-form obtained from the connection forms above satisfies (22.2). Use the second structure equation (7.117) to compute $\bar{\rho}^{ab}$.*

Exercise 22.2 *Show that the volume of S^n is*

$$\Omega_n \equiv \int \prod_{a=1}^{n} \mathrm{d}\theta^a \sqrt{g} = \int \bar{e}^1 \wedge \ldots \wedge \bar{e}^n = \frac{2\pi^{(n+1)/2}}{\Gamma((n+1)/2)}\,. \tag{22.9}$$

22.1.2 Anti-de Sitter space

The first spacetime we discuss is called anti-de Sitter space [276, 277], which we will abbreviate as AdS$_D$ for the D-dimensional case. It was studied in the 1970s and 1980s as a simple solution of gauged supergravity [278, 183]. It also appears in simple Kaluza–Klein reductions of higher dimensional gravity theories [279]. Since 1997, it has enjoyed a special status in the very important AdS/CFT correspondence [280, 281, 282]. AdS has a spherically symmetric metric, as we will see in (22.17), but it is more instructive to approach it from a different standpoint. Then we derive the AdS–Schwarzschild metrics, which describe a black hole in AdS spacetime.

These metrics are solutions of the equations of motion of the gravitational action

$$S = \frac{1}{2\kappa^2} \int \mathrm{d}^D x \sqrt{-g}\,(R - \Lambda)\,. \tag{22.10}$$

The second term is the cosmological constant, which is negative for AdS spacetimes. We write it as $\Lambda = -(D-1)(D-2)/L^2$ where L has dimensions of length. We already encountered this in (9.51) and (16.53).

Exercise 22.3 *Show that the equation of motion for the action (22.10) is equivalent to*

$$R_{\mu\nu} = -\frac{D-1}{L^2} g_{\mu\nu}\,. \tag{22.11}$$

AdS is an example of a maximally symmetric spacetime. This means that the Riemann curvature tensor can be written in terms of the metric tensor $g_{\mu\nu}$ as

$$R_{\mu\nu\rho\sigma} = k\left(g_{\mu\rho}g_{\nu\sigma} - g_{\mu\sigma}g_{\nu\rho}\right), \tag{22.12}$$

where k is a constant of dimension $1/(\text{length})^2$. For AdS, this constant is negative and given by $k = -1/L^2$ where L sets the overall scale of the spacetime.

It follows immediately from (22.12) that the Ricci tensor and the curvature scalar of maximally symmetric spacetimes are

$$\begin{aligned} R_{\mu\nu} &= k(D-1)g_{\mu\nu}, \\ R &= k\,D(D-1). \end{aligned} \tag{22.13}$$

A metric that satisfies (22.13) is called an Einstein metric. Any maximally symmetric spacetime satisfies (22.13), but the converse is certainly not true. The AdS–Schwarzschild spacetimes, which we will study in Sec. 22.1.5, are one example. They are also solutions of the equations of motion (22.11). So they are Einstein metrics, but they are not maximally symmetric.

A maximally symmetric spacetime of dimension D can be embedded as a hyperboloid in flat spacetime of dimension $D+1$. (For Euclidean signature, there is an analogous embedding as a sphere or hyperboloid.) The metric $g_{\mu\nu}$ is then the induced metric, which means that the line element is obtained by evaluating the flat metric of the embedding space on the hyperboloid.

22.1.3 AdS$_D$ obtained from its embedding in $\mathbb{R}^{D+1}$

For AdS$_D$ the embedding space is the pseudo-Euclidean space with metric $\eta_{AB} = \text{diag}(-++\ldots+-)$ with Cartesian coordinates Y^A, $A = 0,\ldots,D$. The hyperboloid that defines AdS$_D$ is the surface

$$f(Y) \equiv Y^A\eta_{AB}Y^B = -(Y^0)^2 + \sum_{i=1}^{D-1}(Y^i)^2 - (Y^D)^2 = -L^2. \tag{22.14}$$

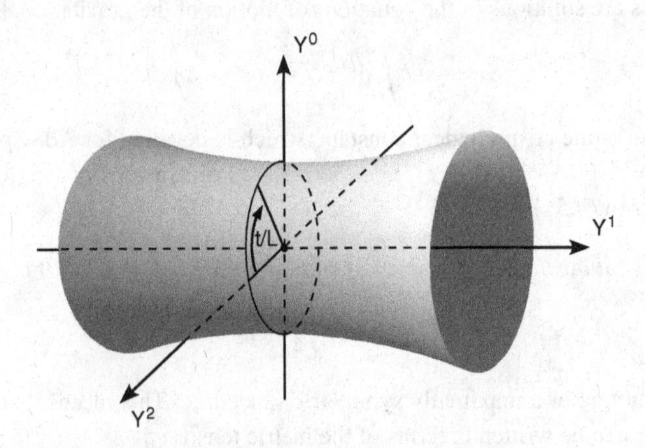

Fig. 22.1 The AdS$_2$ hyperboloid.

A sketch of this hyperboloid, of necessity limited to the case $D = 2$, is presented in Fig. 22.1. The quadratic form in (22.14) is invariant under linear transformations of the coordinates $Y'^A = \Lambda^A{}_B Y^B$ provided that $\Lambda^A{}_B$ is a matrix of the group $SO(D-1,2)$. So the hyperboloid is invariant, and the isometry group of AdS_D is $SO(D-1,2)$. One can repeat the discussion of Ch. 1 for the Lorentz group and show that the differential operators[2]

$$L_{AB} \equiv Y_A \frac{\partial}{\partial Y^B} - Y_B \frac{\partial}{\partial Y^A} \tag{22.15}$$

act as generators of the Lie algebra of $SO(D-1,2)$.

Exercise 22.4 *Show that $L_{AB}\, f(Y) = 0$.*

AdS is a homogeneous space, which means that any two points on the hyperboloid are related by a transformation of the isometry group; see Sec. 12.5. The purpose of the following exercise is to show this explicitly. It is useful to recognize that the maximal compact subgroup of $SO(D-1,2)$ is $SO(D-1) \times SO(2)$. The first factor rotates the Y^i, $i = 1, \ldots, D-1$, and the second factor rotates in the $0D$ plane.

Exercise 22.5 *Show that an arbitrary point Y^A can be brought to the point $(L, 0, \ldots, 0)$ by the following group operations. First transform by an element of the compact subgroup that brings Y^A to the configuration $(Y'^0, Y'^1, 0, \ldots, 0)$ with $Y'^0 > 0$. This is a positive time-like vector, which can be brought to the point $(L, 0, \ldots, 0)$ by a boost in the* 01 *plane.*

We will discuss several commonly used sets of intrinsic coordinates x^μ, $\mu = 0, 1, \ldots, D-1$, on the hyperboloid and obtain the induced metric. The Lie algebra operators L_{AB} may be expressed in terms of the x^μ, and they become Killing vector fields of AdS_D.

The first set of intrinsic coordinates is defined by

$$\begin{aligned} Y^i &= r\bar{x}^i \qquad \text{with} \qquad \sum_{i=1}^{D-1} (\bar{x}^i)^2 = 1\,, \\ Y^0 &= \sqrt{L^2 + r^2}\sin(t/L) \qquad Y^D = \sqrt{L^2 + r^2}\cos(t/L)\,. \end{aligned} \tag{22.16}$$

The unit vector $\bar{x}^i$ parametrizes the unit sphere S^{D-2}; it was written in terms of the $D-2$ angular variables of the hyperspherical coordinate system in (22.5).

The induced metric is obtained from the embedding space line element $ds^2 = \eta_{AB} dY^A dY^B$ by using (22.16) to express each dY^A in terms of the intrinsic coordinates. The result is

[2] Indices are raised and lowered with η_{AB}.

$$ds^2 = -\left(1+\frac{r^2}{L^2}\right)dt^2 + \left(1+\frac{r^2}{L^2}\right)^{-1} dr^2 + r^2\, d\Omega_{D-2}^2, \tag{22.17}$$

where $d\Omega_{D-2}^2$ is the line element on the unit S^{D-2}.

This coordinate system is global. It is easy to see that the entire hyperboloid is covered once as the radial variable varies in the range $0 \le r < \infty$, the angular variables cover S^{D-2}, and the time coordinate t ranges between $0 \le t < 2\pi L$. It is peculiar that time is a periodic variable, and, indeed, the hyperboloid contains closed time-like curves. Any curve on which r and $\bar{x}^i$ are fixed and t varies over a period is a closed time-like curve. This means that AdS has bizarre causal properties, which are usually considered physically unacceptable. In this case the problem is easily solved by unwrapping the time circle and passing to the covering space in which $-\infty < t < +\infty$. Henceforth we assume that we are dealing with this covering space, which we continue to call AdS. It may appear that the origin $r = 0$ of the spatial coordinates in (22.17) is a distinguished point. However, recall Ex. 22.5, which showed that AdS is homogeneous. Thus all points are equivalent and any desired point on the hyperboloid (or its cover) can be made the origin of a global coordinate system.

Exercise 22.6 *Derive (22.17). It is obvious that the shift $t \to t + a$ is an isometry of the metric (22.17), which is generated by the Killing vector $\partial/\partial t$. Show that $L_{0D} = L\,\partial/\partial t$.*

There are other global coordinate systems that are frequently used; they differ from (22.16) by a change of the radial coordinate. For example, if we define the new radial coordinate y by $\cosh(y/L) = \sqrt{1 + r^2/L^2}$, then the line element (22.17) becomes

$$ds^2 = -\cosh^2(y/L)dt^2 + dy^2 + L^2 \sinh^2(y/L)\, d\Omega_{D-2}^2\,. \tag{22.18}$$

Let's introduce another radial coordinate by the change $\cosh(y/L) = 1/\cos\rho$. The range of ρ is $0 \le \rho < \pi/2$. The new coordinate will give some insight into the global structure of (the covering space of) AdS and a useful pictorial representation. It is also convenient to scale the time coordinate by $t = L\tau$. The line element changes to

$$ds^2 = \frac{L^2}{\cos^2\rho}\left[-d\tau^2 + \left(d\rho^2 + \sin^2\rho\, d\Omega_{D-2}^2\right)\right]. \tag{22.19}$$

In this form the AdS metric is a conformal factor $L^2/\cos^2\rho$ times a metric that describes the direct product of the real line $-\infty < \tau < +\infty$ in the time coordinate and a sphere S^{D-1}. The product spacetime, called the Einstein static universe, played a role in the early history of general relativity. It was the way in which Einstein introduced a cosmological constant in his theory of gravitation. The range $0 \le \rho < \pi$ is needed to cover the full S^{D-1}, so the AdS metric includes only the upper hemisphere. It is interesting that AdS is conformal, i.e. related by a Weyl transformation, to half of the Einstein static universe.

The form (22.19) of the metric suggests a useful global picture of AdS spacetime. It is the interior of a cylinder, depicted in Fig. 22.2(a) for the case $D = 3$, in which the time axis is vertical, and the radial and angular coordinates vary horizontally. In the D-dimensional case, each interior point represents an entire S^{D-3}. The cylinder appears to have finite radius $\rho_0 = \pi/2$, but this is a deception, since the 'distance' in the coordinate ρ has no invariant meaning. Because the conformal factor in the metric (22.19) is singular as $\rho \to \pi/2$, the invariant length of any space-like curve, such as a curve on which only ρ varies, becomes infinite as the curve approaches the boundary of the cylinder. So the boundary is at infinite spatial distance from any interior point. After a Weyl scaling of the metric (22.19) by the factor $\cos^2 \rho$, one can take the limit $\rho \to \pi/2$ and find that the rescaled metric on the boundary is the usual metric on the cylinder $\mathbb{R} \times S^{D-2}$.

It is significant that a curve (with fixed angular coordinates) that satisfies $\mathrm{d}\tau = \pm \mathrm{d}\rho$ is a null geodesic. This means that light rays propagate at $\pm 45°$ (with respect to the time axis) in the cylinder. A two-dimensional projection of the light cone through the origin is shown in Fig. 22.2(b). Time-like curves remain within the light cone, and space-like curves lie outside it. Furthermore, information from the origin ($\rho = 0$) propagates to spatial infinity ($\rho = \pi/2$) in finite time $\tau = \pi/2$. Conversely, information from spatial infinity reaches the origin in finite time. This means that the Cauchy problem in AdS is ill defined unless additional boundary conditions are specified at spatial infinity. This feature of AdS is crucial to the AdS/CFT correspondence and to the idea of 'holography' in quantum gravity.

There is a very different set of coordinates widely used in the AdS/CFT correspondence, which are called the Poincaré patch coordinates. The patch covers a region of AdS in which the metric is conformal to half of flat Minkowski spacetime. To introduce these coordinates we return to the Cartesian Y^A of (22.14) and define

$$\begin{aligned}
Y^0 &= L u x^0 \,, \\
Y^i &= L u x^i \,, \qquad i = 1, \ldots, D-2 \,, \\
Y^{D-1} &= \frac{1}{2u}\left(-1 + u^2(L^2 - x^2)\right) \,, \\
Y^D &= \frac{1}{2u}\left(1 + u^2(L^2 + x^2)\right) \,, \\
x^2 &= -(x^0)^2 + \sum_i (x^i)^2 \,.
\end{aligned} \tag{22.20}$$

The time coordinate $x^0 = t$ and the spatial x^i range from $-\infty$ to ∞, and $0 < u < \infty$. The restriction on the range of u is needed to maintain a single-valued map of the hyperboloid coordinate Y^D or Y^i.

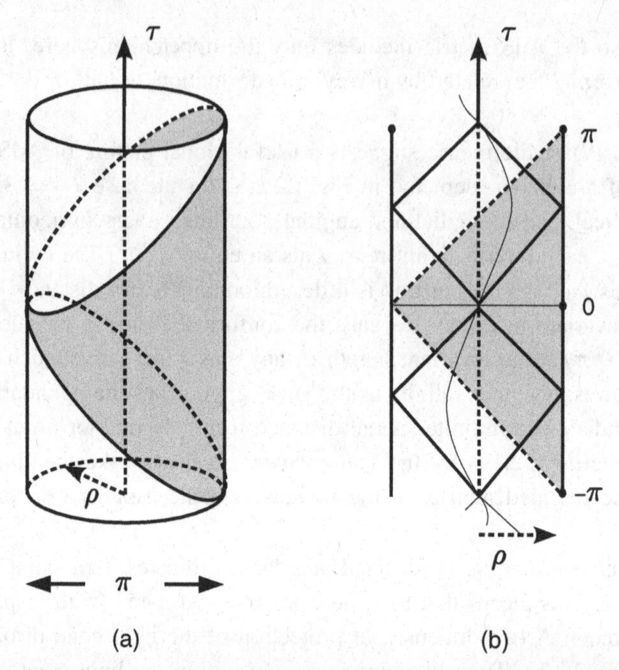

Fig. 22.2 The AdS_3 cylinder. One 'period' of the infinite cylinder is depicted in (a). The Poincaré patch is the interior region bounded by the future and past Cauchy horizons. The boundary of the patch is the shaded 'diamond-shaped' region. A planar section through the center of the cylinder is sketched in (b). The curved line is a typical time-like geodesic through the origin. The solid 45° lines are null geodesics through the origin that reflect from the boundary. The projection of the Poincaré patch on the plane is the shaded triangle.

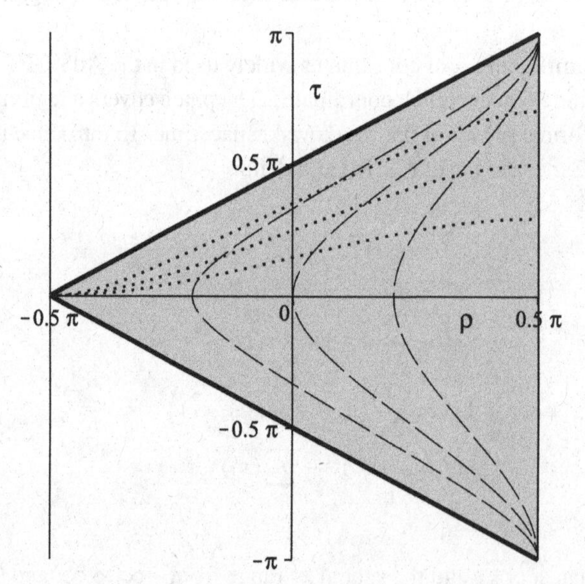

Fig. 22.3 The Poincaré patch of AdS_2: the patch is the shaded triangular region bounded by the past and future horizons. The dashed lines are time-like curves along which patch time t varies for fixed values of z. The dotted lines are the space-like coordinate curves of z at fixed t. The vertical τ-axis is a time-like geodesic that leaves the patch in finite proper time. Other time-like geodesics follow curves similar to those in Fig. 22.2(b).

It is straightforward to show that (22.14) is satisfied by this parametrization, and that the induced metric is

$$ds^2 = L^2 \left[\frac{du^2}{u^2} + u^2 \left(-(dx^0)^2 + \sum_i (dx^i)^2 \right) \right] . \tag{22.21}$$

Exercise 22.7 *Demonstrate these straightforward results.*

Finally we define $z = 1/u$ and rewrite the line element as

$$ds^2 = \frac{L^2}{z^2} \left[dz^2 - (dx^0)^2 + \sum_i (dx^i)^2 \right] . \tag{22.22}$$

This metric is indeed conformal to the positive z region of D-dimensional Minkowski spacetime with its usual coordinates (x^0, x^i, z). In the limit $z \to 0$, one reaches the boundary of the Poincaré patch illustrated in Fig. 22.2. It is the conformal compactification of $(D-1)$-dimensional Minkowski spacetime.

In the simplest case AdS_2, the usual AdS cylinder is replaced by an infinite vertical strip with two distinct boundaries at $\rho = \pm\pi/2$. One segment of this strip is depicted in Fig. 22.3: the Poincaré patch is the shaded triangular wedge with apex at the left boundary. It is instructive to compare global and patch coordinates. From (22.16) (rewritten in terms of the radial coordinate ρ) and (22.20), we find the following relations between global and patch coordinates (we use $x^0 = t$):

$$\begin{aligned} Y^0 &= \frac{L \sin\tau}{\cos\rho} = Lut , \\ Y^1 &= \frac{\sin\rho}{\cos\rho} = \frac{1}{2u}\left(-1 + u^2(L^2 + t^2)\right) , \\ Y^2 &= \frac{L\cos\tau}{\cos\rho} = \frac{1}{2u}\left(1 + u^2(L^2 - t^2)\right) . \end{aligned} \tag{22.23}$$

One can then deduce, using $z = 1/u$, that

$$\begin{aligned} t &= L\frac{\sin\tau}{\cos\tau + \sin\rho} , \\ z &= L\frac{\cos\rho}{\cos\tau + \sin\rho} . \end{aligned} \tag{22.24}$$

Curves of constant patch coordinates are plotted in Fig. 22.3. Note that all 'observers' that follow curves of constant z reach the same boundary point $(\tau = \pi,\ \rho = \pi/2)$ as $t \to \infty$. The lines on which $z = \infty$ are commonly called a horizon. This signifies that the observers cannot receive information from the region above (nor send information to the region below) the horizon. The boundary of the patch is the (disconnected) union of the infinite line $-\infty < t < \infty$ at $\rho = \pi/2$ and the single point $t = 0$ at $\rho = -\pi/2$.

Curves with constant values of z are time-like but non-geodesic. Indeed the time-like geodesics are periodic on the hyperboloid (22.14) and thus best described using global coordinates. One such geodesic is the curve fixed at the spatial origin $r = 0$ of the coordinate chart (22.16). Since any point on the hyperboloid can be brought to the origin, all time-like geodesics can be similarly described, and they all have the same period. Any time-like geodesic crosses the horizon of the Poincaré patch in finite proper time, so the patch is geodesically incomplete.

Because the Poincaré patch metric is conformally flat, it is quite easy to identify all continuous isometries of the patch metric (22.22) of AdS_D. To describe these, we lump x^0, x^i into the $(D-1)$-component vector x^σ, whose indices are contracted with the Minkowski metric. The isometries consist of the following:

(i) translations of the $D - 1$ coordinates x^σ;
(ii) the $(D-1)(D-2)/2$ distinct Lorentz transformations of x^σ;
(iii) a scale transformation of all coordinates, $\delta x^0 = \rho x^0$, $\delta x^i = \rho x^i$, $\delta z = \rho z$, where ρ is an infinitesimal constant parameter;
(iv) $D - 1$ distinct special conformal transformations, $\delta x^\sigma = c^\sigma(x \cdot x + z^2) - 2c \cdot x x^\sigma$, $\delta z = -2c \cdot x\, z$, with constant infinitesimal vector parameter c^σ (with $c^z = 0$).

The total number of independent transformations in this list is $D(D+1)/2$, which is the dimension of the isometry group $\mathrm{SO}(D-1, 2)$. The continuous isometry groups of the global and patch metrics are both the same mathematical group $\mathrm{SO}(D-1, 2)$, but their actions are inequivalent, since the integral curves of the Killing vectors above remain within the patch, while those of the Killing vectors L_{AB} can cross the horizon.

In the limit $z \to 0$ the transformations are those of (15.6). Thus, the isometry group of the Poincaré patch of AdS_D acts as the conformal group of its $\mathrm{Minkowski}_{D-1}$ boundary. This is another fact that is vital for the AdS/CFT correspondence.

Exercise 22.8 *Show that the vector fields for the transformations above are Killing vectors. Show that the Killing vector for the scale transformation (iii) above agrees with the boost* $L_{D(D-1)}$.

22.1.4 Spacetime metrics with spherical symmetry

To see how the geometric 'data' for the sphere S^{D-2} obtained in Sec. 22.1.1 are used in the study of D-dimensional spacetimes, we write a general ansatz for D-dimensional metrics that are static and spherically symmetric:

$$\mathrm{d}s^2 = -\mathrm{e}^{2A(r)}\mathrm{d}t^2 + \mathrm{e}^{2B(r)}\mathrm{d}r^2 + r^2\mathrm{d}\Omega_{D-2}^2\,. \tag{22.25}$$

The spherical metric $\mathrm{d}\Omega_{D-2}^2$ can be written in detail using (22.6) iteratively. However, the schematic form in (22.1) is sufficient for our present purpose, which is to obtain frame, torsion-free connection, and curvature forms for the metric (22.25). We use the Cartan structure equations (7.81) and (7.117). They provide an efficient practical method of computation in many situations. For convenience we rewrite these equations as

$$de^a = -\omega^{ab} \wedge e_b\,, \tag{22.26}$$

$$\rho^{ab} = d\omega^{ab} + \omega^{ac} \wedge \omega_c{}^b\,. \tag{22.27}$$

To begin the process, choose the frame 1-forms $e^t = \exp(A(r))dt$, $e^r = \exp(B(r))dr$, $e^i = r\bar{e}^i$, $i = 1, \ldots, D-2$. The forms $\bar{e}^i$ are any set of 1-forms that depend only on the (angular) coordinates of the sphere, and obey the three properties listed in Sec. 22.1.1. One choice might be the explicit set in (22.7), which was constructed using hyperspherical coordinates, but there are many other possible choices. The metric (22.25) can be rewritten in terms of the frame forms as

$$ds^2 = -e^t\, e^t + e^r\, e^r + \sum_i e^i\, e^i\,. \tag{22.28}$$

The next steps are to determine the (torsion-free) connection 1-forms and curvature 2-forms for the D-dimensional spacetime. The components of the first structure equation (22.26) read

$$\begin{aligned} de^t &= A' \exp(-B)e^r \wedge e^t = -\omega^{tr} \wedge e^r - \omega^{ti} \wedge e^i\,,\\ de^r &= 0 = \omega^{rt} \wedge e^t - \omega^{ri} \wedge e^i\,,\\ de^i &= dr \wedge \bar{e}^i + r d\bar{e}^i = \frac{\exp(-B)}{r} e^r \wedge e^i - \bar{\omega}^{ij} \wedge e^j\\ &= -\omega^{ij} \wedge e^j - \omega^{ir} \wedge e^r + \omega^{it} \wedge e^t\,. \end{aligned} \tag{22.29}$$

All indices are frame indices, which are raised and lowered using the Minkowski metric η_{ab} implicit in (22.28). The connection forms, which are antisymmetric, e.g. $\omega^{tr} = -\omega^{rt}$, can be uniquely determined by solving the equations (22.29). Given the connection 1-forms, it is straightforward to use (7.117) to compute the curvature 2-forms. We leave these tasks as an exercise for readers and summarize the results.

Summary: For the frame 1-forms

$$e^t = \exp(A(r))dt\,, \qquad e^r = \exp(B(r))dr\,, \qquad e^i = r\bar{e}^i\,. \tag{22.30}$$

The connection 1-forms are given by

$$\begin{aligned} \omega^{tr} &= A' \exp(-B)e^t\,, & \omega^{ti} &= 0\,,\\ \omega^{ir} &= (\exp(-B)/r)e^i\,, & \omega^{ij} &= \bar{\omega}^{ij}\,. \end{aligned} \tag{22.31}$$

The components of the curvature 2-form are

$$\begin{aligned} \rho^{tr} &= -\left(A'' - A'B' + A'^2\right) \exp(-2B)e^t \wedge e^r\,,\\ \rho^{ti} &= -\frac{A'}{r} \exp(-2B)e^t \wedge e^i\,,\\ \rho^{ri} &= \frac{B'}{r} \exp(-2B)e^r \wedge e^i\,,\\ \rho^{ij} &= \bar{\rho}^{ij} - \frac{1}{r^2} \exp(-2B)e^i \wedge e^j = (1 - \exp(-2B))\frac{1}{r^2} e^i \wedge e^j\,. \end{aligned} \tag{22.32}$$

Exercise 22.9 *It is useful practice to learn to solve the first Cartan structure equation, so we ask readers to set aside the results (22.31) and try to find them as solutions of the various equations in (22.29). Start with the first equation and find 'reasonable' expressions for ω^{tr} and ω^{ti}. Then go on to the remaining equations. Some guesswork may be involved in the first steps, but there is a unique solution, so one can be assured that the answer is correct once all equations are satisfied.*

In most situations the metric (22.25) must satisfy the Einstein field equations with or without matter sources. Thus we will need the components of the Ricci tensor. We work with the frame components $R^a_b = R^{ac}{}_{bc}$, obtained by contraction with frame components of the curvature tensor. The latter are obtained from the curvature 2-form via

$$\rho^{ab} = \frac{1}{2} R^{ab}{}_{cd} e^c \wedge e^d \,. \tag{22.33}$$

We record the non-vanishing Ricci tensor components

$$\begin{aligned} R^t_t &= -\left[A'' - A'B' + A'^2 + (D-2)\frac{A'}{r}\right]\exp(-2B)\,, \\ R^r_r &= -\left[A'' - A'B' + A'^2 - (D-2)\frac{B'}{r}\right]\exp(-2B)\,, \\ R^i_j &= \delta^i_j\left[(D-3)(1-\exp(-2B))\frac{1}{r^2} + \frac{B'-A'}{r}\exp(-2B)\right]. \end{aligned} \tag{22.34}$$

Exercise 22.10 *This is another computation that will give readers useful practice.*

22.1.5 AdS–Schwarzschild spacetime

The main purpose of this section is to obtain the AdS–Schwarzschild metrics, which are static and spherically symmetric metrics that describe a black hole in AdS space. They have important applications in the AdS/CFT correspondence. These metrics are solutions of the equations (22.11), which means that they are Einstein metrics.

The Einstein space conditions are $R^t_t = R^r_r = \delta^j_i R^i{}_j/(D-2) = -(D-1)/L^2$. The first equality tells us that $A' = -B'$. By choice of scale of the time coordinate, we choose $A(r) = -B(r)$. This gives us the differential equation

$$\frac{\mathrm{d}}{\mathrm{d}r}(1-\exp(-2B)) + \frac{D-3}{r}(1-\exp(-2B)) = -\frac{(D-1)}{L^2}r\,. \tag{22.35}$$

The solution is

$$\exp(-2B) = 1 + r^2/L^2 - (r_0/r)^{D-3} \equiv h(r)\,. \tag{22.36}$$

The reader is invited to check that R^t_t and R^r_r also satisfy the Einstein space conditions.

We have now found the AdS–Schwarzschild metric in its usual presentation:

$$ds^2 = -h(r)dt^2 + \frac{1}{h(r)}dr^2 + r^2 d\Omega_{D-2}^2 \,. \tag{22.37}$$

Notice that the power law $1/r^{(D-3)}$ in $h(r)$ is that of the gravitational potential of a point mass in flat D-dimensional spacetime. The parameter r_0 is proportional to the mass of the black hole. The function $h(r)$ has a single zero at a finite value of r and is singular at $r = 0$. To interpret this one may evaluate the Kretschmann invariant $R_{\mu\nu\rho\sigma}R^{\mu\nu\rho\sigma}$, which is singular at $r = 0$ and otherwise regular. Thus $r = 0$ is a curvature singularity. The zero of $h(r)$ is a regular point, which is the event horizon. A light ray emitted inside the horizon can never escape. The coordinate displacement dr is space-like outside the horizon but time-like inside, while dt switches from time-like to space-like. The point $r = 0$ is called a space-like singularity because the nearby tangent space contains only space-like vectors. We refer readers to general relativity texts for more extensive discussions of the physics of black holes.

In the limit $L \to \infty$ the metric (22.37) becomes asymptotically flat and describes a black hole in Minkowski spacetime. This is the Schwarzschild solution, first found [283] in 1916. In the limit $r_0 \to 0$, we find the AdS$_D$ in the form (22.17).

Exercise 22.11 *Use (7.116) to compute the curvature tensor components $R^{tr}{}_{tr}$ and $R^{ti}{}_{tj}$ for the metric (22.37). Observe that they do* not *satisfy the conditions of maximal symmetry unless the black hole mass vanishes.*

22.1.6 The Reissner–Nordström metric

This metric [284, 285] describes a non-rotating electrically charged black hole. It is a solution to the equations of motion of the Maxwell–Einstein action

$$S = \int d^D x \sqrt{-g}\left(\frac{1}{2\kappa^2}R - \frac{1}{4}F_{\mu\nu}F^{\mu\nu}\right), \tag{22.38}$$

which are

$$\begin{aligned}
\partial_\mu(\sqrt{-g}g^{\mu\rho}g^{\nu\sigma}F_{\rho\sigma}) &= 0\,,\\
R_{\mu\nu} &= \kappa^2\left[T_{\mu\nu} - \frac{1}{D-2}g_{\mu\nu}T^\rho_\rho\right],\\
T_{\mu\nu} &\equiv F_{\mu\rho}F_\nu{}^\rho - \frac{1}{4}g_{\mu\nu}F_{\rho\sigma}F^{\rho\sigma}\,.
\end{aligned} \tag{22.39}$$

The Maxwell–Einstein system is well known in general relativity and in $\mathcal{N} = 2$, $D = 4$ and $D = 5$ supergravity.

The solution is spherically symmetric and static, so we work within the metric ansatz (22.25) and use the frame components of the Ricci tensor in (22.34). The non-vanishing components of the gauge field are $F_{tr} = -F_{rt} = f(r)$, which leads to $F_{\rho\sigma}F^{\rho\sigma} =$

$-2e^{-2(A+B)}f^2$. The non-vanishing components of $F_{\mu\rho}F_\nu{}^\rho$ are $(FF)_{tt} = e^{-2B}f^2$ and $(FF)_{rr} = -e^{-2A}f^2$.

Exercise 22.12 *Reexpress this information in frame components and use (22.34) to derive $A' = -B'$ as in the previous section. Show that the curved space Maxwell equation in (22.39) becomes $\partial_r(r^{D-2}f(r)) = 0$, which has the solution*

$$F_{tr} = f(r) = -\frac{q}{\Omega_{D-2}r^{D-2}}, \tag{22.40}$$

where Ω_{D-2} is the volume of the $(D-2)$-dimensional sphere (see (22.9)), which is 4π in the case of $D = 4$. Check that, with this normalization, q is the electric charge defined as in (7.64).

There remains to solve the (frame-component) equation

$$\begin{aligned} R^i_j &= \delta^i_j\left[(D-3)(1-\exp(-2B))\frac{1}{r^2} + \frac{2B'}{r}\exp(-2B)\right] \\ &= \delta^i_j\frac{\kappa^2 q^2}{(D-2)(\Omega_{D-2}r^{D-2})^2}, \end{aligned} \tag{22.41}$$

and the solution is

$$F(r) \equiv \exp(-2B) = 1 - \left(\frac{r_0}{r}\right)^{D-3} + \frac{\kappa^2 q^2}{(D-2)(D-3)(\Omega_{D-2}r^{D-3})^2}. \tag{22.42}$$

The Reissner–Nordström metric in D-dimensional spacetime is then

$$ds^2 = -F(r)dt^2 + \frac{1}{F(r)}dr^2 + r^2 d\Omega^2_{D-2}. \tag{22.43}$$

Let's focus on the four-dimensional case. The radial function in the line element can then be written in terms of the black hole mass M and Newton's constant $G = \kappa^2/8\pi$ (see (A.11)) as

$$F(r) = 1 - \frac{2MG}{r} + \frac{q^2 G}{4\pi r^2}. \tag{22.44}$$

This function has roots at

$$r = r_\pm = MG \pm \sqrt{M^2G^2 - \frac{q^2G}{4\pi}}, \tag{22.45}$$

and it is infinite at $r = 0$. This point is a physical singularity, since the invariants $R_{\mu\nu\rho\sigma}R^{\mu\nu\rho\sigma}$ as well as $R_{\mu\nu}R^{\mu\nu}$ and $F_{\mu\nu}F^{\mu\nu}$ are singular there.

For small charge $q^2 < 4\pi M^2 G$ the roots $r_\pm$ are real. They are the inner and outer horizons, which surround the singularity. The three invariants are non-singular at the horizons. The outer horizon is the event horizon. A light ray emitted anywhere within this sphere cannot escape. As one moves inwards across the outer horizon the displacement dt changes from time-like to space-like, as in the Schwarzschild case, but it changes back to time-like at the inner horizon. The singularity at $r = 0$ is thus time-like because the tangent space contains a time-like vector.

If $q^2 > 4\pi M^2 G$, the roots $r_\pm$ become imaginary. The singularity is no longer shielded by a horizon and is called a naked singularity. It is common to encounter mathematical solutions with naked singularities, but it is considered to be physically pathological. The cosmic censorship conjecture asserts, with considerable support in theoretical studies, that naked singularities cannot occur as the endpoint of time evolution with generic initial data. Classical solutions with naked singularities are pathological because the arbitrarily strong fields nearby are accessible to an external observer. The classical theory is thus unreliable, so quantum effects must be taken into account; one may hope that they resolve the singularity.

For $q^2 = 4\pi M^2 G$, the inner and outer horizons coincide and we find the extremal Reissner–Nordström solution with metric

$$ds^2 = -(1 - MG/r)^2 dt^2 + (1 - MG/r)^{-2} dr^2 + r^2 d\Omega_2^2 . \tag{22.46}$$

We will study this metric in some detail because it has special significance as a BPS solution of supergravity. The horizon is located at $r = MG$, and it is convenient to shift the radial coordinate by introducing $v = r - MG$. The horizon is now at $v = 0$, and the metric becomes

$$ds^2 = -(1 + MG/v)^{-2} dt^2 + (1 + MG/v)^2 [dv^2 + v^2 d\Omega_2^2] . \tag{22.47}$$

The near-horizon limit of this metric, namely the limiting form as $v \to 0$, is

$$ds^2 \approx -\frac{v^2}{(MG)^2} dt^2 + (MG)^2 \frac{dv^2}{v^2} + (MG)^2 d\Omega_2^2 . \tag{22.48}$$

Finally we define the coordinate $z = (MG)^2/v$ and rewrite the near-horizon metric as

$$ds^2 \approx \frac{(MG)^2}{z^2} [-dt^2 + dz^2] + (MG)^2 d\Omega_2^2 . \tag{22.49}$$

We compare with (22.3) and (22.22) and learn that this line element describes the direct product of two manifolds, namely $\mathrm{AdS}_2 \otimes S^2$. Both the AdS scale L and the radius of the sphere r_S are equal, namely $L = r_S = MG$. The metric (22.49) is known as the Robinson–Bertotti metric.

22.1.7 A more general Reissner–Nordström solution

There is a more general static spherically symmetric solution of the system of gravity coupled to an electromagnetic field. It describes a black hole that carries both electric and magnetic charge. Let's find this solution using the ideas of electromagnetic duality (discussed more generally in Sec. 4.2).

Let us start with a four-dimensional spacetime with metric tensor $g_{\mu\nu}$ and an antisymmetric 2-tensor $F_{\mu\nu}$. For the moment we do not assume that these quantities satisfy any particular equations of motion. We use a gauge field configuration that is the superposition of electric and magnetic fields with real charges q and p, i.e.

$$\mathcal{F}_{\mu\nu} = qF_{\mu\nu} + \mathrm{i}p\tilde{F}_{\mu\nu} = qF_{\mu\nu} + p\,{}^*F_{\mu\nu}\,, \qquad \text{i.e.}\quad \mathcal{F}^+_{\mu\nu} = (q+\mathrm{i}p)F^+_{\mu\nu}\,, \tag{22.50}$$

where we wrote also the expression in terms of the Hodge dual (7.59).

Exercise 22.13 *Show that the electromagnetic stress tensor of $\mathcal{F}$ satisfies*

$$\begin{aligned} T_{\mu\nu}(\mathcal{F}) &\equiv \mathcal{F}_{\mu\rho}\mathcal{F}_\nu{}^\rho - \tfrac{1}{4}g_{\mu\nu}\mathcal{F}_{\rho\sigma}\mathcal{F}^{\rho\sigma} \\ &= (q^2+p^2)\big(F_{\mu\rho}F_\nu{}^\rho - \tfrac{1}{4}g_{\mu\nu}F_{\rho\sigma}F^{\rho\sigma}\big)\,. \end{aligned} \tag{22.51}$$

The exercise shows that the stress tensor of any gauge field configuration is the same as the stress tensor for its dual.[3] *This means that the gravitational effects of the two configurations are identical.*

For any $F^{\mu\nu}$ that satisfies the Maxwell equation in (22.39) and the Bianchi identity, the dual tensor also enjoys the same properties (see Sec. 4.2.2), and so does the superposition $\mathcal{F}_{\mu\nu}$. In the previous section we considered a purely electric $F_{\mu\nu}$ in a static spherically symmetric spacetime. With unit charge we have $F^{tr} = -F_{tr} = 1/(4\pi r^2)$ with other components vanishing. The dual tensor describes a unit point magnetic charge whose only non-vanishing component is $\tilde{F}_{\theta\phi} = -\mathrm{i}\sin\theta/4\pi$.

Exercise 22.14 *Verify that this is the correct dual. Show that for $r > 0$ it satisfies $\partial_\mu(\sqrt{-g}\,{}^*F^{\mu\nu}) = 0$ and $\partial_\mu{}^*F_{\nu\rho} + \partial_\nu{}^*F_{\rho\mu} + \partial_\rho{}^*F_{\mu\nu} = 0$.*

We can now deduce the form of the general Reissner–Nordström metric with no further work. Since the stress tensor (22.51) is proportional to q^2+p^2, the metric (22.43) that we found earlier is still correct if we make the minor modification

$$F(r) = 1 - \frac{2MG}{r} + \frac{(q^2+p^2)G}{4\pi r^2}\,. \tag{22.52}$$

The previous discussion of the curvature singularity and horizons of $F(r)$ also applies to the more general case. The only significant change is that the upper bound on the charges to avoid a naked singularity becomes

[3] Note that this is the transformation discussed already in Ex. 4.15 with $A = q$ and $B = -p$. The result (22.51) is a straightforward extension of (4.61) to curved spacetime.

$$q^2 + p^2 \leq 4\pi M^2 G\,. \tag{22.53}$$

It is interesting that this bound obtained from the cosmic censorship conjecture agrees with the BPS bound of $\mathcal{N} = 2$ supergravity; see Sec. 12.6.2.

22.2 Killing spinors and BPS solutions

Killing spinors are one of the most interesting concepts to emerge from the development of supergravity. Supergravity actions are invariant under SUSY transformations with arbitrary spinor functions $\epsilon(x)$. Killing spinors are associated with special classical solutions of the equations of motion of a supergravity theory. Specifically they are the finite subset of the spinor functions for which the SUSY transformation leaves the solution invariant, i.e. unchanged from its original form. These spinors contain a finite number of *constant parameters*, and they determine the residual *global supersymmetry algebra* of the solution; see Sec. 12.6. In favorable cases it is easier to find classical solutions of the equations of motion by studying the conditions for Killing spinors rather than the equations of motion themselves. The Killing spinor conditions come directly from the fermion transformation rules of supergravity. The conditions determine the spinors themselves, and they give information on the spacetime geometries that support them. Many interesting solutions have been discovered by studying these conditions. There is also an interesting relation between Killing spinors and Killing vectors, which reflects the structure of the supersymmetry algebra.

There are similarities between Killing spinor analysis and the approach of Bogomol'nyi, Prasad and Sommerfield (BPS) [286, 287] to magnetic monopoles. In both formalisms, components of bosonic fields are related by first order differential equations. For this reason a supergravity solution that has Killing spinors is frequently called a BPS solution.

To motivate our discussion let's set supersymmetry and supergravity temporarily aside and consider a purely bosonic theory of gravity coupled to scalar or vector fields. One example is the Maxwell–Einstein theory with action (22.38). The action is invariant under general coordinate transformations with arbitrary infinitesimal parameters $\xi^\mu(x)$. The simplest classical solution of the system is D-dimensional Minkowski spacetime with metric $g_{\mu\nu} = \eta_{\mu\nu}$ and vanishing gauge field $F_{\mu\nu} = 0$. The solution is invariant, unchanged from

Killing spinors and residual global supersymmetry algebra — **Box 22.1**

Killing spinors are the parameters of preserved supersymmetry of a solution. When they are characterized by n_Q constants, we say that there are n_Q preserved supersymmetries. Bilinears of the Killing vectors are Killing vectors, which determine bosonic preserved symmetries.

the form just specified, only for the well-known Killing vectors of Minkowski spacetime, which are a finite dimensional subset of the $\xi^\mu(x)$; see (11.46). These Killing vectors determine the isometry group of the spacetime; in this case it is the Poincaré group in D dimensions. In Sec. 22.1.4 we wrote the ansatz (22.25) for a D-dimensional static and spherically symmetric spacetime metric. Again it is invariant only under a subset of coordinate transformations, which contains the Killing vectors for time translations and spatial rotations. In many situations one is interested in the fluctuations about a particular classical background solution. The dynamics of these fluctuations is covariant under the isometry group.

The situation in supergravity is analogous. Let's consider the basic $\mathcal{N} = 1$, $D = 4$ supergravity theory of Ch. 9. Again the simplest classical solution is Minkowski spacetime with $(g_{\mu\nu})_0 = \eta_{\mu\nu}$ and vanishing gravitino field $(\psi_\mu)_0 \equiv 0$. The action is invariant under the transformation rules (9.3) and (9.4), in which $\epsilon(x)$ is arbitrary, and the spin connection in D_μ includes a torsion correction quadratic in ψ_μ. The residual global SUSY algebra is determined by the conditions

$$\delta e^a_\mu = \tfrac{1}{2}\bar{\epsilon}\gamma^a\psi_\mu = 0\,, \qquad \delta\psi_\mu = D_\mu\epsilon = 0\,, \tag{22.54}$$

in which we insert the background values of the frame field $(e^a_\mu)_0 = \delta^a_\mu$ and gravitino $(\psi_\mu)_0 = 0$. These conditions express the requirement that the residual SUSY transformations do not change the background fields. Since $(\psi_\mu)_0 = 0$, the first condition of (22.54) is automatically satisfied. In the second condition the spin connection (including torsion) vanishes in the background, so the condition becomes simply $\partial_\mu\epsilon = 0$. We conclude that the Killing spinors of the Minkowski background are the set of four independent *constant* Majorana spinors ϵ_α.

At this point we recall that the parameters of global SUSY transformations are constant Majorana spinors. Further we know that there is a particle representation of the SUSY algebra consisting of massless spin-2 and spin-3/2 particles. With this evidence we infer that the symmetry algebra of fluctuations about the Minkowski background is the $D = 4$ Poincaré SUSY algebra. The Minkowski background, with vanishing matter fields, frequently occurs as the simplest classical solution of supergravity coupled to $\mathcal{N} = 1$ gauge and chiral multiplets. It is no surprise that the Poincaré SUSY algebra governs these systems. The Minkowski background is called maximally supersymmetric because there are four independent Killing spinors and thus four conserved supercharges.

Finally we note that, for any pair of Killing spinors ϵ and ϵ', the bilinear $(\bar{\epsilon}'\gamma^\mu\epsilon)$ is a constant vector field and thus a Killing vector for spacetime translations. It is the translation Killing vectors that occur, rather than rotations, because the anti-commutator of two global SUSY transformations closes on spacetime translations.

BPS solutions can be quite complex. They include multiple charged and multi-center black holes, intersecting D-branes and more. But the pattern is the same in all examples. One searches for a preserved supersymmetry of a classical solution involving bosonic fields such as the metric, gauge fields and scalars, while the fermion fields in the same theory, gravitinos and spin-1/2 fields, vanish. Let's denote the bosonic fields collectively by $B(x)$ and the fermions by $F(x)$. The local SUSY transformations of the theory may be written in the schematic form

$$\delta B(x) = \bar{\epsilon}(x)\, f_1(B(x))F(x) + \mathcal{O}(F^3), \qquad \delta F(x) = f_2(B(x))\epsilon(x) + \mathcal{O}(F^2). \quad (22.55)$$

The functions $f_1(B)$, $f_2(B)$ typically include Dirac γ-matrices and spacetime derivatives. For each preserved supersymmetry there is a non-trivial configuration $\epsilon(x)$ for which both $\delta B(x)$ and $\delta F(x)$ *vanish* when the classical solution is substituted in (22.55). Since the fermions vanish classically, the condition $\delta B(x) = 0$ is trivially satisfied, so we need to study only the *linearized* fermion variations

$$\delta F(x)|_{\text{lin}} \equiv f_2(B(x))\epsilon(x) = 0. \quad (22.56)$$

These conditions are a set of coupled first order differential equations involving the fields $B(x)$ and the spinors $\epsilon(x)$. When there is a solution, it will determine a set of n_Q linearly independent spinor configurations of $\epsilon(x)$. Each independent $\epsilon(x)$ is a Killing spinor, and we then say that the BPS solution preserves n_Q supercharges. In practice, the conditions (22.56) also determine much of the information on the field configuration $B(x)$, so it is not necessary to solve the full set of Lagrangian equations of motion. Examples of the solution of (22.56) will be presented in Secs. 22.3 and 22.4.

22.2.1 The integrability condition for Killing spinors

Let's look at the Killing spinor condition $D_\mu \epsilon = 0$ more closely. Note that covariant derivatives in non-trivial spacetimes, those with non-vanishing Riemann tensor, generically do not commute. Non-commutativity is expressed by the Ricci identity, which was discussed in Ch. 7; see (7.123). For spinors it reads

$$[D_\mu, D_\nu]\epsilon = \tfrac{1}{4} R_{\mu\nu ab}\gamma^{ab}\epsilon = 0. \quad (22.57)$$

This algebraic condition expresses the mutual compatibility of the vector components of $D_\mu \epsilon = 0$. It is usually called the integrability condition for Killing spinors, and it is a necessary condition for the existence of such spinors. Integrability is a trivial issue for Minkowski spacetime since the curvature tensor vanishes, but it is a very strong constraint on the geometry of any curved spacetime. The issue of integrability and the Killing spinors for pp-wave spacetimes are discussed in Appendix 22A.

22.2.2 Commuting and anti-commuting Killing spinors

Spinor fields in supergravity actions are anti-commuting quantities, and so are the transformation parameters $\epsilon(x)$. A general supergravity theory contains both gravitinos $\psi_\mu(x)$ and ordinary fermions $\chi(x)$. Killing spinors are determined from the fermion transformation rules $\delta\psi_\mu = 0$ and $\delta\chi = 0$, evaluated in a purely bosonic field configuration. The equations are linear in ϵ but contain no other fermionic quantities. There is nothing to determine whether the Killing spinors are commuting or anti-commuting. Thus we can view the Killing spinor conditions as a set of ordinary differential and algebraic conditions whose solution, when it exists, is expressed in terms of ordinary commuting quantities. However, when the Killing spinors are applied to study fluctuations about a BPS background, they must be converted to anti-commuting. To do this one simply assumes that the independent constant parameters in the Killing spinors are Grassmann numbers.

22.3 Killing spinors for anti-de Sitter space

Anti-de Sitter space first appeared in supergravity as the maximally supersymmetric solution (i.e. four independent Killing spinors) of the cosmological extension of the $\mathcal{N}=1$, $D=4$ theory, which was discussed in Sec. 9.6. However it also occurs for $D>4$, for example, in Kaluza–Klein reductions of 10- and 11-dimensional supergravity. The five-dimensional case is essential for the AdS/CFT correspondence. In this section we will obtain the Killing spinors for AdS_D, which is maximally supersymmetric for any spacetime dimension.

The bosonic action that leads to AdS space is given in (22.10); see also Sec. 9.6. Killing spinors are solutions of the equation

$$\hat{D}_\mu \epsilon \equiv \left(D_\mu - \frac{1}{2L}\gamma_\mu \right)\epsilon = 0 . \tag{22.58}$$

In the case $\mathcal{N}=1$, $D=4$ this is exactly the gravitino transformation rule (9.48) (with vanishing background ψ_μ), and ϵ is a Majorana spinor. However, (22.58) applies in every dimension; in each case ϵ is a specific type of spinor, e.g. symplectic-Majorana in $D=5$. The information we need does not require a detailed specification, so we assume generically that ϵ is a complex (Dirac) spinor.

It is informative to examine the integrability condition for (22.58). This reads as in (9.49):

$$[\hat{D}_\mu\,,\,\hat{D}_\nu]\epsilon = \left(\frac{1}{4}R_{\mu\nu ab}\gamma^{ab} + \frac{1}{2L^2}\gamma_{\mu\nu} \right)\epsilon\,. \tag{22.59}$$

Since AdS_D is a maximally symmetric spacetime, we insert the form $R_{\mu\nu ab} = -(e_{a\mu}e_{b\nu} - e_{a\nu}e_{b\mu})/L^2$ and find that the right-hand side of (22.59) vanishes *identically*. This is a strong hint that AdS_D is maximally supersymmetric with a full set of $2^{[D/2]}$ Killing spinors. We will obtain the explicit form of these spinors shortly.

Exercise 22.15 *Derive the integrability condition (22.59). Assume that ϵ and ϵ' are both Killing spinors. Show that $\nabla_\mu\left(\bar{\epsilon}'\gamma_\nu\epsilon\right) = -(1/L)\bar{\epsilon}'\gamma_{\mu\nu}\epsilon$. Therefore the vector $K_\nu = \bar{\epsilon}'\gamma_\nu\epsilon$ satisfies $\nabla_\mu K_\nu + \nabla_\nu K_\mu = 0$. This exercise shows that the γ_ν bilinear of any pair of AdS Killing spinors is a Killing vector.*

We will study Killing spinors in the Poincaré patch of AdS_D. Killing spinors for global coordinate charts are known for $D=4$ (see [184]), and also for general D (see [288]). It is convenient to follow [289] and transform the radial coordinate in the Poincaré metric (22.22) by $z = L\mathrm{e}^{-r/L}$ so that the boundary appears in the limit $r\to\infty$. The metric becomes

$$\mathrm{d}s^2 = \mathrm{e}^{2r/L}\eta_{\mu\nu}\mathrm{d}x^\mu\mathrm{d}x^\nu + \mathrm{d}r^2\,. \tag{22.60}$$

The indices of the transverse coordinates x^μ take values $\mu = 0,1,\ldots,D-2$. The radial coordinate is referred to as r with no assigned numerical value.

To find Killing spinors use frame 1-forms

$$e^{\hat{\mu}} = \mathrm{e}^{r/L}\mathrm{d}x^\mu\,, \qquad e^r = \mathrm{d}r\,. \tag{22.61}$$

The connection forms are

$$\omega^{\hat{\mu}r} = \frac{1}{L}e^{\hat{\mu}}\,, \qquad \omega^{\hat{\mu}\hat{\nu}} = 0\,. \tag{22.62}$$

Exercise 22.16 *As a check on these data show that the curvature 2-form is maximally symmetric, i.e.* $\rho^{ab} = -e^a \wedge e^b/L^2$.

The γ-matrices are the constant γ^a of flat D-dimensional spacetime. We label them $\gamma^{\hat{\mu}}$, $\hat{\mu} = 0, 1, \ldots, D-2$, and γ^r. The Killing spinor conditions split into radial and transverse components:

$$\hat{D}_r\epsilon = \left(\partial_r - \frac{1}{2L}\gamma_r\right)\epsilon = 0\,,$$
$$\hat{D}_\mu\epsilon = \left(\partial_\mu + \frac{1}{2L}\gamma_\mu(\gamma_r - 1)\right)\epsilon = 0\,. \tag{22.63}$$

To solve them we introduce constant spinors $\eta_\pm$ that satisfy $\gamma_r\eta_\pm = \pm\eta_\pm$. It is then not difficult to see that the Killing spinors are

$$\epsilon_+ = \mathrm{e}^{r/2L}\eta_+\,,$$
$$\epsilon_- = \left(\mathrm{e}^{-r/2L} + \frac{1}{L}\mathrm{e}^{r/2L}x^\mu\gamma_{\hat{\mu}}\right)\eta_-\,. \tag{22.64}$$

The last term includes a sum over the transverse indices. Note that ϵ_- is not an eigenspinor of γ_r.

Exercise 22.17 *The purpose of this exercise is to relate the various* AdS_D *isometries listed in Sec. 22.1.3 to the Killing spinor bilinears* $\bar{\epsilon}'\gamma^M\epsilon$, *where* M *refers to the coordinate indices* μ *and* r. *Use the relation* $\delta z = -z\delta r/L$. *Compute all* $\delta x^M \equiv \bar{\epsilon}'\gamma^M\epsilon$ *and show that they correspond as follows to the* AdS_D *isometries:*

(i) translations $\quad \delta x^\mu = a^{\hat{\mu}} = \bar{\eta}'_+\gamma^{\hat{\mu}}\eta_+\,, \qquad \delta z = 0;$

(ii) Lorentz $\quad \delta x^\mu = \lambda^{\hat{\mu}}{}_{\hat{\nu}}x^\nu\,, \qquad \lambda^{\hat{\mu}\hat{\nu}} = \frac{1}{L}\left(\bar{\eta}'_+\gamma^{\hat{\mu}\hat{\nu}}\eta_- + \bar{\eta}'_-\gamma^{\hat{\mu}\hat{\nu}}\eta_+\right)\,, \qquad \delta z = 0;$

(iii) scale $\quad \delta x^\mu = \rho x^\mu\,, \qquad \delta z = \rho z\,, \qquad \rho = \frac{1}{L}\left(\bar{\eta}'_+\eta_- - \bar{\eta}'_-\eta_+\right);$

(iv) special conformal $\quad \delta x^\mu = c^{\hat{\mu}}(x\cdot x + z^2) - 2c\cdot x x^\mu, \quad \delta z = -2c\cdot x\, z,$ *with* $c^{\hat{\mu}} = \frac{1}{L^2}\bar{\eta}'_-\gamma^{\hat{\mu}}\eta_-$.

Hint: since $\bar{\eta}_\pm\gamma_r = \mp\bar{\eta}_\pm$, *the bilinear* $\bar{\eta}'_+\eta_+$ *vanishes. The* δx^M *above satisfy the differential equations for AdS Killing vectors in any dimension* D, *but they are not necessarily real. They are real in spacetime dimensions 2, 3, 4 mod 8, if the spinors* η, η' *are chosen as anti-commuting Majorana spinors. Comparing with the superconformal algebra of Sec. 16.1.1, one can identify* η_+ *and* η_- *with the parameters of* Q- *and* S-*supersymmetry, respectively.*

22.4 Extremal Reissner–Nordström spacetimes as BPS solutions

In this section we will investigate the possibility of BPS solutions of the Maxwell–Einstein system (22.38). We will use coordinates $t, x^i, i = 1, 2, 3$, and assume that the static spacetime metric takes the form

$$\mathrm{d}s^2 = -\mathrm{e}^{2U(x)}\mathrm{d}t^2 + \mathrm{e}^{-2U(x)}\mathrm{d}x^i\mathrm{d}x^i\,. \tag{22.65}$$

Initially we assume a purely electric gauge field with non-vanishing components $F_{ti}(x) = -\partial_i A_t(x)$. The notation $U(x)$ indicates that the function is independent of t.

We will obtain an interesting generalization of the Reissner–Nordström solution called the Papapetrou–Majumdar spacetime in which the scale factor and gauge potential are

$$\begin{aligned}\mathrm{e}^{-U} &= 1 + \sum_{I=1}^{n}\frac{M_I G}{|\vec{x}-\vec{x}_I|}\,,\\ A_t &= \pm\frac{1}{\sqrt{4\pi G}}\mathrm{e}^{U}\,.\end{aligned} \tag{22.66}$$

Note that the Maxwell equation in the metric (22.65) reads

$$\frac{1}{\sqrt{2}}\kappa\,\partial_i\left(\sqrt{-g}g^{ij}g^{tt}\partial_j A_t\right) = \partial_i(\mathrm{e}^{-2U}\partial_i\mathrm{e}^{U}) = -\nabla^2\mathrm{e}^{-U} = 0\,. \tag{22.67}$$

Thus, e^{-U} is a harmonic function (with respect to Cartesian coordinates x^i). If we choose the function $\mathrm{e}^{-U} = 1 + MG/|\vec{x}|$, then the metric (22.65) agrees with extremal Reissner–Nordström in the form (22.47) with radial coordinate $v = |\vec{x}|$, and charge $q = \mp\sqrt{4\pi G}M$. The more general point-singular function in (22.66) describes a 'multi-center' collection of n extremal Reissner–Nordström black holes with horizons located at $\vec{x} = \vec{x}_I$. The system remains stable and thus time independent because gravitational attraction and electric repulsion cancel between each pair of black holes. It can be shown that all mass parameters must have positive sign to avoid naked singularities.

We now want to see how the Papapetrou–Majumdar solution emerges as a BPS solution of $\mathcal{N} = 2$ supergravity and to find its Killing spinors. We follow [290]. The pure supergravity theory contains the graviton and Maxwell fields and two Majorana gravitinos, and the transformation rules involve two Majorana spinor parameters. It is convenient to use the notation of Sec. 6A and Ch. 20 and work with the chiral projections of the two Majorana spinors. We thus use the up/down position of the R-symmetry indices, now denoted by $A, B = 1, 2$, to specify the chirality. Thus, for the SUSY transformation parameters, we have

$$\gamma_*\epsilon^A = P_L\epsilon^A = \epsilon^A\,, \qquad \gamma_*\epsilon_A = -P_R\epsilon_A = -\epsilon_A\,, \tag{22.68}$$

with similar notation for gravitinos. Each Majorana spinor contains two independent complex or four real components, so there are eight independent real parameters for $\mathcal{N}=2$ SUSY.

Killing spinors are determined by the condition that the linearized gravitino transformation rule vanishes. We discussed this model in Ex. 21.3 and obtained the transformation rule (21.49):

$$\delta\psi_{\mu A} = \left(\partial_\mu + \tfrac{1}{4}\gamma^{ab}\omega_{\mu ab}\right)\epsilon_A - \tfrac{1}{8}\sqrt{2}\kappa F_{ab}\gamma^{ab}\gamma_\mu\varepsilon_{AB}\epsilon^B = 0\,. \tag{22.69}$$

Notice that F_{ab} can be replaced above by the self-dual F^+_{ab}. The reason is that $\delta\psi_{\mu A}$ has the chirality of P_R and so the argument of (14.18) applies.

To analyze the condition (22.69) we work in the frame

$$e^{\hat{0}} = \mathrm{e}^U \mathrm{d}t\,, \qquad e^{\hat{i}} = \mathrm{e}^{-U}\mathrm{d}x^i\,, \tag{22.70}$$

in which the connection 1-forms are

$$\omega^{\hat{0}\hat{i}} = \mathrm{e}^U\partial_i U e^{\hat{0}}\,, \qquad \omega^{\hat{i}\hat{j}} = -\mathrm{d}x^i\partial_j U + \mathrm{d}x^j\partial_i U. \tag{22.71}$$

Notice that $F_{\hat{0}\hat{i}} = F_{ti} = -\partial_i A_t$.

Most of the information we need can be found from the time component of (22.69), which reads

$$\delta\psi_{tA} = \partial_t\epsilon_A + \tfrac{1}{2}\mathrm{e}^{2U}\partial_i U\gamma^{\hat{i}}\gamma^{\hat{0}}\epsilon_A - \tfrac{1}{4}\sqrt{2}\kappa\mathrm{e}^U\partial_i A_t\gamma^{\hat{i}}\varepsilon_{AB}\epsilon^B = 0\,. \tag{22.72}$$

Since the solution we seek is static, it is reasonable to assume that the Killing spinors are independent of the time t. The first term in (22.72) then vanishes, and the second and third terms cancel if the following two conditions hold:

$$A_t = \pm\sqrt{2}\kappa^{-1}\mathrm{e}^U\,, \tag{22.73}$$

$$\epsilon_A = \mp\gamma^{\hat{0}}\varepsilon_{AB}\epsilon^B\,. \tag{22.74}$$

The first condition states that the gauge potential A_t and the scale factor e^U are related exactly as in (22.66), and the second is a purely algebraic condition on the Killing spinors, which we will discuss shortly. First, however, we must examine the spatial components of (22.69).

Exercise 22.18 *Show that the condition $\delta\psi_{kA}=0$ reduces to $(\partial_k - \frac{1}{2}\partial_k U)\epsilon_A = 0$ after substitution of (22.71) and (22.73)–(22.74). The solution of this equation is simply*

$$\epsilon^A = \mathrm{e}^{\frac{1}{2}U(x)}\epsilon_0^A\,, \tag{22.75}$$

where the spinors ϵ_0^A are constant and satisfy (22.74). Show that these ϵ^A contain four independent real parameters. The Papapetrou–Majumdar solutions thus preserve four of the eight supercharges of the $\mathcal{N}=2$ algebra. They are therefore called $\frac{1}{2}$-BPS solutions.

Box 22.2 The black hole attractor mechanism

The scalars in extremal black hole solutions of $\mathcal{N}=2$ supergravity exhibit a gradient flow towards fixed points at the horizon. The fixed point values depend only on the charges carried by the black hole and not on the values of the scalars at large distance. The flow is determined by a black hole potential related to the central charge of the $\mathcal{N}=2$ supergravity algebra.

The information obtained from the Killing spinor analysis is not quite all we need because the scale factor e^{U} is not yet determined. However, we see that (22.73) is exactly the relation used to write the Maxwell equation in the form (22.67), and this tells us directly that e^{-U} is a harmonic function.

There is also a more general form of the Papapetrou–Majumdar solution in which each center carries both electric and magnetic charge $q_I,\ p_I$ with mass $M_I = \sqrt{q_I^2+p_I^2}/(4\pi G)^{1/2}$. This form is also $\frac{1}{2}$-BPS. See [291] and [290] for more information.

Exercise 22.19 *Use the prepotential in (21.47) and the value of X^0 found in Ex. 21.3 to find that the central charge defined in (21.52) is*

$$\mathcal{Z}=(4\pi G)^{-1/2}\,(q+\mathrm{i}p)\,. \tag{22.76}$$

This shows that the condition on the electric and magnetic charges for the supersymmetric solution saturates the BPS bound (12.22), i.e. $M^2=|\mathcal{Z}|^2$.

22.5 The black hole attractor mechanism

Attractor black holes are solutions of supergravity theories that contain both gauge and scalar fields and carry electric and magnetic charges. We will study them in the framework of $\mathcal{N}=2$ supergravity coupled to n gauge multiplets, each containing a gauge field and a complex scalar. Although the attractor mechanism is more general, we focus on static, spherically symmetric and supersymmetric solutions. They possess Killing spinors, but we first take a different viewpoint and show that the dynamics of metric and scalars can be derived from a set of first order differential equations in the radial variable.[4] The attractor mechanism was discovered in [294, 295, 296].

The scalar dynamics is particularly rich. The values of the scalars at spatial infinity are continuously variable parameters of the solution of the equations, and the black hole mass depends on these moduli. However, the scalars approach fixed values at the black hole horizon that depend only on the charges, and 'memory' of their asymptotic values is lost. In fact, the 'flow' of the scalars towards the horizon exhibits the simple and elegant feature of gradient flow towards a fixed point, which is the minimum of a function related to the central charge of the $\mathcal{N}=2$ supergravity algebra. Hence the name 'attractor'.

[4] Much of the analysis below actually applies to extremal solutions and there are also first order equations for non-extremal black holes; see [292, 293].

The near-horizon geometry of the solution is determined by the charges. The horizon area of the black hole and thus its entropy depend *only* on the charges. This is particularly significant for supersymmetric (BPS) black holes, because these solutions persist in the quantized string theory of which supergravity is the low energy limit. Consistency of the quantum theory requires that charges take discrete values that lie on a set of points called the 'charge lattice'. The possible values of horizon area of a BPS black hole are thus discrete and can be matched to the number of 'quantum microstates' in the string theory description of the black hole [297].

22.5.1 Example of a black hole attractor

We introduce the attractor mechanism in a simple $\mathcal{N} = 2$ supergravity model containing the gravity multiplet and one gauge multiplet. The bosonic fields are the metric $g_{\mu\nu}$ and the graviphoton, plus the gauge multiplet photon and complex scalar z. The two photons combine into gauge potentials with field strengths $F_{\mu\nu}$ and $F'_{\mu\nu}$. The scalar target space is the Poincaré plane, with $\operatorname{Im} z > 0$.

This system was obtained as an example of special Kähler geometry in Ch. 20. We use the parametrization that cannot be derived from a prepotential, derived in Ex. 20.19. The bosonic part of the action is[5]

$$S = \frac{1}{2\kappa^2}\int \mathrm{d}^4x\left\{\sqrt{-g}\left[R - \frac{\partial_\mu z\partial^\mu\bar{z}}{2(\operatorname{Im} z)^2} - \tfrac{1}{2}\operatorname{Im} z\left(F_{\mu\nu}F^{\mu\nu} + F'_{\mu\nu}F'^{\mu\nu}\right)\right]\right.$$
$$\left. + \tfrac{1}{4}\operatorname{Re} z\,\varepsilon^{\mu\nu\rho\sigma}[F_{\mu\nu}F_{\rho\sigma} + F'_{\mu\nu}F'_{\rho\sigma}]\right\}. \qquad (22.77)$$

Since $\operatorname{Im} z > 0$, the kinetic terms of both gauge fields are positive. In Ex. 20.20 it is proven that this model has $\mathrm{U}(1) \times \mathrm{SU}(1,1)$ duality symmetry.

A very simple example of a black hole solution that displays the attractor mechanism was presented in [295], which we follow closely. It turns out that $\operatorname{Re} z$ vanishes in this attractor solution so we drop it and define a real scalar ϕ, usually called the dilaton, by $\operatorname{Im} z = \mathrm{e}^{-2\phi}$. The action of our system then becomes

$$S = \frac{1}{2\kappa^2}\int \mathrm{d}^4x\sqrt{-g}\left[R - 2\partial^\mu\phi\partial_\mu\phi - \tfrac{1}{2}\mathrm{e}^{-2\phi}(F^{\mu\nu}F_{\mu\nu} + F'^{\mu\nu}F'_{\mu\nu})\right]. \qquad (22.78)$$

In isotropic coordinates the line element takes the form

$$\mathrm{d}s^2 = -\mathrm{e}^{2U(r)}\mathrm{d}t^2 + \mathrm{e}^{-2U(r)}(\mathrm{d}r^2 + r^2(\mathrm{d}\theta^2 + \sin^2\theta\,\mathrm{d}\phi^2)). \qquad (22.79)$$

The solution is reasonably simple, so we take it from [295] and verify in Ex. 22.26 below that it emerges from the field equations. One gauge field is purely electric and the other magnetic, so its dual is electric. Spherical symmetry determines the 2-forms

[5] We rename $F^0_{\mu\nu} \to F_{\mu\nu}$ and $F^1_{\mu\nu} \to F'_{\mu\nu}$.

$$F = \pm \mathrm{d}(1/H_1) \wedge \mathrm{d}t\,, \qquad G' = \pm \mathrm{d}(1/H_2) \wedge \mathrm{d}t\,,$$
$$G'_{\mu\nu} \equiv -\tfrac{1}{2}\sqrt{-g}\mathrm{e}^{-2\phi}\varepsilon_{\mu\nu\rho\sigma}F'^{\rho\sigma} = \kappa^2\varepsilon_{\mu\nu\rho\sigma}\frac{\delta S}{\delta F_{\rho\sigma}}\,, \tag{22.80}$$

where $H_{1,2}$ are functions of r and $G'^{\mu\nu}$ is the tensor defined in (7.63), up to a factor κ^2 to avoid such factors below.[6] The full solution is then

$$\mathrm{e}^{-2U} = H_1 H_2\,, \qquad \mathrm{e}^{-2\phi} = H_1/H_2\,,$$
$$H_1 = \mathrm{e}^{-\phi_0} + \tfrac{|q|}{4\pi r}\,, \qquad H_2 = \mathrm{e}^{\phi_0} + \tfrac{|p'|}{4\pi r}\,. \tag{22.81}$$

The constant ϕ_0 is the value of the dilaton at spatial infinity. The $\pm$ signs in the 2-forms indicate four different possibilities for the signs of the charges, as can be checked with the following exercise.

Exercise 22.20 *Note that $H_{1,2}$ in (22.81) are point-singular harmonic functions. Show that the magnetic 2-forms at infinity are $G = -(q/4\pi)\sin\theta\, \mathrm{d}\theta \wedge \mathrm{d}\phi$ and $F' = -(p'/4\pi)\sin\theta\, \mathrm{d}\theta \wedge \mathrm{d}\phi$, when we put $q = \pm|q|$ and $p' = \mp|p'|$. These are the 'standard' expressions for point monopoles of electric charge q and magnetic charge p'.*

The basic feature of the attractor mechanism may now be seen by examining

$$-g^{tt} = g_{rr} = 1 + \frac{\mathrm{e}^{-\phi_0}|p'| + \mathrm{e}^{\phi_0}|q|}{4\pi r} + \frac{|p'q|}{(4\pi r)^2}\,,$$
$$\mathrm{e}^{-2\phi} = \frac{\mathrm{e}^{-\phi_0} + |q|/(4\pi r)}{\mathrm{e}^{+\phi_0} + |p'|/(4\pi r)}\,. \tag{22.82}$$

The mass of the black hole is determined by the coefficient of the $1/r$ term in g_{rr}. Comparing with (22.47), we identify

$$8\pi MG = \mathrm{e}^{-\phi_0}|p'| + \mathrm{e}^{\phi_0}|q|\,, \tag{22.83}$$

and observe that it depends on the asymptotic value of the dilaton. The horizon is located at $r = 0$ where, as usual, g_{tt} vanishes. The value of the dilaton at the horizon,

$$(\mathrm{e}^{-2\phi})_{\mathrm{hor}} = \left|\frac{q}{p'}\right|\,, \tag{22.84}$$

[6] In the rest of this chapter, we redefine several quantities by absorbing factors of κ. They then carry different dimensions from those listed in Sec. 21.5.3. We rename $\mathcal{N}_{IJ} \to \kappa^{-2}\mathcal{N}_{IJ}$, where the new $\mathcal{N}_{IJ}$ is dimensionless. $G_{\mu\nu,I}$ is defined with this new $\mathcal{N}_{IJ}$ and hence has dimension $[M]$, as does $F^I_{\mu\nu}$. Therefore, the electric charge q_I, also defined with this new $G_{\mu\nu,I}$ according to (7.65), is $\kappa^2 q_I$ in terms of the one used before, and has dimension of length, as does p^I.

depends only on the electric and magnetic charges. For all choices of its value at long distance, the dilaton is driven to a fixed value at the horizon.

We emphasize that in the coordinate system used above the region $r > 0$ covers the exterior of the black hole with the horizon at $r = 0$. As in the Reissner–Nordström solution, the metric of the dilaton black hole approaches the Robinson–Bertotti metric (22.49) near the horizon. We suggest the following simple exercise to support these statements.

Exercise 22.21 *Show that the horizon is a 2-sphere with area*[7]

$$A = 4\pi \lim_{r\to 0} r^2 \mathrm{e}^{-2U(r)} = \frac{|qp'|}{4\pi}\,. \tag{22.85}$$

Calculate the invariants $F^{\mu\nu}F_{\mu\nu}$ and $F'^{\mu\nu}F'_{\mu\nu}$ and verify that they are non-singular and constant on the horizon.

22.5.2 The attractor mechanism – real slow and simple

In this section we discuss a more general family of solutions of the theory with action (22.77) and develop a more general approach to the attractor mechanism. Let's discuss the metric ansatz first. The harmonic functions that determine the solution (22.81) are quite simple, but they would be even simpler as linear functions of the variable $\tau = 1/r$. With this motivation we introduce τ as a new radial variable and rewrite the line element (22.79) in the new form

$$\mathrm{d}s^2 = -\mathrm{e}^{2U(\tau)}\mathrm{d}t^2 + \mathrm{e}^{-2U(\tau)}\left[\frac{\mathrm{d}\tau^2}{\tau^4} + \frac{1}{\tau^2}(\mathrm{d}\theta^2 + \sin^2\theta\,\mathrm{d}\phi^2)\right]. \tag{22.86}$$

This line element is the limit as $c \to 0$ of the more general form

$$\mathrm{d}s^2 = -\mathrm{e}^{2U(\tau)}\mathrm{d}t^2 + \mathrm{e}^{-2U(\tau)}\left[\frac{c^4\mathrm{d}\tau^2}{\sinh^4(c\tau)} + \frac{c^2}{\sinh^2(c\tau)}(\mathrm{d}\theta^2 + \sin^2\theta\,\mathrm{d}\phi^2)\right]. \tag{22.87}$$

The pattern of black hole solutions of the theory is as follows. For each choice of charges (q, p) and (q', p') with non-vanishing symplectic invariant $qp' - q'p$, there is a family of regular non-extremal black hole solutions whose metric takes the form in (22.87). In the limit $c \to 0$ each of these becomes a regular extremal solution. Only the scalar fields of extremal solutions exhibit the attractor phenomenon. Most of our explanation of attractor behavior applies to extremal solutions, but at the end we restrict to the subset of extremal solutions that are supersymmetric.

Although non-extremal solutions are not of direct concern here, it is useful to give some motivation for the form of the spatial 3-metric in (22.87), which first appeared in [291]. Here is an optional exercise for this purpose.

[7] Consider a k-dimensional submanifold embedded in a D-dimensional spacetime with metric $\mathrm{d}s^2 = g_{\mu\nu}\mathrm{d}x^\mu\mathrm{d}x^\nu$. The embedding equations are $x^\mu(\xi^\alpha)$, $\alpha = 1,\ldots,k$. The induced metric on the submanifold is $h_{\alpha\beta} = \dfrac{\partial x^\mu}{\partial\xi^\alpha}\dfrac{\partial x^\mu}{\partial\xi^\alpha}g_{\mu\nu}(x(\xi))$ and its area is $A = \int\sqrt{\det h_{\alpha\beta}}\,\mathrm{d}^k\xi$. In the exercise it is convenient to take $\xi^1 = \theta$ and $\xi^2 = \phi$.

Exercise 22.22 *Show that the non-extremal Reissner–Nordström metric can be brought to the form in (22.87) by the change of radial coordinate:*

$$\frac{c^2}{\sinh^2(c\tau)} = (r - r_+)(r - r_-)\,, \tag{22.88}$$

where $r_\pm$ are the outer/inner horizons of Reissner–Nordström; see (22.45). Show that the non-extremality parameter c is related to the horizon positions by $c = (r_+ - r_-)/2$.

Exercise 22.23 *For applications in this section and in the multi-field extension below, it is useful to prepare several formulas valid for the metric of (22.86). They involve the ideas of Sec. 4.2 but now applied in curved spacetime using the dual tensors in (7.59). (A subtle ingredient is that in order to have the same value for a scalar quantity such as $\sqrt{-g}\varepsilon_{\mu\nu\rho\sigma}H^{\mu\nu\rho\sigma}$ (for any 4-tensor H) in coordinates $\{tr\theta\phi\}$ and coordinates $\{t\tau\theta\phi\}$, one must take $\varepsilon_{t\tau\theta\phi} = -1$ since $\varepsilon_{tr\theta\phi}$ is $+1$.) For any antisymmetric tensor $F_{\mu\nu}$, show that*

$$\tilde{F}_{t\tau} = \mathrm{i}\frac{\mathrm{e}^{2U}}{\sin\theta}F_{\theta\phi}\,, \qquad \tilde{F}_{\theta\phi} = -\mathrm{i}\mathrm{e}^{-2U}\sin\theta\, F_{t\tau}\,. \tag{22.89}$$

The Lagrangian of the multi-component model is given in (21.4), up to factors κ^2; see footnote 6. The definition of $G_{\mu\nu I}$ is

$$G_{\mu\nu I} \equiv \kappa^2\varepsilon_{\mu\nu\rho\sigma}\frac{\delta S}{\delta F^I_{\rho\sigma}} = \mathrm{i}\,\mathrm{Im}\,\mathcal{N}_{IJ}\tilde{F}^J_{\mu\nu} + \mathrm{Re}\,\mathcal{N}_{IJ}F^J_{\mu\nu}\,. \tag{22.90}$$

Write this equation explicitly for the index pairs $[\mu\nu] = [t\tau]$ and $[\theta\phi]$ in (22.89), and solve the equations to obtain

$$\begin{pmatrix} F_{t\tau} \\ G_{t\tau} \end{pmatrix} = -\Omega\mathcal{M}\frac{\mathrm{e}^{2U}}{\sin\theta}\begin{pmatrix} F_{\theta\phi} \\ G_{\theta\phi} \end{pmatrix}, \qquad \frac{\mathrm{e}^{2U}}{\sin\theta}\begin{pmatrix} F_{\theta\phi} \\ G_{\theta\phi} \end{pmatrix} = \Omega\mathcal{M}\begin{pmatrix} F_{t\tau} \\ G_{t\tau} \end{pmatrix}, \tag{22.91}$$

where Ω is the symplectic metric and $\mathcal{M}$ is the symmetric matrix introduced in [298]:

$$\Omega = \begin{pmatrix} 0 & \delta_I{}^J \\ -\delta^I{}_J & 0 \end{pmatrix}, \qquad \mathcal{M} = \begin{pmatrix} -(I + RI^{-1}R)_{JK} & (RI^{-1})_J{}^K \\ (I^{-1}R)^J{}_K & -(I^{-1})^{JK} \end{pmatrix}, \tag{22.92}$$

using $R_{JK} \equiv \mathrm{Re}\,\mathcal{N}_{JK}$ and $I_{JK} = \mathrm{Im}\,\mathcal{N}_{JK}$.

To return to the model of (22.77), we replace $\mathcal{N}_{IJ} \to -z$ (see (20.145)). The dual is then

$$G_{\mu\nu} = -\mathrm{i}\,\mathrm{Im}\,z\,\tilde{F}_{\mu\nu} - \mathrm{Re}\,z F_{\mu\nu}\,, \tag{22.93}$$

and the matrix $\mathcal{M}$ becomes

$$\mathcal{M} = \frac{1}{\mathrm{Im}\,z}\begin{pmatrix} |z|^2 & \mathrm{Re}\,z \\ \mathrm{Re}\,z & 1 \end{pmatrix}. \tag{22.94}$$

It is an element of $\mathrm{SL}(2,\mathbb{R}) = \mathrm{Sp}(2,\mathbb{R})$.

We now analyze the equations of motion that determine extremal black holes. The steps of the analysis are as follows:

1. Analyze and solve the gauge field equations of motion.
2. Calculate the Ricci tensor components for the metric of (22.86) and the stress tensor components for the matter fields in the Lagrangian of (22.77). Write the components of the Einstein equations and substitute the gauge field solutions in them.
3. Write the scalar equations of motion that are sourced by gauge field bilinears.
4. At this point we have a system of equations to determine the scale factor $U(\tau)$ and the real scalars $\operatorname{Re} z(\tau)$ and $\operatorname{Im} z(\tau)$ in terms of the four charges. We observe that these equations are equivalent to the equations of motion of an effective action with one independent variable, namely τ, plus a constraint, which is essentially the 'energy' conservation law for the system.

Even a journey of a thousand miles begins with one small step, so let's go.

1. We solve the field equations for $F_{\mu\nu}$ assuming that they describe a dyon[8] with charges (p, q). Spherical symmetry and the conventional Bianchi identity require that field strength 2-forms conform to the ansatz:

$$\begin{aligned} F &= F_{t\tau}\mathrm{d}t \wedge \mathrm{d}\tau + F_{\theta\phi}\mathrm{d}\theta \wedge \mathrm{d}\phi \\ &= f(\tau)\mathrm{d}t \wedge \mathrm{d}\tau - \frac{p}{4\pi}\sin\theta\, \mathrm{d}\theta \wedge \mathrm{d}\phi\,. \end{aligned} \tag{22.95}$$

The quantity p is the magnetic charge; the function $f(\tau)$ is to be found from the Euler–Lagrange equation $(\delta S/\delta A_t) = 0$, where $F_{\mu\nu} = \partial_\mu A_\nu - \partial_\nu A_\mu$. Inserting the metric components from the line element (22.86) and the field components from (22.95) one finds the equation

$$\partial_\tau \left[-\mathrm{e}^{-2U} f(\tau) \operatorname{Im} z + \frac{p}{4\pi} \operatorname{Re} z \right] = 0\,. \tag{22.96}$$

(The equations of motion for A_τ, A_θ, A_ϕ are trivial identities.) One can immediately integrate this equation with the electric charge q appearing as an integration constant. One obtains using (22.93) and (22.89)

$$(\sin\theta)^{-1} G_{\theta\phi} = -\mathrm{e}^{-2U} f(\tau) \operatorname{Im} z + \frac{p}{4\pi} \operatorname{Re} z = -\frac{q}{4\pi}\,. \tag{22.97}$$

The electric/magnetic charges are as defined in (7.65). The general result (22.91) allows us to summarize results by displaying the 2-forms F, G as a symplectic vector, which transforms under duality as in (4.54):

$$4\pi \begin{pmatrix} F \\ G \end{pmatrix} = \mathrm{e}^{2U} \Omega \mathcal{M} \begin{pmatrix} p \\ q \end{pmatrix} \mathrm{d}t \wedge \mathrm{d}\tau - \begin{pmatrix} p \\ q \end{pmatrix} \sin\theta\, \mathrm{d}\theta \wedge \mathrm{d}\phi\,, \tag{22.98}$$

where $\mathcal{M}$ is given in (22.94). The 2-forms F', G' can be written as similar expressions, obtained from (22.98) by changing the charges $q \to q'$, $p \to p'$.

[8] The charges p, q can now be either positive or negative.

2. The Einstein equations are

$$R_{\mu\nu} = \kappa^2 \left(T_{\mu\nu} - \tfrac{1}{2} g_{\mu\nu} T^\rho{}_\rho \right) ,$$
$$\kappa^2 T_{\mu\nu} = \operatorname{Im} z \left(F_{\mu\rho} F_\nu{}^\rho - \tfrac{1}{4} g_{\mu\nu} F_{\rho\sigma} F^{\rho\sigma} + F'_{\mu\rho} F'^{\rho}_\nu - \tfrac{1}{4} g_{\mu\nu} F'_{\rho\sigma} F'^{\rho\sigma} \right)$$
$$+ \frac{1}{4(\operatorname{Im} z)^2} \left(\partial_\mu z \partial_\nu \bar{z} + \partial_\nu z \partial_\mu \bar{z} - g_{\mu\nu}\, \partial_\rho z \partial^\rho \bar{z} \right) . \qquad (22.99)$$

The only non-vanishing components of $R_{\mu\nu}$ and $T_{\mu\nu}$ are the four diagonal components. We indicate derivatives with respect to τ by $\dot{U}(\tau)$ and $\ddot{U}(\tau)$, and we introduce the important black hole potential [296]

$$(4\pi)^2 V_{\rm BH} = \tfrac{1}{2} (\, p \quad q\,) \mathcal{M} \begin{pmatrix} p \\ q \end{pmatrix} + \tfrac{1}{2} (\, p' \quad q'\,) \mathcal{M} \begin{pmatrix} p' \\ q' \end{pmatrix} . \qquad (22.100)$$

Why is it important? One virtue is that it simplifies the field equations. With the gauge fields from (22.98) inserted they read

$$R_{tt} = \mathrm{e}^{4U} \tau^4 \ddot{U} = \mathrm{e}^{6U} \tau^4 V_{\rm BH} ,$$
$$R_{\tau\tau} = \ddot{U} - 2\dot{U}^2 = -\mathrm{e}^{2U} V_{\rm BH} + |\dot{z}|^2/2(\operatorname{Im} z)^2 , \qquad (22.101)$$
$$R_{\theta\theta} = \tau^2 \ddot{U} = \tau^2 \mathrm{e}^{2U} V_{\rm BH} .$$

The $R_{\phi\phi}$ and $R_{\theta\theta}$ equations differ by the common factor of $\sin^2\theta$, so only the equations above need be considered. Thus the four diagonal components of the Einstein equations produce just two independent conditions:

$$\ddot{U} = \mathrm{e}^{2U} V_{\rm BH} ,$$
$$\dot{U}^2 = \mathrm{e}^{2U} V_{\rm BH} - \frac{|\dot{z}|^2}{(2 \operatorname{Im} z)^2} . \qquad (22.102)$$

Exercise 22.24 *Show that $V_{\rm BH}$ is a positive quantity. Show that $\mathrm{e}^{2U} V_{\rm BH}$ is essentially the electromagnetic energy density $\sqrt{-g}T^0{}_0$ obtained by inserting the gauge fields F (and F') from (22.98) in the electromagnetic terms of the stress tensor in (22.99).*

Exercise 22.25 *As discussed in Sec. 4.2.3, the scalar z and the charges (p, q) transform under the $\mathrm{SL}(2,\mathbb{R})$ duality symmetry as (note that $Z = -z^*$)*

$$z \to \frac{az - b}{d - cz} , \qquad \begin{pmatrix} p \\ q \end{pmatrix} \to \begin{pmatrix} d & c \\ b & a \end{pmatrix} \begin{pmatrix} p \\ q \end{pmatrix} . \qquad (22.103)$$

The charges (p', q') transform in the same way as (p, q). Show that $V_{\rm BH}$ is invariant under these duality transformations. It suffices to analyze the following three cases separately: inversion ($a = d = 0$, $b = -c = 1$), translation ($a = d = 1$, $c = 0$, b arbitrary), and scale ($d = 1/a$, $b = c = 0$).

3. With some care one can show that the scalar equations of motion are

$$\frac{\kappa^2}{\sin\theta}\frac{\delta S}{\delta z} = \frac{\mathrm{d}}{\mathrm{d}\tau}\left(\frac{\dot{\bar{z}}}{(2\,\mathrm{Im}\,z)^2}\right) - \frac{\mathrm{i}|\dot{z}|^2}{4(\mathrm{Im}\,z)^3} - \mathrm{e}^{2U}\frac{\partial V_{\mathrm{BH}}}{\partial z} = 0\,. \tag{22.104}$$

4. Consider the following positive definite action functional:

$$S[U,z] \equiv \int \mathrm{d}\tau \left(\dot{U}^2 + \frac{|\dot{z}|^2}{(2\,\mathrm{Im}\,z)^2} + \mathrm{e}^{2U}V_{\mathrm{BH}}\right)\,. \tag{22.105}$$

   It resembles the action of a system in classical mechanics, with variables $U(\tau)$, $z(\tau)$ except that the potential appears with an unconventional sign. Noether's theorem implies that the 'energy'

$$\mathcal{E} \equiv \dot{U}^2 + \frac{|\dot{z}|^2}{(2\,\mathrm{Im}\,z)^2} - \mathrm{e}^{2U}V_{\mathrm{BH}} \tag{22.106}$$

   is independent of τ. Now observe that the Euler–Lagrange equations of this system agree with the scalar equations of motion of (22.104) and the second order equation of (22.102) , and that the remaining first order equation implies that we consider only solutions of the mechanical system with $\mathcal{E} = 0$.

This completes the program outlined above. Since a spherically symmetric classical solution has 'cohomogeneity 1' it is not a great surprise that the final equations can be rephrased as those of a classical mechanical system. As so far described, our discussion applies to all extremal solutions.

Exercise 22.26 *Check that the explicit solution described in (22.80) and (22.81) is indeed a solution of all field equations above in the special case of charges* $(0,q)$ *and* $(p',0)$ *and* $\mathrm{Re}\,z = 0$, $\mathrm{Im}\,z = \mathrm{e}^{-2\phi}$. *Identify the solutions (22.98) with (22.80).*

The attractor behavior is not yet in sight, but it will come after we digress in the next section to study the supersymmetry properties of black holes. The discussion requires manipulation of the transformation rules of $\mathcal{N} = 2$ supergravity derived in Ch. 20. Readers who are most interested in understanding how the action (22.105) leads to the attractor property can proceed directly to Sec. 22.7. The essential result needed is the central charge $\mathcal{Z}$ defined in (22.125).

22.6 Supersymmetry of the black holes

22.6.1 Killing spinors

The action (22.77) is a particular model of $\mathcal{N} = 2$ supergravity. We will see how the solution discussed in Sec. 22.5.1 can be derived from an analysis of its linearized fermion transformation rules. Then we will consider the algebra of the preserved supersymmetries and obtain the equality of mass and central charge for these supersymmetry-preserving solutions. Indeed, they are BPS solutions.

We thus start from (21.42) without hypermultiplets, gauging or moment maps. Using A, B for the R-symmetry index, the transformations and the (bosonic part of the) auxiliary fields that appear therein are (using the rescaled $\mathcal{N}_{IJ}$ in footnote 6)

$$\begin{aligned}
\delta\psi_\mu^A &= \left(\partial_\mu + \tfrac{1}{4}\omega_\mu{}^{ab}\gamma_{ab} - \tfrac{1}{2}\mathrm{i}\mathcal{A}_\mu\right)\epsilon^A - \tfrac{1}{16}\gamma^{ab}T^-_{ab}\varepsilon^{AB}\gamma_\mu\epsilon_B\,,\\
\delta\chi_A^\alpha &= \partial\!\!\!/ z^\alpha\epsilon_A - \tfrac{1}{2}\kappa^{-2}g^{\alpha\bar\beta}G^-_{ab\,\bar\beta}\gamma^{ab}\varepsilon_{AB}\epsilon^B\,,\\
\mathcal{A}_\mu &= \tfrac{1}{2}\mathrm{i}\kappa^2\left(\partial_\mu z^\alpha\partial_\alpha\mathcal{K} - \partial_\mu\bar z^{\bar\alpha}\partial_{\bar\alpha}\mathcal{K}\right),\\
T^-_{ab} &= -4X^I\,\mathrm{Im}\,\mathcal{N}_{IJ}F^{-J}_{ab}\,,\qquad G^-_{ab\,\bar\beta} = \overline{\nabla}_{\bar\beta}\bar X^I\,\mathrm{Im}\,\mathcal{N}_{IJ}F^{-J}_{ab}\,,\\
X^I &= yZ^I\,,\qquad y = \mathrm{e}^{\kappa^2\mathcal{K}/2}\,,\qquad \nabla_\alpha Z^I = \left[\partial_\alpha + \kappa^2(\partial_\alpha\mathcal{K})\right]Z^I\,.
\end{aligned}\tag{22.107}$$

We now use the data from Ex. 20.19 in which $I = 0, 1$, and $\alpha = z$ (only one value):

$$Z^0 = 1\,,\qquad Z^1 = \mathrm{i}\,,\qquad \mathrm{e}^{-\kappa^2\mathcal{K}} = 4\,\mathrm{Im}\,z\,,\qquad \mathcal{N}_{IJ} = -z\delta_{IJ}\,,\tag{22.108}$$

leading to the action in (22.77), when $F^0_{\mu\nu}$ is identified with $F_{\mu\nu}$ and $F^1_{\mu\nu}$ with $F'_{\mu\nu}$. In this model some relevant quantities in (22.107) are

$$T^-_{ab} = 4y\,\mathrm{Im}\,z\left(F^{-0}_{ab} + \mathrm{i}F^{-1}_{ab}\right),\qquad G^-_{ab\,\bar z} = \tfrac{1}{2}\mathrm{i}\,y\left(F^{-0}_{ab} - \mathrm{i}F^{-1}_{ab}\right).\tag{22.109}$$

We now look for solutions where z is imaginary, $F^0_{\mu\nu}$ is an electric field, and $F^1_{\mu\nu}$ is magnetic:

$$z = \mathrm{i}\mathrm{e}^{-2\phi}\,,\qquad y = \tfrac{1}{2}\mathrm{e}^{\phi}\,,\qquad F^0_{i0} = \partial_i\psi\,,\qquad \tilde F^1_{i0} = \mathrm{i}\mathrm{e}^{2\phi}\partial_i\chi\,.\tag{22.110}$$

The F_{ij} components of $F^0_{\mu\nu}$ and of $\tilde F^1_{\mu\nu}$ vanish. The ansatz for $F^1_{\mu\nu}$ is needed to make $G_{\mu\nu I}$ for $I = 1$ purely electric, i.e.

$$G_{i0,1} = \partial_i\chi\,,\qquad\text{where}\qquad G^+_{i0,1} = \mathcal{N}_{11}F^{1,+}_{i0} = -\tfrac{1}{2}z\tilde F^1_{i0}\,.\tag{22.111}$$

This gives for the quantities in (22.109)

$$T^-_{i0} = \mathrm{e}^{-\phi}\left[\partial_i\psi + \mathrm{e}^{2\phi}\partial_i\chi\right],\qquad G^-_{i0\,\bar z} = \tfrac{1}{8}\mathrm{i}\mathrm{e}^{\phi}\left[\partial_i\psi - \mathrm{e}^{2\phi}\partial_i\chi\right].\tag{22.112}$$

The $[ij]$ components are determined by these using the (anti-)self-duality, and lead in the transformation laws just to a doubling of the contributions of the $[i0]$ components due to relations similar to (14.18).

The spacetime metric is (22.79) with $U(x^i)$ and with the frame fields and connections of (22.70) and (22.71). Since the solution is static, we also look for static Killing spinors. The BPS conditions are then (remember that hatted indices are frame indices)

$$\begin{aligned}
\delta\psi_0^A &= \tfrac{1}{2}\mathrm{e}^{2U}\gamma_{\hat 0}\gamma_{\hat\imath}(\partial_i U)\epsilon^A - \tfrac{1}{4}\gamma^{\hat\imath}\mathrm{e}^{U-\phi}\left[\partial_i\psi + \mathrm{e}^{2\phi}\partial_i\chi\right]\varepsilon^{AB}\epsilon_B = 0\,,\\
\delta\psi_i^A &= \partial_i\epsilon^A - \tfrac{1}{2}\gamma_{\hat\imath\hat\jmath}(\partial_j U)\epsilon^A + \tfrac{1}{4}\gamma^{\hat 0\hat\jmath}\mathrm{e}^{-U-\phi}\left[\partial_j\psi + \mathrm{e}^{2\phi}\partial_j\chi\right]\gamma_{\hat\imath}\varepsilon^{AB}\epsilon_B = 0\,,\\
\delta\chi_A &= \mathrm{i}\gamma^{\hat\imath}\mathrm{e}^U\partial_i\mathrm{e}^{-2\phi}\epsilon_A + \mathrm{i}\mathrm{e}^{-3\phi}\left[\partial_i\psi - \mathrm{e}^{2\phi}\partial_i\chi\right]\gamma^{\hat 0\hat\imath}\varepsilon_{AB}\epsilon^B = 0\,.
\end{aligned}\tag{22.113}$$

The first and the (C-conjugate of the) last conditions can be combined to

$$\gamma^{\hat{i}}\epsilon^A \partial_i e^{U+\phi} = \gamma^{\hat{i}}(\partial_i \psi)\gamma_{\hat{0}}\varepsilon^{AB}\epsilon_B\,, \qquad \gamma^{\hat{i}}\epsilon^A \partial_i e^{U-\phi} = \gamma^{\hat{i}}(\partial_i \chi)\gamma_{\hat{0}}\varepsilon^{AB}\epsilon_B\,. \tag{22.114}$$

We consider solutions that depend only on the spatial variable r, so we can drop the $\gamma^{\hat{i}}$ on both sides. The remaining Killing equations take the same form as in the Papapetrou–Majumdar solution in Sec. 22.4:

$$\delta\psi_i^A = (\partial_i - \tfrac{1}{2}\partial_i U)\epsilon^A = 0\,. \tag{22.115}$$

We solve these equations by

$$\psi = \pm e^{U+\phi}\,, \qquad \chi = \pm e^{U-\phi}\,, \qquad \epsilon^A = \pm\gamma_{\hat{0}}\varepsilon^{AB}\epsilon_B\,, \qquad \epsilon^A = e^{U/2}\epsilon^A_{(0)}\,, \tag{22.116}$$

where $\epsilon^A_{(0)}$ is constant. We can thus identify this with (22.80) choosing $\psi = \pm H_1^{-1}$ and $\chi = \pm H_2^{-1}$. The signs in (22.116) are related. Using the result of Ex. 22.20, this means that the upper signs imply positive electric and negative magnetic charge, while the lower signs imply negative electric and positive magnetic charge. The condition on the supersymmetry parameters implies that only four of the eight real components of $\epsilon^A_{(0)}$ are preserved. Specifically, change basis from $\{\epsilon_1, \epsilon_2\}$ to the combinations

$$\epsilon_\pm = \epsilon_2 \pm \gamma^{\hat{0}}\epsilon^1\,. \tag{22.117}$$

The condition (22.116) implies that one of these should vanish. Note that these are still chiral, and $\epsilon^\pm$ are their C-conjugates. Hence, these supersymmetric solutions are $\frac{1}{2}$-BPS solutions.[9]

22.6.2 The central charge

As mentioned in Sec. 22.3, when supersymmetries are preserved, their bilinears are Killing vectors. This will give rise to central charges. We consider again $\mathcal{N} = 2$ and consider the supersymmetry algebra (20.74) applied to the frame field component $e_0{}^{\hat{i}}$, which is a component that vanishes in the solution. Only the general coordinate and Lorentz transformations contribute to $[\delta_1, \delta_2]e_0{}^{\hat{i}}$. Consider first a general coordinate transformation. We saw in Sec. 11.3.2 how the covariant general coordinate transformations act on the frame field. In this case Killing vectors are independent of time, so we obtain

$$\delta e_0{}^{\hat{i}} = \partial_t \xi^{\hat{i}} + \omega_0{}^{\hat{i}\hat{0}}\xi_{\hat{0}} = \xi^{\hat{0}} e^{2U}\partial_i U\,. \tag{22.118}$$

The long distance behavior $r \to \infty$ of U determines the mass of the solution, specifically

$$\partial_i U \to \frac{x^i}{r^3} MG\,. \tag{22.119}$$

[9] The reader may wonder why positive electric and positive magnetic charges and vice versa are not BPS. In fact, they are! The action (22.77) is also part of an $\mathcal{N} = 4$ supergravity theory. We considered here only two of the four supersymmetries. In the context of $\mathcal{N} = 4$ the solutions with same sign for electric and magnetic charges preserve combinations of the supersymmetries that we did not treat here. In that context these solutions are $\frac{1}{4}$-BPS, while for vanishing q or p they are $\frac{1}{2}$-BPS [299].

Now we consider a Lorentz transformation of $e_0{}^{\hat{i}}$. Since the components $e_0{}^{\hat{j}}$ vanish in the solution, only the Lorentz boost $\lambda^{\hat{i}\hat{0}}$ contributes and gives

$$\delta e_0{}^{\hat{i}} = -\lambda^{\hat{i}}{}_{\hat{0}} e_0^{\hat{0}} = \lambda^{\hat{i}\hat{0}} \mathrm{e}^U . \tag{22.120}$$

The Lorentz parameter in $[\delta_1, \delta_2] e_0{}^{\hat{i}}$ can be obtained from (20.75). Since (22.112) gives a real value for T^-_{i0}, we obtain the pure Lorentz boost

$$\lambda^{\hat{i}\hat{0}} = \tfrac{1}{4}\left(\bar{\epsilon}'^A \epsilon^B \varepsilon_{AB} + \bar{\epsilon}'_A \epsilon_B \varepsilon^{AB}\right)\left(\mathrm{e}^{-\phi}\partial_i \psi + \mathrm{e}^{\phi}\partial_i \chi\right) . \tag{22.121}$$

(To avoid confusion with the R-symmetry index, we use here ϵ^A for ϵ^i_1, and ϵ'^A for ϵ^i_2.) At long distance from the source this becomes

$$\lambda^{\hat{i}\hat{0}} = \frac{x^i}{2r^3}\left(\bar{\epsilon}'^A \epsilon^B \varepsilon_{AB} + \bar{\epsilon}'_A \epsilon_B \varepsilon^{AB}\right) G \mathcal{Z}_\infty ,$$
$$\mathcal{Z}_\infty = \bar{\mathcal{Z}}_\infty = Q - P , \qquad Q = \mathrm{e}^{\phi_0}\frac{q}{8\pi G} , \qquad P = \mathrm{e}^{-\phi_0}\frac{p'}{8\pi G} . \tag{22.122}$$

It is important to compare with (12.23) and note that the Lorentz transformation from (20.74) produces the central charge $\mathcal{Z}$. Below, we will also identify $\mathcal{Z}_\infty$ with the expression (21.52) at $r \to \infty$.

Using the spinor basis (22.117), we can rewrite the parameters of $[\delta_1, \delta_2] e_0{}^{\hat{i}}$ as

$$\xi^0 = \tfrac{1}{2}\left(\bar{\epsilon}'^A \gamma^{\hat{0}} \epsilon_A + \bar{\epsilon}'_A \gamma^{\hat{0}} \epsilon^A\right) = \tfrac{1}{4}\left(\bar{\epsilon}'^+ \gamma^{\hat{0}} \epsilon_+ + \bar{\epsilon}'_+ \gamma^{\hat{0}} \epsilon^+ + \bar{\epsilon}'^- \gamma^{\hat{0}} \epsilon_- + \bar{\epsilon}'_- \gamma^{\hat{0}} \epsilon^-\right) ,$$
$$\tfrac{1}{2}\left(\bar{\epsilon}'^A \epsilon^B \varepsilon_{AB} + \bar{\epsilon}'_A \epsilon_B \varepsilon^{AB}\right) = \tfrac{1}{4}\left(\bar{\epsilon}'^+ \gamma^{\hat{0}} \epsilon_+ + \bar{\epsilon}'_+ \gamma^{\hat{0}} \epsilon^+ - \bar{\epsilon}'^- \gamma^{\hat{0}} \epsilon_- - \bar{\epsilon}'_- \gamma^{\hat{0}} \epsilon^-\right) . \tag{22.123}$$

Since we consider the algebras of the preserved supersymmetries, we must use the same projection for ϵ and ϵ'. Then the unwanted ξ^i vanishes.

For supersymmetric solutions, the total transformation of $e_0{}^{\hat{i}}$, namely the sum of (22.118) and (22.120), vanishes. At large distance this gives the condition

$$0 = G\frac{x^i}{4r^3}\left[\left(\bar{\epsilon}'_+ \gamma^{\hat{0}} \epsilon^+ + \text{h.c.}\right)(M + \mathcal{Z}) + \left(\bar{\epsilon}'_- \gamma^{\hat{0}} \epsilon^- + \text{h.c.}\right)(M - \mathcal{Z})\right] . \tag{22.124}$$

This shows that the extremal solutions satisfy $M = |\mathcal{Z}_\infty|$. In particular the solution with two positive signs in (22.80) preserves the ϵ_- supersymmetry and has $M = \mathcal{Z}_\infty$, and the solution with two negative signs preserves the ϵ_+ supersymmetry and has $M = -\mathcal{Z}_\infty$.

We gave a general definition of $\mathcal{Z}$ in (21.52). Using the values of Z^I, F_I from (20.144) (and taking into account the redefinition of q in footnote 6) we obtain in this case (with $q_0, p^0 \to q, p$ and $q_1, p^1 \to q', p'$)

$$\mathcal{Z} = \frac{\kappa^{-2}}{\sqrt{\operatorname{Im} z}}\left[(q + \mathrm{i}q') + z(p + \mathrm{i}p')\right] . \tag{22.125}$$

In the case when $z = \mathrm{i}\mathrm{e}^{-2\phi}$ and $q' = p = 0$ and in the limit $\phi(r) \to \phi_0$, this agrees with the central charge $\mathcal{Z}_\infty$ defined in (22.122).

22.6.3 The black hole potential

We now make contact between the definition of the central charge and the black hole potential in the multi-component $\mathcal{N} = 2$ model. To do so, we first use (21.23) in the expressions for $\mathcal{Z}$ and $\nabla_\alpha \mathcal{Z}$ in (21.54), to write these as (after the rescaling of footnote 6)

$$\begin{aligned} \mathcal{Z} &= 2\kappa^{-2} e^{-\kappa^2 \mathcal{K}/2} Z^I(z) \left[q_I - \mathcal{N}_{IJ} p^J \right], \\ \nabla_\alpha \mathcal{Z} &= 2\kappa^{-2} e^{-\kappa^2 \mathcal{K}/2} \nabla_\alpha Z^I \left[q_I - \mathcal{N}_{IJ} p^J \right]. \end{aligned} \tag{22.126}$$

This can be used to write [295, 296], using (20.190),

$$\begin{aligned} \kappa^2 \mathcal{Z}\overline{\mathcal{Z}} + \nabla_\alpha \mathcal{Z} g^{\alpha\bar\beta} \nabla_{\bar\beta} \overline{\mathcal{Z}} &= -2\kappa^{-2} \left[q_I - \mathcal{N}_{IK} p^K \right] \left[q_J - \overline{\mathcal{N}}_{JL} p^L \right] (I^{-1})^{IJ} \\ &= 2\kappa^{-2} \begin{pmatrix} p & q \end{pmatrix} \mathcal{M} \begin{pmatrix} p \\ q \end{pmatrix}, \end{aligned} \tag{22.127}$$

where we used the notations from (22.92). Using the definition as in (22.100),

$$(4\pi)^2 V_{\rm BH} = \tfrac{1}{2} \begin{pmatrix} p & q \end{pmatrix} \mathcal{M} \begin{pmatrix} p \\ q \end{pmatrix}, \tag{22.128}$$

this leads to an expression for the 'black hole potential',

$$(4\pi)^2 V_{\rm BH} = \tfrac{1}{4} \kappa^4 \mathcal{Z}\overline{\mathcal{Z}} + \tfrac{1}{4} \kappa^2 \nabla_\alpha \mathcal{Z} g^{\alpha\bar\beta} \nabla_{\bar\beta} \overline{\mathcal{Z}}, \tag{22.129}$$

which we will immediately exploit. This relation between the black hole potential and the central charge is a consequence of the special geometry of $\mathcal{N} = 2$ supergravity.[10]

To apply this formula, it is useful to notice that, since $\nabla_\alpha \overline{\mathcal{Z}} = 0$, and $\nabla_\alpha (\mathcal{Z}\overline{\mathcal{Z}}) = \partial_\alpha (\mathcal{Z}\overline{\mathcal{Z}})$, we have

$$\nabla_\alpha \mathcal{Z} = 2 \sqrt{\frac{\mathcal{Z}}{\overline{\mathcal{Z}}}} \partial_\alpha |\mathcal{Z}|. \tag{22.130}$$

Thus we can write the simpler form

$$G^{-2} V_{\rm BH} = \mathcal{Z}\overline{\mathcal{Z}} + 4\kappa^{-2} g^{\alpha\bar\beta} \partial_\alpha |\mathcal{Z}| \partial_{\bar\beta} |\mathcal{Z}|. \tag{22.131}$$

Central charge, black hole potential and Killing spinors — **Box 22.3**

The extremal charged black solutions discussed before are BPS solutions. The central charge $\mathcal{Z}$ gives an expression for the black hole potential, and its value at infinity $\mathcal{Z}_\infty$ for the extremal solutions satisfies $M = |\mathcal{Z}_\infty|$.

[10] Suitable functions $|\mathcal{Z}|$ in this relation can also be constructed for certain extremal, non-supersymmetric black holes; see [292].

22.7 First order gradient flow equations

Finally, we derive the attractor behavior of the model of Sec. 22.5.2. The black hole potential (22.129) is related to the central charge (22.125) by

$$G^{-2}V_{\rm BH} = |\mathcal{Z}|^2 + 16({\rm Im}\, z)^2\partial_z|\mathcal{Z}|\partial_{\bar z}|\mathcal{Z}|\,. \tag{22.132}$$

Exercise 22.27 *Readers who did not follow the general derivation in the previous section should check this explicitly.*

Using this relation we can complete squares and write the action (22.105) as

$$\begin{aligned} S[U,z] = \int {\rm d}\tau \left[(\dot U + G{\rm e}^U|\mathcal{Z}|)^2 + \frac{1}{4({\rm Im}\, z)^2}\left|\dot z + 8G({\rm Im}\, z)^2{\rm e}^U\partial_{\bar z}|\mathcal{Z}|\right|^2\right] \\ -2\,G{\rm e}^U|\mathcal{Z}|\Big|_{\tau=0}^{\tau=\infty}\,. \end{aligned} \tag{22.133}$$

This form of the action consists of a positive sum of squares plus a boundary term. The conditions for a minimum are the first order differential equations

$$\dot z = -8G{\rm e}^U({\rm Im}\, z)^2\partial_{\bar z}|\mathcal{Z}|\,, \qquad \dot U = -G{\rm e}^U|\mathcal{Z}|\,. \tag{22.134}$$

The derivation guarantees that any solution of this first order system automatically solves the scalar equation of motion (22.104) and the second order equation for U in (22.102). Furthermore the (properly weighted) sum of the squares of the two first order equations together with the relation (22.132) immediately tells us that the 'energy' (22.106) of the solution vanishes, exactly what we need!

It is quite striking that the dynamics of supersymmetric black holes can be reduced to a pair of first order equations that exhibit the structure known as gradient flow.[11] The metric component $-g_{tt} = {\rm e}^{2U(\tau)}$ flows monotonically downwards from spatial infinity ($\tau = 0$) to the horizon $\tau \to \infty$. At large distance the solution becomes flat and we normalize to ${\rm e}^{2U(0)} = 1$. At the horizon ${\rm e}^{2U(\tau)} \to 0$. The scalar field $z(\tau)$ also flows from a value set by a boundary condition at spatial infinity and approaches a constant value at the horizon determined by the fixed-point condition $\partial_z|\mathcal{Z}| = 0$. Indeed the fixed point is a minimum of the function $|\mathcal{Z}|$.

Useful physical information follows quickly from this picture. We know that the mass of a static asymptotically flat solution is determined by the leading correction to g_{tt} at large distance, specifically by

$$-g_{tt} = {\rm e}^{2U} \approx 1 - \frac{2MG}{r} + \ldots = 1 - 2MG\tau + \ldots\,. \tag{22.135}$$

Thus

$$M = -\frac{1}{2G}\frac{\rm d}{{\rm d}\tau}{\rm e}^{2U(\tau)}|_{\tau=0} = |\mathcal{Z}|_\infty\,. \tag{22.136}$$

[11] The same structure is realized in the first order equations (23.127), which describe domain walls in asymptotically AdS spacetimes.

The mass of any black hole solution of the flow equations is equal to the value of $|\mathcal{Z}|$ at spatial infinity. This is nothing but the BPS condition derived from gradient flow. With Ex. 22.21 in view, one can see that the area of a sphere at fixed τ is $A = 4\pi\mathrm{e}^{-2U(\tau)}/\tau^2$. We are most interested in the area of the horizon, and this is governed by the flow equation written in the form

$$\frac{\mathrm{d}}{\mathrm{d}\tau}\mathrm{e}^{-U} = G|\mathcal{Z}| \stackrel{\tau\to\infty}{\longrightarrow} G|\mathcal{Z}|_{\mathrm{min}}\,. \tag{22.137}$$

The area of the horizon is thus determined by the minimum value of $|\mathcal{Z}|$; indeed, the area is

$$A = 4\pi G^2|\mathcal{Z}|^2_{\mathrm{min}} = 4\pi\, V_{\mathrm{BHmin}}\,. \tag{22.138}$$

22.8 The attractor mechanism – fast and furious

We now extend the previous discussion to describe the attractor mechanism in the more general situation in which $\mathcal{N} = 2$ supergravity is coupled to n abelian vector multiplets. There are $n+1$ gauge fields $F^I_{\mu\nu}$ and n complex scalars z^α. The bosonic parts of the action (21.34) for this case (without Fayet–Iliopoulos terms) are[12]

$$S = \frac{1}{2\kappa^2}\int \mathrm{d}^4x\Big\{\sqrt{-g}\Big[R - 2g^{\mu\nu}g_{\alpha\bar\beta}\partial_\mu z^\alpha\partial_\nu\bar z^{\bar\beta} + \tfrac{1}{2}\,\mathrm{Im}\,\mathcal{N}_{IJ}F^I_{\mu\nu}F^{\mu\nu J}\Big] \\ -\tfrac{1}{4}\,\mathrm{Re}\,\mathcal{N}_{IJ}\,\varepsilon^{\mu\nu\rho\sigma}F^I_{\mu\nu}F^J_{\rho\sigma}\Big\}. \tag{22.139}$$

The gauge kinetic matrix $\mathcal{N}_{IJ} = \mathrm{Re}\,\mathcal{N}_{IJ} + \mathrm{i}\,\mathrm{Im}\,\mathcal{N}_{IJ} = \mathcal{N}_{JI}$ depends on the scalars z^α and $\bar z^\beta$ (with $\mathrm{Im}\,\mathcal{N}_{IJ}$ a negative definite quadratic form so that gauge field kinetic energy is positive). The Kähler metric $g_{\alpha\bar\beta}$ is also positive definite. The gauge kinetic function and the Kähler metric are determined from the prepotential $F(X^I)$ of the projective special Kähler manifold from which the physical action (22.139) is obtained. In some cases the prepotential does not exist, but a symplectic section (X^I, F_I) of the projective embedding manifold does exist and is sufficient. These issues are discussed in Secs. 20.3.3 and 21.2.

We will need only minimal changes in the treatment in Sec. 22.5.2 of the simpler model with one vector multiplet. The key features of the derivation of first order gradient flow equations for BPS black holes were already discussed there, and we must simply 'promote' all quantities to the present case of n scalars and $n+1$ vectors.

The gauge fields in (22.95) now have an index $I = 0, 1, \ldots, n$ and similarly the symplectic vectors in (22.98) have $2(n+1)$ components, and the corresponding $2(n+1)\times 2(n+1)$ matrix $\mathcal{M}$ is given in (22.92). The dual field strengths are defined as in (22.90). The charges form a symplectic vector defined as integrals over a 2-sphere Σ^2 that encloses the origin, viz.

[12] Similar to the rescalings in footnote 6, we have rescaled here also the metric $g_{\alpha\bar\beta}$ with κ^2.

$$\begin{pmatrix} p^I \\ q_I \end{pmatrix} = -\frac{1}{2}\int_{\Sigma^2} \begin{pmatrix} F^I_{\mu\nu} \\ G_{\mu\nu I} \end{pmatrix} \mathrm{d}x^\mu \wedge \mathrm{d}x^\nu\,, \tag{22.140}$$

where $G_{\mu\nu I}$ is defined in (22.90). With this notation (suppressing the indices), (22.98) remains valid.

Exercise 22.28 *Check that $\mathcal{M}$ preserves the symplectic form, i.e. $\mathcal{M}\Omega\mathcal{M} = \Omega$, and is therefore an element of the group* $\mathrm{Sp}(2n, \mathbb{R})$.

The stress tensor of (22.99) is modified as follows:

$$\begin{aligned}\kappa^2 T_{\mu\nu} = &-\operatorname{Im}\mathcal{N}_{IJ}\left(F^I_{\mu\rho}F_\nu{}^{\rho J} - \tfrac{1}{4}g_{\mu\nu}F^I_{\rho\sigma}F^{\rho\sigma J}\right)\\ &+ g_{\alpha\bar\beta}\left(\partial_\mu z^\alpha \partial_\nu \bar z^{\bar\beta} + \partial_\nu z^\alpha \partial_\mu \bar z^{\bar\beta} - g_{\mu\nu}\,\partial_\rho z^\alpha \partial^\rho \bar z^{\bar\beta}\right).\end{aligned} \tag{22.141}$$

Again there is the all-important black hole potential (22.128). As in Ex. 22.24, V_{BH} is a positive quantity, which is closely related to the gauge field energy density.

When the gauge fields of (22.98) are inserted, the source terms for various components of the Ricci tensor obey (22.101), with the only modification

$$|\dot z|^2/2(\operatorname{Im} z)^2 \;\rightarrow\; 2g_{\alpha\bar\beta}\dot z^\alpha \dot{\bar z}^{\bar\beta}\,. \tag{22.142}$$

This also applies to (22.102). The scalar equations of motion (22.104) generalize to

$$\frac{\kappa^2}{\sin\theta}\frac{\delta S}{\delta z^\alpha} = \frac{\mathrm{d}}{\mathrm{d}\tau}\left(g_{\alpha\bar\beta}\dot{\bar z}^{\bar\beta}\right) - (\partial_\alpha g_{\gamma\bar\delta})\dot z^\gamma \dot{\bar z}^{\bar\delta} - \mathrm{e}^{2U}\,\partial_\alpha V_{\mathrm{BH}} = 0\,. \tag{22.143}$$

Finally we come to the one-dimensional effective action that is the generalization of (22.105),

$$S[U, z, \bar z] \equiv \int \mathrm{d}\tau\left(\dot U^2 + g_{\alpha\bar\beta}\dot z^\alpha \dot{\bar z}^{\bar\beta} + \mathrm{e}^{2U}V_{\mathrm{BH}}\right), \tag{22.144}$$

and its conserved 'energy',

$$\mathcal{E} \equiv \dot U^2 + g_{\alpha\bar\beta}\dot z^\alpha \dot{\bar z}^{\bar\beta} - \mathrm{e}^{2U}V_{\mathrm{BH}}\,. \tag{22.145}$$

The Euler–Lagrange equations of this action reproduce the scalar equation of motion (22.143) and the second order equation of (22.102) with the replacement (22.142). The remaining first order equation is just the condition $\mathcal{E} = 0$ for solutions of the equivalent one-dimensional system.

With these trivial modifications, all the ideas of Sec. 22.7 can be applied using the central charge $\mathcal{Z}$ in (22.126). In this way one derives the attractor phenomenon for supersymmetric black holes in $\mathcal{N} = 2$ supergravity with n abelian vector multiplets. For further information see [300, 301, 302].

Appendix 22A Killing spinors for pp-waves

We begin this appendix with a guided exercise to explore the relation between the integrability condition (22.57) and the equation of motion $R_{\mu\nu} = 0$ of pure $\mathcal{N} = 1$, $D = 4$ supergravity. Then we study the family of Ricci flat spacetimes known as pp-waves and their Killing spinors. The abbreviation pp-wave stands for plane-fronted waves with parallel propagation.

Exercise 22.29 *Suppose that ϵ and ϵ' are both Killing spinors (possibly with $\epsilon = \epsilon'$). Deduce from (22.57) that*

$$-\tfrac{1}{2}\bar{\epsilon}'\gamma^\rho\gamma^\nu R_{\mu\nu ab}\gamma^{ab}\epsilon = R_{\mu\nu}\bar{\epsilon}'\gamma^\rho\gamma^\nu\epsilon = 0\,. \qquad (22.146)$$

The sum of this relation and its conjugate (with ϵ and ϵ' interchanged) reads (with no torsion)

$$R_{\mu\nu}\bar{\epsilon}'\{\gamma^\rho,\gamma^\nu\}\epsilon = 2R_\mu^\rho\,\bar{\epsilon}'\epsilon = 0\,. \qquad (22.147)$$

We learn that a spacetime with Killing spinors necessarily satisfies $R_{\mu\nu} = 0$ only if there are Killing spinors ϵ and ϵ' such that $\bar{\epsilon}'\epsilon \neq 0$. The relation between Killing spinors and equations of motion is somewhat subtle because the spinor contraction $\bar{\epsilon}'\epsilon$ may vanish.

It is interesting to ask whether there are other classical solutions of pure $\mathcal{N} = 1$, $D = 4$ supergravity, solutions with non-vanishing Riemann tensor, that possess Killing spinors. There are examples within the family of pp-waves with metric

$$\mathrm{d}s^2 = 2H(u,x,y)\mathrm{d}u^2 + 2\mathrm{d}u\mathrm{d}v + \mathrm{d}x^2 + \mathrm{d}y^2\,. \qquad (22.148)$$

The function $H(u,x,y)$ is initially arbitrary. The pp-wave spacetimes are characterized by the existence of a covariantly constant null vector, which is then a Killing vector. In the Brinkmann coordinates above, this vector is simply $K = \partial/\partial v$. As we will see below, the pp-wave metric is Ricci flat if[13] $\nabla^2 H = H_{xx} + H_{yy} = 0$. Then we have a solution of the source-free Einstein equations and thus a classical solution of supergravity.

For $H = 0$ the metric (22.148) reduces to Minkowski spacetime in light-cone coordinates $u = (x-t)/\sqrt{2}$, $v = (x+t)/\sqrt{2}$. We denote the flat metric tensor in these coordinates by $\hat{\eta}_{ab}$, where $a, b = +, -, 1, 2$. Its non-vanishing components are $\hat{\eta}_{+-} = \hat{\eta}_{-+} = \hat{\eta}_{11} = \hat{\eta}_{22} = 1$. To work with spinors we will need constant Dirac matrices that satisfy $\{\gamma_a,\gamma_b\} = 2\hat{\eta}_{ab}$.

We now develop the Killing spinor analysis using the frame 1-forms

$$e^- = \mathrm{d}u\,,\qquad e^+ = \mathrm{d}v + H\mathrm{d}u\,,\qquad e^1 = \mathrm{d}x\,,\qquad e^2 = \mathrm{d}y\,. \qquad (22.149)$$

The metric (22.148) can then be expressed as $\mathrm{d}s^2 = e^a\hat{\eta}_{ab}e^b = e^+e^- + e^-e^+ + e^1e^1 + e^2e^2$. The Cartan structure equations give the connection forms ω^{ab} with non-vanishing components

$$\omega^{+1} = H_x e^-\,,\qquad \omega^{+2} = H_y e^-\,, \qquad (22.150)$$

[13] We use the notation H_x, H_y, H_{xx}, etc. for derivatives with respect to the coordinates x, y.

and the non-vanishing curvature 2-forms

$$R^{+1} = H_{xx}e^1 \wedge e^- + H_{xy}e^2 \wedge e^- \,, \qquad R^{+2} = H_{yx}e^1 \wedge e^- + H_{yy}e^2 \wedge e^- \,. \quad (22.151)$$

We can now write out the Killing spinor conditions $D_\mu \epsilon = 0$ in detail as

$$D_u\epsilon = (\partial_u - \tfrac{1}{2}H_x\gamma^1\gamma^- - \tfrac{1}{2}H_y\gamma^2\gamma^-)\epsilon = 0\,,$$
$$D_v\epsilon = \partial_v\epsilon = 0\,, \qquad D_x\epsilon = \partial_x\epsilon = 0\,, \qquad D_y\epsilon = \partial_y\epsilon = 0\,. \quad (22.152)$$

All conditions are satisfied if we take constant spinors that satisfy the constraint $\gamma^-\epsilon = 0$. There are two independent spinors that satisfy this condition. It is easy to check that the contraction $\bar{\epsilon}'\epsilon = 0$ for any pairing of such spinors, since $2\bar{\epsilon}'\epsilon = \bar{\epsilon}'(\gamma^+\gamma^- + \gamma^-\gamma^+)\epsilon$. So we encounter the special case of Ex. 22.29 above in which the Killing spinors do not imply that the equation of motion is satisfied. Indeed this is why we have found the spinors with no restrictions on $H(u, x, y)$.

To complete the analysis we need the Ricci tensor. From (22.151) we find the non-vanishing frame components of the curvature tensor:

$$R_{1-}{}^{+1} = H_{xx}\,, \qquad R_{2-}{}^{+1} = H_{xy}\,, \qquad R_{1-}{}^{+2} = H_{xy}\,, \qquad R_{2-}{}^{+2} = H_{yy}\,. \quad (22.153)$$

The only non-vanishing component of the Ricci tensor is R_{--}. It is given by the index contraction $R_{--} = R_{-1-}{}^1 + R_{-2-}{}^2 = -(H_{xx} + H_{yy})$. Thus the pp-wave is a solution of $R_{\mu\nu} = 0$ if and only if H is harmonic in the variables x, y. The dependence on u is arbitrary.

In summary we have obtained two independent Killing spinors for pp-wave spacetimes, and we have shown that it is a classical solution of $\mathcal{N} = 1$, $D = 4$ supergravity if $H_{xx} + H_{yy} = 0$. Since the maximum number of independent Killing spinors is 4, we say that the pp-wave is a $\frac{1}{2}$-BPS solution. Some exercises are appropriate.

Exercise 22.30 *For any pair ϵ, ϵ' of Killing spinors, show that all components of the bilinear $\bar{\epsilon}'\gamma^a\epsilon$ vanish except $\bar{\epsilon}'\gamma^+\epsilon$. Thus we obtain the unique Killing vector $K = \bar{\epsilon}'\gamma^+\epsilon\partial/\partial v$ from the Killing spinors.*

Exercise 22.31 *Show that the integrability condition (22.57) cannot be satisfied by a Schwarzschild black hole, so this is one example of a Ricci flat spacetime with no Killing spinors. The Riemann tensor can be found from the curvature 2-form given in (22.32).*

23 The AdS/CFT correspondence

The AdS/CFT correspondence arose as a duality between Type IIB string theory in the background geometry $\mathrm{AdS}_5 \otimes S^5$ and the maximally supersymmetric quantum field theory (QFT) in $D = 4$ dimensions, the $\mathcal{N} = 4$ super-Yang–Mills (SYM) theory [280, 281, 282]. Most concrete results have been obtained in a low energy limit in which the string theory is well approximated by classical supergravity, initially $D = 10$ Type IIB supergravity. Applications have broadened greatly; as practiced now, the subject includes a general correspondence between theories of gravity (plus other fields) in $D + 1$ spacetime dimensions and quantum field theories without gravity in D dimensions. Tractable calculations in a classical approximation on the gravity side yield information about quantum systems in a strong coupling limit for which the traditional techniques of quantum field theory are inadequate. It is truly surprising how much information about D-dimensional quantum systems can be captured by classical gravity in $D + 1$ dimensions. Field theory also leads to new insights into gravity.

Two theories related by the correspondence must have the same symmetries. The isometry group of AdS_5, namely SO(4, 2), is a particular case of the SO($D - 2$, 2) symmetry group discussed in Sec. 22.1.3. The isometry group of the sphere S^5 is the compact group SO(6). This group has the same Lie algebra as SU(4), and we use the two symbols interchangeably. The global internal symmetry of $\mathcal{N} = 4$ SYM theory is also SU(4). It is frequently called the R-symmetry group because, as in the case of the $\mathrm{U}(1)_R$ symmetry of the $\mathcal{N} = 1$ SUSY theories discussed in Sec. 6.2.1, fermions and bosons transform in different representations. The spacetime symmetry of the $\mathcal{N} = 4$ theory is more unusual, although well known and understood. All fields of the theory are massless, and the only coupling constant is dimensionless, so the action integral is *formally* invariant under conformal transformations. What is remarkable is that the renormalization effects which spoil the naive conformal symmetry in most massless theories are absent in $\mathcal{N} = 4$ SYM.

The conformal group in four dimensions is SO(4, 2), so it does match the isometry of AdS_5. It is a 15-parameter group which includes the translations and Lorentz transformations of the Poincaré group, the dilatation or scale transformation of all coordinates, and special conformal transformations. The Lie algebras of SO(4, 2) and SU(2, 2) are equivalent. The theory is also invariant under four Poincaré supercharges whose anti-commutator is a spacetime translation and four conformal supercharges whose anti-commutator is a special conformal transformation. All symmetries combine into the superalgebra SU(2, 2|4), which is the $\mathcal{N} = 4$ extension of the SU(2, 2|1) superalgebra discussed in Sec. 16.1.1. The superstring background (in its near-horizon limit) also has the same superalgebra, so the match of symmetries with those of $\mathcal{N} = 4$ SYM is complete.

The AdS/CFT correspondence goes well beyond a match of symmetries and includes a great deal of dynamics. Significant strong coupling results and other surprising features of the gauge theory were suggested by analysis on the gravity side of the correspondence and then confirmed in the $\mathcal{N} = 4$ theory. Some of these results will be discussed in the body of the chapter below. The correspondence has also been applied to theories with less symmetry than $\mathcal{N} = 4$ SYM, including situations without conformal symmetry. This includes field theories that describe a renormalization group (RG) flow between different conformal theories in the short and long distance limits. We will discuss the basics of these 'holographic RG flows'. The physics of finite temperature in quantum field theory can also be approached from the AdS/CFT viewpoint, and some aspects will be discussed. Our goals are to convince readers of the power of the AdS/CFT approach and to engage them in the computations necessary to use it effectively.

There are several review articles and lecture notes devoted entirely to the AdS/CFT correspondence; for example, see [303, 304, 305, 306]. Our treatment is necessarily less thorough. The intricate scaling arguments concerning the physics of $D3$-branes in Type II string theory, which motivated the initial conjecture, will not be discussed. A full explanation of some of the needed features of field theory at the quantum level is also beyond the scope of this chapter. We suggest that readers consult the reviews above and several recent books which contain chapters on the subject [124, 307, 308, 309]. These sources provide further results and different perspectives.

The paradigm that underlies AdS/CFT is that there is a $1:1$ correspondence between local fields in the gravity theory and operators in the quantum field theory, typically gauge invariant composite operators. The mass of a field in the gravity theory is related to the scale dimension of the corresponding operator. AdS$_{d+1}$ has a boundary at spatial infinity. In the coordinates of the Poincaré patch (see Sec. 22.1.3) the boundary is a copy of d-dimensional Minkowski space. Asymptotic conditions on the fields of the gravity theory provide the bridge between the two sides of the correspondence. The boundary data of a classical solution of the gravity theory determine n-point correlation functions of the field theory operators. If the gravity solution is pure AdS$_{d+1}$, these correlators are covariant under conformal transformations. If the bulk solution looks like AdS$_{d+1}$ near the boundary but differs in the interior, then conformal symmetry is broken. The boundary data can then act as sources for operators, which perturb a conformal theory and induce renormalization group flows. An alternative is that conformal symmetry is broken by non-vanishing one-point functions $\langle \mathcal{O}(x) \rangle$, which indicate that the boundary CFT$_d$ is not in its ground state.

Box 23.1 AdS/CFT correspondence

D-dimensional quantum systems can be captured by classical gravity in $D+1$ dimensions. There is a $1:1$ correspondence between local fields in the gravity theory and operators in the quantum field theory. The boundary data of a classical solution of the gravity theory determine n-point correlation functions of the field theory operators. If the bulk spacetime includes a black hole with non-zero Hawking temperature, then the gravity dual is a CFT in a thermal ensemble.

If the bulk spacetime includes a black hole with non-zero Hawking temperature, then the gravity dual is a CFT in a thermal ensemble.

23.1 The $\mathcal{N} = 4$ SYM theory

The $\mathcal{N} = 4$ SYM theory contains the gauge potential A_μ, four chiral fermion fields $P_L \chi^\alpha$ and their anti-chiral conjugates $P_R \chi_\alpha$, plus six real scalars X^i. The indices $\alpha = 1, \ldots, 4$ and $i = 1, \ldots, 6$ describe the fundamental, anti-fundamental, and vector representations of the R-symmetry group SU(4) $\sim$ SO(6). All fields transform in the adjoint representation of the gauge group G. Although the theory can be formulated for any gauge group, the AdS/CFT properties are simplest when $G =$ SU(N). We confine our discussion to this case and use the matrix notation of Sec. 4.3.4. Each component field is an $N \times N$ matrix, for example $X^i = t_A X^{iA}$. The matrices t_A, $A = 1, \ldots, N^2 - 1$, are traceless anti-hermitian generators of the fundamental representation, normalized as in (4.97).

The Lagrangian of the theory can be written in several equivalent forms. In the following form the R-symmetry is manifest:

$$\begin{aligned}\mathcal{L} = \mathrm{tr}\Big\{ & \tfrac{1}{2} F_{\mu\nu} F^{\mu\nu} + 2\bar{\lambda}^\alpha \gamma^\mu D_\mu P_L \lambda_\alpha + D_\mu X^i D^\mu X^i \\ & + \Big(g\, (C^\alpha{}_\beta)^i \bar{\lambda}_\alpha P_L [X^i, \lambda^\beta] + \mathrm{h.c.} \Big) + \tfrac{1}{2} g^2 \sum_{i,j} [X^i, X^j]^2 \Big\}. \end{aligned} \tag{23.1}$$

The 4×4 matrices C^i are given in terms of Pauli matrices by

$$\begin{aligned} C^1 &= \begin{pmatrix} 0 & \sigma_1 \\ -\sigma_1 & 0 \end{pmatrix}, & C^2 &= \begin{pmatrix} 0 & -\sigma_3 \\ \sigma_3 & 0 \end{pmatrix}, & C^3 &= \begin{pmatrix} \mathrm{i}\sigma_2 & 0 \\ 0 & \mathrm{i}\sigma_2 \end{pmatrix}, \\ C^4 &= \begin{pmatrix} 0 & \mathrm{i}\sigma_2 \\ \mathrm{i}\sigma_2 & 0 \end{pmatrix}, & C^5 &= \begin{pmatrix} 0 & 1 \\ -1 & 0 \end{pmatrix}, & C^6 &= \begin{pmatrix} -\mathrm{i}\sigma_2 & 0 \\ 0 & \mathrm{i}\sigma_2 \end{pmatrix}. \end{aligned} \tag{23.2}$$

One can regroup component fields to show that this Lagrangian is a special case of the general $\mathcal{N} = 1$ SUSY gauge theory of Sec. 6.3. There is a gauge multiplet plus three chiral multiplets whose scalar components are $Z^i = (X^i + \mathrm{i}X^{i+3})/\sqrt{2}$, $i = 1, 2, 3$. The superpotential is $W = w\, \mathrm{tr}(Z^1, [Z^2, Z^3])$. The $\mathcal{N} = 1$ form of the Lagrangian suggests that it describes a quite conventional renormalizable theory whose ultraviolet divergences lead to running gauge coupling constants $g(\mu)$, $w(\mu)$ governed by renormalization group equations $(\mu \mathrm{d}/\mathrm{d}\mu) g = \beta_g(g, w)$ and $(\mu \mathrm{d}/\mathrm{d}\mu) w = \beta_w(g, w)$. However, higher loop calculations and general arguments indicate that the β-functions vanish exactly when $w = \sqrt{2}\, g$, so that the $\mathcal{N} = 4$ theory enjoys superconformal SU(2, 2 | 4) symmetry at the quantum level.

Observables in a gauge theory must be gauge invariant. The matrix trace is used to construct gauge invariant operators from products of elementary fields, such as the operators

$$\operatorname{tr} X^k \equiv N^{(1-k)/2} \operatorname{tr}\left(X^{(i_1} X^{i_2} \ldots X^{i_k)}\right) - \ldots. \tag{23.3}$$

The ... indicate the subtraction of all contractions of the SO(6) indices. These operators are rank k, symmetric, traceless tensors, and they transform in representations of SO(6) whose Dynkin designation is $[0, k, 0]$. For $k = 2, 3, 4$, the dimensions of these representations are 20, 50, 105, respectively.

Exercise 23.1 *What is the dimension of the* $[0, 5, 0]$ *representation?*

We will be concerned with n-point correlation functions of these operators, namely $F_n(x_i) = \langle \operatorname{tr} X^{k_1}(x_1) \operatorname{tr} X^{k_2}(x_2) \ldots \operatorname{tr} X^{k_n}(x_n)\rangle$. In the weak coupling limit these correlation functions are expressed as sums of Feynman diagrams. It is advantageous to organize the perturbation series in terms of N and the 't Hooft coupling $\lambda = g^2 N$. Diagrams[1] with Euler number χ in the sense of graph theory carry the factor N^χ. The contribution of diagrams of fixed χ can then be presented as a series expansion in λ, schematically

$$F_n^\chi(x_i) = N^\chi\left(f_0(x_i) + \lambda f_1(x_i) + \lambda^2 f_2(x_i) + \ldots\right). \tag{23.4}$$

Therefore planar diagrams, those with $\chi = 2$, dominate in the large N limit. Diagrams with more complex topologies, i.e. $\chi \leq 0$, are suppressed by powers of $1/N^2$.

Lorentz scalar operators $\mathcal{O}(x)$ in a conformal field theory in Minkowski spacetime are characterized by their behavior under scale transformations $x^\mu \to x'^\mu = \rho^{-1} x^\mu$ of the coordinates. A scalar operator transforms as

$$\mathcal{O}(x) \to \mathcal{O}'(x') = \rho^\Delta \mathcal{O}(x). \tag{23.5}$$

The scale dimension Δ is a real number, which must satisfy the unitarity bound $\Delta \geq 1$ required by the representation theory of SO(4, 2). If $\Delta = 1$, then $\mathcal{O}(x)$ is a free massless scalar field. If $\mathcal{O}(x)$ is a generic scalar field then its scale dimension would be expected to depend on N and λ. However, there are important examples of operators whose scale dimension is fixed, as we discuss below. The scale dimension of an operator determines the form of its two-point function, viz.

$$\langle \mathcal{O}(x)\mathcal{O}(y)\rangle = \frac{c}{(x-y)^{2\Delta}}. \tag{23.6}$$

In a superconformal field theory specific sets of bosonic and fermionic operators are joined in unitary irreducible representations of the relevant superalgebra, SU(2, 2|4) in the case of the $\mathcal{N} = 4$ SYM theory. A typical representation contains a lowest dimension primary operator, often a scalar $\mathcal{O}_\Delta(x)$, plus descendent operators, obtained by applying one or more factors of the (Poincaré) supercharges Q to the primary. The superconformal algebra SU(2, 2|1) discussed in Sec. 16.1.1 contains the commutator $[D, Q] = \frac{1}{2}Q$. This shows that the first descendent of $\mathcal{O}_\Delta(x)$ has dimension $\Delta + 1/2$.

[1] Since fields in the adjoint representation carry a pair of SU(N) indices, diagrams are drawn with double lines. See the review [310].

An important feature of the representation theory of SU(2, 2 | 4) is the existence of 'short representations' – representations in which the scale dimension of the primary operator is locked at a value determined by its R-symmetry properties. The operators tr X^k defined above are the primary operators of the simplest family of short representations. As we saw, the Dynkin labels of the representation are $[0, k, 0]$. The scale dimension of tr X^k is fixed as $\Delta = k$. This relation cannot be renormalized unless the underlying superconformal symmetries are broken. The superalgebra also has long representations in which the primary scale dimension Δ is restricted only by a lower bound due to unitarity. One simple example is the Konishi operator $K = \text{tr}(X^i X^i)$. This operator is an SO(6) singlet. Weak coupling calculations have determined the scale dimension $\Delta_K = 2 + 3\lambda/4\pi^2 + \mathcal{O}(\lambda^2)$. This shows that Δ_K depends continuously on the coupling constant of the $\mathcal{N} = 4$ theory. An early result [281] of the AdS/CFT correspondence showed that, when λ is large, the anomalous dimension is of order $\Delta_K \approx \lambda^{1/4}$, independent of N.

The superalgebra SU(2, 2 |4) contains eight Q supercharges and their eight chiral conjugates. In a typical long representation each of the 2^{16} possible products of the 16 Q's produces a distinct operator of the representation, and the range of spins in a representation is $\Delta s = 4$. Short representations contain far fewer operators because some products of Q's annihilate the operator to which they are applied rather than produce new components of the representation. For this reason the range of spins in the short representations headed by tr X^k is $\Delta s = 2$.

Descendent operators can be very important. In fact the short representation headed by the chiral primary operator tr X^2 contains the symmetry currents of the $\mathcal{N} = 4$ theory as descendent operators. These are the stress tensor $T_{\mu\nu}$, the 15 R-symmetry currents J_μ, and the four supercurrents $\mathcal{J}_\mu$ (and their chiral conjugates).

One counts the independent components of these operators by taking the constraints due to superconformal symmetry into account. For example, the symmetric stress tensor has 10 components, but there are constraints due to the conservation law $\partial^\mu T_{\mu\nu} = 0$ and the traceless property $T_\mu{}^\mu = 0$. This means that $T_{\mu\nu}$ contains five independent components. Similarly each conserved R-symmetry current has three independent components. Each chiral projection of the four vector–spinor supercurrents has eight components and satisfies both a conservation law and the superconformal constraint $\gamma^\mu \mathcal{J}_\mu = 0$. Each constraint imposes two conditions, leaving four independent components. The result is summarized by (the last factor 2 for $\mathcal{J}_\mu$ counts the two chiralities)

$$T_{\mu\nu}\colon 5 \text{ comps.}, \qquad J_\mu\colon 3 \times 15 \text{ comps.}, \qquad \mathcal{J}_\mu\colon 4 \times 4 \times 2 \text{ comps.} \tag{23.7}$$

The first factors in each case agree with the count of on-shell degrees of freedom of the relevant gauge field in five-dimensional spacetime, namely the metric, vector potential, and gravitino; see Table 6.2. The match of these numbers is an important clue to a possible correspondence between gauge field theory in four dimensions and gravity in five dimensions!

In addition to the symmetry currents discussed above, the short representation with chiral primary operator tr X^2 contains other descendent operators. There are a total of 128 bosonic plus 128 fermionic operators in the representation.

23.2 Type IIB string theory and $D3$-branes

The story must start with some bare facts about Type IIB string theory. Length scales in string theory are measured with respect to the scale α' which has dimensions of (length)^2, and there is a dimensionless coupling constant g_s. The $D = 10$ gravitational coupling is the product $\kappa_{10}^2 = 8\pi G_{10} = 64\pi^7 g_s^2 \alpha'^4$.

Originally Type IIB string theory appeared to describe only closed strings. But it was found [311] that it also includes open strings whose endpoints lie on $(p+1)$-dimensional hypersurfaces in spacetime called Dp-branes. The D in Dp-branes indicates that the vibrating open string satisfies a Dirichlet boundary condition on the branes. The Type IIB string theory contains odd-dimensional (charged) branes, i.e. $p = 1, 3, 5, 7, 9$. At low energies, open strings ending on the D-branes decouple from closed strings, and their dynamics is described by a gauge field theory. The specific properties of the gauge theory depend on the configuration of the branes. On a stack of N-coincident Dp-branes, it is a non-abelian theory with gauge group SU(N). For a stack of $D3$-branes the particle content and global symmetries are those of the $\mathcal{N} = 4$ SYM theory. The brane description [17, Sec. 13.3] implies the relation $g_s = g_{YM}^2/4\pi$ between the string coupling and the coupling constant of the $\mathcal{N} = 4$ field theory.

The AdS/CFT correspondence between string theory on AdS$_5 \otimes S^5$ and $\mathcal{N} = 4$ SYM was postulated to hold for all values of the parameters g_s and N. However, it is difficult to derive quantitative tests of this strong form of the conjecture (but see [312]). It is therefore important that there is a limit in which string theory is well approximated by supergravity and especially by classical supergravity in which calculations are tractable.

Roughly speaking, supergravity is a good approximation to string theory when the extended structure of strings is not detectable. This means that the length scale L associated with a field configuration in supergravity must be long compared to the string scale, i.e. $L^2 \gg \alpha'$. It is equivalent to say that the scale of the curvature tensor $R_{MNPQ} \sim 1/L^2$ must be very small so that higher derivative corrections to the basic supergravity action are negligible. (We use upper case M, N, P, Q to denote $D = 10$ coordinate indices.) As we will see below, the classical $D3$-brane solution in Type IIB supergravity has the length scale $L^4 = 4\pi g_s \alpha'^2 N$. Thus supergravity is a valid approximation to string theory if $g_s N \gg 1$.

However, we need another condition to suppress quantum effects, since quantum corrections to supergravity suffer from ultraviolet divergences in many cases, and the problem of quantization of the superstring in AdS$_5 \otimes S^5$ is not yet understood. Thus the condition we need is $g_s \ll 1$. This means that the string is weakly interacting. It also means that κ_{10}^2 is very small, so quantum effects in supergravity can be neglected. It is equivalent to write these conditions as $g_{YM}^2 N \gg 1$ and $g_{YM}^2 \ll 1$. Clearly N must be very large. Further, in the previous section, we presented 't Hooft's argument that the effective coupling in an SU(N) theory is $\lambda = g_{YM}^2 N$. These arguments motivate the possibility of a duality between supergravity and gauge theory in the planar, but strongly coupled, limit of the gauge theory and in a weak coupling, and therefore calculable, limit of supergravity.

23.3 The $D3$-brane solution of Type IIB supergravity

We describe only the essential features necessary to understand how the $D3$-brane solution arises and how $AdS_5 \otimes S^5$ appears as its 'near-horizon limit'. For this purpose we need only a subset of the bosonic fields of the theory, namely the metric g_{MN}, 5-form field strength F_{MNPQR}, and the scalar dilaton ϕ. Other fields consistently decouple from this small subset whose dynamics is governed by the action[2]

$$S_{\mathrm{IIB}} = \frac{1}{8\pi G_{10}} \int \mathrm{d}^{10}z \sqrt{g_{10}} \left\{ \frac{1}{2} R_{10} - \frac{1}{2\times 5!} F_{MNPQR} F^{MNPQR} - \frac{1}{2} \partial_M \phi \partial^M \phi \right\}. \tag{23.8}$$

Using x^μ, $\mu = 0, 1, 2, 3$, as Cartesian coordinates of Minkowski space and y^a, $a = 1, 2, 3, 4, 5, 6$, as coordinates of a flat Euclidean signature 'transverse space', we write the following ansatz for the set of fields above:

$$\begin{aligned} \mathrm{d}s_{10}^2 &= \frac{1}{\sqrt{H(y^a)}} \eta_{\mu\nu}\, \mathrm{d}x^\mu \mathrm{d}x^\nu + \sqrt{H(y^a)}\, \delta_{ab}\, \mathrm{d}y^a \mathrm{d}y^b, \\ F &= \mathrm{d}A + {}^*(\mathrm{d}A), \qquad A = \frac{1}{H(y^a)} \mathrm{d}x^0 \wedge \mathrm{d}x^1 \wedge \mathrm{d}x^2 \wedge \mathrm{d}x^3, \\ \phi &\equiv 0. \end{aligned} \tag{23.9}$$

Remarkably the configuration above is a solution of the equations of motion provided that $H(y^a)$ satisfies the Laplace equation in six dimensions, i.e.

$$\sum_{a=1}^{6} \frac{\partial^2}{\partial y^a \partial y^a} H = 0. \tag{23.10}$$

Exercise 23.2 *Verify that the metric and 5-form of (23.9) is a solution of the equations of motion of S_{IIB} if $H(y^a)$ is a harmonic function. Compute the connection and curvature of the metric as an intermediate step. See the discussion of the Cartan structure equations in Sec. 22.1.5 for some guidance.*

The solution that describes a set of N coincident $D3$-branes located at the origin of the y^a coordinates is

$$H = 1 + \frac{L^4}{r^4}, \qquad L^4 = 4\pi\alpha'^2 g_s N, \qquad r = \sqrt{y^a y^a}. \tag{23.11}$$

Using the radial coordinate r and five angular coordinates of a 5-sphere, we can write the 10-dimensional line element as

[2] The actual equation of motion for the 5-form is the self-dual condition $F_5 = {}^*F_5$. This is compatible with the Euler–Lagrange equations of (23.8), but it must be imposed as an extra condition.

Box 23.2 SU(2, 2|4)

The superalgebra SU(2, 2|4), which is the symmetry of $\mathcal{N} = 4$ super-Yang–Mills theory, is also the symmetry of the metric in the near-horizon limit of the $D3$-brane solution of Type IIB supergravity.

$$\mathrm{d}s_{10}^2 = \frac{1}{\sqrt{H(r)}}\eta_{\mu\nu}\,\mathrm{d}x^\mu \mathrm{d}x^\nu + \sqrt{H(r)}\,(\mathrm{d}r^2 + r^2\mathrm{d}\Omega_5). \tag{23.12}$$

Here $\mathrm{d}\Omega_5^2$ is the SO(6) invariant metric on the unit S^5. The integer N is determined by the flux of the 5-form through the 5-sphere.

The solution describes a spacetime that is asymptotically flat as $r \to \infty$. In the limit as $r \to 0$ the spacetime is non-singular but has a horizon. The region $r \approx 0$ is a geodesically long 'throat'. Near the horizon, the constant 1 in the harmonic function $H(r)$ can be neglected, so the limiting form of the metric is

$$\begin{aligned}\mathrm{d}s_{10}^2 &= \frac{r^2}{L^2}\eta_{\mu\nu}\mathrm{d}x^\mu \mathrm{d}x^\nu + \frac{L^2\mathrm{d}r^2}{r^2} + L^2\mathrm{d}\Omega_5^2 \\ &= \frac{L^2}{z^2}\left[\mathrm{d}z^2 + \eta_{\mu\nu}\mathrm{d}x^\mu \mathrm{d}x^\nu\right] + L^2\mathrm{d}\Omega_5^2.\end{aligned} \tag{23.13}$$

It is easy to recognize that the 10-dimensional spacetime becomes a product space $M_5 \otimes S^5$ in the near-horizon limit. In the second line we have used the new radial coordinate $z = L^2/r$, so that readers can compare with (22.22) and learn that M_5 is just the Poincaré patch of AdS$_5$!

The mathematics of the near-horizon limit is quite straightforward, but the arguments for its physical relevance are rather complex. We simply state that it is the throat region of the geometry that determines the physics of the AdS/CFT correspondence, and we restrict our attention to the metric (23.13).

The obvious symmetry of the near-horizon limit is the product SO(4, 2) $\otimes$ SO(6), which is the isometry group of AdS$_5 \otimes S^5$. Killing spinor analysis (see Ch. 22) of the fermion transformation rules of Type IIB supergravity shows that there are also 32 conserved supercharges. Thus, the superalgebra SU(2, 2|4) operates in the near-horizon limit.

23.4 Kaluza–Klein analysis on AdS$_5 \otimes S^5$

The metric AdS$_5 \otimes S^5$ is the product of a non-compact spacetime with a compact 'internal space'. We will be interested in the dynamics of fluctuations about this 'background geometry'. It is common practice to use the Kaluza–Klein (KK) approach in which the fields of Type IIB supergravity are expressed as series expansions in the spherical harmonics of S^5. Each Kaluza–Klein mode is then viewed as a field on AdS$_5$. The equations of motion of these fields are those of a $D = 5$ supergravity theory expanded about the AdS$_5$

background. Among the infinite number of KK fields are a set of 15 massless vector gauge fields of the SO(6) isometry group of the 5-sphere. Thus the memory of the S^5 factor in the $D = 10$ background metric persists in the compactified theory.

The KK modes interact with the strength of the effective $D = 5$ Newton constant

$$\frac{\kappa_5^2}{8\pi} = G_5 = \frac{G_{10}}{\text{Vol}(S^5)} = \frac{\pi L^3}{2N^2}. \tag{23.14}$$

We used $\text{Vol}(S^5) = \pi^3 L^5$ in the last step. If one substitutes the formula for L from (23.11), one gets $G_5 = \pi\alpha'^{3/2}\lambda^{3/4}/2N^2$. Thus $G_5 \to 0$ in the limit $N \to \infty$ and λ fixed but large. So again we find that gravity is weakly coupled.

The linearized equations of motion of the KK modes were analyzed in [313, 314]. The analysis is complicated because the properties of vector and tensor spherical harmonics are needed, and because the independent $D = 5$ modes are mixtures of modes from different $D = 10$ fields. Nevertheless an elegant result emerges. There is a perfect 1 : 1 correspondence between the modes obtained from the Kaluza–Klein analysis and the primary and descendent operators of the short representations of the superalgebra SU(2, 2|4) discussed in Sec. 23.1. In particular there are $D = 5$ supergravity modes $\phi_k(z, x^\mu)$ that are dual to the chiral primary operators $\text{tr}\, X^k$.

The lowest multiplet is headed by ϕ_2 and also contains the $D = 5$ spacetime metric fluctuation, the 15 SO(6) gauge fields, and gravitinos that transform in the $4 + \bar{4}$ representations of the gauge group. These are the supergravity fields which are dual to the symmetry currents of the $\mathcal{N} = 4$ SYM theory; see (23.7). The dynamics of the full set of these fields is governed by the maximal gauged extended $D = 5$ supergravity theory of [315].

Each field of the lowest multiplet is the base of a KK tower of fields of increasing scale dimension. Scalar fields can be expanded in terms of the rank k spherical harmonics[3] $Y^k(y^\alpha)$ on S^5. Each Y^k is an eigenfunction of the invariant Laplacian on S^5 with eigenvalue $m^2 = k(k + 4)/L^2$. Most scalar fields in the $D = 5$ theory involve mixtures of $D = 10$ fields, and there are several cases to consider. Effectively, after mixing is implemented, there are expansions of the form

$$\phi(r, x^\mu, y^\alpha) = \sum_k \phi_k(r, x^\mu) Y^k(y^\alpha). \tag{23.15}$$

Each KK mode $\phi_k(r, x^\mu)$ is a scalar field on AdS_5.

When the expansions are substituted in the $D = 10$ field equations, we find that each Kaluza–Klein mode ϕ^k satisfies an equation of the form

$$(\Box_{\text{AdS}} - m^2)\phi_k = \text{ nonlinear interaction terms.} \tag{23.16}$$

The symbol $\Box_{\text{AdS}}$ is the invariant Laplacian on AdS_5. The terms on the right-hand side involve products of many KK modes. They are determined by nonlinear terms of the $D = 10$ Type IIB Lagrangian. Owing to mixing, it is very difficult to find them explicitly, and results are known only in special cases. The mass m^2 is a function of the mode index k which depends on which scalars are considered. The KK modes of the $D = 10$ dilaton are simplest because there is no mixing. From the quadratic terms of the action

[3] For simplicity we omit the three subsidiary 'quantum numbers' carried by the Y^k.

Box 23.3 **AdS representation and CFT scale dimension**

In all applications of AdS/CFT, whether or not supersymmetry is present, it is very basic that the same lowest energy eigenvalue Δ, which labels the AdS group representation of ϕ, is also the scale dimension of its dual CFT operator.

(23.8), one can show that the mass $m^2 = k(k+4)/L^2$, with $k \geq 0$, is simply the eigenvalue of the Laplacian on S^5 acting on Y^k. For the scalars dual to chiral primary operators, mixing produces the mass $m^2 = k(k-4)/L^2$, with $k \geq 2$.

It is useful to take a general viewpoint and consider scalar fields $\phi(r, x^\mu)$ in AdS_5 with general mass parameter m^2. In the same way that every scalar field in Minkowski space contains an infinite number of momentum modes, each AdS_5 scalar contains an infinite number of modes classified in unitary irreducible representations of the AdS_5 isometry group SO(4, 2). This group has maximal compact subgroup SO(4) $\otimes$ SO(2). A scalar field transforms in a representation labeled by the lowest eigenvalue Δ of the SO(2) generator.[4] This eigenvalue is related to the mass by

$$\Delta = 2 + \sqrt{4 + m^2 L^2}. \tag{23.17}$$

In general Δ need not be an integer, but Δ's for scalars coming from the KK analysis above are locked at integer values because those scalars are further unified in short representations of the superalgebra SU(2, 2|4); see Box 23.3.

An interesting feature of dynamics in AdS spacetime is that m^2 can be negative without instability. Unitarity of the representation of the AdS group requires that Δ is real, so negative values of the mass squared are allowed provided that the bound [183, 184] $m^2 \geq -4/L^2$ is satisfied. In AdS_{d+1}, the relation between lowest eigenvalue and scalar m^2 becomes

$$\Delta = \tfrac{1}{2}\left(d + \sqrt{d^2 + 4m^2L^2}\right). \tag{23.18}$$

The unitarity/stability bound [185] is then $m^2 \geq -d^2/4L^2$.

23.5 Euclidean AdS and its inversion symmetry

It is simpler to discuss the basic ideas of the AdS/CFT correspondence using the Euclidean continuation of AdS spacetime. For flexibility in applications we keep the dimension $d+1$ arbitrary. Euclidean AdS is more properly called hyperbolic space H_{d+1}. It is the upper sheet ($Y^{d+2} \geq L$) of the hyperboloid

[4] The generator is the Killing vector Ld/dt, where t is the time coordinate of a global coordinate chart; see (22.17). In Poincaré patch coordinates the appropriate generator is the scale transformation Lzd/dz.

$$\sum_{i=1}^{d+1}(Y^i)^2 - (Y^{d+2})^2 = -L^2, \tag{23.19}$$

embedded in a $(d+2)$-dimensional Minkowski spacetime. The symmetry group is $SO(d+1,1)$. By a construction similar to (22.20), one can introduce Cartesian coordinates $z_0, z_1, \dots, z_d$ with $z_0 > 0$ and write the induced metric as

$$ds^2 = \frac{L^2}{z_0^2}\left[(dz_0)^2 + \sum_{i=1}^{d}(dz_i)^2\right], \tag{23.20}$$

in the form of the Lobachevsky upper half-space.[5]

The hypersurface $z_0 = 0$ is at infinite geodesic distance from any interior point. It is the conformal boundary of the manifold. As we will see, it is necessary to specify asymptotic data at this surface to obtain a solution of field equations on Euclidean AdS. This leads to the somewhat unconventional boundary value problems, which are central in the AdS/CFT correspondence.

The Killing vectors that generate the SO(5, 1) isometry group are very similar to those given for SO(4, 2) at the end of Sec. 22.1.2, and we shall not repeat them. Instead we focus on the discrete inversion symmetry of Euclidean AdS, which is given by the coordinate transformation:

$$\text{inversion:} \qquad z'_\mu = \frac{z_\mu}{z^2}, \qquad z'^2 = \frac{1}{z^2}. \tag{23.21}$$

Exercise 23.3 *Show that the inversion leaves the metric (23.20) invariant. Show that a finite special conformal transformation*

$$z'_\mu = \frac{(z_\mu + z^2 a_\mu)}{(1 + 2z \cdot a + z^2 a^2)} \tag{23.22}$$

can be expressed as the product (inversion)(translation by a^μ)(inversion). Since the metric (23.20) is translation invariant only in the coordinates z_i, the vector a_μ is restricted to $a_\mu \to (a_0 = 0, a_i)$. The conformal transformation (23.22) is difficult to work with, so it is quite important that a test of conformal symmetry can be reduced to a test of inversion.

There are many well-known applications of inversion to conformal field theories in flat space (which we will explore below), and it is also useful for field theories in AdS. In fact it works more simply in AdS, where it is an *isometry*, than in flat space, where it acts as a *conformal isometry*.

[5] We will usually be indifferent to upper versus lower coordinate indices, i.e. $z^\mu = z_\mu$, and we will frequently represent the coordinates z_i perpendicular to the radial direction by $\vec{z}$. We define the contraction of the $d+1$ indices as $z^2 = \delta^{\mu\nu} z_\mu z_\nu$ and of the d indices $i = 1, \dots, n$ by $(\vec{z})^2 = \delta^{ij} z_i z_j$.

The Jacobian of the inversion is a very useful object, namely

$$\frac{\partial z'_\mu}{\partial z_\nu} = (z')^2 J_{\mu\nu}(z), \qquad J_{\mu\nu}(z) = J_{\mu\nu}(z') = \delta_{\mu\nu} - 2\frac{z_\mu z_\nu}{z^2}. \tag{23.23}$$

It tells us how vector fields on the manifold transform. For example, the transformation (7.3) of a scalar field under a diffeomorphism, namely $\phi'(z') = \phi(z)$, implies that its gradient transforms under inversion as

$$\partial_\mu \phi'(z') = \frac{\partial z_\nu}{\partial z'_\mu}\,\partial_\nu \phi(z) = z^2\left(\delta_{\mu\nu} - 2\frac{z_\mu z_\nu}{z^2}\right)\partial_\nu \phi(z). \tag{23.24}$$

Several properties of the Jacobian tensor $J_{\mu\nu}(z)$ are very useful in applications to both Euclidean AdS and to conformal field theories in flat Euclidean space.

1. The Jacobian satisfies $J_{\mu\rho}(z)J_{\rho\nu}(z) = \delta_{\mu\nu}$, so it is an orthogonal matrix. However, it has d eigenvalues $+1$ and one eigenvalue -1. Thus, viewed as a matrix, the Jacobian is an element of the group $O(d+1)$ that is not in its proper subgroup $SO(d+1)$. As an isometry, inversion is an improper reflection that cannot be continuously connected to the identity in $SO(d+1,1)$.
2. Let z_μ, w_μ denote two points with z'_μ, w'_μ their images under inversion. Then
$$\frac{1}{(z-w)^2} = \frac{z'^2\, w'^2}{(z'-w')^2}. \tag{23.25}$$
(This property is most useful for conformal field theory since $1/(z-w)^{2\Delta}$ is proportional to the two-point correlation function of a scalar primary operator; see (23.6).)
3. Note the (tedious to prove but important) inversion property
$$J_{\mu\nu}(z-w) = J_{\mu\mu'}(z')J_{\mu'\nu'}(z'-w')J_{\nu'\nu}(w'). \tag{23.26}$$

Exercise 23.4 *Derive the properties discussed above.*

23.6 Inversion and CFT correlation functions

We discussed the properties of inversion in AdS_{d+1}, but they are essentially the same in flat d-dimensional Euclidean space. We simply replace z_μ, w_μ by d-vectors x_i, y_i and take $x_i = x'_i/x'^2$, etc. Inversion is now a conformal isometry and a symmetry of CFT_d. Under the inversion $x_i \to x'_i$, a scalar operator of scale dimension Δ is transformed as $\mathcal{O}_\Delta(x) \to \mathcal{O}'_\Delta(x) = x'^{2\Delta}\mathcal{O}_\Delta(x')$. Correlation functions then transform covariantly under inversion, viz.

$$\begin{aligned}&\langle \mathcal{O}_{\Delta_1}(x_1)\mathcal{O}_{\Delta_2}(x_2)\cdots\mathcal{O}_{\Delta_n}(x_n)\rangle \\ &= (x'_1)^{2\Delta_1}(x'_2)^{2\Delta_2}\cdots(x'_n)^{2\Delta_n}\langle \mathcal{O}_{\Delta_1}(x'_1)\mathcal{O}_{\Delta_2}(x'_2)\cdots\mathcal{O}_{\Delta_n}(x'_n)\rangle.\end{aligned} \tag{23.27}$$

The spacetime forms of two- and three-point functions of scalar operators are unique in any CFT_d, a fact that can be established using the transformation law under inversion. These forms are

$$\langle \mathcal{O}_\Delta(x)\mathcal{O}_{\Delta'}(y)\rangle = \frac{c_2\delta_{\Delta\Delta'}}{(x-y)^{2\Delta}}, \tag{23.28}$$

$$\langle \mathcal{O}_{\Delta_1}(x)\mathcal{O}_{\Delta_2}(y)\mathcal{O}_{\Delta_3}(z)\rangle = \frac{c_3}{(x-y)^{\Delta_{12}}(y-z)^{\Delta_{23}}(z-x)^{\Delta_{31}}}, \tag{23.29}$$
$$\Delta_{12} = \Delta_1 + \Delta_2 - \Delta_3, \text{ and cyclic permutations.}$$

Operators such as conserved currents J_i and the conserved traceless stress tensor T_{ij} are important in a CFT$_d$. Under inversion $J_i(x) \to J_{ij}(x')x'^{2(d-1)}J_j(x')$, indicating that its scale dimension is $d-1$. The stress tensor transforms as $T_{ij}(x) \to J_{ik}(x')J_{jl}(x')x'^{2d}T_{kl}(x')$. Conservation requires that the two-point correlators of these operators be constructed from the transverse 'projector' $\pi_{ij} = \partial_i\partial_j - \delta_{ij}\Box$, but they can then be rewritten using $J_{ij}(x-y)$. The correlators, which are unique, are

$$\langle J_i(x)J_j(y)\rangle = b\pi_{ij}\frac{1}{(x-y)^{(2d-4)}} = -4b(d-1)(d-2)\frac{J_{ij}(x-y)}{(x-y)^{(2d-2)}},$$
$$\langle T_{ij}(x)T_{kl}(y)\rangle = \left[2\pi_{ij}\pi_{kl} - (d-1)(\pi_{ik}\pi_{jl} + \pi_{il}\pi_{jk})\right]\frac{c}{(x-y)^{(2d-4)}}$$
$$\sim \frac{J_{ik}(x-y)J_{jl}(x-y) + k \leftrightarrow l - (2/d)\delta_{ij}\delta_{kl}}{(x-y)^{2d}}. \tag{23.30}$$

We repeat that the 'kinematics', that is the SO$(d+1,1)$ symmetry, of a CFT$-d$ determines the spacetime dependence of the correlators above. The 'dynamics' resides in the value of the constants c_2, c_3, b, c. The determination of these constants is one application of the AdS/CFT correspondence. We will soon put inversion to good use in our study of how this is done, but first we suggest two exercises for the reader.

Exercise 23.5 *Show that the correlators (23.28)–(23.30) transform under inversion as their scale dimension and tensor properties require.*

Exercise 23.6 *Consider a current that transforms as $\mathcal{J}_i(x) \to J_{ij}(x')x'^{2\Delta}\mathcal{J}_j(x')$. Show that only the value $\Delta = d-1$ is compatible with a conservation law $\partial_i\mathcal{J}_i = 0$, which is covariant under inversion. Hint: the behavior of operators under scale transformations (23.5) implies that $x^i\partial_i\mathcal{J}_j = -\Delta\mathcal{J}_j$.*

23.7 The free massive scalar field in Euclidean AdS$_{d+1}$

The (Euclidean) action of a free massive scalar field in the AdS background metric of (23.20) is

$$S[\phi] = \frac{1}{2}\int \mathrm{d}^d z\,\mathrm{d}z_0\,\sqrt{g}\left[g^{\mu\nu}\partial_\mu\phi\partial_\nu\phi + m^2\phi^2\right]. \tag{23.31}$$

Exercise 23.7 *Show that the action is invariant under inversion using the transformations discussed in Sec. 23.5. Show specifically that $S[\phi] = S[\phi']$. Some care is needed to define the transformation of the inverse metric $g^{\mu\nu}$. Note that the measure $\mathrm{d}^d z \mathrm{d}z_0 \sqrt{g}$ is invariant.*

The wave equation satisfied by $\phi(x)$ is

$$\frac{1}{\sqrt{g}}\partial_\mu\left(\sqrt{g}g^{\mu\nu}\partial_\nu\phi\right) - m^2\phi = 0, \tag{23.32}$$

$$z_0^{d+1}\frac{\partial}{\partial z_0}\left[z_0^{-d+1}\frac{\partial}{\partial z_0}\phi(z_0,\vec{z})\right] + z_0^2\frac{\partial}{\partial \vec{z}^2}\phi(z_0,\vec{z}) - m^2L^2\phi(z_0,\vec{z}) = 0. \tag{23.33}$$

The hypersurface $z_0 = 0$, which is the conformal boundary of Euclidean AdS, is isomorphic to $\mathbb{R}^d$. A solution of the wave equation requires the specification of boundary data. We now discuss the elegant solution of this boundary value problem[6] derived in [282]. One can verify without difficulty that the function

$$K_\Delta(z_0,\vec{z},\vec{x}) = c_\Delta\left(\frac{z_0}{z_0^2 + (\vec{z}-\vec{x})^2}\right)^\Delta, \qquad c_\Delta = \frac{\Gamma(\Delta)}{\pi^{\frac{d}{2}}\Gamma(\Delta - d/2)}, \tag{23.34}$$

is a solution of (23.33) if (23.18) is satisfied. Further, K_Δ is 'normalized' so that

$$\lim_{z_0\to 0} z_0^{\Delta-d}K_\Delta(z_0,\vec{z},\vec{x}) = \delta(\vec{z}-\vec{x}). \tag{23.35}$$

Therefore the integral

$$\phi(z_0,\vec{z}) = \int \mathrm{d}^d x\, K_\Delta(z_0,\vec{z},\vec{x})\phi_0(\vec{x}) \tag{23.36}$$

determines the solution of (23.32) that approaches the boundary data $\phi_0(\vec{x})$ at the rate

$$\phi(z_0,\vec{z}) \to z_0^{d-\Delta}\phi_0(\vec{z}) \quad \text{as} \quad z_0 \to 0. \tag{23.37}$$

The function K_Δ is called the bulk-to-boundary propagator. Although the norm $\int \mathrm{d}z_0 \mathrm{d}^d z \phi^2$ of this solution diverges, it is exactly what we need to determine correlation functions of operators in the CFT that effectively 'lives' on the boundary, which is a copy of flat $\mathbb{R}^d$.

In inverted coordinates $z'_\mu = z_\mu/z^2$ and $\vec{x}\,' = \vec{x}/(\vec{x})^2$, the solution of the wave equation is

[6] To simplify the discussion we restrict throughout to the range $\Delta > \frac{1}{2}d$. See [316, 317] for an extension to the interval $\frac{1}{2}d \geq \Delta \geq \frac{1}{2}(d-2)$ close to the unitarity bound.

$$\phi'(z'_0, \vec{z}\,') = \int d^d x' K_\Delta(z'_0, \vec{z}\,', \vec{x}\,')\phi'_0(\vec{x}\,'), \tag{23.38}$$

with new boundary data $\phi'_0(\vec{x}')$. We require that the solutions ϕ' and ϕ are related by the scalar transformation law $\phi'(z'_0, \vec{z}\,') = \phi(z_0, \vec{z})$. For this purpose we reexpress, using (23.25), the integral in terms of the original coordinates. We then have

$$K_\Delta(z'_0, \vec{z}', \vec{x}') = K_\Delta(z_0, \vec{z}, \vec{x})(\vec{x})^{2\Delta}, \qquad d^d x' = d^d x/(\vec{x})^{2d}. \tag{23.39}$$

It is now straightforward to see that ϕ' and ϕ satisfy the desired relation if the boundary data are related by

$$\phi'_0(\vec{x}\,') = \frac{1}{(\vec{x})^{2(\Delta-d)}}\phi_0(\vec{x}). \tag{23.40}$$

Exercise 23.8 *Apply similar reasoning to study two solutions $\phi'(z'_0, \vec{z}\,')$ and $\phi(z_0, \vec{z})$ in which coordinates z'_μ and z_μ are related by the scale transformation $z'_\mu = \rho^{-1} z_\mu$. Show that*

$$K_\Delta(z'_0, \vec{z}', \vec{x}') = K_\Delta(z_0, \vec{z}, \vec{x})\rho^\Delta, \qquad d^d x' = d^d x \rho^{-d}. \tag{23.41}$$

Show that the two solutions are related by $\phi'(z'_0, \vec{z}\,') = \phi(z_0, \vec{z})$, if the boundary data scale as

$$\phi'_0(\vec{x}\,') = \rho^{d-\Delta}\phi_0(\vec{x}). \tag{23.42}$$

Exercise 23.9 *In the AdS/CFT correspondence $\phi_0(x)$ is the source term for the dual operators $\mathcal{O}_\Delta(x)$. The source term that appears in the generating function of correlators in the dual CFT is the exponential of the integral $\int d^d x\, \phi_0(x)\mathcal{O}_\Delta(x)$. Show that this integral is invariant under the scale transformations (23.42) and (23.5).*

23.8 AdS/CFT correlators in a toy model

In this section we show how the AdS/CFT correspondence is used to calculate correlation functions of a composite scalar operator $\mathcal{O}_\Delta(x)$ in a CFT$_d$. We assume that the action integral of its dual bulk field $\phi(z)$ in Euclidean AdS$_{d+1}$ is

$$S = \frac{1}{8\pi G}\int d^{d+1}z\sqrt{g}\left(\frac{1}{2}\partial_\mu\phi\partial^\mu\phi + \frac{1}{2}m^2\phi^2 + \frac{1}{3}b\phi^3 + \cdots\right). \tag{23.43}$$

We will look for a solution of the equation of motion (with the notation $\square$ for the differential operator defined in (23.32))

$$\frac{\delta S}{\delta\phi} = (-\square + m^2)\phi + b\phi^2 + \ldots = 0, \tag{23.44}$$

with the same boundary behavior (23.37) found in the linear case of the previous section. We emphasize that (23.44) is a *classical* partial differential equation, and we will work with its classical solution.

Box 23.4 Basic hypothesis of the AdS/CFT correspondence

The on-shell action $S[\phi_0]$, obtained by substituting the solution for $\phi(z)$ in terms of the boundary data ϕ_0, is the generating functional for (connected) correlation functions in the dual CFT$_d$.

Given the solution $\phi(z)$, the next step is to substitute it into the action (23.43). The action integral then becomes a functional of the boundary data $S[\phi_0]$ which is called the 'on-shell' action. This leads to the basic hypothesis of the AdS/CFT correspondence, written in Box 23.4. The boundary data are interpreted as the source of the composite operator $\mathcal{O}(x)$, and correlation functions are computed by applying variational derivatives, i.e.

$$\langle \mathcal{O}_\Delta(x_1)\ldots\mathcal{O}_\Delta(x_n)\rangle = (-)^{n-1}\frac{\delta}{\delta\phi_0(x_1)}\cdots\frac{\delta}{\delta\phi_0(x_n)}S[\phi_0]\bigg|_{\phi_0=0}. \qquad (23.45)$$

It is difficult to find an exact solution of the nonlinear equation (23.44), but a perturbative solution will serve our purpose. It is valid because we eventually set $\phi_0 = 0$ in the computation of correlators. We write

$$\phi_1(z) = \int \mathrm{d}^d\vec{x}\, K_\Delta(z_0, \vec{z}-\vec{x})\phi_0(\vec{x}),$$
$$\phi(z) = \phi_1(z) + b\int \mathrm{d}^{d+1}w\sqrt{g}G(z,w)\phi_1^2(w) + \cdots. \qquad (23.46)$$

The linear solution ϕ_1 was discussed in the previous section. Interaction terms require the bulk-to-bulk propagator $G(z,w)$ which satisfies $(-\Box_z + m^2)G(z,w) = \delta(z,w)/\sqrt{g}$ and is given by the hypergeometric function

$$G_\Delta(u) = \tilde{C}_\Delta(2u^{-1})^\Delta F\left(\Delta, \Delta-d+\tfrac{1}{2}; 2\Delta-d+1; -2u^{-1}\right), \qquad (23.47)$$
$$\tilde{C}_\Delta = \frac{\Gamma(\Delta)\Gamma(\Delta-\frac{1}{2}d+\frac{1}{2})}{(4\pi)^{(d+1)/2}\Gamma(2\Delta-d+1)L^{d-1}},$$
$$u = \frac{(z-w)^2}{2z_0w_0}.$$

We now encounter a fundamental difficulty. The bulk integrals in the action can diverge as the boundary is approached. This means that correlation functions diverge. This should not be a surprise, since ultraviolet divergences are expected in the computation of

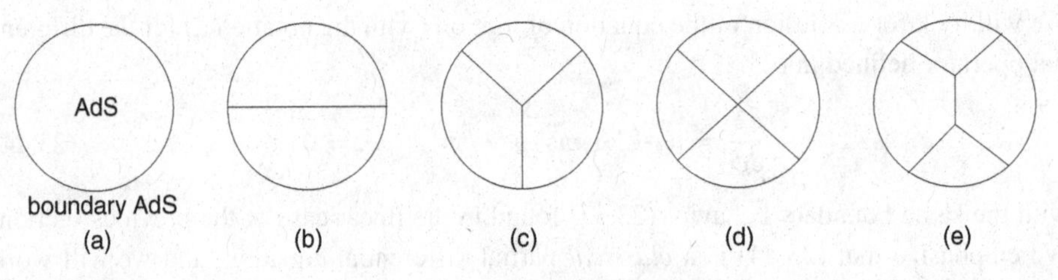

Fig. 23.1 Witten diagrams.

correlators of composite operators in a conformal field theory, although intrinsic couplings of the CFT are not renormalized. For example, divergences occur when operators acquire anomalous dimension. For 'protected' operators, such as tr X^k in $\mathcal{N} = 4$ SYM, there is a divergence in the bulk on-shell action because CFT correlation functions have anomalies. In the method of differential renormalization [318], this is exhibited in anomalous scale dependence of the regulated power laws $1/x^{2k} \sim (\Box)^{k-1}(\ln(M^2x^2)/x^2)$.

To handle the divergences, we insert a cutoff at $z_0 = \epsilon$ in the bulk geometry and restate the boundary value problem for (23.44) as a standard Dirichlet problem at $z_0 = \epsilon$. In most situations a careful treatment of the cutoff is only needed for two-point correlators. We postpone this case and proceed formally to discuss the diagrammatic algorithm proposed in [282], which is valid for n-point correlators with $n \geq 3$.

When the perturbative solution of (23.44) is inserted in (23.43), one finds an expansion of $S[\phi_0]$ in powers of ϕ_0. The various terms can be interpreted as Witten diagrams, such as those shown in Fig. 23.1. In these diagrams the interior and boundary of each disk denote the interior and boundary of the AdS$_{d+1}$ geometry. The boundary points $\vec{x}_i$ are the points in flat Euclidean$_d$ space where field theory operators are inserted. A diagram with n boundary points is interpreted as a contribution to the n-point correlation function. The rules for computation of these diagrams are the following:

1. Bulk points $z, w \in$ AdS$_{d+1}$. They are integrated with the invariant measure $\int d^{d+1}z\sqrt{g(z)}$.
2. Each bulk-to-boundary line carries a factor of K_Δ and each bulk-to-bulk line a factor $G(z, w)$.
3. An n-point vertex carries a coupling factor from the interaction terms of the bulk Lagrangian, e.g. $\mathcal{L} = \frac{1}{3}b\phi^3 + \frac{1}{4}c\phi^4 + \cdots$ with the same combinatoric weights that occur in Feynman–Wick diagrams.
4. For the classical on-shell action, one includes only diagrams without internal loops.

These rules do not apply to two-point correlators which are computed by a more careful procedure to be discussed below.

23.9 Three-point correlation functions

Consider the three-point function for the bulk interaction term $\mathcal{L} = b\phi_1\phi_2\phi_3$ of three scalar fields that are dual to operators $\mathcal{O}_{\Delta_1}$, $\mathcal{O}_{\Delta_2}$, $\mathcal{O}_{\Delta_3}$. The rules for Witten diagrams imply that the correlation function is proportional to the basic integral

$$\begin{aligned} A(\vec{x}, \vec{y}, \vec{z}) &= \int \frac{dw_0 d^d\vec{w}}{w_0^{d+1}} \left(\frac{w_0}{(w-\vec{x})^2}\right)^{\Delta_1} \left(\frac{w_0}{(w-\vec{y})^2}\right)^{\Delta_2} \left(\frac{w_0}{(w-\vec{z})^2}\right)^{\Delta_3}, \\ (w - \vec{x})^2 &\equiv w_0^2 + (\vec{w} - \vec{x})^2. \end{aligned} \tag{23.48}$$

Let's show that the diagrammatic procedure yields a three-point function that transforms correctly under inversion. We change integration variable by $w_\mu = w'_\mu/w'^2$ and at the same time refer boundary points to their inverses, i.e. $\vec{x} = \vec{x}\,'/(\vec{x}\,')^2$ and similar for $\vec{y}, \vec{z}$.

Box 23.5 **CFT_d correlators from AdS_{d+1}**

It is a very general result and a direct consequence of the symmetries that the AdS/CFT procedure produces correlation functions with the transformation properties required by a conformal field theory at the AdS boundary.

Each bulk-to-boundary propagator acquires the inversion factor in (23.39). Since the AdS volume element is invariant, i.e. $\mathrm{d}^{d+1}w/w_0^{d+1} = \mathrm{d}^{d+1}w'/w_0'^{d+1}$, we quickly see that the integral satisfies

$$A(\vec{x}, \vec{y}, \vec{z}) = |\vec{x}\,'|^{2\Delta_1}|\vec{y}\,'|^{2\Delta_2}|\vec{z}\,'|^{2\Delta_3} A(\vec{x}\,', \vec{y}\,', \vec{z}\,'). \tag{23.49}$$

This is the correct inversion property (see (23.27)) of the three-point function $\langle \mathcal{O}_{\Delta_1}(\vec{x})\mathcal{O}_{\Delta_2}(\vec{y})\mathcal{O}_{\Delta_3}(\vec{z})\rangle$ in a CFT_d. Since the integral is clearly invariant under $\mathrm{SO}(d)$ rotations and translations, it is guaranteed that the result of the integral is a correlation function that is proportional to the unique spacetime form (23.29) required by conformal symmetry in CFT_d; see Box 23.5.

We still need to *do* the bulk integral to obtain the constant c_3 in (23.29). The value of this constant has dynamical significance in the boundary CFT. The integral in the original form (23.48) is difficult because it contains three denominators with the restriction $w_0 > 0$. But we can simplify it by using inversion in a somewhat different way [319]. We use translation symmetry to move the point $\vec{z} \longrightarrow 0$, i.e. $A(\vec{x}, \vec{y}, \vec{z}) = A(\vec{x} - \vec{z}, \vec{y} - \vec{z}, 0) \equiv A(\vec{u}, \vec{v}, 0)$. The integral for $A(\vec{u}, \vec{v}, 0)$ is similar to (23.48) except that the third propagator is simplified,

$$\left(\frac{w_0}{(w - \vec{z})^2}\right)^{\Delta_3} \longrightarrow \left(\frac{w_0}{w^2}\right)^{\Delta_3} = (w_0')^{\Delta_3}. \tag{23.50}$$

There is no denominator in the inverted frame since $\vec{z} = 0 \longrightarrow \vec{z}\,' = \infty$. After inversion the integral is

$$A(\vec{u}, \vec{v}, 0) = \frac{1}{|\vec{u}|^{2\Delta_1}|\vec{v}|^{2\Delta_2}} \int \frac{\mathrm{d}^{d+1}w'}{(w_0')^{d+1}} \left(\frac{w_0'}{(w' - \vec{u}\,')^2}\right)^{\Delta_1} \left(\frac{w_0'}{(w' - \vec{v}\,')^2}\right)^{\Delta_2} (w_0')^{\Delta_3}. \tag{23.51}$$

The integral can now be done by conventional Feynman parameter methods, which give

$$\begin{aligned} A(\vec{u}, \vec{v}, 0) &= \frac{1}{|\vec{u}|^{2\Delta_1}|\vec{v}|^{2\Delta_2}} \frac{a(\Delta_1, \Delta_2, \Delta_3)}{|\vec{u}\,' - \vec{v}\,'|^{\Delta_1+\Delta_2-\Delta_3}}, \\ a(\Delta_1, \Delta_2, \Delta_3) &= \frac{\pi^{d/2}}{2} \frac{\Gamma\left[\frac{1}{2}(\Delta_1 + \Delta_2 + \Delta_3 - d)\right]}{\Gamma(\Delta_1)\Gamma(\Delta_2)\Gamma(\Delta_3)} \\ &\quad \times \Gamma\left[\tfrac{1}{2}(\Delta_1 + \Delta_2 - \Delta_3)\right]\Gamma\left[\tfrac{1}{2}(\Delta_2 + \Delta_3 - \Delta_1)\right]\Gamma\left[\tfrac{1}{2}(\Delta_3 + \Delta_1 - \Delta_2)\right]. \end{aligned} \tag{23.52}$$

Exercise 23.10 *Repristinate the original variables $\vec{x}, \vec{y}, \vec{z}$ to obtain the standard form (23.29) with $c_3 = a(\Delta_1, \Delta_2, \Delta_3)$.*

When applied to the $AdS_5 \otimes S^5$ solution of Type IIB supergravity, the diagrammatic prescription discussed in Sec. 23.8 computes correlators in the $\mathcal{N} = 4$ SYM theory in the limit of large N and large 't Hooft coupling $\lambda = g^2 N$. With this in mind we discuss the application of the results for three-point functions to the correlators of the chiral primary operators $\mathrm{tr} X^k$. The cubic couplings b_{klm} of the dual AdS_5 scalars were obtained in [320]. The analysis required mastering the intricate mixing of Kaluza–Klein modes at the non-linear level in the $D = 10$ theory. When combined with the Witten integral above, it was observed that the AdS/CFT prediction

$$\langle \mathrm{tr}\, X^k(\vec{x})\, \mathrm{tr}\, X^l(\vec{y})\, \mathrm{tr}\, X^m(\vec{z})\rangle = b_{klm} a(k, l, m) A(\vec{x}, \vec{y}, \vec{z}) \tag{23.53}$$

for the large N, large λ supergravity limit agreed with the *free field* Feynman amplitude for these correlators. The operators $\mathrm{tr} X^k$ are in short representations of the superconformal algebra, so their scale dimensions are unaffected by interactions, but it was expected that the coefficients would not be 'protected' in this way. The results, however, suggested the strong conjecture [320] that the radiative corrections vanish for all g^2 and N. This conjecture was subsequently confirmed in studies of weak coupling [321, 322, 323] and instanton [324] contributions. General all-orders arguments for non-renormalization have also been developed [323, 325]. The non-renormalization of 3-point functions of chiral primaries (and their descendents) was a surprise and the first of several new results about $\mathcal{N} = 4$ SYM obtained from AdS/CFT.

23.10 Two-point correlation functions

In many applications of AdS/CFT the bulk geometry is not pure AdS_{d+1}. The metric differs from (23.20) in the interior, although it is similar near the boundary. The isometry group is smaller – typically the group of rotations and translations in d dimensions. Because of the reduced symmetry only two-point correlators can be calculated analytically. They are an important case because they contain useful physical information, for example on the spectrum of states in the boundary field theory. They are also a delicate case because it is necessary to introduce a cutoff near the boundary to extract the correct result.

Our technical discussion is restricted to the AdS_{d+1} background, and we present two methods of calculation. The first method is indirect. We extract the two-point function from the Ward identity for the three-point correlator of a conserved current $\mathcal{J}_i(x)$ with a scalar operator $\mathcal{O}_\Delta(x)$ and its conjugate. Since three-point functions do not require a cutoff, this method is conceptually simpler, but the full symmetry of the AdS background is needed. The second method is a direct calculation of the two-point function, and we must use a cutoff. Here the AdS geometry simplifies the calculation, but the method can be generalized to bulk geometries of lower symmetry. It is reassuring that the two methods agree on the final result.

To implement the first method we first note that the unique form of the CFT_d correlator $\langle \mathcal{J}_i \mathcal{O}_\Delta \mathcal{O}^*_\Delta \rangle$ with the correct inversion property is

$$\langle J_i(z)\mathcal{O}_\Delta(x)\mathcal{O}^*_\Delta(y)\rangle \tag{23.54}$$
$$= -\mathrm{i}\xi \frac{1}{(x-y)^{2\Delta-d+2}}\frac{1}{(x-z)^{d-2}(y-z)^{d-2}}\left[\frac{(x-z)_i}{(x-z)^2}-\frac{(y-z)_i}{(y-z)^2}\right].$$

This correlator satisfies

$$\begin{aligned}\frac{\partial}{\partial z_i}\langle J_i(z)\mathcal{O}_\Delta(x)\mathcal{O}^*_\Delta(y)\rangle &= \mathrm{i}\xi[\delta(x-z)-\delta(y-z)]\frac{2\pi^{d/2}}{\Gamma(d/2)}\frac{1}{(x-y)^{2\Delta}}\\ &= \mathrm{i}[\delta(x-z)-\delta(y-z)]\langle\mathcal{O}_\Delta(x)\mathcal{O}^*_\Delta(y)\rangle.\end{aligned} \tag{23.55}$$

The reader is invited to prove the first line in the exercise below. The current is conserved, and thus satisfies the Ward identity in the second line.

Exercise 23.11 *Derive the first line of (23.55) from (23.54). Hints:*

$$\begin{aligned}\frac{\partial}{\partial z_i}\frac{1}{(x-z)^{d-2}} &= (d-2)\frac{(x-z)_i}{(x-z)^d},\\ \Box_z\frac{1}{(x-z)^{d-2}} &= -\frac{(d-2)2\pi^{d/2}}{\Gamma(d/2)}\delta(x-z).\end{aligned} \tag{23.56}$$

To obtain this correlator from the gravity side of the correspondence we extend the toy model of (23.43) to include a massless gauge field A_μ and a pair of charged scalars $\phi,\ \phi^*$ in the bulk. The action of this system is (the bulk coordinate is w_μ)

$$S = \frac{1}{8\pi G}\int \mathrm{d}^{d+1}w\sqrt{g}\left[\frac{1}{4}F_{\mu\nu}F^{\mu\nu} + g^{\mu\nu}(\partial_\mu + iA_\mu)\phi^*(\partial_\nu - iA_\mu)\phi + m^2\phi^*\phi\right]. \tag{23.57}$$

We need the three-point Witten diagram formed with the cubic vertex from (23.57), and bulk-to-boundary propagators for the scalars, given in (23.34), and for the gauge field. For the latter we use the form discussed in [319], namely

$$K_{\mu i}(w,\vec{x}) = C_d\frac{w_0^{d-2}}{[w_0^2+(\vec{w}-\vec{x})^2]^{d-1}}J_{\mu i}(w-\vec{x}), \qquad C^d = \frac{\Gamma(d)}{2\pi^{d/2}\Gamma(d/2)}, \tag{23.58}$$

where $J_{\mu i}$ comes from (23.23). (This form is gauge equivalent to the form first found in [282].) With these ingredients the Witten diagram can be expressed as the AdS integral

$$\langle\mathcal{J}_i(z)\Omega_\Delta(x)\Omega^*_\Delta(y)\rangle = -\mathrm{i}\int\frac{\mathrm{d}^{d+1}w}{w_0^{d+1}}K_{\mu i}(w,\vec{z})w_0^2K_\Delta(w,\vec{x})\overleftrightarrow{\frac{\partial}{\partial w_\mu}}K_\Delta(w,\vec{y}). \tag{23.59}$$

The integral can be done by the inversion technique. This gives a correlator of the form in (23.54) with the coefficient

$$\xi = \frac{(\Delta-d/2)\Gamma(d/2)\Gamma(\Delta)}{\pi^d\Gamma(\Delta-d/2)}. \tag{23.60}$$

Using the Ward identity, we obtain the two-point function

$$\langle \mathcal{O}_\Delta(x)\mathcal{O}^*_\Delta(y)\rangle = \frac{(2\Delta - d)\Gamma(\Delta)}{\pi^{d/2}\Gamma(\Delta - d/2)(x-y)^{2\Delta}}. \tag{23.61}$$

To implement the second method of calculation we place a cutoff at $z_0 = \epsilon$ in the AdS bulk geometry, and we consider a Dirichlet boundary value problem with boundary at this position. We must solve the linear boundary value problem:

$$(\Box - m^2)\phi(z_0, \vec{z}) = 0, \qquad \phi(\epsilon, \vec{z}) = \bar{\phi}(\vec{z}). \tag{23.62}$$

The solution is substituted in the bilinear part of the toy model action (23.43) to obtain the on-shell action. The bulk integral cancels by partial integration, and we are left with the contribution from the (lower) boundary,

$$S[\bar{\phi}] = -\frac{1}{2\epsilon^{d-1}}\int \mathrm{d}^d\vec{z}\,\bar{\phi}(\vec{z})\,\partial_0\phi(\epsilon, \vec{z}). \tag{23.63}$$

The on-shell action is quadratic in the 'source' $\bar{\phi}$ and we will obtain the two-point correlation function of the dual CFT$_d$ operator $\mathcal{O}_\Delta$ as the second functional derivative with respect to the source. Thus the two-point function depends only on the *free field* limit of the bulk action (23.43).

The next step is to solve the boundary value problem (23.62). Since the cutoff region $z_0 \geq \epsilon$ does not have the full symmetry of AdS, an exact solution is difficult in position space, so we work in momentum space. Using the Fourier transform

$$\phi(z_0, \vec{z}) = \int \frac{\mathrm{d}^d p}{(2\pi)^d}\,\mathrm{e}^{\mathrm{i}\vec{p}\cdot\vec{z}}\phi(z_0, \vec{p}), \tag{23.64}$$

we find the boundary value problem

$$\left[z_0^2\partial_0^2 - (d-1)z_0\partial_0 - (p^2z_0^2 + m^2L^2)\right]\phi(z_0, \vec{p}) = 0, \qquad \phi(\epsilon, \vec{p}) = \bar{\phi}(\vec{p}), \tag{23.65}$$

where $\bar{\phi}(\vec{p})$ is the transform of the boundary data.

The differential equation is essentially Bessel's equation, and we choose the solution $z_0^{d/2}K_\nu(pz_0)$, where $\nu = \Delta - d/2$, $p = |\vec{p}|$. This solution is exponentially damped as $z_0 \to \infty$ and behaves as $z_0^{d-\Delta}$ as $z_0 \to 0$. The independent solution $z_0^{d/2}I_\nu(pz_0)$ has subleading behavior z^Δ at the boundary, but it is rejected because it increases exponentially in the deep interior. The normalized solution of the boundary value problem is then

$$\phi(z_0, \vec{p}) = \frac{z_0^{d/2}K_\nu(pz_0)}{\epsilon^{d/2}K_\nu(p\epsilon)}\bar{\phi}(\vec{p}). \tag{23.66}$$

The on-shell action in p-space is

$$S[\bar{\phi}] = -\frac{1}{2\epsilon^{d-1}}\int \mathrm{d}^d p\mathrm{d}^d q(2\pi)^d\delta(\vec{p}+\vec{q})\bar{\phi}(\vec{p})\partial_0\phi(\epsilon, \vec{q}), \tag{23.67}$$

which leads to the cutoff correlation function

$$\langle \mathcal{O}_\Delta(\vec{p})\mathcal{O}_\Delta(\vec{q})\rangle_\epsilon = -\frac{\delta^2 S}{\delta\bar{\phi}(\vec{p})\delta\bar{\phi}(\vec{q})} \equiv (2\pi)^d\delta(\vec{p}+\vec{q})F_\epsilon(p),$$
$$F_\epsilon(p) = \frac{1}{\epsilon^{d-1}}\frac{\mathrm{d}}{\mathrm{d}\epsilon}\ln\left[\epsilon^{d/2}K_\nu(p\epsilon)\right]. \tag{23.68}$$

The final step is to extract the physical correlation function in the limit as $\epsilon \to 0$. The procedure is slightly different in the two cases of integer or non-integer ν, and we discuss the non-integer case here; see [319]. For integer ν, see [304]. We need the near-boundary behavior of the Bessel function (which can be obtained from [326]). For non-integer ν, this is a sum of two series whose leading terms $u^{-\nu}$ and u^ν involve the two Frobenius exponents for the differential equation (23.65):

$$K_\nu(u) = u^{-\nu}(a_0 + a_1u^2 + \ldots) + u^\nu(b_0 + b_2u^2 + \ldots),$$
$$a_0 = 2^{\nu-1}\Gamma(\nu), \qquad b_0 = -2^{-\nu-1}\Gamma(1-\nu)/\nu. \tag{23.69}$$

We will not need the values of the higher coefficients $a_i,\ b_i,\ i \geq 1$. By careful computation of the logarithmic derivative in (23.68) in the boundary limit, we find

$$F_\epsilon(p) = \epsilon^{-d}\left[\frac{d}{2} - \nu(1 + c_2(\epsilon p)^2 + \ldots)\right.$$
$$\left. + \frac{2\nu b_0}{a_0}(\epsilon p)^{2\nu}(1 + d_2(\epsilon p)^2 + \ldots)\right]. \tag{23.70}$$

There is a notable difference between the two series. In the upper line we have a series of *integer* powers of p^2. Upon Fourier transformation to x-space, these would produce singular contact terms in the correlator, e.g. $\delta(x-y),\ \Box\delta(x-y)$, etc. In quantum field theory such contact terms are usually infinite, scheme dependent and unobservable. Indeed the lower order contact terms in (23.70) are infinite. This signals a rather general phenomenon in the AdS/CFT correspondence; the cutoff ϵ at long distance in AdS space plays the role of a short distance cutoff in the dual field theory [327]. We will drop these unphysical terms. Indeed we will see that it is the lower line that produces the two-point function with the required spacetime dependence shown in (23.28).

The lower line contains *non-integer* powers of p. These terms are non-analytic so they contribute to the absorptive part of the two-point function in momentum space. The physical correlator in p-space is determined by the non-analytic terms, in fact only by the first of these, since subsequent terms are subdominant as $\epsilon \to 0$. We therefore write[7]

$$F_\epsilon(p) = \epsilon^{2(\Delta-d)}\frac{2\nu b_0}{a_0}p^{2\nu} = -\epsilon^{2(\Delta-d)}(2\nu)\frac{\Gamma(1-\nu)}{\Gamma(1+\nu)}\left(\frac{p}{2}\right)^{2\nu}. \tag{23.71}$$

Now we face a glaring but mercifully minor issue. The physical correlation function should not have the residual cutoff dependent factor $\epsilon^{2(\Delta-d)}$ that appears in (23.71). The

[7] When ν is an integer, the factor $p^{2\nu}$ in (23.71) is replaced by $p^{2\nu}\ln p$, which is also non-analytic, and the coefficient becomes $2\nu b_0/a_0 = (-)^{(\nu-1)}2^{(2-2\nu)}/\Gamma(\nu)^2$. See [304].

problem arises because the boundary data $\bar{\phi}(\vec{x})$ or $\bar{\phi}(\vec{p})$ must be rescaled as $\epsilon \to 0$ to be compatible with the boundary behavior (23.37) of the exact solution as $z_0 \to 0$. Compatibility requires that we renormalize the boundary data by defining $\bar{\phi}(\vec{x}) \equiv \epsilon^{(d-\Delta)}\phi_0(\vec{x})$. The Fourier transform scales in the same way, namely $\bar{\phi}(\vec{p}) \equiv \epsilon^{(d-\Delta)}\phi_0(\vec{p})$. This rescaling could have (and should have) been done when we computed the two-point function in (23.68). We should differentiate with respect to the renormalized source to obtain the correlator. The chain rule then gives

$$\langle \mathcal{O}(\vec{p})\mathcal{O}(\vec{q})\rangle \equiv -\lim_{\epsilon\to 0}\frac{\delta^2 S}{\delta\phi_0(\vec{p})\delta\phi_0(\vec{q})} = -\epsilon^{2(d-\Delta)}\lim_{\epsilon\to 0}\frac{\delta^2 S}{\delta\bar{\phi}(\vec{p})\delta\bar{\phi}(\vec{q})}. \tag{23.72}$$

The extra factor neatly cancels the unwanted ϵ dependence in (23.71) and produces the physical correlator in p-space:

$$F(p) = -(2\nu)\frac{\Gamma(1-\nu)}{\Gamma(1+\nu)}\left(\frac{p}{2}\right)^{2\nu}. \tag{23.73}$$

It is time to return to x-space. This requires the Fourier transform

$$\langle \mathcal{O}_\Delta(x)\mathcal{O}_\Delta(y)\rangle = \frac{1}{(2\pi)^d}\int \mathrm{d}^d p\, \mathrm{e}^{\mathrm{i}p\cdot(x-y)}F(p). \tag{23.74}$$

We need the basic transform of a power law, namely[8]

$$f(x) \equiv \frac{1}{(2\pi)^d}\int \mathrm{d}^d p\, \mathrm{e}^{\mathrm{i}p\cdot x}p^{2\nu} = \frac{2^{2\nu}}{\pi^{d/2}}\frac{\Gamma(\Delta)}{\Gamma(-\nu)}\frac{1}{x^{2\Delta}}, \tag{23.75}$$

where $\nu = \Delta - d/2$ is used. This result can be combined with (23.73) to give the physical two-point function

$$\langle \mathcal{O}_\Delta(x)\mathcal{O}_\Delta(y)\rangle = \frac{1}{\pi^{d/2}}\frac{(2\Delta-d)\Gamma(\Delta)}{\Gamma(\Delta-d/2)(x-y)^{2\Delta}}, \tag{23.76}$$

which agrees perfectly with the result (23.61). In many applications of the AdS/CFT correspondence the background geometry is not AdS_{d+1}. Instead it is described by a metric of the form

$$\mathrm{d}s^2 = \frac{L^2}{z^2}\left[\mathrm{d}z^2 + F^2(z)\delta_{ij}\mathrm{d}x^i\mathrm{d}x^j\right], \tag{23.77}$$

in which z is the 'radial' coordinate and the x^i, $i = 1, 2, \ldots, d$, are 'transverse' coordinates. The function $F(z)$ is regular as $z \to 0$. The case $F(z) \equiv 1$ reproduces the (Euclidean) AdS metric (23.20), so it is clear that the more general metric has the same boundary structure as AdS, but differs in the interior. The isometry group is now the Poincaré group in d dimensions, so the dual field theory is not conformal, but it does have

[8] This computation requires hyperspherical coordinates in d dimensions, an integral over the principal angle in $\vec{p}\cdot\vec{x} = px\cos\theta$, which produces the Bessel function $J_{(d-2)/2}(px)$, and a final integral over the magnitude p (which requires analytic continuation in Δ). The two relevant integrals in [326] are Sec. 3.915, #5 and Sec. 6.561, #14.

Poincaré$_d$ symmetry. In Sec. 23.12 we will study a situation in which such background geometries appear.

To compute correlation functions of field theory operators dual to massive bulk scalars in the more general background geometry, one considers the 'wave equation' (23.32) with metric components taken from (23.77). Because of the Poincaré symmetry, the Fourier transform is applicable and (23.32) reduces to an ordinary differential equation in z whose structure is similar to (23.65),

$$\left[z^2\partial_z^2 + \left(d\,z^2\partial_z \ln F - (d-1)z\right)\partial_z - \left(\frac{z^2}{F^2}p^2 + m^2L^2\right)\right]\phi(z,p) = 0. \qquad (23.78)$$

Under favorable conditions, which are not guaranteed, the exact solution of this equation can be expressed in terms of special functions (such as the hypergeometric function, for example, see [328]). After inserting a cutoff at $z = \epsilon$, the boundary asymptotics produces series similar to (23.70), which contains singular contact terms plus terms non-analytic in ϵ. The physical correlation function in p-space is again given by the least singular non-analytic term.

23.11 Holographic renormalization

The cutoff method of the previous section may seem rather arbitrary, a rough kludge. In this section we discuss (all too) briefly a related procedure, both systematic and comprehensive, for treating the divergences that occur in AdS/CFT computations. This is the formalism of holographic renormalization [329, 330]. It is a subtle formalism, and we present only a simplified introduction with the recommendation to consult dedicated reviews [331, 332], which include detailed examples. The field theory information that can be obtained using holographic renormalization includes renormalized correlation functions, the Ward identities satisfied by correlators related by symmetries, and certain quantum anomalies. The first success of the method [329] was the computation of the anomalous trace of the stress tensor of a CFT$_d$ using the AdS/CFT correspondence.

The formalism applies to asymptotically anti-de Sitter (AAdS) spacetimes. Near the boundary the metric of a $(d+1)$-dimensional Euclidean signature AAdS spacetime can be brought by coordinate transformation to the form (with $z > 0$)

$$\mathrm{d}s^2 = \frac{L^2}{z^2}\left[\mathrm{d}z^2 + g_{ij}(x,z)\mathrm{d}x^i\mathrm{d}x^j\right]. \qquad (23.79)$$

The limit $z \to 0$ of the 'transverse' metric $g_{ij}(x,z)$ must define a non-degenerate d-dimensional Riemannian metric $g_{(0)ij}(x)$. The metric (23.77) is an example of AAdS with the d-dimensional Poincaré symmetry group. In the general form above there are no isometries. In the AdS/CFT correspondence, the transverse bulk metric g_{ij} is dual to the stress tensor T_{ij} of the boundary QFT$_d$.

AAdS spacetimes occur as the geometries determined by classical solutions of the equations of motion of gravity either alone or coupled to other fields, such as vectors and scalars. We consider a system of gravity and a scalar field with the Euclidean action

$$S = \frac{1}{2\kappa^2}\left\{\int_{z>\epsilon} \mathrm{d}z\mathrm{d}^d x\sqrt{g}[-R + g^{\mu\nu}\partial_\mu\phi\partial_\nu\phi + 2V(\phi)] - 2\int_{z=\epsilon} \mathrm{d}^d x\sqrt{\gamma}K\right\}. \tag{23.80}$$

The second term is the Gibbons–Hawking boundary term for a boundary at $z = \epsilon$. The induced metric on the boundary, and K, the trace of the second fundamental form, are

$$\begin{aligned}\gamma_{ij}(x) \equiv \gamma_{ij}(x,z)\big|_{z=\epsilon}, \qquad \gamma_{ij}(x,z) \equiv \frac{L^2}{z^2}g_{ij}(x,z),\\ K = \nabla_\mu n^\mu = -\frac{z}{L}\partial_z \ln\sqrt{\gamma(x,z)}\Big|_{z=\epsilon},\end{aligned} \tag{23.81}$$

where n^μ is the unit normal to the boundary. Eventually we will need the limit as $\epsilon \to 0$.

The potential $V(\phi)$ must satisfy certain conditions so that the theory has generic AAdS solutions with scale L. First $V(\phi)$ must have a stationary point, assumed without loss of generality to be at $\phi = 0$, at which $V(0) = -d(d-1)/2L^2$. Thus it takes the form

$$V(\phi) = -\frac{d(d-1)}{2L^2} + \tfrac{1}{2}m^2\phi^2 + \tfrac{1}{3}b\phi^3 + \ldots. \tag{23.82}$$

The mass must satisfy the inequalities $-d^2/4 \le m^2L^2 \le 0$. The lower limit is the stability bound for general d, which was discussed at the end of Sec. 23.4. For positive m^2 the back reaction of the scalar fields is too strong near the boundary to allow a generic AAdS solution.[9] Therefore the stationary point is a local *maximum*! Scalar fields with masses in the above range are dual to QFT$_d$ operators $\mathcal{O}_\Delta(x)$ with scale dimension (23.18) in the range $d/2 \le \Delta \le d$. They are relevant or marginal operators.

The first step in the implementation of the holographic renormalization procedure is called near-boundary analysis. This gives the detailed behavior of $g_{ij}(x,z)$ and $\phi(x,z)$ near the boundary at $z = 0$. The simplest asymptotic behavior occurs when the scale dimension Δ is restricted to be an *integer* in the range $d/2 < \Delta \le d$. These restrictions can be removed with no difficulties of principle; we make them only to simplify our discussion.

It is convenient to use the radial coordinate $\rho = z^2$ to discuss the asymptotic solutions for $g_{ij}(x,z)$ and $\phi(x,z)$, which are

$$\begin{aligned}g_{ij}(x,\rho) &= g_{(0)ij}(x) + \rho g_{(2)ij}(x) + \cdots + \rho^{d/2}\left[g_{(d)ij}(x) + \ln\rho\, h_{(d)ij}(x)\right] + \cdots,\\ \phi(x,\rho) &= \rho^{(d-\Delta)/2}(\phi_{(0)}(x) + \rho\phi_{(2)}(x) + \cdots)\\ &\quad + \rho^{\Delta/2}\left[\phi_{(2\Delta-d)}(x) + \ln\rho\,\psi_{(2\Delta-d)}(x)\right] + \cdots.\end{aligned} \tag{23.83}$$

All terms except the transverse traceless part of $g_{(d)ij}$ and $\phi_{(2\Delta-d)}$ can be expressed as *local functions of the leading terms $g_{(0)ij}$ and $\phi_{(0)}$*. One can interpret $g_{(0)ij}$ as the boundary metric, and the locally determined higher order terms involve the curvature tensor and

[9] However, perturbative solutions with positive m^2 scalars can be included; see [332]. They provide the CFT$_d$ correlation functions for operators with $\Delta > d$ discussed in Secs. 23.9 and 23.10.

covariant derivatives for this metric. The logarithmic term $h_{(d)ij}$ appears when d is an even integer and/or $\Delta - d/2$ is an integer, while $\psi_{(2\Delta-d)}$ occur only when $\Delta - d/2$ is an integer. These terms are related to quantum anomalies in the dual quantum field theory, as we discuss below.

An example of the local corrections in (23.83) is

$$g_{(2)ij} = -\frac{1}{d-2}\left[R_{(0)ij} - \frac{1}{2(d-1)}R_{(0)}g_{(0)ij}\right], \tag{23.84}$$

where the subscript (0) indicates that the Ricci tensor and scalar curvature are those of the boundary metric $g_{(0)}$. For some potentials there is an additional term proportional to the scalar source $\phi^2_{(0)}$. It is quite straightforward to obtain these terms by substituting the expansions in the field equations of the action (23.80). Note that the local terms are universal. They are valid for all classical solutions of the field equations for a given potential $V(\phi)$ in (23.80), but they do depend on the details of the potential.

At the order $\rho^{d/2}$ for g_{ij} and $\rho^{\Delta/2}$ for ϕ, near-boundary analysis does *not* determine the solution, and one needs information from the full solution. The reason is that the field equations are second order in the radial variable ρ, so one expects that there are two independent boundary rates. The onset of the subleading rate signals a possible independent asymptotic solution. Both $\phi_{(2\Delta-d)}$ and the undetermined parts of $g_{(d)ij}$ are *non-local functions of the sources* $\phi_{(0)}$ *and* $g_{(0)ij}$. As we will see, these quantities contain the important information on correlation functions.

The next step is to use the asymptotic solution to define a set of local counterterms, which are then added to the on-shell action to cancel the divergent terms that appear as the cutoff $\epsilon \to 0$. These divergences are entirely determined by the asymptotic solution (23.83). Substituting the asymptotic solution in the action (23.80) with boundary at $\rho = \epsilon$, one finds the regulated action in which the divergent terms are displayed explicitly

$$S_{\text{reg}}[g_{(0)}, \phi_{(0)}, \epsilon] = \int \mathrm{d}^d x \sqrt{g_{(0)}}\left[a_0\epsilon^{-d/2} + a_2\epsilon^{(-d/2+1)} + \ldots - a_d \ln\epsilon + \mathcal{O}(\epsilon^0)\right]. \tag{23.85}$$

All coefficients a_ν are local functions of the sources $g_{(0)ij}$ and $\phi_{(0)}$ and do not depend on the undetermined quantities $g_{(d)ij}$ and $\phi_{(2\Delta-d)}$. These infinities can be canceled by adding local covariant counterterms. The counterterm action $S_{\text{ct}}[g_{ij}(x,\epsilon), \phi(x,\epsilon)]$ can be uniquely obtained from (23.83), but to simplify the discussion we omit this technical step and refer readers to [331, 332].

We then define the renormalized action as the limit $\epsilon \to 0$ of the on-shell action (23.80), evaluated after substitution of the full solution of the field equations and with $\rho \geq \epsilon$, plus S_{ct}:

$$S_{\text{ren}}[g_{(0)}, \phi_{(0)}] = \lim_{\epsilon\to 0}\left(S_{\text{on-shell}} + S_{\text{ct}}\right). \tag{23.86}$$

Renormalized correlation functions are computed as functional derivatives of S_{ren} with respect to the sources. It is most useful to define one-point functions in the presence of sources as the completely finite expressions

$$\langle T_{ij}(x)\rangle_s \equiv \frac{2}{\sqrt{g_0}}\frac{\delta S_{\text{ren}}}{\delta g^{ij}_{(0)}(x)} = \frac{d}{\kappa^2}g_{(d)ij}(x) + C_{ij}(g_{(0)},\phi_{(0)}), \tag{23.87}$$

$$\langle \mathcal{O}_\Delta(x)\rangle_s \equiv \frac{1}{\sqrt{g_0}}\frac{\delta S_{\text{ren}}}{\delta \phi_{(0)}(x)} = (d-2\Delta)\phi_{(2\Delta-d)}(x) + C(g_{(0)},\phi_{(0)}). \tag{23.88}$$

The quantities C_{ij} and C are local functions of the sources. An n-point correlation function with $n \geq 2$ is obtained by applying $n-1$ derivatives with respect to the sources. Thus C_{ij} and C contribute only to contact terms, which are usually of little interest. However, C_{ij} and C are important for one-point functions. It is a striking and non-intuitive result of the holographic renormalization procedure that the non-local information in correlators is compactly encoded in the quantities $g_{(d)ij}$ and $\phi_{(2\Delta-d)}$.

The sources $g_{(0)ij}$ and $\phi_{(0)}$ in (23.87)–(23.88) (and in the steps that led to these equations) are completely general functions. However, the goals of the computation are correlation functions in the boundary QFT dual to a specific nonlinear solution of the bulk gravity theory. We refer to this classical solution as the 'background'. This may be a black brane or domain wall with metric of the form (23.77) together with an accompanying scalar field. In most cases the scalar approaches the leading rate $\rho^{(d-\Delta)/2}$, but for some backgrounds the leading term may be absent and the scalar vanishes at the faster rate $\rho^{\Delta/2}$.

To obtain the physical correlators one computes functional derivatives for arbitrary sources and then sets the sources to their values in the background solution. The metric source $g_{(0)ij}$ is interpreted as the external metric to which the boundary QFT is coupled. If the leading scalar $\phi_{(0)}$ in the background is non-vanishing, then it is interpreted as a source for the dual operator. The effective Lagrangian of the boundary theory becomes

$$\mathcal{L}_{\text{QFT}} = \mathcal{L}_{\text{CFT}} + \phi_{(0)}\mathcal{O}_\Delta, \tag{23.89}$$

which describes a relevant or marginal deformation of a CFT. If the leading scalar term vanishes, then the one-point function obtained directly (but carefully) from (23.88) need not vanish, and the result is interpreted as the vacuum expectation value $\langle \mathcal{O}_\Delta(x)\rangle$ of the dual operator. The stress tensor of the boundary theory computed from (23.87) can also acquire a non-zero vacuum expectation value, $\langle T_{ij}(x)\rangle \neq 0$. One situation in which this happens is when the background is a black brane. The dual boundary theory is then a CFT

Holographic renormalization **Box 23.6**

In asymptotically anti-de Sitter spacetimes a near-boundary analysis of the fields reveals divergences dependent on the cutoff distance to the boundary. These are canceled in the renormalized action by local covariant counterterms determined by the asymptotic solution. The physical correlators are obtained by functional derivatives of the renormalized action with respect to arbitrary sources, and then setting the sources equal to their values in the background solution.

in a thermal ensemble at the Hawking temperature. The components of $\langle T_{ij}(x)\rangle$ contain the AdS/CFT results for the energy density and pressure in the ensemble.

We will conclude this section by describing two explicit applications of holographic renormalization. We will not be able to include all details, but we hope that the discussion will help readers to understand how the formalism is actually used.

23.11.1 The scalar two-point function in a CFT$_d$

In this section we consider the bulk theory of a scalar field of mass $m^2L^2 = 1 - d^2/4$. The dual scalar operator $\mathcal{O}_\Delta$ then has scale dimension $\Delta = d/2 + 1$, according to (23.18). We have chosen the lowest dimension relevant operator that satisfies the restriction made above (23.83) because the relevant asymptotic expansion is quite simple, yet illustrative of the ideas involved. To compute the two-point correlation function $\langle \mathcal{O}_\Delta \mathcal{O}_\Delta\rangle$, it is sufficient to consider a *free* scalar field with the mass squared given above in a *fixed* AdS$_{(d+1)}$ background geometry. The metric is given by (23.77) with $F(z) \equiv 1$, which we now write with radial coordinate ρ as

$$\mathrm{d}s^2 = \frac{L^2}{\rho}\left[\frac{\mathrm{d}\rho^2}{4\rho} + \delta_{ij}\mathrm{d}x^i\mathrm{d}x^j\right]. \tag{23.90}$$

The fixed background approximation is justified in this case. We follow [332], which contains more details. The two-point correlator we compute is a special case of the result (23.61), but the discussion below is an opportunity to see the method of holographic renormalization in action.

The scalar equation of motion is

$$(-\Box + m^2)\phi(x,\rho) = \frac{1}{L^2}\left[-4\rho^{1+d/2}\partial_\rho\rho^{1-d/2}\partial_\rho - \rho\delta^{ij}\partial_i\partial_j + (1 - \tfrac{1}{4}d^2)\right]\phi(x,\rho) = 0. \tag{23.91}$$

The most general asymptotic solution is

$$\phi(x,\rho) = \rho^{(d-2)/4}[\phi_{(0)} + \rho\phi_{(2)} + \rho\ln\rho\,\psi_{(2)} + \ldots]. \tag{23.92}$$

It is quite straightforward to substitute this series in (23.91) and learn that $\psi_{(2)} = -\delta^{ij}\partial_i\partial_j\phi_{(0)}/4$, but $\phi_{(2)}$ is not determined. It is not fixed because there is a second linearly independent solution of (23.91), which vanishes at the rate $\rho^{(d+2)/4}$ as $\rho \to 0$. The one-point function (23.88) is

$$\langle \mathcal{O}_{(1+d/2)}\rangle_s = -2(\phi_{(2)} + \psi_{(2)}). \tag{23.93}$$

It contains the undetermined $\phi_{(2)}$ plus the local term C of (23.88), which is proportional to $\psi_{(2)}$ in this case.

At this point we need to examine the equation of motion (23.91) more closely. It is simply the equation (23.65) studied in Sec. 23.10 (with a different radial coordinate $\rho = z_0^2$). We choose the solution that vanishes at large ρ, namely

$$\phi(\vec{p},\rho) = \phi_{(0)}(\vec{p})\rho^{d/4} p K_1(p\rho^{1/2})$$
$$= \phi_{(0)}(\vec{p})\rho^{(d-2)/4}\left[1 + \rho\, p^2\left(b + \tfrac{1}{4}\ln(p^2\rho)\right) + \ldots\right]. \qquad (23.94)$$

The Bessel function now appears with integer order, and we have used its boundary asymptotics (which contains a $\ln\rho$ term) to write the second line above. Note that $p = |\vec{p}|$ and that b is an irrelevant numerical constant. From this equation we identify

$$\phi_{(2)}(\vec{p}) = p^2(b + \tfrac{1}{4}\ln p^2)\phi_{(0)}(\vec{p}), \qquad \psi_{(2)}(\vec{p}) = \tfrac{1}{4}p^2\phi_{(0)}(\vec{p}). \qquad (23.95)$$

The $\psi_{(2)}$ term agrees with the result obtained above from near-boundary analysis in x-space. The $\phi_{(2)}(\vec{p})$ was left undetermined by near-boundary analysis, but it is now fixed by the requirement that the full solution vanishes as $\rho \to \infty$. Note that $\phi_{(2)}$ is non-analytic in p. Its Fourier transform is non-local in x.

Inserting the results (23.95) in (23.93), we find the one-point function with source

$$\langle \mathcal{O}_{(1+d/2)}(\vec{p})\rangle_s = -2p^2\left[(b + \tfrac{1}{4}) + \tfrac{1}{4}\ln p^2\right]\phi_{(0)}(\vec{p}). \qquad (23.96)$$

A further functional derivative produces the two-point function in p-space. We drop the contact term due to the monomial $(b + \frac{1}{4})p^2$ and write

$$\langle \mathcal{O}_{(1+d/2)}(\vec{p})\mathcal{O}_{(1+d/2)}(\vec{q})\rangle = (2\pi)^d\delta(\vec{p}+\vec{q})\tfrac{1}{2}p^2\ln p^2. \qquad (23.97)$$

The final step is the inverse Fourier transform back to x-space. For this we need an extension of the calculation (23.75). We will be content to use the results for $d = 4$ in Appendix A of [318] which give

$$\langle \mathcal{O}_{(3)}(x)\mathcal{O}_{(3)}(0)\rangle = -\frac{1}{8\pi^2}\Box\Box\frac{1}{x^2}\ln(x^2m^2) = \frac{4}{\pi^2}\left(\frac{1}{x^6}\right)_{\text{ren}}. \qquad (23.98)$$

The final result agrees with (23.76). It contains the renormalized power law $1/x^6$. In the method of holographic renormalization the scale m is related to the 'matter conformal anomaly' discussed in [332].

23.11.2 The holographic trace anomaly

The generator of scale transformations in a CFT$_d$ in Minkowski spacetime is given by

$$D = \int \mathrm{d}^{d-1}x\, x_i T^{i0}. \qquad (23.99)$$

It is conserved in time if and only if the stress tensor is conserved, $\partial_i T^{ij} = 0$ and traceless $T_i^i = 0$. Special conformal invariance also requires tracelessness. It is clear from (23.30) that the stress tensor is traceless. Indeed a traceless stress tensor is the *sine qua non* of a CFT$_d$. However, the situation is different if the CFT is coupled to a background metric $g_{(0)ij}$ that is not flat. Then the ultraviolet regularization, which is necessary to define the quantum theory properly, leads to a trace anomaly. In this section we study only spacetime dimension $d = 4$.

In the field theory one can define a renormalized effective action $S_{\rm eff}[g_0]$, which depends on the background metric. The trace anomaly is defined as the change in $S_{\rm eff}[g_0]$ due to an infinitesimal local scale (or Weyl) transformation of the background metric, defined as

$$\delta g_{(0)ij}(x) = 2\sigma(x) g_{(0)ij}(x). \tag{23.100}$$

Thus

$$\delta S_{\rm eff}[g_0] \equiv \tfrac{1}{2}\int \mathrm{d}^4x\, \sqrt{g_{(0)}}\langle T_{ij}\rangle \delta g^{ij}_{(0)} \equiv -\int \mathrm{d}^4x \sqrt{g_{(0)}}\sigma(x)\langle T^i_i\rangle. \tag{23.101}$$

The trace anomaly is a local function of curvature invariants of the background of scale dimension 4. One can show that the general form is

$$\langle T^i_i(x)\rangle = -\frac{1}{\sqrt{g_{(0)}}}\frac{\delta S_{\rm eff}}{\delta\sigma(x)} = \frac{1}{16\pi^2}[c\, W^2 - a\, E^2], \tag{23.102}$$

where the Weyl tensor and Euler densities are

$$\begin{aligned} W^2 &\equiv R^2_{ijkl} - 2R^2_{ij} + \tfrac{1}{3}R^2, \\ E^2 &\equiv {}^*R_{ij}{}^{kl}\, {}^*R_{kl}{}^{ij} = 6R_{ij}{}^{[ij}\, R_{kl}{}^{kl]} = R^2_{ijkl} - 4R^2_{ij} + R^2. \end{aligned} \tag{23.103}$$

The constants c and a are central charges, which are basic characteristics of the CFT. The central charge c also appears in the two-point function (23.30), while a appears in the three-point function of the stress tensor; see [333].

In a superconformal CFT that includes N_g gauge multiplets and N_χ chiral multiplets of $\mathcal{N} = 1$ SUSY, the central charges are (see [334])

$$c = \frac{1}{24}(3N_g + N_\chi), \qquad a = \frac{1}{48}(9N_g + N_\chi). \tag{23.104}$$

The $\mathcal{N} = 4$ SYM theory contains one gauge multiplet and three chiral multiplets in the adjoint representation of the gauge group SU(N). There is also an $\mathcal{N} = 2$ superconformal gauge theory with one gauge multiplet (containing one gauge and one chiral $\mathcal{N} = 1$ multiplet in the adjoint representation) plus $2N$ hypermultiplets, each in the fundamental representation (a total of $4N^2$ single chiral multiplets in $\mathcal{N} = 1$). Thus the central charges are

$$\begin{aligned} \mathcal{N} = 4: &\qquad c = \tfrac{1}{4}(N^2 - 1), &\qquad a = \tfrac{1}{4}(N^2 - 1), \\ \mathcal{N} = 2: &\qquad c = \tfrac{1}{3}N^2 - \tfrac{1}{6}, &\qquad a = \frac{7}{24}N^2 - \frac{5}{24}. \end{aligned} \tag{23.105}$$

If a CFT has a gravity dual we would hope to reproduce the values of c and a, at least for large N, from the AdS/CFT correspondence. We now outline how the trace anomaly is computed using holographic renormalization. Since we are concerned only with a local property of the stress tensor it is sufficient to consider AAdS solutions of a purely gravitational bulk theory. We will factor out the AdS scale L and write the bulk metric as

$$\mathrm{d}s^2 = L^2 G_{\mu\nu}\mathrm{d}x^\mu \mathrm{d}x^\nu = L^2\left(\frac{\mathrm{d}\rho^2}{4\rho^2} + \frac{g_{ij}(x,\rho)}{\rho}\mathrm{d}x^i \mathrm{d}x^j\right). \qquad (23.106)$$

The action is

$$S = -\frac{L^3}{16\pi G_5}\left\{\int_{\rho>\epsilon} \mathrm{d}\rho \mathrm{d}^d x\,\sqrt{G}[R+12] + 2\int_{\rho=\epsilon} \mathrm{d}^4 x\,\sqrt{\gamma}K\right\}, \qquad (23.107)$$

in which G_5 is Newton's constant in five dimensions and $\gamma_{ij} = g_{ij}/\epsilon$ is the induced metric at the cutoff.

It is useful to consider a particular type of infinitesimal diffeomorphism, used in [335], namely

$$\rho = \rho'(1-2\sigma(x')), \qquad x^i = x'^i + a^i(x',\rho'), \qquad (23.108)$$

with

$$a^i(x,\rho) = \tfrac{1}{2}\int_0^\rho \mathrm{d}\hat{\rho}\; g^{ij}(x,\hat{\rho})\partial_j\sigma(x). \qquad (23.109)$$

Exercise 23.12 *To first order in $\sigma(x)$, show that the transformed line element (23.106) takes the form*

$$\mathrm{d}s^2 = L^2\left(\frac{\mathrm{d}\rho^2}{4\rho^2} + \frac{1}{\rho}[g_{ij} + 2\sigma(1-\rho\partial_\rho)g_{ij} + \nabla_i a_j + \nabla_j a_i]\mathrm{d}x^i\mathrm{d}x^j\right). \qquad (23.110)$$

Hint: remember (7.143).

In the limit $\rho \to 0$, both a_i and $\rho \mathrm{d}g_{ij}/\mathrm{d}\rho$ vanish, so that

$$g_{(0)ij}(x) \to (1+2\sigma(x))g_{(0)ij}(x). \qquad (23.111)$$

The net effect of the five-dimensional diffeomorphism is an infinitesimal Weyl transformation of the boundary metric $g_{(0)ij}$. This result shows explicitly that an AAdS spacetime determines the conformal class of its boundary metric, but not the metric itself. This observation will be useful shortly.

We would expect to obtain the holographic trace anomaly from the variation of the renormalized on-shell action obtained from (23.107) due to a Weyl transform of the source $g_{(0)ij}$. Specifically

$$\langle T^i_i\rangle = -\frac{1}{\sqrt{g_{(0)}}}\frac{\delta}{\delta\sigma}S_{\mathrm{ren}}[g_{(0)}] = -\lim_{\epsilon\to 0}\frac{1}{\sqrt{g_{(0)}}}\frac{\delta}{\delta\sigma}(S_{\mathrm{on-shell}} + S_{\mathrm{ct}}). \qquad (23.112)$$

From the near-boundary expansion of the metric in (23.87), one can obtain, by long and delicate analysis, the explicit form of the counterterm action [329, 330]:

$$S_{\mathrm{ct}}[g_{(0)}] = \frac{L^3}{4\pi G_5}\int \mathrm{d}^4 x\sqrt{\gamma}\left(\frac{3}{2} + \frac{R_{(\gamma)}}{8} - \frac{\ln\epsilon}{32}(R^{ij}_{(\gamma)}R_{(\gamma)ij} - \tfrac{1}{3}R^2_{(\gamma)})\right). \qquad (23.113)$$

The subscript γ indicates that $R_{(\gamma)ij}$, $R_{(\gamma)}$ are the curvatures of the induced metric $\gamma_{ij} = g_{ij}(x,\epsilon)/\epsilon$. Thus the second and third terms in the integral carry 'hidden factors' of ϵ and ϵ^2, respectively.

To calculate the variational derivative one must vary the boundary data, $\delta g_{(0)ij} = 2\sigma g_{(0)ij}$, while maintaining the fact that the interior solution corresponds to that variation. Thus one is really carrying out the diffeomorphism (23.108), with the shift $\delta\epsilon = 2\epsilon\sigma(x)$ of the position of the boundary. The on-shell action is invariant under the combined change of coordinates and reparametrization of the boundary. The first two terms in $S_{\rm ct}$ are also invariant (since they contain scalars under the diffeomorphism). Thus the only term that contributes in (23.112) is the explicit variation $\delta \ln \epsilon = 2\sigma(x)$, and this produces the anomalous trace

$$\langle T_i^i \rangle = \frac{L^3}{64\pi G_5}(R_{(0)}^{ij} R_{(0)ij} - \tfrac{1}{3} R_{(0)}^2). \tag{23.114}$$

The tensor structure of this expression can be compared to that of the field theory form (23.102) of the trace anomaly. One can see that the invariant R_{ijkl}^2 is absent, and this means that $c = a$ for any CFT_4 that has a gravity dual. The $\mathcal{N} = 2$ superconformal gauge theory is therefore excluded, but $\mathcal{N} = 4$ SYM is allowed. What remains to check is the AdS/CFT prediction for the central charge, which is

$$c = a = \frac{\pi L^3}{8 G_5} = \frac{1}{4} N^2. \tag{23.115}$$

The final result and success(!) was obtained by inserting the value of G_5 from (23.14).

We conclude this section by referring readers to the papers [336, 337] in which an alternative version of the holographic renormalization procedure is developed. This procedure is based on the Hamiltonian formulation of the bulk equations of motion in which the radial variable of AAdS solutions plays the role of time. This method incorporates symmetries more efficiently and simplifies the calculation of correlation functions.

23.12 Holographic RG flows

This term is used to describe the gravity dual of a CFT deformed by a relevant operator, an operator with scale dimension $\Delta < d$. The Lagrangian describing this situation is given in (23.89). Effects of the relevant perturbation disappear at short distances, so correlation functions of the perturbed theory approach those of the original CFT, which we call the $\mathrm{CFT}_{\mathrm{UV}}$ at short distances. The relevant perturbation does change the low energy behavior of the theory, and we assume that the correlation functions approach those of another CFT, called the $\mathrm{CFT}_{\mathrm{IR}}$ at long distance.

The two CFTs are characterized by different central charges, c_{UV} and a_{UV} at short distance and c_{IR} and a_{IR} at long distance. The intuition, derived from Zamolodchikov's c-theorem for two-dimensional theories [338], is that the central charges measure the number of degrees of freedom of a theory and that this number decreases in the flow from the UV to IR since massive particles drop out at long distance. There is considerable evidence [339, 36, 340] that the a central charge does satisfy $a_{\mathrm{UV}} > a_{\mathrm{IR}}$ in the many models which have been examined, although there is no general proof. As discussed in the previous section, the central charges of a CFT with a gravity dual satisfy $a = c$, so we won't be concerned with the question of a versus c. However, we will give a very simple proof that $a_{\text{UV}} > a_{\text{IR}}$ in any holographic RG flow.

The holographic dual of a particular relevant perturbation [341] of $\mathcal{N} = 4$ SYM theory has been studied [342] and makes remarkably detailed contact with the field theory. Rather than discuss this fairly complex example, we will illustrate the basic dynamics of holographic RG flows in the simple bulk theory model with action (23.80).

23.12.1 AAdS domain wall solutions

We seek gravity solutions that are dual to quantum field theories with the symmetry group of the Poincaré group in d dimensions. (We speak of the Poincaré group although we work in Euclidean signature.) Since the symmetries must match on both sides of the duality, we look for solutions of the $D = (d+1)$-dimensional bulk system with the same symmetry group. The most general such configuration is

$$ds^2 = e^{2A(r)}\delta_{ij}dx^i dx^j + dr^2, \qquad \phi = \phi(r). \tag{23.116}$$

This configuration is called a domain wall.

We assume that the scalar potential $V(\phi)$ has at least two critical points, i.e. $V'(\phi_k) = 0$, at which $V(\phi_k) = -d(d-1)/2L_k^2 < 0$. The potential is sketched in Fig. 23.2. The equations of motion of the action (23.80) have special solutions for which $\phi(x)$ is *fixed* at a critical point, i.e. $\phi(x) \equiv \phi_k$ and the spacetime geometry is $\text{AdS}_{(d+1)}$ with scale L_k. We refer to these solutions as critical solutions and write the metric as

$$ds^2 = e^{2r/L_k}\delta_{ij}dx^i dx^j + dr^2. \tag{23.117}$$

This differs from the previous description (23.20) of the AdS metric by a change of the radial coordinate: $r = -L\ln(z_0/L)$.

The AAdS domain wall we want is a solution of the field equations that conforms to the ansatz (23.116) and interpolates between the critical solutions for two adjacent critical points of the potential, which we label as UV and IR. Specifically the metric has the boundary structure of AdS with scale L_{UV} as $r \to +\infty$ and has the deep interior behavior of (Euclidean) AdS with scale L_{IR} as $r \to -\infty$.

The Euler variation of the action (23.80) yields coupled equations of motion for the scalar ϕ and the metric $g_{\mu\nu}$. However, the domain wall ansatz contains just two unknown functions $A(r)$, $\phi(r)$, and it turns out that all equations of motion are satisfied if $A(r)$, $\phi(r)$ obey the coupled ordinary differential equations[10]

[10] The reader can use (8.8) to check the gravitational equations. The only non-vanishing curvature components are $R_{ij}{}^{k\ell} = -2A'^2\delta_{ij}^{k\ell}$, $R_{ri}{}^{rj} = -(A'' + A'^2)\delta_i^j$ and those related by the symmetry properties. There are two resulting equations from the (rr) and from the (ij) components of (8.8). One linear combination of these gives the first result in (23.118), and the second gives the result in Ex. 23.13.

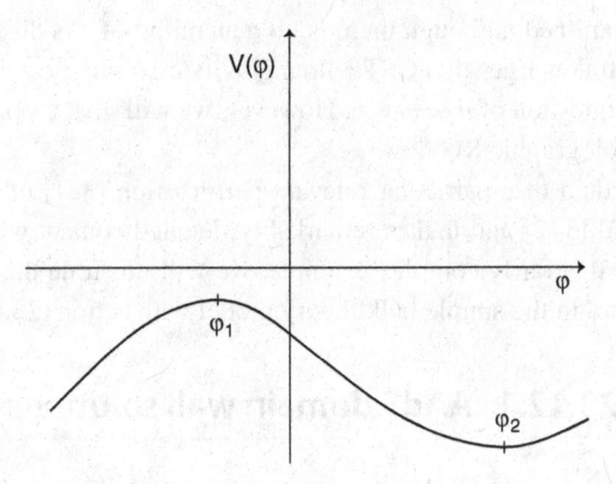

Fig. 23.2 Sketch of a scalar potential that supports domain wall solutions.

$$A'^2 = \frac{1}{d(d-1)}[\phi'^2 - 2V(\phi)], \qquad \phi'' + dA'\phi' = \frac{dV(\phi)}{d\phi}. \tag{23.118}$$

It is easy to see how the critical solutions appear. At each critical point ϕ_k of the potential, the scalar equation is satisfied by $\phi(r) \equiv \phi_k$. The A' equation then yields $A(r) = \pm(r + r_0)/L_k$. The integration constant r_0 and the sign have no significance since they can be changed by scaling of the coordinates x^i and changing $r \to -r$. Thus we can choose $A(r) = r/L_k$ which brings us exactly to the form in (23.117).

Exercise 23.13 *Show that the simple result $A'' = -\phi'^2/(d-1)$ follows from the pair of equations in (23.118). Since the sign is negative the function $A(r)$ is concave downwards. The result $A'' \leq 0$ also turns out to be exactly what is needed to establish the holographic c-theorem, as we show below.*

Exact solutions of the nonlinear second order system (23.118) are difficult (although we will outline an interesting method in Sec. 23.12.3). However, we can learn a lot by linearizing about each critical point. Near the stationary point ϕ_i we express the fields as $\phi(r) = \phi_i + h(r)$ and $A'(r) = 1/L_i + a'(r)$ and work to lowest order in the deviations $h(r)$ and $a(r)$. Following (23.82) we parametrize the potential near ϕ_i as

$$V(\phi) = \frac{1}{2}\left[-\frac{d(d-1)}{L_i^2} + m_i^2 h^2\right]. \tag{23.119}$$

After substituting these expressions in (23.118) we find that $a'(r)$ is of order h^2 and can be neglected, while $h(r)$ satisfies the linear equation

$$h'' + \frac{d}{L_i}h' - m_i^2 h = 0. \tag{23.120}$$

The general solution is

$$h(r) = Be^{(\Delta_i - d)r/L_i} + Ce^{-\Delta_i r/L} \qquad \text{with} \qquad \Delta_i = \tfrac{1}{2}\left(d + \sqrt{d^2 + 4m_i^2 L_i^2}\right). \tag{23.121}$$

The basic idea of linearization theory is that there is an exact solution of the nonlinear equations of motion that is well approximated by a linear solution near a critical point. Thus as $r \to +\infty$, we assume that the exact solution behaves as

$$\phi(r) \approx_{r \gg 0} \phi_1 + B_1 e^{(\Delta_1 - d)r/L_1} + C_1 e^{-\Delta_1 r/L_1}. \tag{23.122}$$

The fluctuation must disappear as $r \to +\infty$. For a generic situation in which the dominant B-term is present, this requires $d/2 < \Delta_1 < d$ or $m_1^2 < 0$. Hence the critical point associated with the boundary region of the domain wall must be a local maximum, and everything is consistent with an interpretation as the dual of a QFT$_d$ that is a relevant deformation of an ultraviolet CFT$_d$. Thus we can set $\phi_1 = \phi_{\rm UV}$ and $L_1 = L_{\rm UV}$.

Near the critical point ϕ_2, which is a minimum, we have $m_2^2 > 0$ so $\Delta_2 > d$. This critical point must be approached at large negative r, where the exact solution is approximated by

$$\phi(r) \approx_{r \ll 0} \phi_2 + B_2 e^{(\Delta_2 - d)r/L_2} + C_2 e^{-\Delta_2 r/L_2}. \tag{23.123}$$

The second term diverges, so we must choose the solution with $C_2 = 0$. Thus the domain wall approaches the deep interior region with the scaling rate of an irrelevant operator of scale dimension $\Delta_2 > d$ exactly as required for infrared fixed points by RG ideas on field theory. We can now set $\phi_2 = \phi_{\rm IR}$ and $L_2 = L_{\rm IR}$.

The nonlinear equation of motion for $\phi(r)$ has two integration constants. We must fix one of them to ensure $C = 0$ as $r \to -\infty$. The remaining freedom is just the shift $r \to r + r_0$ and has no effect on the physical picture. A generic solution with $C = 0$ in the IR would be expected to approach the UV critical point at the dominant rate $Be^{(\Delta_{\rm UV} - d)r/L_1}$, which we have seen to be dual to a relevant operator deformation of the CFT$_{\rm UV}$. It is possible (but exceptional) that the $C = 0$ solution in the IR would have vanishing B-term in the UV and approach the boundary as $C_{\rm UV}e^{-\Delta_{\rm UV} r/L_1}$. In this case the physical interpretation is that of the deformation of the CFT$_{\rm UV}$ by a vacuum expectation value; $\langle \mathcal{O}_{\Delta_{\rm UV}} \rangle \sim C_{\rm UV} \neq 0$; see [343, 316].

The domain wall flow 'sees' the AdS$_{\rm IR}$ geometry only in the deep interior limit. To discuss the CFT$_{\rm IR}$ and its operator perturbations in themselves, we must think of extending this interior region out to a complete AdS$_{d+1}$ geometry with scale $L_{\rm IR} = L_2$.

The interpolating solution we are discussing is plotted in Fig. 23.3. The scale factor $A(r)$ is concave downwards since $A''(r) < 0$ from Ex. 23.13. This means that the slopes of the linear regions in the deep interior and near the boundary are related by $1/L_{\rm IR} > 1/L_{\rm UV}$. Hence,

$$V_{\rm IR} = -\frac{d(d-1)}{2L_{\rm IR}^2} < V_{\rm UV} = -\frac{-d(d-1)}{2L_{\rm UV}^2}. \tag{23.124}$$

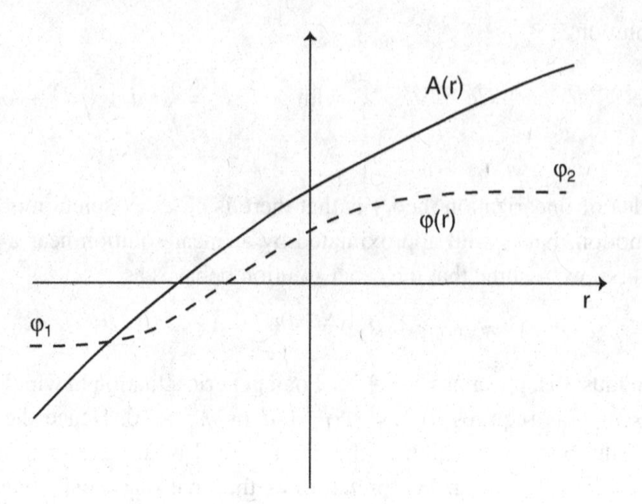

Fig. 23.3 Profiles of the scale factor and scalar field of a domain wall.

Thus the flow from the boundary to the interior necessarily goes to a deeper critical point of $V(\phi)$. Recall that the condition $A''(r) < 0$ is very general and holds in any physically reasonable bulk theory, e.g. a system of many scalars ϕ^I and potential $V(\phi^I)$. Thus any Poincaré invariant domain wall interpolating between AdS geometries is *irreversible*.

23.12.2 The holographic c-theorem

Let's apply the result (23.115) for the central charges obtained from the trace anomaly in AdS_5 to a flow between two CFT_4s. The method of the previous section is applicable to the CFT_{UV} since its trace anomaly is determined by the boundary behavior of the metric. One can show that the anomaly is not affected by the coupling of gravity to matter fields. Hence $c_{UV} = a_{UV} = \pi L_{UV}^3/8G_5$. As explained above, to obtain the anomaly of the CFT_{IR} one must consider the extension of the deep interior metric of the AAdS domain wall to a complete AdS_5 spacetime with scale L_{IR}. It is the boundary behavior of this extended metric that determines the trace anomaly, which is $c_{IR} = a_{IR} = \pi L_{IR}^3/8G_5$. Using the inequality $L_{UV} > L_{IR}$ we can see immediately that $c_{UV} > c_{IR}$. The conclusion is in Box 23.7.

An issue that is discussed in the literature on the c-theorem in quantum field theory is the existence of c-functions. A c-function is a continuous function of the RG scale that decreases monotonically along the flow from UV $\rightarrow$ IR and interpolates between the central charges c_{UV} and c_{IR}. In the AdS/CFT correspondence, the radial coordinate of the bulk

Box 23.7 The c-theorem for the central charges

The central charges at the UV and IR fixed points of any RG flow with a gravity dual necessarily obey the c-theorem: $c_{UV} > c_{IR}$. The radial coordinate plays the role of the RG scale in the bulk solution.

solution plays the role of the RG scale. For any AAdS domain wall one can consider the following scale dependent function (and its radial derivative):

$$C(r) = \frac{\pi}{8G}\frac{1}{A'^3}, \qquad C'(r) = \frac{\pi}{8G}\frac{-3A''}{A'^4} \geq 0. \tag{23.125}$$

We have $C'(r) \geq 0$ as a consequence of the condition $A'' \leq 0$ derived from the domain wall equations (23.118). Thus $C(r)$ is an essentially perfect holographic c-function.

23.12.3 First order flow equations

The domain wall equations (23.118) are a nonlinear second order system with no apparent method of analytic solution. Nevertheless there is an interesting procedure [344] that does give exact solutions in a number of cases. Given the potential $V(\phi)$, suppose that we could solve the following differential equation in field space and obtain an auxiliary quantity, the superpotential $W(\phi)$:

$$\frac{1}{2}\left(\frac{\mathrm{d}W}{\mathrm{d}\phi}\right)^2 - \frac{d}{2(d-1)}W^2 = V(\phi). \tag{23.126}$$

We then consider the pair of first order differential equations

$$\phi'(r) = \frac{\mathrm{d}W}{\mathrm{d}\phi}, \qquad A'(r) = -\frac{1}{d-1}W(\phi). \tag{23.127}$$

It is easy to show that any solution of the first order system (23.127) is also a solution of the original second order set (23.118). The uncoupled system (23.127) has a trivial structure; the two equations can be solved sequentially. If the two required integrals are tractable, one finds an explicit analytic solution for the domain wall.

Exercise 23.14 *It is easy to see that any critical point of $W(\phi)$ is also a critical point of $V(\phi)$, although not conversely. In this exercise we study what happens when the critical point ϕ_i is common. Suppose that $W(\phi)$ is approximated by $W(\phi) \approx -(\lambda_i + \frac{1}{2}\mu_i h^2)/L_i$ near $\phi = \phi_i$. Show that λ_i, μ_i are related to the parameters of the approximate potential (23.119) by $\lambda_i = d-1$ and $m_i^2 L_i^2 = \mu_i(\mu_i - d)$. Show that this requires $m_i^2 \geq -d^2/4L_i^2$. This condition is the perturbative stability bound for general d discussed for $d = 4$ at the end of Sec. 23.4. Thus any critical point common to both W and V must describe a stable AdS solution. Show that the solution $\phi(r)$ of the flow equation (23.127) approaches a common UV critical point at the rate $h \approx \mathrm{e}^{-\mu_i r/L_i}$ with $\mu_i = d - \Delta_i$, the source rate, or $\mu_i = \Delta_i$, the vacuum expectation value rate.*

This structure generalizes to bulk theories with several scalars ϕ^I and Lagrangian

$$L \sim -\tfrac{1}{2}R + \tfrac{1}{2}\sum_I g^{\mu\nu}\partial_\mu\phi^I\partial_\nu\phi^I + V(\phi). \tag{23.128}$$

Given a superpotential $W(\phi^I)$ that satisfies the partial differential equation

$$\frac{1}{2}\sum_I \left(\frac{\partial W}{\partial \phi^I}\right)^2 - \frac{d}{2(d-1)} W^2 = V(\phi), \tag{23.129}$$

the first order flow equations

$$\frac{d\phi^I}{dr} = \frac{dW}{d\phi^I}, \qquad A'(r) = -\frac{1}{d-1} W(\phi) \tag{23.130}$$

automatically give a solution of the second order equations of motion of (23.128) for Poincaré invariant domain walls.

Exercise 23.15 *Prove this and derive first order flow equations with the same property for the case where the scalar kinetic term of (23.130) is replaced by that of a nonlinear σ-model, namely $\frac{1}{2}G_{IJ}(\phi)\partial_\mu\phi^I\partial^\mu\phi^J$.*

The equations (23.130) are conventional gradient flow equations. The solutions are paths of steepest descent for $W(\phi)$, everywhere perpendicular to the contours $W(\phi) = \text{const}$. In applications to RG flows, the $\phi^I(r)$ represent scale dependent couplings of relevant operators in a QFT Lagrangian, so what we have is gradient flow in the space of couplings – an idea frequently discussed in the RG literature!

There are two interesting reasons why there are first order flow equations that reproduce the conventional second order dynamics of domain walls.

1. They emerge as BPS conditions for supersymmetric domain walls in supergravity theories. The superpotential appears directly in the fermion transformation rules. There is no need to solve a differential equation to find it. See [324] for examples.
2. They are the Hamilton–Jacobi equations for the dynamical system of gravity and scalars [345]. The superpotential is the classical Hamilton–Jacobi function, and one must solve (23.126) or (23.129) to obtain it from the potential $V(\phi^I)$.

23.13 AdS/CFT and hydrodynamics

The AdS/CFT correspondence is frequently applied to the strong coupling behavior of a CFT in a thermal ensemble at temperature T. The gravitational dual of this situation is an AdS black hole or black brane at Hawking temperature $T = k/2\pi$, where k is the surface gravity of the black hole. The surface gravity is the acceleration of a test mass at the horizon. The terminology we use is that an AdS black hole has the boundary structure of anti-de Sitter space in global coordinates. For bulk dimension $D = d + 1$ its isometry group is $\mathbb{R} \otimes \mathrm{SO}(d)$ where $\mathbb{R}$ is the group of time translations. An AdS black brane has the boundary structure of AdS in Poincaré patch coordinates. Its isometry group is the direct product of time translations with rotations and translations in $d-1$ directions.

In this section[11] we study the AdS/CFT dual of the Lorentzian signature black brane spacetime

$$ds^2 = \frac{L^2}{z^2}\left[-f(z)\mathrm{d}t^2 + \frac{1}{f(z)}\mathrm{d}z^2 + \delta_{ij}\mathrm{d}x^i\mathrm{d}x^j\right], \qquad f(z) = 1 - \frac{z^d}{z_h^d}, \tag{23.131}$$

with $i, j = 1, \ldots, d-1$. The horizon is at $z = z_h$. The metric has the same boundary behavior as AdS_{d+1}, although it is not written in the standard presentation (23.77) of AAdS metrics. This spacetime is an Einstein space; it satisfies $R_{\mu\nu} = -(d/L^2)g_{\mu\nu}$, but it is not conformally flat.

To investigate the metric near the horizon we introduce the coordinate $\tilde{z} = (2z_h/d)\sqrt{f(z)}$. To lowest order $\tilde{z}$, the near-horizon metric is given by

$$ds^2 \approx \frac{L^2}{z_h^2}\left[-k^2\tilde{z}^2\mathrm{d}t^2 + d\tilde{z}^2 + \delta_{ij}\mathrm{d}x^i\mathrm{d}x^j\right], \tag{23.132}$$

where $k = d/2z_h$. To identify the Hawking temperature we write the analytically continued, i.e. $t \to i\tau$, metric of the Euclidean section,

$$ds^2 \approx \frac{L^2}{z_h^2}\left[k^2\tilde{z}^2\mathrm{d}\tau^2 + \mathrm{d}\tilde{z}^2 + \delta_{ij}\mathrm{d}x^i\mathrm{d}x^j\right]. \tag{23.133}$$

This metric describes a product manifold, namely the 2-plane in polar coordinates $\tilde{z}$, τ with Euclidean $\mathbb{R}^{d-1}$. However, there is a conical singularity, a source of unwanted curvature, unless we identify the angular coordinate as $\tau \equiv \tau + 2\pi/k$. The Hawking temperature is the inverse of the period of the Euclidean time coordinate; hence $T = k/2\pi$.

The Bekenstein–Hawking entropy is the area of the horizon in the Lorentzian metric (23.131). For the black brane the area becomes finite if we assume that the transverse coordinates x^i span a box of volume V in $\mathbb{R}^{d-1}$. Then

$$S = \frac{1}{4G_{d+1}}A = \frac{1}{4G_{d+1}}\int_V \mathrm{d}^{d-1}x\,\sqrt{g} = \frac{1}{4G_{d+1}}\left(\frac{L}{z_h}\right)^{d-1} V. \tag{23.134}$$

The integral in the intermediate step is computed at fixed $z = z_h$, $t = 0$. Later we will need the entropy density, defined as $s = S/V$.

The dual situation in quantum field theory that we are concerned with is the hydrodynamics of the thermal ensemble of temperature T in the $\mathcal{N} = 4$ SYM theory. This is the

[11] Our discussion is based on [346, 306] to which we refer readers for further details. The lecture notes [347] are also very useful.

physics of long wavelength excitations in the ensemble, which propagate in spacetime. Thus we consider finite temperature in Lorentzian signature quantum field theory.

We are interested in the limit in which hydrodynamic phenomena are governed by linear response theory. We will present a very heuristic derivation of the basic linear response formula. For more information, see [348, 349]. Suppose there is an operator $\mathcal{O}$ with external classical source ϕ_0. Real-time correlation functions of $\mathcal{O}$ in the canonical ensemble are given by variational derivatives of the generating function

$$\langle e^{i\int d^dx\, \mathcal{O}(x)\phi_0(x)}\rangle \equiv Tr\left(e^{-\beta H} e^{i\int d^dx\, \mathcal{O}(x)\phi_0(x)}\right) \Big/ Tr\left(e^{-\beta H}\right), \tag{23.135}$$

with $\beta = 1/T$. The one-point function (in the presence of the source) is then

$$\langle \mathcal{O}(x)\rangle_{\phi_0} = -i\frac{\delta}{\delta\phi_0(x)}\langle e^{i\int d^dx\, \mathcal{O}(x)\phi_0(x)}\rangle = \left\langle \mathcal{O}(x) \exp i\int d^dy\, \mathcal{O}(y)\phi_0(y)\right\rangle. \tag{23.136}$$

In most cases, $\langle \mathcal{O}(x)\rangle$ vanishes when the source vanishes. To first order in the source, the ensemble expectation value of an operator is thus given by the convolution of its two-point correlator with the source, i.e.

$$\langle \mathcal{O}(x)\rangle = i\int d^dy \langle \mathcal{O}(x)\mathcal{O}(y)\rangle\phi_0(y). \tag{23.137}$$

There are several types of correlation functions in real time quantum field theory, which differ in the time order of the operators involved. Our heuristic procedure has not treated time order correctly. We simply state that more thorough arguments in finite temperature field theory, given in [348, 349], show that we should express (23.137) in momentum space, i.e.

$$\mathcal{O}(\omega,\vec{k}) = G_R(\omega,\vec{k})\phi_0(\omega,\vec{k}), \tag{23.138}$$

where G_R is the retarded correlator

$$G_R(\omega,\vec{k}) = i\int dt\, d^{d-1}x\, e^{i(\omega t-\vec{k}\cdot\vec{x})}\theta(t)\langle[\phi(t,\vec{x}),\phi(0)]\rangle. \tag{23.139}$$

The expectation value is again taken in the canonical ensemble. Equation (23.138) is the linear response formula.

For a conserved current the low frequency, long wavelength limit of the correlation function is an important physical quantity, the transport coefficient χ, defined by the Kubo formula

$$\chi = \lim_{\omega\to 0}\lim_{\vec{k}\to 0}\frac{1}{\omega}\,\mathrm{Im}\, G_R(\omega,\vec{k}). \tag{23.140}$$

Transport coefficients are important because they are the physical parameters of the hydrodynamic equations of the medium, which determine its macroscopic behavior. Two examples are the DC electric conductivity associated with the operator $\mathcal{O} = J_i$, a component of the electric current, and the shear viscosity η associated with the off-diagonal component $\mathcal{O} = T_1^2$ of the stress tensor perpendicular to a wave in the z-direction. Although $\mathcal{N} = 4$, $d = 4$ SYM theory is our prime concern, we will phrase the discussion for general boundary dimension d.

Let us see how to access information about hydrodynamic phenomena using the AdS/CFT correspondence in Lorentzian signature. As a proxy for physically more interesting cases, we will study a massless scalar field ϕ in the black brane background. We will find the AdS/CFT prediction for its hypothetical transport coefficient. The ideas can then be applied quite easily to the calculation of the shear viscosity. The scalar field action is

$$S = -\frac{1}{2q}\int \mathrm{d}z\, \mathrm{d}t\, \mathrm{d}^{d-1}x\, \sqrt{-g}g^{\mu\nu}\partial_\mu\phi\partial_\nu\phi. \tag{23.141}$$

The parameter q governs the strength of coupling of ϕ to gravity. It is determined by information in the bulk theory, which might come, for example, from the dimensional reduction of Type IIB supergravity on $\mathrm{AdS}_5 \otimes S^5$. As $z \to 0$ a generic solution of the equation of motion for ϕ has the boundary behavior $\phi(z, t, \vec{x}) \to \phi_0(t, \vec{x})$.

The massless scalar is dual to a marginal operator $\mathcal{O}_d$, an operator with scale dimension $\Delta = d$. As in Euclidean signature, the boundary configuration $\phi_0(t, \vec{x})$ is the source of the operator $\mathcal{O}_d$, and its correlation functions are determined as derivatives of the basic formula

$$\langle \mathrm{e}^{\mathrm{i}\int \mathrm{d}t\, \mathrm{d}\vec{x}\, \mathcal{O}_d(t,\vec{x})\phi_0(t,\vec{x})}\rangle = \mathrm{e}^{\mathrm{i}S[\phi_0]}. \tag{23.142}$$

This formula relates the generating function (23.135) to the exponential of i times the on-shell value of the action (23.141), which is a functional of the boundary data ϕ_0.

The equation of motion is

$$\begin{aligned} 0 &= \partial_\mu(\sqrt{-g}g^{\mu\nu}\partial_\nu\phi), \\ 0 &= q\partial_z\Pi(z,t,\vec{x}) - \left(\frac{L}{z}\right)^{d-1}\left[\frac{1}{f(z)}\partial_t^2 - \nabla^2\right]\phi(z,t,\vec{x}), \\ \Pi(z,t,\vec{x}) &\equiv \frac{1}{q}\sqrt{-g}g^{zz}\partial_z\phi. \end{aligned} \tag{23.143}$$

In the last line we defined the canonical momentum for the 'radial evolution' of the solution. After partial integration we can write the variation of the on-shell action as

$$\delta S[\phi_0] = \int \mathrm{d}t\, \mathrm{d}^{d-1}x\, \delta\phi_0(t,\vec{x})\, \Pi(0,t,\vec{x}). \tag{23.144}$$

Although the classical solution $\phi(z, t, \xi)$ is well defined, the canonical momentum Π diverges at the boundary because of the background metric factors in its definition. This is another guise of the same divergence encountered in Sec. 23.11. For a massless bulk scalar it is simply resolved by subtracting a boundary term to make the on-shell action finite and eliminate an unobservable contact term in the correlator G_R to be calculated. We will be precise about this below. For the moment, we simply indicate the subtraction in the variational derivative, which defines the one-point function

$$\langle \mathcal{O}(t,\vec{x})\rangle_{\phi_0} = \frac{\delta S[\phi_0]}{\delta \phi_0(t,\vec{x})} = \Pi_s(z=0,t,\vec{x}). \tag{23.145}$$

To make contact with (23.138) and (23.140), we move to momentum space. The Fourier transform of $\phi(z,t,\vec{x})$ and other quantities is defined by

$$\phi(z,t,\vec{x}) = \int \frac{\mathrm{d}\omega\, \mathrm{d}^{d-1}k}{(2\pi)^d}\, \mathrm{e}^{\mathrm{i}(\vec{k}\cdot\vec{x}-\omega t)}\phi(z,\omega,\vec{k}), \tag{23.146}$$

so that (23.145) becomes

$$\langle \mathcal{O}(\omega,\vec{k})\rangle = \frac{\delta S_{\text{on-shell}}}{\delta \phi_0(\omega,\vec{k})} = \Pi_s(z=0,\omega,\vec{k}). \tag{23.147}$$

As we discuss immediately below, the solution of the equation of motion requires boundary conditions on $\phi(z,\omega,\vec{k})$ at the boundary and at the horizon (where the Fourier transform of (23.143) has regular singular points). Thus, $\Pi_s(z=0,\omega,\vec{k})$ depends non-locally on the source and it is clearly linear in the source. So we identify the AdS/CFT one-point function (23.147) with the field theory linear response (23.138). Therefore the Green's function for the CFT in the thermal ensemble is

$$G_R(\omega,\vec{k}) = \frac{\Pi_s(z=0,\omega,\vec{k})}{\phi_0(\omega,\vec{k})}. \tag{23.148}$$

Actually, we have jumped too far ahead. We must discuss how to obtain the retarded Green's function from the bulk physics. The only 'freedom' that remains resides in the function $\Pi_s(z=0,\omega,\vec{k})$ and thus in the boundary conditions that it satisfies at the horizon. In momentum space the equation (23.143) reads

$$q\partial_z \Pi(z,\omega,\vec{k}) = \left(\frac{L}{z}\right)^{d-1}\left[\frac{-1}{f(z)}\omega^2 + \vec{k}^2\right]\phi(z,\omega,\vec{k}). \tag{23.149}$$

Using $q\Pi(z,\omega,\vec{k}) = (L/z)^{d-1} f(z)\partial_z\phi(z,\omega,\vec{k})$, it is straightforward to find that there are two possible forms of the near-horizon solution of (23.149), namely

$$\phi_\pm(z,\omega,\vec{k}) = (z_h - z)^{\pm(\mathrm{i}\, z_h\, \omega/d)}. \tag{23.150}$$

When the time dependence $\mathrm{e}^{-\mathrm{i}\omega t}$ is included, we see that ϕ_+ and ϕ_- respectively describe outgoing and infalling waves at the horizon. The retarded Green's function produces *causal* wave propagation in field theory, and causal behavior of waves in the exterior of a black hole certainly requires infalling boundary conditions. This heuristic argument in favor of the ϕ_- solution was first given in [350, 351], and it was shown in [352, 353] that the infalling prescription produces the retarded Green's function via (23.148).

We therefore assume that $\phi(z,\omega,\vec{k})$ is a solution of (23.149) that approaches a multiple of $\phi_-(z,\omega,\vec{k})$ at the horizon. The solution is proportional to the boundary value so we write

$$\phi(z,\omega,\vec{k}) \equiv F(z,\omega,\vec{k})\phi_0(\omega,\vec{k}). \tag{23.151}$$

We can now determine the subtraction that is necessary to make the on-shell action finite. We work in ω, $\vec{k}$ space. We write the series expansion $\phi(z,\omega,\vec{k}) = \phi_0(\omega,\vec{k}) + z^2\phi_2 + \ldots + z^d(\phi_d + \ln z\psi_d) + \ldots$, which is similar to those used in Sec. 23.11.[12] Substituting the series in the differential equation (23.149), we find that all terms ϕ_j with $j < d$ and also ψ_d are determined as polynomials in ω, $\vec{k}$ times ϕ_0, while ϕ_d is not determined by the boundary analysis. (It requires information about the behavior at the horizon to obtain ϕ_d.) The first correction is

$$\phi_2 = \frac{1}{2(d-2)}(\omega^2 - \vec{k}^2)\phi_0, \tag{23.152}$$

which is the only term needed in the application to the calculation of the shear viscosity in $\mathcal{N}=4$, $d=4$ SYM theory below. We then define

$$\Pi_s(z,\omega,\vec{k}) \equiv \left(\frac{L}{z}\right)^{d-1} \partial_z \left[\phi(z,\omega,\vec{k}) - \frac{z^2}{2(d-2)}(\omega^2 - \vec{k}^2)\phi_0(\omega,\vec{k}) + \ldots\right], \tag{23.153}$$

where ... includes all terms up to order z^{d-2} in the boundary asymptotics. (The first subtracted term formally contributes a contact term proportional to $\Box\delta(x-y)$ in the Green's function.)

The retarded Green's function (23.148) emerges, after the subtraction, as the boundary limit of the ratio $\Pi(z,\omega,\vec{k})/\phi(z,\omega,\vec{k})$. This quantity obeys a first order differential equation that follows from (23.149). To derive it, note that the general second order linear equation

$$\partial_z\left[A(z)\partial_z\phi(z)\right] = B(z)\phi(z) \tag{23.154}$$

can be recast as the Hamiltonian system

$$\Pi(z) = A(z)\partial_z\phi(z), \qquad \partial_z\Pi(z) = B(z)\phi(z). \tag{23.155}$$

It is then straightforward to derive the Riccati equation [355, 346]

$$\partial_z\left(\frac{\Pi(z)}{\phi(z)}\right) = B(z) - \frac{1}{A(z)}\left(\frac{\Pi(z)}{\phi(z)}\right)^2. \tag{23.156}$$

From (23.149) we identify $A(z) = (L/z)^{d-1}f(z)/q$ and $B(z) = (L/z)^{d-1}(-\omega^2/f(z) + \vec{k}^2)/q$. The equation we need is then [346]

$$\partial_z\left(\frac{\Pi(z)}{\omega\phi(z)}\right) = -\frac{\omega}{f(z)}\left[\left(\frac{qz^{d-1}}{L^{d-1}}\right)\left(\frac{\Pi(z)}{\omega\phi(z)}\right)^2 + \left(\frac{L^{d-1}}{qz^{d-1}}\right)\left(1 - f(z)\frac{\vec{k}^2}{\omega^2}\right)\right]. \tag{23.157}$$

Now note that the right-hand side vanishes in the ordered limit $\lim_{\omega\to 0}\lim_{\vec{k}\to 0}$, so the ratio $\Pi(z)/\omega\phi(z)$ is *independent* of z in this limit. This is the same limit in which the transport coefficient is obtained; see (23.140). Further the subtraction term in Π_s also vanishes

[12] The holographic renormalization procedure discussed in Sec. 23.11 has been extended to Lorentzian signature [354]. However, detailed analysis is not necessary for correlators of massless scalars.

in this limit. Thus we can obtain the desired transport coefficient by combining (23.140) and (23.148) and evaluating the ratio at the horizon, rather than the boundary. We thus get

$$\chi = \lim_{\omega \to 0} \lim_{\vec{k} \to 0} \mathrm{Im}\left(\frac{\Pi(z_h, \omega, \vec{k})}{\omega\phi(z_h, \omega, \vec{k})} \right). \tag{23.158}$$

The infalling wave-form ϕ_- in (23.150) determines the value at the horizon. Indeed, for general ω, $\vec{k}$,

$$\lim_{z \to z_h} \Pi(z, \omega, \vec{k}) = \left(\frac{L}{z_h} \right)^{d-1} \frac{\mathrm{i}\omega}{q} \lim_{z \to z_h} \phi(z, \omega, \vec{k}). \tag{23.159}$$

Equivalently,

$$\lim_{z \to z_h} \mathrm{Im}\left(\frac{\Pi(z, \omega, \vec{k})}{\omega\,\phi(z, \omega, \vec{k})} \right) = \frac{1}{q}\left(\frac{L}{z_h} \right)^{d-1}. \tag{23.160}$$

Inserting this result in (23.158), we obtain the basic AdS/CFT prediction for the transport coefficient:

$$\chi = \frac{1}{q}\left(\frac{L}{z_h} \right)^{d-1}. \tag{23.161}$$

Exercise 23.16 *Verify (23.159) and (23.161).*

Let's comment briefly on the role of the *massless* bulk field. The point is that, for a massive scalar, there is an additional term in the equation of motion (23.149), and it is no longer true that $\partial_z(\Pi/\omega\phi) \to 0$ in the limit ω, $\vec{k} \to 0$. However, only the massless case is relevant because physical transport coefficients are defined for conserved currents in the boundary theory, and the gravity duals of conserved currents are massless bulk gauge fields.

Now we can focus on the specific goal of this section, which is to obtain the shear viscosity η of the $\mathcal{N} = 4$, $d = 4$ SYM theory in a thermal ensemble at temperature T [356, 351]. This is determined from the retarded correlator of the T_1^2 component of the stress tensor,

$$G_R(\omega, k) = \mathrm{i} \int \mathrm{d}t\, \mathrm{d}^3x\, \mathrm{e}^{\mathrm{i}(\omega t - kx^3)} \theta(t) \langle [T_1^2(t, \vec{x}), T_1^2(0)] \rangle. \tag{23.162}$$

Note that we have chosen the wave direction $\vec{k} = (0, 0, k)$ and components of T^μ_ν that are transverse to $\vec{k}$. It is the *shear* viscosity that will then be produced by the formula (23.140).

The stress tensor T^μ_ν is dual to a fluctuation of the bulk metric about the black brane spacetime, so we write $g_{\mu\nu}(z, t, \vec{x}) = \bar{g}_{\mu\nu}(z) + h_{\mu\nu}(z, t, \vec{x})$ where $\bar{g}$ now denotes the black brane background metric (23.131) and h is the fluctuation. The Lorentzian signature bulk action is

$$S = \frac{1}{16\pi G_5} \left\{ \int \mathrm{d}z\, \mathrm{d}t\, \mathrm{d}^3x \sqrt{-g} \left(g^{\mu\nu} R_{\mu\nu} + \frac{12}{L^2} \right) - 2 \int \mathrm{d}t\, \mathrm{d}^3x \sqrt{\gamma} K \right\}. \tag{23.163}$$

As usual, for the two-point correlator (23.162) we need only the linearized equation of motion for $h_{\mu\nu}$, namely

$$R^{\mathrm{Lin}}_{\mu\nu} = -\tfrac{1}{2}\left(D^\rho D_\rho h_{\mu\nu} - D^\rho D_\mu h_{\rho\nu} - D^\rho D_\nu h_{\mu\rho} + D_\mu \partial_\nu h^\rho_\rho\right) = -\frac{4}{L^2} h_{\mu\nu}. \quad (23.164)$$

The expression for the linearized Ricci tensor in a general background geometry was obtained from Ch. 14 of [357]. It is the covariantization of (8.16). Covariant derivatives carry the connection coefficients of the background metric.

We now state several properties of this equation, which will enable us to apply our previous work on the massless bulk scalar field to tensorial fluctuations $h_1{}^2$ of the black brane metric that are proportional to the plane wave $e^{i(kx^3-\omega t)}$.

1. The Fourier mode $h_1{}^2$ decouples from other components of $h_{\mu\nu}$ in the linearized equation (23.164) and to quadratic order in the action (23.163).
2. The uncoupled equation (23.164) reduces to the equation of motion for a massless scalar given in (23.143) or (23.149). It is quite common that transverse traceless perturbations of background metrics in the AdS/CFT correspondence satisfy the same linear equation as a massless scalar in the same background.
3. The effective coupling of $h_1{}^2$ obtained from (23.163) corresponds to $q = 16\pi G_5$ in the scalar action (23.141).

We will discuss these properties further below, but let's now simply note that they allow us to apply (23.158) directly and write the shear viscosity as

$$\eta = \frac{1}{16\pi G_5}\left(\frac{L}{z_h}\right)^3. \quad (23.165)$$

Since this quantity is dimensionful, it is common to divide by the entropy density (23.134) which gives the dimensionless ratio

$$\frac{\eta}{s} = \frac{1}{4\pi}. \quad (23.166)$$

We derived this result as the AdS/CFT prediction obtained from a particularly simple bulk configuration, the purely gravitational black brane. It appears to hold more generally for other solutions, for example a charged black brane, in which Einstein gravity is coupled to other fields in five dimensions. On the other hand the result does change if higher curvature terms are added to the action for gravity in the bulk.

The numerical value $1/4\pi$ is markedly less than experimental values of η/s for common liquids such as water, although liquid helium and ultracold atoms are within one order of magnitude. This led to the initial speculation that the result (23.166) is a universal lower bound. This is not entirely true since a small negative correction is found when 'Gauss–Bonnet terms' are added to the bulk action [358, 359, 360]. The most important application of this result is to the quark–gluon plasma that is momentarily formed in the collision of two heavy nuclei at Brookhaven's Relativistic Heavy Ion Collider (RHIC). Analysis of the data suggests a value close to the AdS/CFT result. It is an astonishing example of cross-fertilization in physics that string theory, as embodied in the AdS/CFT correspondence, may well apply to experiments in nuclear physics.

Let's return to the three properties used above to reduce the calculation of the transport coefficient for the tensor mode ${h_1}^2$ to the simpler case of a massless scalar.

1. Decoupling of ${h_1}^2$ follows from its symmetry properties under a rotation about the x^3 axis [351]. The metric fluctuation components $\frac{1}{2}(h_{11}-h_{22}\pm 2ih_{12})$ carry the maximum two units of angular momentum under this symmetry. The black brane metric is invariant under the full Euclidean group acting on x^1, x^2, x^3, as well as under time translation and reflection. So $\bar{g}_{\mu\nu}$ and $\bar{\Gamma}^\rho_{\mu\nu}$ are invariant under the rotation, and only a small subset of the connection coefficients are non-vanishing, This is enough to conclude that the 12-component of the equation of motion (23.164) reduces to an uncoupled linear differential equation for $h_{12}(z,\omega,k)$.
2. After detailed computation of the various terms in R^{Lin}_{12} one finds that the equation for ${h_1}^2(z,\omega,k)$ reduces to
$$\left[\partial_z^2+\left(\frac{f'(z)}{f}-\frac{3}{z}\right)\partial_z+\frac{\omega^2}{f(z)}-k^2\right]{h_1}^2(z,\omega,k)=0, \tag{23.167}$$
with $f(z)=1-(z/z_h)^4$. This equation is equivalent to (23.149).
3. To obtain the equivalent scalar coupling q, compute the variation of the action (23.163),
$$\delta S=\frac{1}{16\pi G_5}\int \mathrm{d}^5z\,\sqrt{-g}\delta g^{\mu\nu}\left[R_{\mu\nu}-\frac{1}{2}g_{\mu\nu}\left(R+\frac{12}{L^2}\right)\right]. \tag{23.168}$$
To isolate terms involving ${h_1}^2$ we require that the only non-vanishing components of $\delta g^{\mu\nu}$ are $(\delta g)^1{}_2=(\delta g)^2{}_1=-\bar{g}^{11}\bar{g}_{22}\delta {h_1}^2$. There is a factor of 2 because the metric is symmetric, and one can write
$$\begin{aligned}\delta S&=\frac{1}{16\pi G_5}\int \mathrm{d}^5x\,\sqrt{-g}\,2(\delta g)^1{}_2\,R^2{}_1\\&=\frac{1}{8\pi G_5}\int \mathrm{d}^5z\,\sqrt{-g}(\delta g)^1{}_2\,(-\tfrac{1}{2}D^\rho D_\rho {h_1}^2+\ldots).\end{aligned} \tag{23.169}$$
We have linearized in the last step, noting that the only role of the terms ... is to reduce $D^\rho D_\rho {h_1}^2$ to the scalar d'Alembertian, i.e. $D^\rho D_\rho {h_1}^2 \to (1/\sqrt{-\bar{g}})\partial_\rho(\sqrt{-\bar{g}}\bar{g}^{\rho\sigma}\partial_\sigma {h_1}^2)$. This integral can be compared with the variation of the scalar action (23.141), i.e.
$$\delta S=\frac{1}{q}\int \mathrm{d}^5x\sqrt{-\bar{g}}\delta\phi D^\rho\partial_\rho\phi, \tag{23.170}$$
and we can identify the effective coupling $q=16\pi G_5$.

Comparison of notation A

The fact that different sign conventions are used for the same quantities in the physics literature can make the comparison of results difficult. In this appendix, we discuss these conventions. In particular we will define 10 distinct sign choices s_i. In this book, we take $s_i = +1$ for all cases.

A.1 Spacetime and gravity

The first set of conventions is related to the spacetime metric and curvature tensor. On the inside cover of the classic text of Misner, Thorne, and Wheeler [361], three signs are defined to discuss these basic conventions for general relativity. We need additional signs for quantities defined in local Lorentz frames and for spinors. The first sign s_1 defines the metric signature. For the Minkowski metric in coordinates t, x, y, z we write

$$\eta_{ab} = s_1 \operatorname{diag}(-+++). \tag{A.1}$$

The second sign choice appears in the Riemann tensor,

$$R_{\mu\nu}{}^{\rho}{}_{\sigma} = s_2 \left(\partial_\mu \Gamma^\rho_{\nu\sigma} - \partial_\nu \Gamma^\rho_{\mu\sigma} + \Gamma^\rho_{\mu\tau}\Gamma^\tau_{\nu\sigma} - \Gamma^\rho_{\nu\tau}\Gamma^\tau_{\mu\sigma}\right). \tag{A.2}$$

Usually $s_2 = \pm 1$, but exceptionally $s_2 = -\frac{1}{2}$ occurs due to a normalization of p-form components different from the $1/p!$ in (A.19). For the same reason, exceptions to the standard field strength $F_{\mu\nu} = \partial_\mu A_\nu - \partial_\nu A_\mu$ can occur.

The third sign s_3 appears in the Einstein equation $s_3 \left(R_{\mu\nu} - \frac{1}{2} g_{\mu\nu} R\right) = \kappa^2 T_{\mu\nu}$, where T_{00} is always positive, or in the definition of the Ricci tensor:

$$s_2 s_3 R_{\mu\nu} = R^\rho{}_{\nu\rho\mu}. \tag{A.3}$$

We assume $R \equiv g^{\mu\nu} R_{\mu\nu}$, but a rare factor of 2 does occur.

The signs s_1 and s_3 determine the sign of kinetic energies of the scalars and graviton. Positive kinetic terms for scalars, vectors and gravitons require

$$\mathcal{L} = -s_1 \frac{1}{2} \partial_\mu \phi \partial^\mu \phi - \frac{1}{4} F_{\mu\nu} F^{\mu\nu} + s_1 s_3 \frac{1}{2\kappa^2} R. \tag{A.4}$$

Comparison of the kinetic terms is an easy way to recognize the values of s_1 and s_3 used in a given paper. The sign $s_1 s_3$ is also the sign of the scalar curvature of compact spaces.

The curvature is also obtained from the spin connection $\omega_\mu{}^{ab}$. The usual convention is that the two forms $R(\Gamma)$ and $R(\omega)$ are related by contraction with frame fields:

$$R_{\rho\sigma}{}^\mu{}_\nu(\Gamma) = R_{\rho\sigma}{}^{ab}(\omega)e^\mu_a e_{\nu b}. \tag{A.5}$$

There is an independent fourth sign in

$$R_{\mu\nu}{}^{ab} = s_4\left[\partial_\mu\omega_\nu{}^{ab} - \partial_\nu\omega_\mu{}^{ab} + \omega_\mu{}^{ac}\omega_{\nu c}{}^b - \omega_\nu{}^{ac}\omega_{\mu c}{}^b(e)\right]. \tag{A.6}$$

This sign is relevant in covariant derivatives of fermions and vectors, which are of the form

$$\left(\partial_\mu + s_2 s_4 \tfrac{1}{4}\omega_\mu{}^{ab}\gamma_{ab}\right)\psi, \qquad \partial_\mu V^a + s_2 s_4 \omega_\mu{}^{ab} V_b. \tag{A.7}$$

We (anti)symmetrize indices with ‘weight 1’, as in

$$A_{[ab]} = \tfrac{1}{2}\left(A_{ab} - A_{ba}\right) \quad \text{and} \quad A_{(ab)} = \tfrac{1}{2}\left(A_{ab} + A_{ba}\right). \tag{A.8}$$

In some papers $[ab]$ is used as $(ab - ba)$ without the factor $\frac{1}{2}$.

Here are some useful formulas, which illustrate dependences on the choice of signs:

$$\begin{aligned}
0 = \nabla_\mu e_\nu{}^a &= \partial_\mu e_\nu{}^a + s_2 s_4 \omega_\mu{}^{ab}(e) e_{\nu b} - \Gamma^\rho_{\mu\nu} e_\rho{}^a, \qquad g_{\mu\nu} = e_\mu{}^a \eta_{ab} e_\nu{}^b, \\
\omega_\mu{}^{ab}(e) &= s_2 s_4\left[2e^{\nu[a}\partial_{[\mu}e_{\nu]}{}^{b]} - e^{\nu[a}e^{b]\sigma}e_{\mu c}\partial_\nu e_\sigma{}^c\right], \\
\Gamma^\rho_{\mu\nu} &= \tfrac{1}{2}g^{\rho\lambda}\left(2\partial_{(\mu}g_{\nu)\lambda} - \partial_\lambda g_{\mu\nu}\right), \qquad \Gamma^\nu_{\mu\nu} = \tfrac{1}{2}\partial_\nu \ln g, \\
\left[\nabla_\mu, \nabla_\nu\right] V_\rho &= -s_2 R^\sigma{}_{\rho\mu\nu} V_\sigma.
\end{aligned} \tag{A.9}$$

There is another sign choice for the Levi-Civita tensor:

$$\varepsilon_{0123} = s_5, \qquad \varepsilon^{0123} = -s_5. \tag{A.10}$$

(Note that some papers, e.g. [28], use an imaginary Levi-Civita tensor ($s_5 = \pm\mathrm{i}$).)

We write commutators of Lie algebras with real structure constants, i.e. $[T_A, T_B] = f_{AB}{}^C T_C$. This means that matrix generators are anti-hermitian for compact Lie algebras. This differs from much of the physics literature in which $[\hat{T}_A, \hat{T}_B] = \mathrm{i} f_{AB}{}^C \hat{T}_C$ and compact generators are hermitian. The relation between generators is $\hat{T}_A = \mathrm{i}T_A$.

For electromagnetism, we use Heaviside–Lorentz units (with $c = 1$), which differ by a factor 4π from the Gaussian units that are also common in the literature.

The gravitational coupling κ is related to Newton's constant G by $\kappa^2 = 8\pi G$ in all dimensions D. For $D = 4$ the Planck mass is defined by

$$\begin{aligned}
M_{\text{Planck}} &= G^{-1/2} = 1.22 \times 10^{19}\ \text{GeV}, \\
m_{\text{P}} &= \kappa^{-1} = \frac{M_{\text{Planck}}}{\sqrt{8\pi}} = 2.4 \times 10^{18}\ \text{GeV}, \qquad \kappa^2 = 8\pi G.
\end{aligned} \tag{A.11}$$

A.2 Spinor conventions

The basic anti-commutator of two γ-matrices is

$$\gamma_\mu \gamma_\nu + \gamma_\nu \gamma_\mu = 2 s_6 g_{\mu\nu}. \tag{A.12}$$

We use $s_6 = 1$, as is most common in the physics literature. However, there are exceptions as in [120, 362]. Further there is a choice for the matrix γ_*, which is usually called γ_5 for $D = 4$:

$$\gamma_* = \gamma_5 = s_7 \mathrm{i} \gamma_0 \gamma_1 \gamma_2 \gamma_3. \tag{A.13}$$

With $s_7 = \pm 1$, one has $\gamma_* \gamma_* = \mathbb{1}$. This sign, together with s_5, determines the relation

$$\varepsilon_{abcd} \gamma^d = \frac{s_5}{s_7} \mathrm{i} \gamma_* \gamma_{abc}. \tag{A.14}$$

There are two factors, α and β, related to spinor conjugation [12]. The first one appears in the Majorana condition:

$$\bar{\lambda} = \mathrm{i}\alpha^{-1} \lambda^\dagger \gamma^0 = \lambda^T C, \qquad \alpha = \pm 1, \pm \mathrm{i}. \tag{A.15}$$

The second one determines whether complex conjugation changes the order of anti-commuting fermions:

$$(\chi_1 \chi_2)^* = -\beta \chi_1^* \chi_2^* = \beta \chi_2^* \chi_1^*, \qquad \beta = \pm 1. \tag{A.16}$$

In this equation χ_1 and χ_2 are two independent fermionic quantities, e.g. components of spinors (there is no spinor contraction in this equation). The two choices together determine whether bilinears of Majorana spinors such as $\bar{\psi}\lambda$ are real or imaginary. We have (compare with (3.79)) another sign s_* in

$$\left(\bar{\psi}\lambda\right)^* = s_8 (-t_0 t_1) \bar{\psi}\lambda, \qquad s_8 = \beta \alpha^2 s_*, \qquad \gamma_0^\dagger = -s_* \gamma_0. \tag{A.17}$$

Remember that $(-t_0 t_1) = 1$ for spacetime dimensions that allow Majorana spinors. We have $s_8 = 1$ (and where it was explicitly derived in Ch. 3, we took $\alpha = \beta = 1$) and $s_* = 1$. The bilinears $\bar{\psi}\lambda$ and also $\bar{\psi}\gamma_\mu \lambda$ are formally real for Majorana ψ and λ.

The normalization of the supersymmetry algebra varies in the literature. The SUSY commutator on scalars is

$$[\delta(\epsilon_1), \delta(\epsilon_2)]\phi = s_9 \tfrac{1}{2} \bar{\epsilon}_2 \gamma^\mu \epsilon_1 \partial_\mu \phi. \tag{A.18}$$

This is the normalization used in [28], and in the papers on general matter couplings in $N = 1$ supergravity; see [152, 155, 156]. In other books and papers the normalization factor $\frac{1}{2}$ is replaced by 2 or 1. The sign s_9 can be changed by modifying the sign of α. Changing the sign of α is equivalent to changing the sign of the charge conjugation matrix. By changing that sign one changes the sign of all barred spinors, thus flipping the sign in (A.18) and in the kinetic terms of all fermions.

A.3 Components of differential forms

We define the components of a generic p-form, ϕ_p, by

$$\phi_p = \frac{1}{p!}\phi_{\mu_1\cdots\mu_p}\mathrm{d}x^{\mu_1} \wedge \cdots \wedge \mathrm{d}x^{\mu_p}. \qquad \text{(A.19)}$$

In some papers the coefficient is taken as $(1/p!)\phi_{\mu_p\cdots\mu_1}$, i.e. with indices in the other order, or without the factor $p!$. (The other order is convenient for superforms.) The differential of a p-form works from the left as in the example of 1-forms:

$$F = \mathrm{d}A = \partial_\mu A_\nu \, \mathrm{d}x^\mu \wedge \mathrm{d}x^\nu \qquad \Rightarrow \qquad F_{\mu\nu} = 2\partial_{[\mu}A_{\nu]}. \qquad \text{(A.20)}$$

Working from the right would mean $\mathrm{d}A = \partial_\mu A_\nu \mathrm{d}x^\nu \wedge \mathrm{d}x^\mu$, which is often used in papers where the components of forms are defined in reverse order from (A.19).

A.4 Covariant derivatives

We use various covariant derivatives in this book. To help the reader distinguish them, we explain the notation here. There are derivatives that 'covariantize' the spacetime derivatives ∂_μ and derivatives ∂_i with respect to scalars ϕ^i in nonlinear σ-models.

D_μ denotes derivatives with Yang–Mills gauge connection, realized either with matrix generators that act linearly on fields or with Killing vectors. For gravity theories D_μ includes the Lorentz connection. For supergravity theories, the Lorentz connection includes the gravitino torsion, i.e. $\omega_\mu{}^{ab}(e, \psi)$. In some formulas, we exclude gravitino terms and use $\omega_\mu{}^{ab}(e)$. The latter are then indicated by $D^{(0)}_\mu$.

∇_μ further includes Levi-Civita (or affine) connections $\Gamma^\rho_{\mu\nu}$. The similar notation ∇_i is used in σ-models that have an affine connection Γ^k_{ij}. It is denoted by ∇_α in Kähler geometry. In supergravity this also includes the connection for Kähler transformations; see (17.73) and following formula.

$\mathcal{D}_\mu$ is a full covariant derivative including all the gauge fields of the theory under consideration. In supergravity theories it includes the terms with the gravitino. This derivative is introduced in Sec. 11.2 and for gravity theories in Sec. 11.3.2.

$\hat{D}_\mu$ is similar to D_μ but augmented by a σ-model connection of the embedding space pulled back to spacetime. See (17.27), where this occurs for the first time. (An exception occurs in Sec. 9.6, where the notation $\hat{D}$ is used for a covariant derivative that does not conform to the definition above.)

$\hat{\mathcal{D}}_\mu$ is similar: it is a $\mathcal{D}_\mu$ with σ-model connection.

$\hat{\partial}$ has been introduced in (17.101) to denote a covariant derivative with only Yang–Mills gauge connection. For $\mathcal{N} = 2$ see (20.104).

Lie algebras and superalgebras

B

Lie groups, Lie algebras, and superalgebras play an important role in supergravity. In this appendix we review the basic ideas with a focus on the classification of these objects. There are many textbooks available for readers who desire more information.

B.1 Groups and representations

All groups have composition laws. If g_1 and g_2 are any two elements, then their product $g_1 \circ g_2 = g_3$ where g_3 is a third element. We consider continuous groups (Lie groups) here. This means that the g_i are functions of parameters that vary continuously. The dimension of the group is the number of parameters. The most important groups are the *classical groups*. $\mathrm{GL}(n)$ is the group of $n \times n$ matrices, and one can distinguish whether the entries are real or complex (or even quaternionic) numbers. When we further require that the matrices have determinant 1, we obtain the definitions of the groups $\mathrm{SL}(n, \mathbb{R})$ or $\mathrm{SL}(n, \mathbb{C})$. Other groups are determined by the preservation of a metric. When we have a non-degenerate $n \times n$ matrix η that we identify with the metric, we define inner products of vectors V and W as $V^T \eta W$ for the real case, or $V^\dagger \eta W$ in the complex case. The transformations of the vectors that preserve these inner products define the orthogonal, symplectic or unitary groups. These are defined by matrices M such that

$$\begin{array}{llll}
M^T \eta M = \eta, & \eta \text{ symmetric:} & \text{orthogonal matrices,} & \\
M^T \eta M = \eta, & \eta \text{ antisymmetric:} & \text{symplectic matrices,} & \\
M^\dagger \eta M = \eta, & \eta \text{ hermitian:} & \text{unitary matrices.} & \text{(B.1)}
\end{array}$$

This defines the groups $\mathrm{SO}(n) \equiv \mathrm{SO}(n, \mathbb{R})$ or $\mathrm{Sp}(n) \equiv \mathrm{Sp}(n, \mathbb{R})$ of real matrices M with determinant 1, or $\mathrm{SU}(n) \equiv \mathrm{SU}(n, \mathbb{C})$ of complex matrices with $\det M = 1$ in the last case (where $\mathrm{U}(n)$ would appear if unit determinant is not required). When the metric η has indefinite signature, the terminology 'pseudo-orthogonal' is commonly used. This defines the groups $\mathrm{SO}(p, q)$ or $\mathrm{SU}(p, q)$. Some groups can be defined by quaternions; see

$$\begin{aligned}
\mathrm{SO}^*(2N) &\equiv \mathrm{O}(N,\mathbb{H}),\\
\mathrm{SU}^*(2N) &\equiv \mathrm{SL}(N,\mathbb{H}),\\
\mathrm{USp}(2N_+,2N_-) &\equiv \mathrm{U}(N_+,N_-,\mathbb{H}) = \mathrm{U}(2N_+,2N_-,\mathbb{C})\cap \mathrm{Sp}(2N_+ + 2N_-,\mathbb{C}),\\
\mathrm{USp}(2N) &\equiv \mathrm{U}(N,\mathbb{H}) = \mathrm{Sp}(N,\mathbb{H}).
\end{aligned} \tag{B.2}$$

They are obtained by first embedding quaternionic matrices in complex ones, and then imposing the conditions (B.1).

A representation of a group is a map $g \to D(g)$ where $D(g)$ is an $n\times n$ matrix. This map must be a homomorphism, which means that the matrices $D(g)$ have the same multiplication law as the group elements. So if $g_1 g_2 = g_3$, then $D(g_1)D(g_2) = D(g_3)$ for any set of three elements g_i. Representations are important because the fields of a theory transform with the matrices $D(g)$, viz. $\phi^i \to \phi^{i'} = D(g)^i{}_j\phi^j$.

B.2 Lie algebras

In this book, we are mostly concerned with infinitesimal transformations. This means that we consider group transformations close to the identity. We parametrize them with coordinates a^A where A runs over the independent deformations of the identity that belong to the group. We then express the group element by[1]

$$g = \exp(a^A T_A), \tag{B.3}$$

and T_A are the independent 'generators'. The *Lie algebra* is the vector space whose elements are superpositions $a^A T_A$ of the matrices T_A. We distinguish between real algebras, in which the a^A are real, and complex algebras, in which the a^A are complex numbers.

Exercise B.1 *Explain the statement that the group of $n\times n$ complex matrices* $\mathrm{GL}(n,\mathbb{C})$ *can be seen as generated by a complex algebra with n^2 generators, as well as by a real algebra with $2n^2$ generators.*

The requirement that the composition of the group elements (B.3) should lead to a new element of the same form leads to the requirement that the commutator of the generators should lead to a new generator.

Exercise B.2 *Check that* $\exp(a_1^A T_A)\exp(a_2^B T_B)$ *leads up to second order in the coordinates a^A to a group element*

$$\exp\left(a_1^A T_A + a_2^A T_A + \tfrac{1}{2}a_1^A a_2^B [T_A, T_B]\right). \tag{B.4}$$

This commutator defines the structure constants $f_{AB}{}^C = -f_{BA}{}^C$ in

$$[T_A, T_B] = f_{AB}{}^C T_C. \tag{B.5}$$

[1] This defines the 'connected component' of a group, which means that g can be continuously deformed to the identity.

For consistency the structure constants of 'real algebras' must be real, but it turns out that for complex algebras there is also a basis of the T_A such that they are real. Lie algebras can be defined independently of their origin in a group, as the linear spans $a^A T_A$ where there is a composition law, written as commutator in (B.5), that satisfies the Jacobi identity.

A useful quantity is the Cartan–Killing metric

$$g_{AB} = f_{AC}{}^D f_{BD}{}^C. \tag{B.6}$$

One property is that $f_{ABC} = f_{AB}{}^D g_{DC}$ is completely antisymmetric in its three indices. When this metric is non-degenerate, the algebra is 'semisimple'. Semisimple algebras are direct sums of simple algebras, where the latter are defined by having no non-trivial ideals. An ideal is a subset of the algebra that is invariant. This means that, for h in that subset, $[h, a]$ belongs to the subset for any a in the algebra. We will further concentrate on simple algebras. But note that the one-dimensional algebra U(1) is not semisimple since g_{AB} vanishes.

The simple complex algebras are classified in series A_n, B_n, C_n, D_n, where n is any positive integer, plus the exceptional algebras E_6, E_7, E_8, F_4 and G_2. The subscript is the 'rank of the algebra'. The first four series are identified as algebras belonging to classical matrix groups:

A_n: algebra of $\mathfrak{su}(n+1)$,
B_n: algebra of $\mathfrak{so}(2n+1)$,
C_n: algebra of $\mathfrak{sp}(2n)$,
D_n: algebra of $\mathfrak{so}(2n)$ for $n \geq 2$,

where, for example, $\mathfrak{su}(n)$ stands for the algebra of the generators of SU(n).

In the real algebras, one can distinguish compact and non-compact generators, by diagonalizing the Cartan–Killing metric. The generators that have a negative eigenvalue for the Cartan–Killing metric are 'compact generators', those with a positive eigenvalue are 'non-compact generators'. The number of non-compact generators minus the number of compact generators is called the character of the real form. Two real forms are always present. The 'normal form' has character equal to the rank of the algebra. The other one is the compact real form, where all generators are compact, and the character is thus minus the dimension of the algebra. The other real forms have characters between these two extremes. The compact generators define a 'maximal compact subalgebra' of the real algebra. Note that we can make the distinction between compact and non-compact algebras only for algebras over the reals, as multiplying generators with i would change the signs of the entries in the Cartan–Killing metric.

Finite-dimensional representations of a compact algebra in a space with a positive definite metric are anti-hermitian. You can understand this simply by considering the finite transformation in a fixed direction. For a compact transformation to be compact, this should be of the form $\exp(\mathrm{i}\theta)$, rather than $\exp\theta$. Indeed, the first expression defines the transformation on a (compact) circle, and is generated by the 1×1 matrix i, the simplest example of an anti-hermitian matrix. On the other hand, $\exp\theta$ generates the transformations on a (non-compact) line, and is generated by the hermitian 1×1 matrix 1.

Table B.1 Real forms of simple bosonic Lie algebras. The first one in each block is the compact real form, the last one is the normal real form.

algebra	real form	maximal compact subgroup	
A_{n-1}	$\mathrm{SU}(n)$	$\mathrm{SU}(n)$	
A_{n-1}	$\mathrm{SU}(p, n-p)$	$\mathrm{SU}(p) \times \mathrm{SU}(n-p) \times \mathrm{U}(1)$,	$1 \leq p \leq n/2$
A_{2n-1}	$\mathrm{SU}^*(2n)$	$\mathrm{USp}(2n)$	
A_{n-1}	$\mathrm{SL}(n)$	$\mathrm{SO}(n)$	
B_n	$\mathrm{SO}(2n+1)$	$\mathrm{SO}(2n+1)$	
B_n	$\mathrm{SO}(2n+1-p, p)$	$\mathrm{SO}(2n+1-p) \times \mathrm{SO}(p)$,	$0 \leq p \leq n$
B_n	$\mathrm{SO}(n, n+1)$	$\mathrm{SO}(n) \times \mathrm{SO}(n+1)$	
D_n	$\mathrm{SO}(2n)$	$\mathrm{SO}(2n)$	
D_n	$\mathrm{SO}(2n-p, p)$	$\mathrm{SO}(2n-p) \times \mathrm{SO}(p)$,	$0 \leq p \leq n$
D_n	$\mathrm{SO}^*(2n)$	$\mathrm{U}(n)$	
D_n	$\mathrm{SO}(n, n)$	$\mathrm{SO}(n) \times \mathrm{SO}(n)$	
C_n	$\mathrm{USp}(2n)$	$\mathrm{USp}(2n)$	
C_n	$\mathrm{USp}(2p, 2n-2p)$	$\mathrm{USp}(2p) \times \mathrm{USp}(2n-2p)$,	$1 \leq p \leq n/2$
C_n	$\mathrm{Sp}(2n)$	$\mathrm{U}(n)$	
G_2	$G_{2,-14}$	G_2	
G_2	$G_{2,2}$	$\mathrm{SU}(2) \times \mathrm{SU}(2)$	
F_4	$F_{4,-52}$	$F_{4,-52}$	
F_4	$F_{4,-20}$	$\mathrm{SO}(9)$	
F_4	$F_{4,4}$	$\mathrm{USp}(6) \times \mathrm{SU}(2)$	
E_6	$E_{6,-78}$	$E_{6,-78}$	
E_6	$E_{6,-26}$	$F_{4,-52}$	
E_6	$E_{6,-14}$	$\mathrm{SO}(10) \times \mathrm{SO}(2)$	
E_6	$E_{6,2}$	$\mathrm{SU}(6) \times \mathrm{SU}(2)$	
E_6	$E_{6,6}$	$\mathrm{USp}(8)$	
E_7	$E_{7,-133}$	$E_{7,-133}$	
E_7	$E_{7,-25}$	$E_{6,-78} \times \mathrm{SO}(2)$	
E_7	$E_{7,-5}$	$\mathrm{SO}(12) \times \mathrm{SU}(2)$	
E_7	$E_{7,7}$	$\mathrm{SU}(8)$	
E_8	$E_{8,-248}$	$E_{8,-248}$	
E_8	$E_{8,-24}$	$E_{7,-133} \times \mathrm{SU}(2)$	
E_8	$E_{8,8}$	$\mathrm{SO}(16)$	

The list of real forms is given in Table B.1. The second number in the notation for the real forms of exceptional algebras is the 'character'. The conventions that we use for groups are such that $\mathrm{Sp}(2n) = \mathrm{Sp}(2n, \mathbb{R})$ (always even entry), and $\mathrm{USp}(2m, 2n) = \mathrm{U}(m, n, \mathbb{H})$. $\mathrm{SL}(n)$ is $\mathrm{SL}(n, \mathbb{R})$. Note that for the algebras there are the following isomorphisms:[2]

[2] These isomorphism are for the algebras. The covering groups of the orthogonal groups are equal to the groups mentioned at the right-hand sides. For example, the covering group of SO(3) is SU(2) and the correct relation for the groups is $\mathrm{SO}(3) = \mathrm{SU}(2)/\mathbb{Z}_2$.

$$\begin{aligned}
&\mathfrak{so}(3)=\mathfrak{su}(2)=\mathfrak{su}^*(2), \qquad \mathfrak{so}(2,1)=\mathfrak{sl}(2)=\mathfrak{su}(1,1)=\mathfrak{sp}(2),\\
&\mathfrak{so}(4)=\mathfrak{su}(2)\times\mathfrak{su}(2), \qquad \mathfrak{so}(3,1)=\mathfrak{sp}(2,\mathbb{C})=\mathfrak{sl}(2,\mathbb{C}),\\
&\mathfrak{so}(2,2)=\mathfrak{sl}(2)\times\mathfrak{sl}(2), \qquad \mathfrak{so}^*(4)=\mathfrak{su}(1,1)\times\mathfrak{su}(2),\\
&\mathfrak{so}(5)=\mathfrak{usp}(4), \qquad \mathfrak{so}(4,1)=\mathfrak{usp}(2,2), \qquad \mathfrak{so}(3,2)=\mathfrak{sp}(4),\\
&\mathfrak{so}(6)=\mathfrak{su}(4), \qquad \mathfrak{so}(5,1)=\mathfrak{su}^*(4), \qquad \mathfrak{so}(4,2)=\mathfrak{su}(2,2),\\
&\mathfrak{so}(3,3)=\mathfrak{sl}(4), \qquad \mathfrak{so}^*(6)=\mathfrak{su}(3,1),\\
&\mathfrak{so}^*(8)=\mathfrak{so}(6,2).
\end{aligned} \tag{B.7}$$

B.3 Superalgebras

Lie superalgebras are structures with bosonic operators, say B_A, and fermionic ones, say F_α. Between the fermionic ones there are anti-commutation relations, and between the others there are commutation relations. Just as an ordinary Lie algebra must satisfy Jacobi identities, the superalgebra should satisfy a graded version thereof, super-Jacobi identities. We write these as

$$\begin{aligned}
[[B_A, B_B], \mathcal{O}] &= [B_A, [B_B, \mathcal{O}]] - [B_B, [B_A, \mathcal{O}]],\\
[\{F_\alpha, F_\beta\}, \mathcal{O}] &= [F_\alpha, [F_\beta, \mathcal{O}\}] - [F_\beta, [F_\alpha, \mathcal{O}\}],
\end{aligned} \tag{B.8}$$

where $\mathcal{O}$ is either a bosonic or a fermionic operator, and $[F, \mathcal{O}\}$ is a commutator when it is bosonic, or an anti-commutator when it is fermionic. The super-Jacobi identities written in this form can be easily remembered: Write one double (anti-)commutator in the left-hand side. Then write at the right-hand side the operators in the same order, but write the inner bracket with the other adjacent operators. Then write another term in the right-hand side with the two operators of the inner bracket of the left-hand side interchanged, and take the sign such that the symmetry between them in the left-hand side is respected.

Exercise B.3 *Check these identities by writing them out in full.*

The definition of the classical superalgebras starts from a graded space (there are odd and even elements) that are mixed by the transformations of the superalgebra. The matrices are thus of the form

$$M = \begin{pmatrix} A & B \\ C & D \end{pmatrix}, \tag{B.9}$$

where A and D are matrices that do not change the type in the vector space and are thus bosonic transformations, while B and C transform bosonic entries to fermionic or vice versa and are thus 'fermionic transformations'. The simplest superalgebras involve the preservation of a metric. For example, for 'orthosymplectic algebras' this is a symmetric matrix in the upper part (and thus A is an element of an orthogonal algebra) and antisymmetric in the lower part (and thus D is an element of a symplectic algebra).

General superalgebras have been classified in [363, 364, 365, 58], and a convenient table is given in [12]. We give in Table B.2 a list of the superalgebras used in this book with the

Table B.2 Some main Lie superalgebras of classical type.

name	range	bosonic algebra	fermionic
$\mathrm{SU}(m-p,p\vert n-q,q)$	$m\neq n$	$\mathrm{SU}(m-p,p)\oplus$ $\mathrm{SU}(n-q,q)\oplus\mathrm{U}(1)$	$(m,\bar{n})$ $\oplus(\bar{m},n)$
$\mathrm{OSp}(m^*\vert n-q,q)$	m,n,q even	$\mathrm{SO}^*(m)\oplus\mathrm{USp}(n-q,q)$	(m,n)
$F^2(4)$		$\overline{\mathrm{SO}(5,2)}\oplus\mathrm{SU}(2)$	$(8,2)$

bosonic subalgebra (the $A+D$ part of (B.9)) and the representation of the latter formed by the fermionic generators. The superalgebra $\mathrm{SU}(m|m)$ has the same structure as $\mathrm{SU}(m|n)$ except that the U(1) generator can be omitted. Many authors denote the superalgebra without the U(1) by $\mathrm{PSU}(m|m)$.

Exercise B.4 *Use Table B.2 to show that* $\mathrm{SU}(2,2|\mathcal{N})$ *can be the superconformal group for different* $\mathcal{N}$ *in four dimensions. Identify the bosonic part as the direct product of the conformal group (with the identifications in (B.7)) and the R-symmetry group. Check that the fermionic generators are spinors and have the right number of generators for ordinary and special supersymmetries.*

References

[1] Y. Gol'fand and E. Likhtman, *Extension of the algebra of Poincaré group generators and violation of P invariance*, JETP Lett. **13** (1971) 323–326. See also http://www.jetpletters.ac.ru/ps/717/article_11110.shtml (Russian version)

[2] D. Volkov and V. Akulov, *Is the neutrino a Goldstone particle?*, Phys. Lett. **B46** (1973) 109–110

[3] J. Wess and B. Zumino, *Supergauge transformations in four-dimensions*, Nucl. Phys. **B70** (1974) 39–50

[4] D. Z. Freedman, P. van Nieuwenhuizen and S. Ferrara, *Progress toward a theory of supergravity*, Phys. Rev. **D13** (1976) 3214–3218

[5] S. Deser and B. Zumino, *Consistent supergravity*, Phys. Lett. **B62** (1976) 335

[6] É. Cartan, *Sur les groupes projectifs qui ne laissent invariante aucune multiplicité plane*, Bull. Soc. Math. **41** (1913) 53–96

[7] É. Cartan, *The theory of spinors*. Reprinted, Dover Publications, 1981

[8] S. Sternberg, *Group theory and physics*. Cambridge University Press, 1994

[9] M. E. Peskin and D. V. Schroeder, *An introduction to quantum field theory*. Addison-Wesley, 1995

[10] J. Scherk, *Extended supersymmetry and extended supergravity theories*, in *Recent developments in gravitation*, ed. M. Lévy and S. Deser. Plenum Press, 1979

[11] T. Kugo and P. K. Townsend, *Supersymmetry and the division algebras*, Nucl. Phys. **B221** (1983) 357

[12] A. Van Proeyen, *Tools for supersymmetry*, Ann. Univ. Craiova, Phys. **AUC 9 (part I)** (1999) 1–48, `arXiv:hep-th/9910030`

[13] J.-P. Serre, *Representations lineaires des groupes finis*. Hermann, 1967. English translation: *Linear representations of finite groups*, Graduate Texts in Mathematics, vol. 42. Springer, 1977

[14] W. Miller, *Symmetry groups and their applications*. Academic Press, 1972

[15] P. Van Nieuwenhuizen, *An introduction to simple supergravity and the Kaluza–Klein program*, in *Relativity, groups and topology II,* Proceedings of Les Houches 1983, ed. B. S. DeWitt and R. Stora. North-Holland, 1984

[16] T. Ortín, *Gravity and strings*. Cambridge University Press, 2004

[17] J. Polchinski, *String theory*, vol. 2, *Superstring theory and beyond*. Cambridge University Press, 1998

[18] C. Fronsdal, *Massless fields with integer spin*, Phys. Rev. **D18** (1978) 3624

[19] E. S. Fradkin and M. A. Vasiliev, *On the gravitational interaction of massless higher spin fields*, Phys. Lett. **B189** (1987) 89–95

[20] X. Bekaert, S. Cnockaert, C. Iazeolla and M. A. Vasiliev, *Nonlinear higher spin theories in various dimensions*, Lectures given at Workshop on Higher Spin Gauge Theories, Brussels, 12–14 May 2004, `arXiv:hep-th/0503128`

[21] M. A. Vasiliev, *Higher-spin gauge theories in four, three and two dimensions*, Int. J. Mod. Phys. **D5** (1996) 763–797, `arXiv:hep-th/9611024`

[22] J. Bjorken and S. Drell, *Relativistic quantum fields*. McGraw-Hill, 1965

[23] N. Birrell and P. Davies, *Quantum fields in curved space*, Cambridge Monographs on Mathematical Physics. Cambridge University Press, 1982

[24] J. A. Harvey, *Magnetic monopoles, duality, and supersymmetry*, in Proc. of the 1995 Summer School in High-Energy Physics and Cosmology, ed. E. Gava et al., p. 66. World Scientific, 1997, `arXiv:hep-th/9603086`

[25] D. I. Olive, *Exact electromagnetic duality*, Nucl. Phys. Proc. Suppl. **45A** (1996) 88–102, `arXiv:hep-th/9508089`

[26] M. K. Gaillard and B. Zumino, *Duality rotations for interacting fields*, Nucl. Phys. **B193** (1981) 221

[27] P. van Nieuwenhuizen, A. Rebhan, D. V. Vassilevich and R. Wimmer, *Boundary terms in supergravity and supersymmetry*, Int. J. Mod. Phys. **D15** (2006) 1643–1658, `arXiv:hep-th/0606075`

[28] P. Van Nieuwenhuizen, *Supergravity*, Phys. Rep. **68** (1981) 189–398

[29] S. Coleman and J. Mandula, *All possible symmetries of the S matrix*, Phys. Rev. **159** (1967) 1251–1256

[30] S. Weinberg, *The quantum theory of fields*, vol. III, *Supersymmetry*. Cambridge University Press, 2000

[31] R. Haag, J. T. Łopuszański and M. Sohnius, *All possible generators of supersymmetries of the S-matrix*, Nucl. Phys. **B88** (1975) 257

[32] M. B. Green, J. H. Schwarz and E. Witten, *Superstring theory*, vol. 1, *Introduction*, Cambridge Monographs on Mathematical Physics, Appendix 4.A. Cambridge University Press, 1987

[33] J. Polchinski, *String theory*, vol. 1, *An introduction to the bosonic string*. Cambridge University Press, 1998

[34] P. Binétruy, *Supersymmetry: theory, experiment and cosmology*. Oxford University Press, 2006

[35] B. de Wit and D. Z. Freedman, *Combined supersymmetric and gauge-invariant field theories*, Phys. Rev. **D12** (1975) 2286

[36] D. Anselmi, J. Erlich, D. Z. Freedman and A. A. Johansen, *Positivity constraints on anomalies in supersymmetric gauge theories*, Phys. Rev. **D57** (1998) 7570–7588, `arXiv:hep-th/9711035`

[37] M. F. Sohnius, *Introducing supersymmetry*, Phys. Rep. **128** (1985) 39–204

[38] A. Salam and J. A. Strathdee, *Unitary representations of supergauge symmetries*, Nucl. Phys. **B80** (1974) 499–505

[39] T. Eguchi, P. B. Gilkey and A. J. Hanson, *Gravitation, gauge theories and differential geometry*, Phys. Rep. **66** (1980) 213

[40] B. Schutz, *Geometrical methods of mathematical physics*. Cambridge University Press, 1980

[41] Y. Choquet-Bruhat and C. DeWitt-Morette, *Analysis, manifolds, and physics*. North Holland, 1982

[42] C. Nash and S. Sen, *Topology and geometry for physicists*. Academic Press, 1983

[43] T. Frankel, *The geometry of physics: An introduction*. Cambridge University Press, 1997

[44] M. Nakahara, *Geometry, topology and physics*. Taylor & Francis, 2003

[45] H. Flanders, *Differential forms*. Academic Press, 1963

[46] R. M. Wald, *General relativity*. University of Chicago Press, 1984

[47] S. M. Carroll, *Spacetime and geometry: An introduction to general relativity*. Addison-Wesley, 2004

[48] S. Weinberg, *Gravitation and cosmology*. John Wiley, 1972

[49] E. D'Hoker, D. Z. Freedman, S. D. Mathur, A. Matusis and L. Rastelli, *Graviton and gauge boson propagators in $AdS_{(d+1)}$*, Nucl. Phys. **B562** (1999) 330–352, arXiv:hep-th/9902042

[50] D. Volkov and V. Soroka, *Higgs effect for Goldstone particles with spin 1/2*, JETP Lett. **18** (1973) 312–314

[51] P. K. Townsend and P. van Nieuwenhuizen, *Geometrical interpretation of extended supergravity*, Phys. Lett. **B67** (1977) 439

[52] A. H. Chamseddine and P. C. West, *Supergravity as a gauge theory of supersymmetry*, Nucl. Phys. **B129** (1977) 39

[53] D. Z. Freedman and P. van Nieuwenhuizen, *Properties of supergravity theory*, Phys. Rev. **D14** (1976) 912

[54] P. K. Townsend, *Cosmological constant in supergravity*, Phys. Rev. **D15** (1977) 2802–2804

[55] S. Deser and B. Zumino, *Broken supersymmetry and supergravity*, Phys. Rev. Lett. **38** (1977) 1433

[56] M. Cvetič, H. Lü, C. N. Pope, A. Sadrzadeh and T. A. Tran, *Consistent SO(6) reduction of type IIB supergravity on S^5*, Nucl. Phys. **B586** (2000) 275–286, arXiv:hep-th/0003103

[57] M. Cvetič, G. W. Gibbons, H. Lü and C. N. Pope, *Consistent group and coset reductions of the bosonic string*, Class. Quant. Grav. **20** (2003) 5161–5194, arXiv:hep-th/0306043

[58] W. Nahm, *Supersymmetries and their representations*, Nucl. Phys. **B135** (1978) 149

[59] E. Cremmer, B. Julia and J. Scherk, *Supergravity theory in 11 dimensions*, Phys. Lett. **B76** (1978) 409–412

[60] S. Naito, K. Osada and T. Fukui, *Fierz identities and invariance of eleven-dimensional supergravity action*, Phys. Rev. **D34** (1986) 536–552

[61] R. D'Auria and P. Frè, *Geometric supergravity in $d = 11$ and its hidden supergroup*, Nucl. Phys. **B201** (1982) 101–140

[62] I. A. Batalin and G. A. Vilkovisky, *Quantization of gauge theories with linearly dependent generators*, Phys. Rev. **D28** (1983) 2567–2582; erratum **D30** (1984) 508

[63] M. Henneaux, *Lectures on the antifield–BRST formalism for gauge theories*, Nucl. Phys. Proc. Suppl. **18A** (1990) 47–106

[64] J. Gomis, J. París and S. Samuel, *Antibracket, antifields and gauge theory quantization*, Phys. Rep. **259** (1995) 1–145, arXiv:hep-th/9412228

[65] R. Jackiw, *Gauge-covariant conformal transformations*, Phys. Rev. Lett. **41** (1978) 1635

[66] S. Mukhi, *Massive vector multiplet coupled to supergravity*, Phys. Rev. **D20** (1979) 1839

[67] A. Van Proeyen, *Massive vector multiplets in supergravity*, Nucl. Phys. **B162** (1980) 376

[68] J. Strathdee, *Extended Poincaré supersymmetry*, Int. J. Mod. Phys. **A2** (1987) 273

[69] F. Giani and M. Pernici, $N = 2$ *supergravity in ten-dimensions*, Phys. Rev. **D30** (1984) 325–333

[70] I. C. G. Campbell and P. C. West, $N = 2$ $D = 10$ *nonchiral supergravity and its spontaneous compactification*, Nucl. Phys. **B243** (1984) 112

[71] M. Huq and M. A. Namazie, *Kaluza–Klein supergravity in ten-dimensions*, Class. Quant. Grav. **2** (1985) 293

[72] J. H. Schwarz and P. C. West, *Symmetries and transformations of chiral* $N = 2$ $D = 10$ *supergravity*, Phys. Lett. **B126** (1983) 301

[73] J. H. Schwarz, *Covariant field equations of chiral* $N = 2$ $D = 10$ *supergravity*, Nucl. Phys. **B226** (1983) 269

[74] P. S. Howe and P. C. West, *The complete* $N = 2$, $d = 10$ *supergravity*, Nucl. Phys. **B238** (1984) 181

[75] E. Bergshoeff, R. Kallosh, T. Ortín, D. Roest and A. Van Proeyen, *New formulations of* $D = 10$ *supersymmetry and D8–O8 domain walls*, Class. Quant. Grav. **18** (2001) 3359–3382, arXiv:hep-th/0103233

[76] E. A. Bergshoeff, J. Hartong, P. S. Howe, T. Ortín and F. Riccioni, *IIA/IIB supergravity and ten-forms*, JHEP **05** (2010) 061, arXiv:1004.1348 [hep-th]

[77] P. K. Townsend, *A new anomaly free chiral supergravity theory from compactification on K3*, Phys. Lett. **B139** (1984) 283

[78] C. M. Hull, *Strongly coupled gravity and duality*, Nucl. Phys. **B583** (2000) 237–259, arXiv:hep-th/0004195

[79] L. J. Romans, *Massive* $N = 2a$ *supergravity in ten dimensions*, Phys. Lett. **B169** (1986) 374

[80] D. Z. Freedman and A. K. Das, *Gauge internal symmetry in extended supergravity*, Nucl. Phys. **B120** (1977) 221

[81] B. de Wit and H. Nicolai, $N = 8$ *supergravity with local* $SO(8)\times SU(8)$ *invariance*, Phys. Lett. **B108** (1982) 285

[82] F. Cordaro, P. Frè, L. Gualtieri, P. Termonia and M. Trigiante, $N = 8$ *gaugings revisited: An exhaustive classification*, Nucl. Phys. **B532** (1998) 245–279, arXiv:hep-th/9804056

[83] H. Nicolai and H. Samtleben, *Compact and noncompact gauged maximal supergravities in three dimensions*, JHEP **04** (2001) 022, arXiv:hep-th/0103032

[84] B. de Wit, H. Samtleben and M. Trigiante, *Magnetic charges in local field theory*, JHEP **09** (2005) 016, arXiv:hep-th/0507289

[85] T. Damour, M. Henneaux, B. Julia and H. Nicolai, *Hyperbolic Kac–Moody algebras and chaos in Kaluza–Klein models*, Phys. Lett. **B509** (2001) 323–330, `arXiv:hep-th/0103094`

[86] P. C. West, *E_{11} and M theory*, Class. Quant. Grav. **18** (2001) 4443–4460, `arXiv:hep-th/0104081`

[87] M. R. Gaberdiel, D. I. Olive and P. C. West, *A class of Lorentzian Kac–Moody algebras*, Nucl. Phys. **B645** (2002) 403–437, `arXiv:hep-th/0205068`

[88] A. Kleinschmidt, I. Schnakenburg and P. West, *Very-extended Kac–Moody algebras and their interpretation at low levels*, Class. Quant. Grav. **21** (2004) 2493–2525, `arXiv:hep-th/0309198`

[89] F. Riccioni and P. West, *The E_{11} origin of all maximal supergravities*, JHEP **07** (2007) 063, `arXiv:0705.0752 [hep-th]`

[90] E. A. Bergshoeff, I. De Baetselier and T. A. Nutma, *E_{11} and the embedding tensor*, JHEP **09** (2007) 047, `arXiv:0705.1304 [hep-th]`

[91] A. Strominger, *Special geometry*, Commun. Math. Phys. **133** (1990) 163–180

[92] B. de Wit and A. Van Proeyen, *Broken sigma model isometries in very special geometry*, Phys. Lett. **B293** (1992) 94–99, `arXiv:hep-th/9207091`

[93] M. Günaydin, G. Sierra and P. K. Townsend, *The geometry of $N = 2$ Maxwell–Einstein supergravity and Jordan algebras*, Nucl. Phys. **B242** (1984) 244

[94] B. de Wit and A. Van Proeyen, *Potentials and symmetries of general gauged $N = 2$ supergravity – Yang–Mills models*, Nucl. Phys. **B245** (1984) 89

[95] J. Bagger and E. Witten, *Matter couplings in $N = 2$ supergravity*, Nucl. Phys. **B222** (1983) 1

[96] G. Sierra and P. K. Townsend, *An introduction to $N = 2$ rigid supersymmetry*, in *Supersymmetry and supergravity 1983*, ed. B. Milewski. World Scientific, 1983

[97] S. J. Gates, Jr., *Superspace formulation of new non-linear sigma models*, Nucl. Phys. **B238** (1984) 349

[98] S. Ferrara, C. A. Savoy and B. Zumino, *General massive multiplets in extended supersymmetry*, Phys. Lett. **B100** (1981) 393

[99] J. W. van Holten and A. Van Proeyen, *$N = 1$ supersymmetry algebras in $d = 2, 3, 4$ mod 8*, J. Phys. **A15** (1982) 3763

[100] S. Kobayashi and K. Nomizu, *Foundations of differential geometry*, vol. II. John Wiley, 1963

[101] M. Nakahara, *Geometry, topology and physics*. Graduate Student Series in Physics. Hilger, 1990

[102] L. Alvarez-Gaumé and D. Z. Freedman, *A simple introduction to complex manifolds*, in *Unification of the fundamental particle interactions: proceedings*, ed. S. Ferrara, J. Ellis and P. van Nieuwenhuizen. Plenum Press, 1980

[103] K. Yano, *Differential geometry on complex and almost complex manifolds*. Macmillan, 1965

[104] V. I. Arnold, *Mathematical methods of classical mechanics*. Springer, 1978. English translation: K. Vogtmann and A. Weinstein, *Mathematical methods of classical mechanics*, 2nd edn, Graduate Texts in Mathematics, vol. 60. Springer, 2000

[105] S. Ferrara, J. Wess and B. Zumino, *Supergauge multiplets and superfields*, Phys. Lett. **B51** (1974) 239

[106] W. Siegel, *Gauge spinor superfield as a scalar multiplet*, Phys. Lett. **B85** (1979) 333

[107] P. Fayet and J. Iliopoulos, *Spontaneously broken supergauge symmetries and Goldstone spinors*, Phys. Lett. **B51** (1974) 461–464

[108] P. Anastasopoulos, M. Bianchi, E. Dudas and E. Kiritsis, *Anomalies, anomalous $U(1)$'s and generalized Chern–Simons terms*, JHEP **11** (2006) 057, `arXiv:hep-th/0605225`

[109] J. De Rydt, J. Rosseel, T. T. Schmidt, A. Van Proeyen and M. Zagermann, *Symplectic structure of $\mathcal{N} = 1$ supergravity with anomalies and Chern–Simons terms*, Class. Quant. Grav. **24** (2007) 5201–5220, `arXiv:0705.4216 [hep-th]`

[110] S. R. Coleman, *Secret symmetry: An introduction to spontaneous symmetry breakdown and gauge fields*, Subnucl. Ser. **11** (1975) 139. Lectures in Int. Summer School of Physics Ettore Majorana, Erice, Sicily, 1973

[111] M. T. Grisaru, W. Siegel and M. Rocek, *Improved methods for supergraphs*, Nucl. Phys. **B159** (1979) 429

[112] I. Affleck, M. Dine and N. Seiberg, *Dynamical supersymmetry breaking in four-dimensions and its phenomenological implications*, Nucl. Phys. **B256** (1985) 557

[113] K. A. Intriligator and N. Seiberg, *Lectures on supersymmetry breaking*, Class. Quant. Grav. **24** (2007) S741–S772, `arXiv:hep-ph/0702069`

[114] L. O'Raifeartaigh, *Spontaneous symmetry breaking for chiral scalar superfields*, Nucl. Phys. **B96** (1975) 331

[115] A. E. Nelson and N. Seiberg, *R symmetry breaking versus supersymmetry breaking*, Nucl. Phys. **B416** (1994) 46–62, `arXiv:hep-ph/9309299`

[116] P. C. Argyres, *Supersymmetric effective actions in four dimensions*, in *Nonperturbative aspects of string theory and supersymmetric gauge theories*, Proc. ICTP Spring School, Trieste, 1998, ed. M. Duff et al. World Scientific, 1999.

[117] S. Ferrara, L. Girardello and F. Palumbo, *A general mass formula in broken supersymmetry*, Phys. Rev. **D20** (1979) 403

[118] M. T. Grisaru, M. Roček and A. Karlhede, *The superhiggs effect in superspace*, Phys. Lett. **B120** (1983) 110

[119] A. Salam and J. A. Strathdee, *Super-gauge transformations*, Nucl. Phys. **B76** (1974) 477–482

[120] J. Wess and J. Bagger, *Supersymmetry and supergravity*. Princeton University Press, 1992

[121] S. J. Gates, Jr., M. T. Grisaru, M. Roček and W. Siegel, *Superspace, or one thousand and one lessons in supersymmetry*, Front. Phys. **58** (1983) 1–548, `arXiv:hep-th/0108200`

[122] P. C. West, *Introduction to supersymmetry and supergravity*. World Scientific, 1990

[123] I. L. Buchbinder and S. M. Kuzenko, *Ideas and methods of supersymmetry and supergravity: Or a walk through superspace*. IOP Publishing, 1998

[124] J. Terning, *Modern supersymmetry dynamics and duality*, International Series of Monographs on Physics, vol. 132. Oxford Science Publications, 2005

[125] M. Dine, *Supersymmetry and string theory: Beyond the standard model.* Cambridge University Press, 2007

[126] J. Wess and B. Zumino, *Superspace formulation of supergravity*, Phys. Lett. **B66** (1977) 361–364

[127] J. Wess and B. Zumino, *Superfield lagrangian for supergravity*, Phys. Lett. **B74** (1978) 51

[128] W. Siegel, *Solution to constraints in Wess–Zumino supergravity formalism*, Nucl. Phys. **B142** (1978) 301

[129] W. Siegel and S. J. Gates, Jr., *Superfield supergravity*, Nucl. Phys. **B147** (1979) 77

[130] R. Grimm, J. Wess and B. Zumino, *A complete solution of the Bianchi identities in superspace*, Nucl. Phys. **B152** (1979) 255

[131] M. Kaku, P. K. Townsend and P. van Nieuwenhuizen, *Gauge theory of the conformal and superconformal group*, Phys. Lett. **B69** (1977) 304–308

[132] S. Ferrara, M. Kaku, P. K. Townsend and P. van Nieuwenhuizen, *Unified field theories with $U(N)$ internal symmetries: Gauging the superconformal group*, Nucl. Phys. **B129** (1977) 125

[133] M. Kaku and P. K. Townsend, *Poincaré supergravity as broken superconformal gravity*, Phys. Lett. **B76** (1978) 54

[134] M. Kaku, P. K. Townsend and P. van Nieuwenhuizen, *Properties of conformal supergravity*, Phys. Rev. **D17** (1978) 3179

[135] D. Butter, *$\mathcal{N} = 1$ conformal superspace in four dimensions*, Ann. Phys. **325** (2010) 1026–1080, `arXiv:0906.4399 [hep-th]`

[136] E. Sezgin and Y. Tanii, *Superconformal sigma models in higher than two dimensions*, Nucl. Phys. **B443** (1995) 70–84, `arXiv:hep-th/9412163`

[137] B. de Wit, B. Kleijn and S. Vandoren, *Rigid $N = 2$ superconformal hypermultiplets*, in *Supersymmetries and quantum symmetries*, Proc. Int. Sem., Dubna (1997), ed. J. Wess and E.A. Ivanov, Lecture Notes in Physics, vol. 524. Springer, 1999, `arXiv:hep-th/9808160`,

[138] G. Gibbons and P. Rychenkova, *Cones, tri-Sasakian structures and superconformal invariance*, Phys. Lett. **B443** (1998) 138–142, `arXiv:hep-th/9809158`

[139] A. Van Proeyen, *Superconformal tensor calculus in $N = 1$ and $N = 2$ supergravity*, in *Supersymmetry and supergravity 1983*, XIXth Winter School and Workshop of Theoretical Physics, Karpacz, Poland, ed. B. Milewski. World Scientific, 1983

[140] D. V. Belyaev and P. van Nieuwenhuizen, *Simple $d = 4$ supergravity with a boundary*, JHEP **09** (2008) 069, `arXiv:0806.4723 [hep-th]`

[141] S. Ferrara and P. van Nieuwenhuizen, *The auxiliary fields of supergravity*, Phys. Lett. **B74** (1978) 333

[142] K. S. Stelle and P. C. West, *Minimal auxiliary fields for supergravity*, Phys. Lett. **B74** (1978) 330

[143] E. S. Fradkin and M. A. Vasiliev, *S matrix for theories that admit closure of the algebra with the aid of auxiliary fields: the auxiliary fields in supergravity*, Nuovo Cim. Lett. **22** (1978) 651

[144] M. F. Sohnius and P. C. West, *An alternative minimal off-shell version of* $N = 1$ *supergravity*, Phys. Lett. **B105** (1981) 353

[145] P. Breitenlohner, *A geometric interpretation of local supersymmetry*, Phys. Lett. **B67** (1977) 49

[146] P. Breitenlohner, *Some invariant Lagrangians for local supersymmetry*, Nucl. Phys. **B124** (1977) 500

[147] G. Girardi, R. Grimm, M. Müller and J. Wess, *Antisymmetric tensor gauge potential in curved superspace and a* (16+16) *supergravity multiplet*, Phys. Lett. **B147** (1984) 81

[148] W. Lang, J. Louis and B. A. Ovrut, $(16 + 16)$ *supergravity coupled to matter: the low-energy limit of the superstring*, Phys. Lett. **B158** (1985) 40

[149] W. Siegel, *16/16 supergravity*, Class. Quant. Grav. **3** (1986) L47–L48

[150] S. Ferrara, L. Girardello, T. Kugo and A. Van Proeyen, *Relation between different auxiliary field formulations of* $N = 1$ *supergravity coupled to matter*, Nucl. Phys. **B223** (1983) 191

[151] P. Binétruy, G. Girardi and R. Grimm, *Supergravity couplings: a geometric formulation*, Phys. Rep. **343** (2001) 255–462, `arXiv:hep-th/0005225`

[152] E. Cremmer, S. Ferrara, L. Girardello and A. Van Proeyen, *Yang–Mills theories with local supersymmetry: Lagrangian, transformation laws and superhiggs effect*, Nucl. Phys. **B212** (1983) 413

[153] E. Cremmer, B. Julia, J. Scherk, S. Ferrara, L. Girardello and P. van Nieuwenhuizen, *Spontaneous symmetry breaking and Higgs effect in supergravity without cosmological constant*, Nucl. Phys. **B147** (1979) 105

[154] T. Kugo and S. Uehara, *Improved superconformal gauge conditions in the* $N = 1$ *supergravity Yang–Mills matter system*, Nucl. Phys. **B222** (1983) 125

[155] R. Kallosh, L. Kofman, A. D. Linde and A. Van Proeyen, *Superconformal symmetry, supergravity and cosmology*, Class. Quant. Grav. **17** (2000) 4269–4338; erratum **21** (2004) 5017, `arXiv:hep-th/0006179`

[156] P. Binétruy, G. Dvali, R. Kallosh and A. Van Proeyen, *Fayet–Iliopoulos terms in supergravity and cosmology*, Class. Quant. Grav. **21** (2004) 3137–3170, `arXiv:hep-th/0402046`

[157] K. S. Stelle and P. C. West, *Relation between vector and scalar multiplets and gauge invariance in supergravity*, Nucl. Phys. **B145** (1978) 175

[158] T. Kugo and S. Uehara, *Conformal and Poincaré tensor calculi in* $N = 1$ *supergravity*, Nucl. Phys. **B226** (1983) 49

[159] M. A. Lledó, Ó. Maciá, A. Van Proeyen and V. S. Varadarajan, *Special geometry for arbitrary signatures*, in *Handbook on pseudo-Riemannian geometry and supersymmetry*, ed. V. Cortés, IRMA Lectures in Mathematics and Theoretical Physics, vol. 16 chap. 5. European Mathematical Society, 2010, `arXiv:hep-th/0612210`

[160] O. Griffiths and J. Harris, *Principles of algebraic geometry*. John Wiley, 1978

[161] S. J. Gates, Jr., H. Nishino and E. Sezgin, *Supergravity in* $d = 9$ *and its coupling to non-compact* σ*-model*, Class. Quant. Grav. **3** (1986) 21

[162] E. Witten and J. Bagger, *Quantization of Newton's constant in certain supergravity theories*, Phys. Lett. **B115** (1982) 202

[163] S.-S. Chern, *Complex manifolds without potential theory.* Springer, 1979

[164] R. O. Wells, *Differential analysis on complex manifolds.* Springer, 1980

[165] J. Bagger, *Supersymmetric sigma models*, in *Supersymmetry*, ed. K. Dietz et al., NATO Advanced Study Institute, Series B, Physics, vol. 125. Plenum Press, 1985

[166] N. Seiberg, *Modifying the sum over topological sectors and constraints on supergravity*, JHEP **07** (2010) 070, `arXiv:1005.0002 [hep-th]`

[167] S. Cecotti, L. Girardello and M. Porrati, *Ward identities of local supersymmetry and spontaneous breaking of extended supergravity*, in *New trends in particle theory*, Proc. of the 9th Johns Hopkins Workshop, Firenze, ed. L. Lusanna. World Scientific, 1985

[168] S. Ferrara and L. Maiani, *An introduction to supersymmetry breaking in extended supergravity*, in *Relativity, supersymmetry and cosmology*, Proc. of SILARG V, 5th Latin American Symp. on Relativity and Gravitation, Bariloche, Argentina, ed. O. Bressan, M. Castagnino and V. H. Hamity. World Scientific, 1985

[169] S. Cecotti, L. Girardello and M. Porrati, *Constraints on partial superhiggs*, Nucl. Phys. **B268** (1986) 295–316

[170] R. D'Auria and S. Ferrara, *On fermion masses, gradient flows and potential in supersymmetric theories*, JHEP **05** (2001) 034, `arXiv:hep-th/0103153`

[171] Z. Komargodski and N. Seiberg, *Comments on the Fayet–Iliopoulos term in field theory and supergravity*, JHEP **06** (2009) 007, `arXiv:0904.1159 [hep-th]`

[172] T. Banks and N. Seiberg, *Symmetries and strings in field theory and gravity*, Phys. Rev. **D83** (2011) 084019, `arXiv:1011.5120 [hep-th]`

[173] S. Ferrara, C. Kounnas and F. Zwirner, *Mass formulae and natural hierarchy in string effective supergravities*, Nucl. Phys. **B429** (1994) 589–625; erratum **B433** (1995) 255, `arXiv:hep-th/9405188`

[174] J. Polónyi, *Generalization of the massive scalar multiplet coupling to the supergravity*, Hungary Central Inst. Res. - KFKI-77-93

[175] E. Cremmer, S. Ferrara, L. Girardello and A. Van Proeyen, *Coupling supersymmetric Yang–Mills theories to supergravity*, Phys. Lett. **B116** (1982) 231

[176] L. Girardello and M. T. Grisaru, *Soft breaking of supersymmetry*, Nucl. Phys. **B194** (1982) 65

[177] P. Fayet, *Supersymmetric theories of particles and interactions*, Phys. Rep. **105** (1984) 21

[178] H. P. Nilles, *Supersymmetry, supergravity and particle physics*, Phys. Rep. **110** (1984) 1

[179] E. Cremmer, S. Ferrara, C. Kounnas and D. V. Nanopoulos, *Naturally vanishing cosmological constant in $N = 1$ supergravity*, Phys. Lett. **B133** (1983) 61

[180] J. R. Ellis, C. Kounnas and D. V. Nanopoulos, *No-scale supersymmetric GUTs*, Nucl. Phys. **B247** (1984) 373–395

[181] A. Lahanas and D. V. Nanopoulos, *The road to no-scale supergravity*, Phys. Rep. **145** (1987) 1

[182] D. Bailin and A. Love, *Cosmology in gauge field theory and string theory.* IOP Publishing, 2004

[183] P. Breitenlohner and D. Z. Freedman, *Positive energy in anti-de Sitter backgrounds and gauged extended supergravity*, Phys. Lett. **B115** (1982) 197

[184] P. Breitenlohner and D. Z. Freedman, *Stability in gauged extended supergravity*, Ann. Phys. **144** (1982) 249

[185] L. Mezincescu and P. K. Townsend, *Stability at a local maximum in higher dimensional anti-de Sitter space and applications to supergravity*, Ann. Phys. **160** (1985) 406

[186] W. Boucher, *Positive energy without supersymmetry*, Nucl. Phys. **B242** (1984) 282

[187] P. K. Townsend, *Positive energy and the scalar potential in higher dimensional (super)gravity theories*, Phys. Lett. **B148** (1984) 55

[188] J. Distler and E. Sharpe, *Quantization of Fayet–Iliopoulos parameters in supergravity*, Phys. Rev. **D83** (2011) 085010, `arXiv:1008.0419 [hep-th]`

[189] K. R. Dienes and B. Thomas, *On the inconsistency of Fayet–Iliopoulos terms in supergravity theories*, Phys. Rev. **D81** (2010) 065023, `arXiv:0911.0677 [hep-th]`

[190] A. Galperin, E. Ivanov, S. Kalitsyn, V. Ogievetsky and E. Sokatchev, *Unconstrained $N = 2$ matter, Yang–Mills and supergravity theories in harmonic superspace*, Class. Quant. Grav. **1** (1984) 469–498

[191] A. S. Galperin, E. A. Ivanov, V. I. Ogievetsky and E. S. Sokatchev, *Harmonic superspace*. Cambridge University Press, 2001

[192] A. Karlhede, U. Lindström and M. Roček, *Selfinteracting tensor multiplets in $N = 2$ superspace*, Phys. Lett. **B147** (1984) 297

[193] S. J. Gates, Jr., C. M. Hull and M. Roček, *Twisted multiplets and new supersymmetric non-linear σ-models*, Nucl. Phys. **B248** (1984) 157

[194] U. Lindström and M. Roček, *New hyperkähler metrics and new supermultiplets*, Commun. Math. Phys. **115** (1988) 21

[195] U. Lindström and M. Roček, *$N=2$ super Yang–Mills theory in projective superspace*, Commun. Math. Phys. **128** (1990) 191

[196] U. Lindström and M. Roček, *Properties of hyperkähler manifolds and their twistor spaces*, Commun. Math. Phys. **293** (2010) 257–278, `arXiv:0807.1366 [hep-th]`

[197] P. S. Howe, *A superspace approach to extended conformal supergravity*, Phys. Lett. **B100** (1981) 389

[198] P. S. Howe, *Supergravity in superspace*, Nucl. Phys. **B199** (1982) 309

[199] S. M. Kuzenko, U. Lindström, M. Roček and G. Tartaglino-Mazzucchelli, *On conformal supergravity and projective superspace*, JHEP **08** (2009) 023, `arXiv:0905.0063 [hep-th]`

[200] P. Fré and P. Soriani, *The $N = 2$ wonderland: From Calabi–Yau manifolds to topological field theories*. World Scientific, 1995.

[201] L. Andrianopoli, M. Bertolini, A. Ceresole, R. D'Auria, S. Ferrara, P. Frè and T. Magri, *$N = 2$ supergravity and $N = 2$ super Yang–Mills theory on general scalar manifolds: Symplectic covariance, gaugings and the momentum map*, J. Geom. Phys. **23** (1997) 111–189, `arXiv:hep-th/9605032`

[202] R. D'Auria, L. Sommovigo and S. Vaulà, *$N = 2$ supergravity Lagrangian coupled to tensor multiplets with electric and magnetic fluxes*, JHEP **0411** (2004) 028, arXiv:hep-th/0409097 [hep-th]

[203] L. Andrianopoli, R. D'Auria, L. Sommovigo and M. Trigiante, *$D = 4$, $N = 2$ gauged supergravity coupled to vector–tensor multiplets*, Nucl. Phys. **B851** (2011) 1–29, arXiv:1103.4813 [hep-th]

[204] M. Günaydin and M. Zagermann, *The gauging of five-dimensional, $N = 2$ Maxwell–Einstein supergravity theories coupled to tensor multiplets*, Nucl. Phys. **B572** (2000) 131–150, arXiv:hep-th/9912027

[205] A. Ceresole and G. Dall'Agata, *General matter-coupled $\mathcal{N} = 2$, $D = 5$ gauged supergravity*, Nucl. Phys. **B585** (2000) 143–170, arXiv:hep-th/0004111

[206] E. Bergshoeff, S. Cucu, T. de Wit, J. Gheerardyn, R. Halbersma, S. Vandoren and A. Van Proeyen, *Superconformal $N = 2$, $D = 5$ matter with and without actions*, JHEP **10** (2002) 045, arXiv:hep-th/0205230

[207] E. Bergshoeff, S. Cucu, T. de Wit, J. Gheerardyn, S. Vandoren and A. Van Proeyen, *$N = 2$ supergravity in five dimensions revisited*, Class. Quant. Grav. **21** (2004) 3015–3041; erratum **23** (2006) 7149, arXiv:hep-th/0403045

[208] E. Bergshoeff, S. Cucu, M. Derix, T. de Wit, R. Halbersma and A. Van Proeyen, *Weyl multiplets of $N = 2$ conformal supergravity in five dimensions*, JHEP **06** (2001) 051, arXiv:hep-th/0104113

[209] T. Fujita and K. Ohashi, *Superconformal tensor calculus in five dimensions*, Prog. Theor. Phys. **106** (2001) 221–247, arXiv:hep-th/0104130

[210] T. Kugo and K. Ohashi, *Gauge and non-gauge tensor multiplets in 5D conformal supergravity*, Prog. Theor. Phys. **108** (2003) 1143–1164, arXiv:hep-th/0208082

[211] R. Grimm, M. Sohnius and J. Wess, *Extended supersymmetry and gauge theories*, Nucl. Phys. **B133** (1978) 275

[212] R. J. Firth and J. D. Jenkins, *Super-symmetry with isospin*, Nucl. Phys. **B85** (1975) 525

[213] P. Fayet, *Fermi–Bose hypersymmetry*, Nucl. Phys. **B113** (1976) 135

[214] P. Fayet, *Spontaneous generation of massive multiplets and central charges in extended supersymmetric theories*, Nucl. Phys. **B149** (1979) 137

[215] B. de Wit, P. G. Lauwers and A. Van Proeyen, *Lagrangians of $N = 2$ supergravity–matter systems*, Nucl. Phys. **B255** (1985) 569

[216] B. de Wit, C. M. Hull and M. Roček, *New topological terms in gauge invariant actions*, Phys. Lett. **B184** (1987) 233

[217] M. Obata, *Affine connections on manifolds with almost complex, quaternionic or hermitian structure*, Japan J. Math. **26** (1956) 43–79

[218] J. De Rydt and B. Vercnocke, *De Lagrangiaan van vector- en hypermultipletten in $N = 2$ supergravitatie*, Thesis Licenciaat, K.U. Leuven, 2006

[219] J. De Jaegher, B. de Wit, B. Kleijn and S. Vandoren, *Special geometry in hypermultiplets*, Nucl. Phys. **B514** (1998) 553–582, arXiv:hep-th/9707262

[220] B. Zumino, *Normal forms of complex matrices*, J. Math. Phys. **3** (1962) 1055–1057

[221] T. Kugo and K. Ohashi, *Supergravity tensor calculus in 5D from 6D*, Prog. Theor. Phys. **104** (2000) 835–865, arXiv:hep-ph/0006231

[222] E. Bergshoeff, E. Sezgin and A. Van Proeyen, *Superconformal tensor calculus and matter couplings in six dimensions*, Nucl. Phys. **B264** (1986) 653

[223] B. de Wit, J. W. van Holten and A. Van Proeyen, *Transformation rules of $N = 2$ supergravity multiplets*, Nucl. Phys. **B167** (1980) 186–204; erratum **B172** (1980) 543–544

[224] B. de Wit, B. Kleijn and S. Vandoren, *Superconformal hypermultiplets*, Nucl. Phys. **B568** (2000) 475–502, arXiv:hep-th/9909228

[225] P. Breitenlohner and A. Kabelschacht, *The auxiliary fields of $N = 2$ extended supergravity in 5 and 6 space–time dimensions*, Nucl. Phys. **B148** (1979) 96

[226] E. S. Fradkin and M. A. Vasiliev, *Minimal set of auxiliary fields and S matrix for extended supergravity*, Lett. Nuovo Cim. **25** (1979) 79–90

[227] B. de Wit and J. W. van Holten, *Multiplets of linearized* SO(2) *supergravity*, Nucl. Phys. **B155** (1979) 530

[228] E. S. Fradkin and M. A. Vasiliev, *Minimal set of auxiliary fields in* SO(2) *extended supergravity*, Phys. Lett. **B85** (1979) 47–51

[229] P. Breitenlohner and M. F. Sohnius, *Superfields, auxiliary fields, and tensor calculus for $N = 2$ extended supergravity*, Nucl. Phys. **B165** (1980) 483

[230] B. de Wit, J. W. van Holten and A. Van Proeyen, *Structure of $N = 2$ supergravity*, Nucl. Phys. **B184** (1981) 77–108; erratum **B222** (1983) 516–517

[231] B. de Wit, R. Philippe and A. Van Proeyen, *The improved tensor multiplet in $N = 2$ supergravity*, Nucl. Phys. **B219** (1983) 143

[232] T. Kugo and K. Ohashi, *Off-shell $d = 5$ supergravity coupled to matter–Yang–Mills system*, Prog. Theor. Phys. **105** (2001) 323–353, arXiv:hep-ph/0010288

[233] T. Fujita, T. Kugo and K. Ohashi, *Off-shell formulation of supergravity on orbifold*, Prog. Theor. Phys. **106** (2001) 671–690, arXiv:hep-th/0106051

[234] M. F. Sohnius, *Supersymmetry and central charges*, Nucl. Phys. **B138** (1978) 109–121

[235] B. de Wit, J. W. van Holten and A. Van Proeyen, *Central charges and conformal supergravity*, Phys. Lett. **B95** (1980) 51

[236] A. Van Proeyen, *$N = 2$ matter couplings in $d = 4$ and 6 from superconformal tensor calculus*, in *Superunification and extra dimensions,* Proceedings of the 1st Torino meeting, ed. R. D'Auria and P. Fré, World Scientific, 1986, pp. 97–125.

[237] E. Bergshoeff, *Superconformal invariance and the tensor multiplet in six dimensions*, in *Superunification and extra dimensions,* Proceedings of the 1st Torino meeting, ed. R. D'Auria and P. Fré, World Scientific, 1986, pp. 126–137.

[238] F. Coomans and A. Van Proeyen, *Off-shell $\mathcal{N} = (1, 0)$, $D = 6$ supergravity from superconformal methods*, JHEP **1102** (2011) 049, arXiv:1101.2403 [hep-th]

[239] F. Riccioni, *All couplings of minimal six-dimensional supergravity*, Nucl. Phys. **B605** (2001) 245–265, arXiv:hep-th/0101074

[240] L. J. Romans, *Selfduality for interacting fields: covariant field equations for six-dimensional chiral supergravities*, Nucl. Phys. **B276** (1986) 71

[241] H. Nishino and E. Sezgin, *New couplings of six-dimensional supergravity*, Nucl. Phys. **B505** (1997) 497–516, arXiv:hep-th/9703075

[242] S. Ferrara, F. Riccioni and A. Sagnotti, *Tensor and vector multiplets in six-dimensional supergravity*, Nucl. Phys. **B519** (1998) 115–140, arXiv:hep-th/9711059

[243] D. S. Freed, *Special Kähler manifolds*, Commun. Math. Phys. **203** (1999) 31–52, arXiv:hep-th/9712042

[244] D. V. Alekseevsky, V. Cortés and C. Devchand, *Special complex manifolds*, J. Geom. Phys. **42** (2002) 85–105, arXiv:math.dg/9910091

[245] B. de Wit and A. Van Proeyen, *Special geometry, cubic polynomials and homogeneous quaternionic spaces*, Commun. Math. Phys. **149** (1992) 307–334, arXiv:hep-th/9112027

[246] S. Cecotti, S. Ferrara and L. Girardello, *Geometry of type II superstrings and the moduli of superconformal field theories*, Int. J. Mod. Phys. **A4** (1989) 2475

[247] B. Craps, F. Roose, W. Troost and A. Van Proeyen, *What is special Kähler geometry?*, Nucl. Phys. **B503** (1997) 565–613, arXiv:hep-th/9703082

[248] A. Ceresole, R. D'Auria, S. Ferrara and A. Van Proeyen, *Duality transformations in supersymmetric Yang–Mills theories coupled to supergravity*, Nucl. Phys. **B444** (1995) 92–124, arXiv:hep-th/9502072

[249] A. Das, *SO(4) invariant extended supergravity*, Phys. Rev. **D15** (1977) 2805

[250] E. Cremmer, J. Scherk and S. Ferrara, *U(N) invariance in extended supergravity*, Phys. Lett. **B68** (1977) 234

[251] E. Cremmer and J. Scherk, *Algebraic simplifications in supergravity theories*, Nucl. Phys. **B127** (1977) 259

[252] E. Cremmer, J. Scherk and S. Ferrara, *SU(4) invariant supergravity theory*, Phys. Lett. **B74** (1978) 61

[253] E. Cremmer and A. Van Proeyen, *Classification of Kähler manifolds in $N = 2$ vector multiplet–supergravity couplings*, Class. Quant. Grav. **2** (1985) 445

[254] S. Ferrara and A. Van Proeyen, *A theorem on $N = 2$ special Kähler product manifolds*, Class. Quant. Grav. **6** (1989) L243

[255] A. Sen, *Macroscopic charged heterotic string*, Nucl. Phys. **B388** (1992) 457–473, arXiv:hep-th/9206016

[256] J. H. Schwarz and A. Sen, *Duality symmetric actions*, Nucl. Phys. **B411** (1994) 35–63, arXiv:hep-th/9304154

[257] L. Alvarez-Gaumé and D. Z. Freedman, *Geometrical structure and ultraviolet finiteness in the supersymmetric σ-model*, Commun. Math. Phys. **80** (1981) 443

[258] K. Galicki, *A generalization of the momentum mapping construction for quaternionic Kähler manifolds*, Commun. Math. Phys. **108** (1987) 117

[259] P. Frè, *Gaugings and other supergravity tools of p-brane physics*, Proceedings of the Workshop on Latest Development in M-Theory, Paris, France, 1–9 February 2001, arXiv:hep-th/0102114

[260] E. Bergshoeff, S. Cucu, T. de Wit, J. Gheerardyn, S. Vandoren and A. Van Proeyen, *The map between conformal hypercomplex/hyper-Kähler and*

quaternionic(-Kähler) geometry, Commun. Math. Phys. **262** (2006) 411–457, arXiv:hep-th/0411209

[261] N. Ambrose and J. Singer, *A theorem on holonomy*, Trans. Am. Math. Soc. **75** (1953) 428–443

[262] D. V. Alekseevsky, *Classification of quaternionic spaces with a transitive solvable group of motions*, Math. USSR Izv. **9** (1975) 297–339

[263] B. de Wit, M. Roček and S. Vandoren, *Hypermultiplets, hyperkähler cones and quaternion-Kähler geometry*, JHEP **02** (2001) 039, arXiv:hep-th/0101161

[264] L. Sommovigo and S. Vaulà, $D = 4$, $N = 2$ *supergravity with abelian electric and magnetic charge*, Phys. Lett. **B602** (2004) 130–136, arXiv:hep-th/0407205 [hep-th]

[265] M. de Vroome and B. de Wit, *Lagrangians with electric and magnetic charges of* $N = 2$ *supersymmetric gauge theories*, JHEP **08** (2007) 064, arXiv:0707.2717 [hep-th]

[266] S. Cecotti, L. Girardello and M. Porrati, *Two into one won't go*, Phys. Lett. **B145** (1984) 61

[267] S. Ferrara, L. Girardello and M. Porrati, *Minimal Higgs branch for the breaking of half of the supersymmetries in* $N = 2$ *supergravity*, Phys. Lett. **B366** (1996) 155–159, arXiv:hep-th/9510074

[268] J. Louis, P. Smyth and H. Triendl, *The* $N = 1$ *low-energy effective action of spontaneously broken* $N = 2$ *supergravities*, JHEP **1010** (2010) 017, arXiv:1008.1214 [hep-th]

[269] J. Louis, P. Smyth and H. Triendl, *Spontaneous* $N = 2 \to N = 1$ *supersymmetry breaking in supergravity and type II string theory*, JHEP **1002** (2010) 103, arXiv:0911.5077 [hep-th]

[270] N. Seiberg, *Observations on the moduli space of superconformal field theories*, Nucl. Phys. **B303** (1988) 286

[271] S. Ferrara and A. Strominger, $N = 2$ *space–time supersymmetry and Calabi–Yau moduli space*, in *Strings '89*, ed. R. Arnowitt, R. Bryan, M. J. Duff, D. V. Nanopoulos and C. N. Pope, p. 245. World Scientific, 1989

[272] P. Candelas and X. de la Ossa, *Moduli space of Calabi–Yau manifolds*, Nucl. Phys. **B355** (1991) 455–481

[273] P. Candelas, X. C. de la Ossa, P. S. Green and L. Parkes, *A pair of Calabi–Yau manifolds as an exactly soluble superconformal theory*, Nucl. Phys. **B359** (1991) 21–74

[274] P. Candelas, X. C. de la Ossa, P. S. Green and L. Parkes, *An exactly soluble superconformal theory from a mirror pair of Calabi–Yau manifolds*, Phys. Lett. **B258** (1991) 118–126

[275] H. Lü, C. N. Pope and J. Rahmfeld, *A construction of Killing spinors on* S^n, J. Math. Phys. **40** (1999) 4518–4526, arXiv:hep-th/9805151

[276] W. de Sitter, *On the relativity of inertia: Remarks concerning Einstein's latest hypothesis*, Proc. Kon. Ned. Acad. Wet. **19** (1917) 1217–1225

[277] W. de Sitter, *On the curvature of space*, Proc. Kon. Ned. Acad. Wet. **20** (1917) 229–243

[278] S. J. Avis, C. J. Isham and D. Storey, *Quantum field theory in anti-de Sitter spacetime*, Phys. Rev. **D18** (1978) 3565

[279] P. G. O. Freund and M. A. Rubin, *Dynamics of dimensional reduction*, Phys. Lett. **B97** (1980) 233–235

[280] J. M. Maldacena, *The large N limit of superconformal field theories and supergravity*, Adv. Theor. Math. Phys. **2** (1998) 231–252, arXiv:hep-th/9711200

[281] S. S. Gubser, I. R. Klebanov and A. M. Polyakov, *Gauge theory correlators from non-critical string theory*, Phys. Lett. **B428** (1998) 105–114, arXiv:hep-th/9802109

[282] E. Witten, *Anti-de Sitter space and holography*, Adv. Theor. Math. Phys. **2** (1998) 253–291, arXiv:hep-th/9802150

[283] K. Schwarzschild, *Uber das Gravitationfeld eines Massenpunktes nach der Einsteinschen Theorie*, Sitzungsber. Preuss. Akad. Wiss. Berlin (Math. Phys.) **1916** (1916) 189–196. Translated as *On the gravitational field of a mass point according to Einstein's theory* by S. Antoci, arXiv:physics/9905030

[284] H. Reissner, *Uber die Eigengravitation des elektrischen Feldes nach der Einstein'schen Theorie*, Annalen der Physik **50** (1916) 106

[285] G. Nordström, *On the energy of the gravitational field in Einstein's theory*, Verhandl. Koninkl. Ned. Akad. Wetenschap., Afdel. Natuurk., Amsterdam **26** (1918) 1201

[286] E. Bogomol'nyi, *Stability of classical solutions*, Sov. J. Nucl. Phys. **24** (1976) 449–454. Translated from Jadernaja Fiz. **24** (1976), no. 4, 861–870 (Russian)

[287] M. K. Prasad and C. M. Sommerfield, *An exact classical solution for the 't Hooft monopole and the Julia–Zee dyon*, Phys. Rev. Lett. **35** (1975) 760–762

[288] M. C. N. Cheng and K. Skenderis, *Positivity of energy for asymptotically locally AdS spacetimes*, JHEP **08** (2005) 107, arXiv:hep-th/0506123

[289] H. Lü, C. N. Pope and P. K. Townsend, *Domain walls from anti-de Sitter spacetime*, Phys. Lett. **B391** (1997) 39–46, arXiv:hep-th/9607164

[290] T. Mohaupt, *Black hole entropy, special geometry and strings*, Fortsch. Phys. **49** (2001) 3–161, arXiv:hep-th/0007195

[291] G. W. Gibbons and C. M. Hull, *A Bogomolny bound for general relativity and solitons in $N = 2$ supergravity*, Phys. Lett. **B109** (1982) 190

[292] A. Ceresole and G. Dall'Agata, *Flow equations for non-BPS extremal black holes*, JHEP **03** (2007) 110, arXiv:hep-th/0702088

[293] J. Perz, P. Smyth, T. Van Riet and B. Vercnocke, *First-order flow equations for extremal and non-extremal black holes*, JHEP **03** (2009) 150, arXiv:0810.1528 [hep-th]

[294] S. Ferrara, R. Kallosh and A. Strominger, *$N = 2$ extremal black holes*, Phys. Rev. **D52** (1995) 5412–5416, arXiv:hep-th/9508072

[295] S. Ferrara and R. Kallosh, *Supersymmetry and attractors*, Phys. Rev. **D54** (1996) 1514–1524, arXiv:hep-th/9602136

[296] S. Ferrara, G. W. Gibbons and R. Kallosh, *Black holes and critical points in moduli space*, Nucl. Phys. **B500** (1997) 75–93, arXiv:hep-th/9702103

[297] A. Strominger and C. Vafa, *Microscopic origin of the Bekenstein–Hawking entropy*, Phys. Lett. **B379** (1996) 99–104, arXiv:hep-th/9601029

[298] A. Ceresole, R. D'Auria and S. Ferrara, *The symplectic structure of* $N = 2$ *supergravity and its central extension*, Nucl. Phys. Proc. Suppl. **46** (1996) 67–74, `arXiv:hep-th/9509160`

[299] R. Kallosh, A. D. Linde, T. Ortín, A. Peet and A. Van Proeyen, *Supersymmetry as a cosmic censor*, Phys. Rev. **D46** (1992) 5278–5302, `arXiv:hep-th/9205027`

[300] L. Andrianopoli, R. D'Auria, E. Orazi and M. Trigiante, *First order description of black holes in moduli space*, JHEP **11** (2007) 032, `arXiv:0706.0712 [hep--th]`

[301] G. Lopes Cardoso, A. Ceresole, G. Dall'Agata, J. M. Oberreuter and J. Perz, *First-order flow equations for extremal black holes in very special geometry*, JHEP **0710** (2007) 063, `arXiv:0706.3373 [hep-th]`

[302] S. Ferrara, K. Hayakawa and A. Marrani, *Lectures on attractors and black holes*, Fortsch. Phys. **56** (2008) 993–1046; Erice workshop 'Totally unexpected in the LHC era', 2007, `arXiv:0805.2498 [hep-th]`

[303] O. Aharony, S. S. Gubser, J. M. Maldacena, H. Ooguri and Y. Oz, *Large N field theories, string theory and gravity*, Phys. Rep. **323** (2000) 183–386, `arXiv:hep-th/9905111`

[304] E. D'Hoker and D. Z. Freedman, *Supersymmetric gauge theories and the AdS/CFT correspondence*, in *TASI 2001: strings, branes and extra dimensions*, Boulder, Colorado, 3–29 June 2001, ed. S. S. Gubser and J. D. Lykken, pp. 3–158, World Scientific, 2004, `arXiv:hep-th/0201253`

[305] J. M. Maldacena, *Lectures on AdS/CFT*, in *TASI 2003: recent trends in string theory*, Boulder, Colorado, pp. 155–203. World Scientific, 2005, `arXiv:hep-th/0309246`

[306] J. McGreevy, *Holographic duality with a view toward many-body physics*, Adv. High Energy Phys. **2010** (2010) 723105, `arXiv:0909.0518 [hep-th]`

[307] K. Becker, M. Becker and J. H. Schwarz, *String theory and M-theory: A modern introduction*. Cambridge University Press, 2007

[308] E. Kiritsis, *String theory in a nutshell*. Princeton University Press, 2007

[309] B. Zwiebach, *A first course in string theory*, 2nd edn. Cambridge University Press, 2008

[310] S. Coleman, *Aspects of symmetry: Selected Erice lectures of Sidney Coleman*. Cambridge University Press, 1985

[311] J. Polchinski, *Dirichlet-branes and Ramond–Ramond charges*, Phys. Rev. Lett. **75** (1995) 4724–4727, `arXiv:hep-th/9510017`

[312] M. K. Benna, S. Benvenuti, I. R. Klebanov and A. Scardicchio, *A test of the AdS/CFT correspondence using high-spin operators*, Phys. Rev. Lett. **98** (2007) 131603, `arXiv:hep-th/0611135`

[313] M. Günaydin and N. Marcus, *The spectrum of the* S^5 *compactification of the chiral* $N = 2$, $D = 10$ *supergravity and the unitary supermultiplets of* $U(2, 2|4)$, Class. Quant. Grav. **2** (1985) L11

[314] H. J. Kim, L. J. Romans and P. van Nieuwenhuizen, *The mass spectrum of chiral* $N = 2$ $D = 10$ *supergravity on* S^5, Phys. Rev. **D32** (1985) 389

[315] M. Günaydin, L. J. Romans and N. P. Warner, *Compact and non-compact gauged supergravity theories in five dimensions*, Nucl. Phys. **B272** (1986) 598

[316] I. R. Klebanov and E. Witten, *AdS/CFT correspondence and symmetry breaking*, Nucl. Phys. **B556** (1999) 89–114, arXiv:hep-th/9905104

[317] P. Minces and V. O. Rivelles, *Energy and the AdS/CFT correspondence*, JHEP **12** (2001) 010, arXiv:hep-th/0110189

[318] D. Z. Freedman, K. Johnson and J. I. Latorre, *Differential regularization and renormalization: A new method of calculation in quantum field theory*, Nucl. Phys. **B371** (1992) 353–414

[319] D. Z. Freedman, S. D. Mathur, A. Matusis and L. Rastelli, *Correlation functions in the CFT_d/AdS_{d+1} correspondence*, Nucl. Phys. **B546** (1999) 96–118, arXiv:hep-th/9804058

[320] S. Lee, S. Minwalla, M. Rangamani and N. Seiberg, *Three-point functions of chiral operators in $D = 4$, $N = 4$ SYM at large N*, Adv. Theor. Math. Phys. **2** (1998) 697–718, arXiv:hep-th/9806074

[321] E. D'Hoker, D. Z. Freedman and W. Skiba, *Field theory tests for correlators in the AdS/CFT correspondence*, Phys. Rev. **D59** (1999) 045008, arXiv:hep-th/9807098

[322] S. Penati, A. Santambrogio and D. Zanon, *Two-point functions of chiral operators in $N = 4$ SYM at order g^4*, JHEP **12** (1999) 006, arXiv:hep-th/9910197

[323] K. A. Intriligator, *Bonus symmetries of $N = 4$ super-Yang–Mills correlation functions via AdS duality*, Nucl. Phys. **B551** (1999) 575–600, arXiv:hep-th/9811047

[324] M. Bianchi and S. Kovacs, *Non-renormalization of extremal correlators in* $\mathcal{N} = 4$ *SYM theory*, Phys. Lett. **B468** (1999) 102–110, arXiv:hep-th/9910016

[325] P. S. Howe, E. Sokatchev and P. C. West, *3-point functions in* $\mathcal{N} = 4$ *Yang–Mills*, Phys. Lett. **B444** (1998) 341–351, arXiv:hep-th/9808162

[326] I. Gradshteyn and I. Ryzhik, *Table of integrals, series, and products*. Academic Press, 1965

[327] A. W. Peet and J. Polchinski, *UV/IR relations in AdS dynamics*, Phys. Rev. **D59** (1999) 065011, arXiv:hep-th/9809022

[328] M. Bianchi, O. DeWolfe, D. Z. Freedman and K. Pilch, *Anatomy of two holographic renormalization group flows*, JHEP **01** (2001) 021, arXiv:hep-th/0009156

[329] M. Henningson and K. Skenderis, *Holography and the Weyl anomaly*, Fortsch. Phys. **48** (2000) 125–128, arXiv:hep-th/9812032

[330] S. de Haro, S. N. Solodukhin and K. Skenderis, *Holographic reconstruction of spacetime and renormalization in the AdS/CFT correspondence*, Commun. Math. Phys. **217** (2001) 595–622, arXiv:hep-th/0002230

[331] M. Bianchi, D. Z. Freedman and K. Skenderis, *Holographic renormalization*, Nucl. Phys. **B631** (2002) 159–194, arXiv:hep-th/0112119

[332] K. Skenderis, *Lecture notes on holographic renormalization*, Class. Quant. Grav. **19** (2002) 5849–5876, arXiv:hep-th/0209067

[333] J. Erdmenger and H. Osborn, *Conserved currents and the energy–momentum tensor in conformally invariant theories for general dimensions*, Nucl. Phys. **B483** (1997) 431–474, `arXiv:hep-th/9605009`

[334] S. M. Christensen and M. J. Duff, *Axial and conformal anomalies for arbitrary spin in gravity and supergravity*, Phys. Lett. **B76** (1978) 571

[335] C. Imbimbo, A. Schwimmer, S. Theisen and S. Yankielowicz, *Diffeomorphisms and holographic anomalies*, Class. Quant. Grav. **17** (2000) 1129–1138, `arXiv:hep-th/9910267`

[336] I. Papadimitriou and K. Skenderis, *Correlation functions in holographic RG flows*, JHEP **10** (2004) 075, `arXiv:hep-th/0407071`

[337] I. Papadimitriou and K. Skenderis, *AdS/CFT correspondence and geometry*, in *73rd meeting between theoretical physicists and mathematicians: (A)dS-CFT correspondence*, ed. O. Biquard, IRMA Lectures in Mathematics and Theoretical Physics, vol. 8, pp. 73–101. European Mathematical Society, 2005, `arXiv:hep-th/0404176`

[338] A. B. Zamolodchikov, *Irreversibility of the flux of the renormalization group in a 2D field theory*, JETP Lett. **43** (1986) 730–732

[339] J. L. Cardy, *Is there a c theorem in four-dimensions?*, Phys. Lett. **B215** (1988) 749–752

[340] K. A. Intriligator and B. Wecht, *The exact superconformal R-symmetry maximizes a*, Nucl. Phys. **B667** (2003) 183–200, `arXiv:hep-th/0304128`

[341] R. G. Leigh and M. J. Strassler, *Exactly marginal operators and duality in four-dimensional $N = 1$ supersymmetric gauge theory*, Nucl. Phys. **B447** (1995) 95–136, `arXiv:hep-th/9503121`

[342] D. Z. Freedman, S. S. Gubser, K. Pilch and N. P. Warner, *Renormalization group flows from holography supersymmetry and a c-theorem*, Adv. Theor. Math. Phys. **3** (1999) 363–417, `arXiv:hep-th/9904017`

[343] V. Balasubramanian, P. Kraus and A. E. Lawrence, *Bulk vs. boundary dynamics in anti-de Sitter spacetime*, Phys. Rev. **D59** (1999) 046003, `arXiv:hep-th/9805171`

[344] O. DeWolfe, D. Freedman, S. Gubser and A. Karch, *Modeling the fifth-dimension with scalars and gravity*, Phys. Rev. **D62** (2000) 046008, `arXiv:hep-th/9909134 [hep-th]`

[345] J. de Boer, E. P. Verlinde and H. L. Verlinde, *On the holographic renormalization group*, JHEP **08** (2000) 003, `arXiv:hep-th/9912012`

[346] N. Iqbal and H. Liu, *Universality of the hydrodynamic limit in AdS/CFT and the membrane paradigm*, Phys. Rev. **D79** (2009) 025023, `arXiv:0809.3808 [hep--th]`

[347] S. A. Hartnoll, *Lectures on holographic methods for condensed matter physics*, Class. Quant. Grav. **26** (2009) 224002, `arXiv:0903.3246 [hep-th]`

[348] J. I. Kapusta, *Finite-temperature field theory*. Cambridge University Press, 1989

[349] M. Le Bellac, *Thermal field theory*. Cambridge University Press, 2000

[350] D. T. Son and A. O. Starinets, *Minkowski-space correlators in AdS/CFT correspondence: Recipe and applications*, JHEP **09** (2002) 042, `arXiv:hep-th/0205051`

[351] G. Policastro, D. T. Son and A. O. Starinets, *From AdS/CFT correspondence to hydrodynamics*, JHEP **09** (2002) 043, arXiv:hep-th/0205052
[352] S. S. Gubser, S. S. Pufu and F. D. Rocha, *Bulk viscosity of strongly coupled plasmas with holographic duals*, JHEP **08** (2008) 085, arXiv:0806.0407 [hep-th]
[353] N. Iqbal and H. Liu, *Real-time response in AdS/CFT with application to spinors*, Fortsch. Phys. **57** (2009) 367–384, arXiv:0903.2596 [hep-th]
[354] K. Skenderis and B. C. van Rees, *Real-time gauge/gravity duality: Prescription, renormalization and examples*, JHEP **05** (2009) 085, arXiv:0812.2909 [hep--th]
[355] G. Birkhoff and G.-C. Rota, *Ordinary differential equations*, 2nd edn. Blaisdell, 1969
[356] G. Policastro, D. T. Son and A. O. Starinets, *The shear viscosity of strongly coupled* $N = 4$ *supersymmetric Yang–Mills plasma*, Phys. Rev. Lett. **87** (2001) 081601, arXiv:hep-th/0104066
[357] L. D. Landau and E. M. Lifshitz, *The classical theory of fields*, 4th rev. English edn. Pergamon Press, 1975
[358] M. Brigante, H. Liu, R. C. Myers, S. Shenker and S. Yaida, *Viscosity bound violation in higher derivative gravity*, Phys. Rev. **D77** (2008) 126006, arXiv:0712.0805 [hep-th]
[359] M. Brigante, H. Liu, R. C. Myers, S. Shenker and S. Yaida, *The viscosity bound and causality violation*, Phys. Rev. Lett. **100** (2008) 191601, arXiv:0802.3318 [hep-th]
[360] Y. Kats and P. Petrov, *Effect of curvature squared corrections in AdS on the viscosity of the dual gauge theory*, JHEP **01** (2009) 044, arXiv:0712.0743 [hep-th]
[361] C. W. Misner, K. S. Thorne and J. A. Wheeler, *Gravitation*. W.H. Freeman 1970
[362] M. Srednicki, *Quantum field theory*. Cambridge University Press, 2007
[363] V. G. Kac, *A sketch of Lie superalgebra theory*, Commun. Math. Phys. **53** (1977) 31–64
[364] V. G. Kac, *Lie superalgebras*, Adv. Math. **26** (1977) 8–96
[365] M. Parker, *Classification of real simple Lie superalgebras of classical type*, J. Math. Phys. **21** (1980) 689–697

Index

Printed in the United States
By Bookmasters